ANNUAL REVIEW OF PLANT PHYSIOLOGY AND PLANT MOLECULAR BIOLOGY

EDITORIAL COMMITTEE (1989)

ROGER N. BEACHY
WINSLOW R. BRIGGS
ROBERT E. CLELAND
RUSSELL L. JONES
WILLIAM J. LUCAS
WILLIAM L. OGREN
JAMES N. SIEDOW
VIRGINIA WALBOT

Responsible for organization of Volume 40
(Editorial Committee, 1987)

WINSLOW R. BRIGGS
ROBERT E. CLELAND
DEBORAH DELMER
DAVID W. KROGMANN
RUSSELL L. JONES
WILLIAM J. LUCAS
JAMES N. SIEDOW
VIRGINIA WALBOT
MAARTEN CHRISPEELS (GUEST)
STEPHEN MAYFIELD (GUEST)

Production Editor   IKE BURKE
Indexing Coordinator   MARY A. GLASS
Subject Indexer   DONALD C. HILDEBRAND

ANNUAL REVIEW OF PLANT PHYSIOLOGY AND PLANT MOLECULAR BIOLOGY

VOLUME 40, 1989

WINSLOW R. BRIGGS, *Editor*

Carnegie Institution of Washington, Stanford, California

RUSSELL L. JONES, *Associate Editor*

University of California, Berkeley

VIRGINIA WALBOT, *Associate Editor*

Stanford University

ANNUAL REVIEWS INC.   4139 EL CAMINO WAY   PO BOX 10139 PALO ALTO, CALIFORNIA 94303-0897 USA

Ⓡ ANNUAL REVIEWS INC.
Palo Alto, California, USA

International Standard Serial Number: 1040-2519
International Standard Book Number: 0-8243-0640-6
Library of Congress Catalog Card Number: A-51-1660

Annual Review and publication titles are registered trademarks of Annual Reviews
Inc.

∞ The paper used in this publication meets the minimum requirements of American National Standard for Information Sciences—Permanence of Paper for Printed
Library Materials, ANSI Z39.48-1984.

Annual Reviews Inc. and the Editors of its publications assume no responsibility
for the statements expressed by the contributors to this *Review*.

Typesetting by Kachina Typesetting Inc., Tempe, Arizona; John Olson, President
Typesetting coordinator, Janis Hoffman

For the convenience of readers, a detachable order form envelope is bound into the back of this volume.

*Annual Review of Plant Physiology
and Plant Molecular Biology
Volume 40 (1989)*

CONTENTS

TISSUES, ORGANS, AND WHOLE PARTS

INDEXES

ARTICLES OF INTEREST FROM OTHER ANNUAL REVIEWS

From the *Annual Review of Biochemistry*, Volume 58 (1989)

ATP Synthase (H$^+$−ATPase): Results by Combined Biochemical and Molecular Biological Approaches, *M. Futai, T. Noumi, and M. Maeda*
Multiple Isotope Effects on Enzyme-Catalyzed Reactions, *M. H. O'Leary*
Eukaryotic Transcriptional Regulatory Proteins, *P. F. Johnson and S. L. McKnight*
Topography of Membrane Proteins, *M. L. Jennings*
The Bacterial Photosynthetic Reaction Center as a Model for Membrane Proteins, *D. C. Rees, H. Komiya, T. O. Yeates, J. P. Allen, and G. Feher*
Crystal Structures of the Helix-Loop-Helix Calcium-Binding Proteins, *N. C. J. Strynadka and M. N. G. James*
The Mechanism of Biotin-Dependent Enzymes, *J. R. Knowles*

From the *Annual Review of Cell Biology*, Volume 4 (1988)

Microtubule Dynamics and Kinetochore Function in Mitosis, *T. J. Mitchison*
Fatty Acylation of Proteins, *A. M. Schultz, L. E. Henderson, and S. Oroszlan*
Molecular Aspects of Fertilization in Flowering Plants, *E. C. Cornish, M. A. Anderson, and A. E. Clarke*
Assembly of Phospholipids into Cellular Membranes: Biosynthesis, Transmembrane Movement, and Intracellular Translocation, *W. R. Bishop and R. M. Bell*
Biogenesis of Mitochondria, *G. Attardi and G. Schatz*

From the *Annual Review of Genetics*, Volume 22 (1988)

Foreign Genes in Plants: Transfer, Structure, Expression and Applications, *K. Weising, G. Kahl, and J. Schell*
Maize Developmental Genetics: Genes of Morphogenesis, *W. F. Sheridan*
Classical and Molecular Genetics of Tomato: Highlights and Perspectives, *C. M. Rick and J. I. Yoder*
Basic Processes Underlying Agrobacterium-Mediated DNA Transfer to Plant Cells, *P. Zambryski*
Phylogenies from Molecular Sequences: Inference and Reliability, *J. Felsenstein*
The Heat-Shock Proteins, *S. Lindquist and E. A. Craig*

From the *Annual Review of Microbiology*, Volume 42 (1988)

Cell Biology of *Agrobacterium* Infection and Transformation of Plants, *A. N. Binns and M. F. Thomashow*

Host Range Determinants in Plant Pathogens and Symbionts, *N. T. Keen and B. Staskawicz*

From the *Annual Review of Phytopathology*, Volume 27 (1989)

Agrobacterium tumefaciens and Interkingdom Genetic Exchange, *W. Ream*

Xylella fastidiosa: Xylem-Limited Bacterial Pathogen of Plants, *D. L. Hopkins*

The Movement of Viruses in Plants, *R. Hull*

Perspectives on Wound Healing in Resistance to Pathogens, *R. M. Bostock and B. Stermer*

Molecular Genetic Approaches to the Study of Fungal Pathogenesis, *S. A. Leong and D. W. Holden*

Cytoplasmic Male Sterility and Maternal Inheritance of Disease Susceptibility in Maize, *D. R. Pring and D. M. Lonsdale*

Annu. Rev. Plant Physiol. Plant Mol. Biol. 1989. 40:1–18
Copyright © 1989 by Annual Reviews Inc. All rights reserved

MY EARLY CAREER AND THE INVOLVEMENT OF WORLD WAR II

Noburô Kamiya

National Institute for Basic Biology, Okazaki, 444 Japan

CONTENTS

It is an honor to be asked to contribute a prefatory article to the *Annual Review of Plant Physiology and Plant Molecular Biology,* and I accepted the invitation without deeply considering what I should write. Since, however, I underwent unforgetable experiences in my youth—especially before and during World War II—I allow myself to record here my early personal history, remembering especially my teachers and others to whom I owe much.

CHILDHOOD

Born on July 23, 1913 as the third son of Ichiro Kamiya in the uptown district of Tokyo, I spent a calm childhood in a peaceful family. The event most shocking to me during my childhood was the great earthquake that hit the

1

1040-2519/89/0601-0001$02.00

Tokyo and Yokohama area on September 1, 1923. At the onset of the sudden quake around noon of that hot day, I jumped out into the garden, but owing to the violent horizontal and vertical vibrations of the ground I could hardly walk. Fortunately our single-story wooden house did not collapse, but our fire-proof two-story storehouse with its thick mortar wall was severely damaged. A few hours after the earthquake, gigantic cloud columns rose up in the southeastern sky: Fires that lasted until the next day reduced practically all of downtown Tokyo to ashes.

After a few days, accompanied by my father I took a look at the fire-ravaged areas. I could hardly bear to look at the disaster's consequences. Thousands of charred bodies were piled up along the roadsides; drowned bodies were scattered in moats and riverbeds from which victims, surrounded by the fire and smoke, had been unable to escape. To a boy of 10 years old this sight was ghastly. I understood that if our house had been in the downtown distict, I might have encountered the same fate. Rome was not built in a day, but Tokyo was destroyed overnight by a natural calamity. This miserable experience must have changed my view of life. It brought home to me firmly what a blessing it is simply to be alive and how wonderful is a peaceful existence, a keen awareness that has stayed with me ever since.

This terrible blow aside, I spent a happy boyhood playing with brothers and friends. I was absorbed in making various scientific toys and models such as small electric motors and steam engines, devoting less effort to the homework assigned by my schoolteachers. Some boys were enthusiastic about collecting insects or plants, but I was not. Probably I was not a good naturalist. As a high school junior and senior I thought there was no choice for me but to become an engineer. Thus I took a preparatory curriculum in which less weight was laid on biological sciences and in which English was the foreign language emphasized.

BOTANY AS A MAJOR

Being intrigued by the mystery of life, however, I had a latent interest in biology. What dissuaded me from majoring in basic biology was the fact that it was difficult to get a proper job in the university or elsewhere that permitted research in pure biology. The jobs available to biology graduates were mostly posts as high school teachers, whose main duty was the education of junior or senior high school pupils. The graduates from the engineering school had no problem finding good jobs and salaries.

After repeated consideration, however, I concluded that one should select one's profession not according to the convenience and profit but in response to an awareness of one's calling. Getting a good job is of secondary importance. My teacher of botany, the late Dr. Yoshitaka Imai, an internationally known geneticist at what is now Tokyo Metropolitan University, encouraged me in my resolution. To study cell biology or general physiology, however, we had to select either plant science or zoological science—there were at that time no departments of what we now call cell biology, biophysics, molecular biology, and so on. Without hesitation I selected plant sciences, because I did not like to kill animals, not even small insects.

Before World War II, the systems of education in the United States and Japan were somewhat different. In Japan, we had adopted the so-called 6-5-3-3 system [6 years of primary school, 5 years of middle school (junior high, or lower-secondary school), 3 years of high school (senior high, or upper-secondary school), and 3 years of undergraduate study in a university]. I was allowed to enter the Department of Botany, Faculty of Science, University of Tokyo in 1933. Although I found myself in a highly traditional environment, I gleaned enjoyment from each lecture. Plant taxonomy, morphology, genetics, physiology, ecology, microbiology, biochemistry and their class works—everything was fresh to me.

The lectures, however, were disconnected from each other. Morphology, for instance, was pure morphology, including fossil plants; it did not take into consideration the physiological function or meaning of plant structures. I, on the other hand, was keener on interdisciplinary fields, especially the functional aspects of the plant cell. But cytology was a part of morphology, and the class work of cytology was concerned mostly with the morphology of a variety of plant cells. To study chromosomes, whether in connection with taxonomy or genetics, we had to fix the cell with an appropriate fixative and stain with a dye. Few plant physiologists in Japan at this time were interested in either the physical chemistry of the living cell or its dynamic activities. Soon, before the third year of the undergraduate course, I asked my teacher of cytogenetics, Dr. Yoshito Sinotô, to let me work on living cells for a Bachelor's thesis. Dr. Sinotô was broad-minded enough to accept me in his laboratory. Although he was not himself a specialist in this field and could not guide my research, he encouraged me to open up this potentially important field in Japan through my own efforts.

As a happy new member of Dr. Sinotô's laboratory, I immediately faced a serious problem. Most of the important papers in my field of interest were written in German. I was only poorly educated in German, because German was not then as important to prospective engineers as it was to medical and biology students, for whom it seemed almost a prerequisite. I made up my mind to learn the German language all over again. I went to a German-training school and took private lessons from a German woman while I was a university student. In a year or two, I made fairly good progress. I could read German technical papers, and I wrote my Bachelor's thesis in (not very expressive) German. The thesis was on the effect of direct current on mitosis in the stamen hair of *Tradescantia* (1). It reported that the mitotic apparatus moved as a unit mass toward the anode, reversibly in the presence of a comparatively strong but short-lasting electric current. The cell was not killed by this treatment. I did not regard this phenomenon as electrophoresis of the chromosomes having a negative charge but believed it was caused by a temporary hydration and dehydration of the cytoplasm brought about by the ionic imbalance on the anodal and cathodal sides of the cell. I believe this interpretation is still thought to be correct.

In 1937, the year I entered the Graduate School of the University of Tokyo, the International Congress of Education was held on its campus. At that time, not only English (as in recent years) but also German and French were the official languages of the Congress. The office of the Congress therefore recruited translators of German and French from the University of Tokyo to help participants in the Congress from non-English-speaking countries. Although my German was still not fluent enough, I applied for the job and was accepted as one of the German translators. The mission of the translators was not to translate lectures presented at the Congress, a task too difficult for a student from a different discipline, but to accompany distinguished guests from abroad. I was assigned to guide a world-famous German philosopher, Eduard Spranger, who stayed in Japan as a guest professor for some time before and after the Congress. Regardless of his high scholarship and academic reputation, he was always courteous and friendly to me. When the Congress was over, he gave me a beautiful German photo album with a dedication in token of his gratitude. I think his influence helped me to be selected as a Germany-Japan exchange student the next year, providing an opportunity to study in Germany what it was then difficult to study in Japan.

VISITING ERNST KÜSTER AT THE UNIVERSITY OF GIESSEN

My destination in Germany was the Botanical Institute of the University of Giessen, presided over by Professor Ernst Küster, then the world authority on plant cell biology and plant cell pathology. Giessen was a pretty, quiet, garden city in western Germany with a population of only 35,000. It was situated on the east side of the Lahn river, a branch of the Rhein.

I left Japan for Germany in the early part of April, 1938, on board a Japanese freighter. Cherry blossoms were then at their best in Japan. When I reached Giessen two months later, in early June, this cosy city was basking in the fresh verdure of linden trees. A dignified, scholarly looking old gentleman greeted me warmly. Küster was then 63 years old and I was 24. He might well have been curious about me, because there were no Oriental people in Giessen in those days. I was the first (and last) Japanese student ever to work in his laboratory. On meeting me for the first time, he told me he would like to make my stay in his laboratory as fruitful and effective as possible. The professor soon guided me to a chair before his grand desk, just as he would have done had he been giving an oral examination to one of his own, German students. He needed to determine how far my studies had progressed in Japan. I think I answered the first questions pretty well, because they concerned osmotic phenomena in plant cells; but I was shaky on the second group of questions. He asked me if I knew the names of several algae and fungi and their characteristic features. This was the weakest point in my botanical knowledge. What I had learned in systematic botany at the University of Tokyo primarily concerned plants higher than ferns. I knew popular class-work materials such as *Spirogyra, Nitella, Phycomyces,* etc, but otherwise my answers were very unsatisfactory. Professor Küster, however, did not dwell on my ignorance in the matter, but said "Here we are doing experimental work with special attention to lowly plant cells and protophytes. Are you not interested in research along this line?"

I was delighted to hear that I could study there, from that time on, the subjects about which I had little knowledge but that interested me keenly. My life in Giessen thus began. Looking back upon my student days, when I was not yet an independent worker, I think in Giessen I spent the most enriching time in training myself. My stay there was to be cut short at one year and three months by the outbreak of the second European War, a subject to which I shall return below.

DAYS SPENT IN GIESSEN

In the summer semester when I arrived in Giessen, Küster's lecture on General Botany was to be heard every day. I decided to attend this class not only to revise my knowledge in General Botany but also to train my ear for German. This lecture was intended primarily for premedical students, and its level was about that of the famous Strasburger's *Allgemeine Botanik*. Küster spoke without a manuscript (a tiny memo if anything at all), yet the content was condensed and perfect. He said everything necessary and important but nothing unnecessary or trivial for the students, demonstrating simple experiments and showing many specimens arranged on the large lecture table. The lecture was so perfect that one thought the stenographed transcript would make an excellent textbook.

One thing, however, disturbed me. For some months after I had reached Giessen, I stayed in a student dormitory, where the dining hall opened at 7:30 a.m. Küster's lecture on General Botany started punctually as early as 7:00 a.m. every morning and ended at 8:00. The lecture was followed immediately by research work, usually until 12:30. Therefore, I had no chance to take breakfast and had to work the whole morning on an empty stomach! I gradually became accustomed to this habit, and I rather enjoyed every minute of my laboratory work.

It was not Küster's way to use sophisticated new measuring apparatuses or to try to develop a new gadget for a specific purpose. Instead he observed the variety of living plant cells in nature, mostly those of lower plants, and induced abnormal cell structure or cell behavior experimentally. During my sojourn in Giessen, I wrote five papers, of which four were on protophytes [*Melosira* (a diatom)(2), *Euglena* (4), *Spirogyra* (5), and *Lyngbya* (a blue-green alga)(6)] and only one was on onion scale epidermis cells (3). I followed no predetermined plan of study and therefore sought no particular connection among my various observations. My main purpose was to familiarize myself thoroughly with as wide a variety of lower-plant materials as possible. We did not use complicated methods; as a means of experimental study we frequently applied plasmolysis, vital staining, centrifugation, etc.

There were 10 or so research people working in the Botanical Institute, including three postdoctoral assistants, several doctoral candidates, and several guest research workers from other universities. Just before I arrived at Giessen, Walter Michel (19) had succeeded in bringing homospecific as well as heterospecific plant protoplasts into fusion. At that time there were no suitable enzymes strong enough to dissolve the cell wall quickly, and we hence had to resort to plasmolysis followed by the opening of cells in order to obtain the intact plant protoplasts. Though the efficiency was low, it was nevertheless possible to get naked plant protoplasts by this means. After repeated trials and errors, Michel finally fused two protoplasts, a memorable achievement in experimental plant cell research. Who could then have expected that protoplast fusion would become such a popular and important method in plant biotechnology?

It was Küster's daily habit to visit each research worker every other hour, asking "Now, do you have anything new?" Having to listen to such a "greeting" several times a day was certainly too much for us. But if someone was able to exhibit noteworthy phenomena under the microscope, Küster became excited, saying "Sehr interessant! Sehr interessant! Kolossal!" Occasionally Küster would unceremoniously displace a young researcher at his bench and proceed to describe what he observed, while the host of the preparation, sitting to one side, wrote down quickly what Küster dictated. The transcript would be a part of the paper to be published later. It was a good idea to write such descriptions on the spot, especially for phenomena that occurred only rarely. Our memories may so soon fade.

While Küster paid close attention to the research progress of each of his students, he himself wrote more than ten original papers each year. He was the author not only of *Die Pflanzenzelle* (1st ed. 1935; 2nd ed. 1951; 3rd ed. 1956, after his death), his magnum opus, but also of many important monographs. He also acted as a chief-editor of *Zeitschrift Für Wissenschaftliche Mikroskopie und für mikroskopische Technik*. I wondered how so much could be achieved by only one man.

It was not long after I had arrived at his laboratory that I saw a typist entering his big office and taking a seat in front of a large table set at the center of the room. Then the door was closed. Before long, I heard Küster's footsteps as he walked slowly around the central table; he then began dictating to the typist, and there followed the sounds of continual typewriting. Though of course we were not allowed to enter the room then, I learned from one of the professor's assistants that Küster used nothing but a small memo when he dictated. Simply through concentration, he found the proper layout, context, and sentences for a scientific paper while walking at a slow pace around the table. After a few hours, bunches of typewritten sheets, say 30 or so, were finished. Küster took a look at them and made corrections, insertions, or deletions with a pen; but the manuscript was not retyped; it was sent out to a publisher in a day. It must have required very special talent to write a scientific paper in such a way, even in one's mother language.

These recollections of my Giessener days are already half a century old. One of my German colleagues noted that Küster was among the last university professors of an old German type.

Another memorable experience occurred one day when we took dinner together with a few colleagues in a Giessen restaurant. At a table next to ours sat Robert Feulgen, who had established the Feulgen method of staining DNA. Since I thought Feulgen a great figure in the history of cytochemistry, it delighted and surprised me to see him near at hand, taking dinner with a glass of beer just as any other man might.

Things have been much changed since World War II. Nowadays many scientists whose mother tongues are not English (including German scientists) write their papers in English, the most popular and generally accepted language internationally. Therefore, learning English and having to write papers in English are inevitable demands upon internationally minded research workers. This is a big handicap for Japanese scientists. Not a few of them believe that they are the people poorest at learning Western languages, probably because of the basic difference in construction between their mother language and the Western ones. As a matter of fact, it is not rare for us, and I don't hesitate to include myself, to spend more time and energy in preparing the manuscript in English than in performing experiments. Preparing manuscripts of scientific papers by dictation, as Küster did, is difficult enough in one's mother language, let alone in a foreign one. Such expertise in English is at present a dream for most Japanese scientists.

HIDDENSEE

A small marine biological laboratory stood on a tiny island called Hiddensee, just off Rugen Island in the Baltic Sea. The island belongs now to East Germany. It was a custom for Küster and his students, about 10 altogether, to spend three summer weeks at Hiddensee. I had the good fortune to spend two summers there, in 1938 and 1939. It was an enjoyable time for all of us, combining research and recreation. We collected various algae there, finding interesting materials for observation and experiments. If our experiments appeared promising or needed longer to finish up, we took the material back to Giessen. In 1938 we were invited by Professor Boysen-Jensen, a great figure in the field of plant growth physiology, to visit his laboratory in Copenhagen on the way back to Giessen. This was a pleasant and instructive experience for me.

The next summer, 1939, Hiddensee Island was as quiet and peaceful as it had been the year before, but European international relationships were growing more and more strained. On August 25, close to noon, as I was working at the microscope as usual, I got an unexpected long-distance telephone call from Berlin. A Japanese friend of mine who knew my whereabouts transmitted urgent advice from the Japanese Embassy. I was to prepare immediately to depart Germany. I was to board a Japanese boat, the *M. S. Yasukunimaru*, that happened to be in Hamburg and was scheduled to leave the next evening at 8:00 p.m.

I immediately informed Professor Küster of this urgent message from the Japanese Embassy. He was astonished to hear it and told me earnestly, "I feel very sorry to have you leave us. If the reason for leaving Germany is simply a financial problem, don't be bothered with it. As long as I am alive, I think I can help you financially, even if war should break out. When you go to Berlin and meet the officials of the Embassy, tell this to them so that you will be allowed to stay on in Giessen. Giessen is a comparatively safe place if the war begins. If you are allowed to stay in Germany, you need not come back here again, as we shall not stay here long anyway. We will see you in Giessen again. We will take your material (the blue-green alga *Lyngbya*) with us to Giessen. . . ."

Mrs. Küster gave me as a souvenir a photo album of Hiddensee together with a sheet of paper on which she wrote a beautiful poem by the German poet Isolde Kurz entitled "The sun going down in Hiddensee." Underneath this poem she wrote "To the memory of pleasant hours on Hiddensee and Giessen absorbed in the wonder of protists," signing it "Frau Dr. Gertrud Küster." Mrs. Küster and a colleague made me sandwiches, each wrapped separately, and added fruits and chocolate decorated with a branchlet of the beautiful wild flowers of Hiddensee. This would be my snack on the voyage from Hiddensee to Stralsund and in the train from Stralsund to Berlin. The time hurried past. I had to catch a boat bound for Stralsund at 4:30 p.m. The summer sun was still high in the sky. The sight of Professor and Mrs. Küster seeing me off on the pier of Kloster harbor on Hiddensee still lives in my mind. Two of my colleagues accompanied me to the next harbor, Vitte, on the same island. I arrived in Berlin late that evening.

Next morning I met in Berlin several Japanese people and an official of the D.A.A.D. (German Academic Exchange Agency). Although we had received not an evacuation order but only advice, all the Japanese I met in Berlin strongly opposed my going to stay in Giessen. They urged me to go immediately on board *M. S. Yasukunimaru,* a refugee boat. This boat would sail first for Bergen, Norway, where she would stay for some time to see how the international situation developed; if war did not break out, she would return to Hamburg.

It was difficult to anticipate the fate of a foreign student or the possibility of his performing research work in wartime, even if he was not in an enemy country. I decided finally to go to Hamburg, in spite of Küster's kind advice, to board the refugee boat with faint expectation of the ship's returning to Hamburg. If I was to catch the boat, I could not go back to Giessen to gather my belongings from the laboratory and the boarding house. What I had with me was only a small suitcase taken from Giessen to Hiddensee, containing mainly the clothing necessary for a three weeks' stay.

With about 180 refugees aboard, the boat left on time at 8:00 p.m. on the moonlit night of the 26th of August, 1939. Hearing Japanese spoken all around me, I experienced an illusion of being suddenly back in Japan. Most of the refugees were businessmen and their families. Among the few scientists were two physicists, Drs. H. Yukawa and S. Tomonaga. Their names were not so well known then, but they both won the Nobel Prize later. Since Tomonaga was also one of the exchange students of the D.A.A.D., he and I became good roommates in the same second-class cabin.

Two days after embarkation, the refugee boat arrived at Bergen and anchored there for one week, watching for changes in the international situation. While we were in Bergen, Germany invaded Poland and World War II broke out. This deprived me of my last hope of going back to Giessen. I wrote a letter to Küster thanking him for his willingness to have me continue my research and for all he had done for my sake. It was announced that the boat would leave Bergen at noon on the 4th of September, 1939, for Yokohama by way of New York, the Panama Canal, and San Francisco.

In this situation, it occurred to me that it might be possible to continue my work in the United States, since I earnestly hoped to study further before returning home. I knew the name of William Seifriz at the University of Pennsylvania, as I had been a devoted reader of his book *Protoplasm* (1936) when I was in Japan. But I was of course not prepared for landing in the New World. I had no proper visa, no letter of introduction, no acquaintances in the United States. I could read and write English to some extent, but I had had no practice in speaking it. What I had was barely enough money to transit the American continent by train.

When the refugee ship was still in Bergen, I got a "transit" visa, good only for one month. With it I was able to land in the United States. In Bergen I bought a book on English conversation. It was written in Norwegian, but I could pick up English expressions in it and try to memorize them during the voyage to New York. Such hasty efforts did not help much toward speaking English, however. Before the boat reached New York, I wrote a long letter in German to Professor William Seifriz explaining the situation and asking about the possibility of working in his laboratory for a time. Upon my arrival in New York I posted the letter, appending my curriculum vitae and a list of my publications.

VISITING WILLIAM SEIFRIZ AT THE UNIVERSITY OF PENNSYLVANIA AT THE OUTBREAK OF THE EUROPEAN WAR

Taking a far northern route in the Atlantic Ocean, *M. S. Yasukunimaru* reached New York ten days after its departure from Bergen. I think it was I alone who got off the ship at New York.

The first thing I did in New York was visit the Japan Institute. Utterly independent of politics and economy, it had been founded for the purpose of introducing Japanese culture to Americans. Since I had no letter of introduction, I introduced myself to a young official there. Soon another official joined us, and eventually I was guided to the office of the director of the Institute, Tamon Mayeda. (Mayeda was to become the first Minister of Education after World War II.) He must have wondered at the queer young Japanese refugee who had appeared on his Institute's doorstep. I told him my whole story and of my desire to continue to study with Professor Seifriz at the University of Pennsylvania. Mayeda kindly tried to contact Professor Seifriz by phone: "There is a young Japanese scientist now in New York evacuated from Germany on account of the European war. He earnestly hopes to see you and, if possible, to work with you for some time. Would you be good enough to meet him?" Thus Tamon Mayeda made me an appointment with Seifriz on the 18th morning of September, 1939.

It was a clear autumn day with deep blue skies, but my innermost feeling was a complex mixture of aspiration and anxiety. If Seifriz did not speak German, what could I do? Still, I had no choice but to visit him. The letter I had written from the ship must already have been in his hands. Having bought a city map of Philadelphia at the terminal station, I walked to the University campus. I asked the way several times and finally got to the laboratory of Seifriz. It was a rather old fashioned, detached house called Botany Annex.

To my great surprise, the person I met at the entrance of the laboratory was a young Japanese woman, a graduate student of Seifriz. She said "Are you Mr. Kamiya? I am so glad to see you. Dr. Seifriz is waiting for you. He loves to speak in German!" It was a further surprise that she already knew my name.

Guided by her, I stepped upstairs. When I came to the entrance of Seifriz's office, the door of which was open, he walked toward me and shook hands, saying in German that he was very pleased to have me there. He had read my letter already. The young lady, whose name was Masa Uraguchi, seemed to have been waiting downstairs for my coming. It was indeed very fortunate for me that Seifriz spoke German as fluently as his native language. Although he had been born in Washington D.C., near the Capitol, his parents were German and he had studied colloid chemistry for some time with Freundlich in Berlin.

At that time Seifriz was interested in the slime mold *Physarum polycephalum*. It was an excellent experimental material for the study of protoplasm, especially regarding the rhythmicity of its protoplasmic streaming. He was studying the pulsation of the slime mold by taking time-lapse cinefilm. My small paper on the rhythm of euglenoid movement (4) was a hastily written report of a simple observation done in Giessen, but Seifriz seemed to have paid some attention to it, probably because it dealt with a biological rhythm. Since we shared a close interest in this topic, our conversation was lively. A room in front of Seifriz's office, just then vacant, was allotted to me for my laboratory.

It was unusual for Professor Seifriz to have lunch off-campus, but on this day he took Masa Uraguchi and me in his car to a fancy restaurant somewhere near the university. It was funny that we had no common language among us. Seifriz and Masa spoke in English, Seifriz and I spoke in German, and Masa and I spoke in Japanese. But we shared a delightful mood.

On the same day, a modest room was found for me to stay in near the campus. Thus it became possible for me to settle down in Philadelphia at least for a short period. I returned to New York to fetch my suitcase and to report the matter to Tamon Mayeda.

DAYS SPENT AT SEIFRIZ'S LABORATORY

The extension of my transit visa was kindly taken care of by the Chairman of the Department, Professor J. R. Schramm. Although the problem of monetary support was still pending, it had thus become possible for me to stay in Philadelphia and work on *Physarum* with Seifriz as far as my finances would permit. There in Seifriz's laboratory I saw for the first time the beautiful culture of the plasmodium of *Physarum polycephalum*. An important merit of this material for experimental work was that it could be easily cultivated in the laboratory simply by feeding it with oats. Unlike Küster, Seifriz did not suggest what theme I should study or what material I should use. There were a variety of interesting experimental materials in the well-tended greenhouse, but I was attracted to this golden yellow slime mold.

Seifriz's lecture on "The Physics and Chemistry of Protoplasm" began on the 1st of October 1939. However, this was the first rainy day after my arrival in Philadelphia. I had neither umbrella nor raincoat, and I had to run to the laboratory in the rain. Looking at my wet clothes, Seifriz gave me his own English raincoat, saying that he had no need of it because he commuted in his own car.

In his lecture Seifriz gave not merely dry textbook facts, but also the background of the subject and lines of thought by which the discoveries had come about. As was not the case at Küster's lecture, students here had ample opportunity to ask questions or discuss whatever subjects lay within the scope of the lecture. The atmosphere of the class was quite different from that in Giessen. As lecturers the only point Seifriz and Küster had in common was that neither needed elaborate notes.

Seifriz, who was also known as a plant geographer, gave another lecture on "Plants and Climate," with many beautiful slides he had taken during botanical trips all over the world, from the tropical to the arctic zones. These slides, shown more than 50 years ago, were large glass monochrome plates.

Seifriz's main objective in the laboratory was to gain insight into the submicroscopic structure of the protoplasm. Because electron microscopes were not yet available, he used physical properties as a means of looking into the structure of protoplasm. To investigate its physical properties, he liked to resort to the delicate technique of microdissection, which had hitherto not been in general use. In the early days of Seifriz's scientific career, when the now popular apparatus known as the micromanipulator was not available, he used the Barber pipette holder, following Kite & Chambers for microdissection. He tried to estimate such physical properties as the viscosity and elasticity of protoplasm with a glass needle.

Seifriz emphasized that protoplasm is non-Newtonian and generally elastic and insisted that these physical properties are explained by the existence of long-chain molecules attached to one another in one way or another. At a time when the idea of the structure of protoplasm was a confused mixture of fact and theory, Seifriz took the lead in the modern view. The image he had 50 years ago matches well our contemporary knowledge about the cytoskeleton. As to the unique scholarship and personality of Seifriz, I have already written a memorial article elsewhere (9).

For some time after I came to Seifriz's laboratory, I continued to observe the behavior of the slime mold under various experimental conditions, especially its response to mechanical stress. A slight, locally applied mechanical pressure modified the speed and direction of flow of the endoplasm sensitively, and yet produced no injury to the plasmodium. This behavior suggested that the normal flow of the endoplasm, too, was caused by the internal local difference in pressure. Curiosity about such behavior of endoplasmic flow led me to perform further experiments in which air pressure was used instead of mechanical pressure. The double-chamber method for measuring the difference in local internal pressure (i.e. the motive force responsible for protoplasmic streaming) was thus developed (7, 8). The greatest difficulty in preparing the double-chamber lay in separating the two protoplasmic blobs connected by a single vein into two airtight compartments. This was because the vein in which endoplasm flows back and forth was so soft and delicate that it was readily collapsed by mechanical disturbances. The problem was solved using semi-fluid, lukewarm agar.

It took about three months after I first saw the *Physarum* plasmodium to establish this simple technique—until the end of 1939. When we plotted the motive force responsible for streaming against time using this method, we got undulating curves with various wave patterns. It was found, by means of waveform analysis, that all the curves thus obtained could be reconstructed neatly with only two or three sine waves having different periods and amplitudes. Seifriz was very pleased with this result, because it measured the motive force responsible for protoplasmic streaming for the first time and showed that the visible pulsation was composed of more than a single rhythm.

Although I was blessed with good luck in my work, the problem of a source of living expenses had still not been solved. Not only Professor Seifriz but also the staff of the Department were good enough to seek funds for me. However, public opinion in the United States was then set severely against Japan. To the United States at that time Japan was already a so to speak quasi-enemy country. There were not a few foreign students and post-docs, but I met no Japanese students supported by American fellowships. Masa Uraguchi's was an exceptional case: She had come to the United States on a Friends (Quaker) fellowship.

One day in April, Seifriz asked me to accompany him to the dinner of Mrs. Curtin Winsor. I did not know it then, but later I learned that she was a daughter of the President of DuPont and had once been the wife of a son of President Roosevelt. Unaware in what connection Seifriz knew her, I supposed she was his former student. At any rate, Seifriz took me, along with a cine projector and a few reels of cinefilm, to Winsor's home in the Main Line, the most prestigious residential area west of Philadelphia. Mrs. Winsor seemed to be still young, probably in her 30's, a refined and gracious lady. Seifriz introduced us. After dinner was over, Seifriz persuaded her of the importance of basic science and showed the film of the slime mold, which he had taken himself. Seifriz asked her if it would be possible for her to help with my living expenses for one year for the sake of progress in basic science. Mrs. Winsor gave her consent without hesitation. Thus thanks to Seifriz's enthusiasm and Mrs. Winsor's philanthropic support, my financial problems were solved for a year to come. As we rode back toward my boarding house in Seifriz's car, the city of Philadelphia shrouded in spring rain appeared to me a dreamland.

Owing to the efforts of Professor J. R. Schramm, the chairman of the

Department of Botany, I obtained my formal student visa. The problems that had been pending since I landed in New York were thus all solved through the goodwill and efforts of the people around me.

During the Christmas season of the same year, 1940, the annual meeting of the AAAS was held in Philadelphia. As part of the program of the American Society of Plant Physiologists, Seifriz organized a symposium on "The Structure of Protoplasm." Invited to participate, I presented the data so far obtained on *Physarum*. This was my first lecture in English to an audience of about 300. Seifriz chaired the symposium with great charm and humor. The record of the event was published as a book (20) in the early part of 1942—that is, soon after the outbreak of the Pacific War.

MRS. WILLIAM H. COLLINS

Shortly before Miss Masa Uraguchi left for Japan (having finished her thesis), a pious elderly Quaker lady, Mrs. William H. Collins, invited Masa, her close friend Miss Miyeko Mayeda (the first daughter of Tamon Mayeda), and me to a dinner at her home on the campus of Haverford College not far from Philadelphia. Soon after that, Mrs. Collins asked me whether I might be interested in staying in her home, where everything would be taken care of by herself and her able American-Indian butler. My only duty would be to take breakfast and dinner with Mrs. Collins and to accompany her to the Friends Meeting House every Sunday morning. She was the only daughter of a world-renowned 19th-century paleontologist, Edward D. Cope, and the wife of Dr. W. H. Collins, Professor of Astronomy at Haverford College, who had died a couple of years before. She was a graceful Japanophilic lady more than 70 years old and without children.

I gladly accepted her kind invitation. There was a cosy bedroom and a bathroom for my own use. Further, her late husband's study was allotted to me for my work. Delicious meals were served by her butler, an excellent cook. She treated me as if I were her son-in-law. I not only paid no board, but she gave me a coat to wear at home, ties, underwear, etc.—even a golden wristwatch when I somehow lost mine. Occasionally I told her that I felt sorry to be given so much by her while I had nothing to give in return. Then her answer was always "Oh no, Noburô! You give me friendship!" The only inconvenience to me was that I had to return to the house by 6:00 p.m. to take dinner with her, when it had been my habit to work late in the laboratory. At any rate, the blessed encounter with Mrs. Collins changed my life-style.

OUTBREAK OF THE PACIFIC WAR

The international relationship between the United States and Japan worsened day by day, and many people in the States had a foreboding of war with Japan. At the very beginning of December, 1941, the Japanese Government sent a refugee boat, *M. S. Tatsutamaru,* to repatriate Japanese citizens in the

United States. I prepared to go home on board that boat and packed personal articles so that I would be ready to leave Haverford for San Francisco by train on the 8th of December. But alas, when everything was ready for departure and I waited, sad at the thought of bidding Mrs. Collins farewell soon, the radio announced in the early afternoon of December 7th, 1941, the shocking news of a sudden attack on Pearl Harbor by the Japanese Navy. My plans for going back to Japan had all been in vain. From that moment I became an enemy alien. Mrs. Collins was of course much shocked by this drastic news, but her attitude toward me was as good as before; and seemed even gentler, reading my mind.

After that I was visited often by FBI and Naval officers. Once I was asked to appear in person at the Attorney General's Office in downtown Philadelphia. Mrs. Collins, in spite of her age, accompanied me, testifying that I was not a dangerous person at all. Japanese in important positions were sent to the concentration camp on Ellis Island in New York Harbor. As a student I was left more or less free, except that I was not allowed to go farther than 3 miles from my residence without permission. I remained at Mrs. Collins's home under her protection. Since I could not go to the University of Pennsylvania, Seifriz made it a rule to visit me twice a week, regularly.

Surprisingly enough, I got a message a few months after the outbreak of the Pacific War from the Chairman of the Department of Biology of Haverford College, Professor H. K. Henry, granting me the privilege of using a laboratory and library there. I was touched by this generosity of the Quaker College in that most serious international situation. As the building of the Biology Department was a 10 minutes' walk from the Collins home on the same beautiful campus, it was very convenient for me. I gratefully accepted his offer.

Although there was nothing in the laboratory but a large table, a stool, and a lot of bottled snakes preserved in formalin, it was at least possible to cultivate the slime mold there with oats in soup dishes and wet sheets of paper towel. I inverted other soup dishes on the top of the culture to cover it. Having carefully observed the slime mold culture in the soup dishes with the naked eye, I noticed a tendency of the slime mold strands to twist themselves when they came into contact with the water that half-filled the dish.

I then asked Seifriz to bring several small glass cylinders with bottoms and several glass plates of about 2 in^2 to cover them. Pouring a small amount of water in the bottom of the cylinder to make the air inside moist, I tried to hang a 3–5-cm-long strand (vein) segment cut out from the stock culture at the center of this tube from the underside of the flat glass cover. A delicate inverted T-form glass needle was attached to the lower free end of the strand to see how it rotated. If we placed a handmade 360° protractor underneath the bottle and observed the inverted T-form glass needle from the top, we could follow, approximately, even with the naked eye, how the strand segment twisted itself with time. In the first 20–30 min, torsion of the strand was modest and irregular, but it gradually tended to twist regularly and alternately in clockwise and counterclockwise directions in a period of about 2 min, often with a swing angle of more than 180°. But curiously enough, this torsional movement was asymmetrical. It twisted in one direction more each time than in the other. Statistically, I encountered more often the case in which counterclockwise rotation was dominant. At any rate, by repeating the asymmetric movement, the strand twisted itself substantially in one direction many times, practically endlessly. This simple observation made with this simple setup was repeated after the war was over; but its implication in relation to the cytoskeleton and the mechanism of synchronization of local irregular rhythms into one regular, larger rhythm remain interesting problems today (14).

BACK TO JAPAN DURING THE WAR

While I was spending quiet days at Haverford under the protection of Mrs. Collins, sometime in May of 1942, I suddenly got a telegram from the Department of State in Washington, D.C. notifying me of the possibility of repatriation by means of exchange boats. The arrangement had been made by the Red Cross in Switzerland to exchange American civilians in Japan and Japanese civilians in the United States. The telegram said that if I wished to go back to Japan, I should come to the Hotel Pennsylvania in New York by the 10th of June, 1942. The volume of allowable luggage was less than 30 cubic feet. The telegram requested me to answer as soon as possible whether or not I wished to go back. Of course I wanted to take advantage of this opportunity to return home. Since the volume of my luggage was far less than 30 cubic feet, and no restrictions on the contents of the luggage were noted, I went to the Hotel Pennsylvania on the designated date, taking all my belongings. On parting from Mrs. Collins, I expressed to her my hearty gratitude for all she had done for my sake during the previous two years.

At the Hotel Pennsylvania in New York I found many Japanese of various professions. We were all leniently confined on specified floors of the hotel for about one week, and the meals and services of the hotel were good enough for enemy aliens. During the wait all our luggage was inspected severely. When my turn came, not only all the experimental data, but even newspaper sheets used for wrapping were confiscated. When the inspector found several sheets of yellow sclerotia of *Physarum polycephalum* spread on dried sheets of paper towel and filter paper, he asked me with perplexed eyes what they were. In spite of my desperate efforts to explain that it was a dried specimen of an absolutely harmless organism called slime mold, he confiscated all of them mercilessly, saying nothing. In other words, we were allowed to carry only our clothes. For an experimental scientist, loss of his experimental data is grievous. All my efforts of the past few years had thus come to nothing. I was stunned by this result; I felt heartbroken.

We left New York on board a Swedish-American Liner, *S. S. Gripsholm,* on the 18th of June, 1942, if my memory is correct. The then ambassabor, Normura, extraordinary plenipotentiary ambassador Kurusu, and the director of the Japan Institute, Tamon Mayeda, whom I have already mentioned, were all on board. Since the Panama Canal was closed, *S. S. Gripsholm* took a southern route and touched in at Rio de Janeiro for a few days, probably for refueling and reprovisioning. We were not allowed to disembark but we enjoyed the beautiful landscape surrounding the harbor. Then the boat left for Lourenço Marques on the east coast of Africa, rounding the Cape of Good Hope.

The day after we arrived at Lourenço Marques, a Japanese boat, *M. S. Asamamaru*, with large, white crosses on her sides, reached the same port. She carried American citizens from Japan. During our week's stay there, passengers and luggage were exchanged. We had plenty of time to take walks in the city and the suburbs. As a minor botanist, I observed with special curiosity the flora, which was quite different from that of the northern hemisphere. At last all of us were on board the *M. S. Asamamaru* and we left for Japan across the Indian Ocean. The boat touched in at Singapore for a while and reached Yokohama on the 20th of August, 1942. The long voyage from New York to Yokohama had taken more than two months.

I was not drafted until the end of the war. In April, 1943, I was nominated as lecturer at the Botany Department of the University of Tokyo and taught plant cell physiology to the students until 1949. During the war, Tokyo was severely attacked from the air. My house was reduced to ashes, but fortunately the buildings of the University of Tokyo remained almost intact. After the war ended in August of 1945, I tried to restart my experiments on *Physarum*. Although the laboratories were not in good order right after the war, it was not impossible to do some experiments, because my work on the slime mold did not require much money. I constructed myself a queer gadget that enabled us to record the motive force of protoplasmic streaming semiautomatically.

I also started to work on protoplasmic streaming in giant algal cells, especially in the intermodal cell of *Nitella*. Such work was feasible in a devastated, poorly equipped laboratory after the war. Protoplasmic streaming was an intriguing phenomenon, the mechanism of which was hardly known at that time. Whether it was shuttle streaming in *Physarum* or rotational streaming in *Nitella*, protoplasmic streaming was thought to visualize the mystery of cellular activities. As a matter of fact, neither the nature of the force driving the protoplasm nor the site where it was produced in the cell was known.

In 1949, a new Department of Biology was established in Osaka University and the post of professor in charge of cell physiology was offered to me. The next year, in 1950, I was invited again to the University of Pennsylvania to work with Seifriz for 9 months. I continued then the work on torsion of the protoplasmic strand started at Haverford College during wartime but interrupted by my repatriation (17). The subsequent progess in the field of cytoplasmic movements, mostly in the slime mold and giant algal cells, was reviewed by the author in the *Annual Review of Plant Physiology* (11, 13) and elsewhere (10, 12, 14–16).

Looking back on youth, especially the several years I spent in Germany and the United States before and during the Second World War, I would like to believe that an invisible thread of destiny enabled me to meet Professors Küster and Seifriz, Mrs. Collins, and the many benefactors to whom I owed so much. Who could have expected that Dr. Tamon Mayeda, whom I had met in New York for the first time without any letter of introduction, would become my father-in-law after the end of the war. Both Küster and Seifriz had outstanding (and quite different) personalities. Both were most precious teachers in my scientific career, and I can hardly estimate their influence upon my subsequent life.

I would like to express my cordial thanks to Mrs. Judith Fujimoto for her kind help and advice in overcoming the linguistic difficulties I had to cope with here.

Literature Cited

1. Kamiya, N. 1937. Untersuchungen über die Wirkung des electrischen Stromes auf lebende Zellen. I. Das Verhalten der mitotischen Figur unter der Wirkung des Gleichstromes. *Cytologia* (Fujii-Jubiläumsband):1036–42
2. Kamiya, N. 1938. Über Doppelschalen bei *Melosira. Arch. Protistenkunde* 91: 324–42
3. Kamiya, N. 1939. Zytomorphologische Plasmolysestudien an *Allium*-Epidermen. *Protoplasma* 32:373–96
4. Kamiya, N. 1939. Die Rhythmik des metabolischen Formwechsels der Euglenen. *Ber. Dtsch. Bot. Ges.* 57:231–40
5. Kamiya, N. 1939. Beiträge zur Pathologie der Zellteilung und der Querwandbildung. *Protoplasma* 33:427–39
6. Kamiya, N. 1940. Parasiten in Oscillatoriaceen. Beiträge zur Pathologie der Cyanophyceenzelle). *Arch. Protistenkunde* 94:201–11
7. Kamiya, N. 1940. The control of protoplasmic streaming. *Science* 92:462–63
8. Kamiya, N. 1942. Physical aspects of protoplasmic streaming. See Ref. 20, pp. 199–244
9. Kamiya, N. 1956. In memoriam William Seifriz. *Protoplasma* 45:513–24
10. Kamiya, N. 1959. Protoplasmic streaming. *Protoplasmatologia*, 8, 3 a. Vienna: Springer. 199 pp.
11. Kamiya, N. 1960. Physics and chemistry of protoplasmic streaming. *Annu. Rev. Plant Physiol.* 11:323–41
12. Kamiya, N. 1962. Protoplasmic streaming. *Handb. Pflanzenphysiol.* 17(2): 979-1035
13. Kamiya, N. 1981. Physical and chemical basis of cytoplasmic streaming. *Annu. Rev. Plant Physiol.* 32:205–36
14. Kamiya, N. 1985. Some motility characteristics of living cytoplasm: from observations on *Physarum*. In *Cell Motility: Mechanism and Regulation*, ed. H. Ishikawa, S. Hatano, H. Sato, pp. 577–85. Tokyo: Univ. Tokyo Press, 629 pp.
15. Kamiya, N. 1986. Cytoplasmic streaming in giant algal cells: a historical survey of experimental approaches. *Bot. Mag. Tokyo* 99:441–67
16. Kamiya, N., Allen, R. D., Yoshimoto, Y. 1988. Dynamic organization of *Physarum* plasmodium. *Cell Mot. Cytoskel.* 10:107–16
17. Kamiya, N., Seifriz, W. 1954. Torsion in a protoplasmic thread. *Exp. Cell Res.* 6:1–16
18. Kamiya, N., Yoshimoto, Y., Matsumura, F. 1982. Contraction-relaxation cycle of *Physarum* cytoplasm: concomitant changes in intraplasmodial ATP and Ca^{++} concentrations. *Cold Spring Harbor Symp. Quant. Biol.* 46:77–84
19. Michel, W. 1937. Über die experimentelle Fusion pflanzlicher Protoplasten. *Arch. Exp. Zellforsch.* 20:230–52
20. Seifriz, W., ed. 1942. *The Structure of Protoplasm.* Monogr. Am. Soc. Plant Physiol. Ames, Iowa: Iowa State Coll. Press. 283 pp.

Annu. Rev. Plant Phys. Mol. Bio. 1989. 40:19–38

VULNERABILITY OF XYLEM TO CAVITATION AND EMBOLISM

M. T. Tyree and J. S. Sperry

Northeastern Forest Experiment Station, Burlington, Vermont 05402 and Department of Botany, University of Vermont, Burlington, Vermont 05405

CONTENTS

WATER RELATIONS AND THE VULNERABLE PIPELINE

The evolution of cell walls allowed the plant kingdom to solve the problem of osmoregulation in freshwater environments; confining protoplasm inside a rigid exoskeleton prevented cell rupture as a result of osmotic inflow of water. The cost of cell walls for plants was a loss of motility. In contrast, in the

19

1040-2519/89/0601-0019$02.00

animal kingdom osmoregulation involved the evolution of a vascular system that bathed most cells in isosmotic blood plasma; this avoided rigid walls and permitted cell and organismal motilty. Cell walls also placed constraints on the evolution of long-distance transport systems. Tissues were too rigid to evolve a heart pump mechanism. Instead plants evolved two novel transport systems. One is a positive pressure system that moves concentrated, sugar rich sap in the phloem from leaves to growing meristems. Phloem transport uses a standing-gradient osmotic flow mechanism similar to that found in some animal excretory organs, but it is unique in that it occurs at very high pressures (up to 3 MPa) and requires two standing-gradient systems in tandem, one of which works in reverse.

The other transport system distributes water from the soil throughout the plant via the xylem. It is even more extraordinary in that it requires water in the xylem conduits to exist under negative (sub-atmospheric) pressures, typically of -1 and -2 MPa, and sometimes as low as -10 MPa. This means that water must remain liquid at pressures well below its vapor pressure. In this "metastable" state, nucleation of vaporization, or *cavitation,* must be prevented if continuity of the water column in the xylem conduits is to be maintained. Cavitation results in a primarily vapor-filled conduit that eventually fills with air. A conduit in this air-filled state is *embolized* and is not available for water conduction. Thus, plants depend for their water supply on an inherently vulnerable transport system.

Our current understanding of the mechanism of xylem transport is based on the cohesion theory usually ascribed to H. H. Dixon (18). Evaporation from cell wall surfaces in the leaf causes the air-water interface to retreat into the finely porous spaces between cellulose fibers in the wall. Capillarity (a consequence of surface tension) tends to draw the interface back up to the surface of the pores and places the mass of water behind it under negative pressure. This negative pressure is physically equivalent to a tension (a pulling force) that is transmitted to soil water via a continuous water column; any break in the column necessarily disrupts water flow.

Xylem transport can occur by this mechanism because of the special properties of water and the structure of the xylem. Hydrogen bonding promotes cohesion between water molecules and allows water to remain liquid under tension. Calculations of the theoretical tensile strength of water indicate that tensions in excess of 100 MPa would be needed to induce cavitations in the bulk phase (1, 37, 40). Xylem tissue includes a network or matrix of conduits (vessels and tracheids) with rigid walls that do not collapse when the water is under tension. Xylem conduits are water filled from inception and contain no entrapped air bubbles that could nucleate cavitation. Their walls are extremely hydrophilic, decreasing the likelihood of cavitation at the wall-water interface. If a cavitation occurs, or if a conduit is punctured, the resulting vapor or air bubble does not expand beyond the confines of a single conduit because of the surface tension effects at pit membranes between conduits (5). Built-in pathway redundancy insures that water conduction can continue despite limited numbers of cavitations. The optimal design of the xylem argues for strong selection during evolution for resistance to cavitation.

Here we summarize our current knowledge of cavitation and embolism, and evaluate its biological significance in plants. Obviously plants have evolved to survive the threat to their water supply that cavitation imposes. One of the questions we address is how they have done so and what sacrifices were involved. We also consider under what environmental conditions cavitation occurs, and how the consequent disruption of water transport influences the water relations and ecophysiology of plants.

EARLY EFFORTS TO DETECT CAVITATION AND EMBOLISM

The cohesion theory of sap ascent did not initially meet with widespread acceptance. Although there was experimental evidence that negative pressure had to exist in the xylem (e.g. 7), many people argued that cavitation events would be all too common an occurrence if water was under tension for long periods. Paradoxically, however, the demonstration that cavitation events occur can be the most powerful proof of the cohesion theory. From the turn of the century the quest was on to determine the vulnerability of xylem to cavitation events—i.e. to determine how the number of cavitation events increases with tension in plant tissues.

The difficulty of studying the problem in the xylem itself led some to use model systems. In glass vessels, cavitations occurred at tensions of 4–20 MPa and were accompanied by an audible click (6, 18). Because vulnerability to cavitations is determined mostly by the quality of "adhesion" at the wall-water interface (74), many of the earlier attempts to measure the vulnerability of water in artificial vessels are irrelevant to the ultimate issue of when cavitations occur in plants.

In 1915 Renner (41) and Ursprung (70) were the first to watch cavitation in plant cells under the microscope; in the annulus cells of fern sporangia cavitations occur at tensions of ca. 30 MPa. More recently, Milburn (30) observed cavitations in ascospores at tensions of 1.8–7.1 MPa. Peirce in 1936 (39) may have been the first to demonstrate the occurrence of cavitation events in the xylem of plants. He used liquid nitrogen to freeze stem segments of trees in situ. He then excised the stems and while they were still frozen tried to blow air through them. He could not blow air though branches collected in early spring during times of ample water supply, but he could later in the year when transpiration was high. This suggests that some of the xylem conduits must have been embolized prior to freezing and that therefore some cavitation events must have occurred. A similar technique has been used recently (9) to detect embolism in soybean roots.

When negative pressures in plants were actually measured with the thermocouple psychrometer (49) and the pressure bomb (46) in the 1950s and early 1960s, the cohesion theory gained wide acceptance. Given the lack of suitable quantitative studies of cavitation, plant physiologists gradually concluded that if the cohesion theory was correct, cavitation must be relatively rare. However, the validity of the cohesion theory and the commonness of cavitation events are not mutually exclusive.

ACOUSTIC DETECTION OF CAVITATION

Audio (Low-Frequency) Detection

The study of cavitation made an important step forward in 1966 when Milburn & Johnson found that it could be detected in plants by acoustic means (34). A cavitation event causes a rapid relaxation of a liquid tension that produces an acoustic emission (AE) of energy. These AEs can be detected using audio (low-frequency) acoustic transducers and amplifiers. While suspended in air, a *Ricinus* leaf with its petiole attached to an acoustic transducer produced a total of 3000 AEs while it wilted. This number is approximately the number of vessels one might expect to find in such an organ, but an exact count was not made. AE production could be stopped or slowed by adding a drop of water to the cut end of the petiole. The AE production rate could be increased or decreased by any one of several means of increasing or decreasing the rate of transpiration.

Milburn and others have reported audio-range AEs in a wide variety of species (15, 16, 31, 32, 34, 35, 71) and have collected a considerable amount of circumstantial evidence that the audio AEs are indeed correlated with cavitation events. Cavitated vessels soon become embolized and emboli are slow to dissolve (see the section on Embolism Repair, below), so we would expect that once a plant has been stressed to a xylem pressure potential (Ψ_x) of, say, -1 MPa then even if tension were partially released, few new AEs would be detected until Ψ_x drops below -1 MPa again. This appears to be confirmed by Milburn and others (31; 33, p. 165; 38). If emboli were refilled by vacuum infiltration, AE production at Ψ_x above a previous stress level was restored (31).

Ultrasonic Detection

More recently Tyree (62, 64, 65) adapted from engineering a more powerful technique for the acoustic detection of cavitations using ultrasonic frequencies typically between 50 and 1000 kHz. The techniques were well-established methods for the detection of incipient structural failure in solids. As solids begin to fracture under tension forces, there are rapid strain relaxations in the vicinity of the cracks. The potential energy released by this strain relaxation is propagated away from the crack in the form of AEs containing a broad range of frequencies. The energy contained in the lower-audio-range frequencies ($<$ 15 kHz) is usually higher than in the ultrasonic range ($>$ 15 kHz), but engineers choose to work with the ultrasonic frequencies because these can be amplified selectively while lower-range frequencies that might be confounded by audio-range noise can simultaneously be filtered out. Commercial equipment already existed, or circuits could be easily fabricated, that could count individual AE events at rates up to 1000 $\sec^{-1}$ and simultaneously measure the energy and frequency range of the AEs. Automatic monitoring has obvious advantages over Milburn's method, which usually involves aural detection by an observer using earphones. Automatic detection increases the speed of counting, reduces subjectivity in threshold detection, and eliminates the boredom of listening to experiments for many hours. The ultrasonic technique also permits accurate AE detection despite low-frequency noise from various experimental manipulations of the specimen.

Circumstantial evidence comparable to that obtained by Milburn and coworkers for audio AEs relates ultrasonic AEs to cavitation events (20, 27, 38, 44, 62). The following three experiments provide particularly strong evidence for the correspondence:

1. AEs could arise from any structural failure in the liquid or solid phases of stems. When plants dehydrate the xylem cell walls are under radial compressive stress while the water is under tension. In a pressure bomb, however, radial wall compression and water tension can be independently modified. Inside a pressure bomb ultrasonic AEs arise only when xylem water tension exceeds a threshold value of a few tenths of a MPa (64). This suggests that AEs arise at the time of rapid release of tension in xylem water.

2. The ultimate effect of cavitation events in xylem conduits should be a loss of hydraulic conductance of the cavitated stems. Thus the first measurable AEs should correspond in time with the first measurable loss of hydraulic conductance (see the section below on Hydraulic Detection of Embolism). This has been demonstrated in stems of *Thuja, Tsuga,* and *Acer* (63). In principle any cell with walls rigid enough to resist collapse under negative pressure ought to be capable of cavitation. In conifers, where most woody cells are conducting elements, the cessation of AE activity ought to correspond in time with the loss of the last vestiges of hydraulic conductance, and this has been verified (63; Figure 1). In hardwoods, however, AE activity persists long after the loss of most of the measurable hydraulic conductance (56, 63; Figure 1), presumably because of cavitation events in wood fibers, parenchyma, and ray cells.

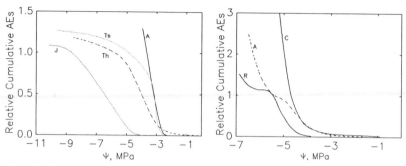

Figure 1   Vulnerability of various species to cavitation measured as relative cumulative acoustic emissions versus water potential. The left-hand graph gives data for conifers: J = *Juniperus virginiana;* Th = *Thuja occidentalis;* Ts = *Tsuga canadensis;* A = *Abies balsamea.* The right-hand graph gives data for hardwood species: R = *Rhizophora mangle;* A = *Acer saccharum;* C = *Cassipourea elliptica.*

3. If AEs correspond to cavitation events then there ought to be a one-to-one correspondence between the number of AEs counted during the dehydration of a sample and the number of rigid-walled cells in that sample, provided (*a*) none of the cells are initially embolized; (*b*) all cells emit sufficiently strong AEs; and (*c*) the sample is small enough so that AEs are not lost by signal attenuation before they reach the transducer. A one-to-one correspondence between AEs and the number of cells in small samples of *Thuja* and *Tsuga* wood has been demonstrated by two independent laboratories (28, 65) using quite different transducers, amplifiers, and counting equipment. Sandford & Grace (44), however, reported only about 16% of the expected AEs from *Chamaecyparis* wood. Either one or more of the above conditions were not met or their custom-built amplifier system had a much lower signal-to-noise ratio, resulting in a less efficient AE counting system.

Physics of Sound Propagation

Initially there was some hope that the energy and frequency composition of individual AEs might be used to determine the origin of AE events (31, 44, 62, 65). For example, it has been suggested that more elastic (potential) energy is stored in large cells deformed by negative sap pressures than in small cells, so AE energy may indicate the size of the cell that has cavitated. It has also been suggested that large cells may emit lower-frequency AEs when they cavitate than do small cells. Ritman & Milburn (42) have recently argued that large conduits produce only low-frequency audio AEs when they cavitate, but we do not agree with their conclusions. Unfortunately, the physics of AE propagation through plant tissues and of the interaction of an AE with a detection system is complex (69), and our level of understanding of these factors prevents reliable interpretation of underlying events.

Because the velocity of sound propagation through water-saturated wood is about the same as in dry wood and greater than in water (8, 69), it follows that sound propagation is through cellulose rather than water. The rate of signal attenuation in cellulose is a function of tissue hardness (lignification?), being approximately 1, 10, and > 20 dB cm^{-1} in hardwoods, softwoods, and herbs, respectively (69 and M. T. Tyree, unpublished). But even in samples cut very small so that attenuation ought to be negligible and so that all cell sizes are about equal, there is an inexplicably wide range of AE amplitudes (65 and M. T. Tyree, unpublished). Consequently, AE amplitude or energy cannot be used to determine what kinds of cells have cavitated.

There is also good reason to believe that the frequency composition of AEs as received by the transducer contains little relevant information, because the transducers commonly used tend to ring on after being struck by an AE, much as a bell does when struck by a clapper. AEs of known duration (e.g. a 0.5 μsec pulse caused by a pencil-lead break) will cause the transducers commonly used to ring on for 50–500 μsec. In addition, the wood through which the sound travels enhances or reduces selected frequencies contained in complex AE waveforms (69).

The properties of the transducer probably dominate. For example, Sandford & Grace (44) reported beats in amplitude of AE signals; similar beats were observed by us in 1983 (M. T. Tyree, unpublished observations). Sandford & Grace suggest that these beats could represent interference of coincident AEs of slightly different frequency. In our opinion they are more likely to arise as waves bounce around inside the transducer, alternately interfering and reinforcing, especially if the transducer has closely spaced harmonics for sound propagation along different axes. Similarly, subtle increases in the average frequency of AEs from woods as they dry probably reflect the effects of water content on the mode of AE transmission through the wood rather than the change in vibrational mode at the source.

HYDRAULIC DETECTION OF EMBOLISM

Cavitation is important biologically because embolized conduits reduce the hydraulic conductivity of the xylem. Sperry (50) introduced a method that quantifies embolism by how much it reduces hydraulic conductivity. In essence, the hydraulic conductivity of the xylem of excised plant segments is measured before and after the removal of air embolisms by a high-pressure treatment (see 52). In this way the cumulative effect of all cavitations that have occurred (and have not been repaired) is measured. Although this method is destructive, it has advantages over the acoustic technique. In addition to directly assessing the impact of cavitation on water transport, it can measure cavitation caused by a variety of factors (such as winter freezing) over long periods (e.g. months). It is well adapted to long-term studies of cavitation in the field.

CAVITATION AND EMBOLISM IN NATURE

Much effort to date has been devoted to the development of reliable methods for detecting cavitation and embolism. Most of this work has involved artificially stressed plants and has not directly addressed the important issue of how common these processes are in nature and what environmental conditions cause them to occur.

The limited information available indicates that cavitation is common in nature, occurring as a result of water stress and winter freezing. Extensive water stress–induced cavitation apparently occurs daily in some herbs and in corn. Using the acoustic method, Milburn & McLaughlin (35) and Tyree et al (66) studied cavitation in *Plantago major* and *Zea mays,* respectively. In both cases, enough AEs occurred during a single day to cause a significant disruption of water flow (perhaps by half for corn). In both species, nightly root pressure apparently served to refill embolized vessels. Using the hydraulic method, Sperry (50, 51) found that embolism occurred in the small palm *Rhapis excelsa* during natural drought. Embolism was confined to the leaves (petioles) and accounted for up to an 84% reduction in hydraulic conductivity. Circumstantial evidence indicated that embolized vessels could be refilled during a prolonged rain when xylem pressures approached atmospheric pressure. Root pressure was not observed.

The only long-term study of embolism in the field has recently been completed, using the hydraulic method, for a stand of sugar maple saplings growing in northern Vermont (53). Embolism during the growing season was confined to the main trunk and had reduced hydraulic conductance 31% by summer's end. The summer was wet; during a dry year this value could be much higher. During the winter, embolism increased until there was over 80% loss of conductance in the upper twigs, with many twigs 100% embolized. Main trunk xylem suffered 60% loss of conductance. Gradual spring recovery from these high winter levels was associated with positive xylem pressures generated in the root and stem. This phenomenon of winter embolism and spring recovery by positive root pressures has also been documented for wild grapevines (54); it may be general for many temperate woody species.

Few as these studies are, they indicate that embolism and recovery may be important processes in the water relations and ecophysiology of plants. It is reasonable to speculate that if nightly recovery did not occur in herbs, their growth and productivity would diminish. Similarly, if spring recovery from winter embolism did not occur in woody plants, the health and vigor of the plants would suffer. The varying abilities of species either to withstand embolism or to recover from it may significantly affect the ecological distribution of some plants. As the severity of winter increases along altitude or latitude gradients, those woody plants that avoid or reverse winter embolism may be at an advantage. Future research on embolism must concentrate on its occurrence in nature and its impact on the growth and productivity of plants.

MECHANISMS OF EMBOLISM FORMATION

Water Stress–Induced Embolism

What is the mechanism linking water stress to embolism? A common hypothesis is that large conduits are more vulnerable to cavitation than small ones. This correlation comes from extensive anatomical surveys using both systematic and floristic approaches. Although some detailed interpretations have been questioned (2–4), it has been concluded that arid zone plants (hypothetically less vulnerable to cavitation) also tend to have smaller conduit sizes (10–14). Recent experimental work indicates that although this correlation exists within an individual, it does not necessarily hold among species (19, 28, 63; see the section on Vulnerability of Xylem to Water Stress–Induced Embolism, below). Thus, the mechanism of embolism formation is not directly related to conduit diameter.

Pickard (40) reviewed two possible mechanisms of cavitation: bubble formation in bulk liquid (homogeneous nucleation) and bubble formation at an interface between water and a solid (heterogeneous nucleation). There is ample theoretical and experimental evidence to indicate that homogeneous nucleation cannot occur in plants in the range of Ψ_xs in xylem lumina (1, 37, 40). Pickard discussed two kinds of heterogeneous nucleation that could occur at the range of Ψ_xs found in plants: (*a*) nucleation at hydrophobic cracks and (*b*) meniscal failure at a pore.

A submicroscopic air bubble can remain in a stable state at the base of a hydrophobic crevice. As the xylem pressure becomes progressively negative the shape of the air-water interface changes until it reaches an unstable volume and it buds off into the bulk solution, nucleating a cavitation. For a crevice of conical shape, Pickard has calculated that this will occur at -1 MPa when the radius of curvature is 0.14 μm.

The walls of all xylem lumina are porous, and when the pore vents onto an air space a concave meniscus must form to balance the negative pressure in the fluid. The radius of curvature needed to sustain a negative pressure, P, is given approximately by $r = -2T/P$ where T is the surface tension of the solution. As P declines so does r, and when r drops below the radius of the pore, r_p, then an air bubble will be sucked into the lumen of the xylem conduit, nucleating a cavitation. The pore radius needed to cause nucleation at

-1 MPa is also about 0.14 μm, a value indistinguishable from that of the hydrophobic crack mechanism.

Although Pickard discounted the meniscal failure mechanism in favor of the other, Zimmermann has recently resurrected the concept (74) under the name of the "air seeding hypothesis." There is a simple experimental method of distinguishing the two mechanisms. If the meniscal failure model is correct then the positive air pressure needed to blow air though the largest water-filled pores should be the same in magnitude but opposite in sign to the Ψ_x needed to cause cavitation. If cavitations are caused by seeding from hydrophobic cracks then there should be no such correspondence.

There is considerable evidence that water stress–induced embolism occurs by air seeding at pores in the intervessel (or intertracheid) pit membranes. Of course, this mechanism can only occur if some vessels are embolized to begin with, but this happens frequently by such prosaic events as herbivory and mechanical damage to stems and leaves. In part, the air pressure required to force air through hydrated stems longer than the longest vessel (and hence through intervessel pits) is of the same magnitude as the tension required to induce embolism in dehydrating stems (15, 54, 56). Additional evidence is that increasing the permeability of pits to air by changing the surface tension of the xylem sap, or by other means (see the section on Pathogen-Induced Embolism, below) also increases the vulnerability of the xylem to embolism (15, 54). Thus, rather than conduit diameter, it is the pit membrane pore diameter that determines a conduit's vulnerability; the larger the pore, the more vulnerable the conduit.

The fact that larger conduits also tend to be more vulnerable within a species apparently results from a correlation between conduit size and pit-membrane pore size that can be explained on a developmental basis. Pit membranes are primary walls laid down at the time of cell division. The rapid cell expansion following cell division tends to stretch and reorient the cellulose fibers in the wall, increasing the spacing between the fibers and therefore increasing the effective porosity. More primary wall is laid down during the early phases of cell expansion, and this tends to reduce the porosity. Large cells are formed in spring when water is plentiful, and this promotes higher cell turgor and more rapid cell expansion. At the same time the demand for carbohydrate reserves is also highest, so the rate of primary (and secondary) wall formation is lowest. The net result will be bigger pit-membrane pores in early spring than in summer. In summer, water is limited, causing slower cell expansion, and carbohydrate reserves are higher, permitting more rapid primary (and secondary) wall growth; thus the cells have smaller pit-membrane pores and are less vulnerable to cavitation. In addition there is a genetic component to the rates of wall formation and cell expansion. For example, cells of the same size but from different species can have different vulnerabilities to cavitation (see Vulnerability of Xylem to Water Stress–Induced Embolism, below).

Although the air-seeding phenomenon explains how cavitation is nucleated, it is not the complete explanation of embolism formation. Once the vessel cavitates, it is initially filled with water vapor and only a little air. Embolism occurs as the vessel becomes air filled. The time required for a vessel to embolize fully following cavitation is likely to be less than 10^3 sec. The justification for this is as follows: Air will diffuse from surrounding tissue and come out of solution in the water vapor void. From air solubility tables it can be calculated that an annulus of water about 1 mm thick will contain about 14 times as much air as needed to fill the void contained in a 20-μm-diameter vessel in the center of the annulus. From a special solution of Fick's law it can be shown that the average time, t, required for an air molecule to diffuse a distance, x, is given by $t = x^2/2D$ where D = the coefficient of diffusion of air molecules in water (about 2×10^{-9} m^2 sec^{-1}). As the air pressure builds up in the void some gas molecules will diffuse out while others move back in. But after about 4 times the time it takes air molecules to diffuse 1 mm, the void ought to be near equilibrium with the air in surrounding tissue; using the above value of D and $x = 10^{-3}$ m yields $4t = 10^3$ sec.

Embolism Formation by Winter Freezing

The winter embolism observed in sugar maple (53) and grapevine (54) is most easily explained as a result of freeze-thaw cycles; several studies have shown that when xylem is frozen while under tension, extensive embolism develops after thaw as air bubbles forced out of solution during freezing expand and nucleate cavitation (45, 47, 72). A curious exception occurs in certain gymnosperms (24, 58).

Another explanation is that embolism is formed by sublimation of ice from frozen vessels. This would lead to large air bubbles on thawing that would remain stable regardless of whether tension was immediately present. This mechanism may have been most responsible for the embolism observed in the sugar maple study, because it occurred primarily during very cold (-25 to $-30°$C) and sunny weather when the main stems probably remained frozen. In addition, the embolism was localized to the south sides of the trunks exposed to the sun where sublimation would be enhanced (53). In all likelihood, both the sublimation and the freeze-thaw mechanism play a role.

Pathogen-Induced Embolism

The role of embolism in vascular disease has not received much attention. Although it has been known for some time that vascular diseases induce water stress in their host by reducing the hydraulic conductivity of the xylem (17), embolism as a cause for this has been virtually ignored. The blockage has generally been attributed to vascular occlusion by material or structures of pathogen or host origin. In the one study that has addressed the role of embolism, evidence was found that it preceded any occlusion of vessels by other means in Dutch Elm disease (36).

Assuming that the basic process of air seeding would be responsible for pathogen-induced embolism, we can speculate on the one hand that it is caused simply by increased water stress induced by the pathogen. Water stress could arise from modification of stomatal behavior (as in fusicoccin diseases, 59), limited occlusion of the xylem, or interference with root uptake. On the other hand, embolism could be caused by a lowering of the threshold for air seeding. This could happen through pathogen-induced changes in sap chemistry. For instance, compounds could be produced that lower the surface tension of the sap. Millimolar concentrations of oxalic acid drastically reduce the air seeding threshold in sugar maple (53) and balsam fir (J. S. Sperry and M. T. Tyree, unpublished); this compound is produced by many pathogenic fungi (57).

EMBOLISM REPAIR

Assuming that the sometimes drastic reduction of xylem transport caused by embolism poses a serious problem for the continued growth of a plant, the embolism repair mechanisms that exist are critical to plant health. The dissolving of air in embolized vessels requires that the xylem pressure must return to positive values or at most a pressure only slightly below atmospheric. Air bubbles are inherently unstable in water at atmospheric pressure even if the water is already saturated with air, provided the body of water is itself in contact with air. Surface tension puts the air bubble in such a body of water under pressure. This pressure equals $2T/r$, where T = the surface tension and r is the radius of the bubble. If r is in microns then this pressure in kilopascals is $140/r$; a bubble about the radius of a vessel (10 μm) thus contains air at 14 kPa above atmospheric pressure. Because the solubility of air in water depends on the pressure of the air at the air-water interface, this air will dissolve in the water. Bubbles contained in *Sphagnum* hyalocyst cells ($r \sim 10$ μm) have been observed under the microscope to dissolve in < 1 min (29). Bubbles in glass capillary tubes (which can dissolve only from their ends until they are smaller than the glass tube) can dissolve in 10 min to 100 hr for bubble radii of 30 and 375 μm (21). Bubbles under positive pressure will no doubt dissolve at much faster rates.

As previously mentioned, root pressure generates positive xylem pressure that helps to repair embolism. This would be particularly effective for repairing embolism in herbaceous plants. How embolism might be reversed in tall plants is not so clear. Perhaps larger root pressures occur during spring than previously suspected. Springtime reversal of embolism has been documented in maple trees (53), but this species is capable of producing stem pressure as part of a springtime sap flow mechanism unique to the genus (60). Spring reversal of embolism in wild grape is driven by rather large spring root pressures (> 0.1 MPa; 54). In grape the vessel and pit membranes are dry during winter, and vessels are very long (several meters). While in the dry state, root pressure drives up the column of water in these long vessels from the ground and literally expels air through the dry pit membranes and out of the vine at leaf and inflorescence scars. Under laboratory conditions a pressure of just a few kilopascals restored about half the hydraulic capacity of dry grape stems. The remaining bubbles apparently dissolve over the next day or so well in advance of leaf flush. Once the pit membranes are wet they are capable of preventing air seeding at xylem pressures below -1 MPa (54).

VULNERABILITY OF XYLEM TO WATER STRESS–INDUCED EMBOLISM

Of ultimate significance to the cohesion theory of sap ascent and to a general understanding of plant water relations is some measure of the vulnerability of the xylem to water stress–induced embolism. The emphasis in the following studies is not directly on water stress–induced embolism in the field, but on the theoretical limits imposed on water transport by embolism. These studies help answer questions concerning the evolution of the xylem and how it may have constrained the evolution of plant water relations in general.

Vulnerability can be defined by the relationship between Ψ_x and embolism; this is termed a "vulnerability curve." In Figure 1 acoustic vulnerability curves are plotted as relative cumulative AE count versus Ψ_x. Relative cumulative counts have to be computed in order to compare between replications and species. Because of the high AE attenuation rate the maximum number of detectable AEs varies with sample geometry within replications in the same species. Because of the dependence of attenuation rate on wood hardness the maximum number of detectable AEs also varies between species, even if the geometries are identical. All the curves in Figure 1 have been normalized such that a relative count of 1 occurs at the Ψ_x when 95% loss of hydraulic conductance is reached, as determined by parallel experiments.

In conifers (left-hand panel) AEs stop at about the time of 95% loss of maximum hydraulic conductivity. One expects this because most cells in conifer wood are tracheids. In vessel-bearing trees (right-hand panel) many more AEs occur after loss of 95% conductance than before. This is probably the result of cavitation events in the numerous wood fibers, ray cells, and parenchyma cells found in vessel-bearing trees.

Vulnerability curves expressed as percentage loss of hydraulic conductance versus Ψ_x are shown in Figure 2 for selected conifers (left-hand panel) and hardwoods (right-hand panel). Loss of hydraulic conductance by embolism is much more meaningful in terms of plant water relations than knowing how many cavitations (AEs) have occurred. This is because AEs can occur in many cells that contribute little to the capacity of plants to conduct water from roots to leaves.

Examination of Figures 1 and 2 shows first that the cohesion theory clearly survives the test of cavitation. Vulnerability to cavitations will not totally incapacitate xylem tissue in the normal range of water potentials experienced by all tree species examined to date (56, 63). However, this is not the same as saying water stress–induced embolism is not a significant threat to plant survival. We argue in the last section of this review that xylem is a highly vulnerable pipeline and that woody plants operate at their theoretical limit of hydraulic sufficiency.

Second, vulnerability is not correlated with whether a conduit is a hard-wood vessel or a softwood tracheid; nor does it correlate with conduit diameter. Although the conifers shown all have tracheids of about the same size (J. S. Sperry and M. T. Tyree, unpublished), there are significant species variations in vulnerability curves; *Juniperus* is the least vulnerable and *Abies* is the most vulnerable. Among the hardwoods there are also significant species variations in vulnerability. *Cassipourea* is the most vulnerable and *Rhizophora* is the least vulnerable, even though the vessel diameters of the two species are the same (56). As explained in the section on Mechanisms of Embolism Formation, above, these differences in vulnerability result strictly from differences in air seeding at interconduit pit membranes (55, 56; J. S. Sperry and M. T. Tyree, unpublished).

In general, the vulnerability of a species correlates with the xylem pressures it experiences in nature. For instance, the mangrove *Rhizophora* experiences

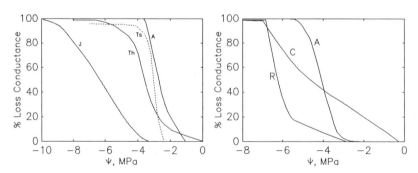

Figure 2 Vulnerability of various species to embolism measured as the percentage loss of hydraulic conductance versus water potential. Species in right- and left-hand curves as in Figure 1.

Ψ_x at least as low as -4.0 MPa and is much less vulnerable than its inland relative *Cassipourea,* which seldom has Ψ_x lower than -1.5 MPa (56). There are trade-offs between vulnerability and hydraulic sufficiency of stem tissue that must relate to the ecology of species. On the one hand, smaller pores confer resistance to cavitation; on the other hand, they may reduce the hydraulic conductivity of the xylem. Thus, the safer the xylem, the less efficient it may be in water conduction.

HYDRAULIC ARCHITECTURE AND SUFFICIENCY OF TREES

Another aspect of the vulnerability of trees to embolism is the degree of redundancy built into the xylem. In other words, how much embolism is too much? Valuable insights can be drawn from recent models of water transport in whole trees based on the vulnerability of their xylem to embolism, and on their "hydraulic architecture" (68).

Hydraulic architecture is a term coined by Zimmerman (73) to describe how the hydraulic conductivity of the xylem in various parts of a tree is related to the leaf area it must supply. This is quantified for a segment of the tree by the Leaf Specific Conductivity (LSC), defined as the absolute conductivity of the segment (k_h = flow rate per unit pressure gradient) divided by the leaf area supplied by the segment (73). The principle utility of this definition of LSC is that it allows a quick estimate of pressure gradients in stems. If the evaporative flux, E, is about the same throughout a tree, then the xylem pressure gradient, dP/dx, in any branch can be estimated from: $dP/dx = E/LSC$.

Generally the LSC for minor branches is 10–1000 times lower than for major branches (22, 23, 51, 67, 73). Thus most of the water potential drop in the xylem occurs in the small branches, twigs, and petioles. In these axes, water potential drops are 10–1000 times greater per unit length than in the main trunk. Prior to Zimmermann's classic work, Hellkvist et al (25) had found, in Sitka spruce, that the main resistance to water transport from soil to leaves resides in the branches. From their data we calculate that on the order of 65% of the total water potential drop occurs in the xylem; roots contribute 20% and leaves 14%. These results explain why trees are able to survive severe, but localized, damage to their trunks—e.g. a double saw-cut (48), or 50% blockage by fire restricted to a few feet of trunk (26). Although such damage to trunks causes a decreased conductivity, it does not cause much of an overall drop in water potential because trunk LSC is very high to begin with, and the damage is localized to a short distance.

Embolism can occur throughout the tree, and by decreasing xylem conductivity over long distances it can substantially influence water status. According to Zimmermann's "segmentation hypothesis" (74), in times of water stress, embolism will preferentially occur in minor branches where LSCs are lowest and consequently xylem tensions are greatest. The effects of the embolism resulting from these cavitation events could drastically worsen the water balance of the tree. As a consequence, peripheral parts of the tree would be sacrificed and the trunk and main branches remain functional. This hypothesis has received support from work on palms (50, 51) and on a woody shrub (43).

We have recently obtained support for the segmentation hypothesis from modeling studies based on woody plants from diverse taxa and environments (61, 68). In this work we examine the "hydraulic sufficiency" of branches of trees to supply water to their leaves. The hydraulic sufficiency depends on both the hydraulic architecture of the tree and the vulnerability of its xylem to water stress–induced embolism. The model is specifically focused on the conditions required to generate "runaway embolism," whereby the blockage of xylem conduits through embolism leads to reduced hydraulic conductance requiring increased tension in the remaining vessel to maintain water flow to leaves and generating more embolism and tension in a vicious circle.

We examined the water relations, hydraulic architecture, and xylem vulnerability of four diverse species; two were tropical (*Rhizophora mangle,* a mangrove, and *Cassipourea elliptica,* a moist forest relative) and two were temperate (*Acer saccharum* and *Thuja occidentalis*). There were great differences among these species in hydraulic architectures, maximum transpiration rates, specific hydraulic conductances of stem tissue, and water relations. Despite these differences the model predicted for all species that: 1. embolism occurs more in minor than in major branches (thus supporting the segmentation hypothesis); 2. xylem tensions could lead to 5–30% loss of transport capacity without adverse effects; 3. if embolism caused more than 5–30% loss of transport capacity, then runaway embolism occurs leading to catastrophic xylem dysfunction (blockage) in a patchwork fashion throughout the crown; and 4. after catastrophic failure of selected minor branches, an improved water balance (less negative Ψ) in surviving minor branches results from dead-shoot leaf loss.

Examples of the predictions of the model are shown in Figure 3. In the vertical axis we have plotted average Ψ of minor shoots bearing leaves in a large branch system. The horizontal axis is the average evaporative flux, $E,$ from leaves required to produce the computed Ψ. The range of Ψs predicted agrees with field observations. The maximum E observed in the field for each species is indicated (∗) near the horizontal axis. Based on the minimum observed Ψ and the vulnerability curves in Figure 2 one might wrongly conclude that xylem embolism might never exceed 5–30%. This model predicts that if stomatal regulation did not limit the maximum E then runaway embolism would occur, leading to shoot dieback. The improved water balance of surviving shoots can be seen in the gradual increase of Ψ versus evaporative flux, E, in the dashed line when E exceeds the threshold value marked (∗). We conclude that these species are operating at the limit of hydraulic sufficiency.

The implication of these results is that xylem structure and vulnerability to embolism place important constraints on the water relations, morphology, and physiology of trees. Specifically the model shows that stomatal regulation and xylem physiology must function and evolve as an integrated unit in order to prevent catastrophic dysfunction. Up until now it has been presumed that the primary role of stomatal regulation was to prevent desiccation damage to the biochemical machinery of the photosynthetic system. It is now clear that another important role of stomatal regulation is to prevent catastrophic xylem embolism while pressing water conduction through stems to their theoretical limit of hydraulic sufficiency. Trees must evolve mechanisms to keep an appropriate balance for carbon allocation between leaves, which increase evaporative demand, and stems, which supply the demand for water evaporated from the leaves.

SUMMARY

Cavitation and embolism occur in plants in response to water and freezing stress. Water stress causes embolism by air seeding at pit membranes between xylem conduits. Freezing of xylem sap forces air out of solution that nucleates cavitation during subsequent thaw, and embolisms can form in frozen vessels

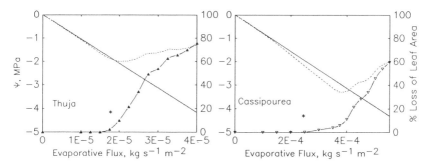

Figure 3   Results of a model that calculates the dynamics between water stress development in two species of trees and the loss of hydraulic conductance of minor stems due to embolism. Solid line: average water potential (Ψ) of all minor shoots versus evaporation rate assuming no embolism occurs. Dashed line: average Ψ of of all surviving minor shoots assuming loss of conductance by embolism. When evaporation rate exceeds the maximum observed in each species under field conditions (given by "∗" near the x-axis), runaway embolism occurs causing loss of leaf area (curves with triangles) in a patchwork fashion. See text for more details.

by sublimation. Repair of embolism by positive xylem pressures has been observed on a nightly basis in herbs, and during the spring in trees and woody vines. Modeling studies imply that vulnerability to water stress–induced embolism has played a major role in the evolution of the xylem and of stomatal regulation. It appears that embolism may be an important factor in the ecophysiology of some plants; specifically in their tolerance of low water availability and freezing temperatures. However, we cannot know the full biological significance of embolism until we know more about its occurrence under natural conditions and how much it reduces growth and productivity.

Literature Cited

1. Apfel, R. E. 1972. The tensile strength of liquids. *Sci. Am.* 227:58–71
2. Baas, P. 1973. The wood anatomical range of *Ilex* (Aquifoliaceae) and its ecological and phylogenetic significance. *Blumea* 21:193–250
3. Baas, P. 1976. Some functional and adaptive aspects of vessel member morphology. *Leiden Bot. Ser.* 3:157–81
4. Baas, P. 1982. Systematic, phylogenetic, and ecological wood anatomy—history and perspectives. In *New Perspectives in Wood Anatomy*, ed. P. Baas, pp. 23–58. The Hague: Martinus Nijhoff
5. Bailey, I. W. 1916. The structure of the bordered pits of conifers and its bearing upon the tension hypothesis of the ascent of sap in plants. *Bot. Gaz.* 62:133–42
6. Berthelot, M. 1850. Sur quelques phénomènes de dilatation forcée des liquides. *Ann. Chim. Phys. 3e Sér.* 30:232–37
7. Böhm, J. 1893. Capillarität und Saftsteigen. *Ber. Dtsch. Bot. Ges.* 11:203–12
8. Bucur, V. 1983. An ultrasonic method for measuring the elastic constants of wood increment cores bored from living trees. *Ultrasonics* 21:116–26
9. Byrne, G. F., Begg, J. E., Hansen, G. K. 1977. Cavitation and resistance to water flow in plant roots. *Agric. Meteorol.* 18:21–25
10. Carlquist, S. 1966. Wood anatomy of compositae: a summary with comment on factors controlling wood evolution. *Aliso* 6:25–44
11. Carlquist, S. 1975. *Ecological Strategies of Xylem Evolution.* Berkeley: Univ. Calif. Press
12. Carlquist, S. 1977. Ecological factors in wood evolution: a floristic approach. *Am. J. Bot.* 64:887–96
13. Carlquist, S. 1977. Wood anatomy of Onagraceae: additional species and concepts. *Ann. Missouri Bot. Gard.* 64:627–37
14. Carlquist, S. 1982. Wood anatomy of *Illicium* (Illiciaceae): phylogenetic, ecological, and functional interpretations. *Am. J. Bot.* 69:1587–98
15. Crombie, D. S., Hipkins, M. F., Milburn, J. A. 1985. Gas penetration of pit membranes in the xylem of *Rhododendron* as the cause of acoustically detectable sap cavitation. *Aust. J. Plant Physiol.* 12:445–53
16. Crombie, D. S., Milburn, J. A., Hipkins, M. F. 1985. Maximum sustainable xylem sap tensions in *Rhododendron* and other species. *Planta* 163:27–33
17. Dimond, A. E. 1970. Biophysics and biochemistry of the vascular wilt syndrome. *Annu. Rev. Phytopathol.* 8:301–322
18. Dixon, H. H. 1914. *Transpiration and the Ascent of Sap in Plants.* London: Macmillan
19. Dixon, M. A., Butt, J. A., Murr, D. P., Tsujita, M. J. 1988. Water relations of cut greenhouse roses: the relationship between stem water potential, hydraulic conductance and cavitation. *Sci. Hort.* 36:109–18
20. Dixon, M. A., Grace, J., Tyree, M. T. 1984. Concurrent measurements of stem density, leaf and stem water potential, stomatal conductance and cavitation on a sapling of *Thuja occidentalis* L. *Plant, Cell Environ.* 7:615–18
21. Ewers, F. W. 1985. Xylem structure and water conduction in conifer trees, dicot trees, and liana. *IAWA Bull.* (n.s.) 6:309–17
22. Ewers, F. W., Zimmermann, M. H. 1984. The hydraulic architecture of balsam fir *(Abies balsamea).* *Physiol. Plant.* 60:453–58
23. Ewers, F. W., Zimmermann, M. H.

1984. The hydraulic architecture of eastern hemlock *(Tsuga canadensis)*. *Can. J. Bot.* 62:940–46

24. Hammel, H. T. 1967. Freezing of xylem sap without cavitation. *Plant Physiol.* 42:55–66

25. Hellkvist, J., Richards, G. P., Jarvis, P. G. 1974. Vertical gradients of water potential and tissue water relations in Sitka spruce trees measured with the pressure chamber. *J. Appl. Ecol.* 7:637–67

26. Jemison, G. M. 1944. The effect of basal wounding by forest fires on the diameter growth of some southern Appalachian hardwoods. *Duke (Univ.) Sch. For. Bull.* 9

27. Jones, H. G., Pena, J. 1986. Relationship between water stress and ultrasound emission in apple *(Malus domestica* Borkh.) *J. Exp. Bot.* 37:1245–54

28. Lewis, A. M. 1987. *Two mechanisms for the initiation of embolism in tracheary elements and other dead plant cells under water stress.* PhD thesis, Harvard Univ.

29. Lewis, A. M. 1988. A test of the air-seeding hypothesis using *Sphagnum* hyalocysts. *Plant Physiol.* 87:577–82

30. Milburn, J. A. 1970. Cavitation and osmotic potentials of *Sordaria* ascospores. *New Phytol.* 69:133–41

31. Milburn, J. A. 1973. Cavitation in *Ricinus* by acoustic detection: induction in excised leaves by various factors. *Planta* 110:253–65

32. Milburn, J. A. 1973. Cavitation studies on whole *Ricinus* by acoustic detection. *Planta* 112:333–42

33. Milburn, J. A. 1979. *Water Flow in Plants.* London/New York: Longman

34. Milburn, J. A., Johnson, R. P. C. 1966. The conduction of sap. II. Detection of vibrations produced by sap cavitation in *Ricinus* xylem. *Planta* 66:43–52

35. Milburn, J. A., McLaughlin, M. E. 1974. Studies of cavitation in isolated vascular bundles and whole leaves of *Plantago major* L. *New Phytol.* 73:861–71

36. Newbanks, D., Bosch, A., Zimmermann, M. H. 1983. Evidence for xylem dysfunction by embolization in Dutch elm disease. *Phytopathology* 73:1060–63

37. Oertli, J. J. 1971. The stability of water under tension in the xylem. *Z. Pflanzenphysiol.* 65:195–209

38. Pena, J., Grace, J. 1986. Water relations and ultrasound emissions of *Pinus sylvestris* L. before, during and after a period of water stress. *New Phytol.* 103:515–24

39. Peirce, G. J. 1936. The state of water in ducts and tracheids. *Plant Physiol.* 11: 623–28

40. Pickard, W. F. 1981. The ascent of sap in plants. *Prog. Biophys. Mol. Biol.* 37:181–229

41. Renner, O. 1915. Theoretisches und Experimentelles zur Kohäsionetheories der Wasserbewegung. *Jahrb. Wiss. Bot.* 56:617–67

42. Ritman, K. T., Milburn, J. A. 1988. Acoustic emissions from plants: ultrasonic and audible compared. *J. Exp. Bot.* 39:1237–48

43. Salleo, S., LoGullo, M. A. 1986. Xylem cavitation in nodes and internodes of whole *Chorisia insignis* H. B. et K. plants subjected to water stress: relations between xylem conduit size and cavitation. *Ann. Bot.* 58:431–41

44. Sandford, A. P., Grace, J. 1985. The measurement and interpretation of ultrasound from woody stems. *J. Exp. Bot.* 36:298–311

45. Sauter, J. J. 1984. Detection of embolization of vessels by a double-staining technique. *J. Plant Physiol.* 116:331–42

46. Scholander, P. F., Hammel, H. T., Bradstreet, E. D., Hemmingsen, E. A. 1965. Sap pressures in vascular plants. *Science* 148:339–46

47. Scholander, P. F., Hemmingsen, W., Garey, W. 1961. Cohesive lift of sap in the rattan vine. *Science* 134:1835–38

48. Scholander, P. F., Ruud, B., Leivestad, H. 1957. The rise of sap in a tropical liana. *Plant Physiol.* 41:529–32

49. Spanner, D. C. 1951. The Peltier effect and its use in the measurement of suction pressure. *J. Exp. Bot.* 2:145–68

50. Sperry, J. S. 1985. Xylem embolism in the palm *Rhapis excelsa. IAWA Bull.* (n.s.) 6:283–92

51. Sperry, J. S. 1986. Relationship of xylem pressure potential, stomatal closure, and shoot morphology in the palm *Rhapis excelsa. Plant Physiol.* 80:110–16

52. Sperry, J. S., Donnelly, J. R., Tyree, M. T. 1987. A method for measuring hydraulic conductivity and embolism in xylem. *Plant, Cell Environ.* 11:35–40

53. Sperry, J. S., Donnelly, J. R., Tyree, M. T. 1988. Seasonal occurrence of xylem embolism in sugar maple *(Acer saccharum). Am. J. Bot.* 75:1212–18

54. Sperry, J. S., Holbrook, N. M., Zimmermann, M. H., Tyree, M. T. 1987. Spring filling of xylem vessels in wild grapevine. *Plant Physiol.* 83:414–17

55. Sperry, J. S., Tyree, M. T. 1988. Mech-

anism of water stress–induced xylem embolism. *Plant Physiol.* 88:581–87
56. Sperry, J. S., Tyree, M. T., Donnelly, J. R. 1988. Vulnerability of xylem to embolism in a mangrove vs an inland species of Rhizophoraceae. *Physiol. Plant.* 74:276–83
57. Stoessl, A. 1981. Structure and biogenetic relations: fungal non host-specific. In *Toxins in Plant Disease,* ed. R. D. Durbin, pp. 109–219. New York: Academic. 515 pp.
58. Sucoff, E. 1969. Freezing of conifer xylem sap and the cohesion-tension theory. *Physiol Plant.* 22:424–31
59. Turner, N. C. 1972. Fusicoccin: a phytotoxin that opens stomata. In *Phytotoxins in Plant Diseases,* ed. R. K. S. Wood, A. Ballio, A. Graniti, pp. 399–401. New York: Academic. 530 pp.
60. Tyree, M. T. 1983. Maple sap uptake, exudation and pressure changes correlated with freezing exotherms and thawing endotherms. *Plant Physiol.* 73:277–85
61. Tyree, M. T. 1988. A dynamic model for water flow in a single tree. *Tree Physiol.* 4:195–217
62. Tyree, M. T., Dixon, M. A. 1983. Cavitation events in *Thuja occidentalis* L.? Ultrasonic acoustic emissions from the sapwood can be measured. *Plant Physiol.* 72:1094–99
63. Tyree, M. T., Dixon, M. A. 1986. Water stress induced cavitation and embolism in some woody plants. *Physiol. Plant.* 66:397–405
64. Tyree, M. T., Dixon, M. A., Thompson, R. G. 1984. Ultrasonic acoustic emissions from the sapwood of *Thuja occidentalis* measured inside a pressure bomb. *Plant Physiol.* 74:1046–49
65. Tyree, M. T., Dixon, M. A., Tyree, E.

L., Johnson, R. 1984. Ultrasonic acoustic emissions from the sapwood of cedar and hemlock: an examination of three hypotheses concerning cavitation. *Plant Physiol.* 75:988–92
66. Tyree, M. T., Fiscus, E. L., Wullschleger, S. D., Dixon, M. A. 1986. Detection of xylem cavitation in corn under field conditions. *Plant Physiol.* 82:597–99
67. Tyree, M. T., Graham, M. E. D., Cooper, K. E., Bazos, L. J. 1983. The hydraulic architecture of *Thuja occidentalis. Can. J. Bot.* 53:1078–84
68. Tyree, M. T., Sperry, J. A. 1988. Do woody plants operate near the point of catastrophic xylem dysfunction caused by dynamic water stress? Answers from a model. *Plant Physiol.* 88:574–80
69. Tyree, M. T., Sperry, J. A. 1989. Characterization and propagation of acoustic emission signals in woody plants: towards an improved acoustic emission counter. *Plant. Cell and Environ.* In press
70. Ursprung, A. 1915. Über die Kohäsion des Wassers im Farnanulus. *Ber. Dtsch. Bot. Ges.* 33:153–62
71. West, D. W., Gaff, D. F. 1976. Xylem cavitation in excised leaves of *Malus sylvestris* Mill. and measurements of leaf water status with the pressure chamber. *Planta* 129:15–18
72. Zimmermann, M. H. 1964. Effect of low temperature on ascent of sap in trees. *Plant Physiol.* 39:568–72
73. Zimmermann, M. H. 1978. Hydraulic architecture of some diffusporous trees. *Can. J. Bot.* 56:2286–95
74. Zimmermann, M. H. 1983. *Xylem Structure and the Ascent of Sap.* Berlin/Heidelberg/New York/Tokyo: Springer-Verlag. 143 pp.

Annu. Rev. Plant Physiol. Plant Mol. Biol. 1989. 40:39–59

THE ROLE OF PLASTIDS IN ISOPRENOID BIOSYNTHESIS

H. Kleinig

Institut für Biologie II, Universität Freiburg, Schänzlestrasse 1, D-7800 Freiburg, Federal Republic of Germany

CONTENTS

INTRODUCTION

Isoprenoids, also known as terpenoids, represent a large family of natural compounds found mainly in higher plants, but also in lower plants, fungi, animals, and prokaryotes. Several thousands of isoprenoid structures have been described, and this number is still increasing. Many biological roles have been attributed to members of this family, including functions as attractants and repellants, hormones, growth inhibitors, pigments, phytoalexins, constituents of electron transport chains, and translocators of sugar residues through membranes. However, it seems unrealistic to seek a distinct role for each isoprenoid found in nature.

In living systems, all isoprenoids are synthesized from isopentenyl diphosphate, which is the universal isoprene unit and isoprenoid precursor (14, 52); it can be considered Ruzicka's hypothetical "active isoprene" (69). Isoprenoids are classified according to the number of isoprene units they

39

1040-2519/89/0601-0039$02.00

contain: monoterpenes (two units), sesquiterpenes (three units), diterpenes (four units), sesterterpenes (five units), triterpenes (six units), tetraterpenes (eight units), and polyterpenes (more than eight units).[1] Some acyclic isoprenoids are known, but most are cyclic, with a wide variety of carbon skeletons. Many oxygenated derivatives occur (e.g. alcohols, epoxides, aldehydes, ketones, acids, and lactones), and even halogen-, sulfur-, and nitrogen-containing forms have been described. Modifications of the carbon skeleton itself have also been observed that exhibit more or fewer carbon atoms than the original structure. Finally, many composite structures—e.g. esters, glycosides, isoprenoid alkaloids, and prenylated aromatics—are known.

In recent years, several biosynthetic pathways have been elucidated in terms of both the sequence of biosynthetic intermediates and their stereochemical properties. Some of the enzymes involved have been purified to homogeneity. Meanwhile, study of the molecular genetics of these compounds has begun. Several recent comprehensive reviews and books cover many aspects of isoprenoid biosynthesis and biochemistry (26, 40, 63, 67) and deal with individual groups of isoprenoids (4, 17, 25, 33, 71). Here I do not discuss reaction mechanisms or stereochemistry extensively but refer the reader to the relevant review literature.

In most publications dealing with isoprenoid biosynthesis, the subcellular compartmentation of the enzymatic reactions is not considered or is confined to the distinction between "membrane-bound" and "soluble" enzymes. Although a compartmentation of some isoprenoid pathways had been proposed by Goodwin & Mercer as early as 1963 (24), reliable studies of intracellular compartmentation have only become possible relatively recently with the development of refined methods of organelle and membrane isolation. Interesting observations are now emerging concerning the assignment of distinct biosynthetic sequences to distinct subcellular compartments.

Within the plant cell, one observes three true plasmic phases: the cytoplasm, the mitoplasm (the mitochondrial matrix), and the plastoplasm (the plastid stroma), which are not directly connected, being separated from each other by the nonplasmic phases between the inner and outer envelope membranes of the mitochondria and plastids.[2] All three plasmic phases of the plant cell are sites of isoprenoid biosynthesis, often in cooperation with membranes. As in many other biosynthetic sequences, distinct steps (enzymes) in

---

[1]The term "prenyl lipid" emphasizes the hydrophobic nature of the compound and is confined to longer-chain isoprenoids.

[2]According to this concept the "plasms" are essentially reserved for soluble compounds. There are, of course, additional compartments embedded within the cytoplasm such as the endoplasmic reticulum, the Golgi apparatus, etc. The interior of these systems inside the membranes, however, is nonplasmic and forms a sort of functional continuum with the exterior of the cell as becomes evident regarding endo- and exocytotic processes.

the pathways are membrane-bound, especially as a consequence of the hydrophobic properties of their substrates and/or products. In most cases, the membrane cooperating with the cytoplasm is the endoplasmic reticulum (ER). In the case of the mitoplasm it is the inner mitochondrial membrane, and in the case of the plastoplasm the inner membrane of the plastid envelope and the thylakoids (in chloroplasts), the prothylakoids (in etioplasts), and the so-called internal membranes (in leucoplasts and chromoplasts).

Here I focus mainly (but not exclusively) on plastids because, as in the case of acyl lipid formation, these organelles appear more and more to represent important sites of isoprenoid/prenyl lipid formation within the plant cell (for a recent review of acyl lipid formation see 32). It should be noted in this context that our knowledge is derived from investigations carried out with relatively few plastid types. These are essentially chloroplasts from spinach and pea leaves and from the green alga *Acetabularia;* etioplasts, etiochloroplasts, and chloroplasts from mustard seedling cotyledons; leucoplasts from castor bean endosperm and calamondin fruits; chromoplasts from tomato and red pepper fruits and daffodil flowers; and perhaps a few others. However, it does not seem likely that fundamental differences exist in the gross organization of isoprenogenic pathways in other plants. On the other hand not every cell (or plastid population of every cell) expresses all the possible isoprenogenic pathways found in the plant in question.

FORMATION OF ISOPENTENYL DIPHOSPHATE/DIMETHYLALLYL DIPHOSPHATE

Isopentenyl diphosphate serves as the universal biosynthetic precursor for all isoprenoid compounds. According to the generally accepted scheme, isopentenyl diphosphate is ultimately synthesized from acetyl-CoA in a sequence including acetyl-CoA acetyltransferase (EC 2.3.1.9), hydroxymethylglutaryl-CoA synthetase (EC 4.1.3.5), hydroxymethylglutaryl-CoA reductase (EC 1.1.1.34), mevalonate kinase (EC 2.7.1.36), phosphomevalonate kinase (EC 2.7.4.2), and diphosphomevalonate decarboxylase (EC 4.1.1.33). Surprisingly, the origin and compartmentation of acetyl-CoA in the plant cell are not completely understood, although several hypotheses exist [see, e.g., the recent review by Liedvogel (41)]. However, the formation of acetyl-CoA within plastids via plastid glycolytic enzymes and a pyruvate dehydrogenase complex seems well established (42, and citations therein).

The strict compartmentation of several isoprenoids such as chlorophylls and carotenoids in the chloroplast gave rise early to the hypothesis that these compounds might also be synthesized within these organelles. A scheme has been proposed for the regulation of isoprenoid biosynthesis by a combination of intracellular enzyme segregation and specific membrane permeability (24). The basis of this scheme is that enzymes for the conversion of mevalonate to prenyl diphosphates must exist both inside and outside the chloroplast (see 67 for a review of the earlier literature). Later, many results based on the utilization of radiolabeled precursors, as well as on enzyme assays using subcellular fractions, supported this concept, although incorporation rates and enzyme activities were often very low and isolation techniques did not fulfil present-day criteria for purified organelle preparations. The failure to detect specific enzyme activities of the mevalonate pathway leading to isopentenyl diphosphate in several plastid types and in mitochondria, on the one hand, along with the high activities of such enzymes in the cytoplasm, on the other (39, 45, 47), led to the alternative suggestion that isopentenyl diphosphate might be synthesized solely in the cytoplasm and then transported into the organelles. However, a translocation system for isopentenyl diphosphate has not yet been detected in either plastid or mitochondrial envelopes. The literature concerning this problem has recently been comprehensively reviewed by Gray (26).

Some recent data are of special interest in the context of isopentenyl diphosphate compartmentation. Some years ago, Moore & Shephard (61) reported the incorporation of appreciable amounts of ^{14}C from $^{14}CO_2$ into carotenoids in chloroplasts isolated from the unicellular green alga *Acetabularia*. These results have now been comfirmed and extended in our laboratory (2). About 5% of the total radioactivity incorporated from $NaH^{14}CO_3$ was recovered in the lipid fraction, wherein the label was distributed almost equally between fatty acids and isoprenoids (chlorophylls, carotenes, and others). If, on the other hand, the label was supplied as [^{14}C]acetate, this precursor was incorporated with a good yield, but exclusively into fatty acids (acetate is converted to acetyl-CoA by the plastid acetyl-CoA synthetase). This pattern of acetate incorporation has also been observed with many other plastid preparations (reviewed in 41) and would be highly unexpected were ^{14}C from $NaH^{14}CO_3$ to be incorporated into both fatty acids and isoprenoids via acetyl-CoA. However, exactly this seems to be the case in *Acetabularia* chloroplasts. Schulze-Siebert & Schultz (72) have also reported an appreciable labeling of β-carotene from $NaH^{14}CO_3$ and [^{14}C]phosphoglycerate in isolated spinach chloroplasts. In these experiments, unlabeled acetate did not dilute β-carotene synthesis. Furthermore, using (for example) [^{14}C]mevalonate as substrate the individual enzymatic steps of the mevalonate pathway could again not be demonstrated in *Acetabularia* plastids, and [^{14}C]isopentenyl diphosphate was only accepted as a precursor for isoprenoid formation in the case of osmotically stressed organelles. Intact *Acetabularia* chloroplasts could not utilize this substrate, implying that their envelope in its natural state is not permeable to isopentenyl diphosphate (2).

These results do not provide a clear picture. If in both the cytoplasmic and plastid compartments principally the same pathway (i.e. the mevalonate path) results in the formation of isopentenyl diphosphate catalyzed by two isoenzyme sets, it would be necessary to postulate that these two sets have quite different properties. The plastid enzymes must be "cryptic" and thus difficult to demonstrate, whereas the corresponding cytoplasmic enzymes are accessible and highly active in vitro. This "cryptic" behavior of the plastid enzymes might be explained by a tight metabolite channeling and regulation, so that exogenously applied intermediates, even at the level of acetate/acetyl-CoA, cannot enter the pathway. Provided that acetyl-CoA is supplied in plastids by the plastid pyruvate dehydrogenase complex—and all evidence now favors this assumption—channeling to the fatty acid and isoprenoid pathways would start at this level. This would also be valid for nonphotosynthetic plastids, since chromoplasts, for example, are provided with dihydroxyacetone phosphate and similar C_3 compounds by the cytoplasm via the phosphate translocator. They then enter the plastid glycolytic pathway to generate acetyl-CoA (36, 43). Additional work is urgently required to clarify these points. This should include work on mitochondria, where a comparable situation may exist.

One must, of course, also consider other possibilities for isopentenyl diphosphate formation in plastids—for example, the bypass via leucine; and the possibility of a completely unknown path should not be entirely ruled out. There is now, however, no experimental evidence for such pathways. In isolated spinach chloroplasts, for example, [U-^{14}C]leucine was not incorporated into isoprenoids (72). In this context it should be noted that, in *Andrographis* tissue cultures, leucine was incorporated into phytosterols via breakdown to acetyl-CoA (1).

Should plastids (and mitochondria) form their own isopentenyl diphosphate, an exchange of the water-soluble isoprenogenic pathway intermediates between different compartments would be unnecessary and would not be expected to occur—i.e. the envelope membranes should be impermeable to these compounds. Indeed, no translocator for isoprenoid compounds has yet been described. However, conflicting reports about the permeability of envelope membranes to isoprenoids have recently been summarized (26). In *Acetabularia*, slight osmotic stress in vitro renders the chloroplast envelope permeable to isopentenyl diphosphate, and this permeability increases with increasing stress (2). This may also be the case for other plastids. With a relatively tight metabolite channeling in vivo, on the other hand, the problem of passive intermediate exchange would not arise (see below).

All three compartments (plastids, mitochondria, and cytoplasm) convert isopentenyl diphosphate into isoprenoids. This implies that an isopentenyl diphosphate isomerase (EC 5.3.3.2; reviewed in 64) is present in each compartment. This enzyme delivers the allylic diphosphate for the first prenyl transfer reaction, a 1'-4 condensation, leading to the C_{10} homologue (Figure 1). The first indication of a compartmentation of the isomerase in plastids and mitochondria was reported by Green et al for the situation in castor bean endosperm (27). An enzyme isolated from *Gossypium* root homogenates, occurring in a multiprotein complex together with a prenyltransferase activity (forming farnesyl diphosphate; 64), is apparently a cytoplasmic enzyme. Isomerases have now been purified and characterized from tomato fruit chromoplasts (77), red pepper chromoplasts (20), and daffodil flower chromoplasts (51). Their molecular masses range between 28,000 and 37,000 Da. According to the plastid type and to the diversity of isoprenoid synthesizing pathways it possesses, the enzymes may be more or less tightly associated with phytoene synthase and a prenyltransferase in the so-called phytoene synthase complex (51). Mitochondrial and cytoplasmic isomerases have not been investigated in detail and thus the relationship among the three isoenzymes is still unknown.

PRENYL TRANSFER REACTIONS AND THE FORMATION OF ISOPRENOIDS

Prenyltransferases (EC 2.5.1.x) catalyze the transfer of an isoprenoid diphosphate to another isoprenoid diphosphate or to a nonisoprenoid acceptor. Three classes of reactions may be distinguished, and these have been extensively reviewed (64, 65): the 1'-4 or head-to-tail condensation between an allylic and a nonallylic diphosphate, leading to linear chain elongation; the 1'-X condensation between an allylic diphosphate and a nonisoprenoid molecule (also called prenylation); and the 1'-2-3, or head-to-middle condensation, which leads, via an asymmetrical intermediate, to squalene or phytoene (Figure 1).

Geranyl diphosphate (C_{10}) is the direct precursor of the large class of monoterpenes. These volatile compounds are constituents of essential oils and are mainly synthesized in specialized anatomical structures of the plant, such as trichomes, secretory cavities, idioblasts, and resin canals. Although many details of the reaction mechanisms involved in monoterpene biosynthesis are known (reviewed in 17), there are only a few reports on the localization of the biosynthetic pathways within the plant cell. The first indication that plastids are the site of monoterpene formation was obtained by Gleizes et al (23, 62), who showed that a leucoplast-enriched fraction from secretory cavities of the calamondin fruit exocarp formed monoterpene hydrocarbons from isopentenyl diphosphate. Recently, it has been found that chromoplasts from daffodil flowers can synthesize acyclic and cyclic monoterpenes (57). They thus contain the corresponding enzyme(s) that result in the establishment of the stereochemical configuration, the loss of the diphosphate group and, where necessary, a cyclization. Daffodil flowers are a particularly useful study object since they contain no special anatomical structures and the chromoplasts of all cells form monoterpenes with a high yield. A geranyl

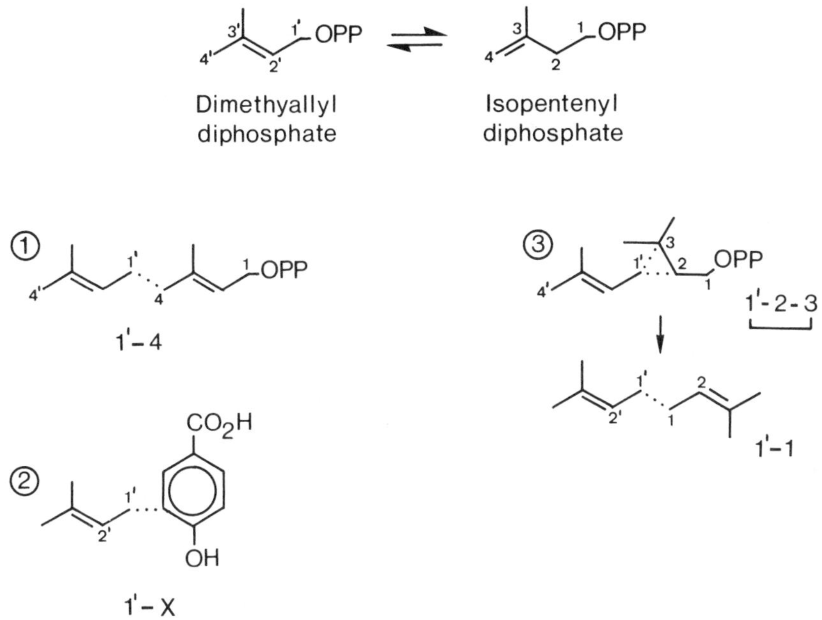

Figure 1 Isopentenyl diphosphate/dimethylallyl diphosphate isomerase reaction (*above*) and prenyl transfer reactions (*below*, according to Ref. 64).

diphosphate synthase (EC 2.5.1.1) is apparently present in the calamondin leucoplasts since monoterpenes are their only isoprenoid products. Daffodil flowers, on the other hand, also synthesize carotenoids. No evidence is yet available for the existence of two individual transferases for the C_{10} and C_{20} moieties, respectively, thus indicating that the geranylgeranyl synthase would furnish a branch point at the intermediate C_{10} level. A geranyl diphosphate synthase was isolated from sage leaf extract by Purkett & Croteau (unpublished; cited in 17). An indication that plastids may be generally the site of monoterpene biosynthesis in plant cells came from the exemplary correlation study carried out by Cheniclet & Carde (15). A comparative analysis of structural and analytical data from 45 plant species revealed a correlation between the presence of leucoplasts in secretory cells and the presence of monoterpenes in the volatile extract. It is interesting that no relationship was observed between the occurrence of these organelles and the presence of sesquiterpenes. Data from older structural work concerning the formation of essential oils are also discussed in this publication.

Sesquiterpenes (C_{15}) are not found in plastids, and the diphosphate precursor farnesyl diphosphate is not formed there by a corresponding synthase (EC 2.5.1.10). On the other hand, it is well established that farnesyl diphosphate serves as precursor for squalene and sterol biosynthesis in the cytoplasm/ER, and there is also evidence assigning sesquiterpene formation, including that of the plant hormone abscisic acid by the direct pathway (30; reviewed in 83), to the cytoplasm/ER (e.g. 3, 9). The absence of squalene formation in vitro from isopentenyl diphosphate in plastid isolates may indeed be regarded as a criterion of purity for these isolates. In vivo studies also suggest that, in secretory cells, distinct sites for mono- and sesquiterpene biosynthesis exist that differ in their accessibility to exogenous radioactively labeled precursors (discussion in 18).

Geranylgeranyl diphosphate (C_{20}) is the major intermediate of isoprenoid biosynthesis in plastids. The synthase (EC 2.5.1.29) is a soluble enzyme and has been isolated from tomato fruit chromoplasts (77), proplastids from castor bean seedlings (22), and red pepper chromoplasts (20). The enzyme has a molecular mass in the range of 72,000 Da and seems to be a dimer (20). Geranylgeranyl diphosphate probably serves also as allylic substrate for a further transferase (EC 2.5.1.11) which forms all-trans-nonaprenol (= solanesol) diphosphate, the side-chain precursor of plastoquinone-45, in chloroplasts (44, 50). A similar activity has been described for plant mitochondria (47). This precursor role of geranylgeranyl diphosphate, however, has not been firmly established. A kinase activity that converts free geranylgeraniol into geranylgeranyl diphosphate in the presence of ATP has been reported to occur in plastids (68).

Although geranylgeranyl diphosphate synthase is obtained as a soluble enzyme upon disintegration of plastids, it may be a membrane-peripheral protein in vivo, since, at the level of the C_{20} diphosphate, many subsequent reactions are membrane-associated (the enzymes being membrane-integral) accompanied by a hydrophobicity of the biosynthetic products and their location and function within membranes. A cooperation between stroma and membranes at this level is evident (8) and, typically, apparently soluble enzymes are greatly stimulated in their activities upon addition of liposomes to the soluble assay (38).

Geranylgeranyl diphosphate serves as a key branching point in plastid isoprenoid metabolism. Delicate mechanisms of regulation and channeling to subsequent pathways must be postulated. These mechanisms are easily disturbed as may be inferred from experiments using, for example, isolated spinach chloroplasts. Although these chloroplasts show the commonly accepted criteria for intactness, the pattern of isoprenoids formed in vitro does not correspond to the in vivo stoichiometry. Synthesis of chlorophyll and especially that of free geranylgeraniol (which does not occur in vivo), apparently due to a released or activated phosphatase activity, predominates, whereas carotenoids, for example, are not formed at all (50). In this respect, "intact chloroplasts" have no advantage over chloroplast homogenates, since comparable results are obtained with both systems. The reason for this apparent impairment of chloroplast biosynthetic capacity when the organelles are removed from their natural cytoplasmic environment is not known. Further investigation of this point would appear fundamental for future research. In contrast to the situation for chloroplasts, in vitro carotenoid biosynthesis with a good yield is observed in etioplasts and chromoplasts (see below).

Geranylgeranyl diphosphate is the direct precursor of diterpenes. Although many details are known about the biosynthesis of this isoprenoid class (reviewed in 81), there are only a few reports from which the site of biosynthesis may be inferred. These reports concern the formation of diterpene hydrocarbons in plastids. The phytoalexin casbene (Figure 2) is synthesized in proplastids of castor bean seedlings (21) and, in several plant species, the formation of *ent*-kaurene, an intermediate of the gibberellin pathway, has been assigned to the plastids (60, 66, 74). In these cases, the so-called B activity of *ent*-kaurene synthase from copalyl diphosphate to *ent*-kaurene (Figure 2) was easier to measure than the utilization of geranylgeranyl diphosphate. A critical discussion of the intracellular site of *ent*-kaurene and gibberellin biosynthesis is provided by Graebe (25). The oxidation reactions of the gibberellin pathway are assigned to the ER (see below).

Geranylgeranyl diphosphate is also the substrate for the esterification of chlorophyllide, catalyzed by chlorophyll synthase (68). The geranylgeranyl moiety is subsequently hydrogenated to the phytyl moiety. The synthase and the NADPH·H$^+$-dependent reductase are firmly bound to the thylakoid membranes in chloroplasts (76). In mustard chloroplast homogenates, a reduction

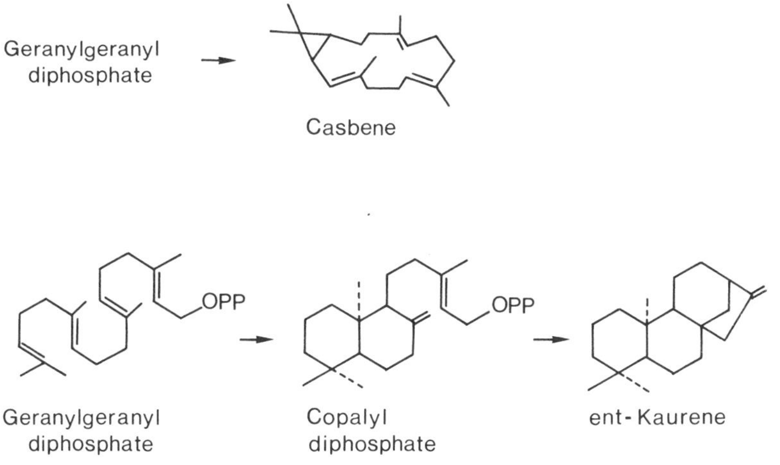

Figure 2   Formation of the phytoalexin casbene (*above*) and the gibberellin intermediate *ent*-kaurene (*below*).

of geranylgeranyl chlorophyll to phytyl chlorophyll has been observed, which is independent of exogenously added NADPH·H$^+$ but dependent on light (44). It is interesting that geranylgeranyl diphosphate can also be directly reduced by another NADPH·H$^+$-dependent reductase located in the chloroplast envelope (76). Phytyl diphosphate is then used for the prenylation of homogentisate and 1,4-dihydroxy-2-naphthoate in the synthesis of tocopherols and phylloquinone, respectively. In a similar reaction, homogentisate is prenylated with nonaprenyl diphosphate, leading to the formation of plastoquinone-45. All these reactions are confined to the inner envelope membrane (75). A review of the synthesis of prenylquinones and tocopherols in chloroplasts has been published recently by Schultz et al (71).

The results presented above show that, in chloroplasts, at least four pathways (chlorophyllide esterification, reduction to phytyl diphosphate, head-to-middle condensation to phytoene (see below), and elongation to nonaprenyl diphosphate) emanate from the geranylgeranyl diphosphate branching point. When diterpene biosynthesis occurs, still more pathways emanate from this point. The different pathways are supplied in vivo with the precursor geranylgeranyl diphosphate in different, but constant, stoichiometries.

The third prenyl transfer reaction, the head-to-middle condensation of two identical diphosphates, is realized twice in the plant cell—namely, in the formation of squalene from farnesyl diphosphates at the ER and in the formation of phytoene from geranylgeranyl diphosphates within plastids. The reaction mechanisms appear to be similar except that in the final step of phytoene synthesis proton loss gives rise to the central double bond, whereas in the squalene pathway a proton is added. As noted above, phytoene (and therefore carotene) synthesis is difficult to achieve with isolated chloroplasts. The corresponding enzyme phytoene synthase is, however, active in chromoplasts from tomato fruits (34), daffodil flowers (6), red pepper fruits (11), and etioplasts from mustard seedling cotyledons (44). The enzyme fulfills the criteria of a membrane-peripheral protein (38) and can be obtained in a soluble form. It catalyzes the two-step process from two molecules of geranylgeranyl diphosphate via prephytoene diphosphate to 15-*cis*-phytoene, which is the predominant isomer in higher plants (Figure 4). In in vitro assays the dephosphorylated alcohol prephytoenol is sometimes obtained as an artefact. The isolation of this alcohol from a natural source has been reported recently (59). The phytoene synthase has as yet been only partly characterized (54, 55, 77, reviewed in 33), so that a comparison of the properties of this enzyme with those of squalene synthase, which has been purified and characterized in detail from yeast (e.g. 70), is not yet possible. This will be an interesting task for future research. The phytoene synthase complex, i.e. phytoene synthase plus geranylgeranyl synthase plus isopentenyl isomerase, has a molecular mass of about 200,000 Da as determined by gel filtration (55).

During the conversion of etioplasts to chloroplasts in mustard seedlings, soluble phytoene synthase activity in isolated plastids gradually decreases and is no longer detectable in chloroplasts from primary leaves (44). The chlorophyll synthase, on the other hand, remains highly active. This contrasts with the in vivo situation in higher plants, where chlorophyll and carotenoid biosynthesis are interdependent processes (53). These results lead to the conclusion that the phytoene synthase may change its topology from a membrane-peripheral protein in nongreen plastids to a more tightly membrane-associated form, with, to some extent, consequently different properties in chloroplasts. A relatively tight correlation of its substrate turnover to that of the chlorophyll synthase (use of geranylgeranyl diphosphate in both cases) in vivo might be the cause for a preponderance of the latter activity in vitro. It is interesting that a phytoene synthase activity could be measured in isolated envelope membranes from spinach chloroplasts, which were essentially free of chlorophyll synthase (48). Meanwhile, an activation of the membrane-bound "silent" phytoene synthase in thylakoid membranes from several plant species could be achieved by the addition of the zwitterinonic detergent Chaps (46). At the moment, however, not much is known about the differences and relationships, respectively, between the soluble and membrane-bound enzymes. A soluble phytoene synthase activity has been found in chloroplasts isolated from young expanding maize leaves (56) and in pepper and pea leaves treated with the detergent Tween-80 (19).

Finally, the formation of fatty acid prenyl esters (C_{20} and C_{45} prenols, C_{16} and C_{18} fatty acids) in isolated chloroplasts has been observed. There are two modes of ester synthesis (50), a transesterification from chlorophyll to a fatty acid moiety from an unknown source occurring within the thylakoids and an acyl-CoA:polyprenol acyltransfer reaction occurring within the envelope membranes. The latter mode was interpreted as a trapping mechanism for escaped free polyprenols, which are slightly amphipathic molecules and, thus, may cause harmful effects. The transesterification may guarantee the elimination of geranylgeranyl chlorophyll that has escaped the reduction mechanism. It may be relevant that both chloroplast membrane systems are equipped with an esterification device that allows the conversion of polyprenols to inert fatty acid esters.

MODIFICATION OF ISOPRENOIDS

In this section I discuss briefly the oxidation of mono- and diterpene hydrocarbons and the reactions leading from phytoene to the naturally occurring colored carotenoids.

Cyclic monoterpene hydrocarbon skeletons are often subject to secondary enzymatic modifications (for review see 17), especially oxidation. Most naturally occurring monoterpenes bear oxygen functions. Two modes of introduction of oxygen functions have been described (Figure 3). In the first case, occurring, for example, in the biosynthesis of borneol and camphor, a cyclic diphosphate is hydrolyzed to the alcohol which is then oxidized to the ketone by a dehydrogenase. Theoretically these reactions could occur in the plastids. However, they have not as yet been assigned to any particular intracellular compartment. The second and more widespread route is the oxidation of a cyclic olefin, occurring, for example, in the synthesis of sabinol/sabinone (Figure 3). In these cases, the hydroxylation reaction has been reported to be NADPH·H^{+}- and O$_2$-dependent and to meet the criteria for a cytochrome P-450–dependent mixed-function oxygenase (35; reviewed in 17). The reaction would thus be expected to be localized at the ER.

Similar arguments may be applied to the pathways and sites of diterpene biosynthesis. Biosynthetic studies with this class of isoprenoids are, however, not as advanced as those with monoterpenes. In the special case of the gibberellin pathway, the enzymes involved in the reactions leading from *ent*-kaurene to gibberellin A$_{12}$ aldehyde and further to the various gibberellin structures are either located at the ER and are cytochrome P-450 dependent or are soluble cytoplasmic enzymes (reviewed in 25). If the site of *ent*-kaurene formation is correctly assigned to the plastid, this would require that *ent*-kaurene be transported from the plastid to the ER for further metabolism.

15-*cis*-Phytoene is the first C$_{40}$ carotenoid in the carotenogenic sequence and the precursor for all those numerous more desaturated and oxygenated carotenoids occurring within lower and higher plant plastids. The dehy-

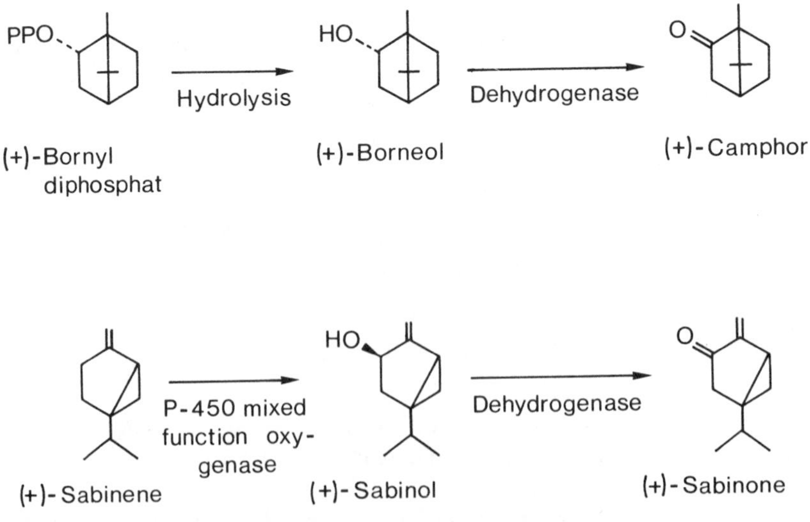

Figure 3   Formation of camphor (*above*) and sabinone (*below*).

drogenation of *cis*-phytoene to lycopene, formally in a four-step mode, and the cyclization reactions leading to the α- and β-ionone rings of the cyclic carotenes (Figure 4) have been investigated to some extent (reviewed in 33). The reactions are membrane bound and can be assigned clearly to the plastid compartment. The solubilization of lycopene cyclase from an acetone powder preparation from red pepper chromoplast membranes (12) and the solubilization and reconstitution of the dehydrogenation/cyclization system from daffodil chromoplast membranes using the zwitterionic detergent Chaps (7) have been achieved. Recently, the dissection of the dehydrogenation sequence in daffodil chromoplast membranes into two independent steps (*cis*-phytoene via phytofluene to ζ-carotene and ζ-carotene via neurosporene to lycopene) has been carried out. Both the dependency of these reactions on molecular oxygen as the possible final electron acceptor and the crucial role of the *cis/trans* state of intermediates have been demonstrated (5). The dependency of carotene dehydrogenation on oxygen is reminiscent of the desaturation of plastid stearoyl-ACP (acyl carrier protein) to oleoyl-ACP, which also requires molecular oxygen (reviewed in 78).

The subsequent transformations leading from carotenes to the wide variety of oxygen-bearing compounds are poorly understood at the enzyme level. It is, however, generally agreed that these reactions are confined to the plastids. For example, the conversion of antheraxanthin to capsanthin and of violaxanthin to capsorubin has been observed in isolated red pepper chromoplasts (10, 13). Furthermore, the so-called violaxanthin cycle is located at the thylakoid membranes. In this cycle, violaxanthin is deepoxidized in the light to zeaxanthin, and the reverse reaction proceeds in the dark (Figure 4; for review see 28, 82). The deepoxidase has been isolated as a soluble enzyme and partially characterized (29), whereas the epoxidase is a membrane-bound enzyme and has not been purified. This epoxidase requires NADPH·H$^+$ and molecular oxygen for its activity (73). Another zeaxanthin epoxidation occurs in the chloroplast envelope. This has been interpreted as a biosynthetic step in the formation of violaxanthin rather than as part of a cycle (16).

There is ample evidence that carotene metabolism takes place exclusively within the plastid compartment and that the end products of the pathway remain there. This means that plastids are indeed equipped with the enzymatic machinery to introduce oxygen functions into isoprenoids. Nevertheless, according to our present knowledge, mono- and diterpenes are not oxygenated within plastids but, as discussed above, at the ER.

CONCLUDING REMARKS

The compartmentation of the isoprenogenic pathways for the plant cell is depicted in Figure 5. This has been carried out in a relatively strict manner,

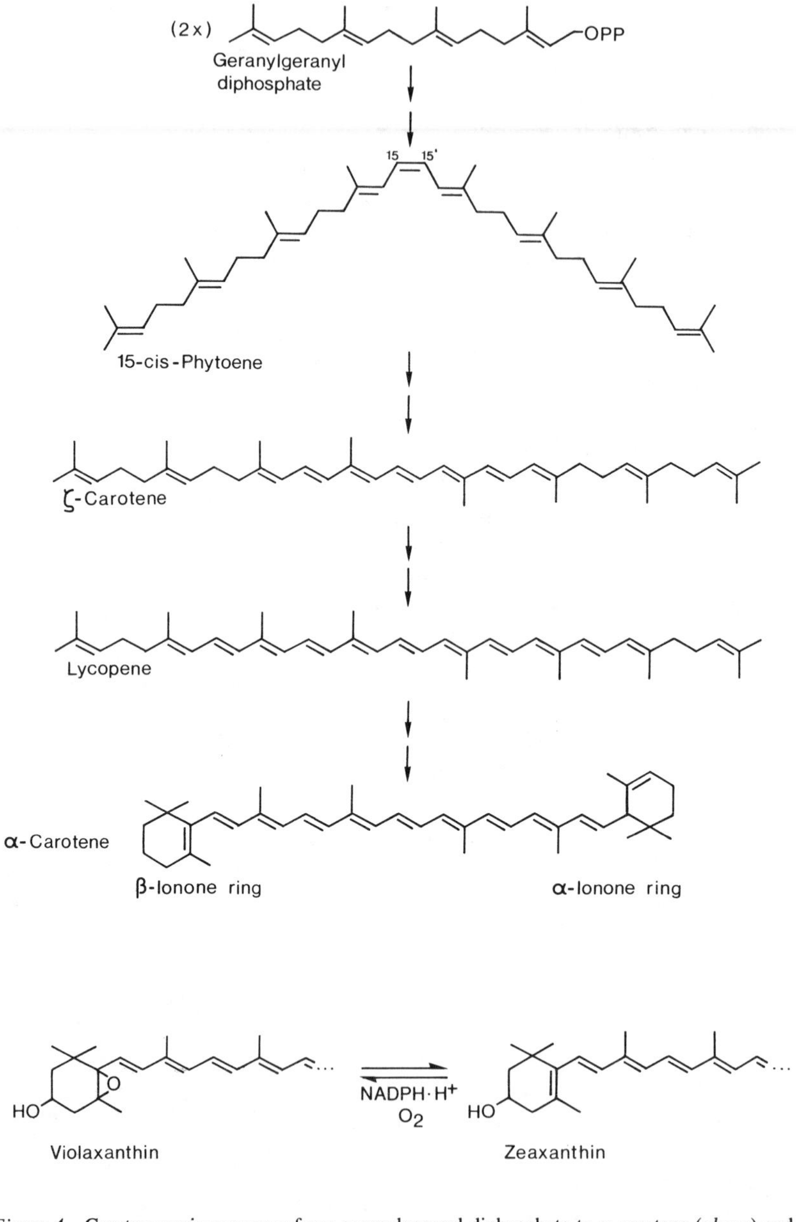

Figure 4   Carotenogenic sequence from geranylgeranyl diphosphate to α-carotene (*above*) and xanthophyll cycle (*below*).

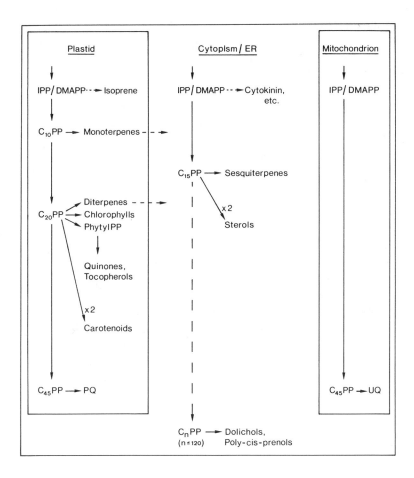

Figure 5 Compartmentation of isoprenogenic pathways in the plant cell. IPP, isopentenyl diphosphate; DMAPP, dimethylallyl diphosphate; PP, diphosphate; PQ, plastoquinone; UQ, ubiquinone; broken lines, tentatively assigned.

assuming that the compartmentation is nearly total. By this I mean that 1'-4 prenyl transfer reactions with distinct chain-length specificities, starting from the precursor isopentenyl diphosphate/dimethylallyl diphosphate, are attributed to distinct compartments and that, with the one exception of the formation of the C_{45} *all-trans* side chains of quinones in both plastids and mitochondria, double equipment does not occur in the cell. In this context, a statement made by West in 1981 should be cited (81): "It seems reasonable to suggest that all aspects of geranylgeranyl diphosphate metabolism, including di-terpene biosynthesis, may be restricted to plastid compartments. . . . although the available evidence at this time is not adequate to permit this to be said with certainty." The synthesis of the ultimate precursor isopentenyl diphos-phate/dimethylallyl diphosphate of the various pathways is, on the other hand, apparently possible in all three compartments involved in iso-prenoid biosynthesis, although the properties of this enzyme sequence from acetyl-CoA in plastids and mitochondria are not understood. The differ-ences between these sequences and the relatively well-known cytoplasmic sequence are also unclear, and an alternative route of formation cannot be ruled out. We hope future research will show whether the concept presented in Figure 5 is correct or whether it will require corrections and/or exten-sions.

It follows from Figure 5 that all geranyl diphosphate and geranylgeranyl diphosphate-derived carbon skeletons are biosynthsized within plastids and all farnesyl diphosphate derivatives within the cytoplasm/ER. In particular, the carotenogenic and the sterol pathways are separated from one another. This is not the case in the carotenoid-containing fungi, which do not possess plastids. Here, both pathways occur in the same compartment, the cytoplasm/ER. As might be expected, in the fungus *Phycomyces* the enzyme providing geranyl-geranyl diphosphate only accepts farnesyl diphosphate as allylic, and isopentenyl diphosphate as nonallylic substrate. This shows that here farnesyl diphosphate is the branching point for sterol and carotenoid formation (49).

The broken lines in Figure 5 leading from mono- and diterpenes in the plastid to the cytoplasmic compartment take into account the current evidence that the modification of these compounds (i.e. the introduction of oxygen functions) takes place in the latter compartment in a sort of "cooperative manner." This point of view, however, awaits further investigation. Another broken line is drawn from farnesyl diphosphate to polyisoprenoid alcohols, which have not as yet been discussed. These alcohols are linear and consist of up to 24 isoprene residues. They are mostly *tri-trans, poly-cis* compounds in contrast to the side chains of plasto- and ubiquinones, which are *all-trans*. These compounds are widespread in plants. Two groups can be distinquished, the so-called *poly-cis* prenols (=ficaprenols) with an α-unsaturated isoprene residue, and the dolichols with an α-saturated residue. A function for the *poly-cis* prenols is not known, whereas the dolichols, in the phosphorylated form, participate in the transport of sugars and oligosaccharides through the ER membrane. For this reason, the synthesis of these compounds is attributed to the cytoplasm/ER in Figure 5. It should be stressed, however, that this assignment is tentative, because polyisoprenoid alcohols (and their fatty acid esters) have been found in several compartments, and the site of their synthesis has not been proven (for review see 31). A polyprenol diphosphate synthase has been partially purified recently from mulberry leaves (37). The activities forming rubber (EC 2.5.1.20, the high-molecular-mass polymer of isopentenyl diphosphate/dimethylallyl diphosphate) in rubber-containing plants also seem to be cytoplasmic (58).

Finally, two further types of reactions should be noted briefly—namely, isoprene emission and the transfer of a C_5 unit from isopentenyl diphosphate/ dimethylallyl diphosphate to nonisoprenoid molecules. Isoprene is emitted from several plants in a light- and temperature-dependent manner, and isoprene formation has been tentatively attributed to the chloroplast (79, and citations therein). There are many cases where a C_5 isoprene residue is attached to other molecules—for example, to 5'-AMP by cytokinin synthase (EC 2.5.1.27) in cytokinin biosynthesis, or during the formation of some phytoalexins, alkaloids, tRNAs etc. These transferases are in part membrane-bound, in part soluble. Although an assignment to a distinct compartment in the cell has not been established in any case, many of them might turn out to be located in the cytoplasm/ER. This is particularly probable for cytokinin synthase, at least when encoded by the T-DNA from the *Agrobacterium* Ti-plasmid (J. Schröder, personal communication).

Figure 5 reveals that the plastid compartment is the most versatile in isoprenoid biosynthesis, followed by the cytoplasm/ER and the mitochondria. Nevertheless, all isoprenogenic enzymes, regardless of the pathway in which they are involved, are—according to present knowledge (for a review of the coding capacities of the sequenced plastid genome from tobacco see e.g. 80)—encoded by the nuclear genome. Different programs of plastid isoprenoid biosynthesis can exist within a given plant—e.g. for the chloroplasts in mesophyll cells, for the leucoplasts in gland cells, or for the chromoplasts in petal cells. In a given cell, on the other hand, the sometimes more than 100 plastids are apparently synchronized, which means that their biosynthetic capacities are roughly identical. Our knowledge about the regulation mechanisms that bring about these different programs is still extremely scanty.

The plastid envelope seems to be impermeable to water-soluble isoprenoids and isoprenoid precursors; at least no special translocator has been described. The question still arises about the situation for the amphiphilic and hydrophobic isoprenoid compounds that may pass a membrane by diffusion. This may partially be the case during the release of volatiles into the surroundings. These aspects, however, have not been investigated in detail. When mono- and diterpenes are synthesized within plastids and metabolized further in the cytoplasm/ER, some channeling device must be postulated. The concept of metabolite channeling, which implies no free diffusion of intermediates, seems to play a major role in the biosynthesis of isoprenoids, as has frequently been mentioned above.

ACKNOWLEDGMENTS

Parts of this review were written during a stay at the Université Pierre et Marie Curie, Paris. I would like to thank my colleague Professeur R. Monéger for his hospitality. I would also like to thank Dr. P. Beyer and Dr. C. Beggs for helpful discussins and for reading the manuscript. Our own research has been supported by the Deutsche Forschungsgemeinschaft.

Literature Cited

1. Anastasis, P., Freer, I., Overton, K., Rycroft, D., Singh, S. B. 1985. The role of leucine in isoprenoid metabolism. Incorporation of [3-^{13}C]leucine and of [2-^{3}H,4-^{14}C]-β,β-dimethylacrylic acid into phytosterols by tissue cultures of *Andrographis paniculata*. *J. Chem. Soc. Chem. Commun.* 1985:148–49

2. Bäuerle, R., Lütke-Brinkhaus, F., Berger, S., Kleinig, H. 1989. Prenyl lipid and fatty acid synthesis in *Acetabularia* chloroplasts. Submitted

3. Belingheri, L., Pauly, G., Gleizes, M., Marpeau, A. 1988. Isolation by an aqueous two-polymer phase system and identification of endomembranes from *Citrofortunella mitis* fruits for sesquiterpene hydrocarbon synthesis. *J. Plant Physiol.* 132:80–85

4. Benveniste, P. 1986. Sterol biosynthesis. *Ann. Rev. Plant Physiol.* 37:275–308

5. Beyer, P., Kleinig, H. 1989. On the role of molecular oxygen and the state of geometrical isomerism of intermediates in the carotene desaturation and cyclization reactions in daffodil chromoplasts. Submitted

6. Beyer, P., Kreuz, K., Kleinig, H. 1980. β-Carotene synthesis in isolated chromoplasts from *Narcissus pseudonarcissus*. *Planta* 150:435–38

7. Beyer, P., Weiss, G., Kleinig, H. 1985. Solubilization and reconstitution of the membrane-bound carotenogenic enzymes from daffodil chromoplasts. *Eur. J. Biochem.* 153:341–46

8. Block, M. A., Joyard, J., Douce, R. 1980. Site of synthesis of geranylgeraniol derivatives in intact spinach chloroplasts. *Biochim. Biophys. Acta* 631:210–19

9. Brindle, P. A., Kuhn, P. J., Threlfall, D. R. 1988. Biosynthesis and metabolism of sesquiterpenoid phytoalexins and triterpenoids in potato cell suspension cultures. *Phytochemistry* 27:133–50

10. Camara, B. 1980. Carotenoid biosynthesis. In vitro conversion of violaxanthin to capsorubin by a chromoplast enriched fraction of *Capsicum* fruits. *Biochem. Biophys. Res. Commun.* 93:113–17

11. Camara, B., Bardat, F., Monéger, R. 1982. Sites of biosynthesis of carotenoids in *Capsicum* chromoplasts. *Eur. J. Biochem.* 127:255–58

12. Camara, B., Dogbo, O. 1986. Demonstration and solubilization of lycopene cyclase from *Capsicum* chromoplast membranes. *Plant Physiol.* 80:172–74

13. Camara, B., Monéger, R. 1981. Carotenoid biosynthesis. In vitro conversion of antheraxanthin to capsanthin by a chromoplast enriched fraction of *Capsicum* fruits. *Biochem. Biophys. Res. Commun.* 99:1117–22

14. Chaykin, S., Law, J., Phillips, A. H., Tchen, T. T., Bloch, K. 1958. Phosphorylated intermediates in the synthesis of squalene. *Proc. Natl. Acad. Sci. USA* 44:998–1004

15. Cheniclet, C., Carde, J. P. 1985. Presence of leucoplasts in secretory cells and of monoterpenes in the essential oil: a correlative study. *Israel J. Bot.* 34:219–38

16. Costes, C., Burghoffer, C., Joyard, J., Block, M., Douce, R. 1979. Occurrence and biosynthesis of violaxanthin in isolated spinach chloroplast envelope. *FEBS Lett.* 103:17–21

17. Croteau, R. 1987. Biosynthesis and catabolism of monoterpenoids. *Chem. Rev.* 87:929–54

18. Croteau, R., Johnson, M. A. 1984. Biosynthesis of terpenoids in glandular trichomes. In *Biology and Chemistry of Plant Trichomes*, ed. E. Rodriguez, P. L. Healy, J. Metha, pp. 133–85. New York: Plenum. 255 pp.

19. Dogbo, O., Bardat, F., Laferriére, A., Quennemet, J., Brangeon, J., Camara, B. 1987. Metabolism of plastid terpenoids. I. Biosynthesis of phytoene in plastid stroma isolated from higher plants. *Plant Sci.* 49:89–101

20. Dogbo, O., Camara, B. 1987. Purification of isopentenyl pyrophosphate isomerase and geranylgeranyl pyrophosphate synthase from *Capsicum* chromo-

plasts by affinity chromatography. *Biochim. Biophys. Acta* 920:140–48

21. Dudley, M. W., Dueber, M. T., West, C. A. 1986. Biosynthesis of the macrocyclic diterpene casbene in castor bean (*Ricinus communis* L.) seedlings. Changes in enzyme levels induced by fungal infection and intracellular localization of the pathway. *Plant Physiol.* 81:335–42

22. Dudley, M. W., Green, T. R., West, C. A. 1986. Biosynthesis of the macrocyclic diterpene casbene in castor bean (*Ricinus communis* L.) seedlings. The purification and properties of farnesyl transferase from elicited seedlings. *Plant Physiol.* 81:343–48

23. Gleizes, M., Pauly, G., Carde, J. P., Marpeau, A., Bernard-Dagan, C. 1983. Monoterpene hydrocarbon biosynthesis by isolated leucoplasts of *Citrofortunella mitis*. *Planta* 159:373–81

24. Goodwin, T. W., Mercer, E. I. 1963. The regulation of sterol and carotenoid metabolism in germinating seedlings. *Biochem. Soc. Symp.* 24:37–41

25. Graebe, J. E. 1987. Gibberellin biosynthesis and control. *Ann. Rev. Plant Physiol.* 38:419–65

26. Gray, J. C. 1987. Control of isoprenoid biosynthesis in higher plants. *Adv. Bot. Res.* 14:25–91

27. Green, T. R., Dennis, D. T., West, C. A. 1975. Compartmentation of isopentenyl pyrophosphate isomerase and prenyl transferase in developing castor bean endosperm. *Biochem. Biophys. Res. Commun.* 64:976–82

28. Hager, A. 1980. The reversible, light-induced conversion of xanthophylls in the chloroplast. In *Pigments in Plants*, ed. F. C. Czygan, pp. 57–79. Stuttgart: Fischer. 447 pp.

29. Hager, A., Perz, H. 1970. Veränderung der Lichtabsorption eines Carotinoids im Enzym (De-Epoxidase)-Substrat (Violaxanthin)-Komplex. *Planta* 93:314–22

30. Hartung, W., Heilmann, B., Gimmler, H. 1981. Do chloroplasts play a role in abscisic acid synthesis? *Plant Sci. Lett.* 22:235–42

31. Hemming, F. W. 1983. Biosynthesis of dolichols and related compounds. In *Biosynthesis in Isoprenoid Compounds*, ed. J. W. Porter, S. L. Spurgeon, 2:305–54. New York: Wiley. 552 pp.

32. Heemskerk, J. W. M., Wintermans, J. F. G. M. 1987. Role of the chloroplast in the leaf acyl-lipid synthesis. *Physiol. Plant.* 70:558–68

33. Jones, B. L., Porter, J. W. 1986. Biosynthesis of carotenes in higher plants. *CRC Crit. Rev. Plant Sci.* 3:295–324

34. Jungalwala, F. B., Porter, J. W. 1967. Biosynthesis of phytoene from isopentenyl and farnesyl pyrophosphates by a partially purified tomato enzyme system. *Arch. Biochem. Biophys.* 119:209–19

35. Karp, F., Harris, J. L., Croteau, R. 1987. Metabolism of monoterpenes: demonstration of the hydroxylation of (+)-sabinene to (+)-*cis*-sabinol by an enzyme preparation from sage (*Salvia officinalis*) leaves. *Arch. Biochem. Biophys.* 256:179–93

36. Kleinig, H., Liedvogel, B. 1980. Fatty acid synthesis by isolated chromoplasts from the daffodil. Energy sources and distribution patterns of acids. *Planta* 150:166–69

37. Koyama, T., Kokubun, Y., Ogura, K. 1988. Polyprenyl diphosphate synthase from mulberry leaves: stereochemistry of hydrogen elimination in the prenyltransferase reaction. *Phytochemistry* 27:2005–9

38. Kreuz, K., Beyer, P., Kleinig, H. 1982. The site of carotenogenic enzymes in chromoplasts from *Narcissus pseudonarcissus* L. *Planta* 154:66–69

39. Kreuz, K., Kleinig, H. 1984. Synthesis of prenyl lipids in cells of spinach leaf. Compartmentation of enzymes for formation of isopentenyl diphosphate. *Eur. J. Biochem.* 141:531–35

40. Law, J. H., Rilling, H. C. 1985. Steroids and isoprenoids. *Methods Enzymol.* 110 and 111 (entire volumes)

41. Liedvogel, B. 1986. Acetyl coenzyme A and isopentenyl-pyrophosphate as lipid precurosrs in plant cells—biosynthesis and compartmentation. *J. Plant Physiol.* 124:211–22

42. Liedvogel, B., Bäuerle, R. 1986. Fatty-acid synthesis in chloroplasts from mustard (*Sinapis alba* L.) cotyledons: formation of acetyl coenzyme A by intraplastid glycolytic enzymes and a pyruvate dehydrogenase complex. *Planta* 169:481–89

43. Liedvogel, B., Kleinig, H. 1980. Phosphate translocator and adenylate translocator in chromoplast membranes. *Planta* 150:170–73

44. Lütke-Brinkhaus, F., Kleinig, H. 1987. Carotenoid and chlorophyll biosynthesis in isolated plastids from mustard seedling cotyledons (*Sinapis alba* L.) during etioplast-chloroplast conversion. *Planta* 170:121–29

45. Lütke-Brinkhaus, F., Kleinig, H. 1987. Formation of isopentenyl diphosphate

via mevalonate does not occur within etioplasts and etiochloroplasts of mustard (*Sinapis alba* L.) seedlings. *Planta* 171:406–11

46. Lütke-Brinkhaus, F., Kleinig, H. 1989. Phytoene synthase in plastids. Submitted

47. Lütke-Brinkhaus, F., Liedvogel, B., Kleinig, H. 1984. On the biosynthesis of ubiquinones in plant mitochondria. *Eur. J. Biochem.* 141:537–41

48. Lütke-Brinkhaus, F., Liedvogel, B., Kreuz, K., Kleinig, H. 1982. Phytoene synthase and phytoene dehydrogenase associated with envelope membranes from spinach chloroplasts. *Planta* 156: 176–80

49. Lütke-Brinkhaus, F., Rilling, H. C. 1988. Purification of geranylgeranyl diphosphate synthase from *Phycomyces blakesleanus*. *Arch. Biochem. Biophys.* 266:607–12

50. Lütke-Brinkhaus, F., Weiss, G., Kleinig, H. 1985. Prenyl lipid formation in spinach chloroplasts and in a cell-free system of *Synechococcus* (Cyanobacteria): polyprenols, chlorophylls, and fatty acid prenyl esters. *Planta* 163: 68–74

51. Lützow, M., Beyer, P. 1988. The isopentenyl-diphosphate Δ-isomerase and its relation to the phytoene synthase complex in daffodil chromoplasts. *Biochim. Biophys. Acta* 959:118–26

52. Lynen, F., Eggerer, H., Henning, U., Kessel, I. 1958. Farnesyl-pyrophosphat und 3-Methyl-Δ^3-butenyl-1-pyrophosphat, die biologischen Vorstufen des Squalens. *Angew. Chem.* 70:738–42

53. Malhotra, K., Oelze-Karow, H., Mohr, H. 1982. Action of light on accumulation of carotenoids and chlorophylls in the milo shoot (*Sorghum vulgare* Pers.) *Planta* 154:654–55

54. Maudinas, B., Bucholtz, M. L., Papastephanou, C., Katiyar, S. S., Briedis, A. V., Porter, J. W. 1975. Adenosine 5'-triphosphate stimulation of the activity of a partially purified phytoene synthetase complex. *Biochem. Biophys. Res. Commun.* 66:430–36

55. Maudinas, B., Bucholtz, M. L., Papastephanou, C., Katiyar, S. S., Briedis, A. V., Porter, J. W. 1977. The partial purification and properties of a phytoene synthesizing enzyme system. *Arch. Biochem. Biophys.* 180:354–62

56. Mayfield, S. P., Nelson, T., Taylor, W. C., Malkin, R. 1986. Carotenoid synthesis and pleiotropic effects in carotenoid-deficient seedlings of maize. *Planta* 169:23–32

57. Mettal, U., Boland, W., Beyer, P., Kleinig, H. 1988. Biosynthesis of

monoterpene hydrocarbons by isolated chromoplasts from daffodil flowers. *Eur. J. Biochem.* 170:613–16

58. Mhadavan, S., Benedict, C. R. 1984. Isopentenyl pyrophosphate *cis*-1,4-polyisoprenyl transferase from guayule (*Parthenium argenteum* Gray). *Plant Physiol.* 75:908–13

59. Monaco, P., Della Greca, M., Onorato, M., Previtera, L. 1988. Prephytoene alcohol from *Myriophyllum verticillatum*. *Phytochemistry* 27:2355–57

60. Moore, T. C., Coolbaugh, R. C. 1976. Conversion of geranylgeranyl pyrophosphate to *ent*-kaurene in enzyme extracts of sonicated chloroplasts. *Phytochemistry* 15:1241–47

61. Moore, F. D., Shephard, D. C. 1977. Biosynthesis in isolated *Acetabularia* chloroplasts II. Plastid pigments. *Protoplasma* 92:167–75

62. Pauly, G., Belingheri, L., Marpeau, A., Gleizes, M. 1986. Monoterpene formation by leucoplasts of *Citrofortunella mitis* and *Citrus unshiu*. Steps and conditions of biosynthesis. *Plant Cell Rep.* 5:19–22

63. Porter, J. W., Spurgeon, S. L. 1981, 1983. *Biosynthesis of Isoprenoid Compounds*, Vols. 1, 2. New York: Wiley. 558, 552 pp.

64. Poulter, C. D., Rilling, H. C. 1981. Prenyl transferases and isomerases. See Ref. 63, 1:161–224.

65. Poulter, C. D., Rilling, H. C. 1981. Conversion of farnesyl pyrophosphate to squalene. See Ref. 63, 1:413–41

66. Railton, I. D., Fellows, B., West, C. A. 1984. *ent*-Kaurene synthesis in chloroplasts from higher plants. *Phytochemistry* 23:1261–67

67. Rogers, L. J., Shah, S. P. J., Goodwin, T. W. 1968. Compartmentation of biosynthesis of terpenoids in green plants. *Photosynthetica* 2:184–207

68. Rüdiger, W., Benz, J., Guthoff, C. 1980. Detection and partial characterization of activity of chlorophyll synthetase in etioplast membranes. *Eur. J. Biochem.* 109:193–200

69. Ruzicka, L. 1953. The isoprene rule and the biosynthesis of terpenic compounds. *Experientia* 9:357–96

70. Sasiak, K., Rilling, H. C. 1988. Purification to homogeneity and some properties of squalene synthetase. *Arch. Biochem. Biophys.* 260:622–27

71. Schultz, G., Soll, J., Fiedler, E., Schulze-Siebert, D. 1985. Synthesis of prenylquinones in chloroplasts. *Physiol. Plant.* 64:123–29

72. Schulze-Siebert, D., Schultz, G. 1987. β-Carotene synthesis in isolated spinach

chloroplasts. Its tight linkage to photosynthetic carbon metabolism. *Plant Physiol.* 84:1233–37

73. Siefermann-Harms, D., Yamamoto, H. Y. 1975. NADPH and oxygen-dependent epoxidation of zeaxanthin in isolated chloroplasts. *Biochem. Biophys. Res. Commun.* 62:456–61

74. Simcox, P. D., Dennis, D. T., West, C. A. 1975. Kaurene synthetase from plastids of developing plant tissues. *Biochem. Biophys. Res. Commun.* 66:166–72

75. Soll, J., Schultz, G., Joyard, J., Douce, R., Block, M. A. 1985. Localization and synthesis of prenylquinones in isolated outer and inner envelope membranes from spinach chloroplasts. *Arch. Biochem. Biophys.* 238:290–99

76. Soll, J., Schultz, G., Rüiger, W., Benz, J. 1983. Hydrogenation of geranylgeraniol. Two pathways exist in spinach chloroplasts. *Plant Physiol.* 71:849–54

77. Spurgeon, S. L., Sathyamoorthy, N.,

Porter, J. W. 1984. Isopentenyl pyrophosphate isomerase and prenyltransferase from tomato fruit plastids. *Arch. Biochem. Biophys.* 230:446–54

78. Stumpf, P. K. 1980. Biosynthesis of saturated and unsaturated fatty acids. In *The Biochemistry of Plants,* ed. P. K. Stumpf, E. E. Conn, 4:177–204. New York: Academic. 693 pp.

79. Tingey, D. T., Evans, R. C., Bates, E. H., Gumpertz, M. C. 1987. Isoprene emission and photosynthesis in three ferns—the influence of light and temperature. *Physiol. Plant.* 69:609–16

80. Weil, J. H. 1987. Organization and expression of the chloroplast genome. *Plant Sci.* 49:149–57

81. West, C. A. 1981. Biosynthesis of diterpenes. See Ref. 63, 1:375–411

82. Yamamoto, H. Y. 1985. Xanthophyll cycles. *Methods Enzymol.* 110:303–12

83. Zeevaart, J. A. D., Creelman, E. A. 1988. Metabolism and physiology of abscisic acid. *Annu. Rev. Plant Physiol. Plant Mol. Biol.* 39:439–73

NOTE ADDED IN PROOF

Phytoene Synthase has recently been isolated and characterized: Dogbo, O., Laferrière, A., D'Harlingue, A., Camara, B. 1988. Carotenoid biosynthesis: Isolation and characterization of a bifunctional enzyme catalyzing the synthesis of phytoene. *Proc. Natl. Acad. Sci. USA* 85:7054–58

Annu. Rev. Plant Physiol. Plant Mol. Biol. 1989. 40:61–94

STRUCTURE AND FUNCTION OF PLASMA MEMBRANE Atpase

Ramon Serrano

Biological Structures Division, European Molecular Biology Laboratory, 6900 Heidelberg West Germany

CONTENTS

INTRODUCTION

The plasma membrane proton pump of plant cells was originally identified by "in vivo" studies on auxin-induced growth (77, 109), active transport (99, 127), and electrical potentials (140). These studies suggested that the proton pump plays a central role in plant physiology. It can be considered as a "master enzyme" that controls many important functions at the cellular and organ level, including cell division and elongation. The pioneering work of Hodges and colleagues (59) demonstrated in isolated plasma membrane vesicles the existence of a distinct ATPase activity. This represented the first step

61

1040–2519/89/0061-0061$02.00

towards the molecular characterization of the proton pump. The plasma membrane ATPase was later characterized at the biochemical level both in vesicles (127, 145) and in purified preparations of enzyme solubilized with detergents (126, 127). These studies demonstrated the similarities between fungal and plant plasma membrane ATPases, which together constitute a novel group of ion-pumping enzymes (126). The code number recently assigned to this group by the Enzyme Nomenclature committee of the International Union of Biochemistry is EC 3.6.1.35 (32).

The yeast *Saccharomyces cerevisiae*, with its well-developed molecular genetics (142), serves as a convenient model system for the study of the plasma membrane H^+-ATPase (128). However, the most complex functions and regulatory aspects can only be studied in plants. It is expected that *Arabidopsis thaliana* will fulfill in the near future the role of a model system susceptible of genetic manipulation (83).

The reviews by Sze (145) and myself (126, 127) constitute the starting point for the present chapter. During recent years the most significant advance has been the cloning and sequencing of fungal and plant[1] plasma membrane H^+-ATPases. This step has clarified the evolutionary origin of the enzyme and has allowed the proposal of testable models for its structure and mechanism. In addition, recent studies on the regulation of the ATPase by protein kinases and the first ATPase mutants obtained in yeast have greatly expanded our understanding of this important enzyme. Recent reviews by Sussman & Surowy (144) and myself (127, 128, 130) discuss the enzymatic and transport properties of purified enzyme preparations, and the methodological aspects of ATPase purification and reconstitution have been described elsewhere (129). Neither of these topics nor the calcium-pumping ATPase recently described in plant plasma membranes (43, 106) is discussed here. The later enzyme has not yet been purified, and its activity is 2–3 orders of magnitude lower than that of the H^+-ATPase. Therefore, the Ca^{2+}-ATPase is a minor component of plant plasma membranes, and this justifies the convention of considering the H^+-ATPase as the genuine plasma membrane ATPase.

PHYSIOLOGICAL ROLE OF THE PROTON PUMP

Suggested Functions

The qualification of the plasma membrane ATPase as a "master enzyme" in fungal and plant physiology needs to be documented by a brief consideration of its proposed physiological roles. The proton pumping activity of the

[1]The plasma membrane ATPase from *Arabidopsis thaliana* has recently been cloned and partially sequenced (J. M. Pardo and R. Serrano, unpublished). These preliminary sequence data have been utilized (Figures 1, 2, and 5) to illustrate its homology with other ATPases.

enzyme directly controls intracellular and extracellular pH, constituting what has been called the "biophysical pH-stat" (138). Intracellular alkalinization activates malic acid synthesis, which by displacing the equilibrium of the malic enzyme reaction increases the $NADP^+/NADPH$ ratio and therefore activates the pentose-phosphate pathway (77). The synthesis and utilization of malic acid in plants constitute a "biochemical pH-stat" acting in concert with the proton pump (28). Nutrient uptake and turgor (mostly determined by intracellular potassium concentration) are dependent on the electrochemical proton gradient generated by the ATPase (77, 99, 127). Cell growth depends on the activity of the ATPase owing to all these effects.

Turgor and nutrients are logical requirements for growth, but pH affects growth by more subtle regulatory mechanisms. Hager et al (47) and Rayle & Cleland (108) independently proposed that the wall-loosening factor induced by auxin is external acidification—the acid-growth theory. Cell enlargement is initiated by wall loosening, and synthesis of new cell wall fixes the expanded wall into its stretched state. Apparently, an acidic pH is required for ill-defined enzymatic loosening reactions (77, 109). In fungi (4, 44) and in animal cells (33), internal alkalinization seems to be required to initiate DNA synthesis by unknown regulatory mechanisms. Probably, both external acidification and internal alkalinization are important for the growth of both plant and fungal cells.

Intracellular pH changes controlled by the plasma membrane H^+-ATPase seems to be involved in the breaking of dormancy during germination of fungal spores (18) and plant seeds (77) and in the differentiation of *Dictyostelium* amoebae into stalk and spore cells (45). At the level of plant organs, the loading of root xylem with inorganic nutrients, the loading of leaf phloem with organic nutrients, and the turgor changes responsible for stomata and pulvini movements seem to depend on active transport processes driven and controlled by the proton pump (73, 77, 127, 158).

A final aspect is the suggested participation of the proton pump in the development of polarity in growing cells and organs such as roots, root hairs, pollen tubes, fungal hyphae, and algal rhizoids. In all these systems polar growth correlates with a spatial segregation of pumps (H^+-ATPase) in nongrowing parts and leaks (K^+ and Ca^{2+} channels, H^+ symports) in growing tips. This results in transcellular or transorgan ion currents (53, 62).

Evidence from Studies with Yeast Mutants

A note of caution, however, must be introduced before considering as established facts all the proposed physiological roles of the plasma membrane ATPase. The experimental evidence for the participation of the enzyme is based on correlations between physiological functions and proton pumping activity and on extrapolations from biochemical studies performed in vitro.

These correlations do not demonstrate a cause-effect relationship between ATPase activity and biological function because the two phenomena may only be coincident. On the other hand, proton pumping is attributed to the plasma membrane ATPase only on the basis of the effects of ATPase inhibitors. As all known ATPase inhibitors have toxic side effects in whole cells (10), this evidence does not constitute rigorous proof. Finally, the in vitro studies only show what the enzyme can do in artificial system but do not address the question of what the enzyme actually does in the cell. Therefore, only a genetic approach that could introduce specific mutations on the ATPase gene may provide conclusive evidence for the physiological role of the plasma membrane ATPase.

This mutational analysis has already been started in yeast, where mutants with no ATPase (131), reduced amounts of ATPase (23, 131), and temperature-sensitive ATPase (24, 132) have already been constructed. The study of these mutants has demonstrated that the plasma membrane H^+-ATPase is essential and rate limiting for yeast growth, supporting the crucial role of the enzyme proposed on the basis of physiological and biochemical studies. In addition, reduction of ATPase activity by specific mutations results in decreased proton pumping activity and decreased amino acid transport. Therefore, the ATPase is the major proton pump energizing the plasma membrane of yeast cells, and any alternative, such as an hypothetical redox pump (26), is not physiologically significant.

These mutants should facilitate studies on the participation of the ATPase in different cellular activities, in particular on the connection between the enzyme and growth. Overproduction of the ATPase has been shown to be toxic for the yeast cell, suggesting that there may be structural constraints for modifying the amount of ATPase without affecting the integrity of the membrane (34). Therefore, a structural role in the membrane should also be considered in future studies. Probably these conclusions can be extrapolated to plant cells, but a mutational analysis of the plant ATPase is required to ascertain the crucial role of the enzyme in plant physiology. Although gene replacement between wild type and in vitro constructed mutant alleles is not yet possible in plants, other classical and molecular-genetic approaches may be useful to manipulate the plant ATPase gene in the near future.

Relationship between H^+P-ATPase and K^+ Transport

Earlier proposals (58, 71) about the plasma membrane ATPase catalyzing ATP-driven potassium transport were based only on the small activation of the enzyme by potassium and did not receive experimental support when the enzyme was purified and reconstituted in proteoliposomes (126, 127). In this defined in vitro system the ATPase was shown to pump protons in the absence of potassium. Potassium transport is probably catalyzed by a different system and driven by the electrical membrane potential generated by the proton pump. Indirect electrical coupling between proton and potassium transport was also suggested by many previous physiological experiments (127) and has received recent support by patch-clamping studies with guard cell protoplasts (6), where the proton pump was shown to function in the absence of potassium.

On the other hand, patch-clamp methodologies have demonstrated the presence in yeast (46) and plant (121, 122) plasma membranes of potassium-selective channels activated by hyperpolarization. Work with the ATPase from *Schizoasaccharomyces pombe* reconstituted into proteoliposomes (152, 153) suggested that the purified, homogeneous ATPase, although not pumping potassium in the course of ATP hydrolysis, contains a voltage-sensitive potassium channel. Recent patch-clamping experiments (J. A. Ramirez, personal communication) have demonstrated that mutants of the yeast ATPase show altered voltage dependence of the potassium channel. This suggests that the ATPase molecule is tightly associated with the K^+ channel and may actually contain it. Further work at the biochemical and genetic level is required to confirm this novel possibility. As discussed below, H^+-pumping ATPases may have evolved from H^+, K^+-exchanging ATPases, and therefore some relationship between the H^+-ATPase and K^+ transport may have persisted.

REGULATION OF PLASMA MEMBRANE ATPase ACTIVITY

Factors Modulating ATPase Activity

The important functions of the plasma membrane H^+-ATPase described above justify the proposition that the activity of the enzyme is modulated by all the important factors controlling fungal and plant physiology. Physiological studies based on the measurements of membrane potentials and external acidification have suggested the following regulatory effects.

The induction of stomata opening by light is based on the stimulation of the proton pump of guard cells, which results in massive potassium uptake and turgor increase (6, 124, 134). Three different classes of light receptors have been implicated in ATPase regulation: the red light receptor (phytochrome), the blue light receptor, and the chloroplast pigments.

Stimulation of the proton pump by a decrease in cell turgor (111, 112, 155) seems to be a key event in osmoregulation. As in the case of guard cells, potassium uptake is secondarily stimulated, resulting in increased turgor.

Practically all plant hormones interact with the proton pump. Auxin activates the enzyme in the growing parts of the shoots, promoting cell elongation (14, 37, 77, 109). The pollen hormone brassinolide rapidly activates the proton pump in roots (115) and shoots (20). Cytokinins activate the pump in cotyledons (77). On the other hand, abscisic acid seems to inhibit the proton pump in guard cells (inducing closing of stomata), germinating seeds (inhibiting germination), and roots (inhibiting xylem loading) (73, 77). Inhibition of the pump is also observed after mechanical (injury) or thermal (chilling or heating) shock of plant tissues (51). Fusicoccin (77) and syringomycin (8, 9) are phytotoxins produced by pathogenic microorganisms which seem to activate the proton pump in all plant tissues. This effect interferes with endogenous modulation of the pump during cell elongation and stomatal opening.

The mechanism of regulation of the H^+-ATPase by these plant-controlling factors has not been elucidated, and this lack is one of the most fundamental knowledge gaps in present plant physiology. In some cases it is not even clear if the observed changes in pump activity reflect modulation of either the amount or the activity of the enzyme. However, the rapid nature of most of the effects suggest that a change in activity is the most likely mechanism.

Two factors are known to activate the yeast proton pump: glucose (125) and acid media (35). Although a change in the amount of enzyme has been discarded (34), the mechanism of modulation is also not known. A covalent modification seems to be involved, resulting in changes on maximal rate, K_m, pH optimum, and vanadate sensitivity. A recently characterized mutation of the yeast ATPase has been shown to result in an enzyme that exists in the activated state in the absence of glucose (24). Apparently this mutation mimics the effects of the covalent modification triggered by glucose. The mutation has been mapped close to the proposed ATP binding site (see below).

Possible Regulation by Protein Kinases

It may be expected that fungal and plant plasma membrane ATPases are regulated by protein kinases, because this is the most frequent regulatory mechanism in eukaryotes (61). Under this basis, the two pertinent questions are the nature of the protein kinase(s) acting on the ATPase and the intracellular messengers that activate these protein kinases and that are generated by the external regulatory factors discussed above.

In the case of yeast cells, it seems that both glucose (owing to intracellular conversion into hexose-phosphates) and acid media (owing to passive proton influx) cause intracellular acidification (19, 102), and this could be the first intracellular messenger of the pathway for ATPase activation. Yeast adenylate cyclase is activated by intracellular acidification (102); therefore an increase in cAMP triggered by glucose may be the second intracellular messenger of this pathway. Accordingly, the product of the CDC25 gene regulates yeast adenylate cyclase (13), and in yeast cells with temperature-sensitive mutations of this gene glucose fails to activate the ATPase at the nonpermissive temperature (100). cAMP also seems to mediate the activation of the proton pump in *Dictyostelium discoideum* (3). However, the fact that the sequence of the yeast plasma membrane ATPase (131) does not contain potential phosphorylation sites recognized by cAMP-dependent protein kinase [clusters of arginines followed by serine, (133)] suggests that this kinase cannot be the direct activator of the ATPase. It is possible that the cAMP-dependent protein kinase phosphorylates and activates another protein kinase acting on the ATPase (see below).

Both the yeast (81) and plant (119) plasma membrane ATPases are phosphorylated in vivo at multiple sites. The fungal enzyme only contains phosphoserine, but both phosphoserine and phosphothreonine are present in the plant ATPase. In vitro, a plasma membrane–bound protein kinase can phosphorylate both ATPases (81, 119), and in the case of plants this kinase was dependent on micromolar calcium (119).

In yeast the only intracellular messenger known to regulate protein kinases during physiological responses is cAMP (50). However, protein kinases activated by this metabolite are not present in plants (104). The nature of the intracellular messengers that could activate plant protein kinases is beginning to be elucidated. The response to mechanical and thermal shocks seems to be mediated by calcium uptake through channels opened by depolarization (113). The observed increase in protein phosphorylation, probably catalyzed by calcium-activated protein kinases, could explain the inhibition of the ATPase observed in shocked plant tissues (160). Both auxin and light seem to decrease intracellular calcium levels (56), and this could release an inhibitory phosphorylation of the ATPase by calcium-dependent protein kinases. However, in the case of auxin other intracellular messengers are currently under discussion. Both auxin and fusicoccin rapidly induce cytosolic acidification (15, 39). On the other hand, artificially induced cytosolic acidification activates both the plasma membrane ATPase and elongation growth (16, 48, 114, 151). Therefore, it has been postulated (17) that acidification is the first intracellular messenger mediating the activation of the ATPase. A simple interpretation of this mechanism could be based on the fact that the pH optimum of the plasma membrane ATPase (about 6.5) is more acidic than normal cytoplasmic pH (17). However, at longer times (more than 30 min) fusicoccin causes cytosolic alkalinization while the proton pump remains activated (79, 110). As described in yeast (19, 102), a transient acidification could trigger the activation of proton pumping and then result in intracellular alkalinization. This stable activation triggered by acidification could again be the result of protein kinase activity; in this respect, the fact that the plasma membrane–bound protein kinase described by Schaller & Sussman (119) has an acidic pH optimum is suggestive. On the other hand, how auxin (and fusicoccin) can acidify the cytosol is not apparent. It has been suggested that auxin could increase intracellular calcium, which would displace protons from common binding sites (17). However, as discussed above, there are indications that auxin may actually reduce the calcium concentration. This point needs to be clarified (56).

A completely different intracellular messenger has recently been proposed based on the measurement of the turnover of phosphatidylinositol lipids. Both light (85) and auxin (36) seem to increase this turnover by activating phospholipase C. This would produce diacylglycerol, a potent activator of protein kinase C in animal (88) and plant (91, 118) cells. However, the phosphorylation of the ATPase by protein kinase C has not been demonstrated.

Stimulation of the ATPase by either fusicoccin (105) or auxin (40, 120, 148) has been reported in in vitro systems, with purified plasma membrane vesicles or extracted proteins. However, these results must be received with caution because similar reports have appeared in the past without later being verified. These in vitro effects on the ATPase must be reproduced in several laboratories and the mechanism of activation must be clarified in order to deserve consideration. In this respect, some recent experimental results (8, 9) provide evidence for the mechanism of the effect of syringomycin on the red beet plasma membrane ATPase. This phytotoxin activates the ATPase both in vivo and in vitro. In isolated membrane vesicles activation correlates with phosphorylation of the ATPase polypeptide by a membrane-bound protein kinase. Extraction of the protein kinase with deoxycholate (which does not solubilize the ATPase) abolishes both the activation and the phosphorylation induced by syringomycin. This strongly suggests that the ATPase is regulated by the syringomycin-activated protein kinase. Syringomycin also seems to activate the yeast proton pump (159), and it is in this organism where the syringomycin-activated protein kinase has been characterized (1, 157). This enzyme is dependent for activity on basic peptides such as histones, polymyxin B, and syringomycin, and it phosphorylates threonine and serine residues surrounded by glutamates. It is possible that this enzyme corresponds to the polyamine-activated protein kinase described in all eukaryotic cells, including plants (104, 150). As cell growth is associated with increased concentrations of polyamines, this protein kinase could provide a link between the activation of both growth and ATPase.

We may conclude this part by formulating the following hypothesis about the regulation of the fungal and plant plasma membrane ATPase. The activity of the enzyme is probably modulated by protein kinases, and the ATPase probably contains multiple phosphorylating sites which may affect its activity in different ways. Protein kinases activated by basic peptides, by calcium, and by diacylglycerol (protein kinase C) may be implicated. The effect of acidification may be mediated by a protein kinase with acidic pH optimum or by the expected increase in calcium after protons displace it from common binding sites. In yeast, acidification seems to increase cAMP levels and activate the cAMP-dependent protein kinase of this organism. Protein kinases acting on the ATPase may be modulated by other protein kinases, such as the cAMP-activated protein kinase of yeast.

Transplasmalemma Redox Activity and the H^+-ATPase

Transplasmalemma redox activity, monitored in the presence of exogenous ferricyanide, stimulates proton efflux in yeast and in plant tissues (26). Although the first interpretation envisaged either a redox proton pump or a direct effect of the redox system on the H^+-ATPase, more recent evidence suggests that the activation of proton efflux results from an indirect effect of the redox system on the ATPase (78, 117, 149). The plasma membrane dehydrogenase seems to transport electrons from intracellular NADPH to extracellular ferricyanide. This transport causes a decrease in the electrical membrane potential because of the efflux of negative charges. In addition, as only electrons seems to cross the membrane, protons resulting from NADPH oxidation acidify the cytosol. The activation of the proton pumping ATPase results from both the depolarization and the cytoplasmic acidification.

Ferricyanide is a nonnatural acceptor for the plasma membrane dehydrogenase, and until a physiological acceptor can be found, the physiological role of this redox system remains obscure.

STRUCTURE AND MECHANISM OF (E-P) ATPases

Evolutionary Relationships

The sequence similarities between different (E-P) ATPases, as determined by the program BESFIT (29), can be clustered into five levels (Table 1). Fungal H^+-ATPases exhibit 85–86% similarity between themselves. The Na^+, K^+-ATPase and the H^+, K^+-ATPase of mammalian cells are 81% similar. The similarity between the H^+-ATPases from plants (see Footnote 1) and protozoa and between any of these and the fungal enzymes is 56–60%. In the case of animal ATPases the Ca^{2+}-ATPase exhibits 53–54% similarity with either the Na^+, K^+-ATPase or the H^+, K^+-ATPase. Finally, pair comparisons of the K^+-ATPases from *Escherichia coli* and *Streptococcus faecalis* and of the H^+-ATPases or K^+-ATPases with any other ATPases indicate a 44–50% similarity. Random comparisons of unrelated sequences by the same algorithm give values in the range of 34–40%, which correspond to the noise of fortuitious similarity. Therefore, all the similarities found between (E-P) ATPases seem significant, and a tentative evolutionary tree can be depicted for the whole family (Figure 1).

The nature of the ancestral cation-pumping ATPases is difficult to guess,

Table 1 Similarities between (E-P) ATPases[a,b]

	Hsc	Hnc	Hsp	Hat	Hld	Ca	NaK	HK	Ksf	Kec
Hsc	100	86.0	85.2	58.2	56.5	49.5	47.1	48.5	48.1	47.8
Hnc		100	86.6	59.7	57.8	47.6	48.8	47.2	49.1	46.3
Hsp			100	56.7	55.6	49.8	47.5	49.4	45.6	43.2
Hat				100	57.8	49.7	50.3	48.9	47.7	44.5
Hld					100	48.0	48.7	50.5	47.8	46.0
Ca						100	53.0	53.8	47.7	45.3
NaK							100	81.1	46.1	43.8
HK								100	44.4	45.1
Ksf									100	47.9
Kec										100

[a] The symbols and references for the different ATPase sequences are: Hsc, H^+-ATPase from *Saccharomyces cerevisiae* (131); Hnc, H^+-ATPase from *Neurospora crassa* (2, 49); Hsp, H^+-ATPase from *Schizosaccharomyces pombe* (42); Hat, H^+-ATPase from *Arabidopsis thaliana* (see Footnote 1); Hld, H^+-ATPase from *Leishmania donovani* (82); Ca, Ca^{2+}-ATPase from sarcoplasmic reticulum (74); NaK, Na^+K^+-ATPase from kidney (137); HK, H^+K^+-ATPase from stomach (136); Ksf, K^+-ATPase from *Streptococcus faecalis* (139); Kec, K^+-ATPase from *Escherichia coli* (57).

[b] Pair comparisons of ATPase sequences were made by the program BESTFIT (29). This uses a symbol comparison table with mismatches scored according to the evolutionary distance between amino acids (123). Basically, the groups of similar amino acids are as follows: (M, I, L, V, F); (F, Y, W); (H, R, K); (D, N, E, Q); (S, T, A, G, P); (C). The percentage similarity between different sequences is given.

mostly because the transport properties of many present ATPases have not been fully characterized. For example, it seems that the K^+-ATPases from bacteria catalyze the exchange of Na^+ or H^+ for K^+ (M. Solioz, personal communication) and that the Ca^{2+}-ATPase exchange Ca^{2+} for H^+ (21, 75, 76, 87, 156). Therefore only H^+-ATPases do not seem to catalyze cation exchanges, and it is possible that cation exchange represents a primitive feature. On this basis, we may speculate that the ancestral pump was a H^+-K^+ exchange ATPase. This fits the proposal that ion pumps arose in primitive anaerobic cells to extrude the acids generated during fermentations (138). Later on, K^+ countertransport was lost to develop the highly electrogenic H^+-ATPases of fungal and plant plasma membranes. The small activation of some fungal and plant ATPases by potassium (127) may reflect binding to a degenerated form of the ancestral potassium transport site. The specialization on Na^+-K^+ exchange evolved as a distinguished feature of the animal plasma membrane ATPase, probably related to life under seawater (high-Na^+) conditions. Present day Na^+, K^+-ATPase still has a marginal H^+-K^+ exchange activity (52) that may reflect this evolutionary origin.

It has recently been suggested that protons are transported as hydrated hydronium (H_3O^+), more related to the other cations in terms of binding sites than anhydrous H^+ (11). The substitution of Na^+-K^+ exchange by Ca^{2+}-H^+ was an important (and complicated) development to achieve efficient Ca^{2+} homeostasis. It is not clear if the electroneutral H^+-K^+ exchange ATPase of

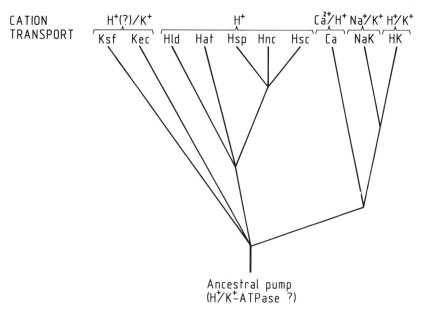

Figure 1 Hypothetical evolutionary tree for (E-P) ATPases. Symbols for different ATPases as in Table 1.

mammalian stomach represents the primitive pump or if it is an adaptation of a primitive electrogenic pump (for example, $2H^+/1K^+$) for massive acidification.

Alignment of ATPase Sequences Based on Structure Prediction

As described above, the overall similarity between distant members of the (E-P) ATPase family is very low. Under these conditions, alignment of the different sequences on the basis of similarity only may be misleading. On the other hand, even distantly related proteins of a given family share a common folding pattern, including the main secondary structural elements and some residues flanking them, which generally include all the active-site peptides (72). Therefore, I have attempted to combine the alignment based on sequence similarity with that based on predicted secondary structure. The result of my analysis (Figure 2) must be received with great reservation because of the uncertainties inherent in the current methodologies of secondary-structure prediction (65) and of identification of transmembrane stretches (92).

The first conclusion of this alignment is that there is strong conservation of only six motifs predicted to be coils or loops connecting elements of secondary structure (Figure 2, Table 2). Because they are highly polar, a surface location can be predicted, as would be expected for their participation in substrate binding and/or catalysis. Motif II includes the aspartate forming the phosphorylated intermediate in eukaryotic ATPases (128). The phosphorylated peptide has not yet been identified in the bacterial enzymes, but it is likely to correspond to this motif. Motif III (KGAP) includes a lysine residue which seems to be essential because its modification by fluorescein isothiocyanate in the $Ca2^+$-, Na^+, K^+-, and H^+, K^+-ATPases inactivates the enzymes (128). ATP at high concentrations protects against inactivation, suggesting that ths lysine is part of the active site. Motifs III and IV are not present in the bacterial ATPases, although related sequences can be found in equivalent positions of my alignment. The other motif for which we have biochemical evidence of function is motif VI. The ATP analogue 5'(p-fluorosulfonyl)benzoyladenosine has been widely used for the affinity labeling of ATP-binding sites (25), and in the Na^+, K^+-ATPase labels the lysine residue at the end of motif VI (89). This lysine is not conserved in the bacterial ATPases. In addition, an ATP analogue with an alkylating group in the gamma-phosphate has been shown to modify the first aspartate of this motif (95). Therefore, motif VI is probably the major component of the ATP binding site of (E-P) ATPases.

Another conclusion of the proposed alignment is that different (E-P) ATPases seem to cross the membrane different numbers of times, ranging from six in the bacterial enzymes, to seven in the Na^+, K^+- and H^+, K^+-ATPases, and to nine in the Ca^{2+}- and H^+-ATPases. Therefore, only the first six hydrophobic stretches (predicted transmembrane helixes) are conserved and may constitute the "core" of the transmembrane cation channel. The extra hydrophobic stretches in eukaryotic ATPases may have a location peripheral to this channel. My earlier (128) proposal of eight transmembrane

---

⟶

Figure 2   Alignment of (E-P) ATPase sequences including structure prediction. A preliminary alignment based on sequence similarity was first made with the program GAP (29), using the *Saccharomyces cerevisiae* H^+-ATPase sequence as template against which all the other sequences were compared. The gapped sequences were put together with the program LINEUP (29) and then a manual correction of the alignment was made on the basis of structure prediction with the program PEPPLOT (29). Hydrophobic stretches predicted to cross the membrane as hydrophobic helixes were deduced from the algorithms of Kyte & Doolittle (68) and of Engelman et al (31). The prediction of secondary structure was based on the empirical method of Chou & Fasman (22) as improved by Garnier et al (41), taking into account the more recent estimations of overall secondary structure by circular dichroism (55) and Raman spectroscopy (93). Symbols for the different ATPases as in Table 1. The *Arabidopsis* sequence is not complete (see Footnote 1), and only hydrophobic and conserved regions are shown. Str: predicted secondary structure (no distinction was made between turns and coils). Con: conserved residues. *: conserved in all the sequences; +: conserved in the eukaryotic ATPases; x: conserved in H^+-ATPases. Groups of similar amino acids as in Table 1.

```
Str   .....coil..............helix..........coil........        1
Con   ....................................................
Hsc   MTDTSSSSSSSSASSVS..AHQPTQEK...........PAKTY......D
Hsp   MADNAG.............EYHDAEKHAPEQQA....PPPQQP..AHAA
Hnc   MADHSASGAPALSTNI...ESGKFDEKAAEAAAYQ...PKP.........
Hat   ....................................................
Hld   MSS....................KKYELDAAAFE..DKP.........
Ca    ....................................................
NaK   MGKGVGRDKYEPAAVSEHGD.....KKKAKKER.................
HK    MGKENYELYSVELGTGPGGD...MAAKMSKKKA.................
Ksf   ....................................................
Kec   ....................................................

Str   ...helix.................coil..........helix..........    2
Con   ....x......................x................x.........
Hsc   DAASESSDDDDIDALIEELQ.....SNHG..........VDDEDSD........
Hsp   APAQDDEPDDDIDALIEELF.....SED............VQEEQED.......
Hnc   ..KVEDDEDEDIDALIEDLE.....SHDG........HDAEEEEEA.......
Hat   ....................................................
Hld   ESHSDAEMT...............PQKP....QRRQSVLSKAVSEHDERA...
Ca    ....................................................
NaK   ...................DMD.......ELKKEV............
HK    ...................GGGGG....KKKEKLENMKKEMEMNDHQL
Ksf   ....................................................
Kec   ....................................................

Str   ...coil........helix....................coil.......      3
Con   ...............................................++x...
Hsc   ..NDGP....VAAGEARPVPEEYLQ...............TDPSYGLTSD
Hsp   ..NDDAP....AAGEAKAVPEELLQ...............TDMNTGLTMS
Hnc   ..TPGGG.........RVVPEDMLQ...............TDTRVGLTSE
Hat   ....................................................
Hld   ............................TGPATDPVPPSKGLTTE
Ca    ..............MENAHTKTVEEVL.........GHFGVNESTGLSL.
NaK   ..SMDD.........HKLSLDELHRKY............GTDLNRGLTTA
HK    ..SVS............ELEQKYQ...............TSATKGLKAS
Ksf   ....................................................
Kec   ....................................................

Str   ...helix.................coil..........stalk.helix.1     4
Con   ..x....................x+.+..............x........
Hsc   ..EVLKRRKK.................YGLN..........QMADEKES....
Hsp   ..EVEERRKK.................YGLN..........QMKEELEN....
Hnc   ..EVVQRRRK.................YGLN..........QMKEEKENH...
Hat   ....................................................
Hld   ..EAEELLKK.................YGRN.........ELPEKKTPS...
Ca    ..EQVKKLKER...............WGSN..........ELPAEEGKTLEE
NaK   ..RAAEILAR...............DGPNALTPPPTTP..EWVKFCRQ....
HK    ..LAAELLLR...............DGPNALRPPRGTP..EYVKFARQ....
Ksf   ......................................MELKQKSP.....
Kec   ..MSRKQLALFEPTLVVQALKEAVK..KLNP.........QAQWRN......

Str   .............transmembrane.helix.1......ext.coil..      5
Con   ................x.x*..*..xx...x..x..x.....xx.x.......
Hsc   .........112 LVVKFVMFFVGPIQFVMEAAAILAA...GLSD.......
Hsp   .........110 PFLKFIMFFVGPIQFVMEMAAALAA...GLRD.......
Hnc   .........113 .FLKFLGFFVGPIQFVMEGAAVLAA...GLED.......
Hat   .............LLKFLGFMWNPLSWVMEAAAIMAI..............
Hld   ..........87 .WLIYVRGLWGPMPAALWIAIIIEF...ALEN.......
Ca    LVIEQFED..60 .LLVRILLLAACISFVLAWF.......EEGEETITA..
NaK   ..........94 .LFGGFSMLLWIGAVLCFLAYGIQAA..TEEEPQNDN..
HK    .........105 .LAGGLQCLMWVAAAICLIAFAIQAS..EGDLTTDDN..
Ksf   ...........9 .AMMTLIAMGITVAYVYSVYSFIANLI..SPHTH....
Kec   ..........34 .PVMFIVWIGSLLTTCISIAMASGAM....PGNA....
```

Figure 2 (continued)

```
Str  ...transmembrane.helix.2......stalk.helix.2.......                6
Con  ....x.x...+..+.*.x+.++x.......*...+x.....+..xx...*.
Hsc  141 WVDFGVICGLLMLNAGVGFV......QEFQAGSIVDELKKTLANTA.
Hsp  139 WVDFGVICALLMLNAVVGFV......QEYQAGSIVDELKKSLALKA.
Hnc  141 WVDFGVICGLLLLNAVVGFV......QEFQAGSIVDELKKTLALKA.
Hat  ....WQDFVGIVCLLVINSTISFV...........................
Hld  115 WPDGAILFAIQIANATIGWY......ETIKAGDAVAALKNSLKPTA.
Ca   .88 FVEPFVILLILVANAIVGVW......QERNAENAIEALKEYEPEMG.
NaK  128 LYLGVVLSAVVIITGCFSYY......QEAKSSKIMESFKNMVPQQA.
HK   139 LYLALALIAVVVVTGCFGYY......QEFKSTNIIASFKNLVPQQA.
Ksf  .40 VMDFFWELATLIVIMLLGHWI.....EMNAVSNASDALQKLAELLPE
Kec  .63 LFSAAISGWLWITVLFANFAEALA..EGRSKAQANSLKGVKKTAFA.


Str  beta.....coil...helix.................coil.........          7
Con  .+.*.....*x.....x..x..x+......x**+*.*..*..*.+*.....
Hsc  VVIR.....DGQ..LVEIPANEVV......PGDILQLEDGTVIPTDGR...
Hsp  VVIR.....EGQ..VHELEANEVV......PGDILKLDEGTIICADGR...
Hnc  VVLR.....DGT..LKEIEAPEVV......PGDILQVEEGTIIPADGR...
Hat  ..................................................
Hld  TVYR.....DS..KWQQIDAAVLV......PGDLVKLASGSAVPADCS...
Ca   KVYR....QDR..KSVQRIKAKDIV.....PGDIVEIAVGDKVPADIR...
NaK  LVIR.....NG...EKMSINAEEVV.....VGDLVEVKGGDRIPADLR...
HK   TVIR.....DGD...KFQINADQLV.....VGDLVEMKGGDRVPADIR...
Ksf  SVKRLK..KDGT....EETVSLKEVH....EGDRLIVRAGDKMPTDGT...
Kec  RKLR.....DA..KYGAAADKVPADQLR..KGDIVLVEAGDIIPCDGE...


Str  ....beta..............coil...beta.....coil....beta          8
Con  ....x.****x*.......****.x.........*.....xx
Hsc  .IVTEDF.LQIDQSAI...231 TGES..LAVDK.....HYGDQ...TFSS
Hsp  .VVTPDVHLQVDQSAI...229 TGES..LAVDK.....HYGDP...TFAS
Hnc  .IVTDDAFLQVDQSAL...231 TGES..LAVDK.....HKGDQ...VFAS
Hat  ...................TGES............................
Hld  ...INEGVIDVDEAAL...203 TGES..LPVTM.....GP.E....HMPK
Ca   LTSIKSTTLRVDQSIL...181 TGES..VSVIKHTD..PVPDP..RAVNQ
NaK  ..IISANGCKVDNSSL...217 TGES..EPQTR......SPD..FTNENP
HK   ..ILSAQGCKVDNSSL...228 TGES..EPQTR......SPE..CTHESP
Ksf  ...IDKGHTIVDESAV...133 TGES..KGVKKQV.....GDS..VIGGS
Kec  ...VIEGGASVDESAI...159 TGES..APVIRES....GGD...FASVT


Str  ..............coil.........beta....coil.....beta.......      9
Con  .x...............*x............+.....**....*x+++xx.xxx...
Hsc  STVK..........RGEG.......FMVVTA...TGDN..TFVGRAAALVNKA.
Hsp  SGVK..........RGEG.......LMVVTA...TGDS..TFVGRAASLVNAA.
Hnc  SAVK..........RGEA.......FVVITA...TGDN..TFVGRAAALVNAA.
Hat  ......................................................
Hld  MGSNVV.........RGEV.......EGTVQY...TGSL..TFFGKTAALLQSVE
Ca   DKKNMLFS..GTNIAAGKA.......MGVVVA...TGVN..TEIGKIRDEMVATE
NaK  LETRNIAFF.STNCVGTA.......RGIVVY...TGDR..TVMGRIATLAS...
HK   LETRNIAFF.STMCVLEGTA.......QGLVVS...TGDR..TIIGRIASLAS...
Ksf  I...........NGDGTIE.....ITVTG.....TGEMV.TCKVMEMVRK....
Kec  ............GGTRILSDW.LVIECSV..NPGE...TFLDRMIAMVE...


Str  ..coil...stalk.helix.3.......transmembrane.helix.3...       10
Con  ..x..........xx...............x..*....+..x
Hsc  ..AGGQG......HFTEVLN......292 GIGIILLVLVIATLLLVWTACFY.
Hsp  ..AGGTG......HFTEVLN......290 GIGTILLVLVLLTLFCIYTAAFY.
Hnc  ..SGGSG......HFTEVLN......292 GIGTILLILVIFTLLIVWVSSFY.
Hat  ........................AIGNFCICSIAVGIAIEIVVMYPI
Hld  ..SDLGN.....IHVILRR.......265 VMFSLCAISFMLCMCCFIYLLA..
Ca   ..QERTP...LQQKLDEFGEQ.....260 LSKVISLICIAVWIINIGHF....
NaK  ..GLEGGQTP.IAAEIEH........289 FIHIITGVAVFLGVSFFILSLIL.
HK   ..GVENEKTP.IAIEIEH........300 FVDIIAGLAILFGATFFVVAMCI.
Ksf  ....AQG...EKSKLEFLSDKVAK..197 WLFYVALVVGIIAFIAWLFLA...
Kec  ...GAQR.....RKTPNE........219 IALTILLIALTIVFLLATATLW..
```

```
Str   ...ext.coil.................transmembrane.helix.4.......    11
Con   .x.........................*.*.*.x++*x**..x**.x+++*+....
Hsc   .RTNGIVRILRYT........327 .....LGITIIGVPVGLPAVVTTTMAVGAAYLA
Hsp   .RSVRLARLLEYT........325 .....LAITIIGVPVGLPAVVTTTMAVGAAYLA
Hnc   .RSNPIVQILEFT........327 .....LAITIIGVPVGLPAVVTTTMAVGAAYLA
Hat   ..........................LLVLLIGGIPIAMPTVLSVTMAIGS....
Hld   .RFYETFRHALQ.........298 ...FAVVVLVVSIPIALEIVVTTTLAVGS....
Ca    NDPVHGGSWIRGAIYYFK...298 ...IAVALAVAAIPEGLPAVITTCLALGT....
NaK   EYTWLE..............318 AVIFLIGIIVANVPEGLLATVTVCLTLTA....
HK    GYTFLR..............329 AMVFFMAIVVAYVPEGLLATVTVCLSLTA....
Ksf   NLPDALER.............226 ....MVTVFIIACPHALGLAIPLVVARSTSIAA
Kec   PFSAWGGNAVS.........252..VTVLVALLVCLIPTTIGGLLSASAVAGM....

Str   ...stalk.helix.4......coil..beta.........coil.....    12
Con   ........*..x+x***.+...+xxx..x*++......*******.+x+.
Hsc   ....KKQAIVQKLSAIESL...AGVE..ILCS..378 DKTGTLTKNKLS
Hsp   ....EKQAIVQKLSAIESL...AGVE..VLCS..376 DKTGTLTKNKLS
Hnc   ....KKKAIVQKLSAIESL...AGVE..ILCS..378 DKTGTLTKNKLS
Hat   ..................................DKTGTLTLNKLS
Hld   KHLSKHKIIVTKLSAIEMM...SGVN..MLCS..351 DRTGTLTLNKME
Ca    RRMAKKNAIVRSLPSVETL...GCTS..VICS..351 DKTGTLTTNQMS
NaK   KRMARKNCLVKNLEAVETL...GSTS..TICS..374 DKTGTLTQNRMT
HK    KRLASKNCVVKNLEAVETL...GSTS..VICS..385 DKTGTLTQNRMT
Ksf   ....KNGLLLKNRNAMEQA...NDLD..VIML..278 DKTGTLTQGKFT
Kec   SRMLGANVIATSGRAVEAA...GDVD..VLLL..307 DKTGTITLGNRQ

Str   ...helix.......................coil..........    13
Con   +.....................................x.........
Hsc   LHEPYTVE......................GVSPDD.........
Hsp   LGEPFTVS......................GVSGDD.........
Hnc   LHDPYTVA......................GVDPED.........
Hat   .............................................
Hld   IQEQCFTFEE....................GNDLKST.......
Ca    VCRMFILDKV....................DGDTCSLN......
NaK   VAHMWFDNQIHEA................DTTENQSGVSFDKTS
HK    VSHLWFDNHLHTA................DTTEDQSGQTFDQSS
Ksf   VTGIEILDEAYQEEEILKYIGALEAH........ANHP..........
Kec   ASEFIPAQGVDEKTLADAAQLASLADE.......TPEGRS.........

Str   .....beta......coil....beta......coil..........    14
Con   ...............................................xx.
Hsc   ....................................LMLT
Hsp   ....................................LVLT
Hnc   ....................................LMLT
Hat   ...................................
Hld   ....................................LVLA
Ca    ......EFTIT.....GSTYAP..IGEVH..KDDKPVKCHQYDG..LVEL
NaK   ATWL..ALSRIA....GLCNRA..VFQA...NQDNLP.........ILKR
HK    ETWR..ALCRVL....TLCNRA..AFKS...GQDAVP.........VPKR
Ksf   .....LAIGIMN...............................YLKE
Kec   ......IVILAK..............................QRFN

Str   .helix...........coil.....helix.................    15
Con   ......x...........++xx.......x...........
Hsc   ACLAASRKKK........GLDAID....KAFLKSLKQY.............
Hsp   ACLAASRKRK........GLDAID....KAFLKALKNY.............
Hnc   ACLAASRKKK........GIDAID....KAFLKSLKYY.............
Hat   .............................................
Hld   ALAAKWRE.........PPRDALDT...MVLGAADLDE.............
Ca    ATICALC...........NDS......ALDYNEAKGVYEKVGEATETALT
NaK   AVA...............GDAS......ESALLKCIEVCC..........
HK    IVI...............GDAS.....ETALLKFSELTL..........
Ksf   K............................................
Kec   LRERDV............QSL.....HATFVPFTAQSRM.........
```

Figure 2 (continued)

```
Str   ...............coil...........helix...........coil         16
Con   ...................................................xxxx
Hsc   ...............PKAKD.........ALTKYKVLEFH......PFDP
Hsp   ...............PGPRS.........MLTKYKVIEFQ......PFDP
Hnc   ...............PRAKS.........VLSKYKVLQFH......PFDP
Hat   ...................................................
Hld   ..................CDNY.........QQLNFV..........PFDP
Ca    CLVEKMNVFDTEL..KGLS.........KIERANAC........NSVI
NaK   ...............GSVK.........EMRERYAKIVEI.....PFNS
HK    ...............GNAMGYRDRFP...KVCEIPFNSTNKFQL..SIHT
Ksf   ...................................................
Kec   ...............SGINIDN.............................

Str   .........helix.........coil......beta...coil......      17
Con   .........xx....x.............x.....................
Hsc   VS.......KKVTAVVE......SPEGE.......................
Hsp   VS.......KKVTAYVQ......APDGT.......................
Hnc   VS.......KKVVAVVE......SPQGE.......................
Hat   ...................................................
Hld   TT.......KRTAATLV......DRRSGE......................
Ca    .........KQLMKKEFTLEF..SRDRKS....MSVYCT..PNKPSRTS.
NaK   TN.......KYQLSIHK......NANAGEP.....................
HK    LEDPRDP............................................
Ksf   ...................................................
Kec   ...................................................

Str   .beta....... coil.....helix.......coil....beta....      18
Con   ..xx........ *+*+.....++...........................
Hsc   ..RIVCV..474 KGAP..LFVLKTVEEDH....PIPED...VHENY...
Hsp   ..RITCV..472 KGAP..LWVLKTVEEDH....PIPED...VLSAY...
Hnc   ..RITCV..474 KGAP..LFVLKTVEEDH....PIPEE...VDQAY...
Hat   ........... KGAP...................................
Hld   ..KFDVT..445 KGAP..HVILQMVYNQ.....DEINDE..VVDII...
Ca    MSKMFV..514 KGAP..EGVIDRCTHIRV...GSTK....VPMTAG..
NaK   RHLLVM...506 KGAP..ERILDRCSSILIH..GKE.....QPLD....
HK    RHLLVM...517 KGAP..ERVLERCSSILIK..GQE.....LPLD....
Ksf   .........332 KITP..YQAQEQKNLA.....GVG.....LEATV...
Kec   ..RMIR...395 KGSV..DAIRRHVEAN.....GGHFP...TDVD....

Str   ...helix.........coil........beta........coil......      19
Con   ..*..x........x*.x.........*xxx...........+x.....
Hsc   .ENKVAELA.......SRGFRA......LGVAR.......KRGEG.....
Hsp   .KDKVGDLA.......SRGYRS......LGVAR.......KIEGQ....
Hnc   .KNKVAEFA.......TRGFRS......LGVARK......RGEGS....
Hat   ...................................................
Hld   ..DSLAA..........RGVRC......LSVA.......KTDQQGR....
Ca    VKQKIMSVIREW......GSGSDT.....LRCLALATH....DNPL....
NaK   .EELKDAFQNAYLEL...GGLGE....RVLGFCHLML....PDEQFPEG..
HK    .EQWREAFQTAYLSL...GGLGE....RVLGFCQLYL....NEKDYPPG..
Ksf   .EDKDVKIINEKEAKRL..GLKIDPE..RLKNYEAQ.....GNTVS.....
Kec   ..QKVDQVARQ........GATP.....LVVVE.......GSR.......

Str   ..helix...........coil.....beta........coil......helix   20
Con   .x.x.xxxx....................................*++*.xx....+*..
Hsc   HWEILGVMPCM....................534 DPPRDDT..AQTVSE
Hsp   HWEIMGIMPCS....................531 DPPRHDT..ARTISE
Hnc   .WEILGIMPCM....................534 DPPRHDT..YKTVCE
Hat   ...............................DPPRHDS........
Hld   .WHMAGILTFL....................501 DPPRPDT..KDTIRR
Ca    .RREEMHLKDSANFIKYE.TNLT..FVGCVGML..600 DPPR..IEVASSVKL
NaK   .FQFDTDDVNF......PVDNLC..FVGLISMI.591 DPPR..AAVPDAVGK
HK    .YTFDVEAMNF......PSSGLC..FAGLVSMI.602 DPPR..ATVPDAVLK
Ksf   .FLVVSDKLVAVIALG..............405 DVIKP..EAKEFIQA
Kec   .VLGVIALK..................447 DIVKGG..IKEAFAQ
```

```
Str     ........beta........coil.....helix.......coil........          21
Con     x...+..*.+x**.......***.....x*xx..x.x....*..xx.........
Hsc     ARHLG..LRVKML..558 TGDAVG..IAKETCRQL...GLGTNIYN.......
Hsp     AKRLG..LRVKML..556 TGDAVD..IAKETARQL...GMGTNIYN.......
Hnc     AKTLG..LSIKML..558 TGDAVG..IARETSRQL...GLGTNIYN.......
Hat     .......VNVKMI......TGDQLA..IAKETGRRL.................
Hld     SKEYG..VDVKML..525 TGDH..LLIAKEMCRML...DLDPNILTADK....
Ca      CRQAG..IRVIMI..624 TGDNKGT..AVAICRRI..GIFGQEED.......
NaK     CRSAG..IKVIMV..615 TGDHP..ITAKAIAK.....GVGIISEGNETVED.
HK      CRTAG..IRVIMV..626 TGDHP..ITAKAIAA.....SVGIISEGSETVED.
Ksf     IKEKN..IIPVML..429 TGDNP..KAAQAVAEYL...GINEYYGG.......
Kec     LRKMG..IKTVMI..471 TGDNR..LTAAAIAAEA...GVDDF..........

Str     ..beta............coil........helix....coil.......        22
Con     ....+................xx......x..............x......
Hsc     ..AERLGL........GGGGDMPGS....ELADFVE...NADG.......
Hsp     ..AERLGL........TGGGNMPGS....EVYDFVE...AADG.......
Hnc     ..AERLGL........GGGGDMPGS....EVYDFVE...AADG.......
Hat     ..................................................
Hld     LPQIK...........DANDLPED..LGEKYGDMML...SVGG.......
Ca      .VTAKAF.........TGR.........EFDEL.....NPSAQRDACLN.
NaK     .IAARLNI........PVSQVNPRD....ARACVVH...GSDLKDMTPEQ
HK      .IAARLRM........PVDQVNKKD....ARACVIN...GMQLKDMDPSE
Ksf     ..................................................
Kec     ..................................................

Str     .............beta...coil...beta.....coil..beta..          23
Con     .............*+.xx..*xx*..x.**..x...xx.....+*..
Hsc     .............FAEVF..PQHK..YRVVEIL..QNRGY..LVAM..
Hsp     .............FGEVF..PQHK..YAVVDIL..QQRGY..LVAM..
Hnc     .............FAEVF..PQHK..YNVVEIL..QQRGY..LVAM..
Hat     ..............................................
Hld     .............FAQVF..PEHK..FMIVETL..RQRGY..TCAM..
Ca      ...........ARCFARVE..PSHK..SKIVEFL..QSFDE..ITAM..
NaK     LDD..ILKYHTEIVFARTS..PQQK..LIIVEGC..QRQGA..IVAV..
HK      LVE..ALRTHPEMVFARTS..PQQK..LVIVESC..QRLGA..IVAV..
Ksf     .............LL.....PDDK..EAIVQRYL.DQGK...KVIM..
Kec     .............LAEAT..PEAK..LALIRQY..QAEGR..LVAM..

Str     ......coil........helix.....coil......helix........        24
Con     ....+***+*******+..+*+.****....**x....x*+x*****+....
Hsc     632 TGDGVNDAPSLK..KADTGIAVE...GATD....AARSAADIVFL...
Hsp     630 TGDGVNDAPSLK..KADTGIAVE...GATD....AARSAADIVFL...
Hnc     632 TGDGVNDAPSLK..KADTGIAVE...GSSD....AARSAADIVFL...
Hat     ....TGDGVNDAPALK..KADIGIAVA...DATD....AARGASDIVLT...
Hld     603 TGDGVNDAPALK..RADVGIAVH...GATD....AARAAADMVLT...
Ca      700 TGDGVNDAPALK..KAEIGIAM...GSGTA....VAKTASEMVLA...
NaK     713 TGDGVNDSPALK..KADIGVAM.GIAGSD....VSKQAADMILL...
HK      724 TGDGVNDSPALK..KADIGVAM.GIAGSD....AAKNAADMILL...
Ksf     474 VGDGINDAPSLA..RATIGMAI...GAGTD....IAIDSADVVLT...
Kec     516 TGDGTNDAPALA..QADVAVAM...NSGTQ....AAKEAGNMVDL...

Str     coil......helix..........coil..........stalk.helix.5       25
Con     .xx++.................................*+.+*..++.++.
Hsc     APGLS..................................AIIDALKTSRQIF.
Hsp     APGLS..................................AIIDALKTSRQIF.
Hnc     APGLG..................................AIIDALKTSRQIF.
Hat     ..................................................
Hld     EPGLS..................................VVVEAMLVSREVF.
Ca      DDNFST.................................IVAAVEEGRAIYN
NaK     DDNFAS.................................IVTGVEEGRLIFD
HK      DDNFAS.................................IVTGVEQGRLIFD
Ksf     NSDPKD.................................ILHFLELAKETRR
Kec     DSNPTK..LIEVVHIGKQMLMTR...GSLTTFSIAND....VAKYFAIIPAAFA
```

Figure 2 (continued)

```
Str     ...............coil.........transmembrane.helix.5...          26
Con     .xx..x......................+..xx..xx.x..xx.........
Hsc     HRMYSY.................692 VVYRIALSLHLEIFLGLWIAIL..
Hsp     HRMYSY.................690 VVYRIALSLHLEIFLGLWLII...
Hnc     HRMYAY.................692 VVYRIALSIHLEIFLGLWIAIL..
Hat     ..........................ITIRIVFGFMLIALIWKF......
Hld     QRMLSF.................663 LTYRISATLQLVCFFFIACFSLT.
Ca      NMKQFIRYLI.....SSNVGE...771 VVCIFLTAALGFPEALIPVQLLWV
NaK     NLKKSIAYTL.....TSNIPE...785 ITPFLIFIIANIPLPLGTVTILCI
HK      NLKKSIAYTL.....TKNIPE...796 LTPYLIYITVSVPLPLGCITILFI
Ksf     KMIQN..................534 .LWWGAGYNIIAIPLAAGIL ...
Kec     ATYPQLNALNIMCLH..SPDS...616 AILSAVIFNALIIVFLIPLAL...

Str     .external..coil.....................................          27
Con     ...................................................
Hsc     DNSLDID............................................
Hsp     RNQLLNLE...........................................
Hnc     .NRSLNIE...........................................
Hat     ...................................................
Hld     PKAYGSVDPHFQFFHLP...................................
Ca      NLVTDGLPATALGFNPPDLDIMNKPPRNPKEPLISGW..............
NaK     DLGTDMVPAISLAYEQAESDIMKRQPRNPQTDKLVNER.............
HK      ELCTDIFPSVSLAYEKAESDIMHLRPRNPRRDRLVNEP.............
Ksf     APIG...............................................
Kec     KGVSYKPLTASAMLRRN..................................

Str     ...transmembrane.helix.6..............coil........          28
Con     ......*xxxxxxx.x...xxx+............x....x..x..x.x..
Hsc     721 ..LIVFIAIFADVATLAIAY..........DNAPYSPKPVKWNLPR
Hsp     719 ..LVVFIAIFADVATLAIAY..........DNAPYSMKPVKWNLPR
Hnc     721 ..LVVFIAIFADVATLAIAY..........DNAPYSQTPVKWNLPK
Hat     .....FMVLIIAILNDGTIMTIS............................
Hld     703 VLMFMLITLLNDGCLMTIGY..........DHVIPSERPQKWNLP.
Ca      832 .LFFRYLAIGCYVGAATVGAAAWWFIAA. .DGGPRVSFYQLSHFLQ
NaK     847 ..LISMAYGQIGMIQALGGFFTYFVIMA.....................
HK      858 ..LAAYSYFQIGAIQSFAGFADYFTAMA....................
Ksf     557 LILSPAVGAVLMSLSTVVVALNALTLK.....................
Kec     654 LWIYGLGGLLVPFIGIKVIDLLLLTVCGLV..................

Str     ........................transmembrane.helix.7.....          29
Con     .........................xx..x.xxx......x.x.......
Hsc     ....................755 .LWGMSIILGIVLAIGSWITLTTMFL
Hsp     ....................753 .LWGLSTVIGIVLAIGTWITNTTMIA
Hnc     ....................755 .LWGMSVLLGVVLAVGTWITVTTMYA
Hat     ........................IFATGVVLGGYMAIMTVVFFWAAY.
Hld     ....................738 VVFVSASILAAVACGSSLMLLWIGL.
Ca      CKEDNPDFEGVDCAIFES.....896 PYPMTMALSVLVTIEMCNAL......
NaK     ...................................................
HK      ...................................................
Ksf     ...................................................
Kec     ...................................................

Str     .coil...........beta......coil....................          30
Con     ........................+..........................
Hsc     .PKGG...........IIQNF.....GAMNG...................
Hsp     QGQNRG..........IVQNF.....GVQDE...................
Hnc     QGENGG..........IVQNF.....GNMDE...................
Hat     ...................................................
Hld     EGYSSQYYENSWFHR..LGLAQL....PQGK...................
Ca      NSLSENQS.........LLRM. ....PPWEN...................
NaK     ENGFLPNH.........LLGIRV....TWDDRWINDVEDSYGQQWTYEQR
HK      QEGWFP ..........LLCVGL....RPQWEDHHLQDLQDSYGQEWTFG
Ksf     ...................................................
Kec     ...................................................
```

```
Str    .........beta......helix.....beta..........coil.        31
Con    .........x...x......xx......x.xxx..........x..x..
Hsc    .........IMFLQI.....SLTEN...WLIFITRAA.......GPFWSS
Hsp    .........VLFLEI.....SLTEN...WLIFVTRCN.......GPFWSS
Hnc    .........VLFLQI.....SLTEN...WLIFITRAN.......GPFWSS
Hat    .................................................
Hld    .........LVTMMYL...KISIS..DFLTLFSSRT......GGHFFFY
Ca     .................................................
NaK    K................IVEFTCHTA..FFVSIVVVQ...WADLVICKT
HK     QRLYQQYTCYT..VFFI.SIEMCQIAD..VLIR..............KT
Ksf    .................................................
Kec    .................................................

Str    coil..............transmembrane..helix.8....coil...      32
Con    ..xx..................+..+.....x.xx....x..........
Hsc    .IPSWQ...........825 .LAGAVFAVDIIATMFTLFGWW...SENWTD
Hsp    .IPSWQ...........825 .LSGAVLAVDILATMFCIFGWF..KGGHQTS
Hnc    .IPSWQ...........827 .LSGAIFLVDILATCFTIWGWF...EHSDTS
Hat    ...................FLLIAFWVAQLIATAIAVYGNW.........
Hld    MPPSP............822 ILFCGAIISLLVSTMAASFW...HKSRPDN.
Ca     .................930 IWLVGSICLSMSLHFLILYV.....EPLP..
NaK    RRNSVFQQGMKNK....951 ILIFGLFEETALAAFLSYCPGMGVAL.....
HK     RRLSAFQQGFFRNR...963 ILVIAIVFQVCIGCFLCYCPGMPNIF.....
Ksf    .................................................
Kec    .................................................

Str    ...................transmembrane..helix.9.......stalk    33
Con    ...................xxxx.x..x..............x
Hsc    ................852 IVTVVRVWIWSIGIFCVLGGFYY......EMSTS
Hsp    ................853 IVAVLRIWMYSFGIFCIMAGTYYIL....SESAG
Hnc    ................854 IVAVVRIWIFSFGIFCIMGGVYYIL....QDSVG
Hat    ................IGWGWAGVIWLYSIVFYFPLDIMKFAI.........
Hld    VLTEGLAWGQTNAEK..864 .LLPLWVWIYCIVWWFVQDVVKVLA...HICMDA
Ca     ................954 .LIFQITPLNVTQWLMVLKISLPVILM..DETLK
NaK    .................................................
HK     .................................................
Ksf    .................................................
Kec    .................................................

Str    .helix....coil..................helix............        34
Con    ...........................x..........x.......
Hsc    EAFDRLM...NGKP.............MKEKKSTRSVEDFMAAMQRVSTQ
Hsp    FDRMM.....NGKP.............KESRNQRSIEDLVVALQRTSTR
Hnc    FDNLMH....GKSP.............KGNQKQRSLEDFVVSLQRVSTQ
Hat    .................................................
Hld    VDLFGCV...SDASGSGPIKPYSDD..MKVNGFEPVKKPAEKS.......
Ca     FVARNYLE..PAILE..................................
NaK    ..........RMYPLKPTWWFCAFPYS..LLIFVYDEVRKLIIR......
HK     ..........NFMPIRFQWWLVPMPFG..LLIFVYDEIRKLGVR......
Ksf    .................................................
Kec    .................................................

Str    ..........coil..........helix...........coil....        35
Con    .xx..............................................
Hsc    HEKET............................................
Hsp    HEKGDA...........................................
Hnc    HEKSQ............................................
Hat    .................................................
Hld    TEKAL..NSSVSSAS...HKALEGLREDTHSPIEEA..SPVNVYVSRDQK
Ca     .................................................
NaK    .......RRPGGWVEKETYY..............................
HK     .......CCPGSWWDQELYY..............................
Ksf    .................................................
Kec    .................................................
```

Table 2 Conserved Motifs of (E-P) ATPases and their Proposed Functions[a]

Motifs	Proposed Functions
I. TGES	Phosphatase activity (E)
II. D(K,R)TGT(L,I)T	Phosphorylation and transduction (D)
III. KGAP	ATP binding and/or kinase activity (K)
(only eukaryotes)	
IV. DPPR	ATP binding (D)
(only eukaryotes)	
V. M(L,I,V)TGD	ATP binding (D)
VI. GDGXND(A,S)P(A,S)LK	ATP binding (two D and K)
(K only in eukaryotes)	

[a] See text for references. X: any amino acid

helixes for all the eukaryotic ATPases must be modified in light of recent experimental evidence. In the Na^+, K^+-ATPase, studies on the release of exposed peptides by trypsin (94, 97) and on the sidedness of binding of antibodies with defined epitopes (96) provide evidence for the seven hydrophobic stretches indicated in Figure 2. Such detailed studies have not yet been performed with other ATPases.

In the case of the yeast enzyme, circumstantial evidence also favors the present model. There is a potential glycosylation site (NWT) after transmembrane helix 8 (Figure 2), but the enzyme has been shown not to be glycosylated, in spite of following the secretory pathway during its biosynthesis (60). In the previous model with only eight transmembrane stretches this site was facing the external side of the membrane and therefore should have been glycosylated. The present model, with nine transmembrane helixes, envisages this site facing the cytosol, thus explaining its lack of glycosylation. In addition, proteolysis by carboxypeptidase Y of membrane-bound ATPase in the presence and absence of detergent suggests that the C-terminus is on the external side of the plasma membrane (27). However, more definitive evidence is needed to confirm the predicted tramsmembrane helixes in these enzymes.

Functional Domains

The predicted structure of (E-P) ATPases described here and in previous publications (12, 93) allows the conceptual division of the enzymes into functional domains. My specific proposal for H^+-ATPases is shown in Figure 3. Most of the hydrophilic regions of the enzyme are predicted to protrude in the cytoplasmic side of the membrane, with little of the polypeptide chain exposed to the external side. This agrees with the low-resolution (2.5 nm) structures of the Ca^{2+}-ATPase (147) and Na^+, K^+-ATPase (84) determined by image reconstruction of two-dimensional membrane crystals.

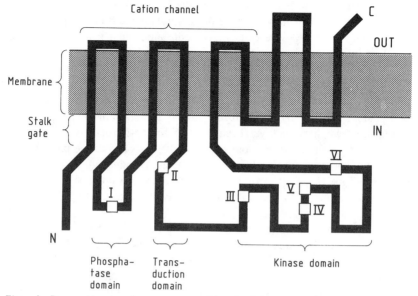

Figure 3   Proposed transmembrane structure and functional domains of H$^+$-ATPases. I to VI are the conserved motifs of Table 2.

The N-terminal domain is the most variable in length and sequence, and it is predicted to consist exclusively of loops and alpha-helixes (between 1 and 6, depending on the enzyme). There are no clues about the function of this domain. In the Na$^+$, K$^+$-ATPase (64) there is a trypsin cutting site affected by the conformational changes experienced by the enzyme during the catalytic cycle (see below). In the yeast H$^+$-ATPase a mutation causing thermolability maps to this domain (24). Therefore, this domain may be structurally important, although its variability and almost absence in the bacterial ATPases suggest that it is not involved in catalysis.

The next cytoplasmic domain is predicted to contain mostly strands of beta-sheet and includes the first highly conserved motif of (E-P) ATPases (TGES, Table 2). Both in the Na$^+$, K$^+$-ATPase (64) and in the Ca^{2+}-ATPase (5) there is a conformation-sensitive tryptic split in this domain. In addition, two mutations of the yeast H$^+$-ATPase that cause thermolability map to this domain (24). Although previous models considered this to be a "transduction domain" (12), the evidence for its participation in coupling ATP hydrolysis to calcium transport has recently been disputed (D. MacLennan, personal communication). A new suggestion has recently been made on the basis of site-directed mutagenesis of the TGES motif (101). Replacement of the glutamate by glutamine does not affect the formation of phosphorylated intermediate but inhibits its hydrolysis. Therefore this motif may be the active site of a phosphatase activity located in the beta-domain. The existence of separate sites for the catalysis of the phosphorylation and dephosphorylation steps of the catalytic cycle could have been anticipated from the fact that covalent inhibitors that probably act on the ATP binding site of the Na^+, K^+-ATPase, such as fluorescein isothiocyanate (66) and 5'-(p-fluorosulfonyl)benzoyladenosine (89) inhibit ATP hydrolysis and the formation of the phosphorylated intermediate but do not affect the hydrolysis of p-nitrophenylphosphate catalyzed by this enzyme in the presence of potassium. Unfortunately, we have not yet been able to find p-nitrophenylphosphatase activity with purified preparations of the yeast H^+-ATPase. the phosphatase inhibitor vanadate should bind to this phosphatase domain and, accordingly, a mutation of the H^+-ATPase from *Schizosaccharomyces pombe* that results in decreased sensitivity to vanadate maps relatively close to the TGES motif (42).

The phosphorylated intermediate is formed in the aspartate of conserved motif II, which is in a small domain separated from the rest of the large central region by tryptic sites, in both the Na^+, K^+-ATPase (64) and Ca^{2+}-ATPase (5). Site-directed mutagenesis of the phosphorylated aspartate to a nonphosphorylated threonine in both the yeast H^+-ATPase (101) and the Ca^{2+}-ATPase (80) inactivates cation transport. This demonstrates the essential role of the phosphorylated intermediate that defines the (E-P) ATPase family. However, the yeast mutant seems to be much more defective in proton transport than in ATP hydrolysis, suggesting that the phosphorylated intermediate may not be essential for catalysis but only for the coupling of ATP hydrolysis to cation transport. The same "uncoupling" between catalysis and transport is observed in the Asp–Glu mutation of the yeast enzyme. Although this mutation does form a phosphorylated intermediate, it is also more impaired in proton transport than in ATP hydrolysis (101). It seems that the formation of the proper intermediate is crucial for coupling. On the basis of these results and the proposed mechanism of the ATPases described below, I consider this phosphorylation domain as the transduction domain of cation pumps.

The ATP binding site and the catalysis of the formation of the phosphorylated intermediate are proposed to occur in the rest of the central region, which includes conserved motifs III, IV, V, and VI. As previously discussed, there is biochemical evidence for the participation of domains III and VI in the catalytic site. Site-directed mutagenesis of the aspartates of motifs IV and V and of the second aspartate of motif VI to asparagine residues (101) results in low-activity enzymes with relaxed nucleotide specificity. It can be speculated that these three aspartates form part of the adenine binding pocket. On the other hand, the first aspartate and last lysine of motif VI would bind the gamma- and alpha-phosphate groups of ATP, in the case of the aspartate probably through a magnesium bridge. In the model of Figure 4, two lysines are present in the catalytic site, but the ATP analogue 5'-(p-fluorosulfonyl)benzoyladenosine only modifies that of motif VI, suggesting that the KGAP lysine does not participate in adenosine binding. This lysine is modified by fluorescein isothiocyanate, which cannot be considered as an ATP analogue. However, as ATP protects against inhibition, I have assumed in Figure 4 that the KGAP motif is close to the phosphates, where it may participate in binding or have a catalytic role. The Lys–Gln mutant obtained in the yeast H^+-ATPase is completely inactive (101). On the other hand, the same mutant in the case of the Ca^{2+}-ATPase (80) retains 25% of wild-type activity. Further studies are needed to clarify the role of this lysine.

The model of Figure 4 is compatible with the known structures of nucleotide binding sites in dehydrogenases (116), phosphoglycerate kinase (7),

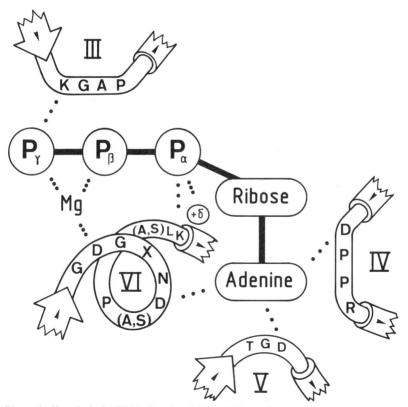

Figure 4   Hypothetical ATP binding site of (E-P) ATPases. Roman numerals III to VI correspond to the conserved motifs of Table 2. Arrows and cylinders represent strands of beta-sheets and alpha-helixes, respectively.

phosphofructokinase (38), elongation factor (69), and the Ras oncogene protein (30). Nucleotide binding sites are generally formed by four surface (polar) loops. Two of them make the purine binding pocket and the other two bind the phosphates. One of the latter ones is rich in glycines to facilitate the proximity of the alpha-phosphate of the nucleotide to the beginning of an alpha-helix, whose positive dipole contributes to phosphate binding. This glycine-rich loop has been proposed to contain the catalytic machinery for phosphate transfer (30). In many nucleotide binding proteins it shows some sequence similarity, basically the degenerated motif GXXXXG followed in many cases by a K (154).

However, even this permissive motif cannot be found in many ATP binding enzymes such as hexokinase, pyruvate kinase, phosphoglycerate kinase, phosphofructokinase, and (E-P) ATPases (128). This suggests that this motif cannot be taken as evidence for nucleotide binding and that it merely reflects one common form the glycine-rich loop may take, not the only possible one. The other phosphate-binding loop contains an aspartate residue which binds to the terminal phosphates of the nucleotide through a magnesium bridge. It has also been noted (154) that this kind of loop shows sequence similarity in nucleotide binding proteins, basically four hydrophobic amino acids followed by an aspartate. However, here too the rule is broken in many kinases, the aspartate being separated from the hydrophobic residues by several amino acids (128). This situation again implies that there are no strict sequence requirements for nucleotide binding and that the small degree of similarity between different nucleotide binding sequences does not support divergence from a common ancestral protein. More likely, convergent evolution of these loops to accomplish nucleotide binding has occurred. In our model of the ATP binding site of (E-P) ATPases (Figure 4) the large loop of motif VI would be equivalent to the two phosphate-binding loops discussed above, with the first aspartate binding to the magnesium connected to the terminal phosphates of the nucleotide and a lysine and alfa-helix close to the alpha-phosphate.

The transmembrane cation channel is proposed to consist of the first six transmembrane stretches (see above). Model studies with synthetic amphiphilic peptides suggest that four (90) to six (70) transmembrane helixs are required to make a cation channel. The highly polar ATPase residues predicted to be buried inside the membrane are shown in Figure 5. In the *Neurospora crassa* ATPase, the glutamate residue of transmembrane helix 1 has been found to be modified by the hydrophobic ATPase inhibitor di-cyclohexylcarbodiimide. This prompted the suggestion that it could be equivalent to the transmembrane carboxylate of $(F/_0 F_1)$ ATPases, which seems to participate in proton transport (143). However, the equivalent residue in the yeast ATPase has been mutagenized to either glutamine or leucine without any detectable effect on either ATPase activity, proton transport, or di-cyclohexylcarbodiimide binding (101). Actually, the position of this glutamate is not fully conserved in H⁺-ATPases because in *Leishmania* the enzyme is located much closer to the surface of the membrane. The four polar residues fully conserved in all H⁺-ATPases are the most likely candidates to participate in proton transport. A transmembrane asparagine in helix 2, a transmembrane aspartate in helix 6, an external aspartate in helix 2, and a cytoplasmic arginine in helix 5 are the candidates, and site-directed mutagenesis could test their role.

There are no transmembrane polar residues conserved in all (E-P) ATPases.

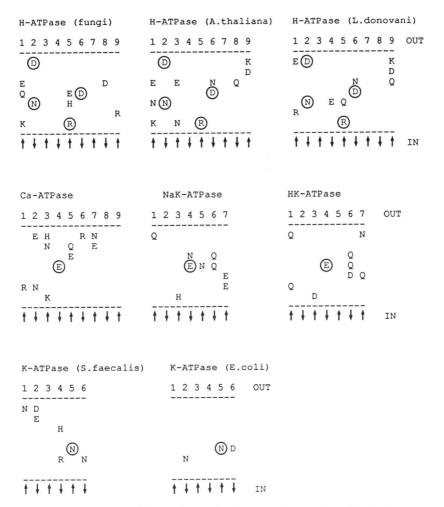

Figure 5 Highly polar residues in the postulated transmembrane helixes of (E-P) ATPases. Circled residues are conserved within subgroups of ATPases (H⁺-ATPases, animal ATPases, bacterial ATPases).

However, a glutamate is conserved in transmembrane helix 4 of all animal ATPases (Ca^{2+}, Na^+-K^+, and H^+-K^+), and an asparagine is conserved in transmembrane helix 5 of bacterial K^+-ATPases. Apparently, the cation channel does not require specific polar amino acids and the conservation of some of them mostly reflects evolutionary proximity, not mechanistic requirements.

The predicted transmembrane helixes 1–5 are extended into the cytoplasm by polar helixes which could constitute the stalk attaching the cytoplasmic domains to the membrane. This stalk could contain the cation binding site, which can be considered as a gate controlling access to the channel.

The carboxy-terminal part of (E-P) ATPases is as variable in sequence and length as the N-terminal domain. It is nonexistent in the bacterial ATPases and may contain one or three transmembrane helixes in the eukaryotic enzymes. Again, there are no clues about the function of this part of the ATPases.

Coupling Mechanism

As discussed elsewhere (128), the most plausible coupling mechanism between ATP hydrolysis and cation transport is that depicted in Figure 6. The enzymes exist in two conformations, differing in catalytic and transport properties. In conformation 1 the transport site is facing the cytoplasmic side and has high affinity. In conformation 2 it is externally oriented and has low affinity. The enzyme is forced to alternate between these two conformations and to bind and release the transported cation, because neither conformation alone can effect the complete catalytic cycle. E_1 is active as a kinase after binding the transported cation, and then it catalyzes the formation of the phosphorylated intermediate. However, it has no hydrolase activity. On the other hand, E_2 has phosphatase activity after releasing the transported cation but is inactive as a kinase. As discussed above, it is likely that the phosphorylation and hydrolytic steps of the reaction cycle are catalyzed by different functional domains. Therefore, the aspartate forming the intermediate must move between the two catalytic domains.

It is assumed that conserved motif II, where this aspartate is located, controls the sidedness and the affinity of the cation-binding site in the channel gate. The movement of motif II between the kinase and phosphatase domains, its effects on the channel gate, and the active/inactive transitions of the kinase and phosphatase domains would depend on the phosphorylation and cation-binding states of the enzyme. It has recently been suggested (146) that cation binding to the ATPases may have an "induced fit" effect on the enzyme-ATP complex, in the same way that glucose binding induces an active conformation of the hexokinase-ATP complex (141). In the later case, crystallographic studies have shown that glucose induces a hinge-closing transition essential

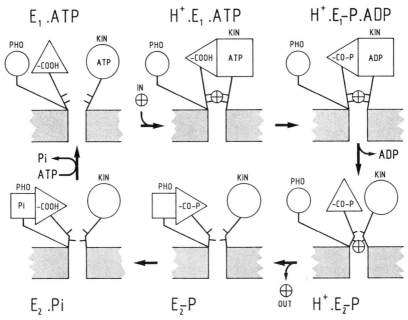

Figure 6 Scheme for the proposed mechanism of H$^+$-ATPases. KIN and PHO are the proposed kinase and phosphatase domains of Figure 3 and the square and circle represent active and inactive conformations, respectively. The triangle is the conserved motif II forming the phosphorylated intermediate. The binding of phosphate to its acyl group is indicated.

for catalysis. In the cation pumps, binding of the transported cation in E_1 may induce both the active conformation of the kinase site and the movement of motif II to approach it. The phosphorylation of motif II would be the trigger for the concerted conformational change affecting the catalytic domains and the channel gate. The release of the cation from E_2 would induce (again as a manifestation of "induced fit") the active conformation of the phosphatase site and the movement of motif II to approach it. Dephosphorylation would again trigger the concerted transition of transport and catalytic sites.

There are few clues to the nature of these crucial conformational changes. Circular dichroism studies indicate that there is no significant change in secondary structure during the E_1/E_2 transition (54, 55, 86). On the other hand, a big conformational change is indicated by changes in accessibility of proteolytic sites, intrinsic tryptophan fluorescence, reactivity of sulfhydryl groups, pKa of acidic groups, labeling by hydrophobic reagents, and the formation of two-dimensional membrane crystals (63). The elucidation of the coupling mechanism of ATPases at the molecular level remains as one of most challenging problems of molecular biology.

SUMMARY AND PERSPECTIVES

After decades of physiological studies, the plant proton pump is beginning to be studied at the molecular and genetic levels. The availability of the cloned gene (see Footnote 1) has opened the way for the in vitro mutagenesis of the ATPase and the analysis of altered pumps in a suitable expression system such as yeast. In addition, studies of yeast mutants deficient in ATPase activity are illuminating the physiological role of the enzyme. A similar approach may be possible for higher plants in the near future. Finally, the study of protein kinases, which may be involved in the regulation of the pump by external factors, is expected to clarify a novel and important regulatory circuit in eukaryotic cells.

Literature Cited

1. Abdel-Ghany, M., Raden, D., Racker, E., Katchalski-Katzir, E. 1988. Phosphorylation of synthetic random polypeptides by protein kinase P and other protein-serine (threonine) kinases and stimulation or inhibition of kinase activities by microbial toxins. *Proc. Natl. Acad. Sci. USA* 85:1408–11
2. Addison, R. 1986. Primary structure of the *Neurospora* plasma membrane H⁺-ATPase deduced from the gene sequence. Homology to Na⁺-/K⁺, Ca²⁺ and K⁺-ATPases. *J. Biol. Chem.* 261:14896–901
3. Aerts, R. J., De Wit, R. J. W., Campagne, M. M. V. L. 1987. Cyclic AMP induces a transient alkalinization in *Dictyostelium*. *FEBS Lett.* 220:366–70
4. Aerts, R. J., Durston, A. J., Moolenaar, W. H. 1985. Cytoplasmic pH and the regulation of the *Dictyostelium* cell cycle. *Cell* 43:653–57
5. Andersen, J. P., Vilsen, B., Collins, J. H., Jorgensen, P. L. 1986. Localization of E_1-E_2 conformational transitions of sarcoplasmic reticulum Ca-ATPase by tryptic cleavage and hydrophobic labeling. *J. Membr. Biol.* 93:85–92
6. Assmann, S. M., Simoncini, L., Schroeder, J. I. 1985. Blue light activates electrogenic ion pumping in guard cell protoplasts of *Vicia faba*. *Nature* 318:285–87
7. Banks, R. D., Blake, C. C. F., Evans, P. R., Haser, R., Rice, D. W., et al. 1979. Sequence, structure and activity of phosphoglycerate kinase: a possible hinge-bending enzyme. *Nature* 279:773–77
8. Bidwai, A. P., Takemoto, J. Y. 1987. Bacterial phytotoxin, syringomycin, in-

duces a protein kinase-mediated phosphorylation of red beet plasma-membrane polypeptides. *Proc. Natl. Acad. Sci. USA* 84:6755–59
9. Bidwai, A. P., Zhang, L., Bachmann, R. C., Takemoto, J. Y. 1987. Mechanism of action of *Pseudomonas syringae* phytotoxin syringomycin. Stimulation of red beet plasma-membrane ATPase activity. *Plant Physiol.* 83:39–43
10. Borst-Pauwels, G. W. F. H., Theuvenet, A. P. R., Stols, A. L. H. 1983. All-or-none interactions of inhibitors of the plasma membrane ATPase with *Saccharomyces cerevisiae*. *Biochim. Biophys. Acta* 732:186–92
11. Boyer, P. D. 1988. Bioenergetic coupling to protonmotive force: Should we be considering hydronium ion coordination and not group protonation? *Trends Biochem. Sci.* 13:5–7
12. Brandl, C. J., Green, N. M., Korczak, B., MacLennan, D. H. 1986. Two Ca²⁺ ATPase genes: homologies and mechanistic implications of deduced amino acid sequences. *Cell* 44:597–607
13. Broek, D., Toda, T., Michaeli, T., Levin, L., Birchmeier, C., et al. 1987. The *Saccharomyces cerevisiae* CDC25 gene product regulates the RAS/adenylate cyclase pathway. *Cell* 48:789–99
14. Brummell, D. A., Hall, J. L. 1987. Rapid cellular responses to auxin and the regulation of growth. *Plant Cell Environ.* 10:523–43
15. Brummer, B., Bertl, A., Potrykus, I., Felle, H., Parish, R. W. 1985. Evidence that fusicoccin and indole-3-acetic acid

induce cytosolic acidification of *Zea mays* cells. *FEBS Lett.* 189:109–14

16. Brummer, B., Felle, H., Parish, R. W. 1984. Evidence that acid solutions induce plant cell elongation by acidifying the cytosol and stimulating the proton pump. *FEBS Lett.* 174:223–27

17. Brummer, B., Parish, R. W. 1983. Mechanism of auxin-induced plant cell elongation. *FEBS Lett.* 161:9–13

18. Busa, W. B. 1982. Cellular dormancy and the scope of pH_i-mediated metabolic regulation. In *Intracellular pH: Its Measurement, Regulation and Utilization in Cellular Functions*, ed. R. Nuccitelli, D. W. Deamer, pp. 417–26. New York: Liss

19. Caspari, G., Tortora, P., Hanozet, G., Guerritore, A. 1985. Glucose-stimulated cAMP increase may be mediated by intracellular acidification in *Saccharomyces cerevisiae*. *FEBS Lett.* 186:75–79

20. Cerana, R., Bonetti, A., Marre, M. T., Romani, G., Lado, P., Marre, E. 1983. Effects of a brassinosteroid on growth and electrogenic proton extrusion in *Azuki* bean epicotyls. *Physiol. Plant.* 59:23–28

21. Chiesi, M., Inesi, G. 1980. ATP-dependent fluxes of manganese and hydrogen ions in sarcoplasmic reticulum. *Biochemistry* 19:2912–18

22. Chou, P. Y., Fasman, G. D. 1978. Empirical predictions of protein conformation. *Ann. Rev. Biochem.* 47:251–76

23. Cid, A., Perona, R., Serrano, R. 1987. Replacement of the promoter of the yeast plasma membrane ATPase gene by a galactose-dependent promoter and its physiological consequences. *Curr. Genet.* 12:105–10

24. Cid, A., Serrano, R. 1988. Mutations of the yeast plasma membrane H^+-ATPase which cause thermosensitivity and altered regulation of the enzyme. *J. Biol. Chem.* In press

25. Colman, R. F. 1983. Affinity labeling of purine nucleotide sites in proteins. *Ann. Rev. Biochem.* 52:67–91

26. Crane, F. L., Sun, I. L., Clark, M. G., Grebing, C., Low, H. 1985. Transplasma membrane redox systems in growth and development. *Biochim. Biophys. Acta* 811:233–65

27. Davis, C. B., Hammes, G. G. 1989. Topology of the yeast plasma membrane proton-translocating. ATPase. *J. Biol. Chem.* In press

28. Davis, D. D. 1973. Control of and by pH. *Symp. Soc. Exp. Biol.* 27:513–33

29. Devereux, J., Haerberli, P., Smithies,

O. 1984. A comprehensive set of sequence analysis programs for the *VAX*. *Nucleic Acids Res.* 12:387–95

30. De Vos, A. M., Tong, L., Milburn, M. V., Matias, P. M., Jancarik, J., et al. 1988. Three-dimensional structure of an oncogene protein: catalytic domain of human c-H-ras p21. *Science* 239:888–93

31. Engelman, D. M., Steitz, T. A., Goldman, A. 1986. Identifying nonpolar transbilayer helices in amino acid sequences of membrane proteins. *Annu. Rev. Biophys. Biophys. Chem.* 15:321–53

32. Enzyme Nomenclature Committee, Union International of Biochemistry. 1984. *Enzyme Nomenclature*. New York: Academic

33. Epel, D., Dube, F. 1987. Intracellular pH and cell proliferation. In *Control of Animal Cell Proliferation*, ed. A. L. Boynton, H. L. Leffert, 2:363–93. New York: Academic

34. Eraso, P., Cid, A., Serrano, R. 1987. Tight control of the amount of yeast plasma membrane ATPase during changes in growth conditions and gene dosage. *FEBS Lett.* 224:193–97

35. Eraso, P., Gancedo, C. 1987. Activation of yeast plasma membrane ATPase by acid pH during growth. *FEBS Lett.* 224:187–92

36. Ettlinger, C., Lehle, L. 1988. Auxin induces rapid changes in phosphatidylinositol metabolites. *Nature* 331:176–78

37. Evans, M. L. 1985. The action of auxin on plant cell elongation. *CRC Crit. Rev. Plant Sci.* 2:317–65

38. Evans, P. R., Farrants, G. W., Hudson, P. J. 1981. Phosphofructokinase: structure and control. *Philos. Trans. R. Soc. London Ser. B* 293:53–62

39. Felle, H., Brummer, B., Bertl, A., Parish, R. W. 1986. Indole-3-acetic acid and fusicoccin cause cytosolic acidification of corn coleoptile cells. *Proc. Natl. Acad. Sci. USA* 83:8992–95

40. Gabathuler, R., Cleland, R. E. 1985. Auxin regulation of a proton translocating ATPase in pea root plasma membrane vesicles. *Plant Physiol.* 79:1080–85

41. Garnier, J., Osguthorpe, D. J., Robson, B. 1978. Analysis of the accuracy and implications of simple methods for predicting the secondary structure of globular proteins. *J. Mol. Biol.* 120:97–120

42. Ghislain, M., Schlesser, A., Goffeau, A. 1987. Mutation of a conserved glycine residue modifies the vanadate sensitivity of the plasma membrane H^+-ATPase from *Schizosaccharomyces pombe*. *J. Biol. Chem.* 262:17549–55

43. Giannini, J. L., Ruiz-Cristin, J., Briskin, D. P. 1987. Calcium transport in sealed vesicles from red beet (*Beta vulgaris* L.) storage tissue. II. Characterization of $^{45}Ca^{2+}$ uptake into plasma membrane vesicles. *Plant Physiol.* 85: 1137–42

44. Gillies, R. J. 1982. Intracellular pH and proliferation in yeast, *Tetrahymena* and sea urchin eggs. In *Intracellular pH: Its Measurement, Regulation and Utilization in Cellular Functions*, ed. R. Nuccitelli, D. W. Deamer, pp. 341–59. New York: Liss

45. Gross, J. D., Bradbury, J., Kay, R. R., Peacey, M. J. 1983. Intracellular pH and the control of cell differentiation in *Dictyostelium discoideum*. *Nature* 303: 244–45

46. Gustin, M. C., Martinac, B., Saimi, Y., Culbertson, M. R., Kung, C. 1986. Ion channels in yeast. *Science* 233:1195–97

47. Hager, A., Menzel, H., Krauss, A. 1971. Versuch und Hypothese zur Primärwirkung des Auxins beim Streckungswachstum. *Planta* 100:47–75

48. Hager, A., Moser, I. 1985. Acetic acid esters and permeable weak acids induce active proton extrusion and extension growth of coleoptile segments by lowering the cytoplasmic pH. *Planta* 163: 391–400

49. Hager, K. M., Mandala, S. M., Davenport, J. W., Speicher, D. W., Benz, E. J., Slayman, C. W. 1986. Amino acid sequence of the plasma membrane H^+-ATPase of *Neurospora crassa*, deduced from its genomic and cDNA sequences. *Proc. Natl. Acad. Sci. USA* 83:7693–97

50. Hanes, S. D., Bostian, K. A. 1987. Control of cell growth and division in *Saccharomyces cerevisiae*. *CRC Crit. Rev. Biochem.* 21:153–223

51. Hanson, J. B., Trewavas, A. J. 1982. Regulation of plant cell growth: the changing perspective. *New Phytol.* 90:1–18

52. Hara, Y., Nakao, M. 1986. ATP-dependent H^+ uptake by proteoliposomes reconstituted with purified Na^+, K^+-ATPase. *J. Biol. Chem.* 261:12655–58

53. Harold, F. M. 1986. *The Vital Force: A Study of Bioenergetics*, pp. 507–18. New York: Freeman

54. Hastings, D. F., Reynold, J. A., Tanford, C. 1986. Circular dichroism of the two conformational states of mammalian $(Na^+ + K^+)$ ATPase. *Biochim. Biophys. Acta* 860:566–69

55. Hennessey, J., Scarborough, G. A. 1988. Secondary structure of the *Neurospora crassa* plasma membrane H^+-ATPase as estimated by circular dichroism. *J. Biol. Chem.* 263:3123–30

56. Hepler, P. K., Wayne, R. O. 1985. Calcium and plant development. *Annu. Rev. Plant Physiol.* 36:397–439

57. Hesse, J. E., Wieczorek, L., Altendorf, K., Reicin, A. S., Dorus, E., Epstein, W. 1984. Sequence homology between two membrane transport ATPases, the Kdp-ATPase of *Escherichia coli* and the Ca^{2+}-ATPase of sarcoplasmic reticulum. *Proc. Natl. Acad. Sci. USA* 81: 4746–50

58. Hodges, T. K. 1976. ATPases associated with membranes of plant cells. In *Encyclopedia of Plant Physiology*, ed. U. Luttge, M. G. Pitman, 2 (Pt. A): 260–83. Berlin: Springer-Verlag

59. Hodges, T. K., Leonard, R. T., Bracker, C. E., Keenan, T. W. 1972. Purification of an ion-stimulated ATPase from plant roots: association with plasma membranes. *Proc. Natl. Acad. Sci. USA* 69:3307–11

60. Holcomb, C. L., Hansen, W. J., Etcheverry, T., Schekman, R. 1988. Secretory vesicles externalize the major plasma membrane ATPase in yeast. *J. Cell Biol.* 106:641–48

61. Hunter, T. 1987. A thousand and one protein kinases. *Cell* 50: 823–29

62. Jaffe, L. F. 1981. The role of ionic currents in establishing developmental patterns. *Philos. Trans. R. Soc. London Ser. B* 295:553–66

63. Jorgensen, P. L. 1982. Mechanism of the Na^+, K^+ pump. Protein structure and conformations of the pure $(Na^+ + K^+)$-ATPase. *Biochim. Biophys. Acta* 694:27–68

64. Jorgensen, P. L., Collins, J. H. 1986. Tryptic and chymotryptic cleavage sites in sequence of alfa-subunit from outer medulla of mammalian kidney. *Biochim. Biophys. Acta* 860:570–76

65. Kabsch, W., Sander, C. 1983. How good are predictions of protein secondary structure? *FEBS Lett.* 155:179–82

66. Karlish, S. J. D. 1980. Characterization of conformational changes in (Na,K) ATPase labeled with fluorescein at the active site. *J. Bioenerg. Biomembr.* 12:111–36

67. Kawakami, K., Noguchi, S., Noda, M., Takahashi, H., Ohta, T., et al. 1985. Primary structure of the alfa-subunit of *Torpedo californica* $(Na^+ + K^+)$ ATPase deduced from cDNA sequence. *Nature* 316:733–36

68. Kyte, J., Doolittle, R. F. 1982. A simple method for displaying the hydropathic character of a protein. *J. Mol. Biol.* 157:105–32

69. La Cour, T. F., Nyborg, J., Thirup, S., Clark, B. F. C. 1985. Structural details of the binding of guanosine diphosphate to elongation factor Tu from *E. coli* as studied by X-ray crystallography. *EMBO J.* 4:2385–88

70. Lear, J. D., Wasserman, Z. R., De-Grado, W. F. 1988. Synthetic amphiphilic peptide models for protein ion channels. *Science* 240:1177–81

71. Leonard, R. T. 1983. Potassium transport and the plasma membrane ATPase in plants. In *Metals and Micronutrients: Uptake and Utilization by Plants*, ed. D. A. Robb, W. S. Pierpoint, pp. 71–86. New York: Academic

72. Lesk, A. M., Chothia, C. H. 1986. The response of protein structures to amino-acid sequence changes. *Philos. Trans. R. Soc. London Ser. A* 317:345–56

73. Luttge, U., Higinbotham, N. 1979. *Transport in Plants*. Berlin: Springer-Verlag

74. MacLennan, D. H., Brandl, C. J., Korczak, B., Green, N. M. 1985. Amino-acid sequence of a $Ca^{2+}+Mg^{2+}$-dependent ATPase from rabbit muscle sarcoplasmic reticulum, deduced from its complementary DNA sequence. *Nature* 316:696–700

75. Madeira, V. M. C. 1978. Proton gradient formation during transport of Ca^{2+} by sarcoplasmic reticulum. *Arch. Biochem. Biophys.* 185:316–25

76. Madeira, V. M. C. 1980. Proton movements across the membranes of sarcoplasmic reticulum during the uptake of calcium ions. *Arch. Biochem. Biophys.* 200:319–25

77. Marre, E. 1979. Integration of solute transport in cereals. In *Recent Advances in the Biochemistry of Cereals*, ed. D. L. Laidman, R. G. Wyn Jones, pp. 3–25. New York: Academic

78. Marre, M. T., Moroni, A., Albergoni, F., Marre, E. 1988. Plasmalemma redox activity and H^+ extrusion. I. Activation of the H^+-pump by ferricyanide-induced potential and cytoplasmic acidification. *Plant Physiol.* 87:25–29

79. Marre, M. T., Romani, G., Bellando, M., Marre, E. 1986. Stimulation of weak acid uptake and increase in cell sap pH as evidence for fusicoccin- and K^+-induced cytosol alkalinization. *Plant Physiol.* 82:316–23

80. Maruyama, K., MacLennan, D. H. 1988. Mutation of aspartic acid-351, lysine-352 and lysine-515 alters the Ca^{2+} transport activity of the Ca^{2+}-ATPase expressed in COS-1 cells. *Proc. Natl. Acad. Sci. USA* 85:3314–18

81. McDonough, J. P., Mahler, H. P. 1982. Covalent phosphorylation of the Mg-dependent ATPase of yeast plasma membranes. *J. Biol. Chem.* 257:14579–81

82. Meade, J. C., Shaw, J., Gallagher, G., Lemaster, S., Stringer, J. R. 1987. Structure and expression of a tandem gene pair in *Leishmania donovani* that encodes a protein structurally homologous to eukaryotic cation-transporting ATPases. *Mol. Cell. Biol.* 7:3937–46

83. Meyerowitz, E. M. 1987. *Arabidopsis thaliana. Annu. Rev. Genet.* 21:93–111

84. Mohraz, M., Simpson, M. V., Smith, P. R. 1987. The three-dimensional structure of the Na, K-ATPase from electron microscopy. *J. Cell Biol.* 105:1–8

85. Morse, M. J., Crain, R. C., Satter, R. L. 1987. Light-stimulated inositolphospholipid turnover in *Samanea saman* leaf pulvini. *Proc. Natl. Acad. Sci. USA* 84:7075–78

86. Nakamoto, R. K., Inesi, G. 1986. Retention of ellipticity between enzymatic states of the Ca^{2+}-ATPase of sarcoplasmic reticulum. *FEBS Lett.* 194:258–62

87. Niggli, V., Sigel, E., Carafoli, E. 1982. The purified Ca^{2+} pump of human erythrocyte membranes catalyzes an electroneutral Ca^{2+}-H^+ exchange in reconstituted liposomal systems. *J. Biol. Chem.* 257:2350–56

88. Nishizuka, Y. 1986. Studies and perspectives of protein kinase C. *Science* 233:305–12

89. Ohta, T., Nagano, K., Yoshida, M. 1986. The active site structure of Na^+/K^+-transporting ATPase: location of the 5' - (p - fluorosulfonyl)benzoyladenosine binding site and soluble peptides released by trypsin. *Proc. Natl. Acad. Sci. USA* 83:2071–75

90. Oiki, S., Dauho, W., Montal, M. 1988. Channel protein engineering: synthetic 22-mer peptide from the primary structure of the voltage-sensitive sodium channel forms ionic channels in lipid bilayers. *Proc. Natl. Acad. Sci. USA* 85:2393–97

91. Olah, Z., Kiss, Z. 1986. Occurrence of lipid and phorbol ester activated protein kinase in wheat cells. *FEBS Lett.* 195:33–37

92. Ovchinnikov, Y. A. 1987. Probing the folding of membrane proteins. *Trends Biochem. Sci.* 12:434–38

93. Ovchinnikov, Y. A., Arystarkhova, E. A., Arzamazova, N. M., Dzhandzhugazyan, K. N., Efremov, R. G., Nabiev, I. R., Modyanov, N. N. 1988. Dif-

ferentiated analysis of the secondary
structure of hydrophilic and hydrophobic
regions in alfa- and beta-subunits of
Na$^+$, K$^+$-ATPase by Raman spectroscopy. FEBS Lett. 227:235–39

94. Ovchinnikov, Y. A., Arzamazova, N.
M., Arystarkhova, E. A., Gevondyan,
N. M., Aldanova, N. A., Modyanov, N.
N. 1987. Detailed structural analysis of
exposed domains of membrane-bound
Na$^+$, K$^+$-ATPase. A model of trans-
membrane arrangement. FEBS Lett.
217:269–74

95. Ovchinnikov, Y. A., Dzhandzugazyan,
K. N., Lutsenko, S. V., Mustayev, A.
A., Modyanov, N. N. 1987. Affinity
modification of E$_1$-form of Na$^+$, K$^+$-
ATPase revealed Asp-710 in the catalyt-
ic site. FEBS Lett. 217:111–16

96. Ovchinnikov, Y. A., Luneva, N. M.,
Arystarkhova, E. A., Gevondyan, N.
M., Arzamazova, A. T., et al. 1988.
Topology of Na$^+$, K$^+$-ATPase.
Identification of the extra- and in-
tracellular hydrophilic loops of the
catalytic subunit by specific antibodies.
FEBS Lett. 227:230–34

97. Ovchinnikov, Y. A., Modyanov, N. N.,
Broude, N. E., Petrukhin, K. E.,
Grishin, A. V., et al. 1986. Pig kidney
Na$^+$, K$^+$-ATPase. Primary structure and
spatial organization. FEBS Lett.
201:237–45

98. Deleted in proof

99. Poole, R. J. 1978. Energy coupling for
membrane transport. Annu. Rev. Plant
Physiol. 29:437–60

100. Portillo, F., Mazon, M. J. 1986. The
Saccharomyces cerevisiae start mutant
carrying the cdc25 mutation is defective
in activation of plasma membrane
ATPase by glucose. J. Bacteriol.
168:1254–57

101. Portillo, F., Serrano, R. 1988. Dissection of functional domains of the yeast
proton-pumping ATPase by directed
mutagenesis. EMBO J. 7:1793–98

102. Purwin, C., Nicolay, K., Scheffers, W.
A., Holzer, H. 1986. Mechanism of
control of adenylate cyclase activity in
yeast by fermentable sugars and car-
bonyl cyanide m-chlorophenylhydra-
zone. J. Biol. Chem. 261:8744–49

103. Deleted in proof

104. Ranjeva, R., Boudet, A. M. 1987.
Phosphorylation of proteins in plants:
regulatory effects and potential involvement in stimulus/response coupling.
Annu. Rev. Plant Physiol. 38:73–93

105. Rasi-Caldogno, F., De Michelis, M. I.,
Pugliarello, M. C., Marre, E. 1986. H$^+$-
pumping driven by the plasma membrane ATPase in membrane vesicles

from radish: stimulation by fusicoccin.
Plant Physiol. 82:121–25

106. Rasi-Caldogno, F., Pugliarello, M. C.,
De Michelis, M. I. 1987. The Ca^{2+}-
transport ATPse of plant plasma membrane catalyzes an H$^+$/Ca^{2+} exchange.
Plant Physiol. 83:994–1000

107. Raven, J. A., Smith, F. A. 1976. The
evolution of chemiosmotic energy coupling. J. Theor. Biol. 57:301–12

108. Rayle, D. L., Cleland, R. 1970.
Enhancement of wall loosening and
elongation by acid solutions. Plant
Physiol. 46:250–70

109. Rayle, D. L., Cleland, R. 1977. Control
of plant cell enlargement by hydrogen
ion. Curr. Top. Dev. Biol. 11:187–214

110. Reid, R. J., Field, L. D., Pitman, M. G.
1985. Effects of external pH, fusicoccin
and butyrate on the cytoplasmic pH in
barley root tips measured by ^{31}P-nuclear
magnetic resonance spectroscopy. Planta 166:341–47

111. Reinhold, L., Seiden, A., Volokita, M.
1984. Is modulation of the rate of proton
pumping a key event in osmoregulation?
Plant Physiol. 75:846–49

112. Reuveni, M., Colombo, R., Lerner, H.
R., Pradet, A., Poljakoff-Mayber, A.
1987. Osmotically induced proton extrusion from carrot cells in suspension culture. Plant Physiol. 85:383–88

113. Rincon, M., Hanson, J. B. 1986. Controls on calcium ion fluxes in injured or
shocked corn root cells: importance of
proton pumping and cell membrane
potential. Physiol. Plant. 67:576–83

114. Romani, G., Marre, M. T., Bellando,
M., Alloatti, G., Marre, E. 1985. H$^+$
extrusion and potassium uptake associated with potential hyperpolarization in
maize and wheat root segments treated
with permeant weak acids. Plant Physiol. 79:734–39

115. Romani, G., Marre, M. T., Bonetti, A.,
Cerana, R., Lado, P., Marre, E. 1983.
Effects of a brassinosteroid on growth
and electrogenic proton extrusion in
maize root segments. Physiol. Plant.
59:528–33

116. Rossman, M. G., Liljas, A., Branden,
C., Banaszak, L. J. 1975. Evolutionary
and structural relationships among dehydrogenases. In The Enzymes, ed. P.
D. Boyer, 11:61–102. New York: Academic. 3rd ed.

117. Rubinstein, B., Stern, A. I. 1986.
Relationship of transplasmalemma redox
activity to proton and solute transport by
roots of Zea mays. Plant Physiol.
80:805–11

118. Schafer, A., Bygrave, F., Matzenauer,
S., Marme, D. 1985. Identification of a

calcium and phospholipid-dependent protein kinase in plant tissue. *FEBS Lett.* 187:25–28

119. Schaller, G. E., Sussman, M. R. 1988. Phosphorylation of the plasma membrane H$^+$-ATPase of oat roots by a calcium-stimulated protein kinase. *Planta* 173:509–18

120. Scherer, G. F. E. 1984. Stimulation of ATPase activity by auxin is dependent on ATP concentration. *Planta* 161:394–97

121. Schroeder, J. I., Hedrich, R., Fernandez, J. M. 1984. Potassium-selective channels in guard cell protoplasts of *Vicia faba. Nature* 312:361–62

122. Schroeder, J. I., Raschke, K., Neher, E. 1987. Voltage dependence of K$^+$ channels in guard-cell protoplasts. *Proc. Natl. Acad. Sci. USA* 84:4108–12

123. Schwartz, R. M., Dayhoff, M. O. 1978. Matrices for detecting distant relationships. In *Atlas of Protein Sequence and Structure*, ed. M. O. Dayhoff, 5(Suppl. 3):353–58. Washington, DC: Natl. Biomed. Res. Found.

124. Serrano, E. E., Zeiger, E., Hagiwara, S. 1988. Red light stimulates an electrogenic proton pump in *Vicia* guard cell protoplats. *Proc. Natl. Acad. Sci. USA* 85:436–40

125. Serrano, R. 1983. In vivo glucose activation of the yeast plasma membrane ATPase. *FEBS Lett.* 156:11–14

126. Serrano, R. 1984. Plasma membrane ATPase of fungi and plants as a novel type of proton pump. *Curr. Top. Cell. Regul.* 23:87–126

127. Serrano, R. 1985. *Plasma Membrane ATPase of Plants and Fungi.* Boca Raton, Fla: CRC Press

128. Serrano, R. 1988. Structure and function of proton translocating ATPase in plasma membranes of plants and fungi. *Biochim. Biophys. Acta* 947:1–28

129. Serrano, R. 1988. H$^+$-ATPase from plasma membranes of *Saccharomyces cerevisiae* and *Avena sativa* roots: purification and reconstitution. *Methods Enzymol.* 157:533–44

130. Serrano, R. 1989. Plasma membrane ATPase. In *The Plant Plasma Membrane. Structure, Function and Molecular Biology*, ed. C. Larsson, I. M. Moller, Chap. 6. Berlin: Springer-Verlag. In press

131. Serrano, R., Kielland-Brandt, M. C., Fink, G. R. 1986. Yeast plasma membrane ATPase is essential for growth and has homology with (Na$^+$+K$^+$), K$^+$ and Ca^{2+}-ATPases. *Nature* 319:689–93

132. Serrano, R., Montesinos, C., Cid, A. 1986. A temperature-sensitive mutant of the yeast plasma membrane ATPase obtained by in vitro mutagenesis. *FEBS Lett.* 208:143–46

133. Shenolikar, S., Cohen, P. 1978. The sustrate specificity of cAMP-dependent protein kinase. *FEBS Lett.* 86:92–98

134. Shimazaki, K., Iino, M., Zeiger, E. 1986. Blue light–dependent proton extrusion by guard-cell protoplasts of *Vicia faba. Nature* 319:324–26

135. Shull, G. E., Greeb, J., Lingrel, J. B. 1986. Molecular cloning of three distinct forms of the Na$^+$, K$^+$-ATPase alfa-subunit from rat brain. *Biochemistry* 25:8125–32

136. Shull, G. E., Lingrel, J. B. 1986. Molecular cloning of the rat stomach (H$^+$-K$^+$) ATPase. *J. Biol. Chem.* 261:16788–91

137. Shull, G. E., Schwartz, A., Lingrel, J. B. 1985. Amino acid sequence of the catalytic subunit of the (Na$^+$+K$^+$) ATPase deduced from a complementary DNA. *Nature* 316:691–95

138. Smith, F. A., Raven, J. A. 1979. Intracellular pH and its regulation. *Annu. Rev. Plant Physiol.* 30:289–311

139. Solioz, M., Mathews, S., Furst, P. 1987. Cloning of the K$^+$-ATPase of *Streptococcus faecalis.* Structural and evolutionary implications of its homology to the KdpB-protein of *Escherichia coli. J. Biol. Chem.* 262:7358–62

140. Spanswick, R. M. 1981. Electrogenic ion pumps. *Annu. Rev. Plant Physiol.* 32:267–89

141. Steitz, T. A., Shoham, M., Bennett, W. S. 1981. Structural dynamics of yeast hexokinase during catalysis. *Philos. Trans. R. Soc. London Ser. B* 293:43–52

142. Struhl, K. 1983. The new yeast genetics. *Nature* 305:391–97

143. Sussman, M. R., Strickler, J. E., Hager, K. M., Slayman, C. W. 1987. Location of a dicyclohexylcarbodiimide-reactive glutamate residue in the *Neurospora crassa* plasma membrane H$^+$-ATPase. *J. Biol. Chem.* 262:4569–73

144. Sussman, M. R., Surowy, T. K. 1987. Physiology and molecular biology of membrane ATPases. *Oxford Surv. Plant Mol. Cell Biol.* 4:47–70

145. Sze, H. 1985. H$^+$-translocating ATPase: advances using membrane vesicles. *Annu. Rev. Plant Physiol.* 36:175–208

146. Tanford, C., Reynolds, J. A., Johnson, E. A. 1987. Sarcoplasmic reticulum calcium pump: a model for Ca^{2+}-binding and Ca^{2+}-coupled phosphorylation. *Proc. Natl. Acad. Sci. USA* 84:7094–98

147. Taylor, K. A., Dux, L., Martonosi, A. 1986. Three-dimensional reconstruction of negatively stained crystals of the

Ca^{2+}-ATPase from muscle sarcoplasmic reticulum. *J. Mol. Biol.* 187:417–27

148. Thomson, M., Krull, U. J., Venis, M. A. 1983. A chemoreceptive bilayer lipid membrane based on an auxin-receptor ATPase electrogenic pump. *Biochem. Biophys. Res. Commun.* 110:300–4

149. Trockner, V., Marre, E. 1988. Plasmalemma redox chain and H$^+$ extrusion. II. Respiratory and metabolic changes associated with fusicoccin-induced and with ferricyanide-induced H$^+$ extrusion. *Plant Physiol.* 7:30–35

150. Veluthambi, K., Poovaiah, B. W. 1984. Polyamine-stimulated phosphorylation of proteins from corn (*Zea mays* L.) coleoptiles. *Biochem. Biophys. Res. Commun.* 122:1374–80

151. Vesper, M. J., Evans, M. L. 1979. Nonhormonal induction of H$^+$ efflux from plant tissues and its correlation with growth. *Proc. Natl. Acad. Sci. USA* 76:6366–70

152. Villalobo, A. 1982. Potassium transport coupled to ATP hydrolysis in reconstituted proteoliposomes of yeast plasma membrane ATPase. *J. Biol. Chem.* 257:1824–28

153. Villalobo, A. 1984. Energy-dependent H$^+$ and K$^+$ translocation by reconstituted yeast plasma membrane ATPase. *Can. J. Biochem. Cell Biol.* 62:865–77

154. Walker, J. E., Saraste, M., Runswick, M. J., Gay, N. J. 1982. Distantly related sequences in the alfa- and beta- subunits of ATP synthase, myosin, kinases and other ATP-requiring enzymes and a common nucleotide binding fold. *EMBO J.* 1:945–51

155. Wyse, R. E., Zamski, E., Tomos, A. D. 1986. Turgor regulation of sucrose transport in sugar beet taproot tissue. *Plant Physiol.* 81:478–81

156. Yamaguchi, M., Kanazawa, T. 1985. Coincidence of H$^+$ binding and Ca^{2+} dissociation in the sarcoplasmic reticulum Ca-ATPase during ATP hydrolysis. *J. Biol. Chem.* 260:4896–4900

157. Yanagita, Y., Abdel-Ghany, M., Raden, D., Nelson, N., Racker, E. 1987. Polypeptide-dependent protein kinase from baker's yeast. *Proc. Natl. Acad. Sci. USA* 84:925–29

158. Zeiger, E. 1983. The biology of stomatal guard cells. *Annu. Rev. Plant Physiol.* 34:441–75

159. Zhang, J., Takemoto, J. Y. 1986. Mechanism of action of *Pseudomonas syringae* phytotoxin, syringomycin. Interaction with the plasma membrane of wild type and respiratory-deficient strains of *Saccharomyces cerevisiae*. *Biochim. Biophys. Acta* 861:201–4

160. Zocchi, G., Rogers, S. A., Hanson, J. B. 1983. Inhibition of proton pumping in corn roots is associated with increased phosphorylation of membrane proteins. *Plant Sci. Lett.* 31:215–21

Annu. Rev. Plant Physiol. Plant Mol. Biol. 1989. 40:95–117

BIOSYNTHESIS AND DEGRADATION OF STARCH IN HIGHER PLANTS*

Erwin Beck and Paul Ziegler

Lehrstuhl Pflanzenphysiologie, Universität Bayreuth, Universitätsstrasse, D-8580 Bayreuth, Federal Republic of Germany

CONTENTS

INTRODUCTION

Numerous reviews on starch biosynthesis and degradation have appeared in the 1980s (4, 23, 39, 40, 51, 73, 100, 101, 124, 125). Here we update established concepts and emphasize three topics that we consider to now merit reexamination: the significance of enzyme multiplicity, a comparison of degradation of reserve and transitory starch, and the localization of starch degrading enzymes in starch-free cellular compartments of leaf tissues. We stress the cell physiological aspects of starch metabolizing enzymes.

*Abbreviations: ADP: adenosine diphosphate; ADPG: ADP-glucose; ABA: abscisic acid; DP: degree of polymerization; GA: gibberellic acid; IEF: isoelectric focusing; PGA: 3-phosphoglycerate; Pi; inorganic phosphate; PPi: inorganic pyrophosphate; pI: isoelectric point; SDS-PAGE: denaturing polyacrylamide gel electrophoresis with sodium dodecylsulfate

1040-2519/89/0601-0095$02.00

RECENT KNOWLEDGE SUPPLEMENTING AN
ESTABLISHED BIOCHEMICAL PICTURE

Research into starch metabolism has focused on the nature and regulation of the biochemical reactions involved. α-1 → 4-Linked polyglucans can be formed from glucose 6-P via glucose 1-P, ADP-glucose, and a primer by the cooperation of phosphoglucomutase (EC 2.7.5.1), ADPG-pyrophosphorylase (EC 2.7.7.27), pyrophosphatase (EC 3.6.1.1), and starch synthase (EC 2.4.1.21), and also from glucose 1-P in both the presence and absence of a primer by phosphorylase (EC 2.4.1.1). The further transformation of amylose to amylopectin via branching enzyme (EC 2.4.1.18) is also well understood in principle (101).

Extensive investigations by Preiss's group have shown that the biochemical regulation of starch synthesis is centered almost exclusively on ADPG-pyrophosphorylase. This enzyme is activated by physiological concentrations of PGA and inactivated by Pi. It has been shown for at least the leaf ADPG-pyrophosphorylases that PGA modulates the sensitivity to Pi inhibition (103). Differences between the tetrameric enzymes (4 × 50 kd) of bacteria, spinach leaves, and potato tubers and the allosterically relatively insensitive corresponding form from maize kernels have recently been explained in terms of proteolytic loss during enzyme purification of a 1-kd peptide of particular regulative importance (97). Another pair of substances capable of modulating the activity of this enzyme has also recently been discovered, namely the inhibitor PPi and its antagonist glucose 1-P (103). Rather high concentractions of PPi (0.2–0.3 mM) have been reported in spinach leaf tissue. These were clearly allocated to the cytosol, however, whereas alkaline pyrophosphatase, which "pulls" the pyrophosphorylase reaction, was largely located in the chloroplast (135). Hence, a regulatory effect of PPi on the in vivo activity of the enzyme is not very likely.

Considerable progress has been made in understanding the biochemical basis of starch degradation. The dissolution of insoluble starch granule material to soluble maltodextrins is thought to be the domain of endoamylases (α-amylases: EC 3.2.1.1). These are the only enzymes generally considered able to bind to and attack raw starch [although this ability has also been ascribed to a tuber β-amylase (5) and to tuber (31) and leaf (120) phosphorylases], and their action is responsible for the physical corrosion of the granule. The debranching and degradation of the resultant maltodextrins and the soluble polymers are considered to be completed by further hydrolases in the case of cereal reserve starch mobilization. These include debranching enzymes (R-enzyme, pullulanase: EC 3.2.1.41; isoamylase: EC 3.2.1.68), exoamylase (β-amylase: EC 3.2.1.2), and α-glucosidase (maltase: EC 3.2.1.20). These topics have been thoroughly covered in various reviews (4,

23, 39, 73, 101, 129). The cooperation between initiatory or "pacemaking" hydrolases and phosphorylases is thought to be an important feature of assimlatory chloroplast starch degradation (8, 100, 123, 125). The regulation of starch degradation is not as well understood as that of starch biosynthesis. Only the genesis of starch degrading enzyme activities in seeds, in particular the classical example of the hormonal induction of α-amylase synthesis and secretion in germinating cereal grains, has been thoroughly investigated (e.g. 3, 4). Otherwise, the in situ presence of proteins inhibitory (16, 86) or stimulatory (105) to starch degradative enzymes has been described, as well as product inhibition of activity (see 40, 129) and (in some cases endogenous) activity rhythms (see 40, 120). Speculation as to the light/dark regulation of starch degradative enzymes (101) has, however, found no experimental support.

A dynamic mediatory role between starch synthesis and degradation has been ascribed to phosphorylase (112). This is certainly even truer of enzyme activities that catalyze transfer reactions between oligodextrins and soluble and insoluble polyglucans (D-enzyme: EC 2.4.1.25) and between maltose and maltodextrins (glucosyl-glucan transferase: 64). The maltose-metabolizing enzymes maltose phosphorylase (EC 2.4.1.8: 111) and maltose synthase (110) should also be mentioned in this regard. Since nothing is known of the regulation of the transferases, it must be assumed that the starch degradative system in particular can follow several pathways leading to the same final products: maltose, glucose, or glucose 1-P.

The cell physiological aspects of starch metabolism have been less well investigated than the biochemical aspects. This is due on the one hand to methodological difficulties associated with biochemical investigations on particulate systems, and with the isolation of intact starch-loaded amyloplasts (24, 27) and chloroplasts (123).

Tentatively speaking, amyloplasts have been thought of as nonphotosynthetic chloroplasts, because both organelles appear to be interconvertible (e.g. 113) and share similar morphologies and metabolic functions (97). Comparison of the envelopes of amyloplasts and chloroplasts of cultured sycamore (*Acer pseudoplatanus* L.) cells has revealed strong similarity of the lipid components, but clearly distinct polypeptide patterns (88). Especially the amounts of the 31-kd Pi-translocator differed in the two types of plastids: The amyloplast envelope was estimated to contain only 10% of the content of this polypeptide found in the chloroplast envelope. In agreement with the idea of a relatively small number of triose-P-translocators in the amyloplast envelope, substantial incorporation of uncleaved glucose moieties into the starch of developing wheat grains has recently been shown, and the existence of a hexose-P-translocator in the corresponding envelope has been suggested (55). Remarkably, this hypothetical carrier was termed a hexose-translocator. Translocators of unphosphorylated hexoses (109) and of maltose (8) have been detected in the chloroplast envelope. Hence, the detour via fructose 1,6-bisphosphate and triose-P no longer needs to be seen as obligatory, either for chloroplast starch-unloading or for the deposition of reserve starch in the amyloplast.

Another problem for investigators of the cell physiology of starch metabolism originates from the fact that the major components of native starch—amylose and amylopectin—are not uniform but represent populations of different molecular forms with regard to chain length and branching characteristics (6, 73). Biochemically and physiologically relevant kinetic parameters (e.g. K_m) cannot thus be related to a defined substrate but only to a certain substrate type. It is conceivable and even probable that several differentiated forms of each relevant enzyme are necessary to cope effectively with the variety of chain lengths and branching intensities exhibited by native starch (103).

MULTIPLE FORMS OF ENZYMES CATALYZING BIOSYNTHESIS AND METABOLISM OF STARCH

Enzymes of starch metabolism represent the classical examples of heterogeneity of an enzyme activity composed of several separable proteins that catalyze the apparently same reaction. The variant forms of such an enzyme activity have been termed isozymes or multiple forms and have been compared with isotopes of an element, since they are usually subsumed under the same identifying number in the Enzyme Commission's list (81). Differences among the members of such an enzyme family may be found in their regulatory properties or substrate specificities.

According to the current definition, isozymes are proteins translated from different mRNAs produced either from different genes (as in the case of plant α-amylases: 13, 80, 87) or by differential splicing of mRNA derived from the same DNA (as in mouse salivary and liver α-amylase: 143). By contrast, multiple forms are understood as posttranslational modifications of an enzyme. Experimentally, they can be distinguished from isozymes by their antigenic relationships. Unfortunately, both terms have been used arbitrarily.

It is interesting that multiple forms of starch-metabolizing enzymes have never been detected as intermediates in the pathway from an enzyme precursor protein to the final active enzyme, which includes removal of a signal peptide, glycosylation, and—eventually—transmembrane transport (79). On the contrary, posttranslational heterogeneity consistently represents alteration of already active enzymes. This can take the form of limited proteolysis (phosphorylase: 15, 32, 112; debranching enzyme: 77; glucoamylase: 44), deamination (α-amylase: 54), protein aggregation (phosphorylase: 31; α-amylase: 57), protein association (starch synthase plus branching enzyme: 94;

α-amylase plus pullulanase from bacteria: 76, 128; α-amylase plus inhibitor protein: 137) and protein adsorption onto membranes (phosphorylase: 112) or starch granules (starch synthase: 68, 69; phosphorylase: 112; α-amylase: 14, 92). While random proteolysis could generate enzyme forms of different regulatory behavior which, however, may turn out to be artifacts (e.g. maize endosperm ADPG-pyrophosphorylase: see 97, Ch. 2), nonartificial heterogeneity should bear physiological significance. In addition to substrate specificity and regulatory properties, compartmentation may represent another factor encountered in the investigation of the multiplicity of starch-metabolizing enzymes. Here we discuss a few examples of the physiological significance of enzyme multiplicity.

Multiplicity of phosphorylases has been related to particular developmental stages, e.g. respective of starch synthesis or degradation in maize endosperm (131, 132), banana fruit pulps (117), and growing and sprouting potato tubers and several other plant propagules (33). The low-molecular-weight forms of phosphorylase in stored roots of sweet potato originate by proteolysis from high-molecular-weight phosphorylases present in growing tubers (15). Correlation of special isozyme patterns with stages of development and tissue differentiation was also established with α-amylase of *Araucacia* seeds (106) and β-amylase of germinating rice seeds (91). In the latter case it has been shown that one β-amylase was synthesized de novo in the scutellum at the onset of germination, whereas subsequently a latent β-amylase tightly bound to the periphery of starch granules of the endosperm was activated. However, even in this well-investigated example the physiological significance and requirement of the enzyme multiplicity are not entirely clear.

A biochemical clue to the phenomenon of enzyme heterogeneity is provided by the idea that the substrate(s) of the enzymic reactions may change during the course of development. Starch composition varies not only owing to transitions between particulate (granular) and soluble forms, but also as a result of alterations in the amylase-to-amylopectin ratios that take place—e.g. during starch granule formation (6). In addition, changes in the length of the basic chain (C-chain) and the number of side chains (A- and B-chains) attached to them may lead to amylopectin exhibiting an either more trichitic or more racemose structure, or even becoming phytoglycogen (73). Even in the transitory granules of assimilatory chloroplastic starch two components have been differentiated: a core of amylopectin surrounded by a pasty mantle layer of more highly branched polyglucan (8). Finally, enzymic degradation patterns imply the presence of different types of starch granule surfaces. Granules of sorghum, maize, and wheat show localized corrosion, which implies that only a few areas or pores provide access to solubilizing enzymes. In contrast, rice, potato, and amylomaize starch granules are degraded evenly, indicating that the enzyme spreads along a homogeneous granule surface (60).

Different structural types of starch should require enzymes of individual specificity for biosynthesis (103), and this may also hold for degradation. Hence, multiplicity of starch metabolizing enzymes may be a biochemical response to the heterogeneity of starch molecules and granule structures. In isolated barley aleurone layers, two α-amylase isozymes (each of which exists in several multiple forms) have been shown to arise sequentially owing to differential induction by low and high GA_3 concentrations (50). A similar incidence was reported for wheat grains, where the first α-amylase produced after GA_3 induction was able to degrade insoluble polysaccharides, while a later-appearing enzyme was only active against smaller, soluble starch components (107). Such a temporal sequence of expression of two types of α-amylase may allow efficient digestion of the starch granules stored in the endosperm.

Of three glucoamylases (GAI, GAI', GAII) isolated from *Aspergillus awamori*, two were proteolytic degradation products of the original GAI (44). An acid fungal protease released a 7-kd glycopeptide (amino acid sequence #471-515) from GAI (#1-616) that contains or represents the raw starch-affinity site. The residual enzyme (GAI') degraded gelatinized starch or glycogen but could not adsorb onto raw—i.e. particulate—starch.

Sequence homology studies at the carboxy terminus of *Aspergillus niger* glucoamylase and *Bacillus macerans* cyclodextrin glucanotransferase have also indicated that a raw starch-binding peptide of about 40 amino acid residues is not essential for enzyme action on soluble starch but is essential for the digestion of raw starch (127). Similarly, the α-amylase forms of immature wheat grains are not adsorbed onto the amylopectin of the mature grain, while the isozymes present in the germinating seeds cannot bind to the amyloplasts of the immature seeds (65).

It is noteworthy that four α-amylases from different animal, bacterial, and fungal sources degraded maize-starch granules at quite different rates but that none of these could degrade potato-starch granules (134). Adsorption to the starch granules appeared to be a prerequisite for digestion. However, two types of adsorption were differentiated: one reaction favored by low temperature and high pH that did not lead to hydrolysis, and another that was also effected at high temperature (35°C) and resulted in rapid hydrolysis.

Analysis of the starch synthases of normal maize kernels and those of the waxy *(wx)* mutant has revealed different patterns of enzyme activity that can be correlated with the different types of starch produced (68). Whereas the *wx* mutant accumulates only amylopectin, the kernels of normal maize additionally contain amylose. Two soluble starch synthase isozymes were detected in both systems, one that is presumably granule bound in the normal kernel but not in the *wx* mutant. The predominant granule-bound isozymes of the normal and the waxy maize kernels are, however, quite different enzymes. Interestingly, the soluble starch synthase activity of both types of maize kernels includes one enzyme form showing substantial activity in the absence of primer. In the primed assay it prefers glycogen, and a second isozyme preferring amylopectin as a primer exhibits only low unprimed activity.

That the relative predominance of starch synthesis or degradation is effected by the vicarious increase of different forms of phosphorylase as quoted above has not yet been unequivocally demonstrated. Two of the (nine) potato tuber phosphorylases adsorb onto insoluble starch and amyloplasts and hence represent almost the only phosphorylases cabable of metabolizing raw starch (31; but see also 120). Both isozymes were identified as dimers; upon formation of more highly aggregated multiple forms these enzymes lost their starch-binding capacity, presumably because of self-masking of the raw starch binding sites (30).

Upon electrophoresis in amylopectin-containing gels, the two phosphorylase isozymes separated into three and two bands, respectively, which showed different activities (30). A similar phenomenon has been observed with spinach leaf debranching enzyme, which upon SDS-PAGE or electrophoresis on substrate-free gels yielded only one protein band with a molecular mass of 100 kd, but separated into three bands when subjected to PAGE on amylopectin-containing gels (66). The three bands are antigenically related, and isoelectric focusing of the protein likewise revealed several proteins with isoelectric points between pH 4.9 and 5.1 (A. Henker, unpublished results). Presumably, binding to substrate results in an alteration of the protein's configuration and concomitantly in an amplification of the originally small charge differences.

The formation of three forms of the debranching enzyme obviously did not result from proteolysis, since it could not be prevented by protease inhibitors (A. Henker, unpublished results). However, in other cases proteolysis is responsible for the occurrence of multiple forms. Freezing of sprouting potato tubers, as well as storage of a crude protein extract thereof, results in the partial degradation of the predominant phosphorylase; the proteolyzed form, however, retains the full activity of the original form (30). In vivo, the endogenous protease(s) do not come into contact with the enzyme, probably owing to compartmentation.

The physiological significance of the original and the proteolyzed forms has been elucidated (112). When tuber formation begins, the original form is located in the cytoplasm. Upon differentiation of the storage parenchyma cells it attaches to the membranes of the proplastids and amyloplasts and also to other membranes, such as the plasmalemma. In mature cells it was located inside the amyloplasts, partly attached to the inner face of the envelope and partly dispersed within the starch granules. In senescing tubers the amyloplast membranes are damaged and the enzyme diffuses into the cytoplasm, where it is acted on by proteases. While the plastid form of the enzyme is presumably involved in starch synthesis during the amyloplast-filling stage, the degraded form must be assumed to participate in polyglucan breakdown, even though neither of these forms of the enzyme could bind to raw starch (see 30, 31).

Three groups of α-amylase have been resolved by IEF of a protein extract of germinated barley kernels, namely one low (α-amylase I = α-AI) and two antigenically related high pI-forms (α-AII and III). Heating or kilning resulted in an only small loss of enzyme activity accompanied by a more or less complete conversion of α-AIII to α-AII (70, 71). Later, α-AIII was revealed to be a complex formed between α-AII and an endogenous heat-sensitive inhibitor protein of 19.9 kd, of obviously widespread occurrence in cereals and leguminous crops (38, 137). A stoichiometric binding ratio of 2:1 for inhibitor with α-AII was determined, while partial inhibition of the enzyme was already observed at a ratio of 1:1. Hence, the occurrence of more than one form of the high-pI α-amylase did not result from an alteration of the protein's primary sequence, but from association with a physiologically relevant regulatory protein.

Owing to the lack of appropriate model substrates, the functional pair polyglucan:starch metabolizing enzyme has not yet been investigated to the point that the significance of the multiplicity of the enzyme activities in question has become understandable. Some progress has been made with respect to extra sequences (raw starch-binding sites) of degrading enzymes mediating starch granule attack, and the significance of the heterogeneity of these enzymes is at least partially conceivable. On the other hand, our understanding of the biochemistry of starch granule formation is still predominantly hypothetical. Due to the high DP of amylose ($\sim 10^3$), amylopectin (10^4–10^5), and phytoglycogen ($\sim 10^5$), realistic representations of the macromolecules are not practical. Any diagram therefore expresses only possible arrangements of the constituent chains and can provide but limited information with respect to the processes of granule physiology. It has been shown with several plant species that the amylose content, the degree of amylopectin branching, and the sizes of the amylose and amylopectin molecules increase with amyloplast maturity (6). However, the question as to how an unbranched and a highly branched α-glucan can be synthesized at the same time and site has not been resolved. Further analysis of enzyme multiplicity will provide insight into this phenomenon.

STARCH DEGRADATION

Of the many examples of starch degradation studied in the organs and tissues of a variety of higher plants, the degradation in the endosperm of germinating cereal seeds represents a unidirectional metabolism of preformed "reserve" material deposited during seed ripening to enable initial heterotrophic embryo growth. In contrast, the degradation of assimilatory leaf starch is a component of a highly dynamic continuum of diurnal synthesis and breakdown ("transitory" starch). Other starch degradative systems are considered here only as they apply to the issues at hand, and the reader is otherwise referred to more comprehensive reviews (4, 10, 23, 39, 40, 73, 101, 125).

Cereal Seed Reserve Starch Degradation

Starch in cereal seeds is characteristically stored in a nonliving tissue (the endosperm), which, at the onset of germination, does not contain suitable enzymatic breakdown machinery. The required enzymes are either not present, are present in a predominantly inactive form, or are effectively sequestered from the starch granules. Reserve starch in cereal seeds (in contrast to that in potato tubers) is thought to be degraded only hydrolytically. Since maltose and especially glucose are the almost exclusive breakdown products, phosphorolysis does not appear to play any significant role (4, 40).

DEVELOPMENT OF IN SITU α-AMYLASE ACTIVITY α-Amylase is not present in resting cereal seeds, with the possible exception of some activity packaged in lytic bodies in maize and *Sorghum* endosperm (2). It must therefore be synthesized by living tissues and secreted into the endosperm upon germination. Cereal seed α-amylases are commercially important (e.g. 122), plentiful, and easily purified. In addition, their production incorporates an excellent model system for studying the induction of enzyme synthesis and protein secretion in general (3, 10, 47).

Whereas the scutellum is clearly an important site of α-amylase production in rice and sorghum seeds (see 39), it has long been debated whether this holds true for other cereals (see 3, 75). In barley seeds α-amylase synthesis and secretion have generally been considered to be the primary domain of the aleurone layers (see e.g. 10, 11). Although several recent studies have suggested that barley seed scutellum plays an important and perhaps even primary role in early α-amylase production (see 3), artifacts cannot be completely ruled out (see 39; 95). However, the detection of in vitro translation products of α-amylase-encoding mRNAs shows at least a partial participation of the scutellum in α-amylase production (83).

The classical gibberellic acid (GA)-induced de novo synthesis of α-amylase in barley aleurone tissue is closely associated with the accumulation of newly synthesized mRNA coding for the enzyme as part of a redirection of protein synthesis (see 4, 47). The use of cloned α-amylase cDNA probes for aleurone mRNA has firmly established the control of α-amylase mRNA by GA in barley (see 3), wheat (7), and wild oats (146). Studies with nuclei isolated from barley aleurone protoplasts provide strong evidence that this type of control is at the transcriptional level (48). However, it remains to be determined whether mRNA stability also is involved to a significant extent: α-Amylase mRNA may be more labile than generally assumed (89).

GA has been shown to be the principal regulator of all α-amylase mRNAs in barley aleurone (21, 22). The removal (via drying, temperature, or detachment treatments) of the normal inability of aleurone tissues of immature wheat kernels to produce and secrete α-amylase upon GA application has also been documented at the mRNA level (19).

Despite these impressive insights into the stimulatory effects of GA in barley, the generality and details of the involvement of the hormone in α-amylase production in cereal seeds are still controversial (3, 39, 130). Even where GA effects are clearly operative, it is doubtful whether they really represent a classical hormone-response syndrome (whereby GA is produced in the embryo, diffuses to the aleurone, and acts there as a primary elicitor: see 10, 11). Alternative possibilities as to GA-related inductive mechanisms have been thoroughly discussed recently (3, 39, 130).

Abscisic acid (ABA), which reverses all of the GA-promoted changes in barley aleurone layers (47), appears to exert its inhibitory effect on GA-induced α-amylase synthesis at the level of transcription (48). ABA may affect the stability of α-amylase mRNA, and it preferentially inhibits the production of high-pI-type α-amylase isoenzymes (89).

The synthesis of an α-amylase/subtilisin inhibitor protein, which inhibits barley α-amylase (86), is stimulated in barley aleurone layers by ABA but reduced by GA, a control antagonistic to that for α-amylase (82). The control of the mRNA levels of this inhibitor protein by GA and ABA indicates that endogenous α-amylase activity may be controlled by the hormones at the transcriptional level, both directly and via the inhibitor (84).

Calcium plays an important role in both barley aleurone and rice scutellum by regulating the differential production of α-amylase isoenzymes and the intracellular transport and secretory release of the molecules (see 3). The production of high-pI α-amylase isoenzymes in barley requires calcium, while that of low-pI isoenzymes does not (see 3). Furthermore, the secretion of low-pI α-amylases is also slightly stimulated by calcium (12). Calcium selectively stimulated the secretion of R-type α-amylase molecules (resistant to hydrolytic degradation by endo-β-N-acetylglucosaminidase H), without affecting that of the S form of the enzyme (susceptible to endo-β-H attack) (78).

These findings still leave unanswered questions as to the site at which the ion is required for the synthesis of the high-pI α-amylase isoenzymes (22), whether the suppression of the synthesis of these isoenzymes may result from coupling to an inhibition of purely secretory phenomena (52), and the possibility that calcium regulates secretion via facilitated movement through the cell wall (12).

There has been uncertainty as to whether the secretion of the α-amylase synthesized in aleurone or scutellar tissues takes place via the Golgi apparatus (3), as is the case with all other eukaryotic cells (56). Positive evidence to this effect with respect to barley aleurone has only recently been presented (28), and has subsequently been confirmed by immunofluorescence and gold-labelling experiments (37). It has also been concluded that the release of α-amylase and other secreted proteins from the aleurone is via digested cell wall channels, for which wall hydrolases are directed by plasmodesmata-lined resistant tubes, thus indicating a subtle role of the wall system in influencing and controlling enzyme release (36).

In the case of rice scutellum, the inhibition of the secretion of only R-type (see above) enzyme molecules by the Golgi-disrupting compound monesin indicates that a vesicular transport of S-type α-amylase independent of the Golgi apparatus may indeed be operative (see 3).

DEVELOPMENT OF OTHER STARCH HYDROLYZING ACTIVITIES β-Amylase in barley is fully preformed in the resting seed and exists as such in "free" (salt-soluble) and "bound" or "latent" forms complexed to proteins (4, 73). The active "free" component may be significant in the early phases of germination. The inactive "bound" or "latent" form can be extracted with reducing agents and is considered to be released in an active form in vivo by proteolytic enzymes generated upon germination in a manner similar to that of α-amylase (4, 73). The primary structure of the enzyme has recently been deduced from the nucleotide sequence of a full-length cDNA (58), and the conversion of the largest-molecular-weight isozyme to three other multiple forms via proteolytic modification of this primary structure (67) could be involved in releasing bound forms from associating proteins (58, 67).

Recent immunochemical (59) and biochemical (43) studies have demonstrated that β-amylase is also deposited on starch granules in the subaleurone layers in the latent form during the desiccation phase of barley seed development. Although the levels of β-amylase are typically very high in germinated cereal grains (see 39), and a maltodextrin-degrading function appears evident, the seeds of some mutants of rye (20) and barley (see 58) exhibiting only very low levels of this enzyme germinate well. β-Amylase may hence be considered insignificant for germination (see also 1, with respect to soybean seeds) and may be a storage protein (35).

While the physiological role of β-amylase in starch degradation is still unclear, a two-fold task can be ascribed to the debranching enzymes: (a) to prevent substantial production of isomaltose and isomaltodextrins during the course of starch degradation; and (b) to render access to branched polyglucans to phosphorylase, which otherwise is arrested by any interchain linkage. Intermediate α-dextrins, produced during α-amylolysis of amylopectin or glycogen, represent the natural substrates of debranching enzymes. Two types of these enzymes have been distinguished: (*a*) the direct debranching enzymes—namely, pullulanase (R-enzyme: EC 3.2.1.41) and isoamylase (EC 3.2.1.68), which cleave $\alpha(1 \rightarrow 6)$ interchain linkages; and (*b*) the indirect debranching enzymes, which combine an amylo-$(1 \rightarrow 6)$-glucosidase with an oligodextrin transferase activity (EC 3.2.1.33 + EC 2.4.1.25: 62). R-enzyme and isoamylase can be distinguished by their ability to degrade pullulan and glycogen (phytoglycogen), respectively. The direct debranchers are of plant and microbial origin, whereas the indirect debranchers are typical of mammals and yeasts.

Pullulanases and isoamylases have been found as "isozymes" in sweet corn (74) and potato tubers (45), although there is no special substrate (phytoglycogen) requiring the latter enzyme activity. A debranching enzyme that is active on pullulan and glycogen has been extracted and characterized from germinating rice seed (46). It differs from the pullulanase and isoamylase present in milky stage and mature rice kernels. High activities of pullulanase were found to be synthesized during ripening of rice kernels (139) and pea cotyledons (114). The enzyme is deposited upon maturation in an inactive form, from which it is liberated by proteolysis during germination. Activation can be effected artificially by digestion with endopeptidases (e.g. papain) or disulfide bond–cleaving reductants (140). The reductively activated enzyme was similar to that of the milky kernel stage (141), and no indication of a direct synthesis of an insoluble zymogen was obtained (139). A similar activation of insoluble, inactive pullulanases was shown for other cereals such as oats, wheat, barley, and rye (142).

In addition to the proteolytic activation of cereal pullulanases, debranching enzymes (and α-glucosidase) are considered to be synthesized de novo and secreted from the aleurone of barley (4, 39, 73) and rice (96) seeds in response to GA upon germination in parallel with α-amylase. Protein bodies in the spherosomal fraction in maize and *Sorghum* seeds contain high activities of α-glucosidase, along with α- and β-amylase (2). Especially these spherosomal activities could rapidly be liberated and thus be of significance at the onset of germination.

REGULATION OF IN SITU STARCH HYDROLYSIS   Once starch hydrolyzing enzymes are present in an active form in the endosperm, they are ensured direct access to the starch granules by the action of concomitantly produced proteases and glucanases, which dissolve the cell walls, membranes, and complexing proteins otherwise compartmenting the starch material (4, 39, 40, 73). A suitably acidic environment is also provided by aleurone tissues (41). There are few factors that might otherwise be involved in regulating the activities of the enzymes themselves.

One of these is the high concentrations of the end products maltose and glucose that develop in the dissolving endosperm and can inhibit amylolytic activities (see 39, 40, 129). Maltose is thought to inhibit the characteristic starch granule–binding ability of barley α-amylase (e.g. 136), perhaps by interfering with noncatalytic sugar-binding or surface sites (see 34, 72, 136). An attempt to demonstrate this directly was not successful, however (25). Work with the soybean storage cotyledon enzyme indicates that both glucose and especially maltose inhibit β-amylase activity by binding to different catalytic subsites (90).

The seeds of many cereals have recently been shown to contain α-amylase inhibitor proteins: Some are active against the endogenous α-amylases. Of these, a bifunctional α-amylase/subtilisin inhibitor from barley seeds has been particularly well characterized (86, 126), and similar proteins are present in other cereal seeds (138). The barley protein is synthesized and deposited in developing endosperm tissue (85), where it is loosely associated with the starch granules and gradually disappears during germination (61). It can directly interact with the endogenous α-amylase (38, 137) and may serve to inactivate inappropriately induced α-amylase upon precocious germination.

Transitory Leaf Starch Degradation

The starch metabolism of assimilatory leaves is intimately related to photosynthesis and the biosynthetic and catabolic requirements of the plant as a whole. It can best be regarded as a buffer to the metabolism of sucrose (123, 125). Transitory leaf starch metabolism differs topographically from that of cereal endosperm reserve starch in that it is compartmented within viable plastids (chloroplasts) of living cells, a situation rather similar to that of the intact amyloplasts of seed cotyledon and storage tuber systems. However, its most characteristic feature of diurnal alteration between net synthesis and degradation is in total contrast to any reserve starch situation. Hence, intracellular compartmentation of starch degradative enzymes and control of rapidly alternating fluxes are principal themes in the degradation of transitory starch.

STARCH DEGRADATION AND ITS ENZYMES IN THE CHLOROPLAST   Product analysis studies with isolated chloroplasts from spinach and pea have demonstrated the importance of phosphorolysis for transitory starch degradation. Triose phosphates stemming from glucose 1-P are accumulated to a major extent, and are thought to represent an energetically favorable end product form for transport to the site of sucrose synthesis in the cytoplasm (100, 123, 125).

This agrees well with the detection of substantial activities of starch phosphorylase in all chloroplast preparations tested. Earlier findings to this effect with respect to spinach and pea (100) have been extended to include

Arabidopsis leaf mesophyll (63), *Commelina* leaf guard cells (104), and maize leaf bundle sheath cells (26).

On the other hand, the complementary "pacemaking" function of hydrolases has also been established. These enzymes also account for the accumulation of maltose and/or glucose in chloroplasts: Transport studies and theoretical-energetics calculations have demonstrated that maltose is indeed compatible with starch end product export from the chloroplast (8). Endohydrolytic attack is now generally assumed to initiate transitory starch degradation in analogy to the seed endosperm situation (8, 100, 123, 125). This idea is corroborated by the fact that the chloroplastic phosphorylases from spinach (102, 116, 121) and pea (119) appear to be unsuited to degrade large branched starch molecules. They were effective with starch granules only following hydrolytic digestion (9, 14, 120).

The general presence of endoamylase in chloroplasts has been in doubt, as the enzyme was reported to be lacking in chloroplasts of pea (see 123; 53) and barley (49) leaves. However, other studies have clearly demonstrated endoamylolytic activity in the chloroplasts of pea and wheat leaf mesophyll (144, 145), as well as in those of spinach (92, 145) and *Arabidopsis* (63) leaf mesophyll, *Chenopodium rubrum* suspension culture cells (145), and maize leaf bundle sheath cells (26).

Evidence that chloroplastic endoamylase alone is responsible for the initial attack of transitory starch granules is not yet completely convincing. Studies to this effect were carried out with either a mixture of chloroplastic hydrolases (9), an endoamylase not shown to be free of other hydrolytic activities (14), or an endoamylase from a nonplant source (120). Unequivocal evidence is hindered by the fact that a chloroplastic endoamylase has never been isolated in a pure form.

All known chloroplastic endoamylases appear to be calcium independent and heat labile (see 100, 123, 125; 63, 144). They thus appear to correspond to the low-pI-type cereal seed α-amylases (see 4; 49), the ability of which to effectively attack native starch is questionable (see 39). However, microscopic studies indicated that such "nonclassical" endoamylases effectively and preferentially dissolve the "pasty" outer layer of larger transitory starch granules (8). Chloroplastic endoamylases may thus be most efficient with relatively highly hydrated forms of granular starch. This might also hold for the low-pI-type cereal seed enzymes.

β-Amylase activity, as deduced from studies with the purified *Vicia faba* enzyme (17), would appear predestined to produce both maltose and some glucose for carbohydrate export from the chloroplast. However, although leaf β-amylase has been reported to be located in the chloroplasts of pea leaves (53), other recent work has clearly shown the extrachloroplastic nature of the enzyme (144, 145). In addition, β-amylase activity in whole leaves (17), leaf mesophyll (49, 63, 145), bundle sheath cells (26), and photoautotrophic culture cells (145) of other species was clearly also mainly extrachloroplastic. If β-amylase is predominantly nonchloroplastic, these functions must be taken over by endoamylase and transglycosylases (see 123, 125). D-enzyme activity is indeed plentiful in pea (53) and *Arabidopsis* (63) leaf chloroplasts. It was only present to a minor extent in spinach leaf chloroplasts (92), however, and since it has otherwise not been checked for, the generality of its presence and significance in chloroplasts is uncertain. This is unfortunately even more true of glucosyl-glucan transferase, which has only been described for spinach leaves (64).

In addition, an activity capable of debranching native starch forms is obviously also required. Debranching enzyme activity appears to be often, if not always, associated with chloroplasts. It is amply present in those of spinach leaves (66, 93, 145) and has recently also been detected in chloroplasts of pea and wheat leaf mesophyll (53, 144, 145), photoautotrophic *Chenopodium rubrum* suspension culture cells (145), and maize leaf bundle sheath cells (26).

In summary, the available evidence corroborates the view that transitory starch degradation represents a cooperative action of hydrolytic attack and phosphorolytic as well as hydrolytic final degradation to metabolizable and/or exportable forms of carbohydrates.

REGULATION OF TRANSITORY STARCH DEGRADATION   The relative preponderance of starch synthesis or breakdown in the chloroplast could be solely determined by the rate of starch synthesis in the absence of control mechanisms for starch degrading enzymes (100). However, starch degradation rates in darkened leaves are not necessarily constant, and while net starch mobilization has been shown to proceed even in the light, it is slower than in the dark (124). Hence the activity of chloroplastic starch degrading enzymes must be subject to control (123, 125).

Knowledge of regulatory mechanisms pertinent to starch degradative enzymes is scarce (100, 123, 125). The only physiologically relevant regulator of chloroplastic phosphorylases—Pi—is a substrate rather than an effector. Feedback regulation of hydrolytic enzymes by the relatively low concentrations of accumulated glucose and/or maltose is unlikely, and speculations as to light-associated influences on the activity of starch degradative enzymes (101) have never been substantiated.

Chloroplastic amylases and debranching enzymes may be significantly regulated by pH, as their activities markedly decrease at pH values higher than those of the optimum value of about pH 6. Accordingly, they would be more strongly inhibited at the stromal pH of 8 established in the light than at that in the dark (about 7). A lesser such effect could also apply to phosphorylases and D-enzyme, which show slightly higher pH optima. In addition to this response to stromal pH changes, endogenous diurnal activity fluctuations have been reported for amylases at an optimal test pH (see 40) and demonstrated with spinach chloroplasts (9, 99). Superimposing the pH effect on the endogenous diurnal oscillation of the amylolytic system, amylolytic degradative activity in the dark could amount to about 5-fold that in the light (8).

Other factors could be involved in the regulation of starch degradative leaf enzymes and thus be significant for the regulation of transitory starch degradation. These include proteins inhibitory to endoamylases (see the section on endosperm starch degradation, above) and inhibitory (16) and stimulatory (105) to phosphorylases. Specific proteolytic activity against spinach leaf phosphorylase (42), but also light-induced NH_4Cl stimulation of phosphorylase and amylase in *Chlorella* cells (79) and hormonal regulation of rice stem amylolytic activity (118) have been reported, too. Light induces de novo synthesis of β-amylase in mustard cotyledons. This phytochrome-mediated control is exerted on the level of mRNA synthesis (115). Water stress stimulates starch degrading enzyme activities in *Aloe* leaves (133) and induces the synthesis of an extrachloroplastic barley leaf α-amylase via transcriptional control (49).

EXTRACHLOROPLASTIC STARCH DEGRADING LEAF ENZYMES

Although in some green algae amylases and phosphorylases appear to be restricted to the chloroplast (62a), with the leaves of higher plants comparisons of mechanically isolated chloroplasts and whole leaf preparations have repeatedly demonstrated the extrachloroplastic localization of the bulk of various starch degrading enzymes (see 100; 53).

Fractionation of leaf cell protoplasts has shown that major starch degrading activities are indeed present in the "cytosolic" compartment of cells performing transitory starch metabolism. All hydrolytic and phosphorolytic starch degrading activities tested in spinach leaf mesophyll (92), maize leaf bundle sheath cell (26), and *Commelina* leaf guard cell (104) protoplasts were predominantly extrachloroplastic, and α- and β-amylase activities were exclusively so in barley leaf mesophyll protoplasts (49). β-Amylase activity of pea, wheat, spinach, and *Chenopodium rubrum* protoplasts was exclusively extrachloroplastic, and that of *C. rubrum* R-enzyme was largely so (145). Some D-enzyme activity in *Arabidopsis* leaf mesophyll protoplasts was extrachloroplastic, along with most amylolytic and phosphorylytic activities (63).

The stringent cytosolic (and chloroplastic) localization of specific phos-

phorylase forms in spinach and pea has also been confirmed using immunocytological and -histochemical techniques (18, 108).

Many recent reviews and articles have mentioned the lack of understanding
of this strong degradative potential in regions of assimilatory cells sequestered
from the starch granule, where no appropriate substrate is known to exist. It
has been pointed out, however, that the high activities would indeed preclude
finding any such substrate (8).

It seems ironic that the extrachloroplastic phosphorylases of spinach and
pea appear to be able to attack starch granules and/or to effectively degrade
high-molecular-weight branched polyglucans, in contrast to the corresponding chloroplastic forms that would have ready access to, but are ineffective
against, such substrates (18, 119–121). The cytosolic comparment of at least
spinach leaf mesophyll has also been shown to contain ample amylase and
R-enzyme activities (92, 145), which could also effectively debranch and
depolymerize any large amylopectin-type substrates. This suggests that complex extrachloroplastic starch substrates might indeed occur in leaves (see also
18).

β-Amylase, while apparently confined to the cytosol in some species
(spinach and C. rubrum), is located primarily in the vacuole of others (pea
and wheat: 145). As already discussed, β-amylase cannot always be correlated with reserve starch degradation, and has been concluded to represent
storage protein. A storage function of an assimilatory cell protein could be
consistent with a vacuolar localization. The protein might be mobilized upon
seed filling, reminiscent of proteins stored in the paraveinal mesophyll cells of
soybean leaves (29).

In addition, Jacobsen et al concluded that the extrachloroplastic α-amylase
of barley leaf mesophyll cells was encoded for by the same genes responsible
for a seed α-amylase isoenzyme (49), and urease in soybean leaves may
represent a seed protein (98). Since pea and wheat seeds both contain β-
amylases, it is possible that at least extrachloroplastic β-amylases represent
unspecific expression of genes coding for reserve mobilization processes.

Finally, the significance of large amylolytic potentials that cannot be
ascribed to assimilatory mesophyll cells and are of hitherto unknown localization must be considered. This is especially the case with respect to all of the
major endoamylase and much of the exoamylase of pea leaves (144). Such
enzymes must be preferentially located in leaf regions not primarily concerned with assimilation, and thus presumably fulfill some function other than
the degradation of "classical" transitory starch.

Starch degrading activities found in epidermis or guard cells (e.g. 17, 104)
are unlikely to account for massive nonmesophyllic activities. Other
possibilities include vascular tissue or specialized cells akin to soybean para-

veinal mesophyll (29). It is tempting to speculate that such regions of a leaf would represent temporary depots of polysaccharide material required for massive export events, thus requiring high degradative activities. However, no substantiation for the existence of such starch metabolizing pathways is available at present.

Literature Cited

1. Adams, C. A., Broman, T. H., Rinne, R. W. 1981. Starch metabolism in developing and germinating soya bean seeds is independent of β-amylase activity. *Ann. Bot.* 48:433–39
2. Adams, C. A., Watson, T. G., Novellie, L. 1975. Lytic bodies from cereals hydrolyzing maltose and starch. *Phytochemistry* 14:953–56
3. Akazawa, T., Hara-Nishimura, I. 1985. Topographic aspects of biosynthesis, extracellular secretion, and intracellular storage of proteins in plant cells. *Annu. Rev. Plant Physiol.* 36:441–72
4. Ashford, A. E., Gubler, F. 1984. Mobilization of polysaccharide reserves from endosperm. In *Seed Physiology, Vol. 2, Germination and Reserve Mobilization*, ed. D. R. Murray, pp. 117–62. Sydney: Academic Press Australia
5. Baba, T., Kainuma, K. 1987. Partial hydrolysis of sweet-potato starch with β-amylase. *Agric. Biol. Chem.* 51:1365–71
6. Banks, W., Muir, D. D. 1980. Structure and chemistry of the starch granule. In *The Biochemistry of Plants*, ed. J. Preiss, 3:321–69. New York/London/Toronto: Academic
7. Baulcombe, D. C., Buffard, D. 1983. Gibberellic acid–regulated expression of α-amylase and six other genes in wheat aleurone layers. *Planta* 157:493–501
8. Beck, E. 1985. The degradation of transitory starch granules in chloroplasts. In *Regulation of Carbon Partitioning in Photosynthetic Tissue*, ed. R. Heath, J. Preiss, pp. 27–44. Rockville: Am. Soc. Plant Physiol.
9. Beck, E., Pongratz, P., Reuter, I. 1981. The amylolytic system of isolated spinach chloroplasts and its role in the breakdown of assimilatory starch. In *Proceedings of the Fifth International Congress on Photosynthesis*, Vol. 4, ed. G. Akoyunoglou, pp. 529–38. Philadelphia: Balaban Int. Sci. Serv.
10. Bewley, J. D., Black, M. 1978. *Physiology and Biochemistry of Seeds. I. Development, Germination and*

Growth. Berlin/Heidelberg/New York: Springer-Verlag. 306 pp.
11. Briggs, D. E. 1973. Hormones and carbohydrate metabolism in germinating cereal grains. In *Biosynthesis and Its Control in Plants*, ed. B. V. Millborrow, pp. 219–79. London: Academic
12. Bush, D. S., Cornejo, M.-J., Huang, C.-N., Jones, R. L. 1986. Ca^{2+}-stimulated secretion of α-amylase during development in barley aleurone protoplasts. *Plant Physiol.* 82:566–74
13. Callis, J., Ho, T.-H. D. 1983. Multiple molecular forms of the gibberellin-induced α-amylase from the aleurone layers of barley seeds. *Arch. Biochem. Biophys.* 224:224–34
14. Chang, C. W. 1982. Enzymic degradation of starch in cotton leaves. *Phytochemistry* 21:1263–69
15. Chang, T.-C., Lee, S.-C., Su, J.-C. 1987. Sweet potato starch phosphorylase—purification and characterization. *Agric. Biol. Chem.* 51:187–95
16. Chang, T.-C., Su, J.-C. 1986. Starch phosphorylase inhibitor from sweet potato. *Plant Physiol.* 80:534–38
17. Chapman, G. W., Pallas, J. E., Mendicino, J. 1972. The hydrolysis of maltodextrins by a β-amylase isolated from leaves of *Vicia faba*. *Biochim. Biophys. Acta* 276:491–507
18. Conrads, J., Van Berkel, J., Schächtele, C., Steup, M. 1986. Non-chloroplast α-1,4-glucan phosphorylase from pea leaves: characterization and in situ localization by indirect immunofluorescence. *Biochim. Biophys. Acta* 882:452–63
19. Cornford, C. A., Black, M., Chapman, J. M., Baulcombe, D. C. 1986. Expression of α-amylase and other gibberellin-regulated genes in aleurone tissue of developing wheat grains. *Planta* 169:420–28
20. Daussant, J., Zbaszyniak, B., Sadowski, J., Wiatroszak, I. 1981. Cereal β-amylase: immunochemical study on two enzyme-deficient inbred lines of rye. *Planta* 151:176–79
21. Deikman, J., Jones, R. L. 1985. Control

of α-amylase mRNA accumulation by gibberellic acid and calcium in barley aleurone layers. *Plant Physiol.* 78:192–98

22. Deikman, J., Jones, R. L. 1986. Regulation of the accumulation of mRNA for α-amylase isoenzymes in barley aleurone. *Plant Physiol.* 80:672–75

23. Duffus, C. M. 1984. Metabolism of reserve starch. In *Storage Carbohydrates in Vascular Plants*, ed. D. H. Lewis, pp. 321–52. Cambridge: Cambridge Univ. Press

24. Duffus, C. M., Rosie, R. 1975. Purification and fractionation of potato amyloplasts. *Anal. Biochem.* 65:11–18

25. Dunn, G. 1974. A model for starch breakdown in higher plants. *Phytochemistry* 13:1341–46

26. Echeverria, E., Boyer, C. D. 1986. Localization of starch biosynthetic and degradative enzymes in maize leaves. *Am. J. Bot.* 73:167–71

27. Echeverria, E., Boyer, C. D., Shannon, J. C. 1985. Isolation of amyloplasts from developing maize endosperm. *Plant Physiol.* 77:513–19

28. Fernandez, D. E., Staehelin, L. A. 1985. Structural organization of ultrarapidly frozen barley aleurone cells actively involved in protein secretion. *Planta* 165:455–68

29. Francheschi, V. R., Giaquinta, R. T. 1983. The paraveinal mesophyll of soybean leaves in relation to assimilate transfer and compartmentation. II. Structural, metabolic and compartmental changes during reproductive growth. *Planta* 157:422–31

30. Gerbrandy, S. J. 1974. Glycogen phosphorylase of potatoes. Purification and thermodynamic properties of the adsorption on glycogen. *Biochim. Biophys. Acta* 370:410–18

31. Gerbrandy, S. J., Doorgeest, A. 1972. Potato phosphorylase isoenzymes. *Phytochemistry* 11:2403–7

32. Gerbrandy, S. J., Shankar, V., Shivaram, K. N., Stegemann, H. 1975. Conversion of potato phosphorylase isozymes. *Phytochemistry* 14:2331–33

33. Gerbrandy, S. J., Verleur, J. D. 1971. Phosphorylase isoenzymes: localization and occurrence in different plant organs in relation to starch metabolism. *Phytochemistry* 10:261–66

34. Gibson, R. M., Svensson, B. 1986. Chemical modification of barley malt α-amlyse 2: involvement of tryptophan and tyrosine residues in enzyme activity. *Carlsberg Res. Commun.* 51:295–308

35. Giese, H., Hejgaard, J. 1984. Synthesis

of salt-soluble proteins in barley. Pulse-labeling study of grain filling in liquid-cultured detached spikes. *Planta* 161:172–77

36. Gubler, F., Ashford, A. E., Jacobsen, J. V. 1987. The release of α-amylase through gibberellin-treated cell walls. An immunocytochemical study with Lowicryl K4M. *Planta* 172:155–61

37. Gubler, F., Jacobsen, J. V., Ashford, A. E. 1986. Involvement of the Golgi apparatus in the secretion of α-amylase from gibberellin-treated barley aleurone cells. *Planta* 168:447–52

38. Halayko, A. J., Hill, R. D., Svensson, B. 1986. Characterization of the interaction of barley α-amylase II with an endogenous α-amylase inhibitor from barley seeds. *Biochem. Biophys. Acta* 873:92–101

39. Halmer, P. 1985. The mobilization of storage carbohydrates in germinated seeds. *Physiol. Vég.* 23:107–25

40. Halmer, P., Bewley, J. D. 1982. Control by external and internal factors over the mobilization of reserve carbohydrates in higher plants. In *Encyclopedia of Plant Physiology (NS)*, Vol. 13A: *Plant Carbohydrates I. Intracellular Carbohydrates*, ed. F. A. Loewus, W. Tanner, pp. 748–93. Berlin/Heidelberg/New York: Springer-Verlag

41. Hamabata, A., Garcia-Maya, M., Romero, T., Bernal-Lugo, I. 1988. Kinetics of the acidification capacity of aleurone layer upon solubilization of reserve substances from starchy endosperm of wheat. *Plant Physiol.* 86:643–44

42. Hammond, J. B. W., Preiss, J. 1983. ATP-dependent proteolytic activity from spinach leaves. *Plant Physiol.* 73:902–5

43. Hara-Nishimura, I., Nishimura, M., Daussant, J. 1986. Conversion of free β-amylase to bound β-amylase on starch granules in the barley endosperm during dessication phase of seed development. *Protoplasma* 134:149–53

44. Hayashida, S., Nakahara, K., Kuroda, K., Kamachi, T., Ohta, K. et al. 1988. Evidence for post-translational generation of multiple forms of *Aspergillus awamori* var. *kawachi* glucoamylase. *Agric. Biol. Chem.* 52:273–75

45. Ishizaki, Y., Taniguchi, H., Maruyama, Y., Nakamura, M. 1983. Debranching enzyme of potato tubers (*Solanosum tuberosum* L.). I. Purification and some properties of potato isoamylase. *Agric. Biol. Chem.* 47:771–79

46. Iwaki, K., Fuwa, H. 1981. Purification and some properties of debranching en-

zyme of germinating rice endosperm. *Agric. Biol. Chem.* 45:2683–88

47. Jacobsen, J. V. 1983. Regulation of protein synthesis in aleurone cells by gibberellin and abscisic acid. In *The Biochemistry and Physiology of Gibberellins,* ed. A. Crozier. 2:157–87. New York: Praeger

48. Jacobsen, J. V., Beach, L. R. 1985. Control of transcription of α-amylase and rRNA genes in barley aleurone protoplasts by gibberellin and abscisic acid. *Nature* 316:275–77

49. Jacobsen, J. V., Hanson, A. D., Chandler, P. C. 1986. Water stress enhances expression of an α-amylase gene in barley leaves. *Plant Physiol.* 80:350–59

50. Jacobsen, J. V., Higgins, T. J. V. 1982. Characterization of the α-amylases synthesized by aleurone layers of Himalaya barley in response to gibberellic acid. *Plant Physiol.* 70:1647–53

51. Jenner, C. F. 1982. Storage of starch. See Ref. 40, pp. 700–47

52. Jones, R. L., Carbonell, J. 1984. Regulation of the synthesis of barley aleurone α-amylase by gibberellic acid and calcium ions. *Plant Physiol.* 76:213–18

53. Kakefuda, G., Duke, S. H., Hostak, M. S. 1986. Chloroplast and extrachloroplastic starch-degrading enzymes in *Pisum sativum* L. *Planta* 168:175–82

54. Karn, R. C., Rosenblum, B. B., Ward, J. C., Merritt, A. D., Shulkin, J. D. 1974. Immunological relations and posttranslational modifications of human salivary amylase (Amy_1) and pancreatic (Amy_2) isoenzymes. *Biochem. Genet.* 12:485–99

55. Keeling, P. L., Wood, J. R., Tyson, R. H., Bridges, J. G. 1988. Starch biosynthesis in developing wheat grains. *Plant Physiol.* 87:311–19

56. Kelly, R. B. 1985. Pathways of protein secretion in eucaryotes. *Science* 230:25–32

57. Koshiba, T., Minamikawa, T. 1981. Purification by affinity chromatography of α-amylase—a main amylase in cotelydons of germinating *Vigna mungo* seeds. *Plant Cell Physiol.* 22:979–87

58. Kreis, M., Williamson, M., Buxton, B., Pywell, J., Hejgaard, J., Svendsen, I. 1987. Primary structure and differential expression of β-amylase in normal and mutant barleys. *Eur. J. Biochem.* 169:517–25

59. Lauriere, C., Lauriere, M., Daussant, J. 1986. Immunohistochemical localization of β-amylase in resting barley seeds. *Physiol. Plant.* 67:383–88

60. Leach, H. W., Schoch, T. J. 1962.

Structure of starch granules. III. Solubilities of granular starches in dimethyl sulfoxide. *Cereal Chem.* 39:318–27

61. Lecommandeur, D., Lauriere, C., Daussant, J. 1987. Alpha-amylase inhibitor in barley seeds: localization and quantification. *Plant Physiol. Biochem.* 25:711–15

62. Lee, E. Y. C., Whelan, W. J. 1971. Glycogen and starch debranching enzymes. In *The Enzymes,* ed. P. D. Boyer, 5:191–234. New York: Academic

62a. Levi, C., Gibbs, M. 1984. Starch degradation in synchronously grown *Chlamydomonas reinhardtii* and characterization of the amylase. *Plant Physiol.* 74:459–63

63. Lin, T.-P., Spilatro, S. R., Preiss, J. 1988. Subcellular localization and characterization of amylases in *Arabidopsis* leaf. *Plant Physiol.* 86:251–59

64. Linden, J. C., Tanner, W., Kandler, O. 1974. Properties of glucosyltransferase and glucan transferase from spinach. *Plant Physiol.* 54:752–57

65. Lowy, G. D. A., Sargeant, J. G., Schofield, J. D. 1981. Wheat starch granule protein: the isolation and characterization of a salt-extractable protein from wheat starch granules. *J. Sci. Food Agric.* 32:371–77

66. Ludwig, I., Ziegler, P., Beck, E. 1984. Purification and properties of spinach leaf debranching enzyme. *Plant Physiol.* 74:856–61

67. Lundgard, R., Svensson, B. 1987. The four major forms of barley β-amylase. Purification, characterization and structural relationships. *Carlsberg Res. Commun.* 52:313–26

68. MacDonald, F. D., Preiss, J. 1985. Partial purification and characterization of granule-bound starch synthases from normal and *waxy* maize. *Plant Physiol.* 78:849–52

69. MacDonald, F. D., Preiss, J. 1983. Solubization of the starch-granule-bound starch synthase of normal maize kernels. *Plant Physiol.* 73:175–78

70. MacGregor, A. W., Ballance, D. 1980. Quantitative determination of α-amylase enzymes in germinated barley after separation by electrofocussing. *J. Inst. Brew.* 86:131–33

71. MacGregor, A. W., Daussant, J., Daussant, J. 1981. Isoelectricfocussing and immunochemical analysis of germinated barley α-amylases after freeze-drying and kilning. *J. Inst. Brew.* 87:155–57

72. MacGregor, E. A., MacGregor, A. W. 1985. The action of cereal α-amylases

on solubilized starch and cereal starch granules. In *New Approaches to Research on Cereal Carbohydrates*, ed. R. D. Hill, L. Munck, pp. 149–60. Amsterdam: Elsevier

73. Manners, D. J. 1985. Starch. In *Biochemistry of Storage Carbohydrates*, ed. P. M. Day, R. A. Dixon, pp. 149–204. London: Academic

74. Manners, D. J., Rowe, K. L. 1967. Hydrolysis of the interchain linkages in glycogen-type polysaccharides by a plant enzyme. *Arch. Biochem. Biophys.* 119:585–86

75. McFadden, G. I., Ahluwalia, B., Clarke, A. E., Fincher, G. B. 1988. Expression sites and developmental regulation of genes encoding (1 → 3, 1 → 4)-β-glucanases in germinating barley. *Planta* 173:51–63

76. Melasniemi, H. 1988. Purification and some properties of the extracellular α-amylase-pullulanase produced by *Clostridium thermohydrosulfuricum*. *Biochem. J.* 250:813–18

77. Mercier, C., Frantz, B. M., Whelan, W. J. 1972. An improved purification of cell-bound pullulanase from *Aerobacter aerogenes*. *Eur. J. Biochem.* 26:1–9

78. Mitsui, T., Akazawa, T., Christeller, J. T., Tartakoff, A. M. 1985. Biosynthesis of rice seed α-amylase: two pathways of amylase secretion by the scutellum. *Arch. Biochem. Biophys.* 241:315–28

79. Miyachi, S., Miyachi, S. 1987. Some biochemical changes related to starch breakdown induced by blue light illumination and by addition of ammonia to *Chlorella* cells. *Plant Cell Physiol.* 28:309–14

80. Miyata, S., Akazawa, T. 1982. Enzymic mechanism of starch breakdown in germinating rice seeds. 12. Biosynthesis of α-amylase in relation to protein glycosylation. *Plant Physiol.* 70:147–53

81. Moss, D. W. 1982. *Isoenzymes*, pp. 5–6. New York: Chapman & Hall. 199 pp.

82. Mundy, J. 1984. Hormonal regulation of α-amylase inhibitor synthesis in germinating barley. *Carlsberg Res. Commun.* 49:439–44

83. Mundy, J., Brandt, A., Fincher, G. B. 1985. Messenger RNAs from the scutellum and aleurone of germinating barley encode (1 → 3, 1 → 4)-β-glucanase, α-amylase and carboxypeptidase. *Plant Physiol.* 79:867–71

84. Mundy, J., Hejgaard, J., Hansen, A., Hallgren, L., Jorgensen, K. G., et al. 1986. Differential synthesis in vitro of barley aleurone and starchy endosperm proteins. *Plant Physiol.* 81:630–36

85. Mundy, J., Rogers, J. C. 1986. Selective expression of a probable amylase/protease inhibitor in barley aleurone cells: comparison to the barley amylase/subtilisin inhibitor. *Planta* 169:51–63

86. Mundy, J., Svendsen, I., Hejgaard, J. 1983. Barley α-amylase/subtilisin inhibitor. I. Isolation and characterization. *Carlsberg Res. Commun.* 48:81–90

87. Muthukrishnan, S., Chandra, G. R., Maxwell, E. S. 1983. Hormonal control of α-amylase gene expression in barley. *J. Biol. Chem.* 258:2370–75

88. Ngernprasirtsiri, J., Harinasat, P., Macherel, D., Strzalka, K., Takabe, T., et al. 1988. Isolation and characterization of the amyloplast envelope-membrane from cultured white-wild cells of sycamore (*Acer pseudoplatanus* L.). *Plant Physiol.* 87:371–78

89. Nolan, R. C., Lin, L.-S., Ho, T.-H. D. 1987. The effect of abscisic acid on the differential expression of α-amylase isozymes in barley aleurone layers. *Plant Mol. Biol.* 8:13–22

90. Nomura, K., Mikami, B., Morita, Y. 1986. Interaction of soybean β-amylase with glucose. *J. Biochem.* 100:1175–83

91. Okamoto, K., Akazawa, T. 1980. Enzymic mechanism of starch breakdown in germinating rice seeds. 9. De novo synthesis of β-amylase. *Plant Physiol.* 65:81–84

92. Okita, T. W., Greenberg, E., Kuhn, D. N., Preiss, J. 1979. Subcellular localization of the starch degradative and biosynthetic enzymes of spinach leaves. *Plant Physiol.* 64:187–92

93. Okita, T. W., Preiss, J. 1980. Starch degradation in spinach leaves. Isolation and characterization of the amylases and R-enzyme of spinach leaves. *Plant Physiol.* 66:870–76

94. Ozbun, J. L., Hawker, J. S., Preiss, J. 1972. Soluble adenosine diphosphate glucose-α-1,4-glucan α-glucosyltransferases from spinach leaves. *Biochem. J.* 126:953–63

95. Palmer, G. H., Duffus, J. H. 1986. Aleurone or scutellar hydrolytic enzymes in malting. *J. Inst. Brew.* 92:512–13

96. Palmiano, E. P., Juliano, B. O. 1972. Biochemical changes in the rice grain during germination. *Plant Physiol.* 49:751–56

97. Plaxton, W. C., Preiss, J. 1987. Purification and properties of nonproteolytic degraded ADPglucose pyrophosphorylase from maize endosperm. *Plant Physiol.* 83:105–12

98. Polacco, J. C., Winkler, R. G. 1984. Soybean leaf urease: a seed enzyme? *Plant Physiol.* 74:800–3

116   BECK & ZIEGLER

99. Pongratz, P., Beck, E. 1978. Diurnal oscillation of amylolytic activity in spinach chloroplasts. *Plant Physiol.* 62: 687–89
100. Preiss, J. 1982. Regulation of the biosynthesis and degradation of starch. *Annu. Rev. Plant Physiol.* 33:431–54
101. Preiss, J., Levi, C. 1980. Starch biosynthesis and degradation. See Ref. 9, pp. 271–320
102. Preiss, J., Okita, T. W., Greenberg, E. 1980. Characterization of the spinach leaf phosphorylases. *Plant Physiol.* 66:864–69
103. Preiss, J., Robinson, N., Spilatro, S., McNamara, K. 1985. Starch synthesis and its regulation. See Ref. 8, pp. 1–26
104. Robinson, N. L., Preiss, J. 1987. Localization of carbohydrate metabolizing enzymes in guard cells of *Commelina communis*. *Plant Physiol.* 85: 360–64
105. Rutherford, W. M., Varkey, J. P., McCracken, D. A., Nadakavukaren, M. J. 1986. Lipoprotein as a regulator of starch synthesizing enzyme activity. *Biochem. Biophys. Res. Commun.* 135:701–7
106. Salas, E., Cardemil, L. 1986. The multiple forms of α-amylase of the *Araucaria* species of South America: *A. araucane* (Mol.) Koch and *A. angustifolia* (Bert.) O. Kutz. A comparative study. *Plant Physiol.* 81:1062–68
107. Sargeant, J. G. 1979. The α-amylase isoenzymes of developing and germinating wheat grain. In *The Biochemistry of Cereals*, ed. D. L. Laidman, R. G. Wyn Jones, pp. 339–43. New York: Academic
108. Schächtele, C., Steup, M. 1986. α-1,4-Glucan phosphorylase forms from leaves of spinach (*Spinacia oleracea* L.) I. In situ localization by direct immunofluorescence. *Planta* 167:444–51
109. Schäfer, G., Heber, U., Heldt, H. W. 1977. Glucose transport into spinach chloroplasts. *Plant Physiol.* 60:286–89
110. Schilling, N. 1982. Characterization of maltose biosynthesis from α-D-glucose-1-phosphate in *Spinacia oleracea* L. *Planta* 154:87–93
111. Schilling, N., Kandler, O. 1975. α-Glucose-1-phosphate, a precursor in the biosynthesis of maltose in higher plants. *Biochem. Soc. Trans.* 3:985–87
112. Schneider, E. M., Becker, J.-U., Volkmann, D. 1981. Biochemical properties of potato phosphorylase change with its intracellular localization as revealed by immunological methods. *Planta* 151: 124–34
113. Senser, M., Beck, E. 1979. Kälteresistenz der Fichte. II. Einfluβ von Photoperiode und Temperatur auf die Struktur und photochemischen Reaktionen von Chloroplasten. *Ber. Dtsch. Bot. Ges.* 92:243–59
114. Shain, Y., Mayer, A. M. 1968. Activation of enzymes during germination: amylopectin-1,6-glucosidase in peas. *Physiol. Plant.* 21:765–76
115. Sharma, R., Schopfer, P. 1987. Phytochrome-mediated regulation of β-amylase mRNA level in mustard (*Sinapis alba* L.) cotyledons. *Planta* 171: 313–20
116. Shimomura, S., Nagai, M., Fukui, T. 1982. Comparative glucan specificities of two types of spinach leaf phosphorylase. *J. Biochem.* 91:703–17
117. Singh, S., Sanwal, C. G. 1975. Characterization of multiple forms α-glucan phosphorylase from *Musa paradisiaca* fruits. *Phytochemistry* 14:113–18
118. Smith, M. A., Jacobsen, J. V., Kende, H. 1987. Amylase activity and growth in internodes of deepwater rice. *Planta* 172:114–20
119. Steup, M., Conrads, J., Van Berkel, J. 1987. Compartment-specific phosphorylase forms from higher plants. In *Progress in Photosynthesis Research*, ed. J. Biggins, 3:479–82. Dordrecht: Martinus Nijhoff
120. Steup, M., Robenek, H., Melkonian, M. 1983. In-vitro degradation of starch granules isolated from spinach chloroplasts. *Planta* 158:428–36
121. Steup, M., Schächtele, C. 1981. Mode of glucan degradation by purified phosphorylase forms from spinach leaves. *Planta* 153:351–61
122. Stewart, G. G. 1987. The biotechnological relevance of starch-degrading enzymes. *Crit. Rev. Biotechnol.* 5:89–94
123. Stitt, M. 1984. Degradation of starch in chloroplasts: a buffer to sucrose metabolism. See Ref. 23, pp. 205–29
124. Stitt, M., Heldt, H. W. 1981. Simultaneous synthesis and degradation of starch in spinach chloroplasts in the light. *Biochim. Biophys. Acta* 638:1–11
125. Stitt, M., Steup, M. 1985. Starch and sucrose degradation. In *Encyclopedia of Plant Physiology (NS)*, Vol. 18: *Higher Plant Cell Respiration*, ed. R. Douce, D. A. Day, pp. 347–89. Berlin/Heidelberg/New York/Tokyo: Springer-Verlag
126. Svendsen, I., Hejgaard, J., Mundy, J. 1986. Complete amino acid sequence of the α-amylase/subtilisin inhibitor from barley. *Carlsberg Res. Commun.* 51:43–50

127. Svensson, B. 1988. Regional distance sequence homology between amylases, α-glucosidases and transglucanosylases. *FEBS Lett.* 230:72–76

128. Takasaki, Y. 1987. Pullulanase-amylase complex enzyme from *Bacillus subtilis.* *Agric. Biol. Chem.* 51:9–16

129. Thoma, J. A., Spradlin, J. E., Dygert, S. 1971. Plant and animal amylases. See Ref. 62, pp. 115–89

130. Trewavas, A. J. 1982. Growth substance sensitivity: the limiting factor in plant development. *Physiol. Plant.* 55: 60–72

131. Tsai, C. Y., Nelson, O. E. 1968. Phosphorylases I and II of maize endosperm. *Plant Physiol.* 43:103–12

132. Tsai, C. Y., Nelson, O. E. 1969. Two additional phosphorylases in developing maize seeds. *Plant Physiol.* 44:159–67

133. Verbücheln, O., Steup, M. 1983. Water-stress induced increase of starch degrading enzyme activities in the CAM-plant *Aloe arborescens.* *Plant Physiol.* 72(Suppl. 1):36

134. Walker, G. J., Hope, P. M. 1963. The action of some α-amylases on starch granules. *Biochem. J.* 86:452–62

135. Weiner, H., Stitt, M., Heldt, H. W. 1987. Subcellular compartmentation of pyrophosphate and alkaline pyrophosphatase in leaves. *Biochim. Biophys. Acta* 893:13–21

136. Weselake, R. J., Hill, R. D. 1983. Inhibition of alpha-amylase-catalyzed starch granule hydrolysis by cycloheptaamylose. *Cereal Chem.* 60:98–101

137. Weselake, R. J., MacGregor, A. W.,

Hill, R. D. 1983. An endogenous α-amylase inhibitor in barley kernels. *Plant Physiol.* 72:809–12

138. Weselake, R. J., MacGregor, A. W., Hill, R. D. 1985. Endogenous alpha-amylase inhibitor in various cereals. *Cereal Chem.* 62:120–23

139. Yamada, J. 1981. Conversion of active and inactive debranching enzymes in rice seeds. *Agric. Biol. Chem.* 45:747–50

140. Yamada, J. 1981. Inactive debranching-enzyme in rice seeds, and its activation. *Carbohydrate Res.* 90:153–57

141. Yamada, J. 1981. Purification of debranching enzyme from mature rice seeds. *Agric. Biol. Chem.* 45:1269–70

142. Yamada, J. 1981. Purification of oat debranching enzyme and occurrence of inactive debranching enzyme in cereals. *Agric. Biol. Chem.* 45:1013–15

143. Young, R. A., Hagenbüchle, O., Schibler, U. 1981. A single mouse α-amylase gene specifies two different tissue-specific mRNAs. *Cell* 23:451–58

144. Ziegler, P. 1988. Partial purification and characterization of the major endoamylase of mature pea leaves. *Plant Physiol.* 86:659–66

145. Ziegler, P., Beck, E. 1986. Exoamylase activity in vacuoles isolated from pea and wheat leaf protoplasts. *Plant Physiol.* 82:1119–21

146. Zwar, J. A., Hooley, R. 1986. Hormonal regulation of α-amylase gene transcription in wild oat (*Avena fatua* L.) aleurone protoplasts. *Plant Physiol.* 80:459–63.

Annu. Rev. Plant Physiol. Plant Mol. Biol. 1989. 40:119–38

THE SINK-SOURCE TRANSITION IN LEAVES

Robert Turgeon

Section of Plant Biology, Division of Biological Sciences, Cornell University, Ithaca, New York 14853

CONTENTS

INTRODUCTION

Carbohydrate metabolism in leaves changes profoundly during development. Young leaves are heterotrophic; they depend in part on carbohydrate imported from other regions of the plant. Mature leaves, on the other hand, are autotrophic; they produce an excess of photoassimilate and act as the plant's major sources of transport sugar. This conversion from sink to source status

119

1040-2519/89/0601-0119$02.00

marks a fundamental transition in the physiology of the leaf, and a considerable amount of study has gone into understanding its biochemical and structural characteristics. Apart from the central role it plays in leaf development, the transition is interesting for another reason; it provides a straightforward system with which to compare the characteristics of sink and source tissue without having to deal with the complications introduced by studying different organs.

In early studies, the primary concern of investigators was to describe changes in leaf function and structure as the sink-source transition proceeds. More attention has been devoted recently to understanding the causal mechanisms of the transition. This subject provides the major focus of the present article.

The sink-source transition in leaves has been discussed briefly elsewhere (69, 101), and reviews of the related topics of leaf development (5, 18, 19, 20), phloem unloading (17, 21, 24, 51, 52), and phloem loading (4, 17, 21, 23, 24, 46, 52, 108) are available. A number of reviews on these subjects can also be found in a volume on phloem transport (16).

STRUCTURAL MATURATION OF LEAVES

Basipetal Maturation

In dicotyledonous plants the transition from photoassimilate sink to source status begins shortly after the leaf has begun to unfold. At this point in development the major morphogenetic events that determine leaf shape are over. Maturation of the phloem and xylem in the midrib and higher-order veins, which occurs in the acropetal (lamina base to tip) direction, is largely complete before the transition begins. While the leaf unfolds, structural and functional maturation of the smaller veins proceeds in the basipetal (tip to base) direction (3). Therefore, there is a gradient in degree of leaf maturation from the base to the tip of the lamina at any point in time during the sink-source transition. This representation of different developmental stages in a single leaf is useful in experimental studies.

Timing of the Sink-Source Transition

Leaves of dicotyledonous plants stop importing and begin to export when they are 30–60% fully expanded. Developing leaves continue to import photoassimilate from source leaves for a period after they have begun to export their own products of photosynthesis (1, 35, 54, 55, 59, 60, 68, 96–98, 103, 111). Jones & Eagles (59) first suggested that this is a result of basipetal maturation; the lamina tip stops importing and begins to export while transport of photoassimilate from other leaves into the relatively immature leaf base continues. Jones & Eagles (59) presented autoradiographic evidence in support of this hypothesis, but the heat-drying method they used to prepare the leaves is known to cause severe displacement artifacts and is not reliable. Later, Larson & Dickson (66), using *Populus deltoides,* and Turgeon & Webb (103), using *Cucurbita pepo, Beta vulgaris,* and *Nicotiana tabacum,* demonstrated by whole-leaf autoradiography that import of labeled translocate continues into the base of a transition leaf after it has terminated at the tip. The boundary between the importing and nonimporting regions is usually clear, although label frequently extends into the nonimporting zone in the larger veins. In many published autoradiographs the entire importing zone appears to have approximately the same level of radioactivity. This is an artifact caused by overexposure. When shorter exposure times are used, or activity is measured by scintillation counting (102), a gradual decline in amount of imported photoassimilate from the base of the lamina to the import-termination boundary is seen. The boundary moves progressively in the basipetal direction as development proceeds until the lamina no longer imports (103). By trimming the lamina of *Cucurbita pepo* it was shown that the development of export capacity also proceeds in the basipetal direction following cessation of import (103).

Development of Minor Veins

The minor venation is the distribution network of the leaf. As the veins grow to keep pace with expansion of the lamina they provide first an importing, and later an exporting, conduit. Vein maturation as it relates to import and phloem unloading has been examined carefully only in two studies. In *Beta vulgaris,* imported labeled photoassimilate is carried primarily in fourth-order veins, but at the import-termination boundary the smallest veins (fifth-order) are also used (89). In tobacco, the finest ramifications of the vein network are not functional at all during the import phase (101, 102); the smallest veins that become labeled are those that were termed Class III veins in a later study (29). An extensive network of veins (Classes IV and V) are enclosed within the boundary of Class III veins and are only used for export. The phloem in these veins does not mature until after the import phase is over (29). Together, the studies on *B. vulgaris* and *N. tabacum* indicate that most of the photoassimilate imported by sink leaves is unloaded by moderately large veins and that the smallest veins are relatively or completely unimportant in this regard.

This makes sense if we consider that leaves at this stage of development are still expanding and the minor-vein phloem must retain the capacity for further elongation. During the import phase, when the tissue is compact, the finest minor veins are not needed for distribution of imported assimilate, and since the sieve elements are immature they are capable of further growth to accommodate expansion. It is only during the final stages of growth that the sieve elements mature in time for the initiation of export.

The development of the phloem has been studied in detail to determine how structural maturation of minor veins is involved in the initiation of photoassimilate export. Isebrands & Larson (58) concluded that structural development and export are synchronized in *Populus*. Fellows & Geiger (35) came to the same conclusion but noted that many of the sieve elements in the finest veins (fifth-order) are structurally mature before export begins (and, in fact, are used during the import phase—see above). In their opinion the structural development of veins is a preparatory step but does not itself trigger the onset of export. In *C. pepo* the situation is somewhat more complex because the minor veins are bicollateral (106). The adaxial phloem contains a single, relatively large sieve element and a companion cell of comparable size. These adaxial sieve elements in the veins delimiting the areoles mature before import ceases, well in advance of export and structural maturation of sieve elements in the abaxial phloem (105). Abaxial sieve elements mature just at the beginning of the export period. This could indicate that export is triggered by development of the abaxial phloem; but this is unlikely because the abaxial sieve elements of larger veins, which also load photoassimilate, mature in advance of export. Some other event(s) apparently causes the leaf to begin exporting photoassimilate. Fellows & Geiger (35) suggested that the critical event is the initiation of phloem loading and subsequent increase in solute concentration in the sieve-element companion cell (SE-CC) complex required to drive export.

Bidirectional Transport

Since leaves of intermediate age are capable of bidirectional transport it is of interest to find out how this is accommodated by the vascular system. Tracing pathways of movement by autoradiography is difficult since the labeled compounds are soluble. In a detailed study, Vogelmann et al (110) demonstrated that, in *Populus*, translocation occurs in specific vascular bundles from source to sink leaves. When export begins, the exporting and importing assimilates are confined to different bundles—i.e. bundles used for export from the lamina tip are distinct from those used for import into the base. No direct evidence was obtained for bidirectional transport in a single vascular bundle.

In species with bicollateral phloem there is evidence that there is some specialization of the external phloem for export and internal phloem for import (10, 54, 81), but both phloem types are capable of translocating assimilates in both directions (54, 111). The timing of sieve element maturation in the minor veins of *Cucurbita pepo* suggests that the adaxial (internal) phloem constitutes a preferential pathway for import while the abaxial phloem is used for export (105). An autoradiographic study of *Cucumis melo* transport supported an export role for the abaxial phloem, but no labeling of adaxial phloem occurred in either importing or exporting leaves (91). This subject warrants further attention.

CARBON BALANCE

Demand for Imported Carbon

The early growth of a leaf is supported by imported carbohydrate. As the lamina expands, the rate of photosynthesis increases (13, 25, 26, 35, 53, 55, 67, 68, 93, 104) while the rates of respiration (25, 55, 68, 104) and growth (66, 104) decline. Eventually, positive carbon balance is reached—i.e. the amount of carbon accumulated during the light period by net photosynthesis is sufficient to meet the requirements of growth and night respiration. Curves of carbon balance over the life span of a *Cucumis sativus* (53, 55), *Cucurbita pepo* (104), or tomato (54) leaf illustrate that net import ceases when the lamina is one sixth to one half fully grown. Approximately the same values are obtained when import is measured by translocation of radioactive photoassimilate from older leaves (61, 111). Maximum rate of import occurs earlier, at approximately 20% full leaf length in sugar beet (35) and *C. pepo* (104). These data suggest that import declines considerably while the leaf is still actively growing, but this is a bit misleading; import does not begin to decline until after the *rate* of growth has peaked (104).

Partitioning of Carbon in Sink Leaves

Growing leaves have two sources of carbon: import and photosynthesis. Joy (62) found that carbon for protein synthesis in the growing sugar beet leaf comes preferentially, but not exclusively, from photosynthesis whereas imported sucrose is preferentially used in the synthesis of structural carbohydrates. These results have been confirmed for other species (13, 27, 54). Dickson & Larson (27) pointed out that the proportion of incoming carbon from the two sources changes dramatically as the leaf grows. Although the same allocation pattern exists in young and more mature growing leaves, in the latter the amount of carbon derived from photosynthesis greatly exceeds that from the transport system. This means that, in the final analysis, most carbon for structural carbohydrate comes from photosynthesis when the entire growing period is analyzed.

Several authors have examined levels of sugar-hydrolyzing enzymes in developing leaves to determine whether hydrolysis might be a key event in determining growth rate. In general, the data on sucrose hydrolysis by either invertase or sucrose synthase are contradictory whereas those on enzymes that split other imported compounds such as stachyose, raffinose, or polyols, are more consistent.

In *Phaseolus vulgaris* the major sucrose-hydrolyzing enzyme is a readily soluble acid invertase that is most active during early stages of leaf growth (78). The high activity of the enzyme correlates well with a high hexose/sucrose ratio at the same period of development. A similar correlation of acid invertase activity and growth is found in *Citrus* (87) and *Lolium* (84). In *Citrus*, invertase has an acid pH optimum at early stages, but later a neutral invertase is more important (87). Unlike the *Citrus* and *Phaseolus* enzymes the acid invertase of *Lolium* is in the insoluble fraction, suggesting that it is extracellular (87). It is unclear whether an extracellular invertase could be involved in the hydrolysis of imported sucrose since the preponderance of evidence indicates that phloem unloading in sink leaves is symplastic—i.e. through plasmodesmata (see below); if this is the case, imported sucrose would not be exposed to an extracellular enzyme. Indeed, analysis of the fate of asymmetrically labeled sucrose in sugar beet indicates that there is no extracellular hydrolysis of sucrose during its accumulation into sink leaves (44). In sugar beet, the activities of acid invertase in sink and source leaves are approximately the same (45). Giaquinta suggested that sucrose enters the vacuole of mesophyll cells in sink leaves intact and is hydrolyzed there by soluble acid invertase to hexoses which traverse the tonoplast back into the cytoplasm in response to demand (45). As in sugar beet, the amount of invertase (acid or alkaline) is not correlated with development in the leaves of a variety of other species (15) and actually increases with declining growth in tomato (75). Clearly, the role of invertase in hydrolysis of imported sucrose, and especially in controlling rate of import, is unclear. As Loescher et al (70) point out, the interpretation of results on invertase are complicated by the presence of multiple forms of the enzyme, the fact that reported levels are higher than needed for complete sucrose hydrolysis, and the fact that the timing of substrate and enzyme level changes often do not coincide. A much more sophisticated understanding of the compartmentation of substrate and enzyme forms and the kinetics of unloading and hydrolysis will be needed before the importance of invertase in regulating growth can be assessed.

Similarly, the importance of sucrose synthase (SS), which is generally considered to function in sucrose hydrolysis, is unclear. Claussen et al (15) determined that SS activity declines considerably as leaves age. They also found a lamina base-to-tip gradient, indicating that changes in activity follow the same basipetal pattern as the sink-source transition. There was no such base-to-tip gradient in acid or alkaline invertase activity. SS activity also declines during development of *Citrus* leaves (87); but in sugar beet there is little difference in activity between sink and source leaves, (45) and in celery (22) SS is highest in mature leaves. Suboptimal extraction conditions may account for the low SS activities reported by some workers (see 15 for discussion).

Schmalstig & Hitz (90) took a novel approach to determining the relative contribution of invertase and SS to sucrose hydrolysis. They provided either

[^{14}C]sucrose or [^{14}C]fluorosucrose (FS) to source leaves and monitored incorporation into sink leaf tissue. Since FS is hydrolyzed by invertase and SS at very different rates it was possible to calculate the contribution of each enzyme to hydrolysis in vivo. All cleavage of sucrose in very young leaves was by SS, but the contribution of invertase to metabolism increased to approximately half by the end of the sink period.

Many plants translocate sugars of the raffinose series as well as sucrose (116). An alkaline form of the stachyose- and raffinose-hydrolyzing enzyme α-galactosidase (42) has high activity in young *Cucurbita pepo* (41) and *Cucumis sativus* (82) leaves, and this activity falls considerably as the leaves mature. The acidic forms of the enzyme change little during development in *C. pepo* (95) but decline in *C. sativus* (82).

The polyols sorbitol and mannitol may be the major translocated forms of carbon in some species. Mannitol is oxidized by immature celery leaf tissue but only slowly by fully grown leaves; sucrose, on the other hand, is degraded by both (34). In apple, a NAD-dependent sorbitol dehydrogenase is found in sink leaves, and the level of this enzyme declines during the sink-source transition (70, 80). In fully expanded pear leaves the metabolism of sorbitol is very slow (7).

Synthesis of Transport Compounds

The metabolism of photosynthetically derived, as opposed to imported, carbohydrate in developing leaves changes dramatically during the sink-source transition. A universal feature of this stage of development is the increase in the proportion of fixed carbon allocated to the "soluble" versus "insoluble" fraction and an increase in the amount in the "neutral" fraction, which contains the transport sugars (22, 27, 28, 64). When absolute concentrations of sugar are analyzed it is common to see an increase in transport sugars, especially sucrose in those species that export sucrose exclusively. However, these data can be difficult to interpret, especially if an entire transition leaf, including sink and source regions, is extracted because both imported and synthesized compounds are included. More cogent information is usually obtained by the analysis of labeled sugars. Sucrose appears to be synthesized from $^{14}CO_2$ in small amounts in young leaves before export begins (45, 104). Similarly, in celery, which exports mannitol (22), and apricot, which exports sorbitol (8), these sugar alcohols are first made while the leaf is still in the importing stage. However, sugars of the raffinose group are made from photosynthetically fixed carbon only at the commencement of export, not before (92, 104). Raffinose and stachyose can be detected in younger leaves but are presumably import sugars (82).

Sucrose-phosphate synthase (SPS), which converts triose phosphates to sucrose and inorganic phosphate, is considered to be of primary importance in regulating sucrose synthesis (57, 94). SPS activity has been detected in young, presumably importing, leaves (22, 83). Giaquinta did not find SPS activity in sugar beet sink leaves, but he did record the synthesis of sucrose at this early stage of development (45). Although sucrose could have been synthesized by another enzyme, possibly SS, it should be noted that the conditions for SPS analysis vary both from one species to another and during development (e.g. see 70, 93) and must be optimized in each case. A great deal has been learned about "coarse" and "fine" control of SPS in recent years (63, 94), and it would be worthwhile to reexamine the initiation of sucrose synthesis during development in light of these findings.

Increasing synthesis of mannitol during leaf development correlates with a dramatic rise in activity of mannose 6-phosphate reductase, the enzyme that catalyzes the synthesis of mannitol 1-phosphate (85). Similarly, aldose 6-phosphate reductase (A6PR) activity increases with increasing leaf age in the leaves of apple, a sorbitol translocator (70).

Raffinose and stachyose syntheses coincide precisely with the onset of export in developing leaves of *Cucurbita* (92, 104). The first enzyme unique to the synthetic pathway of this family of sugars, galactinol synthase, appears to play an important role in regulation of partitioning of carbon between sucrose and galactosyl-sucrose synthesis (48, 82). Galactinol synthase activity is apparent in young, importing cucumber leaves, but at very low levels, and rises substantially at the time of the sink-source transition.

The physiology of phloem loading of the raffinose family of sugars appears to differ significantly from that of sucrose loading (see below). In fact, it has been suggested on the basis of sugar and enzyme analysis on isolated mesophyll cells and veins that these sugars are synthesized in the minor veins (92). However, other studies do not support this conclusion (50, 72, 74). Recently, Madore et al (72) conducted elegant experiments in which mesophyll and vascular tissue–enhanced sections were isolated from *Xerosicyos danguyi,* a member of the Cucurbitaceae. They convincingly demonstrated synthesis of the raffinose family of sugars in mesophyll tissue. It is still possible that these sugars are also made in the phloem. The interaction between tissues in sugar synthesis and transport could prove to be complex.

MECHANISM OF THE SINK-SOURCE TRANSITION

Most work to date on the sink-source transition has been descriptive in its analysis of changing physiology and structure. In future, more attention should be devoted to understanding how the transition occurs—i.e. a more mechanistic than descriptive approach should be taken. One fundamental question, only recently addressed, is whether the transition results directly from changes in carbon balance. It is clear that the timing of the transition coincides with net positive accumulation of carbon in the leaf (see above).

However, this correlation is to be expected and says nothing about cause and effect. To address this question experimentally, achlorophyllous (albino) tobacco shoots were grown in culture and grafted to the detopped stems of green tobacco plants (100). The grafted shoots grew for a period of many months using carbohydrate imported from the stock. As individual leaves grew and matured they stopped importing ^{14}C-labeled assimilates from the stock plant at the same stage of development as green leaves and soon died. These results indicate that termination of import is not caused by the achievement of positive carbon balance in tobacco, since albino leaves never attain that condition.

This conclusion is supported by analysis of the literature on import into mature leaves. If carbon balance controls import, the direction of transport should be reversed when an attached green leaf is kept in darkness. Some early investigators believed that reversal of carbohydrate flow made older, shaded leaves parasitic on unshaded leaves (see references in 71). However, it was subsequently pointed out that the respiration rate of older leaves is very low and drops even further when the leaf is darkened, making it highly unlikely that they are net importers (56, 71). In more recent studies, ^{14}C transport from one mature leaf to another has been analyzed, primarily by whole-leaf autoradiography. In a number of studies either no such transport was seen when the mature "recipient" leaf was darkened, or the amount imported was slight and was attributed to movement of label in the transpiration stream (2, 6, 65). However, other results have been interpreted to mean that the direction of assimilate flow can be reversed (31 and references in 101). In a sense this is correct, for label can be seen in the published autoradiographs. However, three important points must be kept in mind: The protocols used to induce import by reducing competing sink strength are often drastic in the extreme (9), the amount of labeled assimilate imported appears to be exceedingly small (96), and almost all the imported label remains in the veins (see Discussion in 101). This indicates that phloem unloading is very restricted. The most obvious explanation of these data is that the phloem of the midrib and larger veins in mature leaves unloads assimilate into surrounding parenchyma cells no matter what the source of that assimilate is whereas the phloem of smaller veins is able to load but will not unload, even if the lamina is unable to fix carbon from the atmosphere. Consistent with this view is the demonstration that mature albino tobacco tissue is able to vein-load exogenous sucrose (100). The polarity of the loading system restricts import and prevents darkened leaves from becoming a serious drain on the carbon resources of the plant. This is not to say that a small amount of sucrose can not escape the phloem of minor veins. "Leakage" of small quantities of radiolabeled sucrose into the apoplast from leaf tissue, presumably including the minor veins, has been detected in several plant materials (77, 99, and references therein). Some egress of sucrose through plasmodesmata might also occur. However, examination of autoradiographs indicates that exit of translocate from veins of sink leaves (11, 89, 102) far exceeds that from darkened source leaves. The latter system seems inappropriate for the study of phloem unloading.

It is also important to note that a lag period may be seen between the end of the import phase and the beginning of export in a given region of the lamina (35, 89, 103). This would not be expected if import is terminated by the initiation of export. The fact that aphids induce import in mature leaves (12, 49, 115) is consistent with the concept that unloading is inhibited, because aphids feed in the phloem itself. Cytokinin application to darkened mature leaves induces import (79). Thus, it would be worthwhile to examine the effects of cytokinin on phloem unloading.

Is it the case that unloading stops entirely as the result of a developmental program, or does carbon balance play a role as well? Cessation of import is not abrupt, as scintillation counting and autoradiography demonstrate (89, 102). It may be that the unloading pathway is gradually restricted in response to a developmental program and this results in a slow decline in import. It is also possible that the progressive decline in amount of assimilate imported is due to changing carbon balance and that the final, irreversible blockage of the unloading pathway that occurs at the import-termination boundary is sudden.

Restricted phloem unloading is apparently not a universal characteristic of mature leaf tissue. Photoassimilate is transported from the green to the albino region of mature variegated *Coleus blumei* leaves (38, 112) where it unloads, apparently without serious hindrance (112). This albino tissue is capable of very little, if any, phloem loading (112).

In summary, the evidence to date indicates that import is terminated primarily, if not exclusively, by interruption of phloem unloading and that the development of phloem loading and export capability is largely irreversible. The data from mature albino leaf tissue indicates that these processes may be linked; in tobacco, albino tissue vein-loads sucrose but cannot import, whereas, in *Coleus,* import and unloading are unimpeded but the tissue does not load sucrose. It is evident, therefore, that a better understanding of the sink-source transition hinges on research into the control of phloem unloading and loading.

PHLOEM UNLOADING IN SINK LEAVES

From available evidence it appears that import is terminated in developing leaves by some restriction of the phloem unloading pathway (see above). To understand how this restriction occurs we have to know what pathway the sugar takes as it is transported from the SE-CC complex to surrounding tissues in sink leaves. If the pathway is entirely symplastic it is possible that transport through plasmodesmata is retarded.

Schmalstig & Geiger (88) used p-chloromercuribenzenesulfonic acid (PCMBS) to block membrane transport of sucrose in sugar beet sink leaves and determined that this treatment did not block import. They argued for symplastic unloading—i.e. through the plasmodesmata—reasoning that PCMBS would block retrieval of translocate from the free space if this compartment were part of the pathway. I made essentially the same argument, finding that anoxia does not inhibit phloem unloading in tobacco sink leaves (102). In both cases very little imported translocate was found in the free space during inhibitory treatments. While suggestive, inhibitor studies are difficult to interpret. One problem is that neither PCMBS nor anoxia inhibits sucrose uptake entirely. If the unloading pathway were partly apoplastic, the inhibitors might cause a buildup in concentration of unloading sugar in the apoplast that would then result in increased uptake by the first-order transport system, which is strictly concentration dependent (76). Although only 3% of imported sugar was found in the free space of sugar beet leaves during PCMBS treatment, this was 10 times the control value (88).

Another approach used by Schmalstig & Geiger (88) was to feed the sink leaf L-[^{14}C]glucose either in the transpiration stream or the phloem; the resulting autoradiographic images were quite different, suggesting that the sugar in the phloem did not unload into the apoplast. However, the two treatments were not strictly comparable since the sugar did not seem to move through the xylem to the smaller veins from which imported translocate unloads. Giaquinta (44) used asymmetrically labeled sucrose to demonstrate that unloading in sugar beet sink leaves does not involve hydrolysis. This result effectively eliminates a mechanism in which sucrose is unloaded into the apoplast and hydrolyzed before retrieval. However, as Giaquinta pointed out, it does not eliminate the possibility of an apoplastic pathway and retrieval of sucrose without hydrolysis. It therefore can not be used to distinguish between symplastic and apoplastic unloading routes. Insofar as quantitation of plasmodesmata can provide evidence for symplastic transport, a morphometric analysis of the importing veins in tobacco leaves is entirely consistent with this route (29). As the tissue passes through the sink-source transition the number of plasmodesmata along the putative unloading route declines significantly (29).

In total, the evidence is most compatible with symplastic unloading in sink leaves. However, much more work needs to be done to corroborate this conclusion. It should be kept in mind that the unloading pathway may not be the same in different species, in veins of different size in the same sink leaf, or in the same vein in the same leaf at different stages of development. Furthermore, both pathways could be operational at the same time.

PHLOEM LOADING IN SOURCE LEAVES

Apoplastic versus Symplastic Phloem Loading

The pathway of phloem loading in source leaves needs to be understood before we can determine what preparatory changes take place as the tissue is readied for export during the sink-source transition. Over a period of almost 20 years, Geiger, his coworkers, and others have formulated the concept that sucrose passes from its site of synthesis from one mesophyll cell to another until, somewhere in the proximity of the phloem, it enters the apoplast and is then loaded, selectively and against a concentration gradient, into the SE-CC complex by cotransport with protons. The evidence for this hypothesis has been reviewed (46). Although the data in favor of apoplastic loading is compelling for the few species studied in detail, the great variation in source-leaf structure and physiology suggests that this route may not be universal. There is growing enthusiasm for the concept that, in some species, transport sugars follow an entirely symplastic route to the SE-CC complex. The evidence for symplastic phloem loading is meager at best. However, the significance of proving that such a mechanism exists is so great that the idea is well worth pursuing. The subject of symplastic phloem loading has been reviewed elsewhere (23, 108). I make only a few selective points here.

Structural Specialization of Companion Cells

The anatomies of several species have been studied carefully for clues to the pathway of sugar movement. The number of plasmodesmata at the critical interface between the SE-CC complex and surrounding cells varies widely (11 and references therein; 114); in some species, such as *Pisum sativum* (114) and *Zea mays* (32, 33), there are few; in others, such as *Populus deltoides* (86) and *Coleus blumei* (37), there are many. There appears to be a correlation between the structural specialization of the companion cell and the number of plasmodesmata that join it to surrounding cells.

   The terminology used to describe the cell types exhibiting these different forms of specialization has become somewhat confused in the literature and it is worthwhile to attempt to clarify the definitions here. The term *companion cell* is used in the usual ontogenetic sense as being derived from the same precursor as the sieve element (30, p. 122), although in mature tissue the identification of companion cells as distinct from other parenchymatic elements of the phloem is sometimes problematic. *Transfer cells* are distinguished by the presence of cell wall ingrowths that presumably amplify uptake from the apoplast by increasing plasmalemma surface area (47). Transfer cells are found in a number of locations in some, but not all, plants; and in minor veins they may, or may not, be companion cells. *Transfer cell* is not a suitable term for all cell types presumed to function in transferring solute from one cell to another or from apoplast to symplast; it should be used only when wall ingrowths are evident. The use of the term *intermediary cell* (Esau's translation of *Übergangszelle*) has a complicated early history (see 30, pp. 159–63). Fischer (36) first used it in his study of the phloem of the Cucurbitaceae. In these species the companion cells of the minor vein abaxial phloem are very distinctive; they are large, densely cytoplasmic, and rather than being buried within the interior of the phloem they broadly contact the bundle sheath. In a later study (106) it was shown that the interface between the intermediary cell and bundle sheath in *Cucurbita pepo* is traversed by many plasmodesmata arranged in clusters, a feature not common to "ordinary" companion cells (see 37 for a quantitative comparison). Since that time, companion cells with virtually identical structure and placement within the vein have been described in the minor vein phloem of *Cucumis melo* (91), *Coleus blumei* (37), and a number of other species (39, 40), and the term *intermediary cell* has been applied to them. While future research may blur this distinction, it appears that the companion cell of the abaxial phloem of *C. pepo* minor veins is typical of a distinct cell type, and the name *intermediary cell,* in conformity with the modern usage of this term (106), should be reserved for this cell type.

Gamalei & Pakhomova (40) and Gamalei (39) have attempted to correlate companion cell type with frequency of plasmodesmata, pathway of loading, type of sugar or sugar alcohol transported, and growth habit. According to these authors, species in which many plasmodesmata join the SE-CC complex to the bundle sheath (Type I) tend to transport galactosyl-sucrose oligosaccharides. They are mostly tree species. Type II have few or no plasmodesmata with the bundle sheath and transport mainly sucrose. Type III cells are transfer cells and again are associated with sucrose transport. Both Type II and III cells are usually found in herbaceous species. Type I cells are presumed to load symplastically and Types II and III to load from the apoplast.

There are major problems with this broad survey. Few quantitative data were provided, plasmodesmatal frequencies between companion cells and phloem parenchyma cells were apparently not considered, and some obvious discrepancies exist. For example, *Cucurbita* and *Coleus* have both intermediary cells (Type I) and "ordinary" companion cells (Type II) but are listed as having only Type I. Also, careful observation of transfer cells reveals that plasmodesmata do connect them to both the bundle sheath and phloem parenchyma (47, 114). Nevertheless, this broad synthesis may be a fruitful approach to the study of phloem loading and translocation; it deserves further attention.

Although analyses of plasmodesmatal frequencies have been, and will continue to be, useful, they cannot tell us whether or not solute loads symplastically. Even in *Pisum sativum*, in which the companion cells are almost totally isolated symplastically from surrounding cells, the few plasmodesmata present could account for observed rates of loading given the more extreme values in the literature for solute flux across plasmodesmata (114). Furthermore, it is entirely possible that the plasmodesmata at this interface are specialized for transport at rates above those reported for other systems. At the other extreme it is not clear that the presence of numerous plasmodesmata indicates symplastic loading; pores could have served their function during the unloading phase in sink leaves and then become nonfunctional during the sink-source transition.

Free Space Analysis and Uptake of Exogenous Sucrose

Another approach to the study of symplastic loading is to analyze free space sugars (73, 91). In the two cucurbits studied, the labeled sugar in the free space following $^{14}CO_2$ application resembled in composition the sugars of the cytoplasm more than those of the translocation stream. These data apparently do not support the hypothesis of an apoplastic loading route in curcurbits; if the pathway were apoplastic there should be a preponderance of transport sugars in the free space. However, free space analysis has many inherent problems—i.e. difficulties in estimating the efficiency of trapping, contamination from damaged cells, efflux not associated with transport, and induction of extracellular hydrolytic enzyme activity. It is possible that transport compounds are present in the free space but cannot be detected against the background of other sugars owing to limitations of the technique. More compelling is the evidence of Madore & Webb (73) that transport of exogenously supplied sugars does not follow the pattern expected from the apoplastic loading model, especially the finding that stachyose appears in the translocation stream only very slowly following application to the free space.

Evidence from autoradiographic and kinetic analyses following [^{14}C]sucrose application to leaf discs has been used extensively to explore transport pathways. In general, autoradiographs demonstrate preferential uptake of sucrose into the veins, an observation obviously in accord with apoplastic loading (46). This interpretation has frequently been criticized on the grounds that mesophyll cells also accumulate considerable quantities of sucrose, with similar kinetics (see 108 for references). This criticism seems beside the point; if transport sugars enter the apoplast near the SE-CC complex, the uptake properties of mesophyll cells at a distance from the veins are irrelevant. Of course, uptake by mesophyll cells will confuse quantitative comparisons of kinetic and export data (43); but such comparison also require a reasonable estimation of apoplastic sugar concentration *at the loading site*, which does not seem to be feasible anyway. The crux of the matter is that the veins (phloem?) appear in autoradiographs to accumulate exogenous sucrose preferentially.

The valuable point has often been made that these autoradiographs are misleading if sugar is actually accumulated from the free space by the mesophyll and transferred to the veins during the experiment. Attempts have been made to address this question by measuring the uptake potential of the different tissues directly by isolating mesophyll cells and minor veins (14, 109, 113). The consensus from these studies is that there is little or no preferential accumulation by the veins. However, proof that the isolated tissues retain their native uptake characteristics is not yet convincing.

Another way to evaluate the amount of transfer from mesophyll cells to veins is to conduct the experiments over short periods that preclude significant intercellular movement of solute. When leaf discs of *Pisum sativum* are exposed to exogenous [^{14}C]sucrose for only 1 min the veins become clearly labeled, indicating direct uptake (107). However, when leaf discs of *Coleus blumei*, a species with intermediary cells in the minor vein phloem (37), are treated in a similar manner, label is seen in the veins only after a lag period of several minutes, indicating that there is indeed transport to the veins from surrounding tissues (107). It may also be significant that accumulation of exogenous [^{14}]sucrose is not inhibited by PCMBS in *Coleus* as it is in other species (112). Perhaps sucrose loading in *Coleus* is symplastic, or partly symplastic. This will only be resolved by further experimentation.

If the phloem is loaded by an apoplastic route one might expect that the plasmodesmata between the SE-CC complex and surrounding cells are closed during the sink-source transition. This would prevent leakage of solute from the SE-CC complex and is also in accord with the hypothesis that phloem unloading is terminated by physically blocking the unloading route. Symplastic phloem loading presents more of a challenge in this regard. However, if symplastic phloem loading occurs, one might expect that the mechanism responsible for loading is capable only of one-way transport; thus the polarity of the translocation system would be maintained.

CONCLUDING REMARKS

The sink-source transition in leaves has not been given the attention it deserves, considering its importance in carbohydrate metabolism and photosynthate partitioning. The small amount of information available indicates that the timing of vein maturation permits import and then export through mature sieve elements in a way that allows continued expansion of the lamina and that can accommodate bidirectional transport in the petiole. A rapid transition from catabolic to anabolic pathways of carbohydrate metabolism results from poorly understood changes in enzyme activity and compartmentation.

The few studies that approach the subject from a mechanistic perspective indicate that the transition is under developmental control and is not simply a response to changing carbon balance. However, the interactions between the developmental program and environmental conditions are certain to be much more profound and subtle than we presently realize. To ask how these developmental programs are controlled is to address one of the larger questions in biology. This is now feasible. For example, one could bring the techniques of molecular biology to bear on the study of enzyme induction at the initiation of the export phase.

Before we can determine how import stops and export begins we must know how phloem unloading and loading are controlled. The scenario proposed is that symplastic phloem unloading is terminated by closure of plasmodesmata, which effectively ends the import phase. This in turn permits accumulation of high levels of photosynthate in the phloem by sucrose-proton cotransport during the export period. While this hypothesis explains many of the data gathered so far, it has not been fully proven, nor does it deal with other potential pathways of unloading and loading. The mechanism(s) of phloem loading of galactosyl-sucrose oligosaccharides and sugar alcohols is especially perplexing. It must be kept in mind that the sink-source transition may occur by different mechanisms in different species. Progress will be more rapid when investigators become less specialized and undertake coordinated studies of anatomy, physiology, and biochemistry.

ACKNOWLEDGMENTS

Work in the author's laboratory is supported by grants from the National Science Foundation, the USDA Competitive Grants Program, and Hatch Program funds.

Literature Cited

1. Anderson, L. S., Dale, J. E. 1983. The sources of carbon for developing leaves of barley. *J. Exp. Bot.* 34:405–14
2. Aronoff, S. 1955. Translocation from soybean leaves II. *Plant Physiol.* 30:184–85
3. Avery, G. S. 1933. Structure and development of the tobacco leaf. *Am. J. Bot.* 20:565–92
4. Baker, D. A. 1985. Regulation of phloem loading. In *Regulation of Sources and Sinks in Crop Plants,* Monogr. 12, ed. B. Jeffcoat, A. Hawkins, A. D. Stead, pp. 163–76. Bristol: Brit. Plant Growth Regul. Group
5. Baker, N. R. 1985. Energy transduction during leaf growth. In *Control of Leaf Growth,* ed. N. R. Baker, W. J. Davies, C. K. Ong, pp. 115–33. New York: Cambridge Univ. Press. 350 pp.
6. Biddulph, O., Cory, R. 1965. Translocation of C^{14} metabolites in the phloem of the bean plant. *Plant Physiol.* 40:119–29
7. Bieleski, R. L. 1982. Sugar alcohols. In *Encyclopedia of Plant Physiology: Plant Carbohydrates I* (New Ser.), ed. F. A. Loewus, W. Tanner, 13A:158–92. Berlin: Springer-Verlag. 918 pp.
8. Bieleski, R. L., Redgwell, R. J. 1985.

Sorbitol versus sucrose as photosynthesis and translocation products in developing apricot leaves. *Aust. J. Plant Physiol.* 12:657–68

9. Blechschmidt-Schneider, S., Eschrich, W. 1985. Microautoradiographic localization of imported ^{14}C-photosynthate in induced sink leaves of two dicotyledonous C$_4$ plants in relation to phloem unloading. *Planta* 163:439–47

10. Bonnemain, J. L. 1969. Transport du ^{14}C assimilé à partir des feuilles de Tomate en voie de croissance et vers celles-ci. *C. R. Acad. Sci. Paris Ser. D* 269:1600–3

11. Botha, C. E. J., Evert, R. F. 1988. Plasmodesmatal distribution and frequency in vascular bundles and contiguous tissues of the leaf of *Themeda triandra. Planta* 173:433–41

12. Canny, M. J., Askham, M. J. 1967. Physiological inferences from the evidence of translocated tracer: a caution. *Ann. Bot.* (New Ser.) 31:409–16

13. Carr, D. J., Pate, J. S. 1967. Ageing in the whole plant. In *Aspects of the Biology of Ageing,* ed. H. W. Woolhouse, 21:559–99. New York: Academic. 634 pp.

14. Cataldo, D. A. 1974. Vein loading: the role of the symplast in intercellular transport of carbohydrate between the mesophyll and minor veins of tobacco leaves. *Plant Physiol.* 53:912–17

15. Claussen, W., Loveys, B. R., Hawker, J. S. 1985. Comparative investigations on the distribution of sucrose synthase activity and invertase activity within growing, mature and old leaves of some C$_3$ and C$_4$ plant species. *Physiol. Plant.* 65:275–80

16. Cronshaw, J., Lucas, W. J., Giaquinta, R. T., eds. 1986. *Plant Biology,* Vol. 1: *Phloem Transport.* New York: Liss. 650 pp.

17. Daie, J. 1985. Carbohydrate partitioning and metabolism in crops. *Hort. Rev.* 7:69–108

18. Dale, J. E. 1976. Cell division in leaves. In *Cell Division in Higher Plants,* ed. M. M. Yeoman, pp. 315–45. New York: Academic. 542 pp.

19. Dale, J. E. 1985. The carbon relations of the developing leaf. See Ref. 5, pp. 135–53

20. Dale, J. E. 1988. The control of leaf expansion. *Ann. Rev. Plant Physiol. Plant Mol. Biol.* 39:267–95

21. Dale, J. E., Sutcliffe, J. E. 1986. Phloem transport. In *Plant Physiology— A Treatise: Water and Solutes in Plants,* ed. F. C. Steward, J. F. Sutcliffe, J. E.

Dale, 9:455–549. New York: Academic. 611 pp.

22. Davis, J. M., Fellman, J. K., Loescher, W. H. 1988. Biosynthesis of sucrose and mannitol as a function of leaf age in celery (*Apium graveolens* L.). *Plant Physiol.* 86:129–33

23. Delrot, S. 1987. Phloem loading: apoplastic or symplastic? *Plant Physiol. Biochem.* 25:667–76

24. Delrot, S., Bonnemain, J. L. 1984. Mechanism and control of phloem transport. *Physiol. Veg.* 23:199–220

25. Dickmann, D. I. 1971. Photosynthesis and respiration by developing leaves of cottonwood (*Populus deltoides* Bartr.). *Bot. Gaz.* 132:253–59

26. Dickmann, D. I. 1971. Chlorophyll, ribulose-1,5-diphosphate carboxylase, and Hill reaction activity in developing leaves of *Populus deltoides. Plant Physiol.* 48:143–45

27. Dickson, R. E., Larson, P. R. 1975. Incorporation of ^{14}C-photosynthate into major chemical fractions of source and sink leaves of cottonwood. *Plant Physiol.* 56:185–93

28. Dickson, R. E., Larson, P. R. 1981. [14]C fixation, metabolic labeling patterns, and translocation profiles during leaf development in *Populus deltoides. Planta* 152:461–70

29. Ding, B., Parthasarathy, M. V., Niklas, K., Turgeon, R. 1988. A morphometric analysis of the phloem unloading pathway in developing tobacco leaves. *Planta.* In press

30. Esau, K. 1969. *The Phloem.* Berlin: Gebrüder Borntraeger. 505 pp.

31. Eschrich, W., Eschrich, B. 1987. Control of phloem unloading by source activities and light. *Plant Physiol. Biochem.* 25:625–34

32. Evert, R. F., Eschrich, W., Heyser, W. 1977. Distribution and structure of the plasmodesmata in mesophyll and bundle sheath cells of *Zea mays* L. *Planta* 136:77–89

33. Evert, R. F., Eschrich, W., Heyser, W. 1978. Leaf structure in relation to solute transport in phloem loading in *Zea mays* L. *Planta* 138:279–94

34. Fellman, J. K., Loescher, W. H. 1987. Comparative studies of sucrose and mannitol utilization in celery (*Apium graveolens*). *Physiol. Plant.* 69:337–41

35. Fellows, R. J., Geiger, D. R. 1974. Structural and physiological changes in sugar beet leaves during sink to source conversion. *Plant Physiol.* 54:877–85

36. Fischer, A. 1884. Untersuchungen über das Siebröhren-System der Cucurbitacean. Berlin: Gebrüder Borntraeger

37. Fisher, D. G. 1986. Ultrastructure, plasmodesmatal frequency, and solute concentration in green areas of variegated *Coleus blumei* Benth. leaves. *Planta* 169:141–52

38. Fisher, D. G., Eschrich, W. 1985. Import and unloading of ^{14}C assimilate into nonphotosynthetic portions of variegated *Coleus blumei* leaves. *Can. J. Bot.* 63:1708–12

39. Gamalei, Y. V. 1985. Characteristics of phloem loading in woody and herbaceous plants. *Sov. Plant Physiol.* 32:656–65

40. Gamalei, Y. V., Pakhomova, M. V. 1981. Distribution of plasmodesmata and parenchyma transport of assimilates in the leaves of several dicots. *Fiziol. Rast.* 28:901–12

41. Gaudreault, P.-R., Webb, J. A. 1982. Alkaline α-galactosidase in leaves of *Cucurbita pepo*. *Plant Sci. Lett.* 24:281–88

42. Gaudreault, P.-R., Webb, J. A. 1983. Partial purification and properties of an alkaline α-galactosidase from mature leaves of *Cucurbita pepo*. *Plant Physiol.* 71:662–68

43. Geiger, D. R., Sovonick, S. A., Shock, T. L., Fellows, R. J. 1974. Role of free space in translocation in sugarbeet. *Plant Physiol.* 54:892–98

44. Giaquinta, R. 1977. Sucrose hydrolysis in relation to phloem translocation in *Beta vulgaris*. *Plant Physiol.* 60:339–43

45. Giaquinta, R. 1978. Source and sink leaf metabolism in relation to phloem translocation—carbon partitioning and enzymology. *Plant Physiol.* 61:380–85

46. Giaquinta, R. 1983. Phloem loading of sucrose. *Annu. Rev. Plant Physiol.* 34:347–87

47. Gunning, B. E. S., Pate, J. S. 1974. Transfer cells. In *Dynamic Aspects of Plant Ultrastructure*, ed. A. W. Robards, pp. 441–80. Maidenhead: McGraw-Hill

48. Handley, L. W., Pharr, D. M., McFeeters, R. F. 1983. Relationship between galactinol synthase activity and sugar composition of leaves and seeds of several crop species. *J. Am. Soc. Hort. Sci.* 108:600–5

49. Hawkins, C. D. B., Whitecross, M. I., Aston, M. J. 1987. The effect of short-term aphid feeding on the partitioning of [14]CO_2 photoassimilate in three legume species. *Can. J. Bot.* 65:666–72

50. Hendrix, J. E. 1977. Phloem loading in squash. *Plant Physiol.* 60:567–69

51. Ho, L. C. 1988. Metabolism and compartmentation of imported sugars in sink organs in relation to sink strength. *Annu.*

Rev. Plant Physiol. Plant Mol. Biol. 39:355–78

52. Ho, L. C., Baker, D. A. 1982. Regulation of loading and unloading in long distance transport systems. *Physiol. Plant.* 56:225–30

53. Ho, L. C., Hurd, R. G., Ludwig, L. J., Shaw, A. F., Thornley, J. H. M., Withers, A. C. 1984. Changes in photosynthesis, carbon budget and mineral content during the growth of the first leaf of cucumber. *Ann. Bot.* 54:87–101

54. Ho, L. C., Shaw, A. F. 1977. Carbon economy and translocation of ^{14}C in leaflets of the seventh leaf of tomato during leaf expansion. *Ann. Bot.* 41:833–48

55. Hopkinson, J. M. 1964. Studies on the expansion of the leaf surface. IV. The carbon and phosphorus economy of a leaf. *J. Exp. Bot.* 15:125–37

56. Hopkinson, J. M. 1966. Studies on the expansion of the leaf surface. VI. Senescence and the usefulness of old leaves. *J. Exp. Bot.* 17:762–70

57. Huber, S. C., Kerr, P. S., Kalt-Torres, W. 1986. Biochemical control of allocation of carbon for export and storage in source leaves. See Ref. 16, pp. 355–68

58. Isebrands, J. G., Larson, P. R. 1973. Anatomical changes during leaf ontogeny in *Populus deltoides*. *Am. J. Bot.* 60:199–208

59. Jones, H., Eagles, J. E. 1962. Translocation of [14]Carbon within and between leaves. *Ann. Bot.* (New Ser.) 26:505–10

60. Jones, H., Martin, R. V., Porter, H. K. 1959. Translocation of [14]Carbon in tobacco following assimilation of [14]Carbon dioxide by a single leaf. *Ann. Bot.* (New Ser.) 23:493–508

61. Joy, K. W. 1964. Translocation in sugar beet. I. Assimilation of [14]CO_2 and distribution of material from leaves. *J. Exp. Bot.* 15:485–94

62. Joy, K. W. 1967. Carbon and nitrogen sources for protein synthesis and growth in sugar-beet leaves. *J. Exp. Bot.* 18:140–50

63. Kerr, P. S., Kalt-Torres, W., Huber, S. C. 1987. Resolution of two molecular forms of sucrose-phosphate synthase from maize, soybean and spinach leaves. *Planta* 170:515–19

64. Köcher, H., Leonard, O. A. 1971. Translocation and metabolic conversion of ^{14}C-labeled assimilates in detached and attached leaves of *Phaseolus vulgaris* L. in different phases of leaf expansion. *Plant Physiol.* 47:212–16

65. Kursanov, A. L. 1963. Metabolism and

the transport of organic substances in the phloem. *Adv. Bot. Res.* 1:209–78
66. Larson, P. R., Dickson, R. E. 1973. Distribution of imported ^{14}C in developing leaves of eastern cottonwood according to phyllotaxy. *Planta* 111:95–112
67. Larson, P. R., Gordon, J. C. 1969. Leaf development, photosynthesis, and C^{14} distribution in *Populus deltoides* seedlings. *Am. J. Bot.* 56:1058–66
68. Larson, P. R., Isebrands, J. G., Dickson, R. E. 1972. Fixation patterns of ^{14}C within developing leaves of eastern cottonwood. *Planta* 107:301–14
69. Larson, P. R., Isebrands, J. G., Dickson, R. E. 1980. Sink to source transition of *Populus* leaves. *Ber. Dtsch. Bot. Ges.* 93:79–90
70. Loescher, W. H., Merlow, G. C., Kennedy, R. A. 1982. Sorbitol metabolism and sink-source interconversions in developing apple leaves. *Plant Physiol.* 70:335–39
71. Ludwig, L. J., Saeki, T., Evans, L. T. 1965. Photosynthesis in artificial communities of cotton plants in relation to leaf area. *Aust. J. Biol. Sci.* 18:1103–18
72. Madore, M. A., Mitchell, D. E., Boyd, C. M. 1988. Stachyose synthesis in source leaf tissues of the CAM plant *Xerosicyos danguyi* H. Humb. *Plant Physiol.* 87:588–91
73. Madore, M., Webb, J. A. 1981. Leaf free space analysis and vein loading in *Cucurbita pepo*. *Can. J. Bot.* 59:2550–57
74. Madore, M., Webb, J. A. 1982. Stachyose synthesis in isolated mesophyll cells of *Cucurbita pepo*. *Can. J. Bot.* 60:126–30
75. Manning, K., Maw, G. A. 1975. Distribution of acid invertase in the tomato plant. *Phytochemistry* 14:1965–69
76. Maynard, J. W., Lucas, W. J. 1982. A reanalysis of the two-component phloem loading system in *Beta vulgaris*. *Plant Physiol.* 69:734–39
77. M'Batchi, B., Delrot, S. 1988. Stimulation of sugar exit from leaf tissues of *Vicia faba* L. *Planta* 174:340–48
78. Morris, D. A., Arthur, E. D. 1984. An association between acid invertase activity and cell growth during leaf expansion in *Phaseolus vulgaris* L. *J. Exp. Bot.* 35:1369–79
79. Nakata, S., Leopold, A. C. 1967. Radioautographic study of translocation in bean leaves. *Am. J. Bot.* 54:769–72
80. Negm, F. B., Loescher, W. H. 1981. Characterizations of aldose 6-phosphate reductase (alditol 6-phosphate:NADP 1-oxidoreductase) from apple leaves. *Plant Physiol.* 67:139–42

81. Peterson, C. A., Currier, H. B. 1969. An investigation of bidirectional translocation in the phloem. *Physiol. Plant.* 22:1238–50
82. Pharr, D. M., Sox, H. N. 1984. Changes in carbohydrate and enzyme levels during the sink to source transition of leaves of *Cucumis sativus* L., a stachyose translocator. *Plant. Sci. Lett.* 35:187–93
83. Pollock, C. J. 1976. Changes in the activity of sucrose-synthesizing enzymes in developing leaves of *Lolium temulentum*. *Plant Sci. Lett.* 7:27–31
84. Pollock, C. J., Lloyd, E. J. 1977. The distribution of acid invertase in developing leaves of *Lolium temulentum* L. *Planta* 133:197–200
85. Rumpho, M. E., Edwards, G. E., Loescher, W. H. 1983. A pathway for photosynthetic carbon flow to mannitol in celery leaves. *Plant Physiol.* 73:869–73
86. Russin, W., Evert, R. F. 1985. Studies on the leaf of *Populus deltoides* (Salicaceae): ultrastructure, plasmodesmatal frequency and solute concentrations. *Am. J. Bot.* 72:1232–47
87. Schaffer, A. A., Sagee, O., Goldschmidt, E. E., Goren, R. 1987. Invertase and sucrose synthase activity, carbohydrate status and endogenous IAA levels during *Citrus* leaf development. *Physiol. Plant.* 69:151–55
88. Schmalstig, J. G., Geiger, D. R. 1985. Phloem unloading in developing leaves of sugar beet. I. Evidence for pathway through the symplast. *Plant Physiol.* 79:237–41
89. Schmalstig, J. G., Geiger, D. R. 1987. Phloem unloading in developing leaves of sugar beet. II. Termination of phloem unloading. *Plant Physiol.* 83:49–52
90. Schmalstig, J. G., Hitz, W. D. 1987. Contributions of sucrose synthase and invertase to the metabolism of sucrose in developing leaves. *Plant Physiol.* 85: 407–12
91. Schmitz, K., Cuypers, B., Moll, M. 1987. Pathway of assimilate transfer between mesophyll cells and minor veins in leaves of *Cucumis melo* L. *Planta* 171:19–29
92. Schmitz, K., Holthaus, U. 1986. Are sucrosyl-oligosaccharides synthesized in mesophyll protoplasts of mature leaves of *Cucumis melo?* *Planta* 169:529–35
93. Silvius, J. E., Kremer, D. F., Lee, D. R. 1978. Carbon assimilation and translocation in soybean leaves at different stages of development. *Plant Physiol.* 62:54–58
94. Stitt, M., Wilke, I., Feil, R., Heldt, H.

W. 1988. Coarse control of sucrose-phosphate synthase in leaves: alterations of the kinetic properties in response to the rate of photosynthesis and the accumulation of sucrose. *Planta* 174: 217–30

95. Thomas, B., Webb, J. A. 1978. Distribution of α-galactosidase in *Cucurbita pepo*. *Plant Physiol.* 62:713–17

96. Thrower, S. L. 1962. Translocation of labelled assimilates in the soybean. II. The pattern of translocation in intact and defoliated plants. *Aust. J. Biol. Sci.* 15:629–49

97. Thrower, S. L. 1967. The pattern of translocation during ageing. See Ref. 13, pp. 483–506

98. Turgeon, R. 1980. The import to export transition: experiments on *Coleus blumei*. *Ber. Dtsch. Bot. Ges. Bd.* 93: 91–97

99. Turgeon, R. 1984. Efflux of sucrose from minor veins of tobacco leaves. *Planta* 161:120–28

100. Turgeon, R. 1984. Termination of nutrient import and development of vein loading capacity in albino tobacco leaves. *Plant Physiol.* 76:45–48

101. Turgeon, R. 1986. The import-export transition in dicotyledonous leaves. See Ref. 57, pp. 285–91

102. Turgeon, R. 1987. Phloem unloading in tobacco sink leaves: insensitivity to anoxia indicates a symplastic pathway. *Planta* 171:73–81

103. Turgeon, R., Webb, J. A. 1973. Leaf development and phloem transport in *Cucurbita pepo:* transition from import to export. *Planta* 113:179–91

104. Turgeon, R., Webb, J. A. 1975. Leaf development and phloem transport in *Cucurbita pepo:* carbon economy. *Planta* 123:53–62

105. Turgeon, R., Webb, J. A. 1976. Leaf development and phloem transport in *Cucurbita pepo:* maturation of the minor veins. *Planta* 129:265–69

106. Turgeon, R., Webb, J. A., Evert, R. F.

1975. Ultrastructure of minor veins in *Cucurbita pepo* leaves. *Protoplasma* 83:217–32

107. Turgeon, R., Wimmers, L. E. 1988. Different patterns of vein loading of exogenous [^{14}C]sucrose in leaves of *Pisum sativum* and *Coleus blumei*. *Plant Physiol.* 87:179–82

108. van Bel, A. J. E. 1987. The apoplast concept of phloem loading has no universal validity. *Plant Physiol. Biochem.* 25:677–86

109. van Bel, A. J. E., Koops, A. J. 1985. Uptake of [^{14}C]sucrose in isolated minor-vein networks of *Commelina benghalensis* L. *Planta* 164:362–69

110. Vogelmann, T. C., Larson, P. R., Dickson, R. E. 1982. Translocation pathways in the petioles and stem between source and sink leaves of *Populus deltoides* Bartr. ex Marsh. *Planta* 156: 345–58

111. Webb, J. A., Gorham, P. R. 1964. Translocation of photosynthetically assimilated C^{14} in straight-necked squash. *Plant Physiol.* 39:663–72

112. Weisberg, L. A., Wimmers, L. E., Turgeon, R. 1988. Photoassimilate transport characteristics of nonchlorophyllous and green tissue in variegated leaves of *Coleus blumei*. *Planta* 175:1–8

113. Wilson, C., Oross, J. W., Lucas, W. J. 1985. Sugar uptake into *Allium cepa* leaf tissue: an integrated approach. *Planta* 164:227–40

114. Wimmers, L. E. 1988. *Transfer cell structure and function in pea leaf minor vein phloem*. PhD thesis. Cornell Univ., Ithaca, New York. 252 pp.

115. Wu, A., Thrower, L. B. 1973. Translocation into mature leaves. *Plant Cell Physiol.* 14:1225–28

116. Zimmermann, M. H. 1957. Translocation of organic substances in trees. I. The nature of the sugars in the sieve tube exudate of trees. *Plant Physiol.* 32:288–91

Annu. Rev. Plant Physiol. Plant Mol. Biol. 1989. 40:139–68

XYLOGLUCANS IN THE PRIMARY CELL WALL

Takahisa Hayashi

Basic Research Laboratory, Central Research Laboratories, Ajinomoto Co. Inc., Kawasaki 210, Japan

CONTENTS

INTRODUCTION

Two types of cell walls are found in higher plants: primary walls and secondary walls. Primary walls are produced by growing cells, so they can

139

1040-2519/89/0601-0139$02.00

elongate and/or swell; secondary walls cannot. The primary walls derived from growing plant cells are composed of many complex carbohydrates (approximately 90%) and proteins (10%) which may interact with each other during growth. The complexity of cell-wall metabolism has led to numerous primary-wall models and wall-loosening theories, which raise endless questions. In 1973, Albersheim and his coworkers (9, 76, 135) introduced for the first time a complete model of primary walls using growing sycamore cells. The structure of the walls was characterized by hydrolysis with purified glycanases and subsequent chemical analyses of the released fragments. Reconstruction of the fragments led to the suggestion that rhamnogalacturonan, arabinogalactan, xyloglucan, and hydroxyproline-rich protein were interconnected by covalent bonds, whereas hydrogen bonds interconnected cellulose and xyloglucan. Since then, the walls of growing plant cells have generally been thought to act as a macromolecular complex. Many publications have appeared, including excellent reviews of structure and function of primary walls by McNeil et al (98), Bacic et al (5), Darvill et al (22), Dey & Brinson (28), and of cross-linking by Fry (39). General reviews of biosynthesis include those by Delmer & Stone (27), and Bolwell (10); of cellulose biosynthesis by Delmer (24, 25), arabinogalactan-protein by Fincher et al (38), wall proteins by Cassad & Varner (18), and plant β-glucanases by Verma et al (146); of plant growth and wall turnover by Labavitch (83); of cell-wall mechanical properties by Taiz (134); and of biophysics by Cosgrove (21).

This review focuses on xyloglucan metabolism including occurrence, biosynthesis, biodegradation, organization, and growth regulation. Xyloglucan is obviously an important component of primary cell walls, as indicated by the universality of its occurrence in the primary walls of higher plants and the notable decrease in xyloglucan content during growth. Xyloglucan biodegrades rapidly after acid or auxin treatment (83). This suggests that xyloglucan functions in cell enlargement during growth and is a key component in primary walls. Also notable is the structural similarity between xyloglucan and cellulose. Since xyloglucan has a cellulose-like main chain composed of 1,4-linkedβ-glucopyranosyl residues, the 1,4-β-glucans must share a certain biosynthetic mechanism and/or macromolecular organization.

I was early impressed with Albersheim's review, "The walls of growing plant cells" (2). I knew intuitively that xyloglucan is the key component of the primary walls of growing plant cells. I have therefore concentrated on identifying and defining the metabolic reactions required for the biosynthesis and biodegradation of xyloglucan in higher plants. Many other groups have also contributed to our knowledge about this fascinating polymer. This chapter summarizes our current understanding of xyloglucan metabolism as it relates to the function of primary cell walls in plant growth.

OCCURRENCE AND STRUCTURE

Dicotyledons

Xyloglucans were first found as amyloids in the cell walls of plant seeds (78, 79). The polysaccharides form a complex with iodine, and this complex produces a blue substance. The xyloglucans all contain D-glucose, D-xylose, and D-galactose in a molar ratio of approximately 4:3:1. They all possess a 1,4-β-glucan backbone with α-xylosyl residues attached to the 6-position of β-glucosyl residues, and terminal galactose is attached to the 2-position of the xylosyl residues by the β-linkage (80). Microscopic studies on sections of seeds have shown that the polysaccharides are deposited parallel to the cell walls of *Tamarindus indica* (120). They disappear during germination (32, 43). Because the amyloid polysaccharide is utilized by the embryo, it has been thought to function, like starch, as a reserve carbohydrate.

A fucosylated xyloglucan was first isolated as the key component of the primary wall from the medium and the walls of sycamore cells grown as a suspension culture (4, 9). Generally, xyloglucans comprise as much as 20–25% of the primary walls of dicotyledons. They contain L-fucose in addition to D-glucose, D-xylose, and D-galactose. The L-fucose is attached to the 2-position of the galactosyl residues by the α-linkage. Species-specific differences occur in the distribution of additional branching fucosyl-galactosyl residues. Arabinose may sometimes be present in the polysaccharides, but the content of these sugar residues is very small. Hydrolysis of pea xyloglucan with endo-1,4-β-glucanase gives a nonasaccharide (glucose/xylose/galactose/fucose, 4:3:1:1) and a heptasaccharide (glucose/xylose, 4:3) in a proportion of 1:1 (51). Partial enzymic hydrolysis yields dimers of the oligosaccharide units that are composed mainly of nonasaccharide→heptasaccharide, with the oligosaccharide units in a molar ratio of 1:1 at all stages of hydrolysis. Therefore, it appears that pea xyloglucan contains these units distributed primarily in an alternating sequence (Figure 1).

Terminal α-L-fucosyl-(1→2)-β-D-galactosyl residues as side chains are probably associated with the serological activity of human blood group substance H. In fact, xyloglucans that contain terminal fucosyl-galactose residues complex with α-L-fucose-binding lectin from *Ulex europaeus* and *Lotus tetragonolobus* (51). Although plant lectins are biochemically active with sugar residues, the function of the lectins in nature is not yet known. The fucose-binding lectins might be involved in cell-wall metabolism by binding to xyloglucans, but no fucose-binding lectins have been obtained from primary walls, where xyloglucans occur.

The heterogeneity of xyloglucan results from either *(a)* differences in molecular mass, *(b)* the distribution of additional branching galactosyl residues and fucosyl galactosyl residues, or *(c)* levels of substituted xylosyl

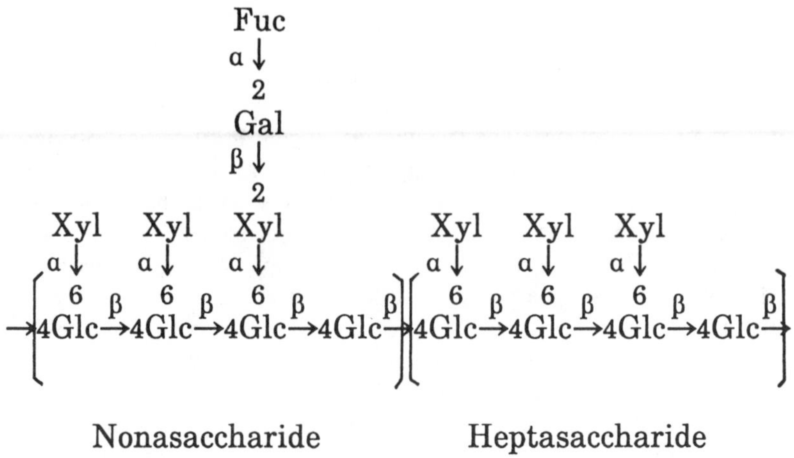

$$
\begin{array}{c}
\text{Fuc} \\
\alpha \downarrow \\
2 \\
\text{Gal} \\
\beta \downarrow \\
2
\end{array}
$$

$$
\begin{array}{ccccccc}
\text{Xyl} & \text{Xyl} & \text{Xyl} & & \text{Xyl} & \text{Xyl} & \text{Xyl} \\
\alpha\downarrow & \alpha\downarrow & \alpha\downarrow & & \alpha\downarrow & \alpha\downarrow & \alpha\downarrow \\
6 & \beta \quad 6 & \beta \quad 6 & \beta \qquad \beta & 6 & \beta \quad 6 & \beta \quad 6 & \beta \qquad \beta
\end{array}
$$

$$
\Big\{ \!\!\to\!\! 4\text{Glc}\!\to\!4\text{Glc}\!\to\!4\text{Glc}\!\to\!4\text{Glc}\!\!\Big[\!\Big] \!\!4\text{Glc}\!\to\!4\text{Glc}\!\to\!4\text{Glc}\!\to\!4\text{Glc}\!\!\to \Big\}
$$

Nonasaccharide   Heptasaccharide

Figure 1   Chemical repeating unit of pea xyloglucan.

residues. Xyloglucans of two different molecular masses were obtained from the alkali-soluble polysaccharide fraction in rapeseed meal (140). The molecular masses were 3100-kd and 9.3kd estimated by chromatography on Sepharose CL-4B, but the polysaccharides had a very similar sugar composition. Cotton xyloglucan (48) is composed of four kinds of subunits: decasaccharide (glucose / xylose / galactose / fucose,4:3:2:1),nonasaccharide (glucose/xylose/galactose/fucose, 4:3:1:1), octasaccharide (glucose/xylose/galactose, 4:3:1), and heptasaccharide (glucose/xylose, 4:3) in a proportion of 2:7:1:3. It appears that all xyloglucans are composed of repeating heptasaccharide units to which variable amounts of galactose and fucose are probably added during synthesis. Three kinds of xyloglucans (with molecular masses of 180 kd, 60 kd, and 30 kd) were obtained from suspension-cultured soybean cells (49), and the molar ratios of nonasaccharide to heptasaccharide varied. These xyloglucans were fractionated by borate anion-exchange chromatography on DEAE-Sephadex. Seven kinds of xyloglucans in the cell walls of apple tissues (126) were also obtained by borate anion-exchange chromatography. Because this anion acts as a chelator of the hydroxyl groups of sugar residues, the separation of xyloglucans probably results from differences in the distribution of the branching sugars. Multiple forms of xyloglucans were also observed in the primary cell walls of *Rosa glauca* cells grown in suspension (67). The multiplicity resulted from the sugar composition, particularly the ratio of glucose to xylose, which ranged from 1.2 to 2 within the different xyloglucans.

York et al (152) determined the location of the O-acetyl substituents on a nonasaccharide repeating unit of sycamore extracellular xyloglucan. The O-2-linked β-galactosyl residues of the nonasaccharide were the dominant site of O-acetyl substitution. These were mono-O-acetylated and di-O-acetylated β-galactosyl residues at O-6, O-4, and O-3 at degrees of 55–60%, 15–20%, and 20–25%, respectively. Pectin is known also to be acetylated (23); the O-acetyl groups are located on the galacturonate residues. It is possible that most plant polysaccharides found in the cell walls are partially acetylated (6) but that acetic esters are hydrolyzed during the alkaline extraction of walls. Recently, it has been found that O-acetylation and O-succination of cell-wall polysaccharides have an important role in biological activities (29, 87), such as *Rhizobium* infection for nitrogen fixation.

There is a structural exception among dicotyledonous xyloglucans in *Solanaceae* species: the xyloglucans from *Nicotiana tabacum* (1, 30, 31) and *Solanum tuberosum* (123). The polysaccharides do not contain fucose, but have arabinose instead, this being linked to xylosyl residues at 2. The glucan backbone contains fewer substituted xylosyl residues, a case similar to that of xyloglucans extracted from monocotyledonous plants, described below.

Monocotyledons

Xyloglucans from monocotyledons contain no terminal fucose, less xylose, and much less galactose than those from dicotyledons. Because the glucan backbone contains fewer substituted xylosyl residues, the substitution pattern of xylosyl residues is very different from that of dicotyledonous xyloglucan. Hydrolysis of monocotyledonous xyloglucan with endo-1,4-β-glucanase yields mainly a pentasaccharide (glucose/xylose, 3 : 2) and glucose, and small amounts of a hexasaccharide (glucose/xylose, 4 : 2), a trisaccharide (glucose/xylose, 2 : 1), and cellobiose (Figure 2) (68, 69, 72, 86, 129). The glucose and cellobiose are derived from the 1,4-β-glucan backbone unsubstituted with xylosyl residues.

Kato & Matsuda (71) obtained large amounts of heptasaccharide (glucose/xylose, 4 : 3) and an octasaccharide (glucose/xylose/galactose, 4 : 3 : 1) (Figure 2) from rice cells grown in suspension culture, although the hepta- and octasaccharide units were not obtained by similar treatment of xyloglucan from cell walls of rice seedlings (69). Because the rice cells were harvested at the middle of exponential growth, the wall materials were certainly derived from the primary walls. The explanation may be that xyloglucans in monocotyledons originally contain many substituted xylosyl and galactosyl residues and that the residues are partially deleted during growth. In fact, plant cell walls contain α-xylosidase and β-galactosidase activities to hydrolyze xylosidic and galactosidic linkages of xyloglucan (81). This possibility is discussed below.

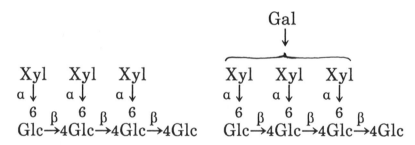

Figure 2   Oligosaccharides obtained from monocotyledon xyloglucans.

There is also a structural exception among xyloglucans of monocotyledonous plants in *Allium cepa* (95, 118). The polysaccharide is probably composed of terminal fucosyl-galactosyl and terminal galactosyl residues, and the backbone contains many substituted xylosyl residues. Therefore, the structural features of the polysaccharide are very similar to those in dicotyledons. In onions, the wall constituents and relative amounts of cell-wall components are similar to those in dicotyledons, and the major hemicellulose is a xyloglucan. Onion thus resembles dicotyledons in its cell-wall composition.

Burke et al (13) examined the cell walls of six different monocotyledonous plants grown in suspension culture by methylation analysis and found 0.3–3.8% 4,6-linked glucosyl residues derived from xyloglucan. By this calculation, monocotyledons contained much less xyloglucan than dicotyledons (9.2% 4,6-linked glucosyl residues in sycamore). However, the real contents should be increased 2 to 5-fold, as 4,6-linked glucosyl residue equivalents by methylation analysis, because monocotyledonous xyloglucans contain fewer substituted xylosyl residues. Furthermore, cell walls in monocotyledons contain less cellulose. Therefore, the ratios of xyloglucan to cellulose in monocotyledons are similar to those in dicotyledons. This suggests that xyloglucans and cellulose microfibrils in monocotyledons function as a macromolecular complex similar to that in dicotyledons.

Conformation of Xyloglucan

The X-ray fiber diffraction pattern of *Tamarindus indica* xyloglucan showed the backbone glucan conformation of an extended two-fold helix similar to cellulose (136). The helixes of 1.4-linked equatorial glucoses generate a straight ribbon-like structure. One pitch of the helix is 1.03 nm. Similar results have been obtained with xyloglucan extracted from pea primary walls.

Ogawa et al (110) analyzed the X-ray fiber pattern and energy calculation of pea xyloglucan prepared from growing cell walls. Interactions between all pairs of nonbonded atoms of the polysaccharide were calculated stepwise by defining the internal van der Waals energy, torsional, and hydrogen-bond factors. Calculations were done with a modified Lennard-Jones "6–12" potential function (128) and with a computer program, PS79 (153). The calculation of interactions in the backbone glucan indicated that a hydrogen bond was present between O-5 and O-3' (distance = 0.260 nm). In addition, interactions between the glucan chain and xylose residues are varied by three rotatable bonds between rigid glucopyranose residues, which have intrinsic torsional potential for rotation around the C-5–C-6, C-6–O-1, and O-1–C-1 bonds (143). These results indicate that, although the conformation of backbone glucan is firmly fixed with the two-fold helix, the xylosyl residues have a very flexible structure. Because the molecular mass of pea xyloglucan is 330 kd (51), the polysaccharide looks like a straight ribbon approximately 500 nm long. The probable conformation of pea xyloglucan is shown in Figure 3.

Based on such studies, most xyloglucans probably have a conformation similar to that of pea xyloglucan. It also appears that the xylosyl, galactosyl, and fucosyl residues substituted do not alter the conformation of the glucan backbone.

BIOSYNTHESIS

Properties of Xyloglucan Synthase

Particulate preparations of xyloglucan synthase were obtained initially from cultured soybean cells (56–58), pea epicotyls (53, 114), cotton fibers (26), and soybean hypocotyls (26, 50). The substrates are UDP-glucose and UDP-xylose, and the stoichiometry of the reaction, as measured with the enzyme, is:

UDP-glucose + (xyloglucan)$_n \rightarrow$ UDP + (xyloglucan)$_{n+1}$

UDP-xylose + (xyloglucan)$_{n+1} \rightarrow$ UDP + (xyloglucan)$_{n+2}$

The enzyme preparation catalyzes the transfer of glucosyl residues and that of xylosyl residues into an endogenous xyloglucan. This in vitro synthesized polysaccharide showed the same molecular mass (180 kd) as soybean xyloglucan prepared from cell walls (60). Extensive efforts to test primers, which are cellodextrins and xyloglucan fragment oligosaccharides, have proved unsuccessful in xyloglucan synthesis in soybean (57). The transfers of glucosyl and xylosyl residues take place at the O-4 and O-6 positions, respectively, of the glucosyl residues at the nonreducing terminal. The incorporation from UDP-xylose was dependent on the presence of UDP-glucose and vice versa, but high levels of UDP-xylose inhibited the incorporation of glucose. Thus, development of the xyloglucan chain proceeds by concurrent transfers of glucose and xylose during xyloglucan synthesis. The synthesis of fucosyl residues from GDP-fucose was enhanced by the addition of UDP-glucose, UDP-xylose, and/or UDP-galactose (16).

It is safe to assume that UDP-glucose and UDP-xylose serve as the precursors for the xylose and glucose groups in a xyloglucan molecule. This notion has been supported by the extraction and isolation of sugar nucleotides

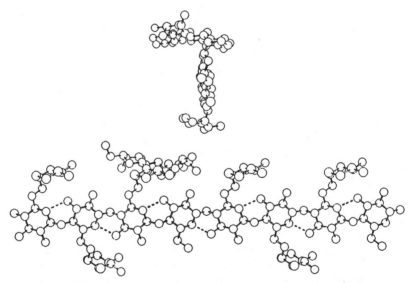

Figure 3 A possible conformation of pea xyloglucan (see Figure 1), projected parallel (upper) and perpendicular (lower) to the chain axis. All the hydrogen atoms are omitted. Covalent bonds are shown as solid lines and hydrogen bonds as dotted lines.

from soybean cells (59). The concentrations of UDP-glucose and UDP-xylose were approximately 102 and 2.4 μmol, respectively, per kg fresh weight of growing cells. In addition, ion-exchange (formic acid system) chromatography of sugar nucleotides showed that galactose was obtained mainly from the UDP-sugar fraction, and fucose was only detected in the fraction containing GDP-sugars. Therefore, it is also safe to assume that UDP-galactose and GDP-fucose serve as galactosyl and fucosyl donors, respectively, for the addition of galactosyl and fucosyl residues.

To date, no evidence has been found for the involvement of a lipid or a protein intermediate in the biosynthesis of xyloglucan. When pea membranes were incubated with UDP-[^{14}C]xylose or UDP-[^{14}C]glucose and sequentially extracted with chloroform/methanol (2:1) and chloroform/methanol/water (10:10:3), charged xylolipid (52) and glucolipid (17), which accumulated in the preparation with time, were formed. Such glycosyl lipids had properties consistent with polyprenyl monophosphoryl-xylose or glucose, but gave no evidence for their further participation in xyloglucan biosynthesis. Pea membranes also contain no transferases required for the formation of charged lipid-linked xylosyl-glucosyl oligosaccharides (93), although xyloglucan is often composed of repeating oligosaccharide units.

Hydrolysis of in vitro synthesized xyloglucan with fungal endo-1,4-β-glucanase gave a nonasaccharide (glucose/xylose/galactose/fucose, 4:3:1:1) (16), a heptasaccharide (glucose/xylose, 4:3), and a pentasaccharide (glucose/xylose, 3:2) (54, 57, 60). The fact that no units smaller than the pentasaccharide unit have been obtained in the enzymatically synthesized polysaccharide strongly suggests that the oligosaccharide units are formed very rapidly up to the five-sugar unit at the site of a multienzyme complex of glycosyltransferases. Xyloglucan synthesized in vitro in the absence of GDP-fucose and UDP-galactose is composed of heptasaccharides and pentasaccharides, and the elongation of the chain proceeds through the transformation of pentasaccharide into heptasaccharide (60). Camirand et al (16) suggest that fucosylation occurs after galactosylation of the heptasaccharide units in peas. Farkas & Maclachlan (35) further showed the transfer of fucose from GDP-fucose into exogenously added unfucosylated xyloglucan from *Tamarindus*. However, it is uncertain whether the fucosylation can occur on all galactosyl residues of the polysaccharide acceptor.

Cotton fibers are single cells that elongate synchronously over a period of almost 3 weeks. The level of xyloglucan synthase activity increased up to 16 days postanthesis and decreased rapidly at the onset of secondary-wall synthesis (26). However, the deposition of secondary-wall cellulose begins at about 16 days postanthesis and continues until about 32 days postanthesis (99). This suggests that the gene expression of xyloglucan and cellulose synthases in cotton fiber cells is developmentally regulated.

In monocotyledons, there are no reports on the biosynthesis of xyloglucan. The biosynthetic mechanism may be different from that in dicotyledons, because xyloglucans in monocotyledons contain many fewer substituted xylosyl residues. The glucosyltransferase activity in the system of xyloglucan synthesis may more closely resemble the activity of "glucan synthetase I" (115, 116), which produces a linear unsubstituted 1,4-β-glucan.

Mechanism of Xyloglucan Synthesis in Dictyosomes

Golgi dictyosomes are known to contain xyloglucan 4-β-glucosyltransferase (15, 58, 114) and 6-α-xylosyltransferase (15, 50, 58, 114). Camirand & Maclachlan (16) showed that total synthesis of xyloglucan occurs in Golgi stacks before transport of the polysaccharide to the cell wall by secretory vesicles. Because fucosylation may be the last step in the completion of xyloglucan unit structure, this also indicates the occurrence of xyloglucan 2-β-galactosyltransferase in the dictyosomes.

UDP-xylose is derived from UDP-glucose by the action of UDP-glucose dehydrogenase (EC 1.1.1.22) (133) and UDP-glucuronate carboxy-lyase (EC 4.1.1.35) (3). John et al (66) reported that the UDP-glucuronate carboxy-lyase of chondrocyte cells is part of a multienzyme complex of glycosyltransferases involved in mucopolysaccharide biosynthesis. Likewise, plant UDP-glucuronate carboxy-lyases are potentially part of the dictyosome-associated xyloglucan glycosyltransferase complex, because UDP-glucuronate carboxy-lyase activity co-migrated with xyloglucan synthase activity in soybean dictyosomes (50). Mung bean membranes (37) also contain the activity, which can be solubilized with 0.5% digitonin. It seems likely that the enzyme is a soluble one localized in the lumen of the membrane vesicles. Hayashi et al (50) showed that added UDP-glucuronate is transported into the Golgi vesicles by a specific carrier and decarboxylated to UDP-xylose within the lumen. On incubation of UDP-[^{14}C]glucuronate with the vesicles in the presence of UDP-glucose, [^{14}C]xylose-labeled xyloglucan was evidently formed. Thus, xyloglucan is subsequently formed from UDP-xylose and UDP-glucose in the Golgi apparatus.

In soybean cells grown in suspension culture (59), the concentration of UDP-xylose (2.5 μM) is far lower than that of UDP-glucose (103 μM). Nevertheless, the primary cell walls of dicotyledons contain a considerable amount of xylose in xyloglucan. Plant cells possess such precise compartmentation of the UDP-xylose biosynthesis in the Golgi apparatus, and thus the cellular pool of this sugar nucleotide is considerably higher at the site of polysaccharide synthesis.

Is "Glucan Synthetase I" Involved in Xyloglucan Synthesis?

Plant membranes may contain at least two 1,4-β-glucan 4-β-glucosyltransferases, one of which is involved in xyloglucan synthesis, and one that forms cellulose. Although the latter should be active in vivo at the plasma membrane (12, 106, 130) and the former in Golgi membranes (124), analyses of plant membranes on sucrose gradients have shown that both glucosyltransferase activities can be identified in association with Golgi membranes (50, 58, 114). However, xyloglucan synthase acts by concurrently transferring glucose and xylose to a nascent xyloglucan acceptor and not by transferring xylose to a preformed 1,4-β-glucan.

When UDP-glucose is supplied to plant membranes, β-glucans are formed at a rate that depends on the level of membrane protein, incubation time, and substrate concentration. The activities of β-glucan synthases produce at least two kinds of linkages. One with a low K_m and a requirement for Mg^{2+} forms 1,4-β-linkages mostly recovered in the alkali-insoluble fraction, and the other with a higher K_m and a requirement for Ca^{2+} and β-glucoside for activity(25) produces 1,3-β-linkages. The latter has evolved as a marker enzyme for plasma membranes (113): UDP-glucose 3-β-glucosyltransferase, known as glucan synthetase II; the former has evolved as a marker enzyme for Golgi membranes (113): UDP-glucose 4-β-glucosyltransferase, known as glucan synthetase I. The open question is whether the UDP-glucose 4-β-glucosyltransferase derives from a proenzyme of cellulose synthase or whether the 4-β-glucosyltransferase is a free form of xyloglucan glucosyltransferase from a multienzyme complex.

Ray (114) proposed that xylosyl residues are transferred to a 1,4-β-glucan core, which is formed from UDP-glucose by the enzyme "glucan synthetase I." The glucan core he mentioned is not a simple 1,4-β-glucan produced by the glucosyl transferase working alone but appears to be a nascent glucan chain. This is in agreement with enzyme properties, in that xylosyl transfer does not occur for preformed 1,4-β-glucan during xyloglucan synthesis; rather, concurrent transfer of both glucose and xylose are obligatory (26, 57, 58, 60). Therefore, he suggested that the Golgi-located 1,4-β-glucan synthase acts in vivo for xyloglucan synthesis, probably indicating that the synthase is a free form of xyloglucan glucosyltransferase.

A common property of the xyloglucan synthase system from many different sources is its stimulation by divalent cations, usually Mn^{2+} but not Mg^{2+}. Although the requirement of glucan synthetase I for Mg^{2+} is well known in higher plants, Mg^{2+} is less effective for the activity of xyloglucan glucosyltransferase (54, 57, 58). In contrast, the biosynthesis of 1,4-β-glucan (alkali-insoluble) is stimulated mainly by Mg^{2+} and much less by Mn^{2+}. In addition, Ca^{2+} does not promote the synthesis of xyloglucan (T. Hayashi, unpublished). It seems that there are indeed two separate synthases for 1,4-β-glucan and xyloglucan in peas (54) as well as soybeans (57). Biosynthesis of both products occurs at maximum rate when the two ions are provided together, but one does not enhance the effect of the other (54). This indicates that xylosyl transfer does not occur on preformed 1,4-β-

glucan, but rather that concurrent transfer of both glucose and xylose are obligatory.

Giddings et al (41) and recently Haigler & Brown (47) found that the Golgi apparatus was responsible for the selective transport and exocytosis of rosettes in mesophyll cells of *Zinnia elegans*. This indicates that the Golgi apparatus functions not only in the synthesis and export of xyloglucan but also in the export of a multienzyme complex of cellulose synthases. Cellulose synthesis usually takes place on the cell surface, whereas xyloglucans are synthesized within the membrane system of the Golgi apparatus. Xyloglucan molecules are probably prepared in the membrane system prior to the synthesis of cellulose, and cellulose formation must begin as soon as the vesicles of the Golgi dictyosomes are fused with the plasma membrane. At that moment, cell membranes begin to achieve complete cell wall thickening. Thus, it also seems possible that the Golgi-located Mg^{2+}-dependent UDP-glucose 4-β-glucosyltransferase, glucan synthetase I, is a latent proenzyme of cellulose synthase.

Coated vesicles containing clathrin (100) are widely distributed in plant cells, particularly in regions of active cell-wall synthesis. The vesicles are considered to function in exocytosis and endocytosis. Griffing (44) found that the coated vesicles in soybean protoplasts were enriched in glucan synthetase I. However, a chase experiment after pulsing with [³H]glucose showed that the radioactivity was not incorporated via the coated vesicles but was transported into plasma membranes from the Golgi. Thus glucan synthetase I may be an enzyme recycled and/or repositioned from the plasma membrane during endocytosis.

The question of the function of Mg^{2+}-dependent 1,4-β-glucan glucosyltransferase, glucan synthetase I, is still unanswerable in biochemical terms. It would seem to be a waste of time to study crude membrane-bound glucosyltransferase activities. Techniques are available for solubilizing and purifying membrane-bound enzymes in an active form. Thelen & Delmer (141) have developed a method for the detection and characterization of membrane-bound glucan synthases of plant and bacterial origin following their separation by electrophoresis in polyacrylamide gels.

BIODEGRADATION

Plant cell walls contain a set of glucanases and glycosidases that hydrolyze xyloglucan into monosaccharides (81, 82). The enzyme activities are extracted from the wall preparation with high-salt buffer (containing 1 M NaCl). The hydrolysis pattern by soybean enzymes showed a two-step degradation of the xyloglucan. It was first endohydrolyzed into large fragments, which were then further hydrolyzed into monosaccharides. This suggests that endo-1,4-β-

glucanase activity is responsible for the first step of degradation, and that the second step requires α-xylosidase, β-glucosidase, α-fucosidase, and β-galactosidase activities.

Properties and Regulation of Auxin-Induced Endo-1,4-β-Glucanases

The growing cells of pea stem generate two distinct endo-1,4-β-glucanases (14). The activities hydrolyze the internal 1,4-β-glucosyl linkages of xyloglucan (62), as well as a variety of (1→4)-β-glucans, (1→4, 1→3)-glucans, and cellodextrins (14, 94, 149). The action patterns of pea endo-1, 4-β-glucanases against pea xyloglucan are essentially the same as those of fungal cellulases; both hydrolyze internal linkages adjacent to unsubstituted glucose residues (51, 62), thereby introducing a free reducing end-group at these points.

The K_m values of the two pea endo-1,4-β-glucanases acting on pea xyloglucan, amyloid xyloglucan, cellohexaose, and CM-cellulose were all remarkably similar (3.5 mg/ml) (62). However, the V_{max} values with the two xyloglucans as substrates were 10–50 times lower for the glucanases. The enzymes possess a binding site that recognizes at least six consecutive 1,4-β-linked glucose units, and random limited substitution in CM-cellulose does not interfere with the binding or hydrolysis of the substrate. However, the 1,4-linked glucose backbone of xyloglucan is only hydrolyzable at every fourth glucose residue. Such structural constraints probably account for a V_{max} lower than that for cellohexaose and CM-cellulose.

Using ferritin conjugate antibody and electron microscopy, Bal et al (7) clearly located the enzymes in ultrathin sections of pea tissues. One of the activities, with a molecular mass of 70 kd, is firmly associated with the inner surface of cell walls. The other activity, with a molecular mass of 15 kd, is also found associated with rough ER where synthesis of the β-glucanases occurs with membrane-bound polysomes. The levels of these activities are further increased about 100-fold after treatment with superoptimal auxin (14, 33, 53, 147). Because the increases are prevented by both puromycin and actinomycin D but not by 5-fluorodeoxyuridine (34), auxin induces the glucanases at the transcriptional level (147). Although auxin evokes both swelling of parenchyma cells and cell division, the glucanases probably occur in expanding parenchyma cells rather than dividing cells.

Is Cellulose a Substrate for Auxin-Induced Endo-1,4-β-Glucoanases?

Auxin induces the biosynthesis of endo-1,4-β-glucanases in vivo (147). The glucanase synthesized is secreted to the wall, where it binds firmly to the inner surface (7). When macromolecular complexes of xyloglucan and cellulose (cell-wall ghosts) were treated in vitro with pea endo-1,4-β-glucanase, the amount of xyloglucan decreased and there was a concomitant release of soluble saccharide (62). There was also a marked increase in the number of free reducing end-groups in the xyloglucan, which remained bound to cellulose microfibrils. In addition, there was a smaller but detectable increase in cellulose chain ends.

Auxin treatment of pea epicotyls in vivo causes increases in net deposition of xyloglucan and cellulose (62) despite marked induction of endo-1,4-β-glucanase activity. However, the average degree of polymerization of the resulting xyloglucan was much lower than in controls, and the decrease in molecular mass led to solubilization of the polysaccharide. Because these events probably occur in vivo mainly on the inner surface of the wall, endohydrolysis of xyloglucan may weaken the cross-linkages that constrain microfibril slippage and creep. Thus, there is a correlation between endo-1,4-β-glucanase activity and xyloglucan degradation. In the primary walls of higher plants, xyloglucan is evidently more accessible to hydrolysis. It is concluded that endo-1,4-β-glucanase undoubtedly functions in vivo primarily as a xyloglucan-degrading enzyme.

Glycosidase Reaction—Is Xyloglucan a Precursor of Cellulose?

Plant cell walls contain α-xylosidase activity, which is probably specific for xyloglucan degradation. The activity does not hydrolyze p-nitrophenyl-α-D-xyloside and isoprimeverose (82, 112), but acts on heptasaccharide (glucose/xylose, 4:3). If α-xylosidase activity can act on a xyloglucan, the resulting hydrolysate may be a cellulose, suggesting that xyloglucan is a precursor of cellulose. Xyloglucans in monocotyledons contain fewer substituted xylosyl residues than those in dicotyledons. As speculated above, xyloglucans in monocotyledons may originally contain many substituted xylosyl residues that are partially deleted during growth. In addition, some xylose residues that remain in the 24% KOH-insoluble fraction (cellulose) in the cell-wall preparations of monocotyledons may derive from terminal xylose, because the residues were confirmed as 2,3,4-tri-O-methyl xylitol by methylation analysis (T. Hayashi, unpublished). This suggests that some 1,4-β-glucan in the cellulose fraction is derived from xyloglucan by deletion of xylosyl residues.

When soybean xyloglucan was incubated with an enzyme preparation from the cell walls of soybean hypocotyls, xylosyl residues were released only from fragmented xyloglucan oligosaccharides and not from high-molecular-mass xyloglucan (81, 82). It is likely that the degradation proceeds by the sequential splitting of the α-xylosidic and β-glucosidic linkages and that the release of xylosyl residues occurs at the nonreducing terminal of the backbone. O'Neill et al (111) partially purified an α-xylosidase from pea stem extracts and used this to hydrolyze xyloglucan heptasaccharide. The result indicated that the xylosidase hydrolyzed the glycosidic linkage of only one of the three xylosyl residues of the heptasaccharide. The mode of action of a bacterial α-xylosidase from *Aspergillus niger* was also the same in the oligosaccharide as that of the plant xylosidases (97). It seems likely that α-xylosidases from soybean and pea cell walls hydrolyze at most one xylosidic linkage of the nonreducing terminal of the glucan backbone. In any case, it does not seem reasonable to conclude that xyloglucan represents a precursor to cellulose, although nothing is known about the occurrence and action of the α-xylosidase in monocotyledons nor has evidence been published to date that the glycosidase acts on a xyloglucan polymer in monocotyledons during growth.

MACROMOLECULAR ORGANIZATION

Localization of Xyloglucan

Localization of xyloglucan on thin sections of sycamore cells grown in suspension culture was carried out by Moore et al (102) using a polyclonal antibody against the polysaccharide. Visualization using immunogold labelling showed that xyloglucan was localized throughout the entire wall and middle lamella, although rhamnogalacturonan (pectin) was restricted to the middle lamella. It has been suggested that xyloglucan is continuously secreted during both cell-plate formation and cell-wall expansion. Moore et al (103) further indicated that a xyloglucan is present within the Golgi apparatus, predominantly in the vesicles surrounding the central stack of cisternae. In addition, it was suggested that a single Golgi stack contained both xyloglucan and rhamnogalacturonan, but that these two matrix components are packaged into separate vesicles.

Misaki et al (101) recently prepared anti-octasaccharide antibody IgG, which interacted strongly with octasaccharide (glucose/xylose/galactose, $4:3:1$)- and nonassaccharide (glucose/xylose/galactose/fucose, $4:3:1:1$)-BSA (bovine serum albumin) but slightly with heptasaccharide (glucose/xylose, $4:3$)- and isoprimeverose-BSA. The antibody tends to recognize β-galactosyl-α-xylosyl groups attached to the $1,4$-β-glucan chain. Because the antibody strongly reacted with soybean xyloglucan, soybean tissues were stained indirectly with the antibody and fluorescein isothiocyanate (FITC)- or peroxidase-labeled antisera against IgG. The antibody bound to the inner layer of the primary walls and not to the middle lamella, although another antibody prepared by Moore et al (102) bound to the middle lamella.

A macromolecular complex (wall ghost) composed of xyloglucan and cellulose (51) was obtained from elongating regions of pea stems and examined by light microscopy using iodine staining, by radioautography after labeling with [^{3}H]fucose, by fluorescence microscopy using a fluorescein-lectin (fucose-binding lectin from *Ulex europaeus*) as a probe, and by electron microscopy after shadowing. The results indicated that xyloglucan is located both on and between cellulose microfibrils. The organization of this macromolecular complex was such that forming cellulose microfibrils continued to be coated with xyloglucan throughout cell elongation (55).

Interaction between Xyloglucan and Cellulose

Binding of pea xyloglucan to cellulose prepared from the elongating region of pea epicotyls proceeded and reached equilibrium within 4 hr at 40°C (55). The interconnection between native pea xyloglucan and cellulose is stable at acidic pH (< 6) and unstable above this pH. This indicates that creep between the two kinds of β-glucans does not occur at acidic pH (55, 144), although acidic pH stimulates cell-wall loosening (117). The binding of xyloglucan to cellulose was very specific and was not affected by the presence of a 10-fold excess of $(1\rightarrow2)$-β-glucan, $(1\rightarrow3)$-β-glucan, $(1\rightarrow6)$-β-glucan, $(1\rightarrow3,1\rightarrow4)$-$\beta$-glucan, arabinogalactan, or pectin (55). This suggests that the binding occurs specifically with structures containing $(1\rightarrow4)$-β-glucosyl linkages in polymers having a complementary conformation to cellulose.

Many xyloglucans were also isolated using a cellulose column that specifically bound xyloglucan from other components (4, 9). This procedure is very useful for obtaining highly purified xyloglucan. Fluorescein-labeled pea xyloglucan (55) was also used to stain cellulose microfibrils. This kind of cellulose visualization contributes to an understanding of the morphology of plant cell walls, and to the localization of cellulose. Thus, xyloglucans resemble an antibody for cellulose.

Organization of β-Glucan Complex

Bauer et al (9) proposed the occurrence of hydrogen bonds between xyloglucan and cellulose, and Chambat et al (19) confirmed this by solubilizaton and fractionation of *Rosa* cell walls. Hayashi et al (55) recently showed the interactions between xyloglucan and cellulose in vitro and in vivo consist of hydrogen bonds. Analysis of the binding capacity for cellulose microfibrils of different surface area showed that the capacity was dependent on the surface area of the microfibrils (55). However, the native xyloglucan:cellulose complex (51) obtained from pea stems contains 8-fold higher levels of xyloglucan than those in the reconstituted complex (Table 1). In the primary cell walls of pea stems, xyloglucan probably not only binds to the surface of cellulose microfibrils but is also woven into the amorphous parts of the microfibrils. This probably explains why concentrated alkali, which causes microfibrils to swell, is required for the extraction of xyloglucan. Mild alkali, which does not cause microfibril swelling, does not dissociate the complex. The native association may form close to, or directly at, the site of cellulose synthesis.

Table 1   Properties of pea xyloglucan-cellulose complex

Binding levels	xyloglucan-cellulose complex formed	
	in vitro	in vivo
Capacity	low	high
xyloglucan content	5%	41%
Interconnection	mild	strong
solvent required for dissociation	4% KOH	24% KOH

Xyloglucan may then be localized in both the inner and outer surfaces of microfibrils, possibly enhancing binding in vivo.

A possible role of the xyloglucans that cover cellulose and bind to and weave into microfibrils is the prevention of hydrogen bonding between cellulose microfibrils. This may help each microfibril to slide during cell enlargement, because xyloglucans have no mutual affinity (55). During the biogenesis of cellulose, xyloglucans probably control the size of microfibrils and prevent their fasciation. In fact, pea xyloglucan interfered with the ribbon assembly of cellulose microfibrils produced by *Acetobacter xylinum* (55): It appears that a xyloglucan associates with the subunits of the ribbon and prevents fasciation into larger bundles. This interaction may also stimulate the rate of cellulose synthesis, as was found in *Acetobacter xylinum* cells in the presence of Calcofluor and carboxymethylcellulose (46), resulting not only in spreading of the fiber network, but also in giving the flexibility necessary for microfibrils to slide.

Xyloglucans may also contribute to the cross-linking of each cellulose microfibril network. This probably gives a rigidity to cell walls by the binding of adjacent microfibrils. The cross-linking between perpendicular fibrils may function as a bracket, and that between parallel fibrils as a beam (Figure 4). When auxin-induced endo-1,4-β-glucanases hydrolyze the cross-linking xyloglucans (62), each cellulose microfibril could become loosened. Cellulose microfibrils in the primary walls of growing pea epicotyl cells are oriented primarily in a transverse direction on the inner surfaces of walls, and some change direction to travel in longitudinally oriented "ribs" on the outer surface (91). Therefore, the endohydrolysis of xyloglucan may be required for cell enlargement during growth. The proposed linkages between xyloglucan and cellulose are summarized in Figure 4.

Xyloglucans are not the main component in the primary walls of monocotyledons and may not be the only form of hemicellulose that binds to cellulose by hydrogen bonds. In monocotyledons, xylans are the major component of the primary walls (5, 22, 28, 98), and these polysaccharides have been shown in vitro to bind to cellulose. The association between the xylans and cellulose

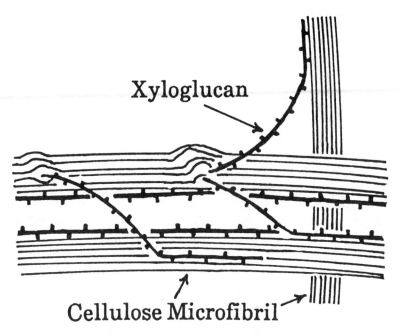

Figure 4 Potential linkages between xyloglucan and cellulose.

is also thought to occur *in vivo* in a manner similar to that between xyloglucan and cellulose in dicotyledons. However, mild alkali (4% KOH) mostly extracts xylans from cell-wall preparations of monocotyledons, and strong alkali (24% KOH) is required for the extraction of xyloglucan even in monocotyledons. This suggests that xyloglucan is more strongly associated with cellulose, and that xylans may bind to cellulose only at the surface of microfibrils, if at all. In addition, xylan is not strong enough to prevent the in vitro association between xyloglucan and cellulose, because xyloglucan has a higher affinity for cellulose (55).

Potential Model of Xyloglucan Exocytosis

A macromolecular complex composed of xyloglucan and cellulose has been obtained from elongating regions of etiolated pea stems. Such macromolecular organization probably occurs at the cell surface during cellulose biosynthesis. Because xyloglucan is synthesized in dictyosomes and secreted to walls via secretory vesicles, the polysaccharide must be prepared for organization with cellulose microfibrils. Because the Golgi apparatus is responsible for the exocytosis of rosettes of particles (41, 47), the terminal complexes are probably also prepared as proterminal complexes for organization. The organization mechanism is unknown, but there is evidence that the exogenous xyloglucan clearly fails to complex with newly formed cellulose microfibrils in the protoplast wall (61). This suggests that the macromolecular organization involves a secretion process on the plasma membranes.

Vian (148) suggested that the secretory products are loose and scattered, like balls of thread. They become ordered later, at the moment of their deposition into the inner wall layers. Staehelin (131) carefully observed the mechanism of secretion and demonstrated fusion of vesicles with plasma membrane necked pores up to about 60 nm in diameter. During discharge, each vesicle was flattened, forming a disc-shaped structure perpendicular to the plane of the membrane. Although the structural appearance of pea xyloglucan resembles a straight ribbon 500 nm long, the polysaccharide is probably folded and packaged in secretory vesicles (20). It appears that xyloglucans bind to each nascent 1,4-β-glucan by a hydrogen bond and that the 1,4-β-glucans bundle themselves up with xyloglucans. This probably explains why the X-ray diffraction patterns of the cellulose of primary walls show a lower degree of crystallinity and are thinner than those of secondary walls. In any event, the microfibrils of native celluloses probably form as unidirectionally oriented 1,4-β-glucans (63). A possible model of xyloglucan exocytosis and macromolecular organization is summarized in Figure 5.

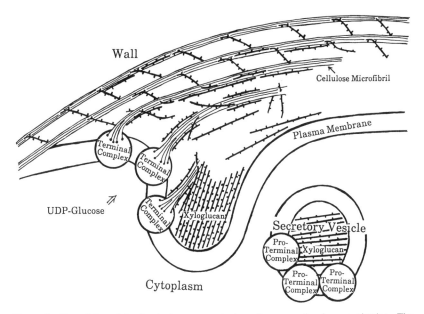

Figure 5 Potential model of xyloglucan exocytosis and macromolecular organization. The moment the xyloglucan synthesized in Golgi is transported to extracellular sites, cellulose synthase (terminal complex) catalyzes cellulose synthesis at the cell surface and the association of cellulose with xyloglucan takes place.

Linkages between Xyloglucan and Other Matrix Components

A primary-wall model proposed by Albersheim (9, 76, 135) was based on the fact that enzymic hydrolysis of sycamore cell walls yielded fragments that could be a linkage point between xyloglucan and galactan, showing the presence of glycosidic bonds between xyloglucan and pectin (and other matrix components). However, recent results (22, 98) cast doubt on the presence of such glycosidic bonds.

Matrix components could bind to xyloglucan because skeleton celluloses are completely covered with xyloglucans. Sequential extraction of pea primary-wall preparations indicated that pectin was mostly extracted with 0.1 M EDTA, galactan and proteins with 4% KOH, and xyloglucan with 24% KOH (51). This suggests that other matrix components might bind to xyloglucan by alkali-labile linkages such as O-esters or proteins. Because xyloglucan and pectin are probably packaged into separate secretory vesicles (103), the linkage between xyloglucan and other components, if any might form at the cell surface.

GROWTH REGULATION

Auxin-Induced Cell Enlargement

Auxin induces cell enlargement. In pea stems, IAA is synthesized in the plumule and transported to epicotyl cells to evoke cell enlargement. Cells in the parenchyma grow about 10 times their length and 1.7 times their width after cell division and, thereby, the cell walls are extended by about 17 times. The walls of enlarged cells continue to receive new wall components, mainly cellulose. Thus cell walls grow with the deposition of the polysaccharide into secondary walls and with a relative decrease in the primary-wall components, but they maintain wall rigidity during cell enlargement. The primary walls are composed of cellulose microfibrils, cross-linking xyloglucans between the microfibrils, and other matrix components. The rigidity of the walls is probably constrained by xyloglucan networks.

Xyloglucan in the primary walls is one of the most probable contributors to the loosening of cellulose microfibril networks, which renders the walls susceptible to turgor-driven expansion (134). Labavitch & Ray (84, 85) showed for the first time that auxin promoted the turnover of xyloglucan. This promotion begins within 15 min after auxin treatment, and the effect not only increases with increased IAA concentration but also corresponds to the elongation rate. Auxin- and/or acid-induced growth has also been proved to accompany xyloglucan degradation in peas (42, 65, 84, 85, 138, 139) *Vigna* (108, 109), *Pinus* (88–90), and even two monocotyledons, *Avena* (64) and *Oryza* (121). Endo-1,4-β-glucanase (cellulase) activities responsible for xyloglucan degradation were also confirmed to be associated with auxin-induced cell expansion in pea (54, 62), cell growth in mung bean (70), and, even in monocotyledons, with developing vessels of barley roots (127) and developing pistils and pollen tubes of *Hemerocallis fulva* (77).

When a bacterial cellulase preparation was incubated with plant tissues, elongation was not induced in *Avena* coleoptiles (125), but some cortical cells in sunflower stems were enlarged (75). To date, the only reported case of plant cellulases acting against plant tissues has been demonstrated in pea. When tissues or cell-wall ghosts prepared from pea epicotyls were incubated with cellulases purified from auxin-treated peas, a large degradation of xyloglucan and a smaller increase in cellulose chain ends were observed (62). Furthermore, β-glucan synthesis from UDP-glucose by pea tissues is increased 2–3-fold after treatment with the purified cellulases (150). The β-glucanases probably act slightly on cellulose in addition to xyloglucan, and this may contribute to the availability of endogenous 1,4-β-glucan chain ends as acceptors for transglucosylation (150), although alkali-insoluble β-glucan synthesis apparently does not need a primer (92). This scheme suggests that both β-glucan degradation and cellulose synthesis are accompanied by cell-wall loosening.

Although xylans, not xyloglucans, are the dominant hemicellulosic component of the cell walls in monocotyledons (22, 98), the xyloglucans are more strongly bound and probably affect the microfibril network more directly. The walls also contain high levels of mixed 1,3-β-1,4-β-linked glucan, which is located in the middle lamella (132). Although exo-1,3-β-glucanases were proposed as wall-loosening enzymes in *Avena* coleoptiles (96), it seems likely that the degradation of the mixed β-glucans or 1,3-β-glucan is only indirectly associated with cell enlargement because of the localization in the middle lamella.

Acid-Induced Growth and Xyloglucan Degradation

Auxin stimulates the hydrogen pump of the plasma membrane, which lowers the pH of the wall in a growing cell. Jacobs & Ray (65) observed promotion of a water-soluble xyloglucan from cell walls on incubation with acidic pH buffer (Na-citrate, pH 4) when the elongation of pea stems was stimulated. Nishitani & Masuda (108) also observed that acid pH induced a decrease in the mass-average molecular weight of xyloglucans in the cell walls of light-grown *Vigna angularis* hypocotyls. They reported that the effect of acid pH on the xyloglucan appeared without a lag period, and that the molecular mass decreased linearly with time. Thus it was observed that not only auxin but also acidic pH caused a similar biochemical effect on xyloglucan degradation during elongation.

Auxin-induced cell elongation occurs in two separate ways (145). It appears that the first response results from acid growth, the second from gene expression. Theologis (142) suggests that rapidly auxin-induced mRNAs mediate H^+ secretion and cell elongation, and that late mRNAs, which are indirectly regulated by auxin, may code for endo-1,4-β-glucanases. Nevertheless, xyloglucan is degraded at an early stage of auxin treatment and during acid growth without a lag time. The rapid response may be caused by *(a)* the noncovalent interactions of cell-wall components that occur by ionic bonds (105, 107, 122, 137) or by lectin-like interactions (8), and/or *(b)* merely by xyloglucan degradation by endo-1,4-β-glucanase activation (14).

Fry (39) noted that peroxidase (which catalyzes the formation of isodityrosine and diferulate cross-links) and pectinesterase (which produces free-acidic pectin for Ca-bridges) require Ca^{2+} and are inhibited by low pH. However, this does not explain xyloglucan degradation by acid-induced growth.

Kauss et al (45, 73, 74) solubilized carbohydrate-binding proteins from the cell walls of nongrowing segments of mung bean hypocotyls. The agglutinating activity is strongly diminished by acidic pH and is inhibited by D-galactose. To date, however, no xyloglucan-binding lectin has been obtained in growing plant cell walls although there are fucose- or glactose-binding lectins from plant seeds that can bind to pea and soybean xyloglucans (51).

Because the activity of endo-1,4-β-glucanase has a pH optimum of 5.5–6.0, acid-induced growth may derive from the activation of the enzyme. However, the level of the activity is increased only 1.3-fold in vitro at the pH optimum compared with neutral pH (14). This suggests that rapid cell enlargement results in a smaller but certain increase in the activity. Although levels of the endo-1,4-β-glucanase activity are further increased about 100-fold after treatment with superoptimal auxin in pea epicotyls, it seems unlikely that cell enlargement requires so much excess activity. Thus, the activity in walls may be controlled in some as-yet-unclear way in vivo by the cellular regulation of wall pH.

Morris & Northcote (104) suggested that the fusion of Golgi-derived vesicles with the plasma membrane was rate-limiting and a potential control point in sycamore cells. Because both cell walls and plasma membrane have a high affinity for Ca^{2+}, it appears that Ca^{2+} causes a decrease in surface potential by directly binding to the cell surface. Auxin and/or low pH may stimulate vesicle fusion with the plasma membrane by depositing cell-wall components and terminal complex plus endo-1/4-β-glucanase, which may then be packaged in a secretory vesicle.

Thus, although the endohydrolysis of xyloglucan must result from the action of endo-1,4-glucanase activity (62), it is still uncertain whether the early response of auxin-induced cell enlargement is caused by changes in the level of this activity.

Inhibition of Cell Elongation

York et al (151) reported that xyloglucan nonasaccharide inhibited auxin-induced elongation of pea stem segments. The inhibitory activity was almost complete with a nonasaccharide concentration of between 10^{-7} and 10^{-8} M. In this assay, 2,4-dichlorophenoxyacetic acid was used at 10^{-6} M, indicating that the concentration of oligosaccharide was much lower than that of the auxin in the incubation mixture. This phenomenon suggests that, although auxin induces endo-1,4-β-glucanase activities that hydrolyze xyloglucans, the fragments formed act as feedback inhibitors of auxin-stimulated growth. The nonasaccharide was also found in the culture medium of spinach cells (40) and accumulated extracellularly to about 4.3×10^{-7} M.

Levels of product inhibition are very high in cellulase-catalyzing systems. Most cellulase activities from bacteria and fungi are inhibited by cellobiose at very low concentrations, whereas cellobiose acts as an inducer for endo-1,4-β-glucanases in lower organisms (119). Although the nonasaccharide is also a kind of product catalyzed by cellulase, the activities of pea cellulases are inhibited by cellodextrins but stimulated by nona- and heptasaccharides (36). Furthermore, xyloglucan heptasaccharide did not inhibit the auxin-stimulated elongation of pea stems (151). These results suggest that the inhibition of stem elongation is not caused by inhibition of endoglucanase activities. A similar physiological phenomenon was recently obtained with α-oligogalacturonides (11), showing that the uronides behave as competitive antagonists of IAA. These oligosaccharides probably act on plant cells and change cell growth by a mechanism unrelated to xyloglucan degradation.

FUTURE STUDIES

I am becoming ever more convinced that xyloglucan is the key component of primary walls and a regulatory factor for cell enlargement in the plant kingdom. The strategy for future studies probably requires three lines of research. 1. We must identify the properties and understand the regulation of cellulose synthase. This would provide the mechanism for the macromolecular organization of xyloglucan and cellulose and probably the function of Golgi-located 1,4-β-glucan 4-β-glucosyltransferase. However, no one has yet succeeded in synthesizing cellulose in a cell-free system in higher plants (25). 2. We must identify the mechanism accountable for the rapid response of cell elongation and xyloglucan degradation by auxin and/or acid. Because the physiological and biochemical changes involved occur within 15 min, it is not clear that the trigger for wall loosening is regulated at the level of gene expression. There may be certain kinds of receptors for the early response, which are associated with biochemical and physiological reactions. 3. We

162   HAYASHI

must elucidate the molecular biology of xyloglucan metabolism. Pea cellulases, endo-1,4-β-glucanases, have been shown to be translated by auxin-induced mRNAs (147), but their genes have not been cloned. Further study should shed light on the gene expression induced by auxin at the transcriptional and translational levels and allow further correlations with known biochemical and physiological changes.

ACKNOWLEDGMENTS

I thank Drs. D. P. Delmer, G. Maclachlan, A. Camirand, K. Matsuda, and Y. Kato for critically reading the manuscript.

Literature Cited

1. Akiyama, Y., Katô, K. 1982. An arabinoxyloglucan from extracellular polysaccharides of suspension-cultured tobacco cells. *Phytochemistry* 21:2112–14
2. Albersheim, P. 1975. The walls of growing plant cells. *Sci. Am.* 232:81–95
3. Ankel, H., Feingold, D. S. 1965. Biosynthesis of uridine diphosphate D-xylose. I. Uridine diphosphate glucuronate carboxy-lyase of wheat germ. *Biochemistry* 4:2468–75
4. Aspinal, G. O., Molloy, J. A., Craig, J. W. T. 1969. Extracellular polysaccharides from suspension-cultured sycamore cells. *Can. J. Biochem.* 47:1063–70
5. Bacic, A., Harris, P. J., Stone, B. A. 1988. Structure and function of plant cell walls. In *The Biochemistry of Plants*, ed. J. Priess, pp. 297–371. New York: Academic
6. Bacon, J. S. D., Gordon, A. H., Morris, E. J. 1975. Acetyl groups in cell-wall preparations from higher plants. *Biochem. J.* 149:485–87
7. Bal, A. K., Verma, D. P. S., Byrne, H., Maclachlan, G. A. 1976. Subcellular localization of cellulases in auxin-treated pea. *J. Cell Biol.* 69:97–105
8. Bates, G. W., Ray, P. M. 1981. pH-Dependent interactions between pea cell wall polymers possibly involved in wall deposition and growth. *Plant Physiol.* 68:158–64
9. Bauer, W. D., Talmadge, K. W., Keegstra, K., Albersheim, P. 1973. The structure of plant cell walls. II. The hemicellulose of the walls of suspension-cultured sycamore cells. *Plant Physiol.* 51:174–87
10. Bolwell, G. P. 1988. Synthesis of cell wall components: aspects of control. *Phytochemistry* 27:1235–53

11. Branca, C., De Lorenzo, G., Cervone, F. 1988. Competitive inhibition of the auxin-induced elongation by α-D-oligogalacturonides in pea stem segments. *Physiol. Plant.* 72:499–504
12. Brown, R. M. Jr., Montezinos, D. 1976. Cellulose microfibrils: visualization of biosynthetic and orienting complexes in association with the plasma membrane. *Proc. Natl. Acad. Sci. USA* 73:143–47
13. Burke, D., Kaufman, P., McNeil, M., Albersheim. P. 1974. The structure of plant cell walls. VI. A survey of the walls of suspension-cultured monocots. *Plant Physiol.* 54:109–15
14. Byrne, H., Christou, N. V., Verma, D. P. S., Maclachlan, G. A. 1975. Purification and characterization of two cellulases from auxin-treated pea epicotyls. *J. Biol. Chem.* 250:1012–18
15. Camirand, A., Brummell, D., Maclachlan, G. 1987. Fucosylation of xyloglucan:localization of the transferase in dictyosomes of pea stem cells. *Plant Physiol.* 84:753–56
16. Camirand, A., Maclachlan, G. 1986. Biosynthesis of the fucose-containing xyloglucan nonasaccharide by pea microsomal membranes. *Plant Physiol.* 82:379–83
17. Camirand, A., Torossian, K., Hayashi, T., Maclachlan, G. 1985. Are charged lipid-linked intermediates involved in the biosynthesis of β-glucans? *Can. J. Bot.* 63:867–71
18. Cassab, G. I., Varner, J. E. 1988. Cell wall proteins. *Annu. Rev. Plant Physiol. Plant Mol. Biol.* 39:321–53
19. Chambat, G., Barnoud, F., Joseleau, J.-P. 1984. Structure of the primary cell walls of suspension-cultured *Rosa*

glauca cells. I. Polysaccharides associated with cellulose. *Plant Physiol.* 74:687–93

20. Chanzy, H. D. 1974. Structural and morphological aspects of cellulose materials. In *Structure of Fibrous Biopolymers*, ed. E. D. T. Atkins, A. Keller, pp. 417–34. London: Butterworths

21. Cosgrove, D. 1986. Biophysical control of plant cell growth. *Annu. Rev. Plant Physiol.* 37:377–405

22. Darvill, A., McNeil, M., Albersheim, P., Delmer, D. P. 1980. The primary cell walls of flowering plants. In *The Biochemistry of Plants*, ed. N. E. Tolbert, pp. 91–162. New York: Academic

23. Dea, I. C. M., Madden, J. K. 1986. Acetylated pectic polysaccharides of sugar beet. *Food Hydrocolloids* 1:71–88

24. Delmer, D. P. 1983. Biosynthesis of cellulose. *Adv. Carbohydr. Chem. Biochem.* 41:105–53

25. Delmer, D. P. 1987. Cellulose biosynthesis. *Annu. Rev. Plant Physiol.* 38:259–90

26. Delmer, D. P., Cooper, G., Alexander, D., Cooper, J., Hayashi, T., et al. 1985. New approaches to the study of cellulose biosynthesis. *J. Cell Sci. Suppl.* 2:33–50

27. Delmer, D. P., Stone, B. A. 1988. Biosynthesis of plant cell walls. See Ref. 5, pp. 373–420

28. Dey, P. M., Brinson, K. 1984. Plant cell-walls. *Adv. Carbohydr. Chem. Biochem.* 42:265–382

29. Djordjevic, S. P., Chen, H., Batley, M., Redmond, J. W., Rolfe, B. G. 1987. Nitrogen fixation ability of exopolysaccharide synthesis mutants of *Rhizobium* sp. strain NGR 234 and *Rhizobium trifolii* is restored by the addition of homologous exopolysaccharides. *J. Bacteriol.* 169:53–60

30. Eda, S., Katô, K. 1978. An arabinoxyloglucan isolated from the midrib of the leaves of *Nicotiana tabacum*. *Agric. Biol. Chem.* 42:351–57

31. Eda, S., Kodama, H., Akiyama, Y., Mori, M., Katô, K., et al. 1983. An arabinoxyloglucan isolated from the cell walls of suspension-cultured tobacco cells. *Agric. Biol. Chem.* 47:1791–97

32. Edwards, M., Dea, I. C. M., Bulpin, P. V., Reid, J. S. G. 1985. Xyloglucan (amyloid) mobilisation in the cotyledons of *Tropaeolum majus* L. seeds following germination. *Planta* 163:133–40

33. Fan, D. F., Maclachlan, G. A. 1966. Control of cellulase activity by indoleacetic acid. *Can. J. Bot.* 44:1025–34

34. Fan, D. F., Maclachlan, G. A. 1967. Massive synthesis of ribonucleic acid and cellulase in the pea epicotyl in response to indoleacetic acid, with and without concurrent cell division. *Plant Physiol.* 42:1114–22

35. Farkas, V., Maclachlan, G. 1988. Fucosylation of exogenous xyloglucans by pea microsomal membranes. *Arch. Biochem. Biophys.* 261:148–53

36. Farkas, V., Maclachlan, G. 1988. Stimulation of pea 1,4-β-glucanase activity by oligosaccharides derived from xyloglucan. *Carbohydr. Res.* In press

37. Feingold, D. S., Neufeld, E. F., Hassid, W. Z. 1960. The 4-epimerization and decarboxylation of uridine diphosphate D-glucuronic acid by extracts from *Phaseolus aureus* seedlings. *J. Biol. Chem.* 235:910–13

38. Fincher, G. B., Stone, B. A., Clarke, A. E. 1983. Arabinogalactan-protein: structure, biosynthesis, and function. *Annu. Rev. Plant Physiol.* 34:47–70

39. Fry, S. C. 1986. Cross-linking of matrix polymers in the growing cell walls of angiosperms. *Annu. Rev. Plant Physiol.* 37:165–86

40. Fry, S. C. 1986. In-vivo formation of xyloglucan nonasaccharide: a possible biologically active cell-wall fragment. *Planta* 169:443–53

41. Giddings, T. H. Jr., Brower, D. L., Staehelin, L. A. 1980. Visualization of particle complexes in the plasma membrane of *Micrasterias denticulata* associated with the formation of cellulose fibrils in primary and secondary cell walls. *J. Cell. Biol.* 84:327–39

42. Gilkes, N. R., Hall, M. A. 1977. The hormonal control of cell wall turnover in *Pisum sativum* L. *New Phytol.* 78:1–15

43. Gould, S. E. B., Rees, D. A., Wight, N. J. 1971. Polysaccharides in germination. Xyloglucans ('amyloids') from the cotyledons of white mustard. *Biochem. J.* 124:47–53

44. Griffing, L. R., Mersey, B. G., Fowke, L. C. 1986. Cell-fractionation analysis of glucan synthase I and II distribution and polysaccharide secretion in soybean protoplasts. Evidence for the involvement of coated vesicles in wall biogenesis. *Planta* 167:175–82

45. Haass, D., Frey, R., Thiesen, M., Kauss, H. 1981. Partial purification of a hemagglutinin associated with the cell walls from hypocotyls of *Vigna radiata*. *Planta* 151:490–96

46. Haigler, C. H., Benziman, M. 1982. Biogenesis of cellulose I microfibrils occurs by cell-directed self-assembly in

164   HAYASHI

Acetobacter xylinum. In Cellulose and Other Natural Polymer Systems, ed. R. M. Brown Jr., pp. 273–97. New York: Plenum

47. Haigler, C. H., Brown, R. M. Jr. 1986. Tranport of rosettes from the Golgi apparatus to the plasma membrane in isolated mesophyll cells of Zinnia elegans during differentiation to tracheary elements in suspension culture. Protoplasma 134:111–20

48. Hayashi, T., Delmer, D. P. 1988. Xyloglucan in the cell walls of cotton fibre. Carbohydr. Res. 181:273–77

49. Hayashi, T., Kato, Y., Matsuda, K. 1980. Xyloglucan from suspension-cultured soybean cells. Plant Cell Physiol. 21:1405–18

50. Hayashi, T., Koyama, T., Matsuda, K. 1988. Formation of UDP-xylose and xyloglucan in soybean Golgi membranes. Plant Physiol. 87:341–45

51. Hayashi, T., Maclachlan, G. 1984. Pea xyloglucan and cellulose. I. Macromolecular organization. Plant Physiol. 75:596–604

52. Hayashi, T., Maclachlan, G. 1984. Biosynthesis of pentosyl lipids by pea membranes. Biochem. J. 217:791–803

53. Hayashi, T., Maclachlan, G. 1984. Pea xyloglucan and cellulose. III. Metabolism during lateral expansion of pea epicotyl cells. Plant Physiol. 76:739–42

54. Hayashi, T., Maclachlan, G. 1986. Pea cellulose and xyloglucan: biosynthesis and biodegradation. In Cellulose: Structure, Modification, and Hydrolysis, ed. R. A. Young, R. M. Rowell, pp. 67–76. New York: Wiley

55. Hayashi, T., Marsden, M. P. F., Delmer, D. P. 1987. Pea xyloglucan and cellulose. V. Xyloglucan-cellulose interactions in vitro and in vivo. Plant Physiol. 83:384–89

56. Hayashi, T., Matsuda, K. 1981. Biosynthesis of xyloglucan in suspension-cultured soybean cells. Synthesis of xyloglucan from UDP-glucose and UDP-xylose in the cell-free system. Plant Cell Physiol. 22:517–23

57. Hayashi, T., Matsuda, K. 1981. Biosynthesis of xyloglucan in suspension-cultured soybean cells. Occurrence and some properties of xyloglucan 4-β-D-glucosyltransferase and 6-α-D-xylosyltransferase. J. Biol. Chem. 256:11117–22

58. Hayashi, T., Matsuda, K. 1981. Biosynthesis of xyloglucan in suspension-cultured soybean cells. Evidence that the enzyme system of xyloglucan synthesis does not contain β-1,4-glucan 4-β-D-glucosyltransferase activ-

ity (EC 2.4.1.12). Plant Cell Physiol. 22:1571–84

59. Hayashi, T., Matsuda, K. 1981. Sugar nucleotides from suspension-cultured soybean cells. Agric. Biol. Chem. 45:2907–8

60. Hayashi, T., Nakajima, T., Matsuda, K. 1984. Biosynthesis of xyloglucan in suspension-cultured soybean cells. Processing of the oligosaccharide building blocks. Agric. Biol. Chem. 48:1023–27

61. Hayashi, T., Polonenko, D. R., Camirand, A., Maclachlan, G. 1986. Pea xyloglucan and cellulose. IV. Assembly of β-glucans by pea protoplasts. Plant Physiol. 82:301–6

62. Hayashi, T., Wong, Y.-S., Maclachlan, G. 1984. Pea xyloglucan and cellulose. II. Hydrolysis by pea endo-1, 4-β-glucanases. Plant Physiol. 75:605–10

63. Hieta, K., Kuga, S., Usuda, M. 1984. Electron staining of reducing ends evidences a parallel-chain structure in Valonia cellulose. Biopolymers 23:1807–10

64. Inouhe, M., Yamamoto, R., Masuda, Y. 1984. Auxin-induced changes in the molecular weight distribution of cell wall xyloglucans in Avena coleoptiles. Plant Cell Physiol. 25:1341–51

65. Jacobs, M., Ray, P. M. 1975. Promotion of xyloglucan metabolism by acid pH. Plant Physiol. 56:373–76

66. John, K. V., Schwartz, N. B., Ankel, H. 1977. UDP-glucuronate carboxylyase in cultured chondrocytes. J. Biol. Chem. 252:6707–10

67. Joseleau, J. P., Chambat, G. 1984. Structure of the primary cell walls of suspension-cultured Rosa glauca cells. II. Multiple forms of xyloglucans. Plant Physiol. 74:694–700

68. Kato, Y., Iki, K., Matsuda, K. 1981. Cell wall polysaccharides in immature barley plants. II. Characterization of a xyloglucan. Agric. Biol. Chem. 45:2745–53

69. Kato, Y., Ito, S., Iki, K., Matsuda, K. 1982. Xyloglucan and β-glucan in cell walls of rice seedlings. Plant Cell Physiol. 23:351–64

70. Kato, Y., Matsuda, K. 1981. Occurrence of soluble and low molecular weight xyloglucan and its origin in etiolated mung bean hypocotyls. Agric. Biol. Chem. 45:1–8

71. Kato, Y., Matsuda, K. 1985. Xyloglucan in the cell walls of suspension-cultured rice cells. Plant Cell Physiol. 26:437–45

72. Kato, Y., Shiozawa, R., Takeda, S., Ito, S., Matsuda, K. 1982. Structural investigation of a β-D-glucan and xylo-

glucan from bamboo-shoot cell-walls. *Carbohydr. Res.* 109:233–48

73. Kauss, H., Bowles, D. J. 1976. Some properties of carbohydrate-binding proteins (lectins) solubilized from cell walls of *Phaseolus aureus*. *Planta* 130:169–74

74. Kauss, H., Glaser, C. 1974. Carbohydrate-binding proteins from plant cell walls and their possible involvement in extension growth. *FEBS Lett.* 45:304–7

75. Kawase, M. 1979. Role of cellulase in aerenchyma development in sunflower. *Am. J. Bot.* 66:183–90

76. Keegstra, K., Talmadge, K. W., Bauer, W. D., Albersheim, P. 1973. The structure of plant cell walls. III. A model of the walls of suspension-cultured sycamore cells based on the interconnections of the macromolecular components. *Plant Physiol.* 51:188–96

77. Konar, R. N., Stanley, R. G. 1969. Wall-softening enzymes in the gynoecium and pollen of *Hemerocallis fulva*. *Planta* 84:304–10

78. Kooiman, P. 1957. Amyloids of plant seeds. *Nature* 179:107–9

79. Kooiman, P. 1960. On the occurrence of amyloids in plant seeds. *Acta Bot. Neerl.* 9:208–219

80. Kooiman, P. 1961. The constitution of *Tamarindus*-amyloid. *Rec. Trav. Chim. Pays-Bas.* 80:849–65

81. Koyama, T., Hayashi, T., Kato, Y., Matsuda, K. 1981. Degradation of xyloglucan by wall-bound enzymes from soybean tissue. I. Occurrence of xyloglucan-degrading enzymes in soybean cell wall. *Plant Cell Physiol.* 22:1191–98

82. Koyama, T., Hayashi, T., Kato, Y., Matsuda, K. 1983. Degradation of xyloglucan by wall-bound enzymes from soybean tissue. II. Degradation of the fragment heptasaccharide from xyloglucan and the characteristic action pattern of the α-D-xylosidase in the enzyme system. *Plant Cell Physiol.* 24:155–62

83. Labavitch, J. M. 1981. Cell wall turnover in plant development. *Annu. Rev. Plant Physiol.* 32:385–406

84. Labavitch, J. M., Ray, P. M. 1974. Turnover of cell wall polysaccharides in elongating pea stem segments. *Plant Physiol.* 53:669–73

85. Labavitch, J. M., Ray, P. M. 1974. Relationship between promotion of xyloglucan metabolism and induction of elongation by indoleacetic acid. *Plant Physiol.* 54:449–502

86. Labavitch, J. M., Ray, P. M. 1978. Structure of hemicellulosic polysaccha-

rides of *Avena sativa* coleoptile cell walls. Phytochemistry 17:933–37

87. Leigh, J. A., Reed, J. W., Hanks, J. F., Hirsch, A. M., Walker, G. C. 1987. *Rhizobium meliloti* mutants that fail to succinylate their Calcofluor-binding exopolysaccharide are defective in nodule invasion. *Cell* 51:579–87

88. Lorences, E. P., Suárez, L., Zarra, I. 1987. Hypocotyl growth of *Pinus pinaster* seedlings. Changes in α-cellulose, and in pectic and hemicellulosic polysaccharides. *Physiol. Plant.* 69:461–65

89. Lorences, E. P., Suárez, L., Zarra, I. 1987. Hypocotyl growth of *Pinus pinaster* seedlings. Changes in the molecular weight distribution of hemicellulosic polysaccharides. *Physiol. Plant.* 69:466–71

90. Lorences, E. P., Zarra, I. 1987. Auxin-induced growth in hypocotyl segments of *Pinus-pinaster* Aiton. Changes in molecular-weight distribution of hemicellulosic polysaccharides. *J. Exp. Bot.* 38:960–67

91. Maclachlan, G. A. 1976. A potential role for endo-cellulase in cellulose biosynthesis. *Appl. Polym. Symp.* 28:645–58

92. Maclachlan, G. A. 1982. Does β-glucan synthesis need a primer? See Ref. 46, pp. 327–39

93. Maclachlan, G. 1985. Are lipid-linked glycosides required for plant polysaccharide biosynthesis? In *Biochemistry of Plant Cell Walls,* ed. C. T. Brett, J. R. Hillman, pp. 199–220. Cambridge: Cambridge Univ. Press

94. Maclachlan, G. A., Wong, Y.-S. 1979. Two pea cellulases display the same catalytic mechanism despite major differences in physical properties. *Adv. Chem. Ser.* 181:347–60

95. Mankarios, A. T., Hall, M. A., Jarvis, M. C., Threlfall, D. R., Friend, J. 1980. Cell wall polysaccharides from onions. *Phytochemistry* 19:1731–33

96. Masuda, Y., Oi, S., Satomura, Y. 1970. Further studies on the role of cell-wall-degrading enzymes in cell-wall loosening in oat coleoptiles. *Plant Cell Physiol.* 11:631–38

97. Matsushita, J., Kato, Y., Matsuda, K. 1987. Characterization of α-D-xylosidase II from *Aspergillus niger*. *Agric. Biol. Chem.* 51:2015–16

98. McNeil, M., Darvill, A. G., Fry, S. C., Albersheim, P. 1984. Structure and function of the primary cell walls of plants. *Annu. Rev. Biochem.* 53:625–63

99. Meinert, M. C., Delmer, D. P. 1977. Changes in biochemical composition of the cell wall of the cotton fiber during

development. *Plant Physiol.* 59:1088–97

100. Mersey, B. G., Griffing, L. R., Rennie, P. J., Fowke, L. C. 1985. The isolation of coated vesicles from protoplasts of soybean. *Planta* 163:317–27

101. Misaki, A., Sone, Y., Kojima, A., Shibata, S. 1988. Immunochemical specificities and histochemical application of the xyloglucan-recognizing antibodies, derived from its oligosaccharides. In *Abstract for XIVth International Carbohydrate Symposium*, Stockholm. 359 pp.

102. Moore, P. J., Darvill, A. G., Albersheim, P., Staehelin, L. A. 1986. Immunogold localization of xyloglucan and rhamnogalacturonan I in the cell walls of suspension-cultured sycamore cells. *Plant Physiol.* 82:787–94

103. Moore, P. J., Staehelin, L. A. 1987. *Cis* and *trans* Golgi cisternae assemble and package different types of secretory polysaccharides. *J. Cell Biol. Suppl.* 105:76

104. Morris, M. R., Northcote, D. H. 1977. Influence of cations at the plasma membrane in controlling polysaccharide secretion from sycamore suspension cells. *Biochem. J.* 166:603–18

105. Moustacas, A. M., Nari, J., Diamantidis, G., Noat, G., Crasnier, M., et al. 1986. Electrostatic effects and the dynamics of enzyme reactions at the surface of plant cells. 2. The role of pectin methyl esterase in the modulation of electrostatic effects in soybean cell walls. *Eur. J. Biochem.* 155:191–97

106. Mueller, S. C., Brown, R. M. Jr. 1980. Evidence for an intramembrane component associated with a cellulose microfibril synthesizing complex in higher plants. *J. Cell Biol.* 84:315–26

107. Nari, J., Noat, G., Diamantidis, G., Woudstra, M., Ricard, J. 1986. Electrostatic effects and the dynamics of enzyme reactions at the surface of plant cells. 3. Interplay between limited cell-wall autolysis, pectin methyl esterase activity and electrostatic effects in soybean cell walls. *Eur. J. Biochem.* 155:199–202

108. Nishitani, K., Masuda, Y. 1982. Acid pH-induced structural changes in cell wall xyloglucans in *Vigna angularis* epicotyl segments. *Plant Sci. Lett.* 28:87–94

109. Nishitani, K., Masuda, Y. 1983. Auxin-induced changes in the cell wall xyloglucans: effects of auxin on the two different subfractions of xyloglucans in the epicotyl cell wall of *Vigna angularis*. *Plant Cell Physiol.* 24:345–55

110. Ogawa, K., Hayashi, T., Okamura, K. 1988. Conformational analysis on xyloglucans. In *The Application of Conformational Analysis to Mono-, Oligo- and Polysaccharides*. Extended Abstract for *XIVth International Carbohydrate Symposium*, ed. K. Bock, pp. 129–34, Stockholm

111. O'Neill, R. A., Darvill, A. G., Albersheim, P. 1987. A xylosidase from pea stems hydrolyzes selected xylosidic linkages of xyloglucan oligosaccharides. *Plant Physiol. Suppl.* 83:106

112. Pierrot, H., Wielink, J. E. V. 1977. Localization of glycosidases in the wall of living cells from cultured *Convolvulus arvensis* tissue. *Planta* 137:235–42

113. Quail, P. H. 1979. Plant cell fractionation. *Annu. Rev. Plant Physiol.* 30:425–84

114. Ray, P. M. 1980. Cooperative action of β-glucan synthetase and UDP-xylose xylosyl transferase of Golgi membranes in the synthesis of xyloglucan-like polysaccharide. *Biochim. Biophys. Acta* 629:431–44

115. Ray, P. M., Eisinger, W. R., Robinson, D. G. 1976. Organelles involved in cell wall polysaccharide formation and transport in pea cells. *Ber. Dtsch. Bot. Ges.* 89:121–46

116. Ray, P. M., Shininger, T. L., Ray, M. M. 1969. Isolation of β-glucan synthetase particles from plant cells and identification with Golgi membranes. *Proc. Natl. Acad. Sci. USA* 64:605–11

117. Rayle, D. L., Cleland, R. 1972. The in vitro acid-growth response: relation to in vivo growth responses and auxin action. *Planta* 104:282–96

118. Redgwell, R. J., Selvendran, R. R. 1986. Structural features of cell-wall polysaccharides of onion *Allium cepa*. *Carbohydr. Res.* 157:183–99

119. Reese, E. T., Mandels, M. 1963. Enzymatic hydrolysis of β-glucans. In *Enzymic Hydrolysis of Cellulose and Related Materials*, ed. E. T. Reese, pp. 197–234. New York: Pergamon

120. Reis, D., Vian, B., Darzens, D., Roland, J.-C. 1987. Sequential patterns of intramural digestion of galactoxyloglucan in tamarind seedlings. *Planta* 170:60–73

121. Revilla, G., Zarra, I. 1987. Changes in the molecular weight distribution of the hemicellulosic polysaccharides from rice coleoptiles growing under different conditions. *J. Exp. Bot.* 38:1818–25

122. Ricard, J., Noat, G. 1986. Electrostatic effects and the dynamics of enzyme reactions at the surface of plant cells. 1. A theory of the ionic control of a com-

plex multi-enzyme system. *Eur. J. Biochem.* 155:183–90

123. Ring, S. G., Selvendran, R. R. 1981. An arabinoxyloglucan from the cell wall of *Solanum tuberosum. Phytochemistry* 20:2511–19

124. Robinson, D. G., Eisinger, W. R., Ray, P. M. 1976. Dynamics of the Golgi system in wall matrix polysaccharide synthesis and secretion by pea cells. *Ber. Dtsch. Bot. Ges.* 89:147–61

125. Ruesink, A. W. 1969. Polysaccharidases and the control of cell wall elongation. *Planta* 89:95–107.

126. Ruperez, P., Selvendran, R. R., Stevens, B. J. H. 1985. Investigation of the heterogeneity of xyloglucans from the cell walls of apple. *Carbohydr. Res.* 142:107–13

127. Sassen, M. M. A. 1965. Breakdown of the plant cell wall during the cell-fusion process. *Acta Bot. Neerl.* 14:165–96

128. Scott, R. A., Scheraga, H. A. 1966. Conformational analysis of macromolecules. III. Helical structures of polyglycine and poly-L-alanine. *J. Chem. Phys.* 45:2091–101

129. Shibuya, N., Misaki, A. 1978. Structure of hemicellulose isolated from rice endosperm cell wall: mode of linkages and sequences in xyloglucan, β-glucan and arabinoxylan. *Agric. Biol. Chem.* 42: 2267–74

130. Shore, G., Maclachlan, G. A. 1975. The site of cellulose synthesis. Hormone treatment alters the intracellular location of alkali-insoluble β-1,4-glucan (cellulose) synthetase activities. *J. Cell Biol.* 64:557–71

131. Staehelin, L. A. 1987. Secretion in plant cells differs from animal cells: intermediate plasma membrane configurations visualized in ultrarapidly frozen cells. *J. Cell Biol. Suppl.* 105:56

132. Stone, B. A. 1984. Noncellulosic β-glucans in cell walls. In *Structure, Function, and Biosynthesis of Plant Cell Walls,* ed. W. M. Dugger, S. Bartnicki-Garcia, pp. 52–74. Rockville: Am. Soc. Plant Physiol.

133. Strominger, J. L., Mapson, L. W. 1957. Uridine diphosphoglucose dehydrogenase of pea seedlings. *Biochem. J.* 66:567–72

134. Taiz, L. 1984. Plant cell expansion: regulation of cell wall mechanical properties. *Annu. Rev. Plant Physiol.* 35: 585–657

135. Talmadge, K. W., Keegstra, K., Bauer, W. D., Albersheim, P. 1973. The structure of plant cell walls. I. The macromolecular components of the walls of suspension-cultured sycamore cells with

a detailed analysis of the pectic polysaccharides. *Plant Physiol.* 51:158–73

136. Taylor, I. E. P., Atkins, E. D. T. 1985. X-ray diffraction studies on the xyloglucan from tamarind *(Tamarindus indica)* seed. *FEBS Lett.* 181:300–2

137. Tepfer, M., Taylor, I. E. P. 1981. The interaction of divalent cations with pectic substances and their influence on acid-induced cell wall loosening. *Can. J. Bot.* 59:1522–25

138. Terry, M. E., Bonner, B. A. 1980. An examination of centrifugation as a method of extracting an extracellular solution from peas, and its use for the study of indoleacetic acid-induced growth. *Plant Physiol.* 66:321–25

139. Terry, M. E., Jones, R. L., Bonner, B. A. 1981. Soluble cell wall polysaccharides released from pea stems by centrifugation. I. Effect of auxin. *Plant Physiol.* 68:531–37

140. Theander, O., Åman, P. 1978. Fractionation and characterization of alkali-soluble polysaccharides in rapeseed *(Brassica napus)* meal. *Swed. J. Agric. Res.* 8:3–10

141. Thelen, M. P., Delmer, D. P. 1986. Gel-electrophoretic separation, detection, and characterization of plant and bacterial UDP-glucose glucosyltransferases. *Plant Physiol.* 81:913–18

142. Theologis, A. 1986. Rapid gene regulation by auxin. *Annu. Rev. Plant Physiol.* 37:407–38

143. Tvaroska, I., Pérez, S., Marchessault, R. H. 1978. Conformational analysis of (1→6)-α-D-glucan. *Carbohydr. Res.* 61: 97–106.

144. Valent, B. S., Albersheim, P. 1974. The structure of plant cell walls. V. On the binding of xyloglucan to cellulose fibers. *Plant Physiol.* 54:105–8

145. Vanderhoef, L. N. 1979. Auxin regulated cell enlargement: Is there action at the level of gene expression? In *NATO Adv. Study Inst. Ser. A: Life Sciences, Genome Organization and Expression in Plants: Proc. Edinburgh, Scotland, July,* ed. C. J. Leaver, 29:159–74. New York: Plenum

146. Verma, D. P. S., Kumar, V., Maclachlan, G. A. 1982. β-Glucanases in higher plants: localization, potential functions, and regulation. See Ref. 46, pp. 459–88

147. Verma, D. P. S., Maclachlan, G. A., Byrne, H., Ewings, D. 1975. Regulation and in vitro translation of messenger ribonucleic acid for cellulase from auxin-treated pea epicotyls. *J. Biol. Chem.* 250:1019–26

148. Vian, B. 1982. Organized microfibril

assembly in higher plant cells. See Ref. 46, pp. 23–43

149. Wong, Y.-S., Fincher, G. B., Maclachlan, G. A. 1977. Kinetic properties and substrate specificities of two cellulases from auxin-treated pea epicotyls. *J. Biol. Chem.* 252:1402–7

150. Wong, Y.-S., Fincher, G. B., Maclachlan, G. A. 1977. Cellulase can enhance β-glucan synthesis. *Science* 195:679–81

151. York, W. S., Darvill, A. G., Albersheim. P. 1984. Inhibition of 2,4-dichlorophenoxyacetic acid–stimulated elongation of pea stem segments by a xyloglucan oligosaccharide. *Plant Physiol.* 75:295–97

152. York, W. S., Oates, J. E., van Halbeek, H., Darvill, A. G., Albersheim, P., et al. 1988. Location of the O-acetyl substituents on a nonasaccharide repeating unit of sycamore extracellular xyloglucan. *Carbohydr. Res.* 173:113–32

153. Zugenmaier, P., Sarko, A. 1980. The variable virtual bond. Modeling technique for solving polymer crystal structures. In *Fiber Diffraction Methods. Am. Chem. Soc. Symp. Ser.*, ed. A. D. French, K. H. Gardner, 141:225–37. Washington DC: Am. Chem. Soc.

Annu. Rev. Plant Physiol. Plant Mol. Biol.. 1989. 40:169–91

PHOTOMORPHOGENESIS IN LOWER GREEN PLANTS[1]

Masamitsu Wada and Akeo Kadota

Department of Biology, Tokyo Metropolitan University, Fukazawa, Tokyo 158, Japan

CONTENTS

INTRODUCTION

Many aspects of plant development are controlled by light (32, 87, 142, 146). Various steps, from the germination of spores (31) and seeds (15, 30) to the formation of reproductive organs (24, 166), require light through various transductive responses such as ion flux (11, 86, 127), metabolism (81, 99), and gene expression (109, 118, 129, 162). Compared with other environmen-

---

[1]Abbreviations: BLP, blue-light-absorbing pigment(s); LGP, lower green plant; Pfr, far-red-light-absorbing form of phytochrome; Pr, red-light-absorbing form of phytochrome

169

1040-2519/89/0601-0169$02.00

tal factors that control plant development, light is an excellent tool for analytical studies of morphogenesis, not only because wavelength, fluence rate, direction of electrical vector, and direction of incident ray can be specified, but also because light irradiation can be controlled spatially and temporally by using microbeams and short pulses aimed at specific photoreceptors.

Here we analyze recent information on plant morphogenesis and summarize current knowledge about photomorphogenesis in lower green plants.

The phrase "lower green plants" (LGPs) usually refers to nonvascular plants that contain chlorophyll a and b. Although ferns (Pteridophyta) are obviously higher plants in the sense that they are vascular, they are often included among lower green plants because they do not bear seeds. Fern gametophytes have no vascular system and are in many ways similar to mosses and other "true" lower plants rather than to higher plants. Thus we include studies of ferns in this article, and we define green plants as eukaryotes that have chlorophyll without a major contribution of accessory pigments—i.e. excluding chromophytes and rhodophytes.

Most LGPs are simple at the cell level, including structures such as single cells, linear cells, and two-dimensional cell sheets. However, they have a rather more complicated fine structure at the organelle level compared with higher plant cells because of their ability to live autonomously. They are highly diverse, both morphologically and physiologically. This diversity makes LGPs model systems for the study of morphogenesis (Tables 1 and 2).

Higher plants also possess homogeneous cells with simple organization, such as stamen hair cells, root hairs, pollen tubes and endosperm cells. However, no photoreaction has been shown in these differentiated cells so far. Recently, a number of studies have used protoplasts as models of simple, homogeneous cells. Several photoresponses have been reported with higher-plant protoplasts (3, 8, 9, 20, 21, 88), and these results are promising; but so far, analytical studies have been few.

One approach to photomorphogenesis has emphasized the biochemistry and spectrophotometry of the photoreceptor itself. The other has investigated the physiological phenomena resulting from final photoresponses. The former information has been obtained mostly from higher plants (114), the latter from LGPs (as seen in this review) as well as higher plants (87, 146). So far the transduction chains connecting photoreceptors and the resulting phenomena have not been identified.

Here we focus on advancements made after 1980 in the study of photomorphogenesis at the cell level in LGPs, mostly in ferns and green algae. We include various physiological responses controlled by light, even if it is not known whether or not the responses are in the transduction chain leading to

Table 1 Phytochrome-dependent phenomena

Organisms	Phenomena	Action mode	References[a]
Pteridophyte	germination	induction	18, 64, 106, 122,
(haplophase)			124, 152, 181
	cell elongation	promotion	72, 73, 102
	phototropism	induction	79
	polarotropism	induction	26, 77
	antheridium formation	inhibition	40, 139
	cell cycle (G1 phase)	keep at beginning	173, (68)
	cell cycle (G2 phase)	lengthen (Pr)	105
	chloroplast movement	positive	187
	membrane potential	depolarization	120
Bryophyte	germination	induction	5, 167
	elongation (chloronema)	promotion	6, (110)
	phototropism	induction	46, 70, (111)
	polarotropism	induction	6, (70)
	branching (chloronema)	promotion	95
	bud formation	induction	147
Charophyte	growth	promotion	126
		inhibition	126
	membrane potential	depolarization	184
Chlorophyte	rhizoid formation	induction	107
	chloroplast movement	positive	47, 58
	membrane potential		
	(surface charge)	depolarization	151
	water permeability	promotion	185
	calcium vesicular		
	fluorescence	inhibition	178

[a] Parentheses around reference numbers denote the studies in which R/FR reversibility has not been examined.

photomorphogenesis. We also mention chloroplast photo-orientation (reviewed in 51) when necessary, because the sensory transduction of photo-orientation may be similar to that of photomorphogenesis. These two processes share the same pigment systems.

PHYTOCHROME STUDIES

Physiological Responses

Cells require photoreceptors before they can make photoresponses. Does phytochrome exist in plant cells through all stages of plant life? Fern spores require some period of time [from 5 min in *Dryopteris* (52) to 4 days in *Lygodium* (165)] for imbibition (31) before becoming photosensitive. In *Lygodium*, phytochrome concentration increases during imbibition (164), and the increase is inhibited by gabaculine (98), an inhibitor of chlorophyll and

Table 2   Phenomena that are dependent upon blue-light-absorbing pigment(s)[a]

Organisms	Phenomena	Action mode	References[b]
Pteridophyte	germination	inhibition	153, 154
(haplophase)	elongation	inhibition	78
	phototropism	induction	(59)
	polarotropism	induction	149, 150, (76)
	apical swelling	induction	174
	cell cycle (G1 phase)	shortening	171, (105)
	chloroplast movement	positive	187
		negative	(187)
	membrane potential	hyperpolarization	(120)
Bryophyte	phototropism	induction	(70)
	polarotropism	induction	70, 149
Chlorophyte	hair whorl formation	induction	135
	cap formation	stimulation	(159)
	chloroplast movement	positive	36, 112
		regulation	138
Vaucher-	growth	promotion	(83)
iophyte	phototropism	induction	82
	apical swelling	induction	84
	branching	induction	(83)
	chloroplast movement	positive	10, 28
		negative	28, 57
	cortical fiber		
	reticulation	induction	12
	electric current	promotion	(13), (85)

[a] Phototaxis and photoregulation of metabolism are excluded.
[b] Parentheses around reference numbers denote the studies in which the action spectra have not been determined.

phytochrome chromophore synthesis. This result suggests that phytochrome is synthesized during imbibition. In *Anemia,* however, the light requirement for germination depends upon light conditions during sporogenesis (140). Moreover gabaculine has no effect on spore germination (141), indicating that phytochrome exists in the dry spore. In addition, the onset of photosensitivity within 5 min after sowing in *Dryopteris* (52) is too rapid for gene expression to have occurred at the level of transcription (160), suggesting again that phytochrome is present in the dry spore.

Phytochrome-mediated photoresponses in LGPs are summarized in Table 1. Although the explanation of each phenomenon is beyond the scope of this review, most of these responses are of two kinds. The first kind comprises very rapid or rather rapid responses with a short lag period, such as ion flux [0.4–3.5 sec in a characean alga, *Nitella* (184)], or chloroplast photo-orientation [15–45 min in a green alga, *Mougeotia* (51), and 30 min in the fern *Adiantum* (187)], and may be mediated by membrane regulation (127).

The second is a long-term phenomenon mediated possibly by gene expression. Currently, the regulation of gene expression by phytochrome has been studied extensively at the molecular level in higher plants (109, 118, 129, 162). In LGPs, however, no evidence of gene expression has been shown. Analytical studies now emphasize the photomorphogenetic responses observable under a light microscope at the cellular or subcellular level. Biochemical studies of the molecular basis for these phenomena, including gene expression, are still few, possibly because it is difficult to obtain enough cells for biochemical studies.

Lifetime of Pfr and the Escape Reaction

If the protonema of the moss *Ceratodon* is precultured under darkness, the apical part swells within 5 min of exposure to red light, and then shows phototropic responses under continuous or intermittent irradiation with red light. To inhibit the swelling, the durations of red/far-red pulses given repeatedly must be shorter than 10 sec (45); but to inhibit the tropic response, the repeated cycles of 5 min red and 2 min far-red light are still effective (46). These results indicate that although the responses appear at first to be a single, sequential phenomenon, the phytochrome molecules responsible for each phenomenon and its transduction chain are different.

In *Adiantum* protonemata, phytochrome-mediated phenomena induced with 5–10 min of red light can usually be reversed by subsequent irradiation by far-red light (105, 170, 173), but some exceptions have been demonstrated. When nongrowing, two-celled protonemata were irradiated with red light for 4 sec or less, the effect, induction of apical growth, was reversible by a subsequent irradiation with far-red light even after 2 min of intervening darkness. However, when red light was given for 16 sec or more, photoreversibility became only partial (73). The loss of the reversibility may not be attributed to an escape reaction in this case, because the intervening dark period does not influence the photoreversibility (73).

The time required for 50% escape of red light–induced fern spore germination from the inhibiton of far-red light is about 6–18 hr (65, 106, 152, 165), which is very long compared with other phenomena; and the spectrophotometrically assayable far-red light–absorbing form of phytochrome (Pfr) disappears before the escape reaction occurs (165). In chloroplast photoorientation, the Pfr gradient of *Mougeotia* was stable for a long time, whereas that of the other green alga *Mesotaenium* was quite short (53), as was that of *Adiantum* (H. Yatsuhashi, personal communication). Since the behavior of phytochrome can vary, even for the same reaction in different species and for the different phenomena in the same species, phytochrome may function in multiple modes.

Molecular Properties

Phytochrome in LGPs has been identified photometrically (41, 44, 158, 164), but its molecular properties have not yet been determined in any species. Phytochrome purification or isolation from LGPs has not progressed since the pioneering work of Taylor & Bonner (158), possibly because of the low concentration of phytochrome in the cell, the small number of cells available, and the interference by chlorophyll during isolation procedures. In higher plants, two different molecules of phytochrome have been identified immunochemically (1, 145, 163) and by peptide mapping patterns (1, 163). Phytochrome I (89) is mainly detected in etiolated tissue, and phytochrome II (89) in green tissue (1, 145, 163). LGPs are usually green, even when grown in darkness (except mutants), and the phytochrome of LGPs might be expected to have properties resembling those of phytochrome II in higher plants.

The spectral properties of phytochrome in some LGPs have been determined by difference spectrophotometry (41, 44, 158, 164). Although it is difficult to discuss the spectral characteristics in any detail from these limited data, the absorption peaks of the red-light-absorbing (Pr) and far-red-light-absorbing (Pfr) forms of phytochrome are similar to those of phytochrome found in higher plants, except that they tend to shift toward the blue spectral region (41, 158).

Monoclonal antibodies specific to phytochrome from etiolated pea *(Pisum sativum)* (19) and maize *(Zea mays)* (136) bind to a polypeptide extracted from ferns, liverworts, and mosses, and from the algae *Mougeotia, Mesotaenium,* and *Chlamydomonas*. Although the polypeptides have not been identified as phytochrome spectrophotometrically, it is evident that LGPs have a phytochrome-like polypeptide with a size similar to that of native oat phytochrome and a part immunologically identical to that found in higher-plant phytochrome.

The dichroic orientation of phytochrome in LGP cells (see below) indicates that phytochrome may be tightly bound to the plasma membrane, either with or without intermediate proteins. Analyses of the amino acid sequence of the phytochrome of LGPs and its hydropathy index are necessary before it will be known whether the phytochrome is a transmembrane or membrane bound protein, although evidence suggests that higher-plant phytochrome is not likely to be transmembranous (63).

Localization and Intracellular Arrangement

Intracellular localization of LGP phytochrome has been studied by microbeam analyses (49, 56, 60, 78, 79, 80, 108, 175, 176) but not by immunocytochemistry, microspectrophotometry, or cell fractionation. Partial irradiation with microbeams has shown that photoreceptive sites were always located at the cell periphery (56, 175, 176). In addition, action dichroism induced by polarized light (see below) suggests strongly that phytochrome localizes on or close to the plasma membrane, although in *Avena* coleoptiles no immunocytochemical localization of phytochrome I was detected on the plasma membrane (101), even after sequestering induced by red light irradiation (101, 148). In this connection, immunocytochemical studies using anti-phytochrome II antibodies would be highly desirable, both for LGPs and for higher plants.

The intracellular arrangement of phytochrome in LGPs might differ from that of higher plants. In LGPs, most of the phytochrome effects show action dichroism, whereas no action dichroism has been observed in higher plants. [The exception is one report of photometric dichroism in *Avena* (100), later questioned by one of the study's two authors (128).] However, failure of action dichroism in higher-plant tissues may simply be the consequence of heavy scattering and resulting depolarization.

Soon after Jaffe (69) pointed out the usefulness of linearly polarized light in studying the intracellular arrangement of photoreceptors, LGPs phytochrome began to attract study (14, 26, 48, 58). Etzold's (26) model of the dichroic orientation of Pr and Pfr, parallel and perpendicular to the cell surface, respectively, was proved by polarized microbeam irradiation (56, 79). This orientation of phytochrome is likely to be common in filamentous cells (6, 46), although different dichroic orientations of Pr and Pfr could not be found in a moss, *Physcomitrium* (111). A similar change in absorption caused by differences in the dipole moments between Pr and Pfr was photometrically detected in phytochrome purified from higher plants and immobilized on agarose beads (156, 157). From light-induced changes in linear dichroism at 730 and 660 nm, a rotational angle of about $31°$ (or its complement, $149°$) was calculated (25). Although the value is small compared with the correlative physiological datum (a difference of $90°$), it is more or less compatible with the angle deduced from calculations based on physiological data for *Mougeotia* and for fern protonema (7, 155).

Laser flash photolysis of purified phytochrome of higher plants revealed that at least two intermediates, $I_{692}(I_{700})$ and I_{bl} exist during the phototransformation from Pr to Pfr, and that the intermediates can be reversed by light to Pr (115). These intermediates and their photoreversibility were also detected in fern spores (132, 133) and protonemata (75) and *Mougeotia* cells (92, 131) by examining their physiological responses. Furthermore, the orientations of the intermediates were analyzed in *Mougeotia* (92, 131) and in *Adiantum* (75), with irradiations with polarized double flashes. An experiment using a polarized far-red flash at an appropriate period after an inductive red flash revealed that the orientation of the dipole moment of each intermediate was not very different from that of Pr and that the orientation changed primarily in the final step (I_{bl} to Pfr) of the pathway (75). In *Adiantum,* the transition moments of Pr, I_{692}, I_{bl}, and Pfr were calculated to be inclined $18°$, $9°$, $15°$, and $72°$, respectively, with respect to the membrane

(155). The rate of the reorientation of phytochrome is sufficiently slow that the change can be ascribed either to the conformational change of the protein moiety (but not of the chromophore) or to a change in the interaction between phytochrome and the receptor connecting phytochrome and the membrane, if it exists (75).

Almost all information on the transition dipole moments of phytochrome are for the main absorption peaks at longer wavelengths, both in Pr and Pfr. A knowledge of the directions of the dipole moments of the second peaks in shorter wavelengths may be essential to understanding the structure of phytochrome in vivo. The only information available is that the orientation of the blue absorption band of Pr is parallel to the cell surface as well, indicating that the transition moments of long- and short-wavelength bands of Pr are not so different from each other when viewed from the edge of plasma membrane (60).

Based on the data of chloroplast photo-orientation, a spiral arrangement of the dipole moments of Pr in the cell surface had been shown by Haupt & Bock (55) in *Mougeotia,* but was not detected in *Adiantum* protonemata (186). Considering the membrane fluidity, a random orientation of Pr in the plane of the cell surface would be easy to understand. However, to maintain a specific orientation of Pr of *Mougeotia* in this plane, there must be some specific structure and/or perhaps some mechanism that holds Pr in a fixed position in the ectoplasm.

LGP phytochrome is not always arranged dichroically. In phytochrome-mediated spore germination of the fern *Dryopteris,* no action dichroism could be demonstrated (54), although a protonema of this species showed typical dichroism in polarotropism (26). The dichroic arrangement of phytochrome or even the existence of phytochrome molecules in the plasma membrane and/or ectoplasm (176) was questioned in the protonemata of a fern, *Pteris,* because no phytochrome response was detected in either polaro- or phototropism or in chloroplast photo-orientation by means of irradiation with polarized red light or with red microbeams (76). However, the responses could be induced by blue light, suggesting that the signal transduction chains of both phenomena are normal. Since spore germination of this species is mediated by phytochrome (152), the phytochrome responsible for this germination might be localized somewhere other than in the ectoplasm and/or plasma membrane. The Pr of a green alga *(Spirogyra)* phytochrome that mediates rhizoid differentiation is distributed throughout the cell periphery and remains in the centripetal end part after centrifugation; but no dichroic orientation for these molecules has been detected (108).

Sensory Transduction

Microbeam experiments using *Mougeotia* and *Adiantum* have shown that phytochrome acts locally in those areas irradiated with red light, and the signal is not transduced a long distance (49, 56, 79, 187). In *Adiantum* protonemata, however, the signal was transduced for short distances within the cell in some responses (72, 73, 172) but no further than the boundary of the cell that was irradiated (23, 73).

How does phytochrome act? Induction of chloroplast movement in *Mougeotia* with a single millisecond flash indicates that cycling between Pr and Pfr is not involved in these phenomena (93). In contrast, chloroplast movement of *Mesotaenium* and *Adiantum* is not a trigger reaction but needs rather continuous or intermittent irradiation (58, 187). However, if the lifetime and the "memory" of Pfr in *Mesotaenium* and *Adiantum* are short relative to those of *Mougeotia*, and if their chloroplast movements are slow, continuous irradiation is necessary even if the primary action of phytochrome is the same as that of *Mougeotia*. The kinetics of phytochrome phototransformation from Pr to Pfr in *Avena* with double flashes of 1 msec is identical with the kinetics obtained from the physiological experiments with *Mougeotia* chloroplast movement. This observation suggests that the physiological response corresponds directly to the amount of Pfr produced and not to phototransformation of intermediates (93).

The germination rate of fern spores shows a positive correlation with %Pfr (Pfr/total phytochrome × 100) (34, 133, 165), indicating that the threshold of the %Pfr required for germination is different in each spore, ranging from a few percent to 80% (34). However, when two adjacent areas of a protonema of *Adiantum* were irradiated for 2 hr with two beams of red light of different fluence rates, chloroplasts in these beams moved to the area of higher fluence rate (188). This result suggests that %Pfr is not involved, because the photostationary states in the two beams must be the same. But if the dark reaction of Pfr to Pr occurs quickly in fern protonemata, as shown in *Sinapis* (61), %Pfr in the two beams might be different and might be involved in chloroplast movement.

The hypothesis that phytochrome is localized along the plasma membrane (176) in LGPs supports the idea of the change in membrane permeability for the primary action of phytochrome. Red light–induced membrane depolarization was detected in *Nitella* (184) and in a fern, *Onoclea* (120), as well as in oat coleoptile (119). The lag phases were 0.4–3.5 sec in *Nitella* (184) and 30 sec in *Onoclea* (120). The membrane potential was repolarized by far-red light (120, 184), showing the involvement of phytochrome. The surface charge of *Mesotaenium* cells was also controlled by phytochrome (151). However, there is no direct evidence that these electrophysiological responses are involved in the primary action, in strict sense, of the following sensory transduction chains leading to photomorphogenetic responses (116).

The magnitude of these electrical changes depended upon the concentration of calcium ion (Ca^{2+}) in the medium, suggesting that one of the ionic species involved is calcium (151, 184). Free cytoplasmic Ca^{2+} has been thought to be a possible second messenger for phytochrome, and has been well studied in *Mougeotia*. Light-dependent transport of ^{45}Ca across the plasma membrane was shown (23, 177). Local application of calcium ionophores induced local chloroplast rotation (144). Calmodulin has been detected (180). Calmodulin antagonists inhibited phytochrome-dependent chloroplast movement (144, 180). Calcium-containing vesicles ["calcium vesicles" (179)] associated with the edges of chloroplasts rather than the cortical cytoplasm were proposed to be the major intracellular storage site of calcium (179). On the basis of this accumulated evidence, it is obvious that Ca^{2+} is fundamental in some part of the transduction chain for chloroplast movement in *Mougeotia*. However, if the calcium-containing vesicles are the main source of Ca^{2+}, the spatial separation of calcium storage from the photoreceptive site indicates that Ca^{2+} could not be the second messenger of phytochrome in this response.

Phytochrome-mediated germination of *Onoclea* spores was shown to be Ca^{2+}-dependent (181), and the increase of intracellular Ca^{2+} was also shown to be stimulated by red light (182), suggesting that Ca^{2+} may be acting as a second messenger (181, 182). The content of a calmodulin-like protein also increases during the early stages of light-induced *Anemia* spore germination (29). However, the timing of Ca^{2+} requirement during phytochrome-dependent fern spore germination is not related to the escape from far-red reversal (65, 134, 181), again indicating that Ca^{2+} might not be the second messenger. These Ca^{2+} effects have recently been well reviewed by Hepler & Wayne (62) and will not be discussed further to avoid repetition.

Phytochrome-mediated gene expression has not been studied in the photo-morphogenesis of LGPs from a biochemical point of view. Chloroplast photo-orientation in *Adiantum* was revealed not to be controlled by direct gene expression at the level of transcription, because the response was induced in an enucleated protonema (169). In *Onoclea* spore germination, newly synthesized protein and mRNA appeared 8 and 24 hr, respectively, after the onset of light irradiation (125). However nuclear migration in a spore, the first detectable response in germination, was observed within 16 hr (4), indicating again that gene expression at the level of transcription is not involved as a primary reaction.

BLUE LIGHT–ABSORBING PIGMENT(S) STUDIES

Physiological Responses

Blue light effects are well known in the plant kingdom, such as in the phototropic responses of higher plants (22, 27), photomorphogenesis in fungi (39, 42, 94, 143), and the metabolism of green algae (90, 143). Photomorphogeneses controlled by blue light are also found in LGPs, as listed in Table 2. Many of these blue light–induced phenomena have an interaction with red light irradiation (see the next section). The second absorption band of phytochrome in the blue light range also functions as a blue light receptor. Moreover, as no blue light–absorbing pigment(s) (BLP) has yet been identified, the phenomena listed here may be heterogeneous with respect to their photoreceptors and/or phototransduction chains when compared with those mediated by phytochrome.

Blue light–induced responses are of two kinds. As in phytochrome-mediated responses, one is a typical trigger response induced by a short pulse (e.g. 59, 82, 135, 153); the other requires long-term irradiation (e.g. 84, 174, 187).

Action Spectra and Pigment(s)

The BLP has been sought since Thimann & Curry's (161) initial, precise action spectrum. A flavin moiety was also proposed as the photoreceptor in LGPs (17, 154). So far, however, the pigment has not been identified. In plants that have phytochrome, the action spectrum will be influenced by the absorption of BLP and phytochrome. To compare the action spectrum with BLP absorption spectra, one has to compensate the contribution of phytochrome. This is because phytochrome has a second peak of absorption in the blue light region, and also both pigment systems often simultaneously show the same effect in a given respone (59). Moreover, in some cases, phytochrome modifies or potentiates the effects of BLP (see next section). However, action spectra taken under compensation of phytochrome are few (36, 78, 154, 171).

The action spectrum between 250 and 800 nm for the inhibition of red light–induced germination of spores in *Pteris* showed prominent peaks at about 260, 370, 440, and 730 nm (154). A similar action spectrum was obtained in *Adiantum* (153). The peaks at 260, 370, and 440 nm were ascribed to a BLP. This structure suggests that the photoreceptor pigment is a flavin-related compound, because carotenoids have no peak in the UV region whereas flavins do (153, 154).

Localization and Intracellular Arrangement

Intracellular localization of BLP has been studied with polarized light (50, 60, 174, 187) and/or microbeam irradiation (10, 13, 28, 74, 83, 84, 97, 172, 187).

The cell cycle of *Adiantum* protonemata is controlled by phytochrome and BLP (170, 171). BLP mediates a shortening of the G1 phase (105). Local irradiation with a blue microbeam clearly showed that only the irradiation of the region containing the nucleus could induce cell division (172), even after relocation of the nucleus by centrifugation (74). In addition, the absence of an action dichroism in this phenomenon indicates that the pigment is not likely to be localized and arranged in or on the plasma membrane (74). It is probable that the nucleus, or at least some organelle closely attached to the nucleus, is perceiving the blue light; but it is still unknown whether or not the nucleus itself is the photoreceptive site (35). A BLP inhibiting protonemal elongation is also localized in the nuclear region and shows no dichroic response by polarized light irradiation (78). Thus a single pigment may have multiple effects through the action of diverse transduction chains (35).

Polarized blue light (60, 174, 187) and cell centrifugation (60) have been used to show action dichroism in BLP localized on or close to the plasma membrane in *Adiantum* protonemata, and these receptors are involved in apical cell swelling (174), polarotropism (60), and chloroplast photo-orientation (187). Similar action dichroism has been shown in other ferns (76, 149), mosses (70), liverworts (149), and algae (50). In *Adiantum*, a precise study using microbeam irradiation proved the parallel orientation of the dipole moment of BLP to the plasma membrane (60). No spiral orientation in the plane parallel to the cell surface was observed in chloroplast photo-orientation (186). Pigments localized throughout the entire cell mediate chloroplast movement; however, those involved in cell swelling and polarotropism are localized only on the apical part of the protonemata (59, 174, 187). These results might show more than the existence of different pigment species mediating different physiological responses. They might indicate differences in the competence of a cell part in providing a particular response. In other words, if the effector system is lacking, light cannot induce any respone even when the photoreceptor exists. Here again, the results suggest the multiple effects of a pigment. On the other hand, in some cases, as discussed in this and the previous paragraph, the blue light–receptive sites that control different phenomena are dispersed in the cell as "diverse target" sites (35). A similar "diverse target" of the pigments has been reported in *Vaucheria* (Vaucheriophyceae) (28, 82, 84), but no precise studies have been performed comparable to those using *Adiantum*.

Sensory Transduction

Changes of membrane permeability are induced by blue light irradiation with a very short lag period of 5–10 sec in *Vaucheria* (13, 85) and 3 sec in *Onoclea* (17). They precede fiber reticulation and chloroplast aggregation in *Vaucheria* (12) and apical cell expansion in *Onoclea* (121). They are considered to be an early step in signal transductions leading to blue light–mediated responses. Blue light caused hyperpolarization of the membrane in an apical cell of an *Onoclea* protonema (17) that had been impaled with a glass microelectrode. In *Vaucheria*, blue light promoted an inward-directed current at the growing apex (85) and induced an outward-directed current from the irradiated lateral region (13). The currents were monitored with vibrating probes. Protons are proposed to be the charge-carrying species, based on experiments involving the proton ionophore carbonil-cyanide-m-chlorophenylhydrazone (13), pH dependency (13, 85, 121), and ion selective electrodes (121), although direct measurements of pH change within or immediately outside the cell under blue light stimulation were very few (121). Na_3VO_4, a potent inhibitor of the plasma membrane ATPase, either prevents initiation or blocks further increases in the blue light–mediated hyperpolarization during the apical cell swelling in *Onoclea,* suggesting the existence of an ATP-driven proton pump in the initial events of the blue light response (17).

Ca^{2+} is not now thought to play a direct role in current generation (13, 121), although Ca^{2+} is probably necessary for general maintenance of membrane integrity (13). On the other hand, in blue light–induced chloroplast translocation of a green alga, *Eremosphaera,* the presence of low concentrations of Ca^{2+} with ionophores in the medium can mimic the light response in the dark, suggesting that calcium has some function in the transduction chain (183).

INTERACTION BETWEEN THE TWO PIGMENT SYSTEMS

Phenomena

In plants living under natural conditions with a wide spectral range of visible light, both phytochrome and BLP are activated. Consequently, various actions mediated by both pigment systems would be maintained at some levels, and various modes of interaction would be expected (130). Under blue light conditions, phytochrome also absorbs light, and in the case of *Lygodium* spores, photometrically detectable %Pfr was raised to 16–34% (165). Three modes of interaction between the two pigment systems have been analyzed in LGPs.

Modification or Potentiation of Other Transduction Chains

The chloroplast in *Mougeotia* moves from a profile to a face position under red light (51). But strong blue light applied prior to or simultaneously with red light induces face-to-profile movement in response to red light, irrespective of the direction of the blue light (38, 137). This result indicates that blue light switches the direction of response in the Pfr gradient. In this response, the lack of direct coupling between the two photoreceptors at the photochemical level was revealed (38).

The chloroplast photo-orientation of *Mesotaenium* is mediated by phytochrome. The effect can be potentiated strongly by blue light, but only red light is responsible for the directionality (91).

The spore germination of several species of ferns is phytochrome dependent (31). Blue light administered before or after the red light inhibits the red light–induced germination (18, 152). Spores of *Cheilanthes* show a similar tendency, but in addition they display another type of interaction. The red light–induced spore germination is not reversed by a subsequent irradiation with far-red light. But red/far-red photoreversibility was observed when spores were irradiated with a blue light for inhibition and then with a higher intensity of red light (124).

Interactions of these types are often reported at the level of physiological responses, but we have as yet little insight regarding the mechanism of the interaction. We must understand the intracellular localization of both pigment systems and the timing of the effectiveness of the pigments before we can tease apart the mechanisms giving rise to these interactions.

Cooperation Sharing a Common Transduction Chain

In *Adiantum* protonemata and *Mougeotia* cells, low-fluence-rate chloroplast movement is mediated by both BLP and phytochrome systems (37, 187). Direction of the responses (37, 187), pigment localization (187), and response to inhibitors (178) are identical in both pigment systems, suggesting that the same transduction chain is shared. Phototropism in *Adiantum* protonemata is mediated by phytochrome (79) and BLP (59). Both pigments are located on or close to the plasma membrane of the subapical part of the protonemata (60, 176). Under blue light irradiation, the BLP is active in the low-fluence range and phytochrome is active in the high fluence range (59). A direct photochemical interaction of photoreceptors has not been detected in any cooperative systems so far. To find the point from which the transduction chain is shared is crucial.

Coaction through Different Transduction Chains

The timing of cell division of fern protonemata is controlled by phytochrome and BLP (168). Blue light shortens the cell cycle and far-red light lengthens it (33, 170). Cell-cycle analysis revealed that blue light given during the G1 phase (66, 67) shortened the G1 phase (105), and far-red light given at the beginning of the G1 phase lengthened the G2 phase (105). This result indicates that each pigment system controls the different phases of the cell cycle independently. Moreover, the intracellular photoreceptive sites of the two pigment systems were different: The BLP was located in the nuclear region (74, 172), phytochrome in the cell periphery (172). If we observe the total length of the cell cycle as one response, as reflected by cell plate formation, it may appear that the two pigment systems control the response antagonistically. Thus, it is necessary to divide the response into separate components in order to determine whether the response is governed by the two systems interacting or simply co-acting.

Similar situations might be blue light–induced growth retardation (78), apical swelling (174), and initiation of cell cycle (68, 170) in the fern protonemata grown under red light. These blue- and red-light effects can be recognized as antagonistic responses in final physiological phenomena. The transduction chains of these responses, however, should be studied precisely in order to know which type of interaction is taking place in these responses.

OTHER LIGHT EFFECTS

Aside from the effects of near-UV, blue, red, and far-red light, other light effects have been reported in LGPs. P_{580} has been postulated for the effect of yellow light on protonemal cell elongation in the fern *Onoclea* (103), but the yellow/red reversible nature of the response was ascribed to phytochrome (16). In fern gametophytes incubated for a long time in the dark, green, yellow, and red light all induced two-dimensional growth, while far-red light was not effective (43). The effect was again interpreted on the basis of phytochrome and BLP. A specific green light receptor other than phytochrome or BLP was postulated in *Mougeotia* chloroplast movement from the evidence of the interaction between green and far-red light (96).

CONCLUDING REMARKS

The physiological and cell biological studies on photomorphogenesis in LGPs have been successful because these plants exhibit simple organization and evolutionary diversity. In the future, however, studies of the biochemical and molecular aspects of photomorphogenesis are urgently required in order to analyze and explain the results obtained to date.

One important technical problem must be overcome: the deficiency of suitable study material because of the low rate of proliferation of some organisms (e.g. *Mougeotia*) and the poor harvests of others (e.g. the spores of ferns and mosses). Improvement of culture conditions and establishment of new strains or cultured cell lines are needed.

Given good materials, biological research progresses quickly and easily. LGPs can be regarded as a treasury containing a diverse array of species with unique and characteristic photoresponses, and we must search for and study such organisms carefully. Even among taxonomically close species, photoresponses can vary considerably [e.g. the existence or nonexistence of phytochrome-dependent chloroplast movement, polarotropism, and phototropism in *Adiantum* and *Pteris*, respectively (76); or the different modes of phytochrome action on protonemal growth of *Adiantum* (72), *Onoclea* (104), and *Lygodium* (123)].

Studies using mutants, another promising area of research, have been undertaken in fern (18) and moss (71). LGPs have a great advantage for genetic analysis because their haplophase life is long and autonomous. Thus transgenic plants can easily express integrated characters. Mutations of the intracellular localization or the mode of existence of photoreceptors would be more useful for analytical studies than would mutants without photoreceptors. Mutations of the transduction chains are also required. Although there is a tomato mutant without phytochrome (113), and a mutant of cucumber lacking phytochrome function in light-grown seedlings (2), much more specific photomorphogenic mutants should be selected for among these LGPs in which physiological analyses are far advanced.

The results obtained from lower plants may not always be consistent with those from higher plants, and vice versa. Nevertheless, investigation of the photoresponses in LGPs will be critical in developing our understanding of the overall phenomena of photomorphogenesis in the plant kingdom.

ACKNOWLEDGMENTS

We would like to thank Drs. W. Haupt, E. Schäfer, R. Scheuerlein, and N. Pierce for helpful criticisms and discussions of this review. The preparation of the review and the research cited therein from our laboratory have partly been supported by Grants from the Ministry of Education, Science and Culture of Japan.

Literature Cited

1. Abe, H., Yamamoto, K. T., Nagatani, A., Furuya, M. 1985. Characterization of green tissue-specific phytochrome isolated immunochemically from pea seedlings. *Plant Cell Physiol.* 26:1387–99

2. Adamse, P., Jaspers, P. A. P. M., Kendrick, R. E., Koornneef, M. 1987. Photomorphogenetic responses of a long hypocotyl mutant of *Cucumis sativus* L. *J. Plant. Physiol.* 127:481–91

3. Assmann, S. M., Simoncini, L., Schroeder, J. I. 1985. Blue light activates electrogenic ion pumping in guard cell protoplasts of *Vicia faba*. *Nature* 318:285–87

4. Bassel, A. R., Kuehnert, C. C., Miller, J. H. 1981. Nuclear migration and asymmetric cell division in *Onoclea sensibilis* spores: an ultrastructural and cytochemical study. *Am. J. Bot.* 68:350–60

5. Bauer, L., Mohr, H. 1959. Der Nachweis des reversiblen Hellrot-Dunkelrot-Reaktionssystems bei Laubmoosen. *Planta* 54:68–73

6. Bittisnich, D., Williamson, R. E. 1985. Control by phytochrome of extension growth and polarotropism in chloronemata of *Funaria hygrometrica*. *Photochem. Photobiol.* 42:429–36

7. Björn, L. O. 1984. Light-induced linear dichroism in photoreversibly photochromic sensor pigments. V. Reinterpretation of the experiments on in vivo action dichroism of phytochrome. *Physiol. Plant.* 60:369–72

8. Blakeley, S. D., Thomas, B., Hall, J. L., 1987. The role of microsomal ATPase activity in light-induced protoplast swelling in wheat. *J. Plant Physiol.* 127:187–91

9. Blakeley, S. D., Thomas, B., Hall, J. L. Vince-Prue, D. 1983. Regulation of swelling of etiolated-wheat-leaf protoplasts by phytochrome and gibberellic acid. *Planta* 158:416–21

10. Blatt, M. R. 1983. The action spectrum for chloroplast movements and evidence for blue-light-photoreceptor cycling in the alga *Vaucheria*. *Planta* 159:267–76

11. Blatt, M. R. 1987. Toward the link between membrane transport and photoperception in plants. *Photochem. Photobiol.* 45:933–38

12. Blatt, M. R., Briggs, W. R. 1980. Blue-

light-induced cortical fiber reticulation concomitant with chloroplast aggregation in the alga *Vaucheria sessilis*. *Planta* 147:355–62

13. Blatt, M. R., Weisenseel, M. H., Haupt, W. 1981. A light-dependent current associated with chloroplast aggregation in the alga *Vaucheria sessilis*. *Planta* 152:513–26

14. Bünning, E., Etzold, H. 1958. Über die Wirkung von polarisiertem Licht auf keimende Sporen von Pilzen, Moosen und Farnen. *Ber. Dtsch. Bot. Ges.* 71:304–6

15. Cone, J. W., Kendrick, R. E. 1986. Photocontrol of seed germination. See Ref. 87, pp. 443–65

16. Cooke, T. J., Paolillo, D. J. Jr. 1979. The photobiology of fern gametophytes. I. The phenomenon of red/far-red and yellow/far-red photoreversibility. *J. Exp. Bot.* 30:71–80

17. Cooke, T. J., Racusen, R. H., Briggs, W. R. 1983. Initial events in the tip-swelling response of the filamentous gametophyte of *Onoclea sensibilis* L. to blue light. *Planta* 159:300–7

18. Cooke, T. J., Racusen, R. H., Hickok, L. G., Warne, T. R. 1987. The photocontrol of spore germination in the fern *Ceratopteris richardii*. *Plant Cell Physiol.* 28:753–59

19. Cordonnier, M. -M., Greppin, H., Pratt, L. H. 1986. Identification of a highly conserved domain on phytochrome from angiosperms to algae. *Plant Physiol.* 80:982–87

20. Dangl, J. L., Hauffe, K. D., Lipphardt, S., Hahlbrock, K., Scheel, D. 1987. Parsley protoplasts retain differential responsiveness to u.v. light and fungal elicitor. *EMBO J.* 6:2551–56

21. Das, R., Sopory, S. K. 1985. Evidence of regulation of calcium uptake by phytochrome in maize protoplasts. *Biochem. Biophys. Res. Commun.* 128:1455–60

22. Dennison, D. S. 1979. Phototropism. In *Encyclopedia of Plant Physiology, New S.*, ed. W. Haupt, M. E. Feinleib, 7:506–66. Berlin/Heidelberg/New York: Springer. 731 pp.

23. Dreyer, E. M., Weisenseel, M. H. 1979. Phytochrome-mediated uptake of calcium in *Mougeotia* cells. *Planta* 146:31–39

24. Dring, M. J. 1988. Photocontrol of development in algae. *Annu. Rev. Plant Physiol. Plant Mol. Biol.* 39:157–74

25. Ekelund, N. G. A., Sundqvist, C., Quail, P. H., Viestra, R. D. 1985. Chromophore rotation in 124-kdalton *Avena sativa* phytochrome as measured by

light-induced changes in linear dichroism. *Photochem. Photobiol.* 41:221–23

26. Etzold, H. 1965. Der Polarotropismus und Phototropismus der Chloronemen von *Dryopteris filix mas* (L.) Schott. *Planta* 64:254–80

27. Firn, R. D. 1986. Phototropism. See Ref. 87, pp. 367–89

28. Fischer-Arnold, G. 1963. Untersuchungen über die Chloroplastenbewegung bei *Vaucheria sessilis*. *Protoplasma* 56:495–520

29. Föhr, K. J., Enssle, M., Schraudolf, H. 1987. Calmodulin-like protein from the fern *Anemia phyllitidis* L. Sw. *Planta* 171:127–29

30. Frankland, B., Taylorson, R. 1983. Light control of seed germination. See Ref. 146, pp. 428–56

31. Furuya, M. 1983. Photomorphogenesis in ferns. See Ref. 146, pp. 569–600

32. Furuya, M., ed. 1987. *Phytochrome and Photoregulation in Plants*. Tokyo: Academic. 354 pp.

33. Furuya, M., Ito, M., Sugai, M. 1967. Photomorphogenesis in *Pteris vittata*. *Jpn. J. Exp. Morphol.* 21:398–408

34. Furuya, M., Kadota, A., Uematsu-Kaneda, H. 1982. Percent Pfr-dependent germination of spores in *Pteris vittata*. *Plant Cell Physiol.* 23:1213–17

35. Furuya, M., Wada, M., Kadota, A. 1980. Regulation of cell growth and cell cycle by blue light in *Adiantum* gametophytes. See Ref. 142, pp. 119–32

36. Gabryś, H. 1985. Chloroplast movement in *Mougeotia* induced by blue light pulses. *Planta* 166:134–40

37. Gabryś, H., Walczak, T., Haupt, W. 1984. Blue-light-induced chloroplast orientation in *Mougeotia*. Evidence for a separate sensor pigment besides phytochrome. *Planta* 160:21–24

38. Gabryś, H., Walczak, T., Haupt, W. 1985. Interaction between phytochrome and the blue light photoreceptor system in *Mougeotia*. *Photochem. Photobiol.* 42:731–34

39. Galland, P., Lipson, E. D. 1984. Photophysiology of *Phycomyces blakesleeanus*. *Photochem. Photobiol.* 40:795–800

40. Gemmrich, A. R. 1986. Antheridiogenesis in the fern *Pteris vittata*. I. Photocontrol of antheridium formation. *Plant Sci.* 43:135–40

41. Giles, K. L., von Maltzahn, K. E. 1968. Spectrophotometric identification of phytochrome in two species of *Mnium*. *Can. J. Bot.* 46:305–6

42. Gressel, J., Rau, W. 1983. Photocontrol of fungal development. See Ref. 146, pp. 603–39

43. Grill, R. 1987. Induction of two-dimensional growth by red and green light in the fern *Anemia phyllitidis* L. Sw. *J. Plant Physiol.* 131:363–71

44. Grill, R., Schraudolf, H. 1981. *In vivo* phytochrome difference spectrum from dark grown gametophytes of *Anemia phyllitidis* L. Sw. treated with Norflurazon. *Plant Physiol.* 68:1–4

45. Hartmann, E. 1984. Influence of light on phototropic bending of moss protonemata of *Ceratodon purpureus* (Hedw.) Brid. *J. Hattori Bot. Lab.* 55:87–98

46. Hartmann, E., Klingenberg, B., Bauer, L. 1983. Phytochrome-mediated phototropism in protonemata of the moss *Ceratodon purpureus* Brid. *Photochem. Photobiol.* 38:599–603

47. Haupt, W. 1959. Die Chloroplastendrehung bei *Mougeotia* I. Über den quantitativen und qualitativen Lichtbedarf der Schwachlichtbewegung. *Planta* 53:484–501

48. Haupt, W. 1960. Die Chloroplastendrehung bei *Mougeotia* II. Die Induktion der Schwachlichtbewegung durch linear polarisiertes Licht. *Planta* 55:465–79

49. Haupt, W. 1970. Über den Dichroismus von Phytochrom $_{660}$ und Phytochom$_{730}$ bei *Mougeotia*. *Z. Pflanzenphysiol.* 62:287–98

50. Haupt, W. 1971. Schwachlichtbewegung des *Mougeotia*-Chloroplasten im Blaulicht. *Z. Pflanzenphysiol.* 65:248–65

51. Haupt, W. 1982. Light-mediated movement of chloroplasts. *Annu. Rev. Plant Physiol.* 33:205–33

52. Haupt, W. 1985. Effects of nutrients and light pretreatment on phytochrome-mediated fern-spore germination. *Planta* 164:63–68

53. Haupt, W. 1987. Phytochrome control of intracellular movement. See Ref. 32, pp. 225–37

54. Haupt, W., Björn, L. O. 1987. No action dichroism for light-controlled fern-spore germination. *J. Plant. Physiol.* 129:119–28

55. Haupt, W., Bock, G. 1962. Die Chloroplastendrehung bei *Mougeotia*. V. Die Orientierung der Phytochrom-Moleküle im Cytoplasma. *Planta* 59:38–48

56. Haupt, W., Mörtel, G., Winkelnkemper, I. 1969. Demonstration of different dichroic orientation of phytochrome Pr and Pfr. *Planta* 88:183–86

57. Haupt, W., Schönfeld, I. 1962. Über das Wirkungsspektrum der "negative Phototaxis" der *Vaucheria*-Chloroplasten. *Ber. Dtsch. Bot. Ges.* 75:14–23

58. Haupt, W., Thiele, R. 1961. Chloroplastenbewegung bei *Mesotaenium*. *Planta* 56:388–401

59. Hayami, J., Kadota, A., Wada, M. 1986. Blue light–induced phototropic response and the intracellular photoreceptive site in *Adiantum* protonemata. *Plant Cell Physiol.* 27:1571–77

60. Hayami, J., Kadota, A., Wada, M. 1989. Intracellular dichroic orientation of the blue light–absorbing pigment and the blue-absorption band of red-absorbing form of phytochrome responsible for phototropism of the fern *Adiantum* protonemata. Submitted for publication.

61. Heim, B., Schäfer, E. 1982. Light-controlled inhibition of hypocotyl growth in *Sinapis alba* L. seedlings. Fluence rate dependence of hourly light pulses and continuous irradiation. *Planta* 154:150–55

62. Hepler, P. H., Wayne, R. O. 1985. Calcium and plant development. *Annu. Rev. Plant Physiol.* 36:397–439

63. Hershey, H. P., Barker, R. F., Idler, K. B., Lissemore, J. L., Quail, P. H. 1985. Analysis of cloned cDNA and genomic sequences for phytochrome: complete amino acid sequences for two gene products expressed in etiolated *Avena*. *Nuc. Acid Res.* 13:8543–59

64. Huckaby, C. S., Kalantari, K., Miller, J. H. 1982. Inhibition of *Onoclea sensibilis* spore germination by far-red light and *cis*-4-cyclohexene-1,2-dicarboximide. *Z. Pflanzenphysiol.* 105:375–78

65. Iino, M., Endo, M., Wada, M. 1989. The occurrence of a Ca^{2+}-dependent period in the red-light-induced late G1 phase of *Adiantum* spores. Submitted for publication

66. Iino, M., Nakagawa, Y., Wada, M. 1988. Blue light–regulation of cell division in *Adiantum* protonemata: an approach with pulse stimulation. *Plant Cell Environ.* 11:547–54

67. Iino, M., Nakagawa, Y., Wada, M. 1988. Blue light–regulation of cell division in *Adiantum* protonemata: kinetic properties of the photosystem. *Plant Cell Environ.* 11:555–61

68. Ito, M. 1970. Light-induced synchrony of cell division in the protonemata of the fern, *Pteris vittata*. *Planta* 90:22–31

69. Jaffe, L. F. 1958. Tropistic responses of zygotes of the Fucaceae to polarized light. *Exp. Cell Res.* 15:282–99

70. Jenkins, G. I., Cove, D. J. 1983. Phototropism and polarotropism of primary chloronemata of the moss *Phys-*

comitrella patens: responses of the wild-type. Planta 158:357–64

71. Jenkins, G. I., Cove, D. J. 1983. Phototropism and polarotropism of primary chloronemata of the moss Physcomitrella patens: responses of mutant strains. Planta 159:432–38

72. Kadota, A., Furaya, M. 1977. Apical growth of protonemata in Adiantum capillus-veneris. I. Red far-red reversible effect on growth cessation in the dark. Devel. Growth Differ. 19:357–65

73. Kadota, A., Furuya, M. 1981. Apical growth of protonemata in Adiantum capillus-veneris IV. Phytochrome-mediated induction in non-growing cells. Plant Cell Physiol. 22:629–38

74. Kadota, A., Fushimi, Y., Wada, M. 1986. Intracellular photoreceptive site for blue light-induced cell division in protonemata of the fern Adiantum—Further analyses by polarized light irradiation and cell centrifugation. Plant Cell Physiol. 27:989–95

75. Kadota, A., Inoue, Y., Furuya, M. 1986. Dichroic orientation of phytochrome intermediates in the pathway from Pr to Pfr as analyzed by double laser flash irradiations in polarotropism of Adiantum protonemata. Plant Cell Physiol. 27:867–73

76. Kadota, A., Kohyama, I., Wada, M. 1989. Polarotropism and photomovement of chloroplasts in the protonema of the ferns Pteris and Adiantum: Evidence for the possible lack of dichroic phytochrome in Pteris. Submitted for publication

77. Kadota, A., Koyama, M., Wada, M., Furuya, M. 1984. Action spectra for polarotropism and phototropism in protonemata of the fern Adiantum capillus-veneris. Physiol. Plant. 61:327–30

78. Kadota, A., Wada, M., Furuya, M. 1979. Apical growth of protonemata in Adiantum capillus-veneris. III. Action spectra for the light effect on dark cessation of apical growth and the intracellular photoreceptive site. Plant Sci. Letter 15:193–201

79. Kadota, A., Wada, M., Furuya, M. 1982. Phytochrome-mediated phototropism and different dichroic orientation of Pr and Pfr in protonemata of the fern Adiantum capillus-veneris L. Photochem. Photobiol. 35:533–36

80. Kadota, A., Wada, M., Furuya, M. 1985. Phytochrome-mediated polarotropism of Adiantum capillus-veneris L. protonemata as analyzed by microbeam irradiation with polarized light. Planta 165:30–36

81. Kasemir, H. 1983. Light control of chlorophyll accumulation in higher plants. See Ref. 146, pp. 662–86

82. Kataoka, H. 1975. Phototropism in Vaucheria geminata. I. The action spectrum. Plant Cell Physiol. 16:427–37

83. Kataoka, H. 1975. Phototropism in Vaucheria geminata. II. The mechanism of bending and branching. Plant Cell Physiol. 16:439–48

84. Kataoka, H. 1981. Expansion of Vaucheria cell apex caused by blue or red light. Plant Cell Physiol. 22:583–95

85. Kataoka, H., Weisenseel, M. H. 1988. Blue light promotes ionic current influx at the growing apex of Vaucheria terrestris. Planta 173:490–99

86. Kendrick, R. E., Bossen, M. E. 1987. Photocontrol of ion fluxes and membrane properties in plants. See Ref. 32, pp. 215–24

87. Kendrick, R. E., Kronenberg, G. H. M., eds. 1986. Photomorphogenesis in Plants. Dordrecht: Martinus Nijhoff. 580 pp.

88. Kim, Y.-S., Moon, D.-K., Goodin, J. R., Song, P.-S. 1986. Swelling of etiolated oat protoplasts induced by cAMP and red light. Plant Cell Physiol. 27:193–97

89. Konomi, K., Abe, H., Furuya, M. 1987. Changes in the content of phytochrome I and II apoproteins in embryonic axes of pea seeds during imbibition. Plant Cell Physiol. 28:1443–51

90. Kowallik, W. 1982. Blue light effects on respiration. Annu. Rev. Plant Physiol. 33:51–72

91. Kraml, M., Büttner, G., Haupt, W., Herrmann, H. 1988. Chloroplast orientation in Mesotaenium: The phytochrome effect is strongly potentiated by interaction with blue light. Protoplasma. In press

92. Kraml, M., Enders, M., Bürkel, N. 1984. Kinetics of the dichroic reorientation of phytochrome during photoconversion in Mougeotia. Planta 161:216–22

93. Kraml, M., Schäfer, E. 1983. Photoconversion of phytochrome in vivo studied by double flash irradiation in Mougeotia and Avena. Photochem. Photobiol. 38:461–67

94. Kumagai, T. 1988. Photocontrol of fungal development. Photochem. Photobiol. 47:889–96

95. Larpent, M., Jacques, R. 1971. Role du phytochrome dans le development du protonema de Funaria hygrometrica (Hedw). C. R. Acad. Sci., Ser. D 273:162–64

96. Lechowski, Z., Białczyk, J. 1987. Interaction between green and far-red light on the low fluence rate chloroplast orientation in *Mougeotia*. *Plant Physiol.* 85:581–84

97. Maekawa, T., Tsutsui, I., Nagai, R. 1986. Light-regulated translocation of cytoplasm in green alga *Dichotomosiphon*. *Plant Cell Physiol.* 27:837–51

98. Manabe, K., Ibushi, N., Nakayama, A., Takaya, S., Sugai, M. 1987. Spore germination and phytochrome biosynthesis in the fern *Lygodium japonicum* as affected by gabaculine and cycloheximide. *Physiol. Plant.* 70:571–76

99. Mancinelli, A. L. 1983. The photoregulation of anthocyanin synthesis. See Ref. 146, pp. 640–61

100. Marmé, D., Schäfer, E. 1972. On the localization and orientation of phytochrome molecules in corn coleoptiles (*Zea mays* L.). *Z. Pflanzenphysiol.* 67:192–94

101. McCurdy, D. W., Pratt, L. H. 1986. Immunogold electron microscopy of phytochrome in *Avena*: identification of intracellular sites responsible for phytochrome sequestering and enhanced pelletability. *J. Cell Biol.* 103:2541–50

102. Miller, J. H., Miller, P. M. 1963. Effects of red and far-red illumination on the elongation of fern protonemata and rhizoids. *Plant Cell Physiol.* 4:65–72

103. Miller, J. H., Miller, P. M. 1967. Action spectra for light-induced elongation in fern protonemata. *Physiol. Plant.* 20:128–38

104. Miller, J. H., Wright, D. R. 1961. An age-dependent change in the response of fern gametophytes to red light. *Science* 134:1629

105. Miyata, M., Wada, M., Furuya, M. 1979. Effects of phytochrome and blue-near ultraviolet light-absorbing pigment on duration of component phases of the cell cycle in *Adiantum* gametophytes. *Dev. Growth Differ.* 21:577–84

106. Mohr, H. 1956. Die Beeinflussung der Keimung von Farnsporen durch Licht und andere Faktoren. *Planta* 46:534–51

107. Nagata, Y., 1973. Rhizoid differentiation in *Spirogyra* I. Basic features of rhizoid formation. *Plant Cell Physiol.* 14:531–41

108. Nagata, Y. 1979. Rhizoid differentiation in *Spirogyra*. III. Intracellular localization of phytochrome. *Plant Physiol.* 64:9–12

109. Nagy, F., Kay, S. A., Chua, N.-H. 1988. Gene regulation by phytochrome. *Trends Genet.* 4:37–42

110. Nebel, B. J. 1968. Action spectra for photogrowth and phototropism in protonemata of the moss *Physcomitrium turbinatum*. *Planta* 81:287–302

111. Nebel, B. J. 1969. Responses of moss protonemata to red and far-red polarized light: evidence for disc-shaped phytochrome photoreceptors. *Planta* 87: 170–79

112. Paques, M., Brouers, M. 1981. Chloroplast phototaxis in *Acetabularia mediterranea*. *Protoplasma* 105:360–61

113. Parks, B. M., Jones, A. M., Adamse. P., Koornneef, M., Kendrick, R. E., Quail, P. H. 1987. The *aurea* mutant of tomato is deficient in spectrophotometrically and immunochemically detectable phytochrome. *Plant Mol. Biol.* 9:97–107

114. Pratt, L. H. 1982. Phytochrome: the protein moiety. *Annu. Rev. Plant Physiol.* 33:557–82

115. Pratt, L. H., Inoue, Y., Furuya, M. 1984. Photoactivity of transient intermediates in the pathway from the red-absorbing to the far-red-absorbing form of *Avena* phytochrome as observed by a double-flash transient-spectrum analyzer. *Photochem. Photobiol.* 39:241–46

116. Quail, P. H. 1980. Phytochrome: the first five minutes from P$_{fr}$ formation. In *Photoreceptors and Plant Development*, ed. J. De Greef, pp. 449–66. Antwerpen:Antwerpen Univ. Press

117. Deleted in proof

118. Quail, P. H., Colbert, J. T., Peters, N. K., Christensen, A. H., Sharrock, R. A., Lissemore, J. L. 1986. Phytochrome and the regulation of the expression of its genes. *Philos. Trans. R. Soc. Lond. Ser. B* 314:469–80

119. Racusen, R. H. 1976. Phytochrome control of electrical potentials and intercellular coupling in oat-coleoptile tissue. *Planta* 132:25–29

120. Racusen, R. H., Cooke, T. J. 1982. Electrical changes in the apical cell of the fern gametophyte during irradiation with photomorphogenetically active light. *Plant Physiol.* 70:331–34

121. Racusen, R. H., Ketchum, K. A., Cooke, T. J. 1988. Modifications of extracellular electric and ionic gradients preceding the transition from tip growth to isodiametric expansion in the apical cell of the fern gametophyte. *Plant Physiol.* 87:69–77

122. Raghavan, V. 1971. Phytochrome control of germination of the spores of *Asplenium nidus*. *Plant Physiol.* 48: 100–2

123. Raghavan, V. 1973. Photomorphogenesis of the gametophytes of *Lygodium japonicum*. *Am. J. Bot.* 60:313–21

124. Raghavan, V. 1973. Blue light interference in the phytochrome-controlled germination of the spores of *Cheilanthes farinosa*. *Plant Physiol.* 51:306–11
125. Raghavan, V. 1987. Changes in poly(A)$^+$RNA concentrations during germination of spores of the fern, *Onoclea sensibilis*. *Protoplasma* 140:55–66
126. Rethy, R. 1968. Red (R), far-red (FR) photoreversible effects on the growth of *Chara* sporelings. *Z. Pflanzenphysiol.* 59:100–2
127. Roux, S. J. 1986. Phytochrome and membranes. See Ref. 87, pp. 115–34
128. Schäfer, E. 1987. Primary action of phytochrome. See Ref. 32, pp. 279–87
129. Schäfer, E., Apel, K., Batschauer, A., Mösinger, E. 1986. The molecular biology of action. See Ref. 87, pp. 83–98
130. Schäfer, E., Haupt, W. 1983. Blue-light effects in phytochrome-mediated responses. See Ref. 146, pp. 723–44
131. Scheuerlein, R., Braslavsky, S. E. 1987. Induction of chloroplast movement in the alga *Mougeotia* by polarized nanosecond dye-laser pulses. *Photochem. Photobiol.* 46:525–30
132. Scheuerlein, R., Inoue, Y., Furuya, M. 1988. Intermediates in the photoconversion of functional phytochrome in fern spores of *Dryopteris:* II. *In vivo* kinetics of the decay of I'$_{700}$ and of the formation of Pfr studied with a double-laser apparatus. *Photochem. Photobiol.* 48:519–24
133. Scheuerlein, R., Koller, D. 1988. Intermediates in the photoconversion of functional phytochrome in fern spores of *Dryopteris:* I. Demonstration and quantitative characterization of the photochromic system Pr$\rightleftharpoons$I'$_{700}$ using nanosecond-laser pulses. *Photochem. Photobiol.* 48:511–18
134. Scheuerlein, R., Wayne, R., Roux, S. J. 1989. Calcium requirement of phytochrome-mediated fern-spore germination. No direct phytochrome-calcium interaction in the phytochrome-initiated transduction chain. *Planta.* In press
135. Schmid, R., Idziak, E. -M., Tünnermann, M. 1987. Action spectrum for the blue-light-dependent morphogenesis of hair whorls in *Acetabularia mediterranea*. *Planta* 171:96–103
136. Schneider-Poetsch, H. A. W., Schwarz, H., Grimm, R., Rüdiger, W. 1988. Cross-reactivity of monoclonal antibodies against phytochrome from *Zea* and *Avena*. Localization of epitopes, and an epitope common to monocotyledons, dicotyledons, ferns, mosses, and a liverwort. *Planta* 173:61–72

137. Schönbohm, E. 1966. Die Bedeutung des Phytochromsystems für die negative Phototaxis des *Mougeotia*-Chloroplasten. *Ber. Dtsch. Bot. Ges.* 79:131–38
138. Schönbohm, E. 1971. Untersuchungen zum Photoreceptorproblem beim tonischen Blaulicht-Effekt der Starklichtbewegung des *Mougeotia*-Chloroplasten. *Z. Pflanzenphysiol.* 66:20–33
139. Schraudolf, H. 1967. Die Steuerung der Antheridienbildung in *Polypodium crassifolium* L. (*Pessopteris crassifolia* Underw. and Maxon) durch Licht. *Planta* 76:37–46
140. Schraudolf, H. 1987. Parental predetermination of dark germination in spores of *Anemia phyllitidis*. *Naturwissenschaften* 74:138
141. Schraudolf, H. 1987. The effect of gabaculine on germination and gametophyte morphogenesis of *Anemia phyllitidis* L. Sw. *Plant Cell Physiol.* 28:53–60
142. Senger, H., ed. 1980. *The Blue Light Syndrome*. Berlin: Springer. 665 pp.
143. Senger, H. 1982. The effect of blue light on plants and microorganisms. *Photochem. Photobiol.* 35:911–20
144. Serlin, B. S., Roux, S. J. 1984. Modulation of chloroplast movement in the green alga *Mougeotia* by the Ca^{2+} ionophore A23187 and by calmodulin antagonists. *Proc. Natl. Acad. Sci. USA* 81:6368–72
145. Shimazaki, Y., Pratt, L. H. 1985. Immunochemical detection with rabbit polyclonal and mouse monoclonal antibodies of different pools of phytochrome from etiolated and green *Avena* shoots. *Planta* 164:333–44
146. Shropshire, W. Jr., Mohr, H., eds. 1983. *Encyclopedia of Plant Physiology, N. S.* 16 A, B, Photomorphogenesis. Berlin/Heidelberg/New York: Springer. 832 pp.
147. Simon, P. E., Naef, J. B. 1981. Light dependency of the cytokinin-induced bud initiation in protonemata of the moss *Funaria hygrometrica*. *Physiol. Plant.* 53:13–18
148. Speth, V., Otto, V., Schäfer, E. 1986. Intracellular localisation of phytochrome in oat coleoptiles by electron microscopy. *Planta* 168:299–304
149. Steiner, A. M. 1967. Action spectra for polarotropism in germlings of a fern and a liverwort. *Naturwissenschaften* 18: 497–98
150. Steiner, A. M. 1969. Action spectrum for polarotropism in the chloronema of the fern *Dryopteris filix-mas* (L.) Schott. *Photochem. Photobiol.* 9:507–13
151. Stenz, H.-G., Weisenseel, M. H. 1986. Phytochrome mediates a reduction of the

surface charge of *Mesotaenium* cells. *J. Plant Physiol.* 122:159–68

152. Sugai, M., Furuya, M. 1967. Photomorphogenesis in *Pteris vittata*. I. Phytochrome-mediated spore germination and blue light interaction. *Plant Cell Physiol.* 8:737–48

153. Sugai, M., Furuya, M. 1985. Action spectrum in ultraviolet and blue light region for the inhibition of red-light-induced spore germination in *Adiantum capillus-veneris* L. *Plant Cell Physiol.* 26:953–56

154. Sugai, M., Tomizawa, K., Watanabe, M., Furuya, M. 1984. Action spectrum between 250 and 800 nanometers for the photoinduced inhibition of spore germination in *Pteris vittata*. *Plant Cell Physiol.* 25:205–12

155. Sugimoto, T., Ito, E., Suzuki, H. 1987. Interpretation of the 'dichroic orientation' of phytochrome. *Photochem. Photobiol.* 46:517–23

156. Sundqvist, C., Björn, L. O. 1983. Light-induced linear dichroism in photoreversibly photochromic sensor pigments. II. Chromophore rotation in immobilized phytochrome. *Photochem. Photobiol.* 37:69–75

157. Sundqvist, C., Björn, L. O. 1983. Light-induced linear dichroism in photoreversibly photochromic sensor pigments. III. Chromophore rotation estimated by polarized light reversal of dichroism. *Physiol. Plant.* 59:263–69

158. Taylor, A. O., Bonner, B. A. 1967. Isolation of phytochrome from the alga *Mesotaenium* and the liverwort *Sphaerocarpos*. *Plant Physiol.* 42:762–66

159. Terborgh, J. 1965. Effects of red and blue light on the growth and morphogenesis of *Acetabularia crenulata*. *Nature* 207:1360–63

160. Theologis, A. 1986. Rapid gene regulation by auxin. *Annu. Rev. Plant Physiol.* 37:407–38

161. Thimann, K. V., Curry, G. M. 1960. Phototropism and phototaxis. In *Comparative Biochemistry*, ed. M. Florkin, H. S. Mas, 1:243–309. New York: Academic

162. Tobin, E. M., Silverthorne, J. 1985. Light regulation of gene expression in higher plants. *Annu. Rev. Plant Physiol.* 36:569–93

163. Tokuhisa, J. G., Daniels, S. M., Quail, P. H. 1985. Phytochrome in green tissue: spectral and immunochemical evidence for two distinct molecular species of phytochrome in light-grown *Avena sativa* L. *Planta* 164:321–32

164. Tomizawa, K., Manabe, K., Sugai, M. 1982. Changes in phytochrome content

during imbibition in spores of the fern *Lygodium japonicum*. *Plant Cell Physiol.* 23:1305–8

165. Tomizawa, K., Sugai, M., Manabe, K. 1983. Relationship between germination and Pfr level in spores of the fern *Lygodium japonicum*. *Plant Cell Physiol.* 24:1043–48

166. Vince-Prue, D. 1983. Photomorphogenesis and flowering. See Ref. 146, pp. 457–90

167. Wada, K., Hirabayashi, Y., Saito, W. 1984. Light germination of *Anthoceros miyabeanus* spores. *Bot. Mag. Tokyo* 97:369–79

168. Wada, M. 1985. Photoresponses in cell cycle regulation. *Proc. Roy. Soc. Edinburgh* 86B:231–35

169. Wada, M. 1988. Chloroplast photoorientation in enucleated fern protonemata. *Plant Cell Physiol.* 29:227–32

170. Wada, M., Furuya, M. 1972. Phytochrome action on the timing of cell division in *Adiantum* gametophytes. *Plant Physiol.* 49:110–13

171. Wada, M., Furuya, M. 1974. Action spectrum for the timing of photoinduced cell division in *Adiantum* gametophytes. *Physiol. Plant.* 32:377–81

172. Wada, M., Furuya, M. 1978. Effects of narrow-beam irradiations with blue and far-red light on the timing of cell division in *Adiantum* gametophytes. *Planta* 138:85–90

173. Wada, M., Hayami, J., Kadota, A. 1984. Returning dark-induced cell cycle to the beginning of G1 phase by red light irradiation in fern *Adiantum* protonemata. *Plant Cell Physiol.* 25:1053–58

174. Wada, M., Kadota, A., Furuya, M. 1978. Apical growth of protonemata in *Adiantum capillus-veneris*. II. Action spectra for the induction of apical swelling and the intracellular photoreceptive site. *Bot. Mag. Tokyo* 91:113–20

175. Wada, M., Kadota, A., Furuya, M. 1981. Intracellular photoreceptive site for polarotropism in protonema of the fern *Adiantum capillus-veneris* L. *Plant Cell Physiol.* 22:1481–88

176. Wada, M., Kadota, A., Furuya, M. 1983. Intracellular localization and dichroic orientation of phytochrome in plasma membrane and/or ectoplasm of a centrifuged protonema of fern *Adiantum capillus-veneris* L. *Plant Cell Physiol.* 24:1441–47

177. Wagner, G., Bellini, E. 1976. Light-dependent fluxes and compartmentation of calcium in the green alga *Mougeotia*. *Z. Pflanzenphysiol.* 79:283–91

178. Wagner, G., Grolig, F., Altmüller, D. 1987. Transduction chain of low irradiance response of chloroplast reorientation in *Mougeotia* in blue or red light. *Photobiochem. Photobiophys.* Suppl. 183–89

179. Wagner, G., Rossbacher, R. 1980. X-ray microanalysis and chlorotetracycline staining of calcium vesicles in the green alga *Mougeotia. Planta* 149:298–305

180. Wagner, G., Valentin, P., Dieter, P., Marmé, D. 1984. Identification of calmodulin in the green alga *Mougeotia* and its possible function in chloroplast reorientational movement. *Planta* 162: 62–67

181. Wayne, R., Hepler, P. K. 1984. The role of calcium ions in phytochrome-mediated germination of spores of *Onoclea sensibilis* L. *Planta* 160:12–20

182. Wayne, R., Hepler, P. K. 1985. Red light stimulates an increase in intracellular calcium in the spores of *Onoclea sensibilis. Plant Physiol.* 77:8–11

183. Weidinger, M., Ruppel, H. G. 1985. Ca^{2+}-requirement for a blue-light-induced chloroplast translocation in *Eremosphaera viridis. Protoplasma* 124:184–87

184. Weisenseel, M. H., Ruppert, H. K. 1977. Phytochrome and calcium ions are involved in light-induced membrane depolarization in *Nitella. Planta* 137:225–29

185. Weisenseel, M. H., Smeibidl, E. 1973. Phytochrome controls the water permeability in *Mougeotia. Z. Pflanzenphysiol.* 70:420–31

186. Yatsuhashi, H., Hashimoto, T., Wada, M. 1987. Dichroic orientation of photoreceptors for chloroplast movement in *Adiantum* protonemata. Non-helical orientation. *Plant Sci.* 51:165–70

187. Yatsuhashi, H., Kadota, A., Wada, M. 1985. Blue- and red-light action in photoorientation of chloroplasts in *Adiantum* protonemata. *Planta* 165:43–50

188. Yatsuhashi, H., Wada, M., Hashimoto, T. 1987. Dichroic orientation of phytochrome and blue-light photoreceptor in *Adiantum* protonemata as determined by chloroplast movement. *Acta Physiol. Plant.* 9:163–73

Annu. Rev. Plant Physiol. Plant Mol. Biol. 1989. 40:193–210

THE *AZOLLA-ANABAENA* SYMBIOSIS: BASIC BIOLOGY

G. A. Peters

Department of Biology, Virginia Commonwealth University, Richmond, Virginia 23284

J. C. Meeks

Department of Microbiology, University of California, Davis, California 95616

CONTENTS

INTRODUCTION

Azolla Lam. is a genus of heterosporous, aquatic ferns with a global distribution. The sporophytes, which normally float freely on the water surface and

1040-2519/89/0601-0193$02.00

multiply vegetatively, are usually 1–3 cm in diameter. Several species, however, can become appreciably larger. While the taxonomy and species recognition may soon be revised (18), six or seven extant species are commonly recognized (58, 67, 75).

All *Azolla* species normally contain, as a symbiont, an heterocyst-forming, N_2-fixing cyanobacterium known as *Anabaena azollae* Strasburger. Strasburger (105), however, routinely referred to the cyanobacterium as "Nostoc strings" (see also 34, 67), apparently in a general taxonomic sense. The symbionts in all other plant–N_2-fixing cyanobacteria associations are generally assigned to the genus *Nostoc* (65, 87, 103), and the *Azolla* symbiont may well belong in this genus rather than *Anabaena* (61, 87; R. Rippka, personal communication). The primary systematic criterion provisionally distinguishing *Nostoc* from *Anabaena* is the formation of hormogonia filaments by *Nostoc* spp. (97), and the cyanobacterium associated with *Azolla* does in fact form hormogonia-like filaments in specific developmental stages of the symbiosis. We see no current need, however, to depart from the historical precedent and have retained *Anabaena azollae* in reference to the cyanobacterium in symbiotic association with *Azolla*.

A. azollae is intimately associated with the fern throughout its life cycle, such that continuity of the symbiosis is maintained during sexual reproduction (6, 12, 15, 45). In the vegetative sporophyte, *A. azollae* is found in specialized leaf cavities (14, 44, 79, 84) where it can provide the total N requirement of the association via the fixation of atmospheric dinitrogen (84, 85). Under optimized conditions these associations can double their biomass in less than two days with dinitrogen as the only N source and contain 5–6.5% N on a dry weight basis (88). In contrast, *Azolla* species in the field usually contain 3–5% N per unit of dry weight (7, 57, 67, 121). *Azolla-Anabaena* also retain nitrogenase activity when growing in the presence of exogenous combined nitrogen sources (37, 72, 80, 82).

Azolla-Anabaena symbioses have a long history of use as a N fertilizer in conjunction with labor-intensive lowland rice cultivation in the Far East (16, 53, 57, 67). As a consequence of the stimulation of research on biological nitrogen fixation during the 1970s, this symbiosis attracted the attention of the international scientific community. Current investigations encompass disciplines ranging from molecular biology through agronomics. Here we select for review the *Azolla-Anabaena* symbiosis as a biological entity with emphasis on current knowledge regarding their organization, perpetuation, and plant-microbe interaction. We provide a few pertinent references for topics including the taxonomy, occurrence, and application of these associations as a biofertilizer for rice, along with our assessment of some recent developments and unresolved issues.

TAXONOMY, OCCURRENCE, AND AGRONOMIC ASPECTS

The genus *Azolla*, established by J. B. Lamarck in 1783 with a description of *Azolla filiculoides*, is usually placed in a single genus family, the Azollaceae, in the Salviniales (45, 57, 67). Fertile sporophytes of these leptosporangiate, heterosporous ferns produce morphologically distinct female megasporocarps and male microsporocarps. The present evolutionary understanding of the genus has been built primarily on the comparative morphology of the megaspore apparatus of fossil and extant species (32, 59). Based on features of the megaspore apparatus, and the nature of the masulla processes in the microsporangia, extant species are divided into two sections (19, 57, 67, 75). The section Azolla (synonymous with Euazolla) is characterized by three megaspore floats and currently includes *A. caroliniana* Willdenow, *A. filiculoides* Lamarck, *A. mexicana* Presl, and *A. microphylla* Kaulfuss. *A. rubra* R. Brown is included as a fifth species in the section Azolla in some treatments but reduced to a variety of *A. filiculoides* in others. The section Rhizosperma, characterized by nine floats in the megaspore apparatus, includes *A. pinnata* R. Brown and *A. nilotica* De Caisne. The latter is distinctive as it can produce a rhizome of up to 40 cm (58). For a review of the taxonomic history of the extant species see (75).

Several species show vegetative polymorphism and, except for epidermal trichomes on the dorsal leaf lobe (18, 58, 75, 91), vegetative characters are of limited use in separating species, especially those in the section Azolla. Current species delineation is based primarily on features of the megaspore apparatus and especially the architecture of the special covering (wall) of the megaspores, called the perine, perispore, or sporoderm (18, 20, 54, 75). For a comparison of the perine architecture of extant species see Perkins et al (75). A preliminary report indicates that taxonomic changes in the section Azolla may include rendering the name *A. microphylla* invalid, regarding *A. caroliniana* as synonymous with *A. filiculoides*, separating *A. filiculoides* into two subspecies (*filiculoides* and *rubra*), and establishing two or more new species (18).

In tropical and temperate regions *Azolla* species may form thick mats on the relatively placid surface of freshwater ponds, marshes, drainage ditches, and rice paddies. The *Azolla* mats sometimes appear red or purple rather than green owing to the presence of anthocyanins, predominantly luteolinin 5-glucoside with lesser amounts of apigeninidin glucoside (33, 36, 91). The ability to assimilate N_2 often enables *Azolla* to grow where nitrogen is limiting to other aquatic plants, but its growth may be limited by the availability of other nutrients (55). *Azolla* does not colonize large lakes or swiftly moving waters because wave action and other forms of turbulence lead to excessive fragmentation and diminished growth (3). The geographical distribution of the currently recognized extant *Azolla* species and human impact on that distribution have been considered elsewhere (57, 67, 79). The worldwide dissemination of populations of all species for use in field trials during the past decade may have had a further influence that is currently undocumented on the species distribution.

Azolla-Anabaena associations have received attention as weeds (3, 57, 67) and as plants with agronomic importance as a biofertilizer for rice. *A. pinnata* has been grown in parts of Vietnam and China as a green manure for centuries (16, 53, 67), and in 1984 *A. pinnata* and *A. filiculoides* were cultivated on over two million hectares in these countries (56). In 1975–1976 field studies were initiated at the International Rice Research Institute (IRRI) in the Philippines (123), at the Central Rice Research Institute in Cuttack, India (101), and at the University of California, Davis (110). Numerous subsequent studies, including those involving 37 sites in 10 countries conducted between 1979 and 1984 as part of the International Network on Soil Fertility and Fertilizer Evaluation for Rice (INSFFER) (122), have demonstrated the potential and problems associated with use of *Azolla* in rice production systems.

Azolla is effective when grown during the fallow season and incorporated into the paddy soil before transplanting rice, when grown in dual culture as a cover crop with periodic incorporation, or when a combination of the two approaches is used. *Azolla* used as a cover crop also suppresses weed growth (67, 110). Under suitable environmental conditions in California, an *Azolla filiculoides* mat had an average N-accumulation rate of 2 kg N/ha/day over a 46-day period and resulted in a biomass containing 93 kg N/ha (109). Considerable variation occurs, however, in growth rates, N content, and mat density within and among species under field conditions around the world (57, 67, 102, 108, 121). In the INSFFER trials the average 15 t/ha biomass yielded only 30 kg N/ha if one assumes 0.2% N (122), which corresponds to about 3% N on a dry weight basis. Additional literature in this area—as well as information on composition; digestability; and use of *Azolla* as a fodder, as fish food, and in the removal of macro- and microelements from sewage effluent and polluted waters—is available (35, 57, 58, 67, 79, 100, 121).

Although the potential of *Azolla* as a biofertilizer for rice is well documented, a number of factors limit its widespread use. In many rice-growing regions, the lack of good control and/or predictable availability of water is a major limitation. The current need to maintain *Azolla* vegetatively during seasons unfavorable for its growth adds to the labor-intensive aspects of its overall use, and available phosphorus often limits biomass accumulation (102, 106, 109, 110, 124). Efforts to expand its use in the humid tropics have met with limited success owing to a host of environmental problems (55), including the detrimental effect of high air and water temperature and the susceptibility of *Azolla* to attack by fungi and a number of insects (7, 57, 58, 66, 121).

ORGANIZATION OF THE SPOROPHYTE SYMBIOSIS

The organization of the sporophyte is dorsiventral (15, 79). Each prostrate multibranched, floating stem (rhizome) bears two lateral rows of alternately arranged, deeply bilobed leaves on the dorsal surface and adventitious roots at nodes on the ventral surface. The anatomical development of these roots is highly ordered and has been characterized in detail (27). The thin ventral leaf lobe is nearly achlorophyllous and is in contact with the water surface. The chlorophyllous, aerial dorsal leaf lobe is fleshy and at maturity contains an ellipsoid cavity in the proximal half of its lamina. The cavity, an extracellular space formed by an infolding of the adaxial epidermis during development, normally contains *A. azollae* (14, 44, 79). It also is formed in *Azolla* freed of *A. azollae* (3, 77); such plants occur rarely in nature (31, 34), but they can be generated in the laboratory by antibiotic treatment (3, 84), dramatic changes in culture conditions (30, 31), or by manipulation of sporocarps (51, 95, and below). The cavity area of plants with and without *A. azollae* can be isolated as a discrete structure following digestion of the leaf tissue with cellulolytic enzymes (77, 90, 117).

An ontogenetic sequence of leaf development is expressed along every axis of the multibranched stem (14, 79), and the *A. azollae* undergoes a parallel pattern of differentiation and development (30, 31, 40, 41). Stem tips are upcurved, supported above the water surface by the rest of the stem axis, and enclosed by the developing bilobed leaves (44, 90). A colony of un-differentiated, or generative (30, 31), *A. azollae* filaments, morphologically similar to hormogonia, is associated with the apical meristem of each stem. Growth of the apical colony is coordinated with that of the fern (3, 30, 79). As each dorsal lobe differentiates, some of the apical colony filaments are partitioned into the developing cavity. This partitioning is facilitated by a rapidly differentiated multicellular epidermal trichome (11, 79), termed the primary branched hair (PBH) (14), which extends into the apical colony. As leaf development and maturation proceed, the PBH and associated *A. azollae* filaments are completely engulfed by the developing leaf cavity in which another branched hair and a separate population of about 25 simple hairs differentiate (14, 79). The latter are randomly distributed around portions of the cavity bordered by photosynthetic mesophyll, while the two branched hairs are always located on the path of the foliar trace (14). Cells of both types of cavity hairs exhibit transfer cell ultrastructure (TCU) (17, 68, 90), a cellular configuration associated with sites of high solute movement in plants (74). During their ontogeny, both hair types undergo a succession of transfer cell differentiation and senescence which begins in their terminal cells and ends in their stalk cells (11). In the region of each branched hair, TCU also occurs in the leaf trace and in the cells separating the leaf trace from the cavity (79).

As the undifferentiated filaments of *A. azollae* in the apical colonies are partitioned into the forming leaf cavities, they continue to grow but begin to differentiate heterocysts and fix N_2 (30, 31, 40, 41). The subcellular reorganization associated with heterocyst differentiation (9, 49, 68) and the ultrastructure of the vegetative cells (9, 68) is essentially the same as that in free-living *Anabaena* or *Nostoc*. Immunocytological analysis has shown that dinitrogenase reductase, the Fe-protein of nitrogenase, is restricted to heterocysts (9). In mature cavities possessing a full complement of epidermal hairs, *A. azollae* cell division is greatly diminished, its cells have enlarged, heterocysts constitute 20–30% of the cells, and nitrogenase activity is maximal (30, 31, 41, 86). The *A. azollae* filaments are associated with the cavity hairs around the periphery of the cavity (14) and use of a cryochamber attached to the scanning-electron-microscope (SEM), which allows observation of frozen, hydrated specimens, has shown that the center is gaseous (S. K. Perkins, unpublished observation). As a leaf begins to senesce, nitrogenase activity declines, *A. azollae* vegetative cells may differentiate into akinetes or become moribund (9, 30, 76), and there is a marked increase in heterotrophic bacteria (11); for information on the latter, see work by Gates and colleagues (23, 120). The succession of TCU differentiation in the cavity hairs coincides with the pattern of differentiation and development exhibited by *A. azollae*. It has been postulated that the cavity hairs are principals in the exchange of nutrients and metabolites between the plant and *A. azollae* (11).

SPORULATION, SPOROCARPS, AND CONTINUITY OF *ANABAENA AZOLLAE* DURING THE SEXUAL CYCLE

Induction

Sporulation is unpredictable; it has been associated with crowded growth conditions, or mat formation, and attributed to interacting effects of environmental variables (2, 7, 107, 108). There is no evidence for either a specific spore-inducing factor or any uniform causal factor. *A. azollae* is not required for sporulation as symbiont-free plants grown on combined nitrogen produce viable sporocarps (51; G. A. Peters, unpublished observation). Sporulation has been induced consistently in specific strains of *A. caroliniana, A. mexicana, A. nilotica*, and an unidentified *Azolla* species under continuous light in controlled environment studies, and in two populations of *A. pinnata*, but not in other *A. pinnata* strains or other species, using a 12/12, 20°/6°C, light/dark cycle (B. H. Marsh and G. A. Peters, unpublished observations). The reports of an ever-bearing *A. pinnata* strain (58) and an *A. mexicana* strain that sporulates consistently and copiously over a broad range of culture conditions (113) do indicate, however, that the controls governing the induction of sporulation can be altered.

Sporocarps and Continuity of Anabaena azollae

Sporocarps occur in pairs (tetrads in *A. nilotica*) of the same or mixed sex on the ventral surface of the fertile sporophyte. Female megasporocarps are ovoid to fusiform and appreciably smaller than the globular male microsporocarps (12, 45, 58). Earlier accounts of these structures, which have been studied at the level of light microscopy for more than a century, are addressed elsewhere (75) and summarized in standard textbooks on pteridophytes. Recent studies using transmission electron microscopy (TEM) have provided additional insight on microsporogenesis (28) and postmeiotic development in the megasporocarp (29). The distinctive features of both megasporocarps and microsporocarps, and interactions of their components leading to the next sporophyte generation, have been depicted using microdissection sequences and scanning electron microscopy (12). Cytochemistry and TEM studies (54) showed that megaspores contain protein bodies, lipid globules, polysaccharide-rich vacuoles, and amyloplasts, while microspores are packed with saturated lipid. These studies also implied that components of the sporocarps were impregnated with sporopollenin. Subsequently sporopollenin (10) was positively identified and shown to account for 43%, 44%, and 30% of the dry weight of pure preparations of megasporocarps, microsporocarps, and microsporangia, respectively (113, 114).

During sporulation, filaments of *A. azollae* are packaged into the developing sporocarps (6, 12, 15, 45, 79, 105). The inoculation of developing sporocarps with *A. azollae* is facilitated by branched epidermal trichomes that develop near the sporangial initial on the apical meristem and grow into the apical *A. azollae* colony (13). Subsequently the sporangial primodium, and associated trichomes with *A. azollae* filaments entangled around them, are displaced ventrally and the *A. azollae* cells, which rapidly begin to differentiate into akinetes, become encased in the developing sporocarps. As sporocarp gender is determined later in development, both megasporocarps and microsporocarps contain the *A. azollae* at maturity (12, 45). Only the *A. azollae* in the megasporocarp, however, can provide for continuity of the symbiosis.

The area occupied by *A. azollae* akinetes in mature megasporocarps (3, 12, 15, 43, 113) serves as the inoculation chamber for the embryonic *Azolla*. When the megaspore germinates, the first archegonium to differentiate on the female gametophyte is positioned such that if its oocyte is fertilized, the embryonic sporophyte will grow directly into the akinetes, trapped between the apical cap of the megaspore apparatus and the indusium, and reinoculation will be accomplished (12, 45). If fertilization is delayed, embryos arise from archegonia at the perimeter of the gametophyte thallus and emerge outside the inoculation chamber to yield endophyte-free plants (G. A. Peters, unpublished observation). As the nascent sporophyte grows toward the inoculation chamber some of the akinetes of *A. azollae* begin to germinate to yield hormogonia-like filaments. The first sporophyte cells to encounter the filaments are branched trichomes of the first dorsal leaf lobe primordium on the apical meristem (13). These trichomes exhibit TCU but are structurally distinct from those associated with the leaf cavity inoculation in the mature sporophyte. Their primary role appears to be facilitating the reassociation of *A. azollae* and the fern (S. K. Perkins, personal communication). A cup-shaped cotyledonary leaf develops at such a disproportionate rate that it rapidly surrounds the entire meristem, entrapping the *A. azollae* filaments and keeping them associated with the branched trichomes, thereby assuring inoculation of developing leaves as the sporophyte emerges from the megaspore apparatus (12, 45).

ANABAENA AZOLLAE IN CULTURE AND SYMBIOSIS

The mechanisms that maintain the continuity of the symbiosis clearly preclude the need for a free-living form of *A. azollae*. They do not exclude the possibility of its occurrence, but the probability of any free-living form naturally reentering the fern is negligible. Nevertheless, free-living culture of nitrogen-fixing cyanobacteria isolated from several *Azolla* species has consistently been reported (3, 5, 8, 34, 61, 70, 111, 119). The identity of these isolates as a free-living form of the associated *A. azollae* has been assumed but never proven by reconstituting the association using symbiont-free *Azolla* and a cultured *Anabaena* or *Nostoc* isolate. A recent report of reestablishment (52) involved maintaining continuity of the symbiosis after removing the inoculation chamber from one megasporocarp and replacing it with that from another. Efforts to reconstitute the functional symbiosis by removing the inoculation chamber and inoculating the embryonic sporophyte with cultured cyanobacterial isolates were unsuccessful (52).

The use of DNA endonuclease restriction fragment length polymorphisms detected by Southern hybridization, primarily with cloned nitrogenase *(nif)* structural genes from nonsymbiotic *Anabaena* sp. PCC 7120, was introduced recently as a stable marker for *A. azollae* (21). Analysis of sequence divergence based on such hybridization studies with leaf cavity symbionts of four species from the section Azolla and five strains of *A. pinnata* (Rhizosperma) indicated that *A. azollae* in all *Azolla* species derive from a common ancestor and that *A. azollae* of the sections Azolla and Rhizosperma belong to internally similar but slightly divergent evolutionary lines (22). Additional studies with two independently cultured isolates from *A. caroliniana* (61, 70) and one from *A. filiculoides* (111) established that these isolates were identical neither to the associated *A. azollae* in either *Azolla* species nor to one another (21, 61, 71). The *A. azollae* in association with *A. caroliniana* was also shown to have a contiguous *nifHDK* gene organization in all of its cells (61). While not unique to *A. azollae* (98; J. C. Meeks, unpublished observation), this *nif* gene organization is distinctive. In the other heterocyst-forming cyanobacteria examined to date, including a cultured isolate from *A. caroliniana* (61), the *nifD* gene (α subunit of dinitrogenase) in vegetative cells is interrupted by an interval of 11 kb (26) or unknown length (39, 61). During the latter stages of heterocyst maturation, this interval is excised with the formation of the transcribed *nifHDK* operon (26).

   The restriction fragment length polymorphism studies (21, 61) strongly imply that *Azolla* species can harbor minor N_2-fixing cyanobacterial symbionts capable of free-living growth in addition to the dominant *A. azollae,* and that the latter may be an obligate symbiont that has not been cultured in the free-living state. Because the minor symbionts cannot be detected by DNA-DNA hybridization with the symbiotic cyanobacteria immediately separated from the plant, they must be under stringent growth control by the *Azolla.* The physical location of the minor symbionts in the sporophyte is unknown, but their occurrence may explain why there are no antigenic similarities between preparations of *A. azollae* and presumptive free-living cultures (1, 24, 47).

   The continuity of the *A. azollae* with *Azolla* throughout its life cycle, together with the hypothesis that *A. azollae* has not been cultured in the free-living state and the concept that culturable minor symbionts are under stringent growth control by *Azolla,* and thus may not contribute substantially to N_2-dependent growth of the association, imply that there may be little or no selective pressure for postulated lectin-mediated recognition between the fern and the free-living *Anabaena* cultures (8, 43, 48, 60, 64). It is possible, however, that lectins, or similar recognition molecules, are involved in the partitioning of *A. azollae* filaments into leaf cavities and developing sporocarps by specific binding to branched epidermal trichomes.

PHYSIOLOGY, BIOCHEMISTRY, AND INTERACTION

Owing to the synchronous differentiation and development of the leaves and *A. azollae* on every axis of the multibranched rhizome (14, 30), physiological and biochemical studies of the intact association, symbiont-free *Azolla,* and *A. azollae* separated from the entire plant reflect a composite of processes and/or activities. While the use of individual leaves, or sequential leaf segments, on the stem axes allows examination of the dynamics of activities and processes associated with the developmental gradient, relatively few such studies have been attempted.

Photosynthesis and Carbon Exchange

In rapidly growing laboratory cultures of the association, A. azollae accounts for about 16% of the total chlorophyll and protein (94), and it retains a full complement of the phycobiliproteins (PBP) phycocyanin, phycoerythrocyanin, and allophycocyanin (40, 115). PBP and chlorophyll a are present in heterocysts as well as in vegetative cells (40, 116).

A. azollae removed from the Azolla has a rate of photosynthesis comparable to that of free-living cyanobacterial cultures and an action spectrum in which the highest quantum yield occurs in the region of PBP absorption (42, 93). The action spectra for photosynthesis, however, both in the association and in Azolla freed of A. azollae, are similar to those of other green plants; they are also so similar to one another that it is difficult to detect any obvious contribution from the A. azollae (93). It has been estimated that A. azollae in fact contributes no more than 5% of the total CO_2 fixed by the intact association (42). A substantial fraction of the photosynthetic CO_2 fixation in separated A. azollae may be contributed by the generative and relatively undifferentiated filaments from the apical segments of the multibranched rhizome. It seems probable that such A. azollae filaments are shaded by the overarching of the more mature leaves (see 14, 86, 90) so that in effect they are light limited and would not contribute to the action spectrum of the symbiosis. The action spectrum for nitrogenase-catalyzed acetylene reduction (below) demonstrates, however, that photosynthetically active radiation reaches A. azollae in mature leaf cavities where vegetative cells still account for more than half of the total cells. Therefore, it would appear that photosynthetic CO_2 fixation by A. azollae in the symbiosis may be altered by some unknown mechanism(s) affecting either noncyclic electron transport or activity of ribulose 1–5, bisphosphate carboxylase, or both.

Azolla and separated A. azollae form intermediates of the reductive pentose phosphate cycle during CO_2 fixation (93). Sucrose is a primary photosynthetic end product in the association and in symbiont-free Azolla (93). Sucrose is not detected as a product of $^{14}CO_2$ fixation by A. azollae (42, 93), but it is present as a major component of the soluble carbohydrate pool extracted from separated A. azollae with boiling water (83). It has been suggested that A. azollae in mature leaf cavities might be capable of photoheterotrophic (85) or mixotrophic metabolism (76) and use sucrose produced by the fern as a reduced carbon source (93). Transfer of fixed carbon, including sucrose, from Azolla to A. azollae has been demonstrated (42), but nothing is known about the biochemical mechanisms of transport or uptake of Azolla photosynthate by the A. azollae.

N_2 Fixation

Photosynthesis is the ultimate source of all the ATP and reductant required for nitrogenase activity in *Azolla-Anabaena*. Dark, aerobic nitrogenase activity is entirely dependent upon endogenous reserves of photosynthate (5, 76, 77, 79, 85) and, assuming that reductant is generated in the same manner in the light and dark, appears to be ATP limited (79, 86). In the light, the driving forces of nitrogenase activity appear to be ATP from cyclic photophosphorylation and reductant that can be provided by prior photosynthesis (76, 77, 85, 86). The strong dependence of nitrogen fixation upon photosynthetic processes is also demonstrated by the action spectrum for nitrogenase-catalyzed acteylene reduction where the rate per incident quantum in both the association and separated *A. azollae* is as great in the region of PBP absorption as it is in the region of chlorophyll absorption (116). Thus, the light energy least effectively harvested by *Azolla* pigments is absorbed by the heterocysts of *A. azollae* in mature leaf cavities and is assumed to support nitrogenase activity via PSI and cyclic photophosphorylation.

For information about the nitrogenase-catalyzed reduction of acetylene, or $^{15}N_2$ and protons, and their relationships, as well as studies of saturating substrate concentration, electron allocation, effects of uptake hydrogenase, and C_2H_2/N_2 conversion factors, we refer the reader elsewhere (69, 77, 78, 80–82, 89).

Nitrogen Assimilation

Studies using ^{15}N (78, 86) and ^{13}N (62, 63) have established that *A. azollae* freshly separated from *Azolla* releases about 40% of the dinitrogen it fixes as ammonium. Use of these isotopes in pulse-chase experiments with main stem axes has shown that fixed nitrogen is translocated from mature cavities to the stem apexes where the undifferentiated filaments lacking nitrogenase activity are located (41, 83). In contrast to the case in legume-rhizobia associations that involve predominantly amide or ureide transport (99), NH_4^+ and all immediate organic products of N_2 fixation appeared to be translocated in *Azolla-Anabaena* (83). Activities of glutamine synthetase (GS), glutamate synthase (GOGAT), and glutamate dehydrogenase (GDH) have been detected in crude extracts of *Azolla* grown with and without *A. azollae* (94). The fern, therefore, appeared to have the capacity to assimilate the N_2-derived NH_4^+ into glutamate either by the GS-GOGAT pathway (glutamate synthase cycle; 96) or by parallel GS-GOGAT and GDH activities. Based on the kinetics of incorporation of ^{13}N into glutamine and glutamate, as well as the effects of inhibitors of GS and GOGAT activities on that incorporation, it is clear that the fern assimilates both N_2-derived and exogenous NH_4^+ by the glutamate synthase cycle with little or no contribution from biosynthetic GDH (62).

Catalytic activities of GS, GOGAT, and GDH have also been detected in crude extracts of *A. azollae* (83, 92, 94, 104, 118). The majority of the GDH activity, however, appears to be associated with leaf cavity trichomes (83, 118) present in the *A. azollae* preparations. Studies of ^{13}N incorporation by separated *A. azollae*, employing kinetic, inhibitor, and substrate-dilution experiments, confirm a lack of a contribution by GDH to NH_4^+ assimilation and the exclusive operation of the glutamate synthase cycle in the cyanobacterium (63).

The specific catalytic activity of GS in preparations of *A. azollae* is consistently reported to be lower than the activity of various free-living *Anabaena* or *Nostoc* cultures, apparently as a consequence of less GS protein in the symbiont. Relative amounts of GS protein in *A. azollae* obtained from all stages of leaf cavity development have been determined using radioimmune (73), immunoelectrophoresis (50, 104), and enzyme-linked immunosorbent assay (50) techniques, coupled with polyclonal antibodies to GS purified from an *Anabaena* sp. (73) and two different symbiotic *Nostoc* spp. (38, 104). In two studies where both activity and protein concentration were determined, they were lower to the same extent (20–35% of GS activity and 22–39% of GS protein) relative to free-living cultures (50, 104). Measurements of activity alone have given rates somewhat higher (50% of free-living; 94), while those of protein (73) or mRNA for the GS structural gene *gln*A (71) are comparatively lower (5 and 10%, respectively). Because of the evolutionarily conserved nature of GS (4), the source of the original antigen or DNA probe is probably of little importance; the variation more likely results from differences in the physiological state of the experimental *Azolla-Anabaena* cultures.

There is no current experimental evidence identifying a gradient of high GS activity or protein in undifferentiated *A. azollae* filaments in the apical region to low activity or protein in N_2-fixing filaments within mature cavities; such a gradient could account for the release of N_2-derived NH_4^+. The regulatory mechanism(s) responsible for the apparently lower rate of synthesis of GS in the *A. azollae* is not known. It has been suggested that the fern might produce a corepressor of transcription of *gln*A in *A. azollae* (71). The promoter region of *gln*A in an obligatory symbiotic *A. azollae* could also have been modified over evolutionary time such that transcription simply became less efficient (50).

CONCLUDING COMMENTS

Basic Biology

Establishment of a functional symbiotic association between rhizobia and legumes involves a series of chemical signals between the microbial and plant partners that results in differential gene expression. In the rhizobia the targets are called symbiotic genes (*nod* and *fix;* 46), while the plant target genes code for proteins termed nodulins (25). An important consequence of this interaction is the regulation of growth and differentiation of the prokaryotic partner. This interaction also affects tissue development of the plant leading to the formation of nodules. Similar events may occur in the actinorhizal symbioses (112). In the phylogenetically diverse group of plant genera forming symbiotic associations with cyanobacteria, minor changes may occur in structures that facilitate metabolite interchange (104). There are, however, no pronounced morphological alterations of the plant tissues. The leaf cavities of *Azolla,* the thallus cavities of bryophytes, the coralloid roots of cycads, and the glands of *Gunnera* develop normally in the absence of symbiotic cyanobacteria (87). Therefore, it may be difficult to determine whether any symbiotic cyanobacteria, and *A. azollae* in particular, transmit chemical signals that influence developmental processes in the various plant partners. The observation that cellular differentiation of *A. azollae* parallels the ontogenetic sequence of leaf development in *Azolla* implies, however, that differential gene expression occurs in *A. azollae* in response to specific chemical signals within that developmental gradient. Owing to the continuity of *A. azollae* in all stages of the *Azolla* life cycle, any chemical signals from *Azolla* that control growth and differentiation of *A. azollae,* and the minor symbionts, may differ from those of other plant groups that constantly undergo natural reinfection by free-living populations to maintain the cyanobacterial symbiont. With the rapid progress being made in molecular genetics and macromolecular analyses, the mechanisms of regulatory control are now amenable to experimental study in *Azolla-Anabaena* and other plant-cyanobacterial associations. The inability to establish free-living cultures of *A. azollae,* however, is a major limitation that must be resolved if the cyanobacterium is to be subjected to routine genetic manipulation.

Applied Biology

User education, along with ecotype selection programs, the development of improved managerial schemes, and the design of specialized machinery hold some promise for improving and extending *Azolla* utilization. A major limitation, however, is the current inability to control the induction of sporulation in the field. When sporocarps are available, parameters are being defined for their use in the field culture of sporophytes in China, India, and Thailand. Overcoming this limitation will provide material not only for the production of sporocarp inoculants, but also for use in germ plasm preservation and in breeding programs.

ACKNOWLEDGMENTS
The research of both authors has been supported in part by grants from the National Science Foundation and the US Department of Agriculture Competitive Grants Office. Thanks to Elsie Lin Campbell and Alice Henin at UCD for assistance in processing the manuscript.

Literature Cited

1. Arad, H., Keysari, A., Tel-Or, E., Kobilier, D. 1985. A comparison between cell antigens in different isolates of *Anabaena azollae. Symbiosis* 1:195–204

2. Ashton, P. J. 1977. Factors affecting the growth and development of *Azolla filiculoides* Lam. In *Proceedings of the Second National Weed Conference of South Africa*, ed. E. B. H. Oliver, pp. 249–68. Capetown: A. A. Balkema

3. Ashton, P. J., Walmsley, R. D. 1976. The aquatic fern *Azolla* and its *Anabaena* symbiont. *Endeavor* 35:39–43

4. Baumann, L., Baumann, P. 1978. Studies of relationship among terrestrial *Pseudomonas, Alcaligenes,* and enterobacteria by an immunological comparison of glutamine synthetase. *Arch. Microbiol.* 119:25–30

5. Becking, J. H. 1976. Contribution of plant-algal associations. In *Proceedings of the 1st International Symposium on Nitrogen Fixation*, ed. W. E. Newton, C. J. Nyman, pp. 556–80. Pullman: Washington State Univ. Press

6. Becking, J. H. 1978. Ecology and physiological adaptation of *Anabaena* in the *Azolla-Anabaena azollae* symbiosis. *Ecol. Bull. (Stockholm)* 26:266–81

7. Becking, J. H. 1979. Environmental requirements of *Azolla* for use in tropical rice production. In *Nitrogen and Rice*, ed. IRRI, pp. 245–74. Manila: Int. Rice Res. Inst.

8. Berliner, M. D., Fisher, R. W. 1987. Surface lectin binding to *Anabaena variabilis* and to cultured and freshly isolated *Anabaena azollae. Curr. Microbiol.* 16:149–52

9. Braun-Howland, E. B., Lindblad, P., Nierzwicki-Bauer, S. A., Bergman, B. 1988. Dinitrogenase reductase (Fe-protein) of nitrogenase in the cyanobacterial symbionts of three *Azolla* species: localization and sequence of appearance during heterocyst differentiation. *Planta.* 176:319–32

10. Brooks, J. 1971. Some chemical and geochemical studies on sporopollenin. In *Sporopollenin*, ed. J. Brooks, P. R. Grant, M. Muir, P. van Gijzel, G. Shaw, pp. 351–407. New York: Academic

11. Calvert, H. E., Pence, M. K., Peters, G. A. 1985. Ultrastructural ontogeny of leaf cavity trichomes in *Azolla* implies a functional role in metabolite exchange. *Protoplasma* 129:10–27

12. Calvert, H. E., Perkins, S. K., Peters, G. A. 1983. Sporocarp structure in the heterosporous water fern *Azolla mexicana* Presl. *Scanning Electron Microsc.* 3:1499–10

13. Calvert, H. E., Perkins, S. K., Peters, G. A. 1985. Involvement of epidermal trichomes in the continuity of the *Azolla-Anabaena* symbiosis through the *Azolla* life cycle. *Am. J. Bot.* 72:808 (Abstr.)

14. Calvert, H. E., Peters, G. A. 1981. The *Azolla-Anabaena azollae* relationship. IX. Morphological analysis of leaf cavity hair populations. *New Phytol.* 89:327–35

15. Campbell, D. H. 1893. On the development of *Azolla filiculoides*, Lam. *Ann. Bot.* 26:155–87

16. Dao, T. T., Tran, O. T. 1979. Use of *Azolla* in rice production in Vietnam. See Ref. 7, pp. 395–405

17. Duckett, J. G., Toth, R., Soni, S. L. 1975. An ultrastructural study of the *Azolla, Anabaena azollae* relationship. *New Phytol.* 75:111–18

18. Dunham, D. G., Fowler, K. 1987. Taxonomy and species recognition in *Azolla* Lam. See Ref. 35, pp. 7–16

19. Florschutz, F. 1938. Die beiden *Azolla*-arten des Niederlandischen Pleistozans. *Recl. Trav. Bot. Neerl.* 35:932–45

20. Fowler, K., Stennet-Wilson, J. 1978. Sporoderm architecture in modern *Azolla. Fern. Gaz.* 11:405–12

21. Franche, C., Cohen-Bazire, G. 1985. The structural *nif* genes of four symbiotic *Anabaena azollae* show a highly conserved physical arrangement. *Plant Sci.* 39:125–31

22. Franche, C., Cohen-Bazire, G. 1987. Evolutionary divergence in the *nif*H.D.K. gene region among nine symbiotic *Anabaena azollae* and between *Anabaena azollae* and some free-living heterocystous cyanobacteria. *Symbiosis* 3:159–78

23. Gates, J. E., Fisher, R. W., Candler, R. A. 1980. The occurrence of coryneform bacteria in the leaf cavity of *Azolla. Arch. Microbiol.* 127:163–65

24. Gates, J. E., Fisher, R. W., Goggin, T. W., Azrolon, N. I. 1980. Antigenetic differences between *Anabaena azollae* fresh from the *Azolla* fern leaf cavity and free-living cyanobacteria. *Arch. Microbiol.* 128:126–29

25. Gloudemans, T., Moerman, M., Van Beckum, J., Gunderson, J., Van Kammen, A., Bisseling, T. 1988. Identification of plant genes involved in the *Rhizobium leguminosarum*–pea root hair

interaction. In *Nitrogen Fixation: Hundred Years After*, ed. H. Bothe, F. J. de Bruijn, W. E. Newton, pp. 611–16. Stuttgart: Gustav Fisher
26. Golden, J. W., Robinson, S. J., Haselkorn, R. 1985. Rearrangement of nitrogen fixation genes during heterocyst differentiation in the cyanobacterium *Anabaena*. *Nature* 314:419–23
27. Gunning, B. E. S., Hughes, J. E., Hardham, A. R. 1978. Formative and proliferative cell divisions, cell differentiation, and developmental changes in the meristem of *Azolla* roots. *Planta* 143:121–44
28. Herd, Y. R., Cutter, E. G., Watanabe, I. 1985. A light and electron microscopic study of microsporogenesis in *Azolla microphylla*. *Proc. R. Soc. Edinburgh Sect. B* 86:53–58
29. Herd, Y. R., Cutter, E. G., Watanabe, I. 1986. An ultrastructural study of postmeiotic development in the megasporocarp of *Azolla microphylla*. *Can. J. Bot.* 64:822–23
30. Hill, D. J. 1975. The pattern of development of *Anabaena* in the *Azolla-Anabaena* symbiosis. *Planta* 122:179–84
31. Hill, D. J. 1977. The role of *Anabaena* in the *Azolla-Anabaena* symbiosis. *New Phytol.* 78:611–16
32. Hills, L. V., Gopal, B. 1967. *Azolla primaeva* and its phylogenetic significance. *Can. J. Bot.* 45:1179–91
33. Holst, R. W. 1977. Anthocyanins of *Azolla*. *Am. Fern. J.* 67:99–100
34. Huneke, A. 1933. Beiträge zur Kenntnis des Symbiose zwischen *Azolla* und *Anabaena*. *Beitr. Biol. Pflanzen* 20:315–41
35. International Rice Research Institute. 1987. *Azolla Utilization*. Manila: IRRI. 296 pp.
36. Ishikura, N. 1982. 3-Desoxyanthocyanin and other phenolics in the water fern *Azolla*. *Bot. Mag. (Jpn.)* 95:303–6
37. Ito, O., Watanabe, I. 1983. The relationship between combined nitrogen uptake and nitrogen fixation in *Azolla-Anabaena* symbiosis. *New Phytol.* 95:647–54
38. Joseph, C. M., Meeks, J. C. 1987. Regulation of expression of glutamine synthetase in a symbiotic *Nostoc* strain associated with *Anthoceros punctatus*. *J. Bacteriol.* 169:2471–75
39. Kallas, T., Coursin, T., Rippka, R. 1985. Different organization of *nif* genes in nonheterocystous and heterocystous cyanobacteria. *Plant Mol. Biol.* 5:321–29
40. Kaplan, D., Calvert, H. E., Peters, G.

A. 1986. Nitrogenase activity and phycobiliproteins of the endophyte as a function of leaf age and cell type. *Plant Physiol.* 80:884–90
41. Kaplan, D., Peters, G. A. 1981. The *Azolla-Anabaena* relationship. X. $^{15}N_2$ fixation and transport in main stem axes. *New Phytol.* 89:337–46
42. Kaplan, D., Peters, G. A. 1988. Interaction of carbon metabolism in the *Azolla-Anabaena* symbiosis. *Symbiosis* 6:53–68
43. Kobiler, D., Cohen-Sharon, A., Tel-Or, E. 1981. Recognition between the N_2-fixing *Anabaena* and the water fern *Azolla*. *FEBS Lett.* 133:157–60
44. Konar, R. N., Kapoor, R. K. 1972. Anatomical studies on *Azolla pinnata*. *Phytomorphology* 22:211–23
45. Konar, R. N., Kapoor, R. K. 1974. Embryology of *Azolla pinnata*. *Phytomorphology* 24:228–61
46. Kondorosi, A., Kondorosi, E., Gyorgypal, A., Banfalvi, J., Gyuris, P., et al. 1988. *Rhizobium meliloti nod* and *fix* genes controlling the initiation and development of root nodules. See Ref. 25, pp. 399–403
47. Ladha, J. K., Watanabe, I. 1982. Antigenic similarity among *Anabaena azollae* separated from different species of *Azolla*. *Biochem. Biophys. Res. Commun.* 109:675–82
48. Ladha, J. K., Watanabe, I. 1984. Antigenic analysis of *Anabaena azollae* and the role of lectin in the *Azolla-Anabaena* symbiosis. *New Phytol.* 98:295–300
49. Lang, N. J. 1965. Electron microscopic study of heterocyst development in *Anabaena azollae* Strasburger. *J. Phycol.* 1:127–34
50. Lee, K. Y., Joseph, C. M., Meeks, J. C. 1988. Glutamine synthetase specific activity and protein concentration in symbiotic *Anabaena* associated with *Azolla caroliniana*. *Antonie van Leeuwenhoek J. Microbiol.* 54:345–55
51. Lin, C., Watanabe, I. 1988. A new method for obtaining *Anabaena*-free *Azolla*, *New Phytol.* 108:341–44
52. Lin, C., Watanabe, I., Liu, C. C., Zheng, D.-Y., Tang, L. F. 1988. Reestablishment of symbiosis to *Anabaena*-free *Azolla*. See Ref. 25, pp. 223–27
53. Liu, C. C. 1979. Use of *Azolla* in rice production in China. See Ref. 7, pp. 375–94
54. Lucas, R. C., Duckett, J. G. 1980. A cytological study of the male and female sporocarps of the heterosporous fern *Azolla filiculoides* Lam. *New Phytol.* 85:408–18
55. Lumpkin, T. A. 1987. Environmental

requirements for successful *Azolla* growth. See Ref. 35, pp. 89–97
56. Lumpkin, T. A., Bartholomew, D. P. 1986. Predictive models for growth of eight *Azolla* accessions to climatic variables. *Crop Sci.* 26:197–11
57. Lumpkin, T. A., Plucknett, D. L. 1980. *Azolla*: botany, physiology, and use as a green manure. *Econ. Bot.* 34:111–53
58. Lumpkin, T. A., Plucknett, D. L. 1982. *Azolla as a Green Manure: Use and Management in Crop Production.* Boulder: Westview Press. 230 pp.
59. Martin, A. R. H. 1976. Some structures in *Azolla* megaspores, and an anomolous form. *Rev. Palaeobot. Palynol.* 21:141–69
60. McCowen, S. W., MacArthur, L., Gates, J. E. 1987. *Azolla* fern lectins that specifically recognize endosymbiotic cyanobacteria. *Curr. Microbiol.* 14:329–33
61. Meeks, J. C., Joseph, C. M., Haselkorn, R. 1988. Organization of the *nif* genes in cyanobacteria in symbiotic association with *Azolla* and *Anthoceros*. *Arch. Microbiol.* 150:61–71
62. Meeks, J. C., Steinberg, N. A., Enderlin, C. S., Joseph, C. M., Peters, G. A. 1987. *Azolla-Anabaena* relationship. XIII. Fixation of [^{13}N]N$_2$. *Plant Physiol.* 84:883–86
63. Meeks, J. C., Steinberg, N. A., Joseph, C. M., Enderlin, C. S., Jorgensen, P. A., Peters, G. A. 1985. Assimilation of exogenous and dinitrogen-derived ^{13}NH$_4$$^+$ by *Anabaena azollae* separated from *Azolla caroliniana* Willd. *Arch. Microbiol.* 142:229–33
64. Mellor, R. B., Gadd, G. M., Rowell, P., Stewart, W. D. P. 1981. A phytohaemagglutinin from the *Azolla-Anabaena* symbiosis. *Biochem. Biophys. Res. Commun.* 99:1348–53
65. Millbank, J. W. 1974. Associations with blue-green algae. In *The Biology of Nitrogen Fixation*, ed. A. Quispel, pp. 238–64. New York: American Elsevier
66. Mochida, O., Yoshiyasu, Y., Dimaano, D. 1987. Insect pests of *Azolla* in the Philippines. See Ref. 35, pp. 207–21
67. Moore, A. W. 1969. *Azolla*: biology and agronomic significance. *Bot. Rev.* 35:17–34
68. Neumuller, M., Bergman, B. 1981. The ultrastructure of *Anabaena azollae* in *Azolla pinnata*. *Physiol. Plant.* 51:69–76
69. Newton, J. W. 1976. Photoreproduction of molecular hydrogen by a plant-algal symbiotic system. *Science* 191:559–61
70. Newton, J. W., Herman, A. I. 1979. Isolation of cyanobacteria from

the aquatic fern, *Azolla. Arch. Microbiol.* 120:161–65
71. Nierzwicki-Bauer, S. A., Haselkorn, R. 1986. Difference in mRNA levels in *Anabaena* living freely or in symbiotic association with *Azolla. EMBO J.* 5:29–35
72. Okoronkwa, N., Van Hove, C. 1987. Dynamics of *Azolla-Anabaena* nitrogenase activity in the presence and absence of combined nitrogen. *Microbios* 49:39–45
73. Orr, J., Haselkorn, R. 1982. Regulation of glutamine synthetase activity and synthesis in free-living and symbiotic *Anabaena* spp. *J. Bacteriol.* 152:626–35
74. Pate, J. S., Gunning, B. E. S. 1972. Transfer cells. *Annu. Rev. Plant Physiol.* 23:173–96
75. Perkins, S. K., Peters, G. A., Lumpkin, T. A., Calvert, H. E. 1985. Scanning electron microscopy of perine architecture as a taxonomic tool in the genus *Azolla* Lamarck. *Scanning Electron Microsc.* 4:1719–34
76. Peters, G. A. 1975. The *Azolla-Anabaena azollae* relationship. III. Studies on metabolic capabilities and a further characterization of the symbiont. *Arch. Microbiol.* 103:113–22
77. Peters, G. A. 1976. Studies on the *Azolla-Anabaena azollae* symbiosis. See Ref. 5, pp. 592–10
78. Peters, G. A. 1977. The *Azolla-Anabaena* symbiosis. In *Genetic Engineering and Nitrogen Fixation*, ed. A. Hollaender, pp. 231–58. New York: Plenum
79. Peters, G. A., Calvert, H. E. 1983. The *Azolla-Anabaena azollae* symbiosis. In *Algal Symbiosis*, ed. L. J. Goff, pp. 109–45. New York: Cambridge Univ. Press
80. Peters, G. A., Calvert, H. E., Kaplan, D., Ito, O., Toia, R. E. Jr. 1982. The *Azolla-Anabaena* symbioses: morphology, physiology and use. *Israel J. Bot.* 30:305–23
81. Peters, G. A., Evans, W. R., Toia, R. E. Jr. 1976. *Azolla-Anabaena azollae* relationship. IV. Photosynthetically driven, nitrogenase-catalyzed H$_2$ production. *Plant Physiol.* 58:119–26
82. Peters, G. A., Ito, O., Tyagi, V. V. S., Kaplan, D. 1981. Physiological studies on N$_2$ fixing *Azolla*. In *Genetic Engineering of Symbiotic Nitrogen Fixation and Conservation of Fixed Nitrogen*, ed. J. M. Lyons, R. C. Valentine, D. A. Phillips, D. W. Rains, R. C. Huffaker, pp. 343–62. New York: Plenum
83. Peters, G. A., Kaplan, D., Meeks, J.

C., Buzby, K. M., Marsh, B., Corbin, J. L. 1985. Aspects of nitrogen and carbon interchange in the *Azolla-Anabaena* symbiosis. In *Nitrogen Fixation and CO₂ Metabolism*, ed. P. W. Ludden, J. E. Burris, pp. 213–22. New York: Elsevier

84. Peters, G. A., Mayne, B. C. 1974. The *Azolla, Anabaena azollae* relationship. I. Initial characterization of the association. *Plant Physiol.* 53:813–19

85. Peters, G. A., Mayne, B. C. 1974. The *Azolla, Anabaena azollae* relationship. II. Localization of nitrogenase activity as assayed by acetylene reduction. *Plant Physiol.* 53:820–24

86. Peters, G. A., Ray, T. B., Mayne, B. C., Toia, R. E. Jr. 1980. The *Azolla-Anabaena* association: morphological and physiological studies. In *Nitrogen Fixation*, ed. W. E. Newton, W. H. Orme-Johnson, 2:293–309. Baltimore: University Park Press

87. Peters, G. A., Toia, R. E. Jr., Calvert, H. E., Marsh, B. H. 1986. Lichens to *Gunnera*—with emphasis on *Azolla*. *Plant and Soil* 90:17–34

88. Peters, G. A., Toia, R. E. Jr., Evans, W. R., Crist, D. K., Mayne, B. C., Poole, R. E. 1980. Characterization and comparisons of five N₂-fixing *Azolla-Anabaena* associations. I. Optimization of growth conditions for biomass increase and N content in a controlled environment. *Plant. Cell Environ* 3:261–69

89. Peters, G. A., Toia, R. E. Jr., Lough, S. M. 1977. *Azolla-Anabaena azollae* relationship. V. ¹⁵N₂-Assimilation, acetylene reduction and H₂ production. *Plant Physiol.* 59:1021–25

90. Peters, G. A., Toia, R. E. Jr., Raveed, D., Levine, N. J. 1978. The *Azolla-Anabaena azollae* relationship. VI. Morphological aspects of the association. *New Phytol.* 80:583–93

91. Pieterse, A. H., Delange, L., Van Vliet, J. P. 1977. A comparative study of *Azolla* in the Netherlands. *Acta Bot. Neerl.* 26:433–49

92. Rai, A. N., Rowell, P., Stewart, W. D. P. 1981. Glutamate synthase activity in symbiotic cyanobacteria. *J. Gen. Microbiol.* 126:515–18

93. Ray, T. B., Mayne, B. C., Peters, G. A., Toia, R. E. Jr. 1979. *Azolla-Anabaena* relationship. VIII. Photosynthetic characterization of the association and individual partners. *Plant Physiol.* 64:791–95

94. Ray, T. B., Peters, G. A., Toia, R. E. Jr., Mayne, B. C. 1978. *Azolla-Anabaena* relationship. VII. Distribution

of ammonia assimilating enzymes, protein, and chlorophyll between host and symbiont. *Plant Physiol.* 62:463–67

95. Reddy, P. M., Fisher, R. W. 1988. A new simple method to produce *Anabaena*-free *Azolla* by in vitro fertilization of micromanipulated megasporocarps. *Plant Cell Rep.* 7:430–33

96. Rhodes, D., Sims, A. P., Folkes, B. F. 1980. Pathway of ammonium assimilation in illuminted *Lemna minor*. *Phytochem.* 19:357–65

97. Rippka, R., Deruelles, J., Waterbury, J. B., Herdman, M., Stanier, R. Y. 1979. Generic assignments, strain histories and properties of pure cultures of cyanobacteria. *J. Gen. Microbiol.* 111:1–61

98. Saville, B., Straus, N., Coleman, J. R. 1987. Contiguous organization of nitrogenase genes in a heterocystous cyanobacterium. *Plant Physiol.* 85:26–29

99. Schubert, K. R. 1986. Products of nitrogen fixation in higher plants. *Annu. Rev. Plant Physiol.* 37:539–74

100. Silver, W. W., Shroder, E. C. 1984. *Practical Applications of Azolla for Rice Production*. Developments in Plant and Soil Science, Vol. 13. Dordrecht: Martin Nijhoff/Dr. W Junk. 227 pp.

101. Singh, P. K. 1977. Multiplication and utilization of fern *Azolla* containing nitrogen-fixing algal symbiont as green manure in rice cultivation. *Il Riso* 26:125–37

102. Singh, P. K. 1979. Use of *Azolla* in rice production in India. See Ref. 7, pp. 407–18

103. Stewart, W. D. P. 1980. Systems involving blue-green algae (cyanobacteria). In *Methods for Evaluating Biological Nitrogen Fixation*, ed. F. J. Bergersen, pp. 583–680. New York: Wiley

104. Stewart, W. D. P., Rowell, P., Rai, A. N. 1980. Symbiotic nitrogen-fixing cyanobacteria. In *Nitrogen Fixation*, ed. W. D. P. Stewart, J. R. Gallon, pp. 239–77. New York: Academic

105. Strasburger, E. 1873. Über *Azolla*. Jena: Hermann Dabis Verlag. 86 pp.

106. Subudhi, B. P. R., Watanabe, I. 1981. Differential phosphorus requirements of *Azolla* species and strains in phosphorus-limited continuous culture. *Soil Sci. Plant Nutr.* 27:237–47

107. Talley, S. N., Lim, E. 1984. Planning *Azolla* research for the 1980's. See Ref. 100, pp. 98–112

108. Talley, S. N., Lim, E., Rains, D. W. 1981. Application of *Azolla* in rice production. See Ref. 82, pp. 363–84

109. Talley, S. N., Rains, D. W. 1980. *Azolla filiculoides* Lam. as a fallow-season

green manure for rice in temperate regions. *Agron. J.* 72:11–18
110. Talley, S. N., Talley, B. J., Rains, D. W. 1977. Nitrogen fixation by *Azolla* in rice fields. See Ref. 78, pp. 259–81
111. Tel-Or, E., Sandovsky, T., Kobiler, D., Arad, C., Weinberg, R. 1983. The unique symbiotic properties of *Anabaena* in the water fern *Azolla*. In *Photosynthetic Prokaryotes: Cell Differentiation and Function*, ed. G. C. Papageorgiou, L. P. Packer, pp. 303–14. New York: Elsevier Science
112. Tjepkema, J. D., Schwintzer, C. R., Benson, D. R. 1986. Physiology of actinorhizal nodules. *Annu. Rev. Plant Physiol.* 37:209–32
113. Toia, R. E. Jr., Buzby, K. M., Peters, G. A. 1987. *Azolla mexicana* sporocarps. I. Isolation and purification of megasporocarps and microsporangia. *New Phytol.* 106:271–79
114. Toia, R. E. Jr., Marsh, B. H., Perkins, S. K., McDonald, J. W., Peters, G.A. 1985. Sporopollenin content of the spore apparatus of *Azolla*. *Am. Fern J.* 75:38–43
115. Tyagi, V. V. S., Mayne, B. C., Peters, G. A. 1980. Purification and initial characterization of phycobiliproteins from the endophytic cyanobacterium of *Azolla*. *Arch. Microbiol.* 128:41–44
116. Tyagi, V. V. S., Ray, T. B., Mayne, B. C., Peters, G. A. 1981. The *Azolla-Anabaena azollae* relationship. XI. Phycobiliproteins in the action spectrum for nitrogenase-catalyzed acetylene reduction. *Plant Physiol.* 68:1479–84

117. Uheda, E. 1986. Isolation of empty packets from *Anabaena*-free *Azolla*. *Plant Cell Physiol.* 27:1187–90
118. Uheda, E. 1986. Isolation of hair cells from *Azolla filiculoides* var. *japonicum* leaves. *Plant Cell Physiol.* 27:1255–61
119. Venkataraman, G. S. 1962. Studies on nitrogen fixation by blue-green algae. III. Nitrogen fixation by *Anabaena azollae*. *Indian J. Agric. Sci.* 32:22–24
120. Wallace, W. H., Gates, J. E. 1986. Identification of eubacteria isolated from leaf cavities of four species of the N-fixing *Azolla* fern as *Arthrobacter* Conn and Dimick. *Appl. Environ. Microbiol.* 52:425–29
121. Watanabe, I. 1982. *Azolla-Anabaena* symbiosis—its physiology and use in tropical agriculture. In *Microbiology of Tropical Soils and Plant Productivity*, ed. Y. R. Dommerques, H. G. Diem, pp. 169–85. The Hague: Martinus Nijhoff/Dr. W Junk
122. Watanabe, I. 1987. Summary report of the *Azolla* program on the International Network on Soil Fertility and Fertilizer Evaluation for Rice. See Ref. 35, pp. 197–205
123. Watanabe, I., Berja, N. S., Alimagno, B. V. 1977. Utilization of the *Azolla-Anabaena* complex as a nitrogen source for rice. *Int. Rice Res. Inst. Res. Pap. Ser.* No. 11. 15 pp.
124. Watanabe, I., Berja, N. S., Del Rosario, D. C. 1980. Growth of *Azolla* in paddy field as affected by phosphorus fertilizer. *Soil Sci. Plant Nutr.* 26:301–7

Annu. Rev. Plant Physiol. Plant Mol. Biol. 1989. 40:211–33

REGULATORY INTERACTIONS BETWEEN NUCLEAR AND PLASTID GENOMES

William C. Taylor

CSIRO Division of Plant Industry, GPO Box 1600, Canberra, A. C. T. 2601, Australia

CONTENTS

INTRODUCTION

A variety of plastid types are found in different plant organs, exhibiting a wide range in size, ultrastructure, pigmentation, biochemical activities, and functional roles (42). These structural and functional differences involve significant changes in gene expression in both the plastid and nuclear genomes. Thus, a key component of plastid development is the coordination of gene expression between two different genomes that are separated by several membranes and that are vastly different in complexity, structure, gene

211

1040-2519/89/0601-0211$02.00

organization, and regulatory mechanisms. Although the mechanisms responsible for regulatory communication between the two genomes are unknown, there are just enough hints to provoke speculation. Here I review data on regulatory communication, focusing on the experimental systems that should provide future understanding.

Plastid type is a function of the cell type in which the plastid is found. Several arguments indicate that the developmental status of the plastid is determined by the same developmental program that controls the fate of the cell in which it resides. Genetic information in the nucleus, therefore, ultimately regulates the expression of the plastid genome, as well as the expression of nuclear genes encoding plastid proteins.

The plastid genome is too small to contain more than a few regulatory genes. The entire sequences of two plastid genomes have been determined (71, 88). These sequences are occupied primarily by genes with assigned products. All of the plastid genes identified to date encode either components of plastid transcription or translation, or proteins involved in photosynthesis. The apparent prokaryotic nature of the plastid genome also makes it unlikely that it contains genes that can interpret various eukaryotic types of developmental information such as timing during morphogenesis and cellular position within an organ.

Another argument is that cellular morphogenesis does not depend on plastid function. Mutations that block chloroplast development and function, such as those producing albinos in maize and tomato, have no direct effect on leaf cell development (59, 85). Neither herbicide treatments that prevent chloroplast development (76) nor heat treatments that preferentially block chloroplast protein synthesis (27) affect leaf morphogenesis.

Finally, the fact that many nuclear mutants block chloroplast development has been interpreted as an argument for the regulation of chloroplast development by nuclear genes. However, such mutations may occur in nuclear genes encoding important components in chloroplast biogenesis whose absence blocks further development. As discussed in the next section, the challenge is to find among all of the mutants affecting plastid development those that identify regulatory genes.

NUCLEAR REGULATION OF PLASTID DEVELOPMENT—GENES AND MECHANISMS

Plastids in meristematic cells show no evidence of differentiation and are termed proplastids (42). Most of the proplastids in a leaf meristem develop into chloroplasts. Not all leaf cells, though, contain well-developed chloroplasts. Cells of the epidermal or vascular tissue, for example, contain either poorly developed chloroplasts or small, relatively undifferentiated plastids (42, 106). Significant differences may also be seen in the differentiated state of chloroplasts in various leaf cell types. Examples are guard cell chloroplasts and the dimorphic chloroplasts in C4 plants of the NADP–malic enzyme type. (42,106). The proplastids in the meristems of other organs have other potential developmental fates. For a more extensive discussion of chloroplast development and gene expression, see a recent review by Mullet (63).

The developmental status of the plastid is not fixed, as illustrated by the redifferentiation of mature chloroplasts into chromoplasts in flowers and in the ripening of fruits such as tomato and citrus. In some citrus the yellow or orange rind will regreen under the appropriate conditions, with chromoplasts redifferentiating into chloroplasts (55, 101). The plasticity of plastid differentiation is also evident in tissue-culture manipulations in which the cells of one organ give rise to other organs or completely regenerate new plants. The plastids redifferentiate as the developmental status of their cells changes. Thus, plastid differentiation involves organ-specific, cell-specific, and developmental stage–specific gene regulation.

Nuclear Genes Coding for Plastid Proteins

What we know about the expression of genes coding for plastid proteins provides some clues about regulatory mechanisms controlling plastid development. The mechanisms governing the expression of nuclear and plastid genes are significantly different. The expression of nuclear genes coding for plastid proteins appears to be regulated primarily, but not exclusively, at the transcriptional level (cf 8, 31, 46). The use of *Agrobacterium tumefaciens* to introduce chimeric gene constructions into plants has provided a system of rapid in vitro mutational analysis that identifies DNA sequences responsible for transcriptional regulation and helps dissect regulatory mechanisms. Identification of the proteins that bind these DNA sequences should elucidate at least some of the factors regulating the transcription of a given gene.

Transcription of the genes from two prevalent chloroplast proteins, the light-harvesting chlorophyll *a/b* protein *(Cab)* and the small subunit (SSu) of ribulose 1,5-bisphosphate carboxylase, is regulated by light acting through phytochrome and a second blue-light receptor (32, 103). Additional factors are responsible for organ-specific, cell-specific, and developmental stage–specific modulation of transcriptional activity (2, 45, 47). Cab gene transcription also exhibits a dramatic circadian rhythm (65, 73; W. C. Taylor, in preparation). Cytokinins affect the expression of some Cab and SSu genes (31). The rate of transcription of each Cab or SSu gene in a given cell is thus dependent upon light, possibly phytohormones and the time of day, and information that locates the cell to an organ and a position within that organ. Positional information determines the stage of a cell in the developmental program of the organ and the cell type into which it will differentiate.

A number of chloroplast enzymes belong to families of related isoenzymes that are differentially distributed in the plant. Pea glutamine synthetase, for example, is encoded by four nuclear genes. One gene encodes the chloroplastic form, one a cytosolic form found in many organs, and two genes code for the isoforms in root nodules (19). The gene coding for the chloroplastic form is expressed in leaves; its expression is induced by light acting, at least in part, through phytochrome (102). Thus, differential expression of the members of a nuclear gene family is an important component of the regulation of some genes coding for chloroplast proteins.

Plastid Gene Expression

In contrast, posttranscriptional mechanisms play important roles in regulating plastid gene expression. Transcription rates seem to be primarily determined by relative promoter strengths. Gruissem and colleagues have used in vitro run-on transcription assays in tomato and spinach to measure relative transcription rates (21). They found that while overall plastid transcriptional activity changes significantly during development, the relative amount of transcription from each promoter remains fairly constant. The major changes in the ratios of plastid RNAs that occur during chloroplast development (21, 36) and during the differentiation of chromoplasts from chloroplasts (36, 74) result primarily from changes in RNA stability. Mullet & Klein (64) used a similar in vitro transcription assay in barley and also found a lack of coordination between transcription rates and RNA abundance, demonstrating the role of RNA stability. However, they also found several cases in which light affected the transcription of specific genes.

Ngernprasirtsiri et al (67) found no transcripts for most plastid genes in the amyloplasts of cultured sycamore cells; these genes were methylated in amyloplasts but not in chloroplasts (66). However, Deng & Gruissem (21a) have demonstrated transcriptional activity in spinach root amyloplasts using their run-on assay. Although the overall level of transcription was lower than in chloroplasts, the relative rate for each gene was the same.

Much of the plastid genome consists of multigene transcription units. The primary transcript is processed in multiple steps to give rise to a complex array of RNAs, many of which are translationally active (5, 105). Other than the splicing of transcripts to remove introns, it is not yet clear if any of the other RNA processing steps affect translation rates. A. Barkan (submitted) has found that the ratio of spliced to unspliced forms of two plastid multigene transcripts is much lower in root amyloplasts and early-stage chloroplasts than in mature chloroplasts. Although not all plastid genes have introns, decreased RNA splicing will limit the synthesis of a range of chloroplast proteins, thereby preventing chloroplast differentiation. Klein et al (43) found translational regulation of some plastid transcripts to be a component of light-induced greening in barley. Deng & Gruissem (21a) also found evidence of translational regulation in spinach root amyloplasts.

These data demonstrate that plastid gene expression is regulated at a number of levels. It appears that gene expression is optimized in chloroplasts and generally or selectively repressed in other plastid types, but more information is required about other plastid types before any final conclusions can be reached. An attractive hypothesis is that the products of nuclear genes are responsible for regulating each step in plastid gene expression. Gruissem et al (37) have suggested that the proteins binding to the inverted repeats at the 3' ends of plastid transcripts are responsible for controlling transcript turnover rates and that these regulatory proteins are encoded in the nucleus. A. Barkan (submitted) has suggested that nucleus-encoded proteins may be responsible for regulating splicing activity in the plastid.

A number of nuclear mutations that interrupt specific processes in chloroplast biogenesis have been identified. These include mutations that disrupt the structural organization of thylakoid membranes (51), the assembly of multiprotein complexes (6, 49, 78), plastid transcription (40), and plastid RNA processing and translation (6, 40, 44, 95). The large number of additional uncharacterized mutations indicates that there are many more nucleus-encoded activities that play important roles in chloroplast biogenesis. Some of these nuclear gene products will likely regulate the synthesis and assembly of specific chloroplast components and may be responsible for some of the differences between chloroplast types.

A posttranslational mechanism involved in coordinating the accumulation of chloroplast proteins has been described. Multimeric protein complexes often consist of nucleus- and plastid-encoded subunits. When the synthesis of the plastid-encoded large subunit of ribulose-1,5-bisphosphate carboxylase was inhibited, the rate of synthesis of the nucleus-encoded SSu was not affected. But both subunits accumulated in stoichiometric amounts owing to rapid proteolysis of unassembled excess subunits (81, 92). While this may be a significant quantitative regulatory mechanism in chloroplasts, it is not clear that it plays any role in the development of other plastid types.

Chloroplast Development Mutants

In principle, genetic analysis is a powerful way to study complex biological processes. Mutations identify genes that play important roles in the process of interest and help dissect a complex mechanism into smaller parts. Mutations can also identify specific gene products and determine their functions. The challenge in plastid development is to predict the phenotypes of mutations in regulatory genes. The pigmentation of chloroplasts and chromoplasts makes it easy to find mutants that block their development. But how does one distinguish a regulatory mutant from one that eliminates an essential component and is therefore rate limiting in plastid development?

A mutation in the regulatory circuitry that controls the development of chloroplasts from proplastids should have a number of pleiotropic effects, among them a lack of internal membrane development with a consequent lack of chlorophyll and carotenoid pigments. Albino seedling mutants would seem to be good candidates for such mutations. However, the primary effect of many of these mutants is to block carotenoid accumulation. The block in chloroplast development is a pleiotropic consequence of the extensive photo-oxidative damage that occurs in the absence of photoprotective carotenoids. Chloroplast development can proceed in many carotenoid-deficient mutants when grown in limited light conditions (4).

At present, mutants have led us to several genes that may regulate the development of chloroplasts from proplastids. Two of these are the maize mutants *iojap (ij)* and *chloroplast mutator (cm)*. Both are recessive nuclear mutations that affect the differentiation of leaf plastids and cause heritable changes in the ability of plastids to differentiate in subsequent generations. It is the latter characteristic that distinguishes these two mutants from other white-striped or variegated leaf mutants and argues for the involvement of both genes in the developmental transition of proplastids into chloroplasts. *Iojap* has been more extensively studied. Plants homozygous for *ij* show albino stripes of varying sizes and frequencies in leaf, stem, husk, and tassel tissue. The defect in *iojap* plastids is maternally inherited; a cross of female *ij/ij* plants with a non-*ij* pollen parent yields progeny (*ij/+*) that include white and striped plants (77). The reverse cross yields only green progeny.

The plastids in white sectors have almost no internal membrane structures and look like proplastids at all stages of leaf development (100). *Iojap*-affected plastids have no ribosomes (89), no protein synthetic capacity (104), and were shown to be deficient in two chloroplast multimeric protein complexes, ribulose 1,5-bisphosphate carboxylase and chloroplast ATPase (90). Each complex is composed of polypeptides encoded in both genomes. The DNA of *iojap*-affected plastids appeared unaltered, as judged by restriction-endonuclease analysis (104). Recent work by C.-D. Han and E. H. Coe, Jr. (personal communication) has demonstrated that albino sectors are deficient in a range of chloroplast proteins, both nucleus- and plastid-encoded.

Because white stripes do not correspond across nodes, Walbot & Coe (104) concluded that the *iojap* defect acted not in the apical meristem but in cells committed to organ formation. White sectors in leaves and ears are not clonal, but are positional, meaning that sectors are independent of cell lineage (17). The effect of *iojap* is dependent upon genotypic background, the position of a cell within an organ, and the state of differentiation of that cell. The positional nature of *iojap* demonstrates that the activity of the normal allele is required at all stages in chloroplast development. Even if a primary function of the gene is in the developmental switch of the proplastid-to-chloroplast program, this developmental decision must be reinforced by continuing activity of the *iojap* gene. The variability in the striping pattern in *ij/ij* plants demonstrates the variability in the timing of action and the degree of expression of the *iojap* defect. The striping pattern was found by Coe et al (17) to be dependent upon the nuclear background and to show only limited variability within a given background. The expression of *iojap* is also affected by the *Isr* locus (17). These observations indicate that other nuclear genes can correct or modify the *iojap* defect, in a direct or indirect fashion. *Iojap* is therefore likely to be one of a set of nuclear genes whose concerted action plays a critical role in chloroplast development.

Expression of the *iojap* defect in *ij/ij* plants is reversible. Very rarely green sectors are seen on grainy leaves. These green leaf sectors are clonal, resulting from the reversion of one of the mutant alleles to normal (17). Evidence for the nuclear reversion comes from plants with green revertant sectors in the tassels which shed *Ij* pollen. These observations indicate that the expression of the *iojap* defect in *ij/ij* plants can be corrected by the action of the normal allele of the *iojap* gene. Also *Ij/ij* heterozygotes are completely green.

The maternal inheritance of the *iojap* defect is profoundly different. White seedlings of *Ij ij* genotype derived from *ij/ij* mothers do not show revertant green stripes. The occasional striped progeny show clonal sectoring. Coe et al (17) concluded that *iojap* causes a permanent dysfunction in some plastids during or prior to gametogenesis and that this dysfunction cannot be rescued or modified by the action of the normal allele of *iojap*. White seedlings arise from eggs that contain only affected plastids, and sectoring results from the sorting out during shoot development of a mix of normal and affected plastids. The fact that maternally inherited *iojap*-affected plastids are incapable of responding to the normal developmental cues of the chloroplast program suggests that the gene plays a fundamental role in the developmental switch of proplastid to chloroplast.

Much less is known about *chloroplast mutator*. It is a recessive nuclear mutation that causes leaf stripes of varying sizes and degrees; affected plastids are maternally inherited (93). Leaf sectors in *cm/cm* plants appear to be clonal, as are the sectors in maternally inherited plants. The variation in sector size and frequency suggests that *chloroplast mutator* activity can occur at any time during embryo or leaf development. The *chloroplast mutator* effect on chloroplast ultrastructure and gene expression is much less severe than that of *iojap* (100).

From the available evidence it is difficult to propose a model that explains the actions of either mutant. Walbot & Coe (104) initially proposed that *iojap* prevented plastid ribosome assembly. The absence of plastid protein synthesis would prevent the assembly of multisubunit protein complexes, causing proteolytic degradation of unassembled nucleus-encoded subunits. The absence of plastid protein sythesis might also block the proliferation of internal plastid membrane structures. Although such a model is consistent with many observations on *iojap*, the evidence in favor is not compelling. Some aspects, such as revertability of the phenotype in *ij/ij* plants, but stability in maternally inherited plants, are difficult to explain by the ribosome-defect model.

The positional nature of *iojap* sectors suggests that the defect in *iojap*-affected plastids can be rescued at a number of times in leaf development. The clonal nature of *chloroplast mutator* sectors suggests that when the mutant acts to block further development in the plastids of a cell, then the plastids in the daughters of that cell also fail to develop.

Further research would be greatly aided by the molecular cloning of both genes. The large size of the maize genome and our complete lack of knowledge about either gene product make this task daunting. One possible approach is to clone the genes using the technique of transposon tagging. The first step is to isolate a new mutant allele of the gene that has been generated by the insertion of a known transposable element into the gene. If the transposable element is present in only a few copies in the genome, one can then clone all DNA sequences containing the element. Genetic segregation analysis should determine which cloned DNA sequence is the mutant allele of the gene (see 25). This strategy has been used to clone several of the genes involved in anthocyanin biosynthesis (18, 25, 52, 72, 99, 107), regulation of seed development (60, 61, 82), and chloroplast thylakoid membrane assembly (R. A. Martienssen, A. Barkan, M. Freeling, W. C. Taylor, in preparation). Once the mutant allele of the gene has been cloned, the normal allele can be isolated from a nonmutant plant. The amino acid sequence of the gene product can be determined from the gene sequence, allowing one eventually to identify the protein and determine its functional role in chloroplast development. The evidence strongly suggests that both genes normally operate at a basic level in the proplastid-to-chloroplast transition, making the laborious approach just outlined worthwhile. Whatever the mode of action of both mutants may be, the positional action of one compared with the clonal action of the other should provide valuable insights into determination and differentiation events in plant cell development.

Developmental Regulation—A Proposal

Based on the information above, I propose the following model for chloroplast development (Figure 1). Proplastids in the zygote (Figure 1, A) are at an undifferentiated ground state. Plastid transcription and translation probably occur at a low level to maintain low levels of RNA polymerase and ribosomes in replicating proplastids during embryogenesis. As the embryo organizes into root and shoot axes, (Figure 1, B) the proplastids gain their first developmental information. Positional information acts directly on the nucleus,

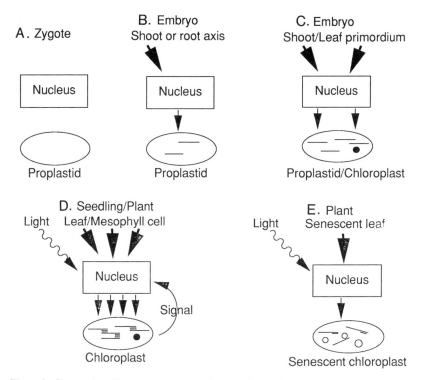

Figure 1 Proposed regulatory communication between the nuclear and plastid genomes during the development and subsequent senescence of a leaf mesophyll cell. Thick arrows represent positional type information received by the cell, thin arrows represent nuclear gene products controlling plastid gene expression or the chloroplast signal controlling nuclear gene expression.

activating the expression of a set of genes whose products play important roles in the shoot apical meristem. The action of one or more of these genes causes a global increase in the transcriptional activity of proplastids in shoot apical cells.

As the apical meristem further organizes and leaf primordia form (Figure 1, C), additional positional information activates additional nuclear genes. The products of this second set of genes are specific for leaf cell development. Some of these gene products further activate proplastid transcription, and most importantly, activate the differentiation of proplastids into chloroplasts. The nucleus-encoded proteins responsible for increased plastid RNA stability, plastid RNA splicing, increased plastid translation, and increased protein stability are among this second set. The same positional information also activates the expression of nuclear genes coding for chloroplast proteins.

Additional cellular organization occurs during the proliferation of each leaf primordium (Figure 1, D). Positional information determines cell fate, which in turn regulates the extent of chloroplast development in each cell. The fine tuning of chloroplast development to give rise to the differences found among mesophyll, epidermal, and stomatal guard cells, for example, may be accomplished by varying the activities of the proteins regulating RNA stability, splicing, or translational activity. In addition, protein stability most likely plays a major role in regulating the extent of chloroplast development. Nuclear genes can control the assmbly of nucleus- and plastid-encoded proteins into multimeric complexes in a number of ways. If the synthesis of thylakoid membranes is limited then the accumulation of thylakoid protein complexes will be limited, and component proteins unable to assemble into membrane complexes will be degraded.

There are significant differences in the expression of nuclear genes in the various cell types in the leaf. The best example is the cell-specific accumulation of nuclear gene transcripts in the mesophyll and bundle sheath cells in C4 plants of the NADP–malic enzyme type, such as maize (13, 54, 84, 87). The positional information that distinguishes a mesophyll cell from a bundle sheath cell therefore acts through different gene regulation mechanisms to cause cell-specific expression of plastid and nuclear genes.

Leaf senescence is accompanied by a shutoff in the expression of plastid genes and nuclear genes coding for chloroplast proteins (Figure 1, E). Although the details of this process have not been determined, we can use the degreening of chloroplasts during fruit ripening as a reasonable analogy. Whether this shutoff is regulated by a new set of nuclear gene products or by a decrease in the gene products responsible for leaf gene expression is not known.

The basic regulatory mechanism involves hierarchies of nuclear gene expression. At the top level is a set of nuclear genes responsible for the proplastid-to-chloroplast developmental switch. These genes interpret positional information, activate the leaf development program, and reinforce decisions during most phases of leaf development. At lower levels in the regulatory hierarchy are nuclear genes that regulate the details of chloroplast development.

The details of this regulatory process are likely to differ among plants. In addition, other organ primordia will have different developmental programs. For example, primordia giving rise to flowers will show the development of proplastids into chromoplasts, sometimes with an intermediate chloroplast stage.

Cytological localization of transcripts in developing maize leaves provides support for positional regulation (47). These data also provide clues about possible interactions among different cell types. The nature of positional information in plants is entirely unknown. Based on the better-understood examples in animal development, one presumes that cells contain greater or lesser concentrations of specific molecules by virtue of their position within an organ or the stage of development of that organ. Furthermore, the *Drosophila* model (86) suggests that there are nuclear genes whose expression is a function of the relative concentration of these molecules. These genes are responsible for interpreting that positional information and reinforcing that interpretation as organ development proceeds.

CHLOROPLAST CONTROL OF NUCLEAR GENE EXPRESSION

Carotenoid Deficiencies Affect Cab Gene Expression

Regulatory information flows in both directions. The state of chloroplast development significantly affects the expression of at least some of the nuclear genes coding for chloroplast proteins. The nuclear mutations *albostrians* and *Saskatoon* in barley block the accumulation of carotenoids and cause complete deficiencies in plastid ribosomes, along with many other plastid components. When Bradbeer et al (12) found both mutants to be deficient in the activities of several nucleus-encoded chloroplast enzymes, they proposed that a plastid-encoded factor, perhaps an RNA, might be required for the cytosolic synthesis of chloroplast proteins. Although there had been arguments for and against chloroplast involvement in developmental regulation (16, 23), this was the first evidence in favor of a chloroplast component of regulation. From their studies of chlorophyll-deficient maize mutants, Harpster et al (38) and Mayfield & Taylor (58) found that the chloroplast affected the level of cytosolic Cab mRNA. Mutants completely deficient in pigmented carotenoids failed to accumulate Cab mRNA (58). Similar results were obtained when the carotenoid deficiency was produced with the herbicide norflurazon (Sandoz 9789) (58, 69).

Carotenoid deficiencies cause a wide range of pleiotropic effects. Chloroplast development is blocked at an early stage, with plastids showing little internal membrane structure (3, 24). Chlorophyll is undetectable (1). Plastid rRNAs (76) and mRNAs (59) are absent, as are all plastid-encoded and most nucleus-encoded chloroplast proteins (28, 38, 39, 57). Peroxisomal enzyme activities are also significantly reduced (26). Despite these major effects on plastids and peroxisomes, other cellular compartments seem unaffected. Most cytosolic proteins (9, 76) and polyadenylated MRNAs (58) are present at normal levels. Leaf morphogenesis proceeds normally (9, 33, 76). Phytochrome is present in normal amounts (59), and many phytochrome-mediated light responses are normal (33, 35, 76).

Light Conditional Photooxidation

Many of these pleiotropic effects are caused by photooxidative damage to the plastid compartment resulting from unquenched triplet chlorophyll and lipid peroxides (29). Pigmented carotenoids normally protect chlorophylls from photooxidation (1). In norflurazon-treated plants and in many albino mutants, chlorophyll is synthesized but owing to rapid photooxidation fails to accumulate. Photooxidation also destroys most other chloroplast components. Because the extent of photooxidative damage depends upon light quantity and quality, many carotenoid deficiencies have light-conditional phenotypes. When grown in either very dim white light (4, 7, 58) or higher-intensity far-red light (69, 76), chloroplast development proceeds in the complete absence of carotenoids. Chloroplasts develop an ultrastructure typical of low light growth (4, 33) and accumulate a normal complement of both nucleus- and plastid-encoded chloroplast proteins (56). Very low levels of chlorophyll are detectable in seedlings grown in dim white light (56), and (what is most important) Cab mRNA accumulates to normal levels (7, 69, 98). The carotenoid deficiency per se has no effect on chloroplast development or gene expression. Rather, photooxidation within the chloroplast is responsible for both the block in chloroplast development and the very low level of Cab mRNA in the cytosol.

A formal proof that photooxidative damage to the chloroplast causes the Cab mRNA deficiency was provided by Burgess & Taylor (15), who made plants doubly deficient in both chlorophyll and carotenoid synthesis. This was accomplished by treating a maize chlorophyll-deficient mutant, *1-Blandy4*, with norflurazon. When grown in dim light these plants had high levels of Cab mRNA, as did norflurazon-treated wild-type plants. Transfer to high light had no effect on Cab mRNA in the norflurazon-treated *1-Blandy4* plants, while norflurazon-treated wild-type plants showed a major decrease in Cab mRNA.

The conditional nature of carotenoid deficiencies has been a useful experimental tool. Light conditions permissive for chloroplast development in maize (59, 98), mustard (69), and barley (7) promoted normal accumulation of Cab and SSu mRNAs. When plants grown in permissive conditions were then transferred to higher-intensity white light, both mRNAs decreased to very low levels (59, 69, 98). The decrease in Cab mRNA was rapid: After 6 hr in high-intensity light about 90% of Cab mRNA had disappeared (59). A chloroplast mRNA, the transcript of the plastid *psaA2* gene, decreased by at least 90% in the first hour of high light (15). The effect of photooxidative damage to the chloroplast is very specific, affecting only chloroplast RNAs and cytosolic mRNAs coding for chloroplast proteins. The mRNA coding for the cytosolic isoform of aldolase is unaffected by photooxidation (15), as is the mRNA coding for phosphoenolpyruvate (PEP) carboxylase (59). Although PEP carboxylase is a key enzyme of C4 photosynthesis, it is located in the cytosol, not the chloroplast.

The decrease in Cab and SSu mRNAs caused by photooxdative damage to the chloroplast could be the consequence of increases in their rates of turnover, inhibition of their syntheses, or a combination of both. Transcription measurements in isolated nuclei demonstrated that photooxidation caused a major decrease in the rate of Cab gene transcription (7, 15, 96). Cab transcription rates were normal in carotenoid-deficient plants grown in very dim light, but they rapidly decreased when plants were transferred to high-intensity light (15). Transcription of the cytosolic aldolase isoform and PEP carboxylase did not change significantly during the course of the experiment. The time course of the decrease in Cab transcription paralleled the decrease in its mRNA level (15). To date, owing to the difficulties in performing pulse labeling measurements in leaves, it has not been determined if photooxidation affects the turnover rate of Cab mRNA. Nevertheless, if we assume a half-life of 1 hr or less for Cab mRNA, then the block in Cab gene transcription could be responsible for the loss of Cab mRNA.

Further evidence for the involvement of photooxidative damage to the chloroplast in blocking transcription comes from experiments with transgenic tobacco plants. The expression of chimeric gene constructions using SSu or Cab promoters was blocked when transgenic plants were treated with norflurazon and grown in normal light conditions (91). However, when these promoters were replaced by the nopaline synthase promoter, expression was unaffected by norflurazon treatment.

It is worth noting that not all albino mutants have light-conditional phenotypes. The maize mutant, *wl-47,* showed no evidence of chloroplast development in limited light (97). The fact that this mutant also showed only low levels of red- or white-light induced Cab or SSu mRNA accumulation, despite normal levels of phytochrome, suggests that a primary defect may be in the light regulation mechanism. The primary defect of *wl-47* may also be to block chloroplast development at a very early stage. The barley mutant, *alb-f17,* used by Batschauer et al (7) in their transcription studies also failed to green in dim white light. These authors concluded that photooxidation does not cause the absence of Cab mRNA in *alb-f17,* given the absence of chlorophyll.

A Chloroplast Signal Regulates Cab Gene Transcription

How does photooxidative damage to the chloroplast affect the transcriptional activity of only a small number of nuclear genes? It is difficult to imagine how a product of photooxidation would have this sort of specificity. An alternative proposal is that photooxidation destroys a factor or signal of chloroplast origin that is a necessary component of optimal Cab gene transcription (7, 10, 15, 59, 68, 69, 91, 98). Although the chloroplast signal has not been identified as a specific molecule, we can make some deductions about its nature. Transferring carotenoid-deficient plants from permissive, dim light conditions to high light allows us to turn off rapidly the chloroplast's capacity for further synthesis of the signal (15, 59). The fact that transcription decreases rapidly tells us that the signal has a short half-life and is continuously necessary for optimal Cab gene transcription. Photooxidation most likely acts only at the site of synthesis of the signal and does not act directly on those signal molecules that are present in the nucleus, actively involved in promoting Cab transcription. Other data, described below, indicate that the signal is not dependent upon photosynthesis or chlorophyll biosynthesis (58), that it originates at an early stage of chloroplast development (53, 68), and that it is most likely not a protein (68).

Oelmuller et al (68) have provided additional support for the chloroplast signal proposal from their work with mustard seedlings. Chloroplast ultrastructure developed in cotyledons between 36 and 120 hr after germination under their growth conditions. Cab and SSu mRNA accumulation occurred during this same period. Chloramphenicol was used to block plastid protein synthesis at various times. Chloramphenicol treatments during the first 36 hr after seed germination blocked both chloroplast development and Cab and SSu mRNA accumulation. However, treatments 48 hr after germination had little effect on Cab or SSu mRNA accumulation. These results strongly support the conclusion that the chloroplast signal is made at an early stage of chloroplast development. They further suggest that the signal is not a plastid-synthesized protein. Inhibition of plastid protein synthesis during the first 36 hr of germination blocked chloroplast development from reaching the critical stage when the chloroplast signal is first synthesized. Once this stage is reached the signal is continuously synthesized even if further chloroplast protein synthesis is blocked. Photooxidative damage to the chloroplast, however, causes an immediate cessation of signal synthesis regardless of the stage of chloroplast development.

Mayfield & Taylor (58) found that Cab mRNA was present in normal amounts in maize mutants blocked in chlorophyll biosynthesis (grown in high light). Chlorophyll-deficient mutants have provided important clues about the chloroplast signal, for they demonstrate it is present in plants incapable of photosynthesis. They further demonstrate that signal synthesis is independent of chlorophyll synthesis or accumulation. Because the maize chlorophyll-deficient mutants used by Mayfield & Taylor (58) affected only the final steps in the chlorophyll biosynthetic pathway, it is still possible that precursors at an earlier step in the pathway are involved (see 41).

The accumulation of Cab and SSu mRNAs occurs in cells with chloroplasts at early stages of development. Martineau & Taylor (53) measured the accumulation of cytosolic and plastid mRNAs as a function of the stage of cellular development in maize leaves. Cab mRNA was present at a low level in the basal 1 cm of leaves. Most chloroplasts in this region are at an early stage of development, exhibiting the initial phases of thylakoid membrane assembly (48).

Does a Chloroplast Signal Regulate Other Nuclear Genes?

Whatever the chloroplast signal may be, its role in Cab gene transcription seems to be universal in higher plants. Photooxidative damage in carotenoid-deficient chloroplasts blocks Cab mRNA accumulation in both monocots (maize; 58; barley; 7) and dicots (mustard, 69; tomato, 34; tobacco, 91; pea, 79; petunia, W. C. Taylor, unpublished data). The cause of the carotenoid deficiency, whether herbicide treatment or a mutation in the nuclear genome, does not seem to affect the Cab mRNA data. The situation with other nucleus-encoded chloroplast proteins is less clear. Ssu mRNA is blocked by photooxidative damage to the chloroplast in maize (59), mustard (69), and tomato (34). Transcription of tomato Ssu genes is blocked by photooxidation (34). Expression of chimeric genes with a pea Ssu promoter in transgenic tobacco is blocked in norflurazon-bleached tissue in normal light (91). However, Ssu mRNA and transcription were unaffected in norflurazon-treated barley and in the carotenoid-deficient mutant *alb-f17* (7). Norflurazon-treated peas showed a slight decrease in Ssu mRNA in white light (79). Although it was difficult to tell if Ssu transcription also decreased in norflurazon-treated pea seedlings, the normal light induction of Ssu transcription was not seen (79). The expression of individual potato Ssu genes was measured in white, fluoridon-treated leaves and found to be only slightly reduced, never more than 50% (P. Schreier, personal communication). Accurate measurements of Ssu transcription are difficult to make in maize because it is difficult to isolate transcriptionally competent nuclei from bundle sheath cells (15).

Several possible explanations may be drawn from these data. The expression of some or all members of the Ssu gene family may be independent of a chloroplast signal in some plants, whereas optimal transcription of all highly expressed Ssu genes may depend upon a chloroplast signal in other plants. Another possibility involves technical differences between laboratories. A disadvantage of using herbicide treatments of seedlings, rather than mutants, to block carotenoid synthesis is that significant levels of carotenoids can accumulate during embryogenesis. This is particularly noticeable in maize, where initiation of the first several leaves occurs during embryogenesis. Even if maize seeds are imbibed in a solution of norflurazon and the seedlings watered with it, the tip of every seedling leaf may be green. While removal of pigmented tips of maize seedling leaves prior to RNA or nuclei isolation is easy, it may be difficult or impossible to distinguish those regions with some carotenoid pigments in the small leaves of some seedlings, especially when grown in dim light or darkness. Any carotenoids present will confer some level of photoprotection and may allow some synthesis of chloroplast signal molecules. If this technical difficulty contributes to any of the differences in Ssu results, it cannot also explain the fact that all experimental systems provide fairly similar results with Cab transcription and mRNA. One must also propose that Cab transcriptional activity is much more sensitive to any change in the signal than is SSu transcription.

Whether a chloroplast signal is also involved in the expression of other nuclear genes coding for chloroplast proteins is not clear. Studies of the effects of norflurazon have noted that the activities of the nucleus-encoded chloroplast proteins nitrate reductase (20) and NADP-glyceraldehyde-3-phosphate dehydrogenase (28, 76) were significantly reduced in herbicide-treated plants grown in white light. Burgess & Taylor (14) measured the levels of seven cytosolic polyadenylated mRNAs coding for chloroplast proteins, in addition to Cab and SSu. All of these mRNAs were found at normal levels in carotenoid-deficient seedlings grown in permissive dim light conditions. Transfer to higher-intensity light for 24 hr caused a major decrease in some, a partial decrease in others, and had little effect on two. In all cases, mRNAs failed to show the normal light-induced increase in the carotenoid-deficient seedlings. One could propose that transcription of all of these mRNAs was blocked by photooxidative damage to the chloroplast and that the differing extents of mRNA decrease reflected different RNA turnover rates. One could also make other proposals. Detailed transcription measurements in isolated nuclei or analyses of gene promoters in carotenoid-deficient transgenic plants will be required before the involvement of a chloroplast signal in gene expression can be demonstrated. The fact that many of the cytosolic mRNAs coding for chloroplast proteins are less abundant than Cab or SSu mRNAs means that in vitro transcription measurements are technically more difficult.

In *Chlamydomonas,* Cab mRNA levels are primarily controlled by light and an intermediate in chlorophyll biosynthesis (41). It is interesting that synthesis of photosystem II polypeptides but not their mRNAs is affected by the stage of chloroplast development (50).

The activity of nitrate reductase, a cytosolic enzyme, was greatly reduced in norflurazon-treated plants (20, 75) and albino mutants (11, 80). When norflurazon-treated mustard cotyledons were allowed to recover from a brief period of photooxidative damage, the recovery of nitrate reductase activity roughly paralleled that of ribulose-1,5-bisphosphate carboxylase (83). On the basis of these observations it has been proposed that the same chloroplast signal that regulates Cab gene expression is also involved in nitrate reductase gene expression (70, 83). However, M. R. Ward, E. H. Coe Jr., and R. C. Huffaker (submitted) have recently found that feeding sugar to albino *iojap* (maternally inherited) seedlings resulted in normal levels of nitrate reductase. While this result suggests that nitrate reductase activity is influenced by carbon metabolism, Finke et al (30) found that inhibition of photosynthesis by herbicide treatment generally did not affect nitrate reductase activity. Whether or not the status of chloroplast development is ultimately shown to affect nitrate reductase activity, the extremely pleiotropic nature of carotenoid deficiencies makes it difficult to conclude that various effects, even similar ones, have the same cause.

The Role of a Chloroplast Signal

Expression of Cab and Ssu genes is always correlated with the presence of chloroplasts (22, 62, 94; 46 and references therein). The obvious role for a chloroplast signal is to provide the nuclear genome with information about the development of the plastids in each cell. This information could be important in the early stages of chloroplast development when high-level transcription is not required until the rapid assembly of thylakoid membranes begins. Low-level Cab transcription seems to be possible in the absence of the signal, suggesting that its primary role is to optimize transcription rates. Light plays a very similar role, making the relationship between the light regulatory mechanism and the chloroplast signal mechanism an important topic for future study. The chloroplast signal may also play an important role in controlling nuclear gene expression when chloroplasts senesce or develop into other plastid types. As senescence or redifferentiation begins, synthesis of the chloroplast signal might cease, causing a major decrease in the transcription of some or all of the nuclear genes coding for chloroplast-specific proteins (Figure 1). When tomato (74) and citrus (S. P. Mayfield, personal communication) fruit degreen, Cab and SSu mRNAs decrease to undetectable levls. When degreened grapefruit was made to regreen, the level of Cab mRNA increased to high levels (S. P. Mayfield, personal communication). In the first part of this review I argued that the developmental program controlling the plastid resides in the nucleus. I therefore suggest that the chloroplast signal does not control the developmental program of nuclear genes but rather provides quantitative modulation of the activity of these genes.

What sort of molecule might the chloroplast signal be? Chloramphenicol treatments of mustard cotyledons suggest that it is not a protein (68). It appears to be a positive regulator, given that Cab is transcriptionally inactive in cells with nongreen plastids or with photooxidatively destroyed chloroplasts. Despite my use of the term synthesis to describe the origin of the signal molecule in the chloroplast, there is no evidence that it is any special type of regulatory molecule. It could easily be a small molecule produced by chloroplast metabolic activity. Such a molecule might bind to a regulatory DNA-binding protein in the nucleus that in turn binds to Cab gene promoters to increase their rates of transcription.

CONCLUDING REMARKS

Recent work in seemingly unrelated areas has provided much needed insight into regulatory communication between the plastid and nuclear genomes. The timing of these reports was perfect for someone attempting to review the area. While there are still few solid answers to long-standing questions about regulatory communication, the new data and new approaches have helped to focus our thinking and speculation.

Perhaps the most important contributions in the long term will be the introduction of new approaches and new technologies. The molecular biological approaches to studies of gene regulation will continue to play fundamental roles in elucidating regulatory communication. Two exciting new approaches will most certainly involve molecular genetics and molecular cell biology. The use of genetic analysis to identify and clone the genes that perform crucial functions in developmental regulation will be a powerful approach, especially when combined with the in vitro genetics of gene-transfer technologies. In situ localization techniques, especialy when combined with appropriate developmental mutants, will provide crucial information about the relationships between cellular and organellar development and gene regulation. It should all make for exciting times ahead.

Exactly ten years ago, Ciferri wrote a short review entitled "The chloroplast DNA mystery" (16). At that time few plastid genes had been identified and little was known about mechanisms of gene expression. Nevertheless, he concluded his review with the still timely comment that "the time is ripe for the search for chloroplast proteins that regulate the activity of chloroplast genes and/or the nuclear genes on which chloroplast 'life' depends."

ACKNOWLEDGMENTS

I wish to thank everyone who provided reprints, preprints, and communications of recent work. I am particularly indebted to Rob Martienssen and Ed Coe for stimulating discussions on *iojap* and *chloroplast mutator*, to David Goodchild for sharing his wisdom on chloroplast development, and to Paul Whitfeld for critical readings of various versions of the manuscript. Work in my laboratory at the University of California at Berkeley was supported by grants from the National Science Foundation, the National Institutes of Health, and the Competitive Grants Program of the US Department of Agriculture.

Literature Cited

1. Anderson, I. C., Robertson, D. S. 1960. Role of carotenoids in protecting chlorophyll from photodestruction. *Plant Physiol.* 35:531–34
2. Aoyagi, K., Kuhlemeier, C., Chua, N.-H. 1988. The pea *rbcS-3A* enhancer-like element directs cell-specific expression in transgenic tobacco. *Mol. Gen. Genet.* 213:179–85
3. Bachmann, M. D., Robertson, D. S., Bowen, C. C., Anderson, I. C. 1967. Chloroplast development in pigment-deficient mutants of maize. *J. Ultrastruct. Res.* 21:41–60
4. Bachmann, M. D., Robertson, D. S., Bowen, C. C., Anderson, I. C. 1973. Chloroplast ultrastructure in pigment-deficient mutants of *Zea mays* under reduced light. *J. Ultrastruct. Res.* 45:384–406
5. Barkan, A. 1988. Proteins encoded by a complex chloroplast transcription unit are each translated from both monocistronic and polycistronic mRNAs. *EMBO J.* 7:2637–44
6. Barkan, A., Miles, D., Taylor, W. C. 1986. Chloroplast gene expression in nuclear, photosynthetic mutants of maize. *EMBO J.* 5:1421–27
7. Batschauer, A., Mosinger, E., Kreuz, K., Dorr, I., Apel, K. 1986. The implication of a plastid-derived factor in the transcriptional control of nuclear genes encoding the light-harvesting chlorophyll *a/b* protein. *Eur. J. Biochem.* 154:625–34
8. Berry, J. O., Carr, J. P., Klessig, D. F. 1988. mRNAs encoding ribulose-1,5-bisphosphate carobxylase remain bound to polysomes but are not translated in amaranth seedlings transferred to darkness. *Proc. Natl. Acad. Sci. USA* 85:4190–94
9. Blume, D. E., McClure, J. 1980. Developmental effects of Sandoz-6707 on activities of enzymes of phenolic and general metabolism in barley shoots grown in the dark or under low or high intensity light. *Plant Physiol.* 65:238–44
10. Borner, T. 1986. Chloroplast control of nuclear gene function. *Endocytobiosis Cell Res.* 3:265–74
11. Borner, T., Mandel, R. R., Schiemann, J. 1986. Nitrate reductase is not accumulated in chloroplast-ribosome-deficient mutants of higher plants. *Planta* 169:202–7
12. Bradbeer, J. W., Atkinson, Y. E., Borner, T., Hagemann, R. 1979. Cytoplasmic synthesis of plastid polypeptides may be controlled by plastid-synthesized RNA. *Nature* 279:816–17
13. Broglie, R., Coruzzi, G., Keith, B., Chua, N.-H. 1984. Molecular biology of C4 photosynthesis in *Zea mays:* differential localization of proteins and mRNAs in the two leaf cell types. *Plant Mol. Biol.* 3:431–44
14. Burgess, D. G., Taylor, W. C. 1987. Chloroplast photooxidation affects the accumulation of cytosolic mRNAs encoding chloroplast proteins in maize. *Planta* 170:520–27
15. Burgess, D. G., Taylor, W. C. 1988. The chloroplast affects the transcription of a nuclear gene family. *Mol. Gen. Genet.* 214:89–96
16. Ciferri, O. 1978. The chloroplast DNA mystery. *Trends Biochem. Sci.* 3:256–58
17. Coe, E. H. Jr., Thompson, D., Walbot, V. 1988. Phenotypes mediated by the *iojap* genotype in maize. *Am. J. Botany* 75:634–44
18. Cone, K. C., Burr, F. A., Burr, B. 1986. Molecular analysis of the maize anthocyanin regulatory locus C1. *Proc. Natl. Acad. Sci. USA* 83:9631–35
19. Coruzzi, G. M., Edwards, J. W., Tingey, S. V., Tsai, F.-Y, Walker, E. L. 1988. Glutamine synthetase: molecular evolution of an eclectic multi-gene family. *UCLA Symp. Mol. Cell. Biol.* (New Ser.) 92: In press
20. Deane-Drummond, C. E., Johnson, C. B. 1980. Absence of nitrate reductase-activity in San-9789 bleached leaves of barley seedings (*Hordeum vulgare* cv. Midas). *Plant Cell Environ.* 3:303–7
21. Deng, X.-W., Gruissem, W. 1987. Control of plastid gene expression during development: the limited role of transcriptional regulation. *Cell* 49:379–87
21a. Deng, X.-W., Gruissem, W. 1988. Constitutive transcription and regulation of gene expression in non-photosynthetic plastids of higher plants. *EMBO J.* In press
22. Eckes, P., Schell, J., Willmitzer, L. 1985. Organ-specific expression of three leaf/stem specific cDNAs from potato is regulated by light and correlated with chloroplast development. *Mol. Gen. Genet.* 199:216–21

23. Ellis, R. J. 1977. Protein synthesis by isolated chloroplasts. *Biochim. Biophys. Acta* 463:185–215

24. Faludi-Daniel, A., Fridvalszky, L., Gyurjan, I. 1968. Pigment composition and plastid structure in leaves of carotenoid mutants of maize. *Planta* 78:184–95

25. Federoff, N. V., Furtek, D., Nelson, O. E. Jr. 1984. Cloning of the *bronze* locus in maize by a simple and generalizable procedure using the transposable controlling element *Activator (Ac)*. *Proc. Natl. Acad. Sci. USA* 81:3825–29

26. Feierabend, J., Kemmerich, P. 1983. Mode of interference of chlorosis-inducing herbicides with peroxisomal enzyme activities. *Physiol. Plant.* 57:346–51

27. Feierabend, J., Schrader-Reichardt, U. 1976. Biochemical differentiation of plastids and other organelles in rye leaves with a high-temperature-induced deficiency of plastid ribosomes. *Planta* 129:133–45

28. Feierabend, J., Schubert, B. 1978. Comparative investigation of the action of several chlorosis-inducing herbicides on the biogenesis of chloroplasts and leaf microbodies. *Plant Physiol.* 61: 1017–22

29. Feierabend, J., Winkelhusener, T. 1982. Nature of photooxidative events in leaves treated with chlorosis-inducing herbicides. *Plant Physiol.* 70:1277–82

30. Finke, A. L., Warner, R. L., Muzik, I. J. 1977. Effect of herbicides on in vivo nitrate and nitrate reduction. *Weed Sci.* 25:18–22

31. Flores, S., Tobin, E. M. 1986. Benzyl-adenine modulation of the expression of two genes for nuclear-encoded chloroplast proteins in *Lemna gibba:* apparent post-transcriptional regulation. *Planta* 168:340–49

32. Fluhr, R., Kuhlemeier, C., Nagy, F., Chua, N.-H. 1986. Organ-specific and light-induced expression of plant genes. *Science* 232:1106–12

33. Frosch, S., Jabben, M., Bergfeld, R., Kleinig, H., Mohr, H. 1979. Inhibition of carotenoid biosynthesis by the herbicide SAN-9789 and its consequences for the action of phytochrome on plastogenesis. *Planta* 145:497–505

34. Giuliano, G., Scolnik, P. A. 1988. Transcription of two photosynthesis-associated nuclear gene families correlates with the presence of chloroplasts in leaves of the variegated tomato *ghost* mutant. *Plant Physiol.* 86:7–9

35. Gorton, H. L., Briggs, W. R. 1980. Phytochrome responses to end-of-day irradiations in light-grown corn grown in the presence and absence of Sandoz-9789. *Plant Physiol.* 66:1024–26

36. Gruissem, W. 1988. Chloroplast RNA: transcription and processing. In *The Biochemistry of Plants: A Comprehensive Treatise,* ed. A. Marcus. New York: Academic. In press

37. Gruissem, W., Barkan, A., Deng, X.-W., Stern, D. 1988. Transcriptional and posttranscriptional control of plastid mRNA levels in higher plants. *Trends Genet.* 4:258–63

38. Harpster, M. H., Mayfield, S. P., Taylor, W. C. 1984. Effects of pigment-deficient mutants on the accumulation of photosynthetic proteins in maize. *Plant Mol. Biol.* 3:59–71

39. Jabben, M., Deitzer, G. F. 1979. Effects of the herbicide Sandoz 9789 on photomorphogenic responses. *Plant Physiol.* 63:481–85

40. Jensen, K. H., Herrin, D. L., Plumley, F. G., Schmidt, G. W. 1986. Biogenesis of photosystem-II complexes: transcriptional, translational, and posttranslational regulation. *J. Cell Biol.* 103:1315–25

41. Johanningmeier, U., Howell, S. H. 1984. Regulation of the light-harvesting chlorophyll-binding protein in *Chlamydomonas reinhardtii*. *J. Biol. Chem.* 259:13541–49

42. Kirk, J. T. O., Tilney-Bassett, R. A. E. 1978. *The Plastids; Their Chemistry, Structure, Growth and Inheritance.* Amsterdam/New York: Elsevier/North Holland Biomedical Press. 650 pp. 2nd ed.

43. Klein, R. R., Mason, H., Mullet, J. E. 1987. Light-regulated translation of chloroplast proteins. Transcripts of psaA-psaB, psbA and rbcL are associated with polysomes in dark-grown and illuminated barley seedlings. *J. Cell Biol.* 106:289–302

44. Kuchka, M. R., Mayfield, S. P., Rochaix, J.-D. 1988. Nuclear mutations specifically affect the synthesis and/or degradation of the chloroplast-encoded D2 polypeptide of photosystem-II in *Chlamydomonas reinhardtii*. *EMBO J.* 7:319–24

45. Kuhlemeier, C., Cuozzo, M., Green, P. J., Goyvaerts, E., Ward, K., Chua, N.-H. 1988. Localization and conditional redundancy of regulatory elements in *rbcS-3A*, a pea gene encoding the small subunit of ribulose-bisphosphate carboxylase. *Proc. Natl. Acad. Sci. USA* 85:4662–66

46. Kuhlemeier, C., Green, P. J., Chua, N.-H. 1987. Regulation of gene expression in higher plants. *Annu. Rev. Plant Physiol.* 38:221–57

47. Langdale, J. A., Rothermel, B. A., Nelson, T. 1988. Cellular pattern of photosynthetic gene expression in developing maize leaves. *Genes Dev.* 2:106–15

48. Leech, R. M., Rumsby, M. G., Thomson, W. W. 1973. Plastid differentiation, acyl lipid, and fatty acid changes in developing green maize leaves. *Plant Physiol.* 52:240–45

49. Leto, K. J., Bell, E., McIntosh, L. 1985. Nuclear mutation leads to an accelerated turnover of chloroplast-encoded 48 kd and 34.5 kd polypeptides in thylakoids lacking photosystem-II. *EMBO J.* 4:1645–53

50. Malnoe, P., Mayfield, S. P., Rochaix, J.-D. 1988. Comparative analysis of the biogenesis of photosystem II in the wild type and y-1 mutant of *Chlamydomonas reinhardtii. J. Cell Biol.* 106:609–16

51. Martienssen, R. A., Barkan, A., Scriven, A., Taylor, W. C. 1987. Identification of a nuclear gene involved in thylakoid structure. *UCLA Symp. Mol. Cell. Biol.* (New Ser.) 63:181–92

52. Martin, C., Carpenter, R., Sommer, H., Saedler, H., Coen, E. S. 1985. Molecular analysis of instability in flower pigmentation of *Antirrhinum-majus* following isolation of the *pallida* locus by transposon tagging. *EMBO J.* 4: 1625–30

53. Martineau, B., Taylor, W. C. 1985. Photosynthetic gene expression and cellular differentiation in developing maize leaves. *Plant Physiol.* 78:399–404

54. Martineau, B., Taylor, W. C. 1986. Cell-specific gene expression in maize determined using cell separation techniques and hybridization *in situ. Plant Physiol.* 82:613–18

55. Mayfield, S. P., Huff, A. 1986. Accumulation of chlorophyll, chloroplastic proteins and thylakoid membranes during reversion of chromoplasts to chloroplasts in *Citrus sinensis* epicarp. *Plant Physiol.* 81:30–35

56. Mayfield, S. P., Nelson, T., Taylor, W. C. 1986. The fate of chloroplast proteins during photooxidation in carotenoid-deficient maize leaves. *Plant Physiol.* 82:760–64

57. Mayfield, S. P., Nelson, T., Taylor, W. C., Malkin, R. 1986. Carotenoid synthesis and pleiotropic effects in carotenoid-deficient seedlings of maize. *Planta* 169:23–32

58. Mayfield, S. P., Taylor, W. C. 1984. Carotenoid-deficient maize seedlings fail to accumulate light-harvesting chlorophyll *a/b* binding protein (LHCP)

mRNA. *Eur. J. Biochem.* 144:79–84

59. Mayfield, S. P., Taylor, W. C. 1987. Chloroplast photooxidation inhibits the expression of a set of nuclear genes. *Mol. Gen. Genet.* 208:309–14

60. McCarty, D. R., Carson, C. 1988. Molecular structure and expression of the viviparous-1 locus in maize. *J. Cell. Biochem.* 12C (Suppl.): 173 (Abstr.)

61. Motto, M., Maddaloni, M., Ponziani, G., Brembilla, M., Marotta, R., et al. 1988. Molecular cloning of the *o2-m5* allele of *Zea mays* using transposon tagging. *Mol. Gen. Genet.* 212:488–94

62. Muller, M., Viro, M., Balke, C., Kloppstech, K. 1980. Polyadenylated mRNA for the light-harvesting chlorophyll a/b protein: its presence in green and absence in chloroplast-free plant cells. *Planta* 148:444–47

63. Mullet, J. E. 1988. Chloroplast development and gene expression. *Annu. Rev. Plant Physiol. Plant Mol. Biol.* 39:475–502

64. Mullet, J. E., Klein, R. R. 1987. Transcription and RNA stability are important determinants of higher plant chloroplast RNA levels. *EMBO J.* 6:1571–79

65. Nagy, F., Kay, S. A., Chua, N.-H. 1988. A circadian clock regulates transcription of the wheat *Cab-1* gene. *Genes Dev.* 2:376–82

66. Ngernprasirtsiri, J., Kobayashi, H., Akazawa, T. 1988. DNA methylation as a mechanism of transcriptional regulation in nonphotosynthetic plastids in plant cells. *Proc. Natl. Acad. Sci. USA* 85:4750–54

67. Ngernprasirtsiri, J., Macherel, D., Kobayashi, H., Akazawa, T. 1988. Expression of amyloplast and chloroplast DNA in suspension-cultured cells of sycamore (*Acer pseudoplatanus* L.). *Plant Physiol.* 86:137–42

68. Oelmuller, R., Levitan, I., Bergfeld, R., Rajasekhar, V. K., Mohr, H. 1986. Expression of nuclear genes as affected by treatments acting on the plastids. *Planta* 168:482–92

69. Oelmuller, R., Mohr, H. 1986. Photo-oxidative destruction of chloroplasts and its consequences for expression of nuclear genes. *Planta* 167:106–13

70. Oelmuller, R., Schuster, C., Mohr, H. 1988. Physiological characterization of a plastidic signal required for nitrate-induced appearance of nitrate and nitrite reductases. *Planta* 174:75–83

71. Ohyama, K., Fukuzama, H., Kohchi, T., Shirai, H., Sano, T. et al. 1986. Chloroplast gene organization deduced

from complete sequence of liverwort *Marchantia-polymorpha* chloroplast DNA. *Nature* 322:572–74

72. O'Reilly, C. N., Shepherd, N. S., Pereira, A., Schwarz-Sommer, Z., Bertram, I., et al. 1985. Molecular cloning of the *al* locus of *Zea mays* using the transposable elements *En* and *Mul*. *EMBO J.* 4:877–82

73. Piechulla, B., Gruissem, W. 1987. Diurnal mRNA fluctuations of nuclear and plastid genes in developing tomato fruits. *EMBO J.* 6:3593–99

74. Piechulla, B., Pichersky, E., Cashmore, A. R., Gruissem, W. 1986. Expression of nuclear and plastid genes for photosynthesis-specific proteins during tomato fruit development and ripening. *Plant Mol. Biol.* 7:367–76

75. Rajasekhar, V. K., Mohr, H. 1986. Appearance of nitrite reductase in cotyledons of the mustard (*Sinapis alba* L.) seedling cotyledons as affected by nitrate, phytochrome and photooxidative damage of plastids. *Planta* 168:369–76

76. Reiss, T., Bergfeld, R., Link, G., Thien, W., Mohr, H. 1983. Photooxidative destruction of chloroplasts and its consequences for cytosolic enzyme levels and plant development. *Planta* 159:518–28

77. Rhoades, M. 1943. Genic induction of an inherited cytoplasmic difference. *Proc. Natl. Acad. Sci. USA* 29:327–29

78. Rochaix, J.-D., Erickson, J. 1988. Function and assembly of photosystem II: genetic and molecular analysis. *Trends Biochem. Sci.* 13:56–59

79. Sagar, A. D., Horwitz, B. A., Elliott, R. C., Thompson, W. F., Briggs, W. R. 1988. Light effects on several chloroplast components in norflurazon-treated pea seedlings. *Plant Physiol.* In press

80. Sawhney, S. K., Prakash V., Naik, M. S. 1972. Nitrate reductase and nitrite reductase activities in induced chlorophyll mutants of barley. *FEBS Lett.* 22:200–2

81. Schmidt, G. W., Mishkind, M. L. 1983. Rapid degradation of unassembled ribulose-1,5-bisphosphate carboxylase small subunits in chloroplasts. *Proc. Natl. Acad. Sci. USA* 80:2632–36

82. Schmidt, R. J., Burr, F. A., Burr, B. 1987. Transposon tagging and molecular analysis of the maize regulatory locus *opaque-2*. *Science* 238:960–63

83. Schuster, C., Oelmuller, R., Bergfeld, R., Mohr, H. 1988. Recovery of plastids from photooxidative damage: significance of a plastidic factor. *Planta* 174: 289–97

84. Schuster, G., Ohad, I., Martineau, B.,

Taylor, W. C. 1985. Differentiation and development of bundle sheath and mesophyll thylakoids in maize: thylakoid polypeptide composition, phosphorylation and organization of photosystem-II. *J. Biol. Chem.* 260:11866–73

85. Scolnik, P. A., Hinton, P., Greenblatt, I. M., Giuliano, G., Delanoy, M. R., et al. 1987. Somatic instability of carotenoid biosynthesis in the tomato *ghost* mutant and its effect on plastid development. *Planta* 171:11–18

86. Scott, M. P., Carroll, S. B. 1987. The segmentation and homeotic gene network in early *Drosophila* development. *Cell* 51:689–98

87. Sheen, J.-Y., Bogorad, L. 1986. Differential expression of six light-harvesting chlorophyll *a/b* binding protein genes in maize leaf cell types. *Proc. Natl. Acad. Sci. USA* 83:7811–15

88. Shinozaki, K., Ohme, M., Tanaka, M., Wakasugi, T., Hayashida, N., et al. 1986. The complete nucleotide sequence of the tobacco chloroplast genome: its gene organization and expression. *EMBO J.* 52043–49

89. Shumway, L. K., Weier, T. E. 1967. The chloroplast structure of *iojap* maize. *Am. J. Bot.* 54:773–80.

90. Siemenroth, A., Borner, T., Metzger, U. 1980. Biochemical studies on the *iojap* mutant of maize. *Plant Physiol.* 65:1108–10

91. Simpson, J., van Montague, M., Herrera-Estrella, L. 1986. Photosynthesis-associated gene families: differences in response to tissue-specific and environmental factors. *Science* 233:34–38

92. Spreitzer, R. J., Goldschmidt-Clermont, M., Rahire, M., Rochaix, J.-D. 1985. Nonsense mutations in the *Chlamydomonas* chloroplast gene that codes for the large subunit of ribulose bisphosphate carboxylase/oxygenase. *Proc. Natl. Acad. Sci. USA* 82:5460–64

93. Stroup, D. 1970. Genic induction and maternal transmission of variegation in *Zea mays*. *J. Hered.* 61:131–41

94. Sugita, M., Gruissem, W. 1987. Developmental, organ-specific, and light-dependent expression of the tomato ribulose-1,5-bisphosphate carboxylase small subunit gene family. *Proc. Natl. Acad. Sci. USA* 84:7104–8

95. Taylor, W. C., Barkan, A., Martienssen, R. A. 1988. Use of nuclear mutants in the analysis of chloroplast development. *Dev. Genet.* 8:305–20

96. Taylor, W. C., Burgess, D. G., Mayfield, S. P. 1986. The use of carotenoid deficiencies to study nuclear-chloroplast

regulatory interactions. In *Current Topics in Plant Biochemistry and Physiology*, ed. D. D. Randall, C. D. Miles, C. J. Nelson, D. G. Blevins, J. A. Miernyk, 5:117–27. Columbia: Univ. Missouri Press

97. Taylor, W. C., Mayfield, S. P. 1985. Ontogenetically regulated photosynthetic genes. In *Molecular Biology of the Photosynthetic Apparatus*, ed. K. Steinback, S. Bonitz, C. J. Arntzen, L. Bogorad, pp. 413–16. New York: Cold Spring Harbor Lab.

98. Taylor, W. C., Mayfield, S. P., Martineau, B. 1984. The role of chloroplast development in nuclear gene expression. *UCLA Symp. Mol. Cell. Biol.* (New Ser.) 19:601–10

99. Theres, B. H., Scheele, T., Starlinger, P. 1987. Cloning of the *Bz2* locus of *Zea mays* using the transposable element *Ds* as a gene tag. *Mol. Gen. Genet.* 209: 193–97

100. Thompson, D., Walbot, V., Coe, E. H. Jr. 1983. Plastid development in *iojap*- and chloroplast mutator-affected maize plants. *Am. J. Bot.* 70:940–50

101. Thomson, W. W., Lewis, L. N., Coggins, C. W. 1967. The reversion of chromoplasts to chloroplasts in valencia oranges. *Cytologia* 32:117–24

102. Tingey, S. V., Tsai, F.-Y., Edwards, J. W., Walker, E. L., Coruzzi, G. M. 1988. Chloroplast and cytosolic glutamine synthetase are encoded by homologous nuclear genes which are expressed *in vivo*. *J. Biol. Chem.* 263:9651–57

103. Tobin, E. M., Silverthorne, J. 1986. Light regulation of gene expression in higher plants. *Annu. Rev. Plant Physiol.* 36:569–93

104. Walbot, V., Coe, E. H. Jr. 1979. Nuclear gene *iojap* conditions a programmed change to ribosomeless plastids in *Zea mays*. *Proc. Natl. Acad. Sci. USA* 76:2760–64

105. Westhoff, P., Herrmann, R. G. 1988. Complex mRNA maturation in chloroplasts: the *psbB* operon from spinach. *Eur. J. Biochem.* 171:551–64

106. Whatley, J. M. 1979. Plastid development in the primary leaf of *Phaseolus vulgaris:* variations between different types of cell. *New Phytol.* 82:1–10

107. Wienand, U., Weydemann, U., Neisbach-Kloesgen, U., Peterson, P. A., Saedler, H. 1986. Molecular cloning of the *c2* locus of *Zea mays*, the gene coding for chalcone synthase. *Mol. Gen. Genet.* 203:202–7

Annu. Rev. Plant Physiol. Plant Mol. Biol. 1989. 40:235–69

DO POLYAMINES HAVE ROLES IN PLANT DEVELOPMENT?

Phillip T. Evans and Russell L. Malmberg

Botany Department, University of Georgia, Athens, Georgia 30602

CONTENTS

INTRODUCTION[1]

How Might Polyamines Be Involved in Plant Development?

Polyamines have been implicated in an overwhelming array of plant growth and developmental processes. It is nearly impossible, however, to say ex-

---

[1]Abbreviations used: ORNdc, ornithine decarboxylase; ARGdc, arginine decarboxylase; SAMdc, s-adenosylmethionine decarboxylase; SAM, s-adenosylmethionine; HPLC, high performance liquid chromatography; TLC, thin layer chromatography; DFMO, difluoromethylornithine; DFMA, difluoromethylarginine; MGBG, methylglyoxal-bis(guanylhydrazone); CHA, cyclohexylamine; ACC, aminocyclopropylcarboxylic acid

1040-2519/89/0601-0235$02.00

plicitly how polyamines function in higher-plant physiology. This dilemma is not unique to plants. Polyamines are found in all organisms, in a wide variety of capacities. They bind to nucleic acids (53) for example, play an essential role in the growth of thermophilic bacteria (111), and are involved in teratogenic mammalian cell cultures (71); but their precise niche in any organism cannot be described simply. They are essential for cell viability, and are correlated with, or required for, a variety of physiological events. In plants, polyamines exist in both free and bound forms (Figure 1). The bound polyamines are conjugated to various phenolic secondary metabolites; these conjugates may have at least as significant a function in plant development as do the free polyamines.

Some authors have postulated that polyamines, and related compounds, are a type of plant growth regulator or hormonal second messenger (40, 62, 64). Also possible are less direct roles. For example, because polyamines are important in cell division (73), they might influence patterns of cell division and the form of the plant. Of course, polyamines might also have superficially interesting developmental correlations and distributions, but only as a late consequence of some primary event, rather than as a causal agent.

The definition of plant hormones, the distinctions of this definition from that of mammalian hormones, and the general nature of plant growth substances have produced some likely theoretical discussions recently (40, 157, 158). A plant hormone should have a significant physiological or developmental effect, and should be active at relatively low concentrations. The extent to which hormones produce concentration-dependent effects is still under discussion. A phytohormone may be synthesized in one location and then transported to another for its function; it may also be active in the location of its synthesis.

How do polyamines fit these criteria? Generally, polyamines are found in plant cells at levels significantly higher than those of the plant hormones. The concentrations necessary for biological effects are often on the order of millimolar, rather than the micromolar levels typical of the traditionally accepted hormones. The question of whether or not polyamines are translocated in the plant is still unsettled, although they are clearly taken up by suspension cultures (115, 116). Davies (40) and Galston & Kaur-Sawhney (64) have argued that polyamines have clear physiological and developmental effects on plants and therefore should be regarded as members of a more loosely defined category, that of plant growth regulators, rather than as hormones per se.

Here we review evidence for polyamine involvement in various growth and developmental phases: cell division, embryogenesis, rooting, flowering, and pollen tube growth. We cover various postulated links between polyamines and plant growth regulators. The interactions of polyamines with

COMMON POLYAMINES

PUTRESCINE $\overset{+}{H_3}N\text{-}CH_2\text{-}CH_2\text{-}CH_2\text{-}CH_2\text{-}\overset{+}{N}H_3$

SPERMIDINE $\overset{+}{H_3}N\text{-}CH_2\text{-}CH_2\text{-}CH_2\text{-}CH_2\text{-}\overset{+}{N}H_2\text{-}CH_2\text{-}CH_2\text{-}CH_2\text{-}\overset{+}{N}H_3$

SPERMINE $\overset{+}{H_3}N\text{-}CH_2\text{-}CH_2\text{-}CH_2\text{-}\overset{+}{H_2}N\text{-}CH_2\text{-}CH_2\text{-}CH_2\text{-}CH_2\text{-}\overset{+}{N}H_2\text{-}CH_2\text{-}CH_2\text{-}CH_2\text{-}\overset{+}{N}H_3$

CADAVERINE $\overset{+}{H_3}N\text{-}CH_2\text{-}CH_2\text{-}CH_2\text{-}CH_2\text{-}CH_2\text{-}\overset{+}{N}H_3$

COMMON POLYAMINE-CONJUGATES

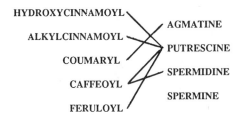

HYDROXYCINNAMOYL

ALKYLCINNAMOYL

COUMARYL

CAFFEOYL

FERULOYL

AGMATINE

PUTRESCINE

SPERMIDINE

SPERMINE

COMMON POLYAMINE SYNTHESIS INHIBITORS

DFMO -- DIFLUOROMETHYLORNITHINE
    inhibits Ornithine decarboxylase, specific, enzyme activated

DFMA -- DIFLUOROMETHYLARGININE
    inhibits Arginine decarboxylase, specific, enzyme activated

MGBG -- METHYLGLYOXAL-BIS(GUANYLHYDRAZONE)
    inhibits SAM decarboxylase, not completely specific, competitive

CHA -- CYCLOHEXYLAMINE
    inhibits spermidine synthase

Figure 1 Polyamines, polyamine-conjugates, and polyamine synthesis inhibitors. Other polyamines, conjugates, and inhibitors can be found, or are available; however, the ones shown are the ones considered or used in most plant polyamine research.

senescence and with ethylene are given their own section because of the abundance of work on these topics. We do not review aspects of polyamine metabolism except where they touch on polyamine functions in development (see 130, 136 for more biochemically oriented reviews). Recent research on polyamines and plant defense also falls outside the scope of this chapter (see 24, 162, 167 for examples).

Overview of Polyamine Biochemistry

The term polyamine has been used in the literature in both a generic sense to include putrescine, spermidine, spermine, other amines, and various derivatives) and a more restricted sense (to mean only those primary amines that have more than two amine groups, such as spermidine and spermine). Putrescine, spermidine, and spermine (diamine, triamine, and tetraamine, respectively) are illustrated in Figure 1. A variety of other primary amine compounds have been found in plants, including cadaverine. Particular attention should be paid the group of compounds referred to as either polyamine-conjugates or hydroxycinnamic acid amides (depending on whether one's first interest is polyamines or phenolics). In some plants, the levels of polyamines in these bound forms will greatly exceed that in the free form. Some of these compounds are diagrammed in Figure 1. The free polyamines are positively charged at intracellular pH. Some of their possible biological functions may be explained by their role as organic carriers of positive charges, and hence their role in many physiological processes. Effects of polyamines based on their charge have been noted in binding both to nucleic acids (53) and to the phospholipids of the plasma membrane (27, 68, 141).

The pathways of polyamine metabolism are shown in Figure 2. Figure 2a shows polyamine synthesis from ornithine and arginine via ornithine decarboxylase (ORNdc) and arginine decarboxylase (ARGdc); not shown is the enzyme lysine decarboxylase, which synthesizes cadaverine from lysine. The dualism of pathways to putrescine synthesis is notable, since fungi and animals get along with just the ornithine decarboxylase pathway. The two paths reportedly have different tissue distributions (reviewed in 136), and different regulation (75). ARGdc has been linked to stress responses (56, 172, 174), while ORNdc is primarily linked to the cell cycle and rapid cell division (34, 72, 73). Repression of the synthetic enzymes by polyamines has been reported in tobacco cell cultures (75). ARGdc and s-adenosylmethionine decarboxylase (SAMdc) were fully repressed by spermidine and spermine, whereas ORNdc is, at best, only partially repressed. Srivenugopal & Adiga (143) have demonstrated in *Lathyrus* that ornithine transcarbamylase, carbamate kinase, putrescine transcarbamylase, and agmatine iminohydrolase are part of a single multi-functional polypeptide, given the name putrescine synthase.

POLYAMINE SYNTHESIS FROM ORNITHINE AND ARGININE

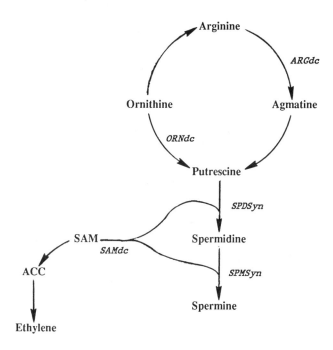

POLYAMINE OXIDASES AND DEGRADATION

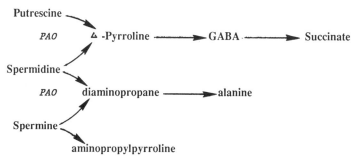

Figure 2 Pathways of polyamine synthesis and degradation. Abbreviations used are: ARGdc—arginine decarboxylase; ORNdc—ornithine decarboxylase; SPDSyn—spermidine synthase; SAM—s-adenosylmethionine; SAMdc—SAM decarboxylase; ACC—aminocyclopropane-1-carboxylic acid; SPMSyn—spermine synthase; PAO—polyamine oxidase; GABA—γ-aminobutyric acid.

Polyamine degradation in plants is less well studied (Figure 2b). Polyamine oxidases have been characterized in some cereals and legumes (83, 144, 147). In tobacco, an unusual putrescine-utilizing cell line has been found with a form of polyamine degradation not found in normal cells (54). The biochemistry of polyamine-phenolic conjugate synthesis and degradation has not yet received the attention it deserves, although a few reports have appeared (13, 15, 17, 19, 21, 137).

Figure 2 also shows the potential involvement of polyamines, through cellular pools of s-adenosylmethionine (SAM), in ethylene biosynthesis. Whereas most cellular metabolism uses SAM as a methyl donor, polyamine synthesis and ethylene biosynthesis make use of the propylamine moiety of SAM. The common use of the propylamine group has led to the speculation that there may be regulatory interconnections between polyamines and ethylene, a topic we explore below.

Finally, Figure 1 shows one of the unique tools available to polyamine investigators—the group of specific or potent inhibitors available for nearly every enzyme in the synthesis pathway. Most biochemical and physiological studies of polyamines in plants and animals have depended on the availability of these inhibitors. The organic chemists who synthesized them deserve a great deal of credit, as does Merrell-Dow Pharmaceuticals for its willingness to share some of these compounds with academic researchers (25, 102). These inhibitors are invaluable, but caution is needed in interpreting data obtained using them. In particular, not all studies have demonstrated that a particular biological effect caused by the inhibitor is reversible by adding the inhibitor plus the appropriate polyamine. Without this control, assigning the biological effect to the polyamine is questionable.

Several HPLC and TLC methods (55, 131) are available for assaying polyamine levels in plant tissues. One indirect means of determining polyamine-conjugate levels is to acid-hydrolyze a given extract and then measure the increase in free polyamines resulting from release of polyamines from the bound conjugates. This provides information on the polyamine half of the polyamine-conjugate. Assaying the phenolic half requires a different set of chromatographic techniques (70, 148). Most polyamine investigators routinely assay ORNdc, ARGdc, and SAMdc enzyme activities, because they represent three branches of the synthesis pathway and because the decarboxylase assay method is relatively easy. Birecka et al (20, 22) have noted that a number of artifacts are possible in measuring enzyme-specific activities in crude plant extracts. They have recommended modifying the enzyme assay in several ways to insure that the decarboxylation seen is really enzyme specific. Their conclusions call into question some results that are based solely on measurement of enzyme activity in crude plant extracts. Smith & Marshall (138) have recently concurred, noting additional potential pitfalls with the use of standard ORNdc, ARGdc, and SAMdc enzyme assays in plants.

POLYAMINES IN DEVELOPMENTAL PROCESSES

Cell Division

Numerous investigations have correlated an increase in polyamine levels with cell division and a drop in polyamines during any subsequent lessening of metabolic activity. Heimer & Mizrahi (73), for example, demonstrated a significant level of ORNdc activity correlated with cell division frequency in tobacco suspension cultures and tomato ovaries. Berlin & Forche (16) reported that low doses of DFMO (an ORNdc inhibitor) caused tobacco cells in culture to stop dividing while cell enlargement continued. The observed cell enlargement without division suggests that DFMO blocks one stage in the cell cycle. Walker et al (161) observed not only a rapid accumulation of polyamines correlated with rapid cell division in *Acer saccharum* seedlings, but also an inhibition of cell division and a drop in polyamine titers with addition of D-arginine, D-arginine + DFMO, MGBG (a SAMdc inhibitor), or CHA (a spermidine synthase inhibitor). Cell elongation was unaffected. In a recent study of maize roots (123), Schwartz and coworkers found a high spermine content in primary root apexes and in decapitated roots as laterals formed, using labeled DFMO and thymidine. They also showed that ORNdc was localized primarily in the meristematic zones (123).

In oat protoplasts the application of the polyamines cadaverine, spermidine, and spermine stimulated both DNA synthesis and a limited amount of mitotic activity (63, 82). More recently, it has been shown that exogenous applications of arginine (1–10 mM) or putrescine, spermidine or spermine (0.1–1 mM) can induce cell division in almond protoplasts on a basal medium. Putrescine was the most effective, but no clear dose dependency was observed for any of the amines (170).

Felix & Harr (51) surveyed the polyamine contents in different organs of seedlings of a number of plant species. Their general conclusions were that although increases in the polyamine contents of cotyledons and endosperms do occur upon germination, increases are frequently not seen in those organs exhibiting rapid cell division. This conclusion is at odds with most other reports.

Embryogenesis

Several studies have investigated the role of polyamines and particularly ARGdc in carrot somatic embryogenesis systems. Montague et al (105) were the first to demonstrate that there was a significant rise in ARGdc activity and in putrescine pools when carrot cultures were shifted from callus medium to embryogenesis medium. Feirer et al (50) found that the inhibitor DFMA (an ARGdc inhibitor) would block the transition from disorganized growth into somatic embryogenesis, and that the addition of putrescine with DFMA would restore the embryogenic potential. DFMA did not block growth of the carrot cells on callusing medium, so the putrescine requirement, demonstrated by DFMA inhibition, seems to be unique to the transition to embryogenic growth. Fienberg et al (52) examined polyamine synthesis in a carrot cell mutant that would not go through embryogenesis; the cell line W001C has high internal levels of auxin that presumably inhibit this transition. When placed in medium without auxin, the mutant line also failed to show the typical increase in polyamine content and in ARGdc and SAMdc activities that wild-type cells display.

A significant genotype difference in polyamine requirement has been noted in a study of somatic embryogenesis in two alfalfa lines (100). Both embryogenic genotypes showed putrescine accumulation during induction on medium with auxin and both exhibited sharp decreases in putrescine concentration upon transfer to the differentiation medium without auxin. The polyamine inhibitors CHA, MGBG, DFMO, and DFMA all reduced polyamine contents in both genotypes tested; however, embryogenesis was only reduced in one of the two genotypes. This result implies that the requirement for polyamines in initiation of somatic embryogenesis may not be universal.

Roots

The possible role of polyamines in root formation and growth has been investigated in several plant systems, most frequently in *Phaseolus* and in *Vigna*. In *Phaseolus,* Jarvis et al (78) found evidence that polyamines are not only correlated with root initiation and early growth, but may be essential for these processes. Treatment of cuttings with indolebutyric acid increased levels of putrescine, spermidine, and spermine in hypocotyls prior to root primordia development, and exogenous application of spermine increased the number of roots. Application of the SAMdc inhibitor of MGBG reduced endogenous levels of spermine and spermidine, raised putrescine levels, inhibited the indolebutyric acid–induced rise in spermine and spermidine, and inhibited root induction and root growth in the presence or absence of indolebutyric acid. Palavan-Unsal (112) determined through the use of canavanine that ARGdc was the major branch of putrescine synthesis involved in root growth in *Phaseolus,* although both ORNdc and ARGdc activities were high in the root apex. Inhibition of ARGdc reduced polyamine titers and growth, both of which could be partially reversed upon addition of putrescine. Applications of α-methyl-ornithine slightly reduced root length and inhibited ORNdc activity. Recently Kakkar & Rai (79) corroborated earlier findings that spermine is associated with enhanced rooting and found that spermidine in combination with indoleacetic acid increased carbohydrate content. It may be significant that the above studies have found spermine effective in enhancing rooting and that Dumortier et al (47) found spermine in maize roots only in the apical region.

In mung beans, Chatterjee et al (28) followed root and nodule growth and found a pattern of polyamine titers similar to patterns of estimates of RNA, DNA, and protein contents; ARGdc activity roughly matched the polyamine pattern. Shyr & Kao (126) found similarly enhanced rooting with ornithine, as well as spermine and spermidine; MGBG inhibited rooting to all but a small extent. Friedman et al (58) failed to find any stimulation of rooting in *Vigna* hypocotyls but did measure an increase in polyamines after indolebutyric acid–induced rooting. Additional studies by Friedman et al (59) with ^{14}C-ornithine and ^{14}C-arginine implied that both pathway branches contributed to the IBA-stimulated increase of polyamines.

Working with apple, Wang & Faust (164) noted a considerable increase in polyamine levels accompanying the induction and growth of roots. But, in contrast to the reports on *Phaseolus,* use of the inhibitor DFMO was more effective than DFMA in reducing fresh weight increase, indicating a major role for ORNdc rather than ARGdc. Tiburcio et al (149), observing root organogenesis in tobacco callus cultures, found root production to be inversely related to putrescine and alkaloid titers. The addition of DFMA or D-arginine decreased the putrescine and alkaloid titers and promoted rooting, implying the involvement of the ARGdc branch of polyamine synthesis in production of putrescine and putrescine-derived alkaloids and an influence of polyamines over in vitro organogenesis. Chriqui et al (33) have provided evidence that there exists a synergistic effect between auxins and ornithine in rhizogenesis in *Datura innoxia* leaf explants. Although addition of polyamines to auxin-containing medium did not enhance rhizogenesis, exogenous putrescine may have promoted growth of the roots (data not shown). The authors suggest that ORNdc may be important in de-differentiation of the leaf explants, a process presumably requiring cell divisions, while ARGdc may support redifferentiation events.

Floral Initiation

Polyamines and/or polyamine-conjugates have been analyzed with respect to floral initiation in intact plants and in in vitro systems. Most studies on initiation in intact plants have been correlations of developmental age with the appearance of a particular class of compounds. Some comparisons between plants grown under different physiological conditions, and of variant plants with wild-type have also been made.

Cabanne et al (26) have studied the accumulation of various hydroxycinnamic acid amides (or polyamine-conjugates, primarily mono- and di-caffeoylputrescine) in *Nicotiana tabacum* cv. Xanthi plants. They found accumulation of the conjugates in apical shoots and leaves of plants that had been grown at 20°C, where they would flower, as well as at 30°C, where they would not. The appearance of the conjugates occurred late in development at both temperatures; however, there clearly was no correlation with flowering per se. On the basis of topping and leaf growth analysis, the authors proposed, however, that the conjugates might be related to the appearance of ripening to flower. Presumably the plants at 30° were ripened to flower but were re-pressed from expressing it.

Dumas and colleagues (45, 46) have examined the levels of polyamine-conjugates in shoot apices of the *Nicotiana* stock RMB7, which is not capable of flowering. This hybrid plant arose from a program of interspecific crossing of *Nicotiana rustica* to *Nicotiana tabacum* cv. Maryland Mammoth, which was subsequently backcrossed to Maryland Mammoth 7 times. Comparisons of RMB7 with its ancestral lines and with cv. Xanthi showed that it has reduced conjugate levels. However, in this sort of interspecific hybrid, it is very difficult to decide what the proper control comparison is; RMB7 is not isogenic with any cultivar. Changes in polyamine or polyamine-conjugate levels might be the result of a genetic factor that is independent of the flowering effect.

Because of the difficulties in doing appropriate experiments with intact plants, several investigators have turned to analyses of floral initiation and development in organogenic cultures. The tobacco thin-layer method of Tran Thanh Van (154) has been used to study the effects of polyamines and their inhibitors on floral initiation in vitro; the system has been reviewed by Tran Thanh Van (155). Briefly, thin surface strips of tissue from internodes in the tobacco inflorescence are placed on a defined medium, and then cultured. Depending upon the hormones, pH, and special additives used, de novo generation of flowers, callus, roots, or vegetative shoots have all been reported from these explants. As all those who use this system note, the thin layers are taken from tissue already committed to flowering (127). Thus the surprising capabilities reported are the generation of roots and vegetative buds from the explants, and the apparent loss of the floral commitment on certain media.

Torrigiani et al (152) measured free and bound polyamines at different times during the culture of thin layers taken from floral-determined tissues and from vegetative stem tissues. The rates of appearance of putrescine, spermidine, and spermine were slightly different between the vegetative and floral bud–forming tissues; but both showed significant increases (10–20-fold) first of spermidine, then of putrescine levels. The polyamine-conjugates, both soluble and insoluble, were found at roughly 10-fold higher levels than the free polyamines, and the levels of the putrescine-conjugates increased significantly during the course of explant development. These results show a correlation of polyamines and conjugates with bud formation but no differences between floral and vegetative programs. These results imply that polyamines may be correlated with bud formation in general, as opposed to floral bud formation in particular.

Kaur-Sawhney et al (87) have reported more dramatic results in switching developmental programs. They used two basic media for culturing floral stem epidermal peels. One led to development of floral buds ("floral medium"), the other to development of vegetative buds ("vegetative medium"); the underlying media difference was a 10-fold higher level of cytokinin in the "vegetative medium." They found that spermidine levels were 4.5-fold higher in explants cultured on floral bud medium than in those cultured on vegetative bud medium. This result led them to test the effects of exogenous spermidine on the buds formed on the vegetative medium and the effects of CHA on the buds formed on the floral medium. Spermidine in the range 0.5–5 mM decreased the number of vegetative buds on vegetative medium by 70% and caused the appearance of a number of floral buds that had not been seen previously on the vegetative medium. CHA at 10–20 mM caused both a 75% reduction in the amount of spermidine in the explants and a significant shift from floral buds to vegetative buds, on what was otherwise the floral medium. The simultaneous addition of 10 mM CHA and 1 mM spermidine provided a partial reversal of the CHA-induced shift, suggesting that the depletion of spermidine really was an important underlying cause of the shift.

The floral thin-layer system has some remarkable properties for the study of in vitro organogenesis, but it is difficult to know precisely how to evaluate the results, even when they seem clear cut. The cyclohexylamine-spermidine-induced shift of Kaur-Sawhney et al (87) could be interpreted as spermidine masking the effects of the high cytokinin in the vegetative medium, or it could imply that spermidine had some completely different role in promoting floral initiation, ideas that were pointed out by the authors. The levels of auxin and cytokinin that affected the floral-vegetative shift in this system were 1–10 μM whereas the levels of polyamines and polyamine inhibitors with similar effects were 1–20 mM. This difference in required exogenous concentration could be related to effects such as poor polyamine transport (172) or large initial pools of polyamines; or it could indicate that the polyamines are less likely than the cytokinins to be natural regulators.

A further complication is that a number of different variables have now been reported that cause shifts in the organogenic program, including pH, cytokinin amount and type (155), and oligosaccharins (156), as well as polyamines. One has the sense that either the system is so plastic that a variety of signals can induce a change, or that the real underlying variable has not yet been found. The implication is either that in the intact plant a variety of signals can be alternatively associated with the floral/vegetative switch, or that the in vitro and in vivo systems respond differently. An additional difficulty is that researchers using the system are not always consistent in how a vegetative shoot is defined. Since any shoot will flower eventually, the operational difference between a regenerated floral bud and vegetative shoot may be only a very small number of nodes. Without a clear definition of a vegetative shoot, based on a significant developmental assay, it may be that the various "vegetative" media are actually only delaying flowering for a short time, as opposed to permitting a change of the tissue to a true vegetative mode of growth.

It seems best to interpret the results from this system cautiously at the moment, although they represent the best evidence to date on the involvement of polyamines in floral initiation. It would be valuable to have comparable information from similar studies done on vegetative bud formation in various in vitro systems, for example from leaf discs.

Floral Development

Initial reports from Heimer and colleagues (72, 73) demonstrated high ORNdc levels specifically in the developing ovaries of the tomato flower. Subsequently, they have demonstrated that feeding tomato ovaries with the inhibitor DFMO will block their development, whereas DFMO plus putrescine allows the ovaries to grow normally (34). Slocum & Galston (128) have characterized the enzyme activities, polyamines, soluble polyamine-conjugates, and insoluble polyamine-conjugates in developing tobacco ovaries. ORNdc-specific activity rose about 3-fold during the course of ovary development and fruit set; this increase was correlated with a doubling in the free putrescine titer but with no significant change in the spermidine and spermine titers. At its peak, the ORNdc activity was 140-fold that of ARGdc. More than 90% of the total content for all three polyamines was found in the bound conjugated form, probably as caffeoyl derivatives. This result reemphasizes the need to consider the interplay of the free and bound forms of the polyamines. Subsequently, Slocum & Galston (129) demonstrated that DFMO would interfere with tobacco ovary development as well; an apparent effect of DFMA on ovary development resulted from cellular conversion of DFMA to DFMO by arginase, after which it could inhibit the ORNdc. A requirement for ARGdc, not ORNdc, has been suggested in avocado fruits (169).

Martin-Tanguy and colleagues (99) have examined the polyamine-conjugates (hydroxycinnamic acid amides) in cytoplasmic male sterile lines of *Zea mays*. These compounds were found to be absent in anthers from plants with the Texas male sterile cytoplasm, and present in plants that also contained the appropriate nuclear restorer gene. Analysis of the postfertilization events of cob and grain development and maturation showed no changes in the hydroxycinnamic acid content in male sterile and male fertile lines. These changes were developmentally restricted to the affected anthers.

Gerats et al (65) began a study with *Petunia* mutants that had been identified as having alterations in floral morphology. Polyamine, polyamine-conjugate, and enzyme activities were measured to see if any changes could be found. A polymorphism was observed in comparing two wild-type lines of *Petunia*, with high and low putrescine levels segregating as a simply inherited trait. This result implied that not all changes in polyamine levels or ratios have to be associated with floral morphology changes. However, one out of the four floral morphology mutations screened showed significantly higher levels of putrescine and ARGdc in flowers and older leaves but not younger leaves. The putrescine elevation co-segregated with the floral morphology change in both of the two wild-type genetic backgrounds. The phenotype of the lesion, *alf*, is quite chaotic, with floral parts turning into other floral parts on a frequent and irregular basis; the vegetative parts formed later in development are also abnormal, with smaller leaves and prolific branching. This result implies that some significant changes in polyamine synthesis may be found by screening plants selected for their unusual floral phenotype.

Our laboratory has used tobacco cell cultures to select variants resistant to two inhibitors of polyamine synthesis, MGBG and DFMO (48, 74, 94–98). Plants have been regenerated from 14 of these lines, while a number of lines have regenerated only into green coral-like calli. Among the 14 regenerated lines, 2 were extreme dwarves and did not flower (*ts4* and *Dfrl*); each of these two lines has multiple pleiotropic effects on the polyamine pathway including low levels of ORNdc (74, 96). A revertant of *ts4*, *Rt1*, showed higher levels of ORNdc, more normal internodes, and flowered with petaloid anthers. Two more lines (*Mgr15* and *Mgr21*) were extreme dwarves and did flower with an aberration in developmental timing referred to as puzzle-box (97). Their pistils were incomplete, forming hollow cylinders, with additional whorls of stamen and carpel primordia inside the hollow pistil cylinder. The polyamine and polyamine-conjugate levels of *Mgr21* are all low, ranging from 25% to 50% of wild type, while the corresponding enzyme activities are nearly normal; the basis for the MGBG resistance phenotype is unclear. Ten plants have been regenerated to flower, without showing the extreme dwarfism trait, although these can have partially shortened internodes (97). The floral phenotypes seen included petaloid anthers, stigmoid anthers, extra petals, and green spaghetti (a transformation of ovules into long slender growths). The biochemical basis for some of these lines has been at least partially worked out: *Mgr12* has an altered SAMdc that is kinetically resistant to the inhibitor MGBG, and it has shrunken anthers in the flower. *Mgr3*, a green spaghetti mutant, has elevated spermidine and spermidine-conjugate levels (95, 98).

Of the cell culture polyamine variants that have been regenerated to flowering plants, 100% show aberrations in floral development. Generally these aberrations are so severe that the plants are sterile, and no genetics can be done. Two lines, *Mgr3* and *Mgr12*, have been partially analyzed genetically, on the basis of a very small number of seeds (4–16 seeds total per cross) obtained in crosses with wild-type pollen. Both of these apparently are nuclear dominant mutations with the floral phenotypes co-segregating with the polyamine phenotype (96, 98). The extent to which somaclonal variation, cell culture artifacts, and multiple mutations arising from mutagenesis contribute to the floral morphology aberrations is unclear in the absence of better genetic analysis. The situation is frustrating, because the 100% correlation observed to date suggests an involvement of polyamines in floral development; however, without further genetic tests, the results are inconclusive.

Although each of the studies discussed has its flaws, the collective evidence implies that polyamines (or their conjugates) significantly affect floral development. Polyamines may be required during crucial differentiation steps, or they may be part of the hormonal regulation of sex development in plants. An alternative explanation is that similar perturbation of many other metabolic pathways would also produce systematic floral aberrations.

Fruit Development

Several lines of evidence have implicated a role for polyamines in fruit development. In apple, Biasi et al (18) observed high free and bound polyamine levels during the early periods of fruit growth, especially bound spermine. In developing "Murcott" manderin fruit, correlations were found between growth rate and levels of ARGdc, ORNdc, putrescine, and spermidine (109). Correlations have also been reported between rates of cell division and putrescine and spermidine levels in avocado pulp (3, 169) and between onset of rapid cell division and ORNdc activities in postfertilization tobacco ovary tissues (128).

Following a report of increased ovary fresh weight in tomato upon putrescine application (34), Costa & Bagni (37) showed in apples that spraying polyamines at mM concentrations on flowers nine days after full bloom increased both fruit set and yield, apparently by increasing fruit growth rate during the stage of rapid cell division. Flower bud formation was also increased. A more recent study of exogenous putrescine applications on apple has reported increased fruit set with one cultivar on one rootstock but no effect on another rootstock or with two other cultivars (160). Increased fruit set has also been reported in olives following application of putrescine at high concentration during flowering (122).

Pollen

The possible role of polyamines in pollen development has seen very limited attention. Using radioactively labeled arginine, Bagni et al (7) demonstrated that synthesis of polyamines precedes the emergence of apple pollen tubes. Speranza et al (140) found little effect of exogenous polyamine on apple pollen germination, although under Ca^{2+} deficiency exogenous spermine did have a promotive effect. Prakash et al (117) studying in vitro pollen tube growth in *Catharanthus* demonstrated a promotive effective of spermidine on pollen germination at 0.01 mM and inhibitory effects at 0.1 mM. The presence of MGBG at 0.5 mM, without spermidine addition, reduced germination percentage, and 1.5 mM totally inhibited germination. Apparently no attempt was made to apply MGBG and spermidine simultaneously.

SENESCENCE AND ETHYLENE

Senescence

Because of the potential metabolic connection between polyamines and ethylene through the propylamine group of SAM, a great deal of attention has been given to the involvement of polyamines in fruit development and senescence. Two sorts of claims have been made for polyamines. One is simply that exogenous application will retard senescence. The second is that the lowering of the polyamine concentration is an important early step in triggering senescence. Also bearing on these two hypotheses is the question of whether exogenous application of polyamines causes novel physiological effects or performs the normal functions of the endogenous pools.

Galston and colleagues (2, 63, 80, 84–86, 125) have reported an anti-senescence effect of polyamines on excised oat leaves incubated in the dark. A short exposure of the excised leaves to a buffer containing a polyamine significantly retarded senescence as measured by retention of chlorophyll and inhibition of RNAse and protease activities. Similar observations have now been made on detached leaves and cell cultures of a variety of monocot and dicot species (84, 107, 121, 142). The effects of the exogenous polyamine applications are thus similar to those of exogenous cytokinin, although cytokinins are typically applied at 0.1 mM and polyamines are applied at roughly 10 mM.

Cheng & Kao (30) noted in soybean leaves that the effects of polyamines on retarding senescence were localized to the cut edges and near the large veins, suggesting that polyamines are not transported well in these leaf cells. Cheng et al (31) have studied senescence of detached rice leaves and have suggested the involvement of diaminopropane in the results obtained with exogenously applied spermidine and spermine. All three appeared to retard senescence in the dark and all promoted chlorophyll degradation in the light; however, the presence of β-hydroxyethylhydrazine, an inhibitor of conversion of polyamines to diaminopropane by polyamine oxidase, reversed the effects of spermine and spermidine in the light, indicating a possible requirement for conversion to diaminopropane. Ca^{2+} was shown to inhibit the observed effects competitively, leading the authors to suggest that an initial attachment to a membrane site may be required (31).

The mechanism of inhibiting senescence by exogenous polyamines may be related to their possible inhibition of ethylene synthesis (4, 5) and to stabilization of membranes (2, 68). An observation that supports an involvement of membrane stabilization is that the addition of Ca^{2+} ions will counteract the ability of polyamines to stabilize chlorophyll levels in senescing leaves (84, 141). Fuhrer et al (61) found in peeled oat leaves that exogenous polyamines repressed ethylene synthesis, particularly at the conversion of ACC to ethylene; they suggested that exogenous polyamines initially attach to membranes and then inhibit ethylene production and retard senescence. In agreement with this, Drolet et al (44) have found significant free radical scavenging by polyamines correlated with the number of amino groups; they suggested this might be part of the membrane stabilization and senescence retardation observed in various systems. They also noted that the conversion of ACC to ethylene is superoxide dependent, and that this reaction was inhibited by polyamines. There is thus a reasonable working model for the inhibition of senescence by exogenous polyamines, involving binding to membranes, prevention of lipid peroxidation, and quenching of the free radicals needed for the ACC-to-ethylene conversion.

Roberts et al (119) presented evidence with microsomal membranes from *Phaseolus* that exogenously applied polyamines associate with membrane lipids and substantially reduce membrane fluidity. The authors caution that some physiological effects attributed to exogenously added polyamines may reflect membrane rigidification rather than true physiological responses; they advise that such experiments be paralleled by use of the polyamine synthesis inhibitors. They also suggest that at physiological concentrations the polyamines might selectively rigidify membrane surfaces and play a role in maintenance of membrane integrity. However, many of their experiments involve 50-mM concentrations of the polyamines, much higher than commonly reported endogenous levels, and administration of putrescine at 1 mM did not cause a significant effect, although that of spermidine and spermine did. Agazio et al (1) found that polyamines could inhibit washing-stimulated K^+ influx and H^+ extrusion without interfering with K^+ uptake and H^+ extrusion stimulated by fusicoccin. They argue that this differential effect is evidence that polyamine effects on membranes are specific at physiological concentrations.

A distinctly different mechanistic hypothesis is that the senescence signal itself acts by decreasing polyamines as one of the critical control steps. Evidence countering the role of polyamines in control of senescence was provided by Smith & Davies (132), who worked with a photoperiod-inducible senescence in the apical bud of peas, a property of two dominant alleles, *Sn* and *Hr*. They found during bud senescence that the amounts of polyamines per organ declined with the decreased size of the bud, but there was no decrease of polyamines prior to the appearance of early symptoms of senescence. They concluded that polyamines were not part of the chain of events leading from the photoperiodic signal to the initiation of senescence.

An important question is thus whether or not exogenous polyamines affect senescence the same way the internal pools of polyamines do. Birecka et al (23) compared chlorophyll and protein degradation of *Avena, Nicotiana,* and *Heliotropium* detached leaves in the dark. They found that older *Heliotropium* leaves with very low polyamine levels exhibited only a weak senescence syndrome, whereas the leaves of the other two species had high levels of polyamines and exhibited pronounced senescence with no significant polyamine decline. These observations led them to suggest that endogenous polyamines may have effects different from the possibly nonspecific effects frequently reported for exogenous polyamines. In contrast, Dibble et al (41) have measured polyamine levels in a landrace of tomato (alcobaca) that ripens more slowly than other tomatoes and has better storage characteristics. This landrace could be compared to a mutation isolated from alcobaca that had lost the storage and ripening characteristics (in a sense a revertant); the mutant thus had the developmental timing seen in most other tomato cultivars. The alcobaca fruit showed a clear increase in putrescine content, especially at later stages of development, compared to normal tomato and to the revertant of alcobaca. These genetic data thus provide significant additional evidence that polyamine levels are inversely correlated with senescence and that large endogenous pools of polyamines may have effects similar to those of exogenous polyamine applications.

Several counterexamples have been reported of systems where polyamines did not seem to retard senescence. Downs & Lovell (43) found that addition of putrescine and spermidine, most significant at 10 mM, to culture solutions of cut carnations could actually result in greater ethylene production and reduced bloom longevity. A previous paper (120) had reported that endogenous putrescine levels rose during senescence of carnations without apparent inhibition of ethylene, but that inhibition of ethylene with aminooxyacetic acid resulted in increases in spermidine in petal tissues. This increase in spermidine would be predicted if there were channeling of SAM to the polyamine pathway because of blockage in the ethylene pathway (49). Srivastava (142) compared endogenous polyamine levels in light- and dark-grown barley seedlings and senescing leaves. Levels were comparable under all conditions with the possible exception of a decline in spermidine with senescence, a decline suppressed by the addition of kinetin. Furthermore, exogenous polyamines did not retard senescence in the dark, and spermine and spermidine increased it in the light.

Although there are a few counterexamples, exogenous polyamines retard senescence in many species and experimental systems. The data of Dibble et al (41) suggest that endogenous and exogenous polyamines are having similar physiological functions during senescence. The mechanism of senescence inhibition is probably related to effects on the membranes, and possibly to inhibition of the ACC-to-ethylene transition by quenching of free radicals. It seems unlikely, however, that a lowering of polyamine levels is actually a significant early part of the response pathway for a senescence signal.

Ethylene

Inhibitors of both polyamine and ethylene synthesis make it possible to probe the interaction of polyamines with ethylene in a variety of experimental procedures. The most common hypothesis tested is that polyamines and ethylene may regulate each other's synthesis, either directly or by metabolic competition for SAM. In addition, Miyazaki & Yang (104) have recently pointed out that polyamines and ethylene biosynthesis must both allow for recycling of methylthioadenosine and that this process may be as significant as the more frequently studied fate of the propylamine group from SAM.

Polyamines inhibit ethylene formation in a number of plant tissues, including apple fruits, bean and tobacco leaf explants (4), *Tradescantia* petals, and mung bean hypocotyls (146). Apelbaum et al (5) have measured ARGdc activity in the apical meristem of peas. In response to ethylene, the levels of ARGdc activity decreased by 90% within 18 hr; ethylene also increased the K_m and decreased the V_{max} of ARGdc. Reducing endogenous ethylene with hyperbaric pressure or treatments with silver thiosulfate and 2,5-norbornadiene led to 30–50% increases in ARGdc activity. Icekson et al (77) similarly showed that ethylene reduced SAMdc activity in pea apexes. Icekson et al (76) subsequently reported in the same experimental system that lysine decarboxylase activity increased and cadaverine levels rose in response to ethylene administration. They speculated that this might be a compensation for the decrease in ARGdc.

In peeled oat leaves, spermidine and diaminopropane were shown to reduce ethylene levels both by inhibition of ACC synthase and by conversion of ACC to ethylene (61). A reduction in chlorophyll loss also occurred that was not caused by lowered ethylene levels. Since Ca^{2+} addition could competitively reduce these polyamine effects it was suggested that a membrane attachment by the polyamines was responsible for the effects. However using apple disks and correlating microsomal membrane microviscosity with ethylene production Ben-Avie et al (14) showed the inhibitory effects of Ca^{2+} and spermine to be temperature dependent and to have very different curves. They concluded that the modes or sites of action of the Ca^{2+} and spermine may differ.

A recent report (88) dealing with induced ethylene production in suspension cultures of pear fruit cells indicated that spermidine at 1 mM had to be applied prior to ethylene induction in order to have an effect on ethylene appearance; the application also resulted in reduced polysome numbers. Putrescine and spermine had similar effects. These results were interpreted to mean that the suppression of ethylene prodution may have been caused by suppression of macromolecular synthesis with subsequent lower levels of ACC synthase and ACC.

The use of the polyamine synthesis inhibitors on cut carnations increased their rate of ethylene production and senescence, while the application of an ACC synthase inhibitor decreased ethylene production and increased polyamine levels (120). Even-Chen et al (49) used 3,4-^{14}C-methionine to trace carbon flow into either ethylene or polyamines in aged orange peel discs. A variety of treatments were applied to inhibit ethylene synthesis, including phosphate/Co^{2+}, aminoethoxyvinylglycine, and exogenous putrescine. These treatments resulted in a 3–4-fold increase in the specific label of spermidine, suggesting that the two pathways are connected by sharing a precursor pool of SAM. Unlabeled spermidine inhibited transfer of label to ACC but did not stimulate label incorporation into spermidine; the spermidine inhibition of label transfer to ACC was partially overcome by the addition of Ca^{2+}, a result consistent with earlier senescence studies (61, 84).

Two alternative forms of interaction between polyamines and ethylene biosynthesis have been proposed recently. Drolet et al (44) demonstrated that spermine, spermidine, putrescine, and cadaverine could effectively act as scavengers of free radicals, although at the relatively high concentrations of 10 and 50 mM, and could inhibit the superoxide-dependent conversion of ACC to ethylene. Winer & Apelbaum (169) suggested that polyamines may inhibit ACC synthase by forming a Schiff base with its cofactor, pyridoxal phosphate. No evidence for this proposal has yet appeared.

Counterexamples exist in which ethylene and polyamines are not mutually antagonistic. A study of chilling effects on cucumber seedlings has shown that prevention of ACC induction by aminooxyacetic acid does not raise polyamine levels and that increases in spermidine and ACC are stimulated simultaneously by the chilling treatment. These observations imply that regulation may not occur at the level of SAM competition (163). Similar situations in which the formation of ACC and the formation of polyamine increase simultaneously occur in apple and cherry buds (165, 166). Kushad et al (90) measured polyamines and ethylene during avocado fruit development and found that polyamines peaked earlier than ethylene. They suggested there was no competition between the two pathways, although their results did not exclude the possibility of metabolic competition during the peak phase of polyamine synthesis.

Cohen & Kende (36) have examined the polyamine synthetic enzymes in deep-water rice. Submergence or treatment with gibberellic acid or ethylene stimulates increased cell division and elongation in the intercalary meristems of these plants. They found a 2-fold increase in ARGdc activity and an 8-fold increase in SAMdc activity peaking at 8 hr after submergence; putrescine levels rose 4-fold and spermidine levels rose 2-fold. Treatment of isolated stem sections in air with either ethylene or GA also resulted in 2–4-fold increases in ARGdc and SAMdc activities, although no change was detected in polyamine levels. Neither treatment affected ORNdc. These results thus contrast with those showing ethylene and polyamines to be mutually inhibitory or antagonistic. In this system, ethylene stimulates both growth of the intercalary meristem and ARGdc and SAMdc activities.

In some experimental systems polyamines and ethylene clearly interact through the intermediate SAM; in others the evidence contradicts this metabolic competition. The link of polyamines to rapid cell divisions seems always to be maintained. That the connection between ethylene and polyamines may be species- and circumstance-dependent suggests that low polyamine levels are not an integral part of ethylene induction or ethylene response. In some species and some tissues the pool of SAM may be limiting, and in these cases there may be a competition. A careful analysis of the metabolic fluxes in systems is needed, comparing systems with and without a clearly defined ethylene/polyamine competition.

LINKS TO GROWTH REGULATORS AND ENVIRONMENTAL STIMULI

Some investigators have suggested that polyamines are part of the signal-response pathway for various plant hormones. In one of the first reports, Bagni (6) demonstrated polyamines could help break dormancy and stimulate cell proliferation in tuber slices of *Helianthus tuberosus,* normally a hormone-dependent process. Since then, polyamines have been implicated in the response of plants to ethylene, to gibberellic acid, and to cytokinins. They may also be regulated by phytochrome. Experiments have been performed (*a*) to test if polyamines can substitute for a defined hormone in a bioassay system, (*b*) to examine the time course of response of polyamines to growth regulators, and (*c*) to analyze polyamine levels in hormone-related mutants.

Regulation of Xylogenesis and Division

The original observations of Bagni (6) on *Helianthus tuberosus* have been followed up by Bagni and colleagues (10, 12, 124). A rapid increase and peak of polyamine titer occurred 24 hr after excision from the tuber tissue; 5 mM DFMO or a combination of 1 mM each of DFMO and DFMA could suppress the cell division response; and they observed some tracheary differentiation in their cultures. Polyamines were substituted for hormones in the media of some cultures. Spermidine and spermine at 0.1 mM produced some of the effects, but more weakly, of 35 μM auxin. D'Orazi & Bagni (42) found apparent high activity of ORNdc in rapidly growing portions of *Helianthus* shoots and tubers; however, strong oxidizing conditions in their extracts may have caused artifactual activity.

Torrigiani et al (153), studying the first synchronous cell cycle induced by auxin in *Helianthus* tuber slices, found that ORNdc, ARGdc, and SAMdc activities as well as polyamine titer increased before and during the S phase, declining during cell division. This finding agrees with other published papers indicating a rise in polyamine content and synthesis before and during nuclear DNA synthesis in plants (34, 124).

Phillips et al (114) used a *Helianthus tuberosus* tuber slice system optimized to give a high frequency of tracheary element differentiation in culture after the initial spurt of cell division. Rather than putrescine, spermidine, and spermine as reported by Serafini-Fracassini et al (124), Phillips et al (114) noted the presence of spermidine, diaminopropane, and cadaverine during the initial 24-hr activation and onset of mitosis. Having tested for variation due to media, the authors attributed these differences to possible differences in cultivar or environment during tuberization. They also examined the effects of MGBG on spermidine levels and tracheary element formation. MGBG addition at 1 mM or 2.5 mM did not alter division rates but did depress xylem differentiation. In the absence of auxin, spermidine at 1 mM did promote cell division to a small extent, but this stimulation was absent in the presence of auxin. These authors failed to find an effect of exogenous polyamines on tracheary element differentiation, either when the polyamines were added alone or when they replaced either the auxin or cytokinin in the medium. Thus both groups of researchers working with *Helianthus tubers* found a correlation of polyamine levels with early rapid cell division and also presented evidence that polyamines could induce a limited pattern of cell division. Phillips et al (114), however, found no evidence that exogenous polyamines could replace auxin or cytokinins in their defined system for tracheary element differentiation.

In a second paper on *Helianthus,* Phillips et al (113) found that the inhibitors DFMO and DFMA did not significantly reduce cell division rates; however, DFMO substantially reduced xylem differentiation while stimulating spermidine accumulation. Exogenous spermidine inhibited xylogenesis much more than cell division when applied at 1 mM or higher concentrations, suggesting an inverse correlation between spermidine accumulation and cytodifferentiation.

Gibberellic Acid

Several laboratories have analyzed gibberellic acid dwarf mutants of *Pisum sativum* (peas) in order to examine the possible role of polyamines in hormone response. Dai et al (39) looked at a dwarf pea after treatment with GA to induce internode elongation. They found that the activity of ARGdc and the levels of putrescine and spermidine all rose in parallel with the promotion of internode elongation by GA. Further, the use of a partial GA antagonist prevented internode elongation and also prevented the rise in ARGdc and polyamine levels. The content of polyamines per gram of material remained constant, but the content per stem section rose after internode elongation. Kaur-Sawhney et al (81) found in 9-day-old pea seedlings that treatment with gibberellin induced a peak of ARGdc at 9 hr, whereas the peak of internode elongation occurred at 12 hr; DFMA treatment with the gibberellin inhibited 70% of the internode elongation, but simultaneous addition of polyamines with the DFMA did not restore the growth. Sufficient polyamines may be a requirement for the induced internode growth, although the lack of reversal of the DFMA inhibition requires further explanation.

Smith et al (133) carried this type of analysis further by extending the measurements of polyamine levels to mutants with four different internode phenotypes *(slender, tall, dwarf, nana)*. The quantity of polyamines measured correlated directly with amount of internode elongation. DFMO and DFMA, when applied exogenously, inhibited both internode elongation and the increase in polyamine titer; the simultaneous application of putrescine and agmatine partially restored elongation. The authors suggested that polyamines might be important in the portion of the GA response that results from cell division but not in the portion resulting from cell elongation.

Kyriakidis (91) reported that gibberellic acid stimulated ORNdc activity in germinating barley seeds. Lin (93) has carefully investigated the polyamine role in the classic barley seed aleurone layer system, where gibberellic acid induces α-amylase activity after an 8-hr lag phase. He found no change in the polyamine levels in the 8 hr before α-amylase appearance. Gibberellic acid did not affect the pattern of label incorporation into polyamines, nor did it affect the uptake or turnover of labeled putrescine. However, the SAMdc inhibitor MGBG reduced GA induction of α-amylase activity by 25–50%, and the combination of MGBG plus spermidine partially restored α-amylase levels. Thus gibberellic acid may not act through polyamines; instead, sufficient spermidine may be required for the α-amylase induction.

Other Growth Regulators

The results of the in vitro floral initiation studies of Kaur-Sawhney et al (87) as well as the anti-senescent properties of polyamines could be interpreted as evidence that polyamines have some relationship to cytokinins. There is no body of evidence for cytokinins comparable to the ethylene-polyamine data, although a few reports show an increase of polyamine synthesis by cytokinins in lettuce (32) and cucumbers (145). The abnormal floral morphology phenotypes of the polyamine variants that we have isolated and regenerated from cell culture (97) suggest a disturbance in gibberellic acid and cytokinin balance, although they could also be consistent with the known involvement of ethylene in sexual differentiation (103).

Phytochrome control of ARGdc activity has been reported. Dai & Galston (38) found that red light both induced and repressed ARGdc activity, depending upon the tissue assayed. Red light increased ARGdc activity 2-fold in buds and decreased ARGdc activity 2-fold in epicotyls. Cycloheximide and actinomycin D treatments showed an incrase in ARGdc activity about 30% smaller in red light–treated buds, suggesting some role of protein synthesis in this response.

Goren et al (66) suggested a link between phytochrome-controlled growth and polyamine titers. Using etiolated pea seedlings, they measured polyamines in buds and internodes before and after red light treatments. Red light inhibited internode growth, stimulated bud development, and increased the titers of putrescine, agmatine, and spermidine in the bud while reducing them in the internode. Goren et al (67) demonstrated that these effects were red light dependent rather than growth dependent, could be reversed by far-red light or inhibited by GA_3 application, and involved increased ARGdc activity through de novo synthesis. Inhibitors of ARGdc activity also exerted an inhibitory influence on internode growth similar to that of red light, but the effect of the inhibitors was reversed only slightly by application of polyamines, leaving the mechanism of action of the inhibitors unclear.

Transport

If polyamines are to be considered a class of growth regulator then the possibility of intercellular transport must be evaluated. Since polyamines are fully protonated at common physiological pH (106) and have been shown to bind to membranes as well as cell walls (171), it might be expected a priori that transport within the plant would be limited.

Young & Galston (173) introduced radioactive carbon–labeled amino acids and polyamines into the cotyledons of etiolated pea seedlings and detected a limited degree of transport of putrescine and spermidine into shoots and roots. Using excised *Saintpaulia* petals, they showed that putrescine, spermidine, and spermine uptake at low concentrations proceeded against a concentration gradient; uptake was pH dependent and was related to membrane potential (11). Further evidence for transport was obtained by tracing the flow of labeled canavanine from roots to leaves in *Nicotiana* and by demonstrating that putrescine synthesized in the young leaves and fruitlets of apple can be translocated in the peduncle (8, 9).

In a search for evidence of transport of endogenous polyamines, Friedman et al (60) performed the seemingly simple experiment of isolating plant sap and measuring polyamine levels. They identified putrescine and trace amounts of spermine in xylem exudate, as well as putrescine and spermidine in presumptive phloem sap. Putrescine concentrations in sunflower xylem exudate rose when plant roots were placed under a salt stress, lending support to the contention that putrescine accumulation in leaves of salt-stressed plants may be caused by translocation from the roots.

Cheng & Kao (30), working with exogenous polyamine applications to soybean primary leaves, measured localized and differential effects on chlorophyll loss. They interpreted the results as indicating that the tetraamine spermine is less readily transported in vivo than the diamine and triamine forms diaminopropane and spermidine.

A study designed to demonstrate anit-ozonant properties of polyamines may indirectly support the possibility of intercellular transport of polyamines. When cut stems of tomato were placed into vials of 0.01 mM putrescine, spermine, or spermidine solutions, ozone damage to leaves could be reduced (110).

Stresses

A variety of reports indicate that polyamine levels and polyamine synthetic enzyme activities may increase in response to various plant stresses. We mention some of these here since plants must respond to environmental challenges by changing their growth and development patterns.

Potassium deficiency was one of the first physiological stresses shown to lead to increased putrescine levels in barley and oats (118, 174). Potassium stress resulted in a 20-fold increase in putrescine levels and a 6-fold increase in ARGdc activity over an 18-day growth period (174). In regenerating buds of tobacco, Klinguer et al (89) found that potassium stress was associated with a 7-fold increase in putrescine and phenethylamine levels; hydroxycinnamic acid amides (polyamine-phenolic conjugates) decreased with lower potassium levels, although they increased with the frequency of bud initiation. Similar associations with putrescine and/or ARGdc activity increases have been reported for magnesium deficiency (134, 135) and for a surplus of ammonium ion (92). A number of other ionic treatments (118, 174) cause no changes in polyamines.

Cadmium treatment of detached oat leaves and intact oat seedlings caused up to a 10-fold increase in the levels of putrescine, with only a small effect on the levels of spermidine and spermine (168). The response was inhibited by the addition of DFMA, implying that ARGdc was the enzyme responsible for the putrescine synthesized in response to the cadmium treatment. However, the levels of ARGdc activity increased only slightly in response to the cadmium, intimating that the resulting putrescine increase may not have resulted entirely from changes in synthesis, even though the synthesis that did exist was through ARGdc. The authors noted that the levels of cadmium they tested were similar to those found in some polluted environments.

Low pH increases putrescine and ARGdc activity in several systems. In excised oat leaves incubated for 3–9 hr on various buffers, low pH stimulated a 5-fold increase in putrescine and a 2-fold increase in ARGdc activity; cycloheximide at least partially reduced the putrescine accumulation, implying some requirement for protein synthesis as part of the response (172). Hiatt & Malmberg (74) examined some low-pH effects on polyamines in tobacco cell cultures. They reported that a tobacco cell culture variant resistant to DFMO, *Dfr1*, had high endogenous putrescine levels. Under low-pH growth conditions, *Dfr1* cell cultures grew significantly faster than did a normal wild type tobacco cell culture. This result suggested that high putrescine levels might have some protective function in the plant response to low pH stress.

As part of their original polyamine and oat protoplast viability experiments, Flores & Galston (56) noted that osmotic treatments could induce high levels of putrescine and ARGdc in detached oat leaves. Tiburcio et al (151) surveyed several genera for polyamine metabolism in response to the osmoticum used during protoplast isolation. Oats, a species difficult to regenerate from protoplasts, responded very differently from the regeneration-competent dicot species tested. Oats responded with increased ARGdc activity and massive accumulations of putrescine, whereas other species showed increases in spermidine and spermine with declines in putrescine and activities of ORNdc and ARGdc. The authors speculated that a high ratio of diamine (putrescine) to polyamines (spermidine and spermine) is detrimental. In a companion paper (150) Tiburcio et al demonstrated that pretreatment of oat leaves with DFMA before an osmotic shock causes a decrease of putrescine and increase of spermine titers. Upon extended osmotic shock only modest increases occurred in putrescine while spermine and spermidine increased several fold. This pretreatment also improved the protoplast viability, supporting the previous contention that a rise in the putrescine/polyamine ratios may contribute to difficulties encountered in cereal protoplast culture. Other work had indicated that high levels of putrescine can be harmful in plant systems (130).

Turner & Stewart (159), using barley leaves, provided evidence that the presence of phosphate ions in buffers used for osmotic stress induction can actually cause an increase in the difference between control and experimental treatments over that caused by osmotic stress alone. This increase is caused by the loss of putrescine in control tissue, inflating the values for apparent induction of putrescine. They also found that putrescine acccumulation was concomitant with decreases in leaf turgor but that accumulation did not occur in leaf sections in which turgor was maintained. Additionally, the authors make a strong case for calculating data for metabolite concentrations during water stress experiments on the basis of dry weight rather than wet weight. The authors indicate that the experimental conditions and presentation of results of earlier studies may have overestimated the effects of osmotic stress on polyamines.

The involvement of polyamines in low-temperature stress has been studied in several plant systems in recent years. In three species of *Citrus* and one of *Capsicum* an accumulation of putrescine was correlated with chilling injury. Thus putrescine accumulation may either be a cause of (130) or result from (56) stress-induced injury. Polyamine content prior to chilling does not correlate with chill tolerance in *Phaseolus*; upon chilling, however, the largest increase in polyamine levels, especially of putrescine, occurs in hardened plants. Furthermore, genotypes with high chill tolerance show a significant increase in putrescine when chilled, but nonhardened chill-sensitive geno-types remain unchanged or exhibit lower putrescine concentrations on expo-sure to chilling (69). In a recent study of cold hardening of alfalfa and wheat, putrescine was found to accumulate in both species during hardening in-duction; but there was a rapid loss of putrescine on dehardening, the decline being much more rapid than deacclimation (108).

Levels of putrescine seem to increase in response to many of the plant stress conditions tested. The physiological rationale for this increase is still unclear. The induced high levels of putrescine may be the cause of the stress injury (130); putrescine may be beneficial for the plants' protection from stress (74); or high levels of putrescine may simply be one of many physiological changes, without any special significance (56).

CONCLUSIONS

There is considerable murkiness about the roles of polyamines in plant development. A number of studies suggest that polyamines do something interesting and important. However, these are suggestions, not clear-cut conclusions. On balance very little can be said about polyamine function in higher-plant physiology for which there is not significant contradictory evi-dence.

There are a few points of agreement: Polyamines are associated with rapid cell division. Polyamines have a function in carrot somatic embryogenesis, although the story in alfalfa may show some genotype-dependent differences. Polyamines are essential to tobacco and tomato ovary and fruit development, and may also play a role in early differentiation events in the floral meristem. Exogenous polyamines can retard senescence in detached leaves, but it isn't clear whether this provides insight into senescence physiology. Under some circumstances there can be a metabolic competition between ethylene synthesis and polyamine synthesis for the propylamine moiety of SAM. Putrescine levels may increase in response to some stresses.

Because researchers agree on only these few items it seems premature to conclude that polyamines are a class of plant growth regulators. The evidence to date is that polyamines are one of perhaps many groups of metabolites required for certain developmental processes, but there is no unequivocal data yet to support their role as hormones, second messengers, or other growth regulators. Similarly, although there are hints that polyamine-conjugates or hydroxycinnamic acid amides may do interesting things in the plant, a role for them in development has not yet been proved. The existing data are not sufficiently reductionistic and mechanistic to be used as the basis for conclusions.

This lack of agreement about the functions of polyamines in plants results partly from a lack of common experimental systems and a lack of experimental treatment standards. In studies of embryogenesis, several authors have chosen to work on the same system, carrot suspension cultures. Although the story is not yet complete, their results can be compared and their studies have been deep enough to convince. On the other extreme, reports on fruit development have included several species in which a "spray and pray" approach toward finding practical applications has been used. Many studies are not comparable because different experimental systems and different polyamine pathway parameters have been measured. Investigators seem to choose randomly (a) whether or not to measure endogenous polyamines; (b) which polyamines to measure; (c) whether or not to use inhibitors; (d) which inhibitors to use; (e) whether or not to use exogenous applications of polyamines and which to use; (f) concentration ranges to use; (g) which enzyme activities to measure, if any; (h) whether or not to measure polyamine conjugate levels; and (i) whether or not to use radioactive tracers.

We suggest the following minimum standards: Polyamine synthesis inhibitors should not be used to impute a function to polyamines without also demonstrating that the inhibitors reduce the internal concentration of the polyamines and that the combination of inhibitor plus appropriate polyamine reverses the effect of the inhibitor alone. Investigators should treat results from use of inhibitors cautiously: At least one inhibitor (MGBG) is not completely specific, and the other inhibitors may be altered or metabolized, regardless of their specificity in in vitro enzyme kinetic studies. Enzyme activity assays based on measurements with crude extracts should be treated with caution since some plant materials have nonspecific decarboxylases and other sources of artifacts. Both free and bound (conjugate) polyamine levels should be measured in each extract, and the experimenter should carefully consider whether it is more appropriate to standardize as polyamines per milligram of protein, per gram of tissue weight, or per cell. We also strongly encourage the measurement of metabolic fluxes by the use of radioactive tracers, since this provides a much better estimate of pathway perturbation than simply contemplating changes in enzyme activity levels and corresponding changes in polyamine levels.

A large portion of the research conducted so far has not been mechanistic. Primarily of a survey nature, it has emphasized correlative evidence. It is extremely difficult to determine cause and effect or identify a signal-response pathway when studies merely measure enzyme activity or metabolite levels in response to a treatment. By contrast connections among polyamines, ethylene biosynthesis, and senescence, have begun to be explored in some depth; biochemical experiments such as those of Even-Chen et al (49) and the genetic approach of Dibble et al (41) are providing significant information.

Polyamines may be required for some portions of plant development, but their precise role remains elusive. We do not yet have sufficient reason to consider them plant growth regulators. The field is ripe for investigators with incisive new approaches.

ACKNOWLEDGMENTS

A special mention should be made of Arthur Galston and his colleague Ravindar Kaur-Sawhney, who have initiated many of the current lines of research on polyamines. They developed analytical techniques for plant polyamines and found provocative correlations of polyamines with many of the areas of plant physiology discussed in this article, as can be seen in the frequency with which we cite and discuss their results. The research on senescence, ethylene, and polyamines can be traced back to their original studies on oat leaf protoplasts. The concept that polyamines play a role in plant development arose largely because their research got everyone else excited about the prospects.

The authors thank Erin Bell, Claudia Kaye, and Debra Mohnen for advice and critical readings of the manuscript. Research in the authors' laboratory has been supported by grants from the National Science Foundation (DMB-85-44021, DCB-85-00172, DMB-87-15799), a McKnight Foundation grant, a CIBA-GEIGY corporation grant, and a University of Georgia Biotechnology grant.

Literature Cited

1. Agazio, M. de, Giardina, M. C., Grego, S. 1988. Effect of exogenous putrescine, spermidine, and spermine on K+ uptake and H+ extrusion through plasma membrane in maize root segments. *Plant Physiol.* 87:176–78

2. Altman, A., Kaur-Sawhney, R., Galston, A. W. 1977. Stabilization of oat leaf protoplasts through polyamine mediated inhibition of senescence. *Plant Physiol.* 60:570–74

3. Apelbaum, A. 1986. Polyamine involvement in the development and ripening of avocado fruit. *Acta Horticult.* 179:779–85

4. Apelbaum, A., Burgoon, A. C., Anderson, J. D., Lieberman, M., Ben-Arie, R., Mattoo, A. K. 1981. Polyamines inhibit biosynthesis of ethylene in higher-plant tissue and fruit protoplasts. *Plant Physiol.* 68:453–56

5. Apelbaum, A., Goldlust, A., Icekson, I. 1985. Control by ethylene of arginine decarboxylase activity in pea seedlings and its implication for hormonal regulation of plant growth. *Plant Physiol.* 79:635–40

6. Bagni, N. 1966. Aliphatic amines and a growth factor of coconut milk stimulate cellular proliferation of *Helianthus tuberosus* in vitro. *Experientia* 22:732–36

7. Bagni, N., Adamo, P., Serafini-Fracassini, D. 1981. RNA, proteins and polyamines during tube growth in germinating apple pollen. *Plant Physiol.* 68:727–30

8. Bagni, N., Baraldi, R., Costa, G. 1984. Translocation and metabolism of aliphatic polyamines in leaves and fruitlets of *Malus domestica* (cv "Ruby Spur"). *Acta Horticult.* 149:173–78

9. Bagni, N., Creus, J., Pistocchi, R. 1986. Distribution of cadaverine and lysine decarboxylase activity in *Nicotiana glauca* plants. *J. Plant Physiol.* 125:9–15

10. Bagni, N., Malucelli, B., Torrigiani, P. 1980. Polyamines, storage substances and abscisic acid-like inhibitors during dormancy and very early activation of *Helianthus tuberosus* tissue slices. *Physiol. Plant.* 49:341–45

11. Bagni, N., Pistocchi, R. 1985. Putrescine uptake in *Saintpaulia* petals. *Plant Physiol.* 77:398–402

12. Bagni, N., Serafini-Fracassini, D. 1985. Involvement of polyamines in the mechanism of break of dormancy in *Helianthus tuberosus*. *Bull. Soc. Bot. France* 132:119–25

13. Balint, R., Cooper, G., Staebell, M., Filner, P. 1987. N-caffeoyl-4-amino-*n*-butyric acid a new flower-specific metabolite in cultured tobacco cells and tobacco plants. *J. Biol. Chem.* 262:11026–31

14. Ben-Arie, R., Lurie, S., Mattoo, A. K. 1982. Temperature-dependent inhibitory effects of calcium and spermine on ethylene biosynthesis in apple discs correlate with changes in microsomal membrane microviscosity. *Plant Sci. Lett.* 24:239–47

15. Berlin, J. 1981. Formation of putrescine and cinnamoyl putrescines in tobacco cell cultures. *Phytochemistry* 20:53–55

16. Berlin, J., Forche, E. 1981. DL-α-difluoromethylornithine causes enlargement of cultured tobacco cells. *Z. Pflanzenphysiol.* 101:272–82

17. Berlin, J., Knobloch, K. H., Hofle, G., Witte, L. 1982. Biochemical characterization of two tobacco cell-lines with different levels of cinnamoyl putrescines. *J. Nat. Prod.* 45:83–87

18. Biasi, R., Bagni, N., Costa, G. 1988. Endogenous polyamines in apple and their relationship to fruit set and fruit growth. *Physiol. Plant.* 73:201–5

19. Bird, C. R., Smith, T. A. 1984. Agmatine metabolism and hordatine formation in barley seedlings. *Ann. Bot.* 53:483–88

20. Birecka, H., Bitonti, A. J., McCann, P. P. 1985. Assaying ornithine and arginine decarboxylases in some plant-species. *Plant Physiol.* 79:509–14

21. Birecka, H., Birecka, M., Cohen, E. J., Bitonti, A. J., McCann, P. P. 1988. Ornithine decarboxylase, polyamines, and pyrrolizidine alkaloids in *Senecio* and *Crotalaria*. *Plant Physiol.* 86:224–30

22. Birecka, H., Bitonti, A. J., McCann, P. P. 1985. Activities of arginine and ornithine decarboxylases in various plant species. *Plant Physiol.* 79:515–19

23. Birecka, H., DiNolfo, T. E., Martin, W. B., Frohlich, M. W. 1984. Polyamines and leaf senescence in pyrrolizidine alkaloid-bearing *Heliotropium* plants. *Phytochemistry* 23:991–97

24. Birecka, H., Garraway, M. O., Baumann, R. J., McCann, P. P. 1986. Inhibition of ornithine decarboxylase and growth of the fungus *Helminthosporium maydis*. *Plant Physiol.* 80:798–800

25. Bitonti, A. J., Casara, P. J., McCann P. P., Bey, P. 1987. Catalytic irreversible inhibition of bacterial and plant arginine

decarboxylase activities by novel substrate and product analogues. *Biochem. J.* 242:69–74

26. Cabanne, F., Dalebroux, M. A., Martin-Tanguy, J., Martin, C. 1981. Hydroxy cinnamic-acid amides and ripening to flower of *Nicotiana tabacum* cultivar Xanthi N. C. *Physiol. Plant.* 53:399–404

27. Chapel, M., Teissie, J., Alibert, G. 1984. Electrofusion of spermine treated plant protoplasts. *FEBS Lett.* 173:331–36

28. Chatterjee, S., Choudhuri, M. M., Ghosh, B. 1983. Changes in polyamine contents during root and nodule growth of *Phaseolus mungo*. *Phytochemistry* 22:1533–56

29. Chen, C.-T., Kao, C.-H. 1986. Localized effect of 1,3-diaminopropane and benzyladenine on chlorophyll loss in soybean primary leaves. *Bot. Bull. Acad. Sin.* 27:97–100

30. Cheng, S.-H., Kao, C.-H. 1983. Localized effect of polyamines on chlorophyll loss. *Plant Cell Physiol.* 24:1463–67

31. Cheng, S.-H., Shyr, Y.-Y., Kao, C.-H. 1984. Senescence of rice leaves. XII. Effects of 1,3-diaminopropane, spermidine and spermine. *Bot. Bull. Acad. Sin.* 25:191–96

32. Cho, S. C. 1983. Effects of cytokinin and several inorganic cations on the polyamine content of lettuce cotyledons. *Plant Cell Physiol.* 24:27–32

33. Chriqui, D., D'Orazi, D., Bagni, N. 1986. Ornithine and arginine decarboxylases and polyamine involvement during in vivo differentiation and in vitro dedifferentiation of *Datura innoxia* leaf explant. *Physiol. Plant.* 68:589–96

34. Cohen, E., Arad, S., Heimer, Y., Mizrahi, Y. 1982. Participation of ornithine decarboxylase in early stages of tomato fruit development. *Plant Physiol.* 70:540–43

35. Cohen, E., Arad, S., Heimer, Y. H., Mizrahi, Y. 1984. Polyamine biosynthetic enzymes in the cell cycle of *Chlorella* correlation between ornithine decarboxylase and DNA synthesis at different light intensities. *Plant Physiol.* 74:385–88

36. Cohen, E., Kende, H. 1986. The effect of submergence ethylene and gibberellin on polyamines and their biosynthetic enzymes in deepwater-rice internodes. *Planta* 169:498–504

37. Costa, G., Bagni, N. 1983. Effects of polyamines on fruit-set of apple. *Hortscience* 18:59–61

38. Dai, Y. R., Galston, A. W. 1981.

Simultaneous phytochrome-controlled promotion and inhibition of arginine decarboxylase activity in buds and epicotyls of etiolated pea. *Plant Physiol.* 67:266–69

39. Dai, Y. R., Kaur-Sawhney, R., Galston, A. W. 1982. Promotion by gibberellic-acid of polyamine biosynthesis in internodes of light-grown dwarf peas. *Plant Physiol.* 69:103–5

40. Davies, P. J. 1987. The plant hormones: their nature, occurrence, and functions. In *Plant Hormones and Their Role in Plant Growth and Development*, ed. P. J. Davies, Ch. A1. Boston: Martinus Nijhoff

41. Dibble, A. R. G., Davies, P. J., Mutschler, M. A. 1988. Polyamine content of long keeping alcobaca tomato fruit. *Plant Physiol.* 86:338–40

42. D'Orazi, D., Bagni, N. 1987. Ornithine decarboxylase activity in *Helianthus tuberosus*. *Physiol. Plant* 71:177–83

43. Downs, C. G., Lovell, P. H. 1986. The effect of spermidine and putrescine on the senescence of cut carnations. *Physiol. Plant.* 66:679–84

44. Drolet, G., Dumbroff, E. B., Legge, R. L., Thompson, J. E. 1986. Radical scavenging properties of polyamines. *Phytochemistry* 25:367–71

45. Dumas, E., Perdrizet, E., Vallee, J. 1981. Évolution quantitative des acides aminés et amines libres au cours du développement de diverses espèces de *Nicotiana. Physiol. Veg.* 19:155–65

46. Dumas, E., Vallee, J., Perdrizet, E. 1982. Étude comparée du métabolisme des acides aminés et amines libres chez deux espèces de Tabac. *Physiol. Veg.* 20:505–14

47. Dumortier, F. M., Flores, H. E., Shekhwat, N. S., Galston, A. W. 1983. Gradients of polyamines and their biosynthetic enzymes in coleoptiles and roots of corn. *Plant Physiol.* 72:915–18

48. Evans, P. T., Holaway, B. L., Malmberg, R. L. 1988. Biochemical differentiation in the tobacco flower probed with monoclonal antibodies. *Planta* 175:259–69

49. Even-Chen, Z., Mattoo, A. K., Goren, R. 1982. Inhibition of ethylene biosynthesis by aminoethoxyvinylglycine and by polyamines shunts label from C14-methionine into spermidine in aged orange peel discs. *Plant Physiol.* 69:385–88

50. Feirer, R., Mignon, G., Litvay, J. 1984. Arginine decarboxylase and polyamines required for embryogenesis in the wild carrot. *Science* 223:1433–35

51. Felix, H., Harr, J. 1987. Association of

polyamines to different parts of various plant species. *Physiol. Plant.* 71:245–50

52. Fienberg, A. A., Choi, J. H., Lubich, W. P., Sung, Z. R. 1984. Developmental regulation of polyamine metabolism in growth and differentiation of carrot culture. *Planta* 162:532–39

53. Flink, L., Pettijohn, D. E. 1975. Polyamines stabilize DNA folds. *Nature* 253:62–63

54. Flores, H. E., Filner, P. 1985. Metabolic relationships of putrescine, GABA, and alkaloids in cell and root cultures of Solanaceae. In *Primary and secondary metabolism of Plant Cell Cultures,* ed. Neumann et al. pp. 174–85. Berlin: Springer-Verlag

55. Flores, H. E., Galston, A. W. 1982. Analysis of polyamines in higher-plants by high performance liquid-chromatography. *Plant Physiol.* 69:701–6

56. Flores, H. E., Galston, A. W. 1982. Polyamines and plant stress: activation of putrescine biosynthesis by osmotic shock. *Science* 217:1259–61

57. Flores, H. E., Galson, A. W. 1984. Osmotic stress-induced polyamine accumulation in cereal leaves. II. Relation to amino-acid pools. *Plant Physiol.* 75:110–13

58. Friedman, R., Altman, A., Bachrach, U. 1982. Polyamines and root formation in mung bean hypocotyl cuttings. I. Effects of exogenous compounds and changes in endogenous polyamine content. *Plant Physiol.* 70:844–48

59. Friedman, R., Altman, A., Bachrach, U. 1985. Polyamines and root formation in mung bean *Vigna radiata* hypocotyl cuttings. II. Incorporation of precursors into polyamines. *Plant Physiol.* 79:80–83

60. Friedman, R., Levin, N., Altman, A. 1986. Presence and identification of polyamines in xylem and phloem exudates of plants. *Plant Physiol.* 82:1154–57

61. Fuhrer, J., Kaur-Sawhney, R., Shih, L.-M., Galston, A. W. 1982. Effects of exogenous 1,3-diaminopropane and spermidine on senescence of oat leaves. *Plant Physiol.* 70:1597–1600

62. Galston, A. W. 1983. Polyamines as modulators of plant development. *Bioscience* 33:382–88

63. Galston, A. W., Altman, A., Kaur-Sawhney, R. 1978. Polyamines, ribonuclease, and the improvement of oat leaf protoplasts. *Plant Sci. Lett.* 11:69–79

64. Galston, A. W., Kaur-Sawhney, R. 1987. Polyamines as endogenous growth substances. See Ref. 41, Ch. E2

65. Gerats, A. G. M., Kaye, C., Collins, C., Malmberg, R. L. 1988. Polyamine levels in *Petunia* genotypes with normal and abnormal floral morphologies. *Plant Physiol.* 86:390–93

66. Goren, R., Palavan, N., Flores, H., Galston, A. W. 1982. Changes in polyamine titer in etiolated pea-seedlings following red-light treatment. *Plant Cell Physiol.* 23:19–26

67. Goren, R., Palavan, N., Galston, A. W. 1982. Separating phytochrome effects on arginine decarboxylase activity from its effect on growth. *J. Plant Growth Regul.* 1:61–73

68. Grimes, H. D., Slocum, R. D., Boss, W. F. 1986. α-Difluoromethylarginine treatment inhibits protoplast fusion in fusogenic wild carrot protoplasts. *Biochim. Biophys. Acta* 886:130–34

69. Guye, M. G., Vigh, L., Wilson, J. M. 1986. Polyamine titre in relation to chill-sensitivity in *Phaseolus sp. J. Exp. Bot.* 37:1036–43

70. Hartley, R. D. 1987. HPLC for the separation and determination of phenolic compounds in plant cell walls. In *High Performance Liquid Chromatography in Plant Sciences,* ed. H. F. Linskens, J. F. Jackson, Berlin: Springer-Verlag pp. 92–103

71. Heby, O. 1981. Role of polyamines in the control of cell proliferation and differentiation. *Differentiation* 19:1–20

72. Heimer, Y., Mizrahi, Y. 1982. Characterization of ornithine decarboxylase of tobacco cells and tomato ovaries. *Biochem. J.* 201:373–76

73. Heimer, Y., Mizrahi, Y., Bachrach, U. 1979. Ornithine decarboxylase activity in rapidly proliferating plant cells. *FEBS Lett.* 104:146–49

74. Hiatt, A. C., Malmberg, R. L. 1988. Utilization of putrescine in tobacco cell lines resistant to inhibitors of polyamine synthesis. *Plant Physiol.* 86:441–46

75. Hiatt, A. C., McIndoo, J., Malmberg, R. L. 1986. Regulation of polyamine synthesis in tobacco. *J. Biol. Chem.* 261:1293–98

76. Icekson, I., Bakhanashvili, M., Apelbaum, A. 1986. Inhibition by ethylene of polyamine biosynthetic enzymes enhanced lysine decarboxylase activity and cadaverine accumulation in pea seedlings. *Plant Physiol.* 82:607–9

77. Icekson, I., Goldlust, A., Apelbaum, A. 1985. Influence of ethylene on s-adenosylmethionine decarboxylase activity in etiolated pea seedlings. *J. Plant Physiol.* 119:335–46

78. Jarvis, B. C., Yasmin, S., Coleman, M. T. 1985. RNA and protein metabolism during adventitious root formation in stem cuttings of Phaseolus aureus cultivar berkin. Physiol. Plant. 64:53–59
79. Kakkar, R. K., Rai, V. R. 1987. Effects of spermine and IAA on carbohydrate metabolism during rhizogenesis in Phaseolus vulgaris. 1. Hypocotyl cuttings. Indian J. Exp. Biol. 25:476–78
80. Kaur-Sawhney, R., Altman, A., Galston, A. W. 1978. Dual mechanism in polyamine mediated control of ribonuclease activity in oat leaf protoplasts. Plant Physiol. 62:158–60
81. Kaur-Sawhney, R., Dai, Y. R., Galston, A. W. 1985. Effect of inhibitors of polyamine biosynthesis on gibberellin-induced internode growth in light-grown dwarf peas. Plant Cell Physiol. 27:253–60
82. Kaur-Sawhney, R., Flores, H. E., Galston, A. W. 1980. Polyamine-induced DNA-synthesis and mitosis in oat leaf protoplasts. Plant Physiol. 65:368–71
83. Kaur-Sawhney, R., Flores, H. E., Galston, A. W. 1981. Polyamine oxidase in oat leaves: a cell wall localized enzyme. Plant Physiol. 68:494–98
84. Kaur-Sawhney, R., Galston, A. W. 1979. Interaction of polyamines and light on biochemical processes involved in leaf senescence. Plant, Cell Environ. 2:189–96
85. Kaur-Sawhney, R., Shih, L. M., Cegielska, T., Galston, A. W. 1982. Inhibition of protease activity by polyamines, relevance for control of leaf senescence. FEBS Lett. 145:345–49
86. Kaur-Sawhney, R., Shih, L. M., Flores, H. E., Galston, A. W. 1982. Relation of polyamine synthesis and titer to aging and senescence in oat leaves. Plant Physiol. 69:405–10
87. Kaur-Sawhney, R., Tiburcio, A. F., Galston, A. W. 1988. Spermidine and flower-bud differentiation in thin-layer explants of tobacco. Planta 173:282–84
88. Ke, D., Romani, R. J. 1988. Effects of spermidine on ethylene production and the senescence of suspension-cultured pear fruit cells. Plant Physiol. Biochem. 26:109–16
89. Klinguer, S., Martin-Tanguy, J., Martin, C. 1986. Potassium nutrition, growth bud formation and amine and hydroxycinnamic-acid amide contents in leaf explants of Nicotiana tabacum cultivar Xanthi N. C. cultivated in vitro. Plant Physiol. 82:561–65
90. Kushad, M. M., Yelenosky, G., Knight, R. 1988. Interrelationship of polyamine and ethylene biosynthesis

during avocado fruit development and ripening. Plant Physiol. 87:463–67
91. Kyriakidis, D. A. 1983. Effect of plant growth hormones and polyamines on ornithine decarboxylase activity during the germination of barley seeds. Physiol. Plant. 57:499–504
92. LeRedulier, D., Goas, G. 1975. Influence des ions ammonium et potassium sur l'accumulation de la putrescine chez les jeunes plantes de Soja hispida Moench, privées de leurs cotyledons. Physiol. Veg. 13:125–36
93. Lin, P. 1984. Polyamine metabolism and its relation to response of the aleurone layers of barley seeds to gibberellic acid. Plant Physiol. 74:975–83
94. Malmberg, R. L. 1980. Biochemical, cellular, and developmental characterization of a temperature sensitive mutant of Nicotiana tabacum and its second site revertant. Cell 22:603–29
95. Malmberg, R. L., McIndoo, J. 1983. Abnormal floral development of a tobacco mutant with elevated polyamine levels. Nature 305:623–25
96. Malmberg, R. L., McIndoo, J. 1984. Ultraviolet mutagenesis and genetic analysis of resistance to methylglyoxal-bis(guanylhydrazone). Mol. Gen. Genet. 196:28–34
97. Malmberg, R. L., McIndoo, J., Hiatt, A. C., Lowe, B. A. 1985. Genetics of polyamine synthesis in tobacco—developmental switches in the flower. Cold Spring Harbor Symp. 50:475–82
98. Malmberg, R. L., Rose, D. J. 1987. Biochemical genetics of resistance to MGBG in tobacco: mutants that alter SAM decarboxylase or polyamine ratios, and floral morphology. Mol. Gen. Genet. 207:9–14
99. Martin-Tanguy, J., Perdrizet, E., Martin, C. 1982. Hydroxycinnamic acid amides in fertile and cytoplasmic male sterile lines of maize. Phytochemistry 21:1939–45
100. Meijer, E. G. M., Simmonds, J. 1988. Polyamine levels in relation to growth and somatic embryogenesis in tissue cultures of Meidcago sativa L. J. Exp. Bot. 203:787–94
101. Mengoli, M., Bagni, N., Biondi, S. 1987. Effect of α-Difluoromethylornithine on carrot cell cultures. J. Plant Physiol. 129:479–86
102. Metcalf, B. W., Bey, P., Danzin, C., Jung, M. J., Casara, P., Vevert, J. P. 1978. Catalytic irreversible inhibition of mammalian ornithine decarboxylase by substrate and product analogues. J. Am. Chem. Soc. 100:2251–53

103. Metzger, J. D. 1987. Hormones and reproductive development. See Ref. 41, Ch. E10
104. Miyazaki, J. H., Yang, S. F. 1987. The methionine salvage pathway in relation to ethylene and polyamine biosynthesis. *Physiol. Plant.* 69:366–70
105. Montague, M., Koppenbrink, J., Jaworski, E. 1978. Polyamine metabolism in embryogenic cells of *Daucus carota*. I. Changes in intracellular content and rates of synthesis. *Plant Physiol.* 62:430–33
106. Morris, D. R., Harada, J. J. 1980. Participation of polyamines in the proliferation of bacterial and animal cells. In *Polyamines in Biomedical Research*, ed. J. M. Gaugas, Ch. 1. NY: Wiley
107. Muhitch, M. J., Edwards, L. A., Fletcher, J. S. 1983. Influence of diamines and polyamines on the senescence of plant suspension cultures. *Plant Cell Rep.* 2:82–84
108. Nadeau, P., Delaney, S., Chouinard, L. 1987. Effects of cold hardening on the regulation of polyamine levels in wheat and alfalfa. *Plant Physiol.* 84:73–77
109. Nathan, R., Altman, A., Monselise, S. P. 1984. Changes in activity of polyamine biosynthetic enzymes and in polyamine contents in developing fruit tissues of "Murcott" mandarin. *Sci. Horticult.* 22:359–64
110. Ormrod, D. P., Beckerson, D. W. 1986. Polyamines as antiozonants for tomato. *Hortscience* 21:1070–71
111. Oshima, T. 1983. Novel polyamines in *Thermus thermophilus. Methods Enzymol.* 94:401–10
112. Palavan-Unsal, N. 1987. Polyamine metabolism in the roots of *Phaseolus vulgaris*—interaction of the inhibitors of polyamine biosynthesis with putrescine in growth and polyamine biosynthesis. *Plant Cell Physiol.* 28:565–72
113. Phillips, R., Press, M. C., Bingham, L., Grimmer, C. 1988. Polyamines in cultured artichoke explants: effects are primarily on xylogenesis rather than cell division. *J. Exp. Bot.* 201:473–80
114. Phillips, R., Press, M. C., Eason, A. 1987. Polyamines in relation to cell division and xylogenesis in cultured explants of *Helianthus tuberosus:* lack of evidence for growth regulatory activity. *J. Exp. Bot.* 38:164–72
115. Pistocchi, R., Bagni, N., Creus, J. A. 1986. Polyamine uptake, kinetics, and competition among polyamines and between polyamines and inorganic cations. *Plant Physiol.* 80:556–60
116. Pistocchi, R., Bagni, N., Creus, J. A.

117. 1987. Polyamine uptake in carrot cell cultures. *Plant Physiol.* 84:374–80
117. Prakash, L., John, P., Nair, G. M., Prathapasenan, G. 1988. Effect of spermidine and methylglyoxal-bis(guanylhydrazone) (MGBG) on in vitro pollen germination and tube growth in *Catharanthus roseus. Ann. Bot.* 61:373–75
118. Richards, F. J., Coleman, R. G. 1952. Occurrence of putrescine in potassium deficient barley. *Nature* 170:460
119. Roberts, D. R., Dumbroff, E. B., Thompson, J. E. 1986. Exogenous polyamines alter membrane fluidity in bean leaves—a basis for potential misinterpretation of their true physiological role. *Planta* 167:395–401
120. Roberts, D. R., Walker, M. A., Thompson, J. E., Dumbroff, E. B. 1983. The effects of inhibitors of polyamine and ethylene biosynthesis on senescence, ethylene production and polyamine levels in cut carnations. *Plant Cell Physiol.* 25:315–22
121. Rodriguez, M. T., Gonzalez, M. P., Linares, J. M. 1987. Degradation of chlorophyll and chlorophyllase activity in senescing barley leaves. *J. Plant Physiol.* 129:369–74
122. Rugini, E., Mencuccini, M. 1985. Increased yield in the olive with putrescine treatment. *Hortscience* 20:102–3
123. Schwartz, M., Altman, A., Cohen, Y., Arzee, T. 1986. Localization of ornithine decarboxylase and changes in polyamine content in root meristems of *Zea mays. Physiol. Plant.* 67:485–92
124. Serafini-Fracassini, D., Bagni, N., Cionini, P. G., Bennici, A. 1980. Polyamines and nucleic acids during the first cell-cycle of *Helianthus tuberosus* tissue after the dormancy break. *Planta* 148:332–37
125. Shih, L. M., Kaur-Sawhney, R., Fuhrer, J., Samanta, S., Galston, A. W. 1982. Effects of exogenous 1,3-diaminopropane and spermidine on senescence of oat leaves. I. Inhibition of protease activity, ethylene production, and chlorophyll loss as related to polyamine content. *Plant Physiol.* 70:1592–96
126. Shyr, Y.-Y., Kao, C.-H. 1985. Polyamines and root formation in mung bean hypocotyl cuttings. *Bot. Bull. Acad. Sin. (Taipei)* 26:179–84
127. Singer, S. R., McDaniel, C. N. 1986. Floral determination in the terminal and axillary buds of *Nicotiana tabacum. Dev. Biol.* 118:587–95
128. Slocum, R. D., Galston, A. W. 1985.

Changes in polyamine biosynthesis associated with postfertilization growth and development in tobacco ovary tissues. *Plant Physiol.* 79:336–43

129. Slocum, R. D., Galston, A. W. 1985. In vivo inhibition of polyamine biosynthesis and growth in tobacco ovary tissues. *Plant Cell Physiol.* 26:1519–26

130. Slocum, R. D., Kaur-Sawhney, R., Galston, A. W. 1984. The physiology and biochemistry of polyamines in plants. *Arch. Bioch. Biophys.* 235:283–303

131. Smith, M. A., Davies, P. J. 1985. Separation and quantitation of polyamines in plant-tissue by high-performance liquid-chromatography of their dansyl derivatives. *Plant Physiol.* 78:89–91

132. Smith, M. A., Davies, P. J. 1985. Effect of photoperiod on polyamine metabolism in apical buds of g-2 peas in relation to the induction of apical senescence. *Plant Physiol.* 79:400–5

133. Smith, M. A., Davies, P. J., Reid, J. B. 1985. Role of polyamines in gibberellin-induced internode growth in peas. *Plant Physiol.* 78:92–99

134. Smith, T. A. 1973. Amine levels in mineral deficient *Hordeum vulgare* leaves. *Phytochemistry* 12:2093–2100

135. Smith, T. A. 1984. Putrescine and inorganic ions. *Adv. Phytochem.* 18:7–54

136. Smith, T. A. 1985. Polyamines. *Ann. Rev. Plant Physiol.* 36:117–43

137. Smith, T. A., Best, G. R. 1978. Distribution of the hordatines in barley. *Phytochemistry* 17:1093–98

138. Smith, T. A., Marshall, J. H. A. 1988. Oxidation of amino acids by manganous ions and pyridoxal phosphate. *Phytochemistry* 27:1611–13

139. Smith, T. A., Sinclair, C. 1967. The effect of acid feeding on amine formation in barley. *Ann. Botany (N.S.)* 31:103–11

140. Speranza, A., Calzoni, G. L., Bagni, N. 1983. Effect of exogenous polyamines on in vitro germination of apple pollen. In *Pollen: Biology and Implications for Plant Breeding*, ed. D. L. Mulcahy, E. Ottavio, New York: Elsevier Biomedical pp. 21–103

141. Srivastava, S. K., Smith, T. A. 1982. The effect of some oligoamines and guanidines on membrane permeability in higher plants. *Phytochemistry* 21:997–1008

142. Srivastava, S. K., Vashi, D. J., Naik, B. I. 1983. Control of senescence by polyamines and guanidines in young and mature barley leaves. *Phytochemistry* 22:2151–54

143. Srivenugopal, K. S., Adiga, P. R. 1983.

Putrescine synthase from *Lathyrus sativus* seedlings. *Methods Enzymol.* 94:335–39

144. Suresh, M. R., Adiga, P. R. 1979. Diamine oxidase of *Lathyrus sativus* seedlings, purification and properties. *J. Biosci.* 1:109–24

145. Suresh, M. R., Ramakrishna, S., Adiga, P. R. 1978. Regulation of arginine decarboxylase and putrescine levels in *Cucumis sativus* cotyledons. *Phytochemistry* 20:1477–88

146. Suttle, J. C. 1981. Effect of polyamines on ethylene production. *Phytochemistry* 20:1477–80

147. Suzuki, Y., Yanagiswa, H. 1980. Purification and properties of maize polyamine oxidase: a flavoprotein. *Plant Cell Physiol.* 21:1085–94

148. Tiburcio, A. F., Galston, A. W. 1987. Analysis of alkaloids in tobacco callus by HPLC. See Ref. 70, pp. 228–42

149. Tiburcio, A. F., Kaur-Sawhney, R., Galston, A. W. 1987. Effect of polyamine biosynthetic inhibitors on alkaloids and organogenesis in tobacco callus cultures. *Plant Cell Tissue Organ Cult.* 9:111–20

150. Tiburcio, A. F., Kaur-Sawhney, R., Galston, A. W. 1986. Polyamine metabolism and osmotic stress. II. Improvement of oat protoplasts by an inhibitor of arginine decarboxylase. *Plant Physiol.* 82:375–78

151. Tiburcio, A. F., Masdeu, M. A., Dumortier, F. M., Galston, A. W. 1986. Polyamine metabolism and osmotic stress. I. Relation to protoplast viability. *Plant Physiol.* 82:369–74

152. Torrigiani, P., Altamura, M. M., Pasqua, G., Monacelli, B., Serafini-Fracassini, D., Bagni, N. 1987. Free and conjugated polyamines during denovo floral and vegetative bud formation in thin cell-layers of tobacco. *Physiol. Plant.* 70:453–60

153. Torrigiani, P., Serafini-Fracassini, D., Bagni, N. 1987. Polyamine biosynthesis and effect of dicyclohexylamine during the cell cycle of *Helianthus tuber*. *Plant Physiol.* 84:148–52

154. Tran Thanh Van, K. M. 1973. Direct flower neoformation from superficial tissues of small explant of *Nicotiana tabacum*. *Planta* 115:87–92

155. Tran Thanh Van, K. M. 1981. Control of morphogenesis in in vitro cultures. *Ann. Rev. Plant Physiol.* 32:291–311

156. Tran Thanh Van, K. M., Toubart, P., Cousson, A., Darvill, A. G., Gollin, D. J., Chelf, P., Albersheim, P. 1985. Manipulation of the morphogenetic

pathways of tobacco explants by oligo-saccharins. *Nature* 314:615–17

157. Trewavas, A. 1981. How do plant growth substances act? *Plant Cell Environ.* 4:203–28

158. Trewavas, A., Cleland, R. E. 1983. Is plant development regulated by changes in the concentration of growth substances or by changes in the sensitivity to growth substances? *Trends Biochem. Sci.* 8:354–57

159. Turner, L. B., Stewart, G. R. 1988. Factors affecting polyamine accumulation in barley (*Hordeum vulgare* L.) leaf sections during osmotic stress. *J. Exp. Bot.* 39:311–16

160. Volz, R. K., Knight, J. N. 1986. The use of growth regulators to increase precocity in apple trees. *J. Hort. Sci.* 61:181–89

161. Walker, M. A., Roberts, D. R., Shih, C. Y., Dumbroff, E. B. 1985. A requirement for polyamines during the cell division phase of radicle emergence in seeds of *Acer saccharum. Plant Cell Physiol.* 26:967–72

162. Walters, D. R., Wylie, M. A. 1986. Polyamines in discrete regions of barley leaves infected with the powdery mildew fungus, *Erysiphe graminis. Physiol. Plant.* 67:630–33

163. Wang, S. Y. 1987. Changes of polyamines and ethylene in cucumber seedlings in response to chilling stress. *Physiol. Plant.* 69:253–57

164. Wang, S. Y., Faust, M. 1986. Effect of growth retardants on root formation and polyamine content in apple seedlings. *J. Am. Hort. Sci.* 111:912–17

165. Wang, S. Y., Faust, M., Steffens, G. L. 1985. Metabolic changes in cherry flower buds associated with breaking of dor-mancy in early and late blooming cultivars. *Physiol. Plant.* 65:89–94

166. Wang, S. Y., Steffens, G. L. 1985. Effect of paclobutrazol on water stress-induced ethylene biosynthesis and polyamine accumulation in apple seedling leaves. *Phytochemistry* 24:2185–90

167. Weinstein, L. H., Galston, A. W. 1985. Prevention of a plant disease by specific inhibition of fungal polyamine biosynthesis. *Proc. Natl. Acad. Sci. USA* 82:6874–78

168. Weinstein, L., Kaur-Sawhney, R., Rajam, M. V., Wettlaufer, S., Galston, A. W. 1986. Cadmium induced accumulation of putrescine in oat and bean leaves. *Plant Physiol.* 82:641–45

169. Winer, L., Apelbaum, A. 1986. Involvement of polyamines in the development and ripening of avocado fruits. *J. Plant Physiol.* 26:223–34

170. Wu, S. C., Kuniyuki, A. H. 1985. Isolation and culture of almond protoplasts. *Plant Sci.* 41:55–60

171. Young, D. H., Kauss, H. 1983. Release of calcium from suspension-cultured *Glycine max* cells by chitosan, other polycations, and polyamines in relation to effects on membrane permeability. *Plant Physiol.* 73:698–702

172. Young, N. D., Galson, A. W. 1983. Putrescine and acid stress. *Plant Physiol.* 71:767–71

173. Young, N. D., Galston, A. W. 1983. Are polyamines transported in etiolated peas? *Plant Physiol.* 73:912–14

174. Young, N. D., Galston, A. W. 1984. Physiological control of arginine decarboxylase activity in potassium deficient oat shoots. *Plant Physiol.* 76:331–35

Annu. Rev. Plant Physiol. Plant Mol. Biol. 1989. 40:271–303

INTRACELLULAR pH: MEASUREMENT AND IMPORTANCE IN CELL ACTIVITY

A. Kurkdjian and J. Guern

Laboratoire de Physiologie Cellulaire Végétale, CNRS-INRA, Avenue de la Terrasse, 91190, Gif sur Yvette, France

CONTENTS

INTRODUCTION

Proteins are sensitive to modifications of the proton concentration in their environment, and pH regulates the activity of key enzymes and metabolic steps (see e.g. 45, 188). The control of intracellular pH in animal cells is

271

1040-2519/89/0601-0271$02.00

fundamental for the achievement of a number of important physiological processes, such as protein synthesis (192), DNA and RNA synthesis (50, 127, 135), control of the cell cycle (22 and references therein; 51, 184), and changes in the conductance of membranes for ions (97 and references therein). The cytoplasmic pH is maintained at a value close to neutrality by powerful ion exchange mechanisms at the plasmalemma and by the high buffering capacity of the cytosol (161).

In plants, despite the scarcity of information about the control exerted by pH on physiological functions, the main features of the distribution of protons in plant cells are now reasonably established (137). Large transmembrane pH differences are built across membranes such as the tonoplast (pH at the cytoplasmic side 7.5; pH at the vacuolar side 5.0) of the thylakoid membrane in the light (pH inside the thylakoid 4.0; pH in the stroma 8.0). The cytoplasmic pH is maintained at relatively constant values close to neutrality (see references in Table 1) despite the existence of pH-perturbing processes which have been extensively analyzed in several reviews (137, 142, 143, 175). The situation is quite different for vacuoles, which generally have a rather variable pH in the range of 5–6.5, with some remarkable exceptional values as low as 1.0 in some plant species (142, 175 and references therein).

Our review concerns pH measurement in the main cell compartments (cytoplasm and vacuole) excluding the one of small organelles such as mitochondria and chloroplasts. We describe the technical improvements brought about since the previously published reviews on techniques for intracellular pH measurement in plants (92, 130, 175). In the second and third parts of the review we discuss some of the recent knowledge concerning the mechanisms of cytoplasmic and vacuolar pH regulation. The fourth section concerns the role played by cytoplasmic pH in the control of development, with a special emphasis on the action of plant hormones. Our last section is devoted to problems and prospects for future studies on the regulation of intracellular pH in plants and on the roles of intracellular pH shifts as signals triggering growth and developmental changes.

MEASUREMENT OF INTRACELLULAR pH

Techniques for Intracellular pH Measurement

Various methods are available for measuring intracellular pH in plant cells. Most were first developed to study animal systems (126, 161) and were adapted afterwards for cytoplasmic and vacuolar pH measurements in plants.

INTRACELLULAR pH MEASUREMENTS WITH CHEMICAL PROBES   The technique is based on the preferential diffusion through the membranes of the undissociated molecules of a radiolabeled or fluorescent probe. This probe dissociates outside and inside the cell according to the pH of the compartments, and a diffusion equilibrium is established between the undissociated molecules in the cells and in the incubation medium. The intracellular and extracellular total concentrations of the probe are measured at equilibrium. The corresponding accumulation ratio is a simple function of the extracellular pH (phe), intracellular pH (phi), and pKa of the probe, as shown first by Waddell & Butler (190).

This technique has been employed to measure cytoplasmic pH using a weak acid such as 5,5-dimethyl-oxazolidine dione (DMO) in cell suspension cultures (88), in protoplasts (131), in organ segments (102, 108, 187), and in algal cells (36, 75, 115, 145, 173, 191).

Weak bases, such as radiolabeled methylamine (MeA), benzylamine (BA), and nicotine (Nic), as well as fluorescent bases, such as 9-aminoacridine (9AA), have been used to measure vacuolar pH in populations of cells (83, 87, 96), isolated vacuoles (2, 29, 125, 177), and organ segments (107). In addition, the technique of 9AA microfluorimetry has been used to estimate the vacuolar pH of individual cells, protoplasts, and isolated vacuoles (85, 92, 98).

Several limitations often render inaccurate the estimation of the cytoplasmic pH (phc) in plant cells when using the DMO probe. They correspond (a) to the difficulty of estimating the relative volume of the cytoplasm, (b) to the fact that the amount of DMO accumulated in the vacuole often represents a significant part of the total amount of DMO absorbed by the cells, (c) to the difficulty of checking that only the undissociated acid diffuses across the plasmalemma (107, 115, 144), (d) to the metabolism of the probe leading to an apparent overaccumulation (and consequently to an overestimation of pH) (7, 86, 115), and (e) to the necessity of using low concentrations of the probe in order to avoid pH clamping of the cytoplasm to the external value by H^+ exchange associated with the transmembrane transfer of the undissociated probe. The amount of plant material needed is 0.1–1g FW, the resolution of the technique is at best 01.–0.2 pH unit, and its response time is too long (86, 88, 107, 172) to allow the detection of short-term pH variations. Several of the limitations described above specifically concern small cells, and the most reliable phc determinations using the DMO technique have been obtained with large Charophyte cells (173).

As regards vacuolar pH (phv) measurements, pH values calculated from the accumulation ratio of a base probe are often lower than the ones obtained with other techniques (92), indicating an "overaccumulation" of the probe molecules compared to the accumulation that can be simply ΔpH driven. Possible reasons for this overaccumulation have been discussed (see 92 for a review). They are, briefly, (a) penetration of the protonated form of the base at the plasmalemma as demonstrated for MeA (87, 96, 189); (b) active transport of the dissociated or undissociated form of the base at the plasmalemma or at the tonoplast, as demonstrated in yeast cells for 9AA (183) and in some plant materials for MeA (87, 141); and (c) accumulation in the diffusive layer of membranes and accumulative adsorption of the probe on proteins (112), DNA (81), polysaccharides, and biological membranes (26), as shown for 9AA. Metabolism of basic probes has not been studied for all bases but appears to be of limited extent, at least for the time needed to reach the diffusion equilibrium (7, 83, 189). Artifacts caused by ion pairing of anions with acridine orange have also been described recently (132). The response time and the resolution of pHv measurements with base probes are comparable to those relative to pHc measurements with DMO.

CYTOPLASMIC AND VACUOLAR pH MEASUREMENTS WITH H^+-SELECTIVE MICROELECTRODES H^+-selective liquid membrane microelectrodes are now preferred over the pH-sensitive glass microelectrodes first described by Thomas in 1974 (185). These microelectrodes are built with an H^+-exchange carrier and allow the measurement of an electron motive force (EMF) related to H^+ ion activity in the compartment where the electrode is inserted. This technique, first developed by Ammann et al (3) for animal cells, was later adapted to plant material (84); its use for measuring cytoplasmic pH in plant cells has been developed considerably by Felle (37–44).

Recent progress has been made in use of these microelectrodes. Double-barreled microelectrodes allow measurement of both voltage (PD) and pH in the same area of the cell (37, 148). Dipping the electrode tip in a solution of PVC (37) prevents the ion exchange resin from being pushed back into the shank of the electrode by the high turgor pressure in the cell. Nitrocellulose can play the same role (148). The fast response of these microelectrodes (a few seconds) makes possible the continuous recording of pHc variations under a variety of experimental conditions (37, 43, 84, 148, 176). The response time is short (a few seconds in principle), but the PD and pH electrodes have different response times, producing artifactual pH transients (37). The relative resolution of the technique for each cell impalement can be as good as 0.02–0.05 pH unit. Artifactual measurements occur in certain conditions. The wounding of the cell around the tip of the electrode, for example, may create transient local pH variations (66, 181).

CYTOPLASMIC AND VACUOLAR pH MEASUREMENT BY ^{31}P NMR SPECTROMETRY   The ^{31}P NMR technique allows simultaneous measurement of the cytoplasmic and vacuolar pH (see 150 for a review). The technique relies on the fact that the chemical shift (i.e. the resonance frequency relative to a standard) of ^{31}P compounds is pH dependent. From adequate calibration curves, the inorganic phosphate (Pi) signals issued from the determination of the cytoplasmic and vacuolar Pi pools and the glucose-6-phosphate cytoplasmic signal allow a simultaneous determination of pH in both compartments. This technique has been applied to various types of plant materials: organ fragments (74, 147, 150–157, 181, 186), algae (35, 117), cell suspension cultures (8, 17, 46, 60, 110, 113, 128, 167, 193), and intact growing seedlings (153). It has recently been adapted to the study of isolated plant cell organelles, namely vacuoles isolated from *Catharanthus roseus* cells (113), using the possibility of loading the vacuolar compartment of plant cells with phosphate (8, 17, 113) prior to vacuole isolation to increase the sensitivity of the technique.

Advantages and limitations of this technique have been analyzed and discussed in several reviews (92, 150, 161). The main limitations concern (*a*) its low sensitivity, requiring large amounts of biological material (1–5 g FW) and resulting in a rather low time resolution (5 min at best), and (*b*) its relative inaccuracy for cytoplasmic pH determinations above pH 7.7 and vacuolar pH measurements under pH 5.0, an inaccuracy resulting from the shape of the calibration curve for Pi (the relative resolution is at best 0.05 pH unit and usually not better than 0.1–0.15 pH unit). Some recent improvements of the technique concern the reduction of the time needed to get one measurement (good spectra may be obtained in less than 5 min) and better conditions for a proper oxygenation of rather large amounts of plant material. Continuous perfusion of medium in the NMR tube is now preferred over air bubbling, which disturbs the homogeneity of the magnetic field and decreases the accuracy of pH determination (113, 158, 181).

CELL SAP pH MEASUREMENTS   The last technique used for pHv measurement consists in the extraction of the cell sap and the measurement of its pH with a glass electrode (23, 92, 99, 109, 125, 145). This technique is easy to use for various types of plant material (tissue culture, organ segments). Its limitations have already been reported (92).

Selection of a Technique According to Specific Objectives

All the techniques reported here can be used to monitor pH variations. The problem is more difficult to solve when absolute pH values are required. This is mainly because each technique has limited accuracy and presents specific sources of artifacts. Among the criteria that must be taken into consideration when selecting a technique are the artifacts or uncertainties that may derive from the type of biological material used (isolated cells, massive organs, highly vacuolated cells etc), the response time when rapid pH variations are expected, the amount of biological material required, and the ease of utilization.

The individual measurement of cytoplasmic and vacuolar pH on isolated cells, protoplasts, and vacuoles can only be achieved by using microfluorimetry or microelectrodes. These individual measurements demonstrated the variability of vacuolar pH values (1–2 pH units between the more and the less acidic vacuoles) within cell populations at the exponential phase of growth (85, 92). Isolating protoplasts and vacuoles has no major effect on the variability and mean pH value. Because of this variability, pHv values determined from a population of cells or in a plant organ represent "mean" values, which, as discussed elsewhere (92), depend on the technique of measurement. For example, the value obtained by cell sap extraction depends not only on the vacuolar H^+ concentration but also on the buffering power of the vacuolar sap of the various subclasses of cells from the population, with a shift of the "mean" towards the subclass having the highest buffering power. The same is true for ^{31}P NMR measurements where the "mean" pH value is a complex function of the vacuolar H^+ and inorganic phosphate concentrations, the "mean" being likely shifted towards the subclass having the highest vacuolar phosphate content. Some variability has also been described for cytoplasmic pH measured in *C. roseus* protoplasts with microelectrodes (H. Barbier-Brygoo and A. Kurkdjian, unpublished results) and in *Chara* cells, with the DMO technique (173). However, the extent of these pHc differences from cell to cell is low compared to the one described for pHv.

When pH has to be continuously recorded and transient pH variations studied, microelectrodes are well suited. Data are then obtained on the simultaneous variations of transmembrane ΔpH and potential difference ($\Delta \Psi$). The NMR technique is recommended for studying pH changes in cell populations or in organ segments. Measurements of pH aside, this technique produces interesting information about changes in the energetic status and phosphate compartmentation of the cells.

As discussed below, the techniques available for intracellular pH measurement in plants are not wholly satisfactory. There is an urgent need to cross-check different techniques by using them on the same biological material. Progress has to be made in improving the time resolution of the techniques and in selecting a good fluorescent pH probe specific for the cytoplasmic compartment. Finally a problem still to be solved concerns the study of intracellular pH in massive plant organs and the topography of proton concentrations at the cell level.

REGULATION OF CYTOPLASMIC pH

Plants face a number of pH-perturbing processes associated either with their metabolism, the transport of solutes, or the effect of various environmental factors. The most important cause of intracellular pH perturbation in plants utilizing CO_2 from the atmosphere is the H^+ imbalance caused by nitrogen nutrition. Nitrate utilization is a factor in intracellular alkalinization, while using ammonium as a nitrogen source is acidifying. This has been excellently analyzed in detail by Raven (137, 138, 140 and references therein). As to the pH perturbations linked to solute uptake, the reader is also referred to the analysis by Raven (137) of what could be a minimum estimate of the rate of H^+ influx linked to proton symports.

In spite of these pH-perturbing factors, the cytoplasmic pH is maintained close to neutrality in most plant species (Table 1) but presents variations in specific physiological circumstances. Here we briefly analyze the effects of anaerobiosis, acidic gases, light/dark transitions, temperature, and external pH.

Modifications of Cytoplasmic pH in Response to Environmental Factors

Anaerobiosis induces cytoplasmic pH decrease (from 0.4–0.8 pH unit) (35, 110, 145, 152, 193) while, in most cases, vacuolar pH is not modified (see however 151). The acidifying effects of anaerobiosis per se must be separated from those due to CO_2 accumulation and can only be measured when CO_2 is eliminated by N_2 bubbling (114, 151, 152). CO_2 at concentrations around 1% and above acidifies the cytoplasm (15, 60, 88, 110, 193). Other acidic gases such as SO_2 or NO_2 solubilize in water to form acids that decrease the cytoplasmic pH of plant cells (see 130 for a review). SO_2, a much stronger acid than CO_2, accumulates as SO_3^- in the cytoplasm of barley mesophyll protoplasts (131), with acidifications as large as 1.5 pH units, and is likely responsible for the damages caused by low concentrations of SO_2.

Light-to-dark transitions reversibly acidify the cytoplasm of photosynthesizing cells by 0.2–0.6 pH unit (36, 43, 117, 145, 148, 176). These light-induced pH variations most likely result from the transport of protons between the stroma and the cytoplasm (43). However, abrupt changes in the rate of CO_2 consumption or other metabolic effects could also play a role in the pH changes observed.

The intracellular pH of plant cells is sensitive to temperature changes as it is in nonhomeothermic animals (161 and references therein). The transfer of maize root tips from $+4°C$ to $+28°C$ induces a pHc decrease of about 0.5 pH unit (1). In *Chara corallina* (144), pHc also decreases when temperature is raised with a slope $dpHc/dT$ of about -0.005 pH unit $°C^{-1}$, lower than the value characteristic of ectothermic animals (-0.017 pH unit $°C^{-1}$) necessary to keep a constant H^+/OH^- ratio with temperature variations.

The sensitivity of the cytoplasmic pH to external pH variations in the range 4.0–9.0 is low (35, 75, 147, 148, 186, 193; see also references in 175). The interpretation of the mechanisms by which the external pH modifies the

Table 1 Cytoplasmic and vacuolar pH values in a variety of plant materials

Plant species	Technique used	Cytoplasmic pH	Vacuolar pH	ΔpH	References
Higher plant cells					
Suspension cultures					
A. pseudoplatanus	^{31}P NMR	7.3–7.5	5.9	1.4–1.6	110
C. roseus	^{31}P NMR	7.5–7.6	5.4	2.1–2.2	113
R. damascena	^{31}P NMR	7.5–7.7	5.9	1.7–1.9	121
L. multiflorum	^{31}P NMR	6.7	4.0	2.7	167
N. tabacum	^{31}P NMR	7.5	5.9	1.6	193
P. vulgaris	^{31}P NMR	7.5	5.3	2.2	128
Organs					
roots					
Z. mays	^{31}P NMR	6.8–7.0	5.5	1.3–1.5	74
	^{31}P NMR	7.0–7.2	5.5	1.5–1.7	1
	^{31}P NMR	7.0–7.2	5.6	1.4–1.6	155
	DMO and cell sap extract	7.7–7.9	5.6	2.1–2.3	108
R. sativus	microelectrodes	7.2	6.0	1.2	177
R. fluitans	microelectrodes	7.3	4.8	2.5	47
S. alba	microelectrodes	7.3	4.6	2.7	37
leaves					
E. densa	DMO and cell sap extract	7.5	5.3	2.2	107
internodes					
P. sativum	^{31}P NMR	7.5	5.8	1.7	181
Algal cells					
N. obtusa	^{31}P NMR	7.3	5.3	2.0	117
L. populosum	DMO and cell sap extract	7.5	4.9–5.1	1.4–1.6	75

proton concentration in the cytoplasm is still a matter of debate. Vacuolar pH is more sensitive to external pH variations (75, 186, 193; see also references in 175). The difference of behavior between the two compartments illustrates the tightness of the homeostasis of the cytoplasmic pH and indicates that exchanges between cytoplasmic and vacuolar compartments are likely involved in the adaptation of plant cells to a large range of external proton concentrations.

Mechanisms for Cytoplasmic pH Regulation

As reported above, cytoplasmic pH modifications can occur through intracellular production or consumption of protons or through the exchange of protons or proton equivalents across the boundary membranes. As discussed in detail by Raven (137, 139, 175), to these two main sources of pH perturbation correspond "biochemical" (internal production or consumption of protons) or "biophysical" (net fluxes of H^+ or OH^- across the plasmalemma) mechanisms of regulation. "Long-term" pH regulation, which is related to pH imbalances linked to the synthesis of living matter from external mineral elements throughout growth, is also thought to involve mechanisms different from those responsible for "short-term" pH regulation, which involves mechanisms of resistance to sudden pH stresses (38).

ACID AND BASE LOADING AS A STRATEGY FOR THE STUDY OF CYTOPLASMIC pH REGULATION   Most of what we know about the mechanisms involved in intracellular pH regulation in animal cells derives from the study of the reactions of cells and organs when subjected to an acid load (for a review see 161). This approach has been used recently in plant cells, where it produced interesting information.

Various lipophilic weak acids (e.g. acetic, propionic, butyric, and benzoic) have been used to induce a cytoplasmic acidification in a variety of plant systems. Cytoplasmic acidifications in the range 0.1–1.0 pH unit, depending on experimental conditions (external pH, concentration and type of acid, plant species, etc) have been measured. The kinetics of cytoplasmic acidification vary according to the type of biological material and to the technique used. An interesting situation has been described in *Acer pseudoplatanus* cells (60, 114), where the strong initial pHc decrease induced by propionate loading is followed by a partial recovery, even when the acid is still present in the external medium. This demonstrates the operation of strong mechanisms able to compensate for the proton load caused by the entry of propionic acid. The same transient acidification has been reported for *Riccia fluitans* rhizoid cells treated with various weak acids (47). Various situations as to the occurrence and extent of the recovery and possible origins of the observed differences have been discussed (60).

Little is known about the reactions of plant cells to alkaline loads, and only a few descriptions of cytoplasmic alkalinizations induced by weak bases are available (43). This is probably linked to the fact that the effect of a weak base treatment is difficult to interpret in terms of direct cytoplasmic pH regulation because in most cases a large vacuolar alkaline loading occurs. Interestingly, acid-loaded *A. pseudoplatanus* cells react immediately by a strong and transient cytoplasmic alkalinization when the acid is washed out (60). This "overshoot," which demonstrates the operation of mechanisms able efficiently to counteract the alkalinization of the cytoplasm of plant cells, is a classic behavior of acid-loaded animal cells (161 and references therein).

CYTOPLASMIC BUFFERS   The buffering capacity of the cytoplasm of a few plant species has been estimated either from acid loading experiments (37, 43, 60, 164), from titrations of tissue homogenates and cell organelles (129), or from calculations based on estimated concentrations of the buffering components in the cytoplasm (143). The corresponding values, reported in Table 2, agree with those published for animal cells (161), where mean buffering capacities between 10 and 50 mM pH unit^{-1} have recently been reported (97). The buffering capacity of the cytoplasm is relatively low compared to the intensity of changes in proton concentration. For example, in *A. pseudoplatanus* cells, when the plasmalemma H^+ pump is fully activated by a fusicoccin treatment, the rate of H^+ excretion can be as high as 15 μEq $H^+hr^{-1}g^{-1}$ FW. With a mean buffering capacity of 30 μEq ml^{-1} pH unit^{-1} (60) and assuming a relative cytoplasmic volume of 10%, the pump has potential to alkalinize the cytoplasm by 1 pH unit in 12 min. Calculations made by Raven (137) also demonstrate that internal cytoplasmic buffers are not efficient enough to compensate for the H^+ imbalances plant cells face. Furthermore, the buffering capacity has no significance for long-term pH regulation (175) because the

Table 2 Buffering capacity of the cytoplasm of various types of plant cells

Type of cells	Buffering power (μEq H^+. ml^{-1})	References
Neurospora crassa[a]	30–35	164
Chlorella fusca[b]	20	143
Riccia fluitans rhizoids[a]	40–80	43
Sinapis alba root hairs[a]	50	37
Acer pseudoplatanus cells[a]	20–40	60
Hordeum vulgare leaves[c]	35–100	129

Buffering capacities have been estimated from acid-loading experiments (*a*), from estimates of the cytoplasmic concentration of the buffering components (*b*) or from titration of leaf homogenate and organelle preparations (*c*). Values correspond to the buffering power around pH 7.0–7.5.

buffer components are themselves synthesized by H^+-generating pathways. The natural conclusion is that more powerful systems must contribute to pHc regulation.

ROLE OF THE PLASMALEMMA H^+-ATPASE   A H^+-pumping ATPase located at the plasmalemma of plant cells, now well characterized, is supposed to play a major role in pHc regulation as the most efficient system used by plant cells to pump out H^+ ions (see 106, 169, 178 for reviews). As a matter of fact, the H^+ pump is activated by acidifying the cytoplasm as revealed by the hyperpolarization induced by acid loads (6, 9, 20, 21, 47, 109) and by the associated stimulation of K^+ uptake (109). Conversely, procaine, which alkalinizes the cytoplasm over pH 8, strongly depolarizes *R. fluitans* rhizoid cells, likely reflecting an inhibition of the H^+ pump (43). The H^+-ATPase of *Streptococcus faecalis* is also assumed to regulate cytoplasmic pH on the basis of its reduced activity as cytoplasm alkalinizes over pH 8.0 (76).

The pump that contributes to the regulation of cytoplasmic pH is also a sensor of the concentration of protons in the cytoplasm. Its activity is strongly pH dependent, with an optimal pH around 6.6 and, as a consequence, a marked stimulation occurs when pHc drops from normal values around pH 7.5 (106, 169, 178 and references therein). The pump is also strongly dependent on the membrane potential as shown in acid-loaded *Neurospora* (11) where the H^+ pump participates in pHc regulation in cells, in cooperation with a high-affinity transport system for K^+ corresponding to a strongly electrogenic H^+/K^+ symport (159). In K^+-depleted cells, the coupled operation of the H^+ excretion by the pump and K^+ uptake by the K^+/H^+ symport results in a net 1 $H^+/1K^+$ exchange. This transport loop is competent in regulating pHc in case of acid loading because the current circulation through the symport increases drastically the efficiency of proton excretion by the pump (11). Micromolar concentrations of K^+ are high enough to release the pump from the voltage inhibition created by its electrogenic activity. If such a loop were also operational in higher plant cells it would provide a coherent unifying interpretation of various data relating H^+ excretion and K^+ uptake. This proposal, however, needs further investigation.

As calculated above, the potential activity of the pump is high enough to remove large numbers of protons from the cytoplasm. However, the contribution of the pump to the resistance to acid stresses is difficult to evaluate quantitatively. This results in part from the fact that erythrosin B, vanadate, DCCD, or conditions where the energetic supply of the pump is depressed have a limited efficiency and a poor (or not so well established) specificity. Furthermore, as stressed by Raven (138, 139), the efficiency of H^+ excretion in cells with a free access to a large extracellular compartment (cells in liquid culture, root hair cells, rhizoids, etc) is likely different from the one that can operate in bulky plant organs. The efficiency of the H^+ pump in counteracting cytoplasmic acidification can be hampered by limitations imposed upon the diffusion of protons from the membrane surface to the bulk medium through the complex cell wall network pathway in compact organs. Limitations in the diffusion of protons have been evidenced recently in corn roots (168).

Despite these difficulties in estimating the part played by the H^+ pump in the regulation of pHc, it is not the sole component of the regulating system. This is evidenced by the fact that, even in conditions of severe inhibition of the H^+ pump (anaerobiosis or erythrosin B treatment), acid-loaded *A. pseudoplatanus* cells display a typical recovery from the initial strong propionate-induced pH drop (114). In normal conditions, it was estimated that H^+ excretion driven by the H^+ pump represents at most 50% of the total H^+ excreted during the recovery, indicating that other systems are involved in pHc regulation. From their study on the effect of fatty acids on intracellular pH of *Riccia fluitans* rhizoid cells, Frachisse et al (47) also concluded that the pump does not account for all the H^+ excreted as a reaction to an acid load. The same conclusion was reached by Sanders & Slayman (164), who demonstrated that pump activity is not the major factor stabilizing pHc in *Neurospora*.

We do not know precisely how sensitive the pump is to small variations of cytoplasmic pH or which other factors (aside from $\Delta\Psi$) could modulate its response to cytoplasmic pH changes. The recent finding that the plasma membrane ATPase from oat roots is phosphorylated by a calcium-stimulated, pH-sensitive protein kinase brings new possibilities for investigation of the regulation of the activity of the pump (166). The most promising future in this field has to be expected from the rapidly increasing knowledge of the molecular properties of the H^+-ATPase at the plasmalemma of fungi and higher plant cells (170). Cloning and sequencing the ATPase gene will offer, through the use of transformation and site-directed mutagenesis, new possibilities for investigation of the role of this pump in the regulation of cytoplasmic pH and the identification of factors regulating its activity.

ROLE OF THE H^+/NA^+ AND H^+/K^+ ANTIPORTS   In animal cells, a H^+/Na^+ antiport is the major pH regulating system involved in counteracting cytoplasmic acidification (48, 54, 97, 161 and references therein). Only a few reports indicate that H^+/Na^+ antiport is present at the plasmalemma of plant cells (16, 27, 69) but it appears not to be a ubiquitous characteristic of higher plants (69). This antiport is assumed to be involved in the extrusion of Na^+ from the cytoplasm (16, 27). However, some evidence has been obtained in *A. pseudoplatanus* cells that at least in special physiological circumstances where the cytoplasm is acidified and the external Na^+ concentration is high, the exchange between cytoplasmic H^+ and external Na^+ could play a role in the regulation of cytoplasmic pH after acid loading (114).

A H^+/K^+ antiport has been demonstrated to contribute to intracellular pH regulation in bacteria (78, 79, 123) and in algal cells (174). In *Vibrio alginolyticus* the H^+/K^+ antiport acts as a regulator of cytoplasmic pH in the alkaline region, catalyzing H^+ influx and K^+ efflux when pHc rises over 7.8 (123). In *Chlorella fusca*, various conditions, including weak acid treatment, that acidify the cytoplasm stimulate the uptake of K^+ (189 and references therein). Conversely, a cytoplasmic alkalinization with a weak base induces a release of K^+ and an uptake of H^+, the system working without an energy source. The H^+/K^+ antiport is also thought to be specifically activated by high cytoplasmic pH values inducing H^+ influx and K^+ efflux (189).

In higher plants, the presence of a H^+/K^+ antiport at the plasmalemma is not demonstrated. There is plentiful evidence that the uptake of K^+ is associated with H^+ excretion, but the type of coupling between these fluxes is still a matter of investigation. Marré et al (104, 107–109, 160) demonstrated in various materials and in a variety of circumstances that the coupling between the K^+ and H^+ fluxes is likely indirect, at least when K^+ is in the millimolar range. The influx of K^+ driven by the transmembrane potential built up by the H^+-pump ATPase limits the extent of hyperpolarization and prevents the pump from being inhibited by high $\Delta\Psi$. Recently, H. Felle (personal communication) demonstrated that acid-loaded *Riccia fluitans* rhizoids realkalinized their cytoplasm, provided external K^+ was present. Such a K^+-dependent restoration also occurred when the H^+ pump was inhibited by NaCN or NaN$_3$. From this evidence, a nH^+/xK^+ antiporter involved in the resistance to acids loads, was proposed as an alternative to the H^+-pump.

ROLE OF ANION EXCHANGES   In animal cells, HCO_3^- exchanges play an important role in pHc regulation. Several bicarbonate-dependent regulatory systems, counter-exchanging chloride and bicarbonate, dependent or not on Na^+, have been identified. Net efflux of bicarbonate acidifies the cytoplasm whereas net influx is alkalinizing (48, 97, 161, and references therein). Such exchangers are able to catalyze acid/base fluxes even greater than those mediated by the Na^+/H^+ antiport itself (24).

There are not many data about the participation of anion exchanges in pHc regulation in plant cells. In *Chlorella* (33) and in *Chara* (165), Cl^- uptake is assumed to be catalyzed by a Cl^-/OH^- antiport. Alkaline bands at the surface of *Chara corallina* cells, associated to HCO_3^- assimilation, likely correspond to local internal pH changes (93). In higher plants, indirect evidence for the involvement of HCO_3^- exchanges in the reaction of *A. pseudoplatanus* cells to acid stresses has been reported in conditions of low external Na^+ and severe inhibition of the H^+ pump by erythrosin B (114). A tentative study of the role of HCO_3^- exchanges in the regulation of pHc in rose cells has also been published (121).

The role of CO_2 in perturbing intracellular pH or regulating it through the biochemical pH stat has been reviewed by Bown (15). CO_2 concentrations around atmospheric levels (0.033% at pH 7.5 and 25°C) have little influence on cytoplasmic pH. However, CO_2 at high concentrations (5%) induces strong cytoplasmic acidifications in A. *pseudoplatanus* cells (88) with a typical pattern of acid loading reaction (60). Such high concentrations can be reached in bulky tissues when the pathway for diffusion of CO_2 to the external atmosphere is limiting or in flooded soils where roots can experience CO_2 concentrations as high as 12% (15).

To conclude, our knowledge about the role of anion exchanges in intracellular pH regulation in higher plants is weak. This is specially true for CO_2–HCO_3^- exchanges. One of the reasons is that the bicarbonate-dependent regulatory systems are difficult to study, even in animal cells, because of the absence of specific inhibitors, the possible coexistence of several systems, and variations in CO_2 concentration obscuring transmembrane bicarbonate transfers.

PRODUCTION AND CONSUMPTION OF ORGANIC ACIDS   The "biochemical pH stat" of Davies (30, 31), i.e. a pH control of the balance between production and consumption of organic acids, is thought to play an important role in pHc regulation (see 143, 175 and references therein) at least in the short-term range. Various experimental evidence support the idea that the balance between production and consumption of malic acid is a key process in the regulation of intracellular pH in plants. When treatments supposed to alkalinize the cytoplasm are given to plant cells (imbalance between K^+ and SO_4^{2-} uptake, fusicoccin treatment, etc), malate synthesis is stimulated and total cell malate increases (68, 91, 104 and references therein). However, the in vivo dependency of the stimulation of malate synthesis with respect to the cytoplasmic alkalinization is not quantitatively characterized. At the opposite, some information is available about the effect of cytoplasmic acidification on malate synthesis. When the cytoplasm of A. *pseudoplatanus* cells is acidified by acid loading, $^{14}CO_2$ incorporation in malate is strongly depressed and a net malate consumption occurs that can account for up to 55% of the total H^+ consumption during the partial recovery following the initial pH drop (114). The subsequent increase of the cytoplasmic pH during the recovery phase is associated with a reactivation of malate synthesis.

PEP carboxylase (PEPc) is classically considered one of the critical enzymes of the "biochemical pH stat." It is markedly activated in vitro by a pH increase in the range pH 6.8–8.0 and is a likely candidate for counteracting cytoplasmic alkalinization (30, 31, 143). The problem is to determine how sensitive to small pH changes is the PEPc activity—i.e. how efficient this enzyme can be in counteracting cytoplasmic alkalinizations. This question was discussed a few years ago stressing the fact that the steepness of the activation by pH of PEPc extracted from *A. pseudoplatanus* cells was strongly dependent on concentrations of the substrate PEP and inhibitor malate (58). In recent years, knowledge of PEPc properties has made substantial progress (5 and references therein). Its complex regulation by a variety of metabolites in corn leaves and the cooperation between malate and pH in modulating the PEPc activity in this material have been extensively described (34, 53). Other factors of complexity in the regulation of PEPc activity have been discovered recently in Crassulacean Acid Metabolism (CAM) plants where two forms of PEPc exist: a night form having a high sensitivity for PEP and being relatively insensitive to malate inhibition, and a form with the opposite properties (80 and references therein). The mechanisms governing the interconversion of the two forms have not yet been identified (19, 80. 194).

These developments call for a reassessment of the pH sensitivity of PEPc. For the moment, the relationship between the phosphorylation state of the enzyme and the sensitivity to malate and pH is not clearly established. Furthermore, as many different factors other than H^+ are able to regulate the activity of the enzyme (5, 34), the possibility should be considered that the increase of PEPc activity observed when the cytoplasm is supposed to be alkalinized could be the result of a complex interaction among various factors, including pH.

However, one must stress that there is no absolute evidence that in vivo the rate of malate synthesis and accumulation is controlled by the activity of PEPc. An alternative that should be investigated in plant cells is that the rate of the glycolytic flux, controlled by the activity of the phosphofructokinase (PFK), could be the limiting factor of malate production. The extreme pH sensitivity reported for PFK of muscle cells (188), which accounts for a fine pH control of the glycolytic flux in these cells (45), agrees with this hypothesis and calls for a reinvestigation of the relationships between the glycolytic activity in plant cells and the regulation of their cytoplasmic pH.

REGULATION OF VACUOLAR pH

Ion exchanges across the tonoplast have a high potential to influence the pH of the vacuolar compartment. Recent progress in this field has concerned the characterization of a H^+-pump ATPase, a H^+-pump pyrophosphatase, and several H^+ antiporters that may modify the proton concentration in the vacuolar and cytoplasmic compartments (12, 146 and references therein). However, little is yet known about the in vivo regulation of the vacuolar pH. Direct evidence for vacuolar pH regulation comes from vacuolar perfusion of internodal cells of *Chara australis*, but the mechanisms involved in this regulation are not known (119). Data in the literature report a broad range of vacuolar pH values with very acidic vacuolar contents in cells accumulating large amounts of organic acids (120, 129). This suggests that, contrary to the requirement for a well-controlled homeostasis of the proton concentration in the cytoplasm, plant cells can sustain large variations of their vacuolar pH.

Vacuolar pH Variations in Different Physiological Circumstances

Vacuolar pH variations are induced when the uptake of external anions and cations is unbalanced (i.e. when ions with very different uptake rates such as SO_4^{2-} and K^+ or Cl^- and Ca^{2+} are offered to the cells). This leads to internal ionic readjustments through the production or consumption of malate (68). The vacuole of CAM plants undergoes dramatic pH shifts, from pH 6.0 in the light to pH 3.5 in the dark, owing to the accumulation of malic acid and protons in the vacuole (94, 95). In *C. roseus* cells, the accumulation of malate is also associated with a decrease of vacuolar pH (99, 101). Conversely, in *A. pseudoplatanus* cells (90–92), the accumulation of malate is associated with a vacuolar pH increase. In *Hordeum vulgare* roots, accumulation of total organic acids and vacuolar pH increase are also linked (68).

Auxin or fusicoccin treatment of plant organs or plant cells also results in vacuolar pH variations associated with changes in the vacuolar malate. This has been demonstrated in *A. pseudoplatanus* cells where vacuoles are alkalinized by 0.3 pH unit following a 90-min treatment with 3-μM FC (57, 91). Such a FC-induced vacuolar alkalinization has also been observed in maize root segments (108) and in *E. densa* leaves (107), but the phenomenon is apparently not of general occurrence (181). The synthetic auxin, 2,4-D (8 μM) also induces a pHv increase (0.2 pH unit after 15 hr), associated with an increase in the malate content of *A. pseudoplatanus* cells (57, 90).

Variety of Systems Potentially Involved in Vacuolar pH Regulation

THE H$^+$ ATPASE   The activity of a tonoplast H$^+$-ATPase was first demonstrated in vacuoles isolated from *Hevea brasiliensis* latex (29) and later characterized in a number of plant species (12 and references therein; 59, 70, 106, and references therein; 146, 178 and references therein; 180). Most of the measurements of the $\Delta\Psi$ and ΔpH established by proton pumping have been performed on tonoplast vesicles, and only a few studies used intact isolated vacuoles (59 and references therein). To get an idea of the capacity of the pump to acidify the vacuolar compartment, isolated vacuoles of *C. roseus* cells have been studied, using ^{31}P NMR to monitor the transtonoplast ΔpH and the ATP hydrolysis (59). Starting from an initial vacuolar pH of 5.7, the ATPase activity results in a vacuolar acidification up to 0.8 pH unit, corresponding to a ΔpH as high as 2.2 pH units, with apparently two protons translocated per ATP molecule hydrolyzed (H^+/ATP = 1.97 ± 0.06; mean ± SE) (59).

THE H^+-PYROPHOSPHATASE   Another system that may contribute to the acidification of vacuoles is the H^+-pyrophosphatase (12, 59 and references therein; 103, 146 and references therein; 180). Pyrophosphate-induced acidifications of about 0.5 pH unit corresponding to a final transtonoplast ΔpH up to 2.1 pH units have been described in isolated vacuoles of *C. roseus* (59).

THE NA^+/H^+ ANTIPORTER   A Na^+/H^+ antiporter has been characterized at the tonoplast of beet tissue (12–14). Like the plasmalemma antiporter of animal cells, the tonoplast antiport is inhibited by amiloride and analogs (13). Recently, a quantitative study of the Na^+/H^+ exchanger has been performed in intact vacuoles of *C. roseus* cells (59) using ^{31}P NMR and 9-aminoacridine microfluorimetry to characterize the vacuolar pH variations associated with the activity of the antiport. The capacity of this Na^+/H^+ is high enough to catalyze the exchange of large amounts of Na^+ and H^+ across the tonoplast and to induce consequently large variations in the concentration of these ions in the cytoplasm. However, direct evidence is lacking to demonstrate that this antiport is involved in the correction of pH imbalances either in the cytoplasm or in the vacuole.

OTHER SYSTEMS EXCHANGING H^+ OR H^+ EQUIVALENTS?   As described above, variations in vacuolar pH are more often associated with modifications of malate concentration in the vacuole. Thus, unraveling the molecular mechanisms and regulation of the transfer of malate through the tonoplast would be an important step toward understanding how the vacuolar pH value is determined. Only a few studies have been devoted to the problem of malate transport across the tonoplast. It has been established that in CAM plants, transport of 1 malate^{2-} is accompanied by 2 H^+ and is coupled to the activity of the tonoplast H^+-ATPase (94, 95). Energy-dependent uptake of malate by vacuoles isolated from barley mesophyll protoplasts (111), *Kalanchoë daigremontiana* leaves (124), and *C. roseus* cells (100) has been reported. Furthermore, the recent finding that the major ionic channel at the tonoplast of plant cells can conduct both cations and anions, including malate (28, 63–65), could explain how malate translocation to the vacuole could be dependent on the electrical activity of the tonoplast ATPase. The important point that remains to be explained is how the selectivity of the cation (H^+ or K^+) accompanying malate is achieved, selectivity that appears of utmost importance for the determination of the direction of pH variations associated with malate accumulation.

The Variability of Vacuolar pH Values in Populations of Cells, Protoplasts, or Isolated Vacuoles

As described above, the variability of vacuolar pH values within cell populations at the exponential phase of growth has been demonstrated using H^+ microelectrodes and 9AA microfluorimetry (84, 85, 92). Variations as large as 1.5 pH unit from one cell to the other have been recorded. This variability is also revealed by the broadness of the vacuolar inorganic phosphate peak of ^{31}P NMR spectra, which contrasts with the rather narrow cytoplasmic Pi peak. Protoplasts prepared from these cells and the corresponding isolated vacuoles display the same pHv variability, without major changes in the mean pH value compared to that of the cells (85). The variability of pHv has been confirmed recently with a totally different method of pHv measurement based on ^{13}C NMR of malate (25).

The origin and the biological significance of this variability, which contrast with the rather constant "mean" values measured in well-defined physiological conditions, raise interesting questions. Paradoxically, only the "mean" pH value appears regulated and not the vacuolar pH of individual cells. In fact, we have shown that this situation likely corresponds to the properties of the vacuolar buffer—i.e. small variations in the relative amounts of the main buffering components (malate, citrate, and K^+) result in considerably amplified variations in the concentration of protons (92). More precisely, a pH difference of 1.9 pH unit between the most acidic and the most alkaline vacuoles (i.e. an 80-fold difference in the concentration of protons) can be accounted for by differences of less than 2-fold in the concentrations of malate, citrate, or K^+. Furthermore, combined but opposite variations of malate and K^+ as low as 20–30% of the mean content can be responsible for the large pH variations observed (92). This suggests that the parameters that are primarily regulated, with a small variability from one cell to the other, are the concentrations of malate and K^+ in the vacuole (and maybe their balance), the vacuolar pH value and its variability being a consequence of this regulation.

Do Vacuole and Cytoplasm Interact in Regulating Their pH?

The exchange systems characterized up to now at the tonoplast (H^+-ATPase, H^+-PPase, Na^+/H^+ antiport) have a rather strong potential for modifying the vacuolar pH and consequently a stronger potential for modifying the concentration of protons in the cytoplasmic compartment. However, it is paradoxical to see that there is not, up to now, clear-cut evidence that vacuoles are involved in the short-term regulation of cytoplasmic pH. This is because the vacuolar and cytoplasmic compartments have similar buffering capacities but differ largely in their relative volumes. If we simply assume that protons are transferred across the tonoplast, from the cytoplasm to the vacuole, the resulting drop of pHv should be associated with an opposite increase of the cytoplasmic pH about 10 times higher than the change in pHv. The vacuolar acidification of about 3 pH units observed in CAM plants should correspond to a fantastic cytoplasmic alkalinization!

These quantitative aspects of the relationships between the cytoplasmic and vacuolar compartments in terms of pH suggest that pH changes in the cytoplasm must be primary to pH changes in the vacuole—i.e. a cytoplasmic alkalinization should be limited by an export to the vacuole of OH^- or OH^- equivalents inducing an associated pHv increase. In agreement with this idea, a survey of the literature shows that in almost all cases studied, vacuolar pH and cytoplasmic pH changes occur in the same direction (1, 47, 60, 75, 108, 121, 128, 145, 157, 186, 193), except in corn and pea root tips under hypoxia, where the cytoplasmic acidification associated with an increase of the vacuolar pH likely corresponds to a release of protons from the vacuole (151).

The interaction in vivo between cytoplasm and vacuole in terms of pH regulation is weakly understood. This feature of plant cells should be actively investigated.

INTRACELLULAR pH AND DEVELOPMENT

Developmentally Associated pH Changes

In animal systems, an acidic cytoplasm appears associated with dormant or quiescent states, whereas an increase in pHc signals cellular reactivation (22, 48, 126, 161). A number of reports indicate that modifications of pHc are important for the control of the cell cycle, division, and growth (51, 52, 56). An increase of pH is necessary to bring the cells from G0 to G1 and into S phase for echinoderm eggs, protozoa, slime molds, and mammalian cells (see 97 for a review). More recently, Ober & Pardee (127) reported that tumorigenic Chinese hamster embryo fibroblast cell lines maintain an internal pH significantly higher than the one of normal cells. All these results suggest that pHc behaves as a signal in the control and regulation of cell division. Evidence that the cytoplasmic pH can determine the choice between alternative pathways of cell differentiation in *Dictyostelium discoideum* has also been provided (56).

Except in the controversial domain of the action of plant hormones on the intracellular pH, there are few reports concerning cytoplasmic pH variations in relation to growth and development in plants. In Japanese artichoke (187), slow-growing tubers and fast-growing sprouts from nondormant tubers display a difference in intracellular pH measured with the DMO technique. In Jerusalem artichoke, dormant and nondormant tubers grown in vitro also display a difference in their intracellular pH values. The cell sap pH and the cytoplasmic pH estimated with the DMO technique are both higher by 0.4 pH unit in dormant tubers than in nondormant ones (49). These results concerning a highly differentiated multicellular system are at variance with the fact that dormant states are associated with acidic cytoplasms in nonplant systems (22 and references therein). Such findings must be confirmed by using different and more elaborate techniques of pH measurement.

Plant Hormones and Intracellular pH

In animal systems, the Na^+/H^+ antiport is regulated by hormones (such as insulin) and several growth factors (such as EGF). Evidence has also been provided that the bicarbonate exchange systems are also under hormonal control, as revealed by their sensitivity to phorbol esters (48 and references therein).

The development of plants is under the control of hormones, and a first type of interaction between hormones and pH concerns the influence of intracellular pH on the distribution of weakly acidic plant hormones between the extracellular medium and the cells, as well as between the different cell compartments. This has been demonstrated to some extent for abscisic acid and auxins (72, 89, 163).

The question of hormone action on intracellular pH in plant cells is raising much debate because contradictory results have been obtained. Auxin stimulates the H^+ pump of the plasmalemma thus increasing the electrical potential difference across this membrane and the uptake of K^+ (6, 105 and references therein). Fusicoccin, which closely mimicks various aspects of auxin action (104, 105), displays the same type of activity. Contradictory results have been obtained as to the action of these two chemicals on pHc. A decrease of pHc of 0.1 pH unit has been measured with microelectrodes in corn coleoptile cells treated with 0.1 μM or 10 μM IAA (21, 41, 44). This cytoplasmic acidification is considered to be the origin of the activation of the proton pump and of the stimulation of growth (9, 20, 21, 61). The effect of IAA on pHc has also been studied in pea internodes using the ^{31}P NMR technique, with a setup allowing good precision and sensitivity in the measurement (181). In conditions where it was checked that tissues actively excreted protons in response to auxins and FC, no significant variation of pHc under the effect of IAA (17 μM) was observed. In FC-treated maize root segments, on the other hand, Marré et al (105, 107, 108) measured with the DMO technique an increase of pHc of 0.14 pH unit. Using the ^{31}P NMR technique on the same material, Reid et al (147) measured a FC-induced pHc increase of 0.1 pH unit.

The discrepancies among these results are difficult to interpret. It is difficult to reconcile the idea that cytoplasm is acidified with the other well-known effect of auxins and FC on malate synthesis. According to Talbott et al (181), the cytoplasmic acidification measured by Felle et al may have resulted from the injury of the cells when introducing the microelectrode. As a matter of fact, local transient pH drops have been described in *Chaos carolinensis*, the intensities of which are related to the degree of damage caused by the electrode (66). However, the fact that the acidification measured in the presence of auxin was dependent on the type of auxin analog used (H. Felle, personal communication) suggests it is not a simple artifact of measurement. Even if artifactual acidifications were caused by microelectrodes, it remains puzzling why some workers measured FC-induced cytoplasmic and vacuolar alkalinizations whereas Roberts et al did not detect any pH variation in either compartment (155, 181). We recently confirmed by using the ^{31}P NMR technique on *E. densa* leaves, in cooperation with E. Marré, the results he obtained with the DMO technique concerning the effect of FC on the cytoplasmic and vacuolar pH. Clear-cut cytoplasmic alkalinizations of about 0.3 pH unit and vacuolar alkalinizations up to 0.6 pH unit were measured (unpublished results). This was apparently the first time that two totally different techniques of pH measurement were used on the same material. There is an urgent need to extend such comparative approaches to other materials, other techniques, and other laboratories.

We propose that the discrepancies among the results described above are not simply linked to the techniques of pH measurement but could result mainly from differences in the biological material and its physiological activity. This hypothesis is based on two sets of arguments. The vacuolar alkalinization caused by FC in *A. pseudoplatanus* cells and *E. densa* leaves is large and can be measured whatever the technique used (cell sap pH, weak bases, ^{31}P NMR). Thus the lack of FC-induced phv increase in corn roots and pea internodes (155, 181) can only be attributed to the properties of this material and/or to the physiological conditions offered to it. Furthermore, when tested with the same technique (^{31}P NMR) *A. pseudoplatanus* cells display a lower cytoplasmic alkalinization than *E. densa* mesophyll cells, demonstrating that the intensity of the response could be related to the biological material. If one assumes that the extent of cytoplasmic alkalinization depends upon the relative intensities of H^+ pumping by the ATPase and of proton reabsorption by the H^+-symports one can imagine a large variability of the net proton excretion according to the biological material and/or the composition of the external medium. An interesting example of the influence of the composition of the external medium on the effect of a growth factor on the cytoplasmic proton concentration is the situation of fibroblasts. They respond to growth factors by an activation of the Na^+/H^+ antiport resulting in a cytoplasmic alkalinization only in the absence of bicarbonate in the external medium (48). Furthermore, the reaction of frog skeletal muscle cells to insulin provides a good example of opposite effects of hormone on the cytoplasmic pH according to the presence or absence of external Na^+ (45). In both cases insulin stimulates the Na^+/H^+ antiport with a cytoplasmic alkalinization when the antiport pumps Na^+ in and with a pHc decrease when the antiport catalyzes a Na^+ excretion in the absence of external sodium. This raises the hypothesis that in plants, as in animals, the cytoplasmic alkalinization linked to the activity of the pump could be more or less compensated or even overcompensated by one or several systems involved in the regulation of pHc and able to compensate for the pH imbalances created by the pump. A variety of H^+-symports operate in plant cells (149). Chloride/H^+ and sugar/H^+ (33) symports could be good candidates for a limitation of auxin or FC-induced cytoplasmic alkalinization, but their role needs to be reinvestigated. Our proposal is also in agreement with the fact that paradoxical equilibrium values between processes acidifying or alkalinizing the cytoplasm can apparently be obtained as reported for *R. fluitans,* where the activity of H^+-amino acid symports depolarizes the cells and activates the pump but induces a cytoplasmic alkalinization (71). Even the effects of FC on plant cells have been recently interpreted as resulting either from a direct activation of the proton pump or more likely from an activation of the pump associated with a reduction of current passage through one or more of the return pathways for protons (10).

Intracellular pH as a Second Messenger

There is increasing evidence that cytoplasmic pH changes play the role of second messenger in animal systems (see e.g. 22, 48 and references therein), but the corresponding information concerning plants is dramatically weak (41 and references therein). As discussed above, the hypothesis that protons act as secondary messengers of plant hormones is the subject of much controversy. Talbott et al recently rejected the hypothesis that protons could act as secondary messengers in auxin-induced cell elongation on the basis of the absence of measurable pHc shift in auxin-treated pea internodes (181). However, the possibility should be considered that local pH variations in the vicinity of the inner surface of the plasmalemma could be functionally significant in terms of growth control. The activity of the H^+ pump could create local intracellular proton domains where the pH is increased in such a way as to induce the events involved in growth control. Such local pH variations cannot be evidenced by using conventional techniques of intracellular pH measurement, but their occurrence is suggested by recent data demonstrating the existence of local pH domains at the external surface of the plasmalemma (136, 168).

The possibility that nonhormonal signals use intracellular pH variations as second messengers should also be investigated. A few recent results suggest such a possibility. In the algae *Eremosphaera viridis*, the acidification of the cytoplasm induces an "action potential like response" (77, 176) corresponding to an increase in the conductivity of K^+ channels. This suggests that in plant cells, as in excitable animal cells, intracellular pH modifies the properties of ionic channels (118) and plays an important role in the signal transduction from photosystems in the chloroplast to K^+-channels in the plasmalemma. In *Phaseolus vulgaris* cells in suspension culture, a glucan treatment that elicits the synthesis of a phytoalexin induces rapid and transient decreases of vacuolar and cytoplasmic pH (128). The mechanisms of these pH variations are not known, and the hypothesis that they could act as secondary messengers of elicitor action remains to be tested. This system offers an interesting parallel with the chemotactic behavior of human neutrophils reacting to a foreign intruder by cytoplasmic acidification (195). The early expression of chemotaxis can be induced or suppressed simply by manipulating the H^+ concentration in the cytosol. This suggests that the interaction of a chemoattractant with its specific receptor is translated into cytosolic acidification, which then triggers the chemotactic transduction cascade.

Changes in cytoplasmic pH and cytosolic Ca^{2+} are often associated in animal cells (22, 48, 126 and references therein). Correlative evidence shows that in some systems the cytoplasm is acidified when cytoplasmic calcium rises, while direct evidence comes from the cytoplasmic acidification induced by injection of Ca^{2+} in snail neurons (48 and references therein). Changes in calcium concentration in response to pHc modifications have been reported for a variety of cell types such as rat lymphocytes, where the hormonal activation of the Na^+/H^+ antiport induces a pHc increase (55) paralleled by an increase of cytoplasmic calcium. Alkalinization of the cytoplasm by a variety of procedures increases Ca^{2+}. This suggests an important role for pHc, which behaves as an intracellular messenger controlling Ca^{2+} level in the cytoplasm. A more complex pattern has been described recently in Swiss 3T3 fibroblasts between the Ca^{2+} and H^+ signals, which can be generated simultaneously but independently in response to mitogen-receptor interaction (67).

In plants, only a few reports on simultaneous variations of cytosolic Ca^{2+} and pH are available (41 and references therein). This is likely due to the fact that cytosolic Ca^{2+} concentrations are even more difficult to measure than cytosolic pH. In coleoptiles of *Zea mays* an increase of free calcium as measured with a specific Ca^{2+} microelectrode is associated with the decrease of pHc induced by IAA (39). Illumination of *Nitellopsis* elicits an increase of cytoplasmic calcium which is reversed by darkness (116). Such a treatment is known to induce a reversible alkalinization of the cytoplasm. These results suggest that in plant cells, as in animal systems, pH and Ca^{2+} messages can be associated. Recent data demonstrate that modulating cytoplasmic pH in *R. fluitans* rhizoids and *Z. mays* roots results in cytosolic free calcium changes: Acidifying the cytoplasm increases Ca^{2+} while a decrease of cytosolic Ca^{2+} is a consequence of cytoplasmic alkalinization (40). Much work is now needed to determine how pH and calcium variations interact in terms of effects.

CONCLUDING REMARKS AND PROSPECTS

Much progress is needed before we will understand not only the mechanisms of intracellular pH regulation in plant cells but also the functional consequences of cytoplasmic pH variations. In plants, as in animals, several systems cooperate to regulate intracellular pH. Viewing intracellular pH changes as the sole consequence of the exchange of protons catalyzed by the proton-pump ATPases at the plasmalemma and tonoplast is a far too simple approach to the problem. A more realistic view is to consider the cytosolic and vacuolar saps as buffer solutions, the pH of which is exclusively determined by the balance between strong cations and anions and by the concentration of weak acid and weak base groups. Thus, a pH change in these two compartments can only result from a change in the intensity of one or several of these four parameters. The role of proton exchanges catalyzed by the H^+ pumps or by H^+ cotransports is only significant when viewed as a part of several coupled systems (through electrical and/or chemical coupling) driving the transmembrane exchange of components of the buffer systems (K^+, Cl^-, malate, etc). Facing this complex multifactorial regulation, we must take new approaches, such as the selection and study of mutants of intracellular pH regulation.

Technical improvements are obviously required. Investigators must cross-check different techniques on the same material in order to clear up the conflicting situations reported in the literature. A more detailed study of the distribution of protons in the cytosol by simultaneous pHc measurement in different regions of the same cell is also needed. The techniques currently used (DMO, ^{31}P NMR) do not detect pH heterogeneity in the cytosol. However, local pHc domains, similar to the local cytoplasmic Ca^{2+} domains recently demonstrated in plant cells (18, 73), can be expected in different parts of the same cell, in the vicinity of organelles exchanging large numbers of protons with the cytosol (4, 32, 62), and in the vicinity of the inner surface of the plasmalemma if H^+ diffusion from the bulk cytosol is rate limiting compared to an intense H^+ secretion by the pump. Local pH measurements performed in large cells of *Amoeba* (66) using pH-sensitive microelectrodes should now be possible on a larger scale by using video image technology (171, 179), which can monitor intracellular pH and calcium (82) and reveal cytoplasmic pH heterogeneity as shown in fibroblasts (182) and yeast cells (171). Cytoplasmic pH heterogeneity has also been demonstrated in protoplasts of *Penicillium cyclopium* using photographic densitometry (162).

These techniques should be adapted to plant cells, but this implies first that a convenient fluorescent pH probe is selected for cytoplasmic pH measurement in plants. A comparative study of a few probes used for animal cells recently published could serve as a guide (122).

The idea that intracellular pH acts as a regulatory signal implies that some critical steps in cell activity are highly sensitive to pH variations. If protons act in a highly cooperative manner, very small changes in pH should be enough to induce drastic shifts in the activity of some critical proteins. It is clear that our knowledge of the "receptors" sensing cytosolic pH changes is weak. Nonexclusive candidates for pH sensing in plants are the H^+-pump ATPase at the plasmalemma, the PFK, and the PEP carboxylase in the cytoplasm; these appear as sensors and regulators of the cytosolic pH. As already discussed, the quantitative aspects of their homeostatic role are difficult (but necessary) to study.

The various systems responsible for intracellular pH regulation most likely constitute a network and can compensate each other. This is exemplified in the animal kingdom by the fact that the role of the Na^+/H^+ antiport was studied in the absence of bicarbonate in the extracellular medium to make silent the other pH-regulating systems. Reciprocally, the bicarbonate-linked regulating systems can be studied in amiloride-treated cells (48). New tools (specific inhibitors, mutants, special physiological conditions) should be built up by plant physiologists in order to identify completely the systems involved in intracellular pH regulation, to evaluate their relative contribution to the overall regulation, and to identify the specific effectors of each system. Much progress has been made toward determining the role of the Na^+/H^+ antiport of animal cells by selecting mutant cells either deficient in the antiport or overexpressing it (133, 134). These mutants also proved to be valuable tools in unraveling the cooperation between the sodium antiport and the Cl^-/HCO_3^- exchange system (134). In bacteria, mutants of *Streptococcus faecalis* with a deficiency in the extrusion system for Na^+ (76) and mutants of *E. coli* that have lost the capacity to resist high external pH and to extrude sodium ions (195 and references therein) have been isolated. Plant cell mutants selected for their inability to use one of the systems involved in pH regulation (or overexpressing it) would be especially useful in analyzing the multifactorial system of pH regulation in plant cells.

ACKNOWLEDGMENTS

We thank H. Felle, Y. Mathieu, G. Ephritkhine, and H. Barbier-Brygoo for helpful criticisms, and our colleagues who have sent preprints. We thank also Mrs. B. Cervoni for typing the manuscript and Mrs. F. Lelievre for her assistance in preparing the bibliography.

Literature Cited

1. Aducci, P., Federico, R., Carpinelli, G., Podo, F. 1982. Temperature dependence of intracellular pH in higher plant cells. A³¹P nuclear magnetic resonance study on maize root tips. *Planta* 156: 579–82
2. Alibert, G., Carrasco, A., Boudet, A. M. 1982. Changes in biochemical composition of vacuoles isolated from *Acer pseudoplatanus* L. during cell culture. *Biochim. Biophys. Acta* 721:22–29
3. Ammann, D., Lanter, F., Steiner, R. A., Schulthess, P., Shijo, Y., Simon, W. 1981. Neutral carrier based hydrogen ion-selective microelectrode for extra and intracellular studies. *Anal. Chem.* 53:2267–69
4. Anderson, W. P., Robertson, R. N., Wright, B. J. 1977. Membrane potentials in carrot cells. *Aust. J. Plant Physiol.* 4:241–52
5. Andreo, C. S., Gonzalez, D. H., Iglesias, A. A. 1987. Higher plant phosphoenolpyruvate carboxylase—structure and regulation. *FEBS Lett.* 213(1):1–8
6. Bates, G. W., Goldsmith, M. H. M. 1983. Rapid response of the plasma-membrane potential in oat coleoptiles to auxin and other weak acids. *Planta* 159:231–37
7. Beffagna, N., Romani, G. 1986. Chromatographic analysis of some probes utilized in the weak acid distribution technique in *Elodea densa*. *Gi. Bot. Ital.* 120(1–2) Suppl.:55–56
8. Ben-Hayyim, G., Navon, G. 1985. Phosphorus-31 NMR studies of wild-type and NaCl-tolerant citrus cultured cells. *J. Exp. Bot.* 36(173):1877–88
9. Bertl, A., Felle, H. 1985. Cytoplasmic pH of root hair cells of *Sinapis alba* recorded by a pH-sensitive micro-electrode. Does fusicoccin stimulate the proton pump by cytoplasmic acidification? *J. Exp. Bot.* 36(168):1142–49
10. Blatt, M. R. 1988. Mechanisms of fusicoccin action: A dominant role for secondary transport in higher plant cells. *Planta* 174:187–200
11. Blatt, M. R., Slayman, C. L. 1987. Role of "active" potassium transport in the regulation of cytoplasmic pH by non animal cells. *Proc. Natl. Acad. Sci. USA* 84:2737–41
12. Blumwald, E. 1987. Tonoplast vesicles as a tool in the study of ion transport at the plant vacuole. *Physiol. Plant.* 69: 731–34
13. Blumwald, E., Cragoe, E. J., Poole, R. J. 1987. Inhibition of Na⁺/H⁺ antiport

14. Blumwald, E., Poole, R. J. 1985. Na⁺/H⁺ antiport in isolated tonoplast vesicles from storage tissue of *Beta Vulgaris*. *Plant Physiol.* 78:163–67
15. Bown, A. W. 1985. CO₂ and intracellular pH. *Plant Cell Environ.* 8:459–65
16. Braun, Y., Hassidim, M., Lerner, H. R., Reinhold, L. 1988. Evidence for a Na⁺/H⁺ antiporter in membrane vesicles isolated from roots of the Halophyte *Atriplex nummularia*. *Plant Physiol.* 87: 104–8
17. Brodelius, P., Vogel, H. J. 1985. A phosphorus-31 nuclear magnetic resonance study of phosphate uptake and storage in cultured *Catharanthus roseus* and *Daucus carota* plant cells. *J. Biol. Chem.* 260:3556–60
18. Brownlee, C., Wood, J. W. 1986. A gradient of cytoplasmic free calcium in growing rhizoid cells of *Fucus serratus*. *Nature* 320:624–26
19. Brulfert, J., Vidal, J., Le Maréchal, P., Gadal, P., Queiroz, O., Kluge, M., Krüger, I. 1986. Phosphorylation-dephosphorylation process as a probable mechanism for the diurnal regulatory changes of phosphoenolpyruvate carboxylase in CAM plants. *Biochem. Biophys. Res. Commun.* 136(1):151–59
20. Brummer, B., Bertl, A., Potrykus, I., Felle, H., Parish, R. W. 1985. Evidence that fusicoccin and indole-3-acetic acid induce cytosolic acidification of *Zea mays* cells. *FEBS Lett.* 189(1):109–14
21. Brummer, B., Felle, H., Parish, R. W. 1984. Evidence that acid solutions induce plant cell elongation by acidifying the cytosol and stimulating the proton pump. *FEBS Lett.* 174(2):223–27
22. Busa, W. B., Nuccitelli, R. 1984. Metabolic regulation via intracellular pH. *Am. J. Physiol.* 246:409–38
23. Caldwell, P. D. 1956. Intracellular pH. *Int. Rev. Cytol.* 5:229–77
24. Chaillet, J. R., Amsler, K., Boron. W. F. 1986. Optical measurements of intracellular pH in single LLC-PK1 cells: Demonstration of Cl-HCO₃ exchange. *Proc. Natl. Acad. Sci. USA* 83:522–26
25. Chang, K., Roberts, J. K. M. 1988. Observation of cytoplasmic and vacuolar malate in corn root tips by ¹³C-NMR spectroscopy. *Plant Physiol.* In press
26. Chow, W. S., Barber, J. 1980. 9-Aminoacridine fluorescence changes as

a measure of surface charge density of the thylakoid membrane. *Biochim. Biophys. Acta* 589:346–52

27. Colombo, R., Bonetti, A., Lado, P. 1979. Promoting effect of fusicoccin on Na⁺ efflux in barley roots: evidence for a Na⁺-H⁺ antiport. *Plant Cell and Environ.* 2:281–85

28. Coyaud, L., Kurkdjian, A., Kado, R., Hedrich, R. 1987. Ion channels and ATP-driven pumps involved in ion transport across the tonoplast of sugarbeet vacuoles. *Biochim. Biophys. Acta* 902:263–68

29. Crétin, H. 1982. The proton gradient across the vacuo-lysosomal membrane of lutoids from the latex of *Hevea brasiliensis*. I. Further evidence for a proton-translocating ATPase on the vacuo-lysosomal membrane of intact lutoids. *J. Membrane Biol.* 65:175–84

30. Davies, D. D. 1973. Control of and by pH. *Symp. Soc. Exp. Biol.* 27:513–29

31. Davies, D. D. 1986. The fine control of cytosolic pH. *Physiol. Plant.* 67:702–6

32. Dilley, R. A., Theg, S. M., Beard, W. A. 1987. Membrane-protons interactions in chloroplast energetics: localized proton domains. *Annu. Rev. Plant Physiol.* 38:348–81

33. Doblinger, R., Tromballa, H. W. 1982. The effect of glucose on chloride uptake by *Chlorella*. II. Effect of intracellular acidification on chloride uptake. *Planta* 156:16–20

34. Doncaster, H. D., Leegood, R. C. 1987. Regulation of phosphoenolpyruvate carboxylase activity in maize leaves. *Plant Physiol.* 84:82–87

35. Enami, I., Akutsu, H., Kyogoku, Y. 1986. Intracellular pH regulation in an acidophilic unicellular alga, *Cyanidium caldarium:* ³¹P NMR determination of intracellular pH. *Plant Cell Physiol.* 27:1351–59

36. Falkner, G., Horner, F., Werdan, K., Heldt, H. W. 1976. pH changes in the cytoplasm of the blue-green alga *Anacystis nidulans* caused by light-dependent proton flux into the thylakoid space. *Plant Physiol.* 58:717–18

37. Felle, H. 1987. Proton transport and pH control in *Sinapis alba* root hairs. A study carried out with double-barrelled pH microelectrodes. *J. Exp. Bot.* 38:340–54

38. Felle, H. 1988. Short-term pH-regulation in plants. *Physiol. Plant.* 74(3):583–91

39. Felle, H. 1988. Auxin causes oscillations of cytosolic free calcium and pH in *Zea mays* coleoptiles. *Planta* 174(4):495–99

40. Felle, H. 1988. Cytoplasmic free calcium in *Riccia fluitans* L. and *Zea mays* L.: Interaction of Ca²⁺ and pH? *Planta* 534:1–8

41. Felle, H. 1989. pH as a second messenger in plants. In *Second Messengers in Plant Growth and Development*, ed. W. F. Boss, D. J. Morré, pp. 145–66. New-York: Liss. In press

42. Felle, H., Bertl, A. 1986. The fabrication of H⁺-selective liquid-membrane microelectrodes for use in plant cells. *J. Exp. Bot.* 37:1416–28

43. Felle, H., Bertl, A. 1986. Light-induced cytoplasmic pH changes and their interrelation to the activity of the electrogenic proton pump in *Riccia fluitans*. *Biochim. Biophys. Acta* 848:176–82

44. Felle, H., Brummer, B., Bertl, A., Parish, R. W. 1986. Indole-3 acetic acid and fusicoccin cause cytosolic acidification of corn coleoptile cells. *Proc. Natl. Acad. Sci. USA* 83:8992–95

45. Fidelman, M. L., Seeholzer, S. H., Walsh, K. B., Moore, R. D. 1982. Intracellular pH mediates action of insulin upon glycolysis in frog skeletal muscle. *Am. J. Physiol.* 242:87–93

46. Foyer, C., Walker, D., Spencer, C., Manu, B. 1982. Observations on the phosphate status and intracellular pH of intact cells, protoplasts and chloroplasts from photosynthetic tissue using phosphorus-31 nuclear magnetic resonance. *Biochem. J.* 202:429–34

47. Frachisse, J. M., Johannes, E., Felle, H. 1988. The use of weak acids as physiological tools: a study of the effects of fatty acids on intracellular pH and electrical plasmalemma properties of *Riccia fluitans* rhizoid cells. *Biochim. Biophys. Acta* 938:199–210

48. Frelin, C., Vigne, P., Ladoux, A., Lazdunski, M. 1988. The regulation of the intracellular pH in cells from vertebrates. *Eur. J. Biochem.* 174:3–14

49. Gendraud, M., Lafleuriel, J. 1983. Caractéristiques de l'absorption de saccharose et de tétraphénylphosphonium par les parenchymes de tubercules de Topinambour dormants et non dormants cultivés in vitro. *Physiol. Vég.* 21:1125–33

50. Gerson, D. F. 1982. See Ref. 126, pp. 375–83

51. Gillies, R. J., Deamer, D. W. 1979. Intracellular pH changes during the cell cycle in *Tetrahymena*. *J. Cell Physiol.* 100:23–32

52. Gillies, R. J., Ugurbil, K., Den Hollander, J. A., Shulman, R. G. 1981. ³¹P NMR studies of intracellular pH and

phosphate metabolism during cell division cycle of *Saccharomyces cerevisiae*. *Proc. Natl. Acad. Sci. USA* 78(4):2125–29

53. Gonzalez, D. H., Iglesias, A. A., Andreo, C. S. 1984. On the regulation of phosphoenolpyruvate carboxylase activity from Maize leaves by L-malate. Effect of pH. *J. Plant Physiol.* 116:405–34

54. Grinstein, S., Cohen, S., Goetz, J. D., Rothstein, A., Gelfand, E. W. 1985. Characterization of the activation of Na⁺/H⁺ exchange in lymphocytes by phorbol esters: change in cytoplasmic pH dependence of the antiport. *Proc. Natl. Acad. Sci. USA* 82:1429–33

55. Grinstein, S., Goetz, J. D. 1985. Control of free cytoplasmic calcium by intracellular pH in rat lymphocytes. *Biochim. Biophys. Acta* 819:267–70

56. Gross, J. D., Bradbury, J., Kay, R. R., Peacey, M. J. 1983. Intracellular pH and the control of cell differentiation in *Dictyostelium discoideum*. *Nature* 303:244–45

57. Guern, J., Kurkdjian, A., Mathieu, Y. 1982. Hormonal regulation of intracellular pH: hypotheses versus facts. In *Plant Growth Substances*, ed. P. F. Wareing, pp. 427–37. New York: Academic. 683 pp.

58. Guern, J., Mathieu, Y., Kurkdjian, A. 1983. Phosphoenolpyruvate carboxylase activity and the regulation of intracellular pH in plant cells. *Physiol. Vég.* 21:855–66

59. Guern, J., Mathieu, Y., Kurkdjian, A., Manigault, P., Manigault, J., Gillet, B., Beloeil, J.-C., Lallemand, J. Y. 1989. Regulation of vacuolar pH of plant cells. II. A³¹P NMR study of major ionic exchanges involved in the control of vacuolar pH in isolated vacuoles. *Plant Physiol.* In press

60. Guern, J., Mathieu, Y., Péan, M., Pasquier, C., Beloeil, J.-C., Lallemand, J. Y. 1986. Cytoplasmic pH regulation in *Acer pseudoplatanus* cells. I. A³¹P NMR description of acid-load effects. *Plant Physiol.* 82:840–45

61. Hager, A., Moser, I. 1985. Acetic acid esters and permeable weak acids induce active proton extrusion and extension growth of coleoptile segments by lowering the cytoplasmic pH. *Planta* 163:391–400

62. Heber, U., Heldt, H. W. 1981. The chloroplast envelope: structure, function and role in leaf metabolism. *Annu. Rev. Plant Physiol.* 32:131–68

63. Hedrich, R., Flügge, U. I., Fernandez, J. M. 1986. Patch-clamp studies of ion transport in isolated plant vacuoles. *FEBS Lett.* 204:228–32

64. Hedrich, R., Neher, E. 1987. Cytoplasmic calcium regulates voltage-dependent ion channels in plant vacuoles. *Nature* 329:833–35

65. Hedrich, R., Schroeder, J. 1989. Physiology of ion channels and ion pumps of higher plant cells. *Annu. Rev. Plant Physiol. Plant Mol. Biol.* 40: XXX-XX

66. Heiple, J. M., Taylor, D. L. 1980. Intracellular pH in single motile cells. *J. Cell Biol.* 86:885–90

67. Hesketh, T. R., Moore, J. P., Morris, J. D. H., Taylor, M. V., Rogers, J., Smith, G. A., Metcalfe, J. C. 1985. A common sequence of calcium and pH signals in the mitogenic stimulation of eukaryotic cells. *Nature* 313:481–84

68. Hiatt, A. J. 1967. Relationship of cell sap pH to organic acid change during ion uptake. *Plant Physiol.* 42:294–98

69. Jacoby, B., Teomy, S. 1988. Assessment of Na⁺/H⁺ antiport in ATP-depleted beet slices and barley roots. *Plant Sci.* 55:103–6

70. Jochem, P., Lüttge, U. 1987. Proton transporting enzymes at the tonoplast of leaf cells of the CAM plant *Kalanchoë daigremontiana*. I. The ATPase. *J. Plant Physiol.* 129:251–68

71. Johannes, E., Felle, H. 1987. Implication for cytoplasmic pH, protonmotive force, and aminoacid transport across the plasmalemma of *Riccia fluitans*. *Planta* 172:53–59

72. Kaiser, W. U., Hartung, W. 1981. Uptake and release of abscisic acid by isolated photoautotrophic mesophyll cells, depending upon pH gradients. *Plant Physiol.* 68:202–6

73. Keith, C. H., Ratan, R., Maxfield, F. R., Bajer, A., Shelanski, M. L. 1985. Local cytoplasmic calcium gradients in living mitotic cells. *Nature* 316:848–50

74. Kime, M. J., Ratcliffe, R. G. 1982. The application of ³¹P nuclear magnetic resonance to higher plant tissue. *J. Exp. Bot.* 33(135):670–81

75. Kirst, G. O., Bisson, M. A. 1982. Vacuolar and cytoplasmic pH, ion composition, and turgor pressure in *Lamprothamnium* as a function of external pH. *Planta* 155:287–95

76. Kobayashi, H., Murakami, N., Unemoto, T. 1982. Regulation of cytoplasmic pH in *Streptococcus faecalis*. *J. Biol. Chem.* 257:1346–52

77. Köhler, K., Steigner, W., Kolbowski, J., Hansen, U. P., Simonis, W., Urbach, W. 1986. Potassium channels in *Eremosphaera viridis*. II. Current-

and voltage-clamp experiments. *Planta* 167:66–75

78. Koyama, N., Nosoh, Y. 1985. Effect of potassium and sodium ions on the cytoplasmic pH of an alkalophilic *Bacillus*. *Biochim. Biophys. Acta* 812:206–12

79. Kroll, R. G., Booth, J. R. 1983. The relationship between intracellular pH, the pH gradient and potassium transport in *Escherichia coli*. *Biochem. J.* 216: 709–16

80. Krüger, I., Kluge, M. 1987. Diurnal changes in the regulatory properties of phosphoenolpyruvate carboxylase in plants: are alterations in the quaternary structure involved? *Bot. Acta* 101:24–27

81. Kubota, Y., Modota, Y. 1980. Nanosecond fluorescence decay studies of the deoxyribonucleic acid-9-aminoacridine and deoxyribonucleic acid-9-amino-10-methyl acridinium complexes. *Biochem.* 19:4189–97

82. Kudo, Y., Ozaki, K., Miyakawa, A., Amano, T., Ogura, A. 1986. Monitoring of intracellular Ca^{2+} elevation in a single neural cell using a fluorescence microscope/video camera system. *Jpn. J. Pharmacol.* 41:345–51

83. Kurkdjian, A. 1982. Absorption and accumulation of nicotine by *Acer pseudoplatanus* and *Nicotiana tabacum* cells. *Physiol. Vég.* 20:73–83

84. Kurkdjian, A., Barbier-Brygoo, H. 1983. A hydrogen ion-selective liquid-membrane microelectrode for measurement of the vacuolar pH of plant cells in suspension culture. *Anal. Biochem.* 132:96–104

85. Kurkdjian, A., Barbier-Brygoo, H., Manigault, J., Manigault, P. 1984. Distribution of vacuolar pH values within populations of cells, protoplasts and vacuoles isolated from suspension cultures and plant tissues. *Physiol. Vég.* 22:193–98

86. Kurkdjian, A., Guern, J. 1978. Intracellular pH in higher plant cells. I. Improvements in the use of the 5,5-dimethyloxazolidine-2[14C], 4-dione distribution technique. *Plant Sci. Lett.* 11:337–44

87. Kurkdjian, A., Guern, J. 1981. Vacuolar pH measurement in higher plant cells. I. Evaluation of the methylamine method. *Plant Physiol.* 67:953–57

88. Kurkdjian, A., Leguay, J. J., Guern, J. 1978. Measurement of intracellular pH and aspects of its control in higher plant cells cultivated in liquid medium. *Respir. Physiol.* 33:75–89

89. Kurkdjian, A., Leguay, J. J., Guern, J. 1979. Influence of fusicoccin on the control of cell division by auxins. *Plant Physiol.* 64:1053–57

90. Kurkdjian, A., Mathieu, Y., Guern, J. 1982. Evidence for an action of 2,4-dichlorophenoxyacetic acid on the vacuolar pH of *Acer pseudoplatanus* cells in suspension culture. *Plant Sci. Lett.* 27:77–86

91. Kurkdjian, A., Morot-Gaudry, J. F., Wuilleme, S., Lamant, A., Jolivet, E., Guern, J. 1981. Evidence for an action of fusicoccin on the vacuolar pH of *Acer pseudoplatanus* cells in suspension culture. *Plant Sci. Lett.* 23:233–43

92. Kurkdjian, A., Quiquampoix, H., Barbier-Brygoo, H., Péan, M., Manigault, P., Guern, J. 1985. Critical evaluation of methods for estimating the vacuolar pH of plant cells. In *Biochemistry and Function of Vacuolar ATPase in Fungi and Plant Cells*, ed. B. P. Marin, pp. 98–113. Berlin: Springer-Verlag. 259 pp.

93. Lucas, W. J. 1975. Analysis of the diffusion symmetry developed by the alkaline and acid bands which form at the surface of *Chara corallina* cells. *J. Exp. Bot.* 26:271–86

94. Lüttge, U., Ball, E. 1979. Electrochemical investigation of active malic acid transport at the tonoplast into the vacuoles of the CAM plant *Kalanchoë daigremontiana*. *J. Membr. Biol.* 47: 401–22

95. Lüttge, U., Smith, J. A. C., Marigo, G., Osmond, C. B. 1981. Energetics of malate accumulation in the vacuoles of *Kalanchoë tubiflora* cells. *FEBS Lett.* 126:81–84

96. MacFarlane, J. J., Smith, F. A. 1982. Uptake of methylamine by *Ulva rigida:* transport of cations and diffusion of free base. *J. Exp. Bot.* 33(133):195–207

97. Madshus, I. H. 1988. Regulation of intracellular pH in eukaryotic cells. *Biochem. J.* 250:1–8

98. Manigault, P., Manigault, J., Kurkdjian, A. 1983. A microfluorimetric method for vacuolar pH measurement in plant cells using 9-aminoacridine. *Physiol. Vég.* 21:129–36

99. Marigo, G., Bouyssou, H., Boudet, A. M. 1986. Accumulation des ions nitrate et malate dans des cellules de *Catharanthus roseus* et incidence sur le pH vacuolaire. *Physiol. Vég.* 24(1):15–23

100. Marigo, G., Bouyssou, H., Laborie, D. 1988. Evidence for a malate transport into vacuoles isolated from *Catharanthus roseus* cells. *Bot. Acta* 101(2):187–91

101. Marigo, G., Delorme, Y. M., Lüttge, U., Boudet, A. M. 1983. Rôle de l'acide malique dans la régulation du pH

vacuoloaire dans des cellules de *Catharanthus roseus* cultivées in vitro. *Physiol. Vég.* 21:1135–44

102. Marigo, G., Lüttge, U., Smith, A. C. 1983. Cytoplasmic pH and the control of Crassulacean acid metabolism. *Z. Pflanzenphysiol.* 109:405–13

103. Marquardt, G., Lüttge, U. 1987. Proton transporting enzymes at the tonoplast of leaf cells of the CAM plant *Kalanchoë daigremontiana*. II. The pyrophosphatase. *J. Plant Physiol.* 129:269–86

104. Marré, E. 1979. Fusicoccin: a tool in plant physiology. *Annu. Rev. Plant Physiol.* 30:273–88

105. Marré, E. 1982. Hormonal regulation of transport: data and perspectives. In *Plant Growth Substances*, ed. P. F. Wareing, pp. 407–18. New-York: Academic. 683 pp.

106. Marré, E., Ballarin-Denti, A. 1985. The proton pumps at the plasmalemma and the tonoplast of higher plants. *J. Bioenerg. Biomembr.* 17:1–21

107. Marré, E., Beffagna, N., Romani, G. 1988. Potassium transport and regulation of intracellular pH in *Elodea densa* leaves. *Bot. Acta* 101:24–31

108. Marré, M. T., Romani, G., Bellando, M., Marré, E. 1986. Stimulation of weak acid uptake and increase in cell sap pH as evidence for fusicoccin and K⁺-induced cytosol alkalinization. *Plant Physiol.* 82:316–23

109. Marré, M. T., Romani, G., Marré, E. 1983. Transmembrane hyperpolarization and increase of K⁺ uptake in maize roots treated with permeant weak acids. *Plant Cell Environ.* 6:617–23

110. Martin, J. B., Bligny, R., Rebeillé, F., Douce, R., Leguay, J. J., Mathieu, Y., Guern, J. 1982. A³¹P NMR study of intracellular pH of plant cells cultivated in liquid medium. *Plant Physiol.* 70: 1156–61

111. Martinoia, E., Flügge, U. I., Kaiser, G., Heber, U., Heldt, H. W. 1985. Energy-dependent uptake of malate into vacuoles isolated from barley mesophyll protoplasts. *Biochim. Biophys. Acta* 806:311–19

112. Marty, A., Viallet, P. 1979. Etude de l'inhibition de la fluorescence de l'amino-9-acridine et de l'amino-9 chloro-6 acridine par la sérumalbumine humaine. *C. R. Acad. Sci. Paris* 288:1715–18

113. Mathieu, Y., Guern, J., Kurkdjian, A., Manigault, P., Manigault, J., Zielinska, T., et al. 1988. Regulation of vacuolar pH of plant cells. I. Isolation and properties of vacuoles suitable for ³¹P NMR studies. *Plant Physiol.* In press

114. Mathieu, Y., Guern, J., Péan, M., Pas-

quier, C., Beloeil, J.-C., Lallemand, J. Y. 1986. Cytoplasmic pH regulation in *Acer pseudoplatanus* cells. II. Possible mechanisms involved in pH regulation during acid-load. *Plant Physiol.* 82:846–52

115. De Michelis, M. I., Raven, J. A., Jayasuriya, H. D. 1979. Measurement of cytoplasmic pH by the DMO technique in *Hydrodictyon africanum*. *J. Exp. Bot.* 30:681–95

116. Miller, A. J., Sanders, D. 1987. Depletion of cytosolic free calcium induced by photosynthesis. *Nature* 326:397–400

117. Mimura, T., Kirino, Y. 1984. Changes in cytoplasmic pH measured by ³¹P-NMR in cells of *Nitellopsis obtusa*. *Plant Cell Physiol.* 25(5):813–20

118. Moody, W. 1984. Effects of intracellular H⁺ on the electrical properties of excitable cells. *Annu. Rev. Neurosci.* 7:257–78

119. Moriyasu, Y., Shimmen, T., Tazawa, M. 1984. Vacuolar pH regulation in *Chara australis*. *Cell Struct. Funct.* 9:225–34

120. Moskowitz, A. H., Hrazdina, G. 1981. Vacuolar contents of fruit subepidermal cells from *Vitis* species. *Plant Physiol.* 68:686–92

121. Murphy, T. M., Matson, G. B., Morrison, S. L. 1983. Ultraviolet-stimulated KHCO₃ efflux from rose cells. *Plant Physiol.* 73:20–24

122. Musgrove, E., Rugg, C., Hedley, D. 1986. Flow cytometric measurement of cytoplasmic pH: a critical evaluation of available fluorochromes. *Cytometry* 7: 347–55

123. Nakamura, T., Tokuda, H., Unemoto, T. 1984. K⁺/H⁺ antiporter functions as a regulator of cytoplasmic pH in a marine bacterium, *Vibrio alginolyticus*. *Biochim. Biophys. Acta* 776:330–36

124. Nishida, K., Tominaga, O. 1987. Energy-dependent uptake of malate into vacuoles isolated from CAM plant. *Kalanchoë daigremontiana*. *J. Plant Physiol.* 127:385–93

125. Nishimura, M. 1982. pH in vacuoles isolated from castor bean endosperm. *Plant Physiol.* 70:742–46

126. Nuccitelli, R., Deamer, D. W., eds. 1982. *Intracellular pH: Its Measurement, Regulation and Utilization in Cellular Functions*. New York: Liss. 594 pp.

127. Ober, S. S., Pardee, A. B. 1987. Intracellular pH is increased after transformation of Chinese hamster embryo fibroblasts. *Proc. Natl. Acad. Sci. USA* 84:2766–70

128. Ojalvo, I., Rokem, S., Navon, G.,

Goldberg, I. 1987. ^{31}P NMR study of elicitor treated *Phaseolus vulgaris* cell suspension cultures. *Plant Physiol.* 85: 716–19

129. Pfanz, H., Heber, U. 1986. Buffer capacities of leaves, leaf cell organelles in relation to fluxes of potentially acidic gases. *Plant Physiol.* 81:597–602

130. Pfanz, H., Heber, U. 1989. Determination of extra- and intracellular pH values in relation to the action of acidic gases on cells. In *Modern Methods of Plant Analysis,* ed. H. F. Liskens, J. F. Jackson, New Ser. 9:322–43. Berlin: Springer Verlag

131. Pfanz, H., Martinoia, E., Lange, O. L., Heber, U. 1987. Flux of SO$_2$ into leaf cells and cellular acidification by SO$_2$. *Plant Physiol.* 85:928–33

132. Pope, A. J., Leigh, R. A. 1988. Dissipation of pH gradients in tonoplast vesicles and liposomes by mixtures of acridine orange and anions. *Plant Physiol.* 86:1315–22

133. Pouyssegur, J. 1985. The growth factor activatable Na$^+$/H$^+$ exchange system: a genetic approach. *TIBS* 11:453–55

134. Pouyssegur, J., Franchi, A., Kohno, M., L'allemain, G., Paris, S. 1985. Na$^+$/H$^+$ exchange and growth control in fibroblasts: a genetic approach. *Curr. Top. Membr. Transp.* 26:201–20

135. Pouyssegur, J., Franchi, A., L'allemain, G. 1985. Cytoplasmic pH a key determinant of growth factor–induced DNA synthesis in quiescent fibroblasts. *FEBS Lett.* 190:115–19

136. Price, G. D., Badger, M. R. 1985. Inhibition by proton buffers of photosynthetic utilization of bicarbonate in *Chara corallina. Austr. J. Plant Physiol.* 12:257–67

137. Raven, J. A. 1985. pH regulation in plants. *Sci. Prog. Oxford* 69:495–509

138. Raven, J. A. 1985. Regulation of pH and generation of osmolarity in vascular plants: a cost benefit analysis in relation to efficiency of use of energy, nitrogen and water. *New Phytol.* 101:25–77

139. Raven, J. A. 1986. Biochemical disposal of excess H$^+$ in growing plants? *New Phytol.* 1041:75–206

140. Raven, J. A. 1988. Acquisition of nitrogen by the shoots of land plants: its occurrence and implications for acid-base regulation. *New Phytol.* 109:1–20

141. Raven, J. A., Farquhar, G. D. 1981. Methylammonium transport in *Phaseolus vulgaris* leaf slices. *Plant Physiol.* 67:859–63

142. Raven, J. A., Smith, F. A. 1974. Significance of hydrogen ion movement in plant cells. *Can. J. Bot.* 52:1035–49

143. Raven, J. A., Smith, F. A. 1976. Cytoplasmic pH regulation and electrogenic H$^+$ extrusion. *Curr. Adv. Plant Sci.* 8:649–60

144. Raven, J. A., Smith, F. A. 1978. Effect of temperature and external pH on the cytoplasmic pH of *Chara corallina. J. Exp. Bot.* 29(111):853–66

145. Raven, J. A., Smith, F. A. 1980. Intracellular pH regulation in the giant-celled marine alga *Chaetomorpha darwinii. J. Exp. Bot.* 31(124):1357–69

146. Rea, P. A., Sanders, D. 1987. Tonoplast energization: two H$^+$ pumps, one membrane. *Physiol. Plant.* 71:131–41

147. Reid, R. J., Field, L. D., Pitman, M. G. 1985. Effects of external pH, fusicoccin and butyrate on the cytoplasmic pH in barley root tips measured by ^{31}P-nuclear magnetic resonance spectroscopy. *Planta* 166:341–47

148. Reid, R. J., Smith, F. A. 1988. Measurements of the cytoplasmic pH of *Chara corallina* using double-barrelled pH microelectrodes. *J. Exp. Bot.* 39(207):1421–33

149. Reinhold, L., Kaplan, A. 1984. Membrane transport of sugars and amino acids. *Annu. Rev. Plant Physiol.* 35:45–83

150. Roberts, J. K. M. 1984. Study of plant metabolism in vivo using NMR spectroscopy. *Annu. Rev. Plant Physiol.* 33:375–86

151. Roberts, J. K. M., Callis, J., Jardetzky, O., Walbot, V., Freeling, M. 1984. Cytoplasmic acidosis as a determinant of flooding intolerance in plants. *Proc. Natl. Acad. Sci. USA* 81:6029–33

152. Roberts, J. K. M., Callis, J., Wemmer, D., Walbot, V., Jardetzky, O. 1984. Mechanisms of cytoplasmic pH regulation in hypoxic maize root tips and its role in survival under hypoxia. *Proc. Natl. Acad. Sci. USA* 81:3379–83

153. Roberts, J. K. M., Pia Testa, M. 1988. ^{31}P NMR spectroscopy of roots of intact corn seedlings. *Plant Physiol.* 86:1127–30

154. Roberts, J. K. M., Ray, P. M., Wade-Jardetzky, N., Jardetzky, O. 1980. Estimation of cytoplasmic and vacuolar pH in higher plant cells by ^{31}P NMR. *Nature* 283:870–72

155. Roberts, J. K. M., Ray, P. M., Wade-Jardetzky, N., Jardetzky, O. 1981. Extent of intracellular pH changes during H$^+$ extrusion by maize root-tip cells. *Planta* 152:74–78

156. Roberts, J. K. M., Wade-Jardetzky, N., Jardetzky, O. 1981. Intracellular pH measurements by ^{31}P nuclear magnetic resonance. Influence of factors other

than pH on ^{31}P chemical shifts. *Biochem.* 20:5389–94
157. Roberts, J. K. M., Wemmer, D., Ray, P. M., Jardetzky, O. 1982. Regulation of cytoplasmic and vacuolar pH in maize root tips under different experimental conditions. *Plant Physiol.* 69:1344–47
158. Roby, C., Martin, J.-B., Bligny, R., Douce, R. 1987. Biochemical changes during sucrose deprivation in higher plant cells. Phosphorus-31 nuclear magnetic resonance studies. *J. Biol. Chem.* 262(11):5000–7
159. Rodriguez-Navarro, A., Blatt, M. R., Slayman, C. L. 1986. A potassium-proton symport in Neurospora crassa. *J. Gen. Physiol.* 87:649–74
160. Romani, G., Marré, M. T., Bellando, M., Alloatti, G., Marré, E. 1985. H$^+$ extrusion and potassium uptake associated with potential hyperpolarization in maize and wheat root segments treated with permeant weak acids. *Plant Physiol.* 79:734–39
161. Roos, A., Boron, W. F. 1981. Intracellular pH. *Physiol. Rev.* 61:296–434
162. Roos, W., Slavik, J. 1987. Intracellular pH topography of *Penicillium cyclopium* protoplasts. Maintenance of ΔpH by both passive and active mechanisms. *Biochim. Biophys. Acta* 899:67–75
163. Rubery, P. H. 1987. Auxin transport. In *Plant Hormones and Their Role in Plant Growth and Development*, ed. P. J. Davies, pp. 341–62. Dordrecht: Martinus Nijhoff. 681 pp.
164. Sanders, D., Slayman, C. L. 1982. Control of intracellular pH. Predominant role of oxidative metabolism on proton transport in the eucaryotic microorganism *Neurospora*. *J. Gen. Physiol.* 80:377–402
165. Sanders, D., Smith, F. A., Walker, N. A. 1985. Proton/chloride cotransport in *Chara*: mechanism of enhanced influx after rapid external acidification. *Planta* 163:411–18
166. Schaller, G. E., Sussman, M. R. 1988. Phosphorylation of the plasma-membrane H$^+$-ATPase of oat roots by a calcium-stimulated protein kinase. *Planta* 173:509–18
167. Schibeci, A., Henry, R. J., Stone, B. A., Brownlee, R. T. C. 1983. ^{31}P NMR measurement of cytoplasmic and vacuolar pH in endosperm cells of ryegrass *(Lolium multiflorum)* grown in suspension culture. *Biochem. Int.* 6:837–44
168. Sentenac, H., Grignon, C. 1987. Effect of H$^+$ excretion on the surface pH of corn root cells evaluated by using weak acid influx as a pH probe. *Plant Physiol.* 84:1367–72
169. Serrano, R. 1984. Plasma membrane ATPase of fungi and plants as a novel type of proton pump. *Curr. Topics Cell. Regul.* 23:87–126
170. Serrano, R. 1988. Structure and function of proton translocating ATPase in plasma membranes of plants and fungi. *Biochim. Biophys. Acta* 947:1–28
171. Sisken, J. E., Barrows, G. H., Grasch, S. D. 1986. The study of fluorescent probes by quantitative video intensification microscopy (QVIM). *J. Histochem. Cytochem.* 34(1):61–66
172. Smith, F. A. 1984. Regulation of the cytoplasmic pH of *Chara corallina* in the absence of external Ca^{2+}: its significance in relation to the activity and control of the H$^+$ pump. *J. Exp. Bot.* 35(159):1525–36
173. Smith, F. A. 1986. Short-term measurements of the cytoplasmic pH of *Chara corallina* derived from the intracellular equilibration of 5,5-dimethyloxazolidine-2,4-dione (DMO). *J. Exp. Bot.* 37(184):1733–45
174. Smith, F. A., Gibson, J. L. 1985. Effect of cations on the cytoplasmic pH of *Chara corallina. J. Exp. Bot.* 36:1331–40
175. Smith, F. A., Raven, J. A. 1979. Intracellular pH and its regulation. *Annu. Rev. Plant Physiol.* 30:289–311
176. Steigner, W., Köhler, K., Simonis, W., Urbach, W. 1988. Transient cytoplasmic pH changes in correlation with opening of potassium channels in *Eremosphaera. J. Exp. Bot.* 39(198):23–26
177. Strack, D., Sharma, V., Felle, H. 1987. Vacuolar pH in radish cotyledonal mesophyll cells. *Planta* 172:563–65
178. Sze, H. 1984. H$^+$-translocating ATPases of the plasma-membrane and tonoplat of plant cells. *Physiol. Plant.* 61:683–91
179. Takamatsu, T., Kitamura, T., Fujita, S. 1986. Quantitative fluorescence image analysis. *Acta Histochem. Cytochem.* 19:61–71
180. Takeshige, K., Tazawa, M., Hager, A. 1988. Characterization of the H$^+$ translocating adenosine triphosphatase and pyrophosphatase of vacuolar membranes isolated by means of a perfusion technique from *Chara corallina. Plant Physiol.* 86:1168–73
181. Talbott, L. D., Ray, P. M., Roberts, J. K. M. 1988. Effect of indoleacetic acid- and fusicoccin-stimulated proton extrusion on internal pH of pea internode cells. *Plant Physiol.* 87:211–16

182. Tanasugarn, L., McNeil, P., Reynolds, G. D., Taylor, D. L. 1984. Microspectrofluorometry by digital image processing: measurement of cytoplasmic pH. *J. Cell. Biol.* 98:717–24

183. Theuvenet, A. P. R., Van der Wijngaard, W. M. H., Borst-Pauwels, G. W. F. H. 1983. 9-aminoacridine, a fluorescent probe of the thiamine carrier in yeast cells. *Biochim. Biophys. Acta* 730:255–62

184. Thomas, K. C. 1981. Stimulation by pH of cell cycle initiation in the yeast *Candida utilis. Can. J. Bot.* 59:2043–48

185. Thomas, R. C. 1974. Intracellular pH of snail neurones measured with pH-sensitive glass microelectrode. *J. Physiol.* 238:159–80

186. Torimitsu, K., Yazaki, Y., Nagasuka, K., Ohta, E., Sakata, M. 1984. Effect of external pH on the cytoplasmic and vacuolar pHs in mung bean root-tip cells: a ^{31}P nuclear magnetic resonance study. *Plant Cell Physiol.* 25(8):1403–9

187. Tort, M., Gendraud, M. 1984. Contribution à l'étude des pH cytoplasmique et vacuolaire en rapport avec la croissance et l'accumulation des réserves chez le Crosne du Japon. *C. R. Acad. Sci. Paris* 299 (10):431–34

188. Trivedi, B., Danforth, W. H. 1966. Effect of pH on the kinetics of frog muscle phosphofructokinase. *J. Biol. Chem.* 241(17):4110–14

189. Tromballa, H. W. 1987. Base uptake, K$^+$ transport and intracellular pH regulation by the green alga *Chlorella fusca. Biochim. Biophys. Acta* 904:216–26

190. Waddell, W. J., Butler, T. C. 1959. Calculation of intracellular pH from the distribution of DMO. Application to skeletal muscle of the dog. *J. Clin. Invest.* 38:720–29

191. Walker, N. A., Smith, F. A. 1975. Intracellular pH in *Chara corallina* measured by DMO distribution. *Plant Sci. Lett.* 4:125–32

192. Winkler, M. M. 1982. See Ref. 126, pp. 325–39

193. Wray, V., Schiel, O., Berlin, J., Witte, L. 1985. Phosphorus-31 nuclear magnetic resonance investigation of the in vivo regulation of intracellular pH in cell suspension cultures of *Nicotiana tabacum:* the effects of oxygen supply, nitrogen, and external pH change. *Arch. Biochem. Biophys.* 236(2):731–40

194. Wu, M. X., Wedding, R. T. 1987. Regulation of phosphoenolpyruvate carboxylase from *Crassula argentea. Plant Physiol.* 84:1080–83

195. Yuli, I., Oplatka, A. 1987. Cytosolic acidification as an early transducing signal of human neutrophil chemotaxis. *Science* 235:340–42

196. Zilberstein, D., Padan, E., Schuldiner, S. 1980. A single locus in *Escherichia coli* governs growth in alkaline pH and on carbon sources whose transport is sodium dependent. *FEBS Lett.* 116(2):177–80

Annu. Rev. Plant Physiol. Plant Mol. Biol. 1989. 40:305–46

MOLECULAR AND CELLULAR BIOLOGY ASSOCIATED WITH ENDOSPERM MOBILIZATION IN GERMINATING CEREAL GRAINS

Geoffrey B. Fincher

Research Centre for Protein and Enzyme Technology, Department of Biochemistry, La Trobe University, Bundoora, Victoria 3083, Australia

CONTENTS

305

1040-2519/89/0601-0305$02.00

INTRODUCTION[1]

Grains of the common cereals are the major source of carbohydrate and protein in human nutrition. They are also used extensively to produce alcoholic beverages and stockfeed. Because germination plays a central role in the propagation of cereal crops, a large research effort has been accorded the regulation of enzymes responsible for endosperm mobilization in germinating grains. The discovery that α-amylase levels are elevated by gibberellic acid (GA) treatment of barley grain (188, 276), coupled with the isolation of viable barley aleurone layers for in vitro experimentation (49), has led to the widespread adoption of barley and other cereal aleurone layers as a model system for studies on plant hormone action, on cellular processes leading to the secretion of plant enzymes, and more recently, on the regulation of plant gene expression.

Here I focus on recent advances in our knowledge of cereal aleurone biology, particularly on the regulation of gene expression during germination. However, the aleurone cannot be considered in isolation. It is now evident that the scutellum also participates in endosperm mobilization, both as a source of hydrolytic enzymes and as an absorptive tissue for the uptake of endosperm degradation products, and that regulation of gene expression in the aleurone and scutellum is subject to highly coordinated temporal and tissue-specific control mechanisms (160). Since gene expression encompasses changes in chromatin structure, gene activation and transcription, mRNA processing and translation, posttranslational modification and correct folding of nascent polypeptides, and the transfer of functional gene products to their sites of action in the grain, the entire process must be considered not only in molecular terms but also in a cellular context. Here, the biological objectives of germination and the molecular and cellular strategies invoked to achieve these objectives are related to the ultrastructure and composition of key tissues in ungerminated grain. Once germination is initiated, the perception and transduction of hormonal signals are important early events. I examine subsequent synthesis of hydrolytic enzymes and their secretion from the aleurone and scutellum in relation to their role in degrading starchy endosperm reserves. Finally, I present recent findings on gene structure as the first clues about how gene expression is regulated in germinating cereal grains.

ULTRASTRUCTURE AND COMPOSITION OF UNGERMINATED GRAIN

During grain development, the outer endosperm differentiates into the aleurone, a morphologically and functionally distinct layer that may be one to

[1]Abbreviations: ER, endoplasmic reticulum; GA, gibberellic acid; ABA, abscisic acid.

several cells in thickness, depending on the species (69). Although the aleurone and the starchy endosperm share a common origin, only the cells of the aleurone layer remain alive after the storage reserves of the endosperm are deposited and the grain matures and dries. At grain maturity the embryonic axis, consisting of the parenchymatous, meristematic cells of the plumule, the embryonic leaf with its ensheathing coleoptile, and the radicle with its protective coleorhiza can be clearly distinguished from the parenchyma cells of the scutellum (69). These are living tissues in the quiescent grain. The scutellar epithelium is a single layer of specialized elongated cells that lies at the interface of the embryo and the endosperm (69). In this section, I compare the ultrastructure and composition of the aleurone, the scutellar epithelium, and the starchy endosperm in ungerminated grain. Most information has been accumulated from studies on wheat, barley, maize, and rice.

Aleurone Layer

CELL CONTENTS   Mature aleurone cells of the common cereals are characteristically packed with specialized protein bodies, known as aleurone grains, which are surrounded by lipid bodies and interspersed with numerous mitochondria and occasional plastids (172) and ER (40). Lipid bodies also line the cell periphery. Starch granules are absent in the aleurone of mature grain. The lipid bodies contain a matrix of storage triacylglycerols encased in a half-unit membrane consisting of a single phospholipid layer and a few major proteins (104).

Although little information on the proteins in aleurone grains is available, their composition differs from the reserves of the starchy endosperm. In particular, they are relatively rich in basic amino acids (14, 72, 234). Within the aleurone grains, both phytin globoids (type I inclusions) and niacytin particles (type II inclusions) can be observed in the protein matrix (24, 172, 195). Phytin consists of the potassium and magnesium salts of myo-inositol hexaphosphate (235) and represents the major phosphate reserve in the grain. The type II inclusions contain bound niacin, o-aminophenol, protein, and carbohydrate. The chemistry of the bound niacin is not yet clear, although in wheat it is probably covalently associated through ester linkages to a β-linked glucose oligomer that is also associated with an unidentified phenolic component; the complex has a molecular weight of approximately 2800 and contains no protein (122; B. A. Stone, personal communication).

Other important components of aleurone cells in quiescent grains include reserve carbohydrates (mostly sucrose, raffinose, and fructans) (35, 130), ribonucleic acids (160), exo- and endopeptidases (3, 59, 163), and vitamins.

Collectively, the reserve proteins, carbohydrates, lipids, phosphate, cations, and cofactors, together with hydrolytic enzymes and mitochondria, permit rapid proliferation of the intracellular membrane system and the protein synthesizing machinery, and provide amino acids necessary for the synthesis of hydrolytic enzymes after the initiation of germination.

CELL WALLS   The aleurone protoplast is surrounded by a cell wall. These walls are deposited during cell enlargement and are therefore classified as primary walls; they contain no secondary thickening or lignin. In wheat and barley, aleurone walls consist of an inner, relatively thin layer and a thicker outer layer (13, 245). There is indirect evidence for a thin cementing layer between the inner and outer wall layers (13). Intercellular wall canals or plasmodesmata connect protoplasts of adjacent aleurone cells and appear to be lined with a material similar to that of the inner wall (245).

Aleurone walls isolated from wheat and barley have been subjected to detailed structural analyses (14). The major polysaccharide components are arabinoxylans (approximately 65–67% by weight) and (1→3, 1→4)-β-glucans (26–29% by weight), with small amounts of glucomannans and cellulose (approximately 2% each) (14). (1→3)-β-Glucans are not structural components of aleurone walls but are restricted to small, discrete deposits that occupy an extracellular location in the subaleurone region of the endosperm (13, 73). Phenolic acids, predominantly ferulic acid, are covalently linked to the walls (71); ferulic acid is esterified to C(O5) atoms of arabinosyl substituents of arabinoxylans (14, 230).

Both barley and wheat aleurone wall preparations contain proteinaceous material, but whether or not it represents an integral structural component of the wall remains uncertain (45, 69). Hydroxyproline-rich glycoproteins of the so-called "extensin" class, which are widely distributed in dicotyledonous walls, are absent (14). However, there is evidence for hydroxyproline-rich wall glycoproteins in rice endosperm and bran, and some of this material may be associated with aleurone walls (69). Similar glycoproteins are found in the pericarp walls of developing maize kernels (101).

The distribution of specific components in cereal aleurone walls has been studied using light, fluorescence, and electron microscopy. Interpretation of these studies relies on the specificity of a variety of reagents. Protein-specific stains indicate that proteinaceous material is concentrated in the inner layer of wheat and barley aleurone walls (11). The fluorochromes Calcofluor and Congo Red have a high affinity for (1→3, 1→4)-β-glucans and preferentially bind to the inner layer of wheat aleurone walls (268). Recent work with fluorescein-labeled (1→3, 1→4)-β-glucanase provides further support for this distribution pattern (B. A. Stone, S. J. Joyner, and F. M. Grave, personal communication). Ferulic acid is distributed evenly across the walls (71). The balance of existing evidence suggests that the inner wall layer is enriched in protein and (1→3, 1→4)-β-glucan, while arabinoxylan and its associated ferulic acid (although distributed across the two wall layers) is probably the major constituent of the outer wall. Unequivocal confirmation of these distribution patterns, together with detailed chemical characterization of constituent macromolecules, is central to our understanding of the processes that lead to the movement of secreted hydrolytic enzymes through aleurone walls into the starchy endosperm—particularly in view of the differences in susceptibility of the two aleurone wall layers to degradative enzymes in the germinating grain.

Scutellum

In the context of starchy endosperm mobilization, the scutellar epithelium is perhaps the most important component of the scutellum, since it is located at the interface of the embryo and the starchy endosperm where it can participate in both enzyme secretion and the absorption of degradation products.

CELL CONTENTS   Cells of the scutellar epithelium in resting grains share many structural features with aleurone cells. They are packed with specialized protein bodies, analogous to aleurone grains, which are surrounded by lipid bodies (255) and contain phytin (7, 183, 241, 246). Mitochondria, free ribosomes, a central nucleus, and some ER are found, but dictyosomes are difficult to identify and starch is usually absent (183, 184, 241). The scutellum in wheat also contains vitamins (153) and high levels (up to 20% on a dry-weight basis) of soluble sugars, predominantly sucrose and raffinose (60). Thus, the scutellar cells are packed with protein, lipid, and phosphate reserves, together with enzymes and a readily available energy source; like resting aleurone cells, they can rapidly initiate protein synthesis early in the germination process.

CELL WALLS   Walls of scutellar epithelial cells also share structural features with aleurone walls. Two distinct layers can be distinguished, particularly in the early stages of germination (242). Individual components of scutellar walls have not been characterized, but both the parenchyma and epithelial walls of maize scutellum consist mainly of arabinoxylan, possibly with low levels of $(1{\to}3, 1{\to}4)$-β-glucan (69). Barley and wheat scutellar walls are intensely autofluorescent, largely owing to associated ferulic acid (229). Lignin and pectin are absent from scutellar walls of wheat, barley, and oats (228).

Starchy Endosperm

The starchy endosperm occupies the bulk of the cereal grain and represents the nutrient store that is mobilized during germination to nourish the growing seedling. Thus the starchy endosperm constitutes the target tissue for enzymes secreted by the aleurone and scutellum. Cells of the starchy endosperm are packed with starch granules in a matrix of storage protein. Although the starchy endosperm cells of mature grains are nonliving, compacted remnants of nuclei, ribosomes, ER and RNA can be detected, particularly in the sub-aleurone region (24, 160). The walls of starchy endosperm cells in wheat, barley, and rice are normally much thinner than those of the aleurone (69).

CELL CONTENTS   Starch granules are synthesized in amyloplasts during endosperm development. Granule morphology and size vary among species. Starch granules are composed predominantly of the linear $(1\rightarrow4)$-α-glucan (amylose) and the branched $(1\rightarrow4, 1\rightarrow6)$-$\alpha$-glucan (amylopectin); the proportion of each varies among species and varieties. Lipids and proteins constitute minor components of starch granules.

Storage protein of the starchy endosperm cells is synthesized by polyribosomes attached to ER and accumulates primarily in membrane-bound protein bodies of ER origin, but also in vacuoles, during grain development (98, 141). In mature tissue, the shapes of protein bodies are distorted by dehydration and tissue compaction. Thus, individual protein bodies in the starchy endosperm of wheat cannot be distinguished and form a continuous matrix in which starch granules are embedded. Cereal storage proteins have been the subject of detailed chemical, genetic, and molecular studies. They can be divided into four classes, based on their solubility: the water-soluble albumins, salt-soluble globulins, aqueous alcohol–soluble prolamins, and acid- or alkali-soluble glutelins (187). Prolamins are usually the most abundant, except in oats and rice which contain significant levels of unrelated legumin-type storage protein (225). Cereal storage proteins are rich in asparagine, glutamine, and proline, as expected for a nitrogen reserve that must be efficiently packaged at high density into starchy endosperm cells.

The prolamins of wheat, rye, barley, and maize are characterized by structural domains that contain tandemly repeated sequences ranging in size from approximately 7 to 20 amino acid residues (29, 76, 93, 126, 194). These repetitive sequences are presumably important determinants of peptide folding during packaging of the storage proteins into protein bodies. Most of the information on the primary structure of prolamins has been obtained from the nucleotide sequences of cloned cDNAs. Sequence analyses and immunological studies have revealed that within a particular cereal, the prolamins constitute polymorphic mixtures of proteins that can be classified into a number of groups, each containing an homologous group of structurally related proteins derived from multigene families of up to 30 members (93, 225). Prolamins from different cereals also exhibit sequence similarities and almost certainly share a common ancestry (66, 125, 227). Cells of the starchy endosperm also contain water- and salt-soluble proteins, which comprise 20–40% of total endosperm nitrogen in barley (78). These proteins include

β-amylase, protease, and amylase inhibitors and other proteins of unknown function (94).

CELL WALLS Walls of the starchy endosperm are similar to those of the aleurone in that they are composed mainly of arabinoxylans and (1→3, 1→4)-β-glucans (69). The walls do not possess distinct internal layering patterns, are not lignified, and, except for rice, have no associated pectin, xyloglucan, or hydroxyproline-rich glycoprotein (69). The absence of secondary thickening and lignin in walls of the starchy endosperm renders them more susceptible to rapid enzymic degradation during endosperm mobilization in the germinating grain.

HORMONE ACTION DURING GERMINATION

Germination of nondormant cereal grains is initiated by water uptake, and its successful completion is signaled by the emergence of the developing roots and shoot. Following the uptake of water, hormonal signals, probably released from the embryo, are believed to result in the synthesis of hydrolytic and other enzymes by the aleurone. In this section, I examined the evidence supporting a role for GA in the process, together with the possible functions of ABA in germinating grains.

Gibberellic Acid

Diffusible factors from the embryo can stimulate the production of hydrolytic enzymes in cereal aleurone cells; the effect can be mimicked in vitro by GA (188, 276), and isolated barley embryos release gibberellins (204). These observations have led to the conclusion that GA released from the embryo during the initial stages of germination diffuses to the aleurone layer and induces reserve mobilization and the synthesis and secretion of hydrolytic enzymes. Such a model is consistent with the overall pattern of endosperm modification (160), but current evidence also suggests that the model is oversimplified. First, it is not certain that the factor diffusing from the embryo in intact grain is, in all cases, GA; direct evidence for the distribution of GA in intact, germinating grains is not yet available. Second, hormonal communication in the germinating grain is significantly more complicated than initially suspected (250). Aleurone cells of maize (92), sorghum (52), and oats (190) are relatively insensitive to added GA_3, as are certain varieties of barley and wheat (182, 208). The yardstick for GA sensitivity has been the aleurone layer of Himalaya barley, usually isolated from grain of selected harvests collected in specific locations. These aleurone layers secrete high levels of α-amylase in the presence of GA_3 and little or no enzyme in its absence, but this response is not typical. In other barley varieties and cereal species, GA_3 acts merely to accelerate rather than to initiate enzyme production and associated changes in aleurone ultrastructure (114, 182, 208).

Third, the site of synthesis of GA in the germinating grain has not been demonstrated convincingly, and this situation has limited our understanding of germination physiology. In barley, the embryonic axis (152), the scutellum (204), and the aleurone (12) have all been proposed as sources of the hormone, although the role of the aleurone as a source has been questioned (79). Furthermore, residual GA in ungerminated grains may be an important source of the hormone during germination (208, 251), and varying levels of residual hormone may account, in part at least, for variability in GA responsiveness within and between cereal species. A fourth complication relates to the chemical variants of GA found in germinating grain (79, 197) and whether these variants originate from different tissues or are responsible for separate functions. Isolated barley aleurone layers secrete α-amylase in response to several GAs (50).

In the decades since the initial observations were documented, the influence of exogenous GA on transcription, translation, and enzyme release in isolated aleurone layers has been confirmed (109), but little progress has been made in defining the molecular mechanism of its action. By analogy with other eukaryotic hormones, protein receptors of GA would be expected to mediate its function. The specific receptor(s) might be located in the plasma membrane, where GA binding could elicit a cascade of intracellular responses, through the action of second-messenger molecules or activated enzymes, in a mechanism similar to that observed for polypeptide-hormone and growth-factor action in mammalian cells. Calcium and phosphate ions, albeit at high concentrations, have been implicated in the GA response in cereal aleurone (114, 221). Protein phosphorylation (206) and protein kinases, including Ca^{2+}-regulated protein kinases, have been identified in plants (62, 199, 220), together with other components of the calcium signal transduction system (95, 189). These systems have not been examined in cereal aleurone. Another possibility is that the putative GA receptor protein is cytosolic and that a GA-receptor complex is translocated to the nucleus where it binds to DNA sequences flanking hormone responsive genes, causing their transcription. This is analogous to the mechanism of steroid-hormone action in mammalian cells. In addition to the specific binding models for GA action, it has been suggested that GA acts nonspecifically to stimulate the expression of all active genes in the cell, rather than exerting its effect on the transcription of specific genes (20). A relatively nonspecific role is indicated by observations that GA forms a complex with phosphatidyl choline, thereby causing increased fluidity of cellular membranes (267). This might lead to increased permeability to ions and to regulatory compounds, or to changes in the activities of membrane-bound enzymes, which could, in turn, alter the interactions between DNA and transcription factors. A high level of inorganic phosphate generated by GA_3 action has been suggested as a "nonspecific" trigger for the enhancement of certain aleurone enzymes (221). Although the weight of evidence supports a role for a diffusible factor (which may or may not be a gibberellin) in the germination of cereal grains, there remain wide gaps in our knowledge of the process. Most information has been obtained for aleurone cells, yet the scutellar epithelium participates in hydrolase secretion and might also be expected to respond to GA. Exogenous GA_3 failed to enhance α-amylase secretion in excised scutella from rice (170), sorghum (5), and barley (149). However, small but significant increases in (1→3, 1→4)-β-glucanase secretion from isolated barley scutella are induced by GA_3, and the effects of exogenous GA on excised scutella may be partly masked by endogenous hormone (236). Moreover, scutellar epithelial cells may respond to GAs other than GA_3, or to other diffusible factors.

Abscisic Acid

During grain development ABA accumulates in the embryo and endosperm and is believed to play a major role both in the control of embryo maturation and in the prevention of precocious germination of the morphologically mature but still unripe grain (119, 203). As the grain dries, however, ABA levels fall and the barrier to germination is lifted (119). The ABA effect in unripe grain is two-fold. First, the hormone suppresses the expression of "germination-specific" genes, namely those that are expressed in the presence of GA (263); this may help prevent precocious germination. Second, ABA induces the expression of "embryogenic" genes, some of which may be necessary for embryo maturation (263). Among the identified proteins that accumulate in the embryo in response to ABA are wheat germ agglutinin (252), rice lectin (196), and a 7S globulin storage protein (263). Neither the precise location of these proteins in the embryo nor the site(s) of action of ABA have been defined.

The dual effect of ABA can also be observed in aleurone layers isolated from mature grains, where it not only suppresses the expression of genes encoding GA-induced proteins, including (1→3, 1→4)-β-glucanase and α-amylase, but also raises the levels of ABA-specific proteins (109, 111, 142, 174, 177). One of the ABA-induced proteins in barley aleurone layers is an inhibitor of endogenous α-amylase (175, 179, 259); this protein may inhibit any α-amylase that may be synthesized during premature germination. Thus, α-amylase activity could be suppressed by ABA both at the transcriptional level and at the enzyme level itself. The identities of other proteins induced by ABA in aleurone layers have not been established. They may include inhibitors of key enzymes in the germination process, and possibly proteins involved in aleurone maturation or in the protection of grain from pathogen invasion.

Retention of ABA sensitivity by aleurone layers of mature cereal grains suggests that ABA plays a regulatory role in germination. Perhaps ABA-inducible genes function to arrest endosperm mobilization transiently in response to unfavorable environmental conditions such as water, temperature, or salt stress. Indeed, dehydration of young barley seedlings is accompanied by large increases in ABA-inducible mRNAs in the aleurone (46) and in the ABA-inducible α-amylase inhibitor protein (R. D. Hill, personal communication). Furthermore, the ABA-inducible proteins in barley aleurone are mostly heat stable and could protect cellular proteins or organelles from heat or moisture stress during the final stages of grain maturation or during germination (J. V. Jacobsen, personal communication).

ENZYME SYNTHESIS AND SECRETION DURING GERMINATION

During germination and early seedling development, cellular activity in the aleurone is directed rapidly to the synthesis and secretion of hydrolytic enzymes that catalyze the depolymerization of storage macromolecules in cells of the starchy endosperm. The scutellar epithelium performs a dual role. First, production and secretion of hydrolases by the scutellar epithelium is an early function, the relative importance of which depends on the cereal species (5, 186). Second, the uptake of endosperm degradation products and their translocation to the developing seedling are additional scutellar functions crucially important for the successful germination of the grain. Enzymes produced by the aleurone and scutellar epithelium during germination can be classified broadly into three groups: (a) "house-keeping" enzymes involved in ongoing intermediary metabolism within the cells, (b) enzymes that mobilize the reserves of the aleurone and scutellar cells themselves, and (c) hydrolytic enzymes, that are synthesized de novo and secreted into the starchy endosperm. In the sections below, I discuss only enzymes involved in internal aleurone and scutellar reserve mobilization and those secreted for the mobilization of starchy endosperm reserves.

Mobilization of Aleurone and Scutellar Reserves

ULTRASTRUCTURAL CHANGES   The protein matrix of aleurone grains and their phytin and niacytin inclusions disappear rapidly on germination; protein bodies coalesce to form vacuoles that eventually occupy most of the cell (254). Lipid bodies decrease in abundance and become smaller, while mitochondria develop well-defined cristae characteristic of high levels of metabolic activity (81, 254). The mobilization of reserves is accompanied by a dramatic proliferation of rough ER, initially associated with the lipid bodies around aleurone grains but later developing into prominent, fenestrated stacks

(40, 51, 113, 219). Abundant Golgi complexes also appear (64, 84, 97). These ultrastructural changes are associated with the mobilization of amino acid, phosphate, and lipid reserves and the elaboration of protein-synthesizing and secretory machinery necessary for the aleurone's function in starchy endosperm degradation.

Similarly, in the scutellar epithelium of germinating cereals, protein bodies and associated phytin inclusions are mobilized rapidly, and their remnants coalesce into large vacuoles. Lipid bodies slowly disappear (63), mitochondria become metabolically active, and a marked development of ER and Golgi is observed (7, 81, 184, 186, 241, 242). In contrast to aleurone cells, both the parenchyma and epithelial cells of the scutellum accumulate starch granules in the early stages of germination (7, 184, 242). Later in germination, starch disappears from epithelial cells but persists longer in parenchyma cells (242). α-Amylases may participate in the removal of the scutellar starch granules.

Ultrastructural changes that occur in the scutellar epithelium but not in the aleurone include partial degradation of the anticlinal walls, resulting in lateral disconnection of individual cells. Their simultaneous elongation leads to the formation of cylindrical papillae (181, 184, 242). A significant increase in surface area, which is undoubtedly related to the absorptive role of the epithelial cells, is thereby achieved. In some oat *(Avena)* species this adaptation is further developed; the apical region of the scutellum elongates to form a papillate, finger-like projection, that extends through the starchy endosperm towards the distal end of the grain (181).

DEGRADATIVE ENZYMES   Protein bodies have been identified as a major repository of hydrolytic enzymes in ungerminated grain (3, 112). These enzymes are prime candidates for a role in the mobilization of aleurone and scutellar reserves during the early stages of germination. Whether GA participates in the mobilization of aleurone or scutellar reserves is unknown. A battery of endo- and exopeptidases, including acid endopeptidases, acid carboxypeptidases, neutral and alkaline aminopeptidases, and dipeptidases, has been detected in ungerminated cereals (163); but the specific roles of individual enzymes in aleurone grain degradation have not been established.

Phosphate reserves associated with the phytin inclusions of aleurone and scutellar protein bodies are released by phytases in the early stages of germination. These enzymes are apparently members of a large group of acid phosphatases. The acid phosphatase isoenzymes involved in phytin hydrolysis are probably those found in aleurone grains of ungerminated grain (11, 74). Other acid phosphatases may function as ATPases (115), protein phosphatases (198), nucleotide phosphatases (48), and in general metabolic reactions. Levels of certain acid phosphatases are enhanced by GA and some are secreted from GA-treated barley aleurone layers; the latter are likely to participate in starchy endosperm mobilization.

Lipases play a central role in the release of fatty acids from the reserve triacylglycerols of lipid bodies in the aleurone and scutellum of germinating cereals. Lipases are synthesized de novo on free ribosomes in the scutellum of germinating maize and are translocated to the membrane of the lipid bodies where they remain bound (258). Lipase synthesis in the aleurone is more complicated. Significant lipase activity is detected in isolated barley aleurone layers before GA application, but preliminary imbibition of half grains may induce its synthesis (65). Thus, it is not clear whether lipase is present in the aleurone of ungerminated grain or whether it is all synthesized de novo. However, a "storage" form of the enzyme is initially associated with protein bodies of imbibed barley aleurone layers, and GA_3 induces its translocation, via a membrane route, to an active form in lipid bodies (65).

Fatty acids released by lipases act as precursors for the extensive phospholipid synthesis necessary to support proliferation of cellular endomembrane systems (253). This may be their primary role early in the germination process (262). After the depletion of soluble sugars stored in the scutellum and aleurone of ungerminated grains, the fatty acids may also function in the provision of metabolic energy or in glucose synthesis. Enzymes of the glyoxylic acid pathway and glyoxysomes have been reported in barley aleurone (114); but in the scutellar epithelium of germinating rice, glyoxysomes are difficult to detect (8), and any synthesis of glucose that might be necessary could be generated by hydrolysis of scutellar starch deposits.

Secretory Pathway and Posttranslational Modification

Investigations of enzyme secretion in germinating cereals have so far been directed mainly towards α-amylase secretion from barley aleurone layers, where the enzymes are synthesized on rough ER and targeted to the lumen of the ER, probably through the interaction of signal-recognition particles and leader peptides (9). Leader peptides have been identified from nucleotide sequence analyses of cDNAs encoding barley α-amylases (47, 213). By analogy with mammalian systems, leader peptides are probably excised from primary translation products in the ER, and the rice scutellar α-amylase leader peptide is excised before translation is complete (169). The participation of both the rough ER and the Golgi apparatus in the secretion of barley α-amylase is now generally accepted (51, 64, 84, 97). Direct immunocytochemical evidence has been obtained by probing tissue sections with affinity-purified polyclonal antibodies against α-amylase (84). High levels of probe were detected over the lumen of the rough ER, and binding was particularly concentrated over both the *cis* and *trans* regions of the Golgi apparatus (84). Golgi-derived vesicles appear to be involved in the delivery of secretory products to the plasma membrane (64). The Golgi has also been implicated in the secretion of α-amylase from the scutellar epithelial cells of barley (81, 184) and rice (9). It is too early to conclude that all α-amylase isoenzymes and other secreted hydrolases follow a secretory route through the Golgi (9, 161), although this seems likely. Experiments with specific, gold-labeled monoclonal antibodies may soon enable investigators to define the secretory pathways of individual isoenzymes.

In eukaryotic cells, the Golgi plays a central role in sorting, modifying, and packaging newly synthesized protein transported from the ER, although the ER-to-Golgi transport process itself is not understood. Rice scutellar α-amylases are N-glycosylated, and oligosaccharide chain elaboration occurs co- and posttranslationally as the enzymes move through the ER and Golgi (9). Wheat α-amylase contains no associated carbohydrate (249), and although glycosylated barley α-amylases have been reported (212), recent experiments have provided no evidence that barley α-amylase isoenzymes are either N- or O-glycosylated (108, 116). Nucleotide sequence analyses of cDNAs show that a barley α-amylase from the low-pI family has no Asn-X-Ser/Thr glycosylation site in the mature peptide but that a high-pI α-amylase does (213). Barley $(1{\rightarrow}3, 1{\rightarrow}4)$-$\beta$-glucanase isoenzyme II, which is secreted predominantly from the aleurone (236), contains one glycosylation site (67) and approximately 4% by weight carbohydrate (269); isoenzyme I, the major form of the enzyme secreted from isolated scutella (236), carries little, if any, associated carbohydrate (269). The nuclease I secreted by barley aleurone layers is also an N-glycosylated protein, containing approximately 6% carbohydrate (39), as are the carboxypeptidases in germinating barley and wheat (34). In addition to N-glycosylation at Asn-X-Ser/Thr sequences, O-glycosylation of Ser or Thr residues may occur. Furthermore, an increasing number of eukaryotic secretory proteins are found to contain sulfated tyrosine residues. This modification occurs in the *trans* Golgi and can affect both the intracellular transport of proteins and their biological activity (106). Whether hydrolytic enzymes secreted from the aleurone or scutellum of germinating cereal grains are modified in these ways is largely unexplored, but it has been shown that the unglycosylated, low-pI group of barley aleurone α-amylases is derived from an intracellular precursor by posttranslational modification that leads to a lowering of the pI of the enzymes without any apparent change in their molecular weight (108, 116). Sulfation and phosphorylation are apparently not involved (116). Changes in pI can also be induced by post-translational acylation. Significant levels of trimethylation of lysyl residues in wheat α-amylases have been reported (173), but these are unlikely to affect pI's in the range 4.5–6.5.

Functional activity of an enzyme depends ultimately on the correct folding of the nascent polypeptide. Formation of disulfide bonds, which are a feature of many secreted proteins, is an early posttranslational event associated with protein folding in the lumen of the rough ER. Evidence is accumulating that the reaction is catalyzed in mammalian systems by the enzyme protein-disulfide isomerase (124, 216). The enzyme has also been reported in developing wheat endosperm (70). In addition, *cis-trans* isomerization of prolyl-peptide bonds appears to be an important and possibly rate-limiting determinant of protein folding, and in some instances it may be enzymically catalyzed (132).

ROLE OF CALCIUM   The levels of hydrolytic enzymes secreted by GA-treated cereal aleurone layers and scutella are strongly influenced by Ca^{2+}. Thus, secretion of rice scutellar α-amylase (168), barley scutellar (1→3, 1→4)-β-glucanase (236) and barley aleurone α-amylase, (1→3, 1→4)-β-glucanases, endopeptidases, nucleases, acid phosphatases, and (1→3)-β-glucanases are stimulated by 10–20 mM concentrations of exogenous Ca^{2+} (49, 117, 236). Furthermore, individual isoenzymes of α-amylase (9, 55) and (1→3, 1→4)-β-glucanase (236) are affected differentially by Ca^{2+} and GA. The important, but as yet unanswered, questions raised by these observations are: Where does Ca^{2+} exert its effect, and what is the molecular mechanism for Ca^{2+} participation in GA-induced responses?

Accumulation of mRNA encoding low-and high-pI α-amylases in barley aleurone layers is unaffected by Ca^{2+}, indicating that Ca^{2+} does not affect transcription of these genes, although it does control high-pI α-amylase synthesis at a later stage in the biosynthetic pathway (55). Experiments with isolated aleurone layers and aleurone protoplasts show that Ca^{2+} regulates secretion of barley α-amylase at the intracellular transport level rather than through an ion exchange process in the cell wall (43). The effects may be explained simply by a Ca^{2+} requirement for intracellular membrane fusion, for example, for the fusion of secretory vesicles with the plasma membrane during exocytosis. Alternatively, Ca^{2+} may play a far more complicated role as a multifunctional regulator of stimulus-secretion coupling (95).

A major concern in assigning a functional role to extracellular Ca^{2+} in the secretion of hydrolytic enzymes from cereal aleurone layers and scutella has been the very high concentrations (10–30 mM) required to elicit the various responses (95). The cytoplasmic Ca^{2+} concentration in aleurone protoplasts has been estimated to be 200–250 nM and increases to 350 nM on addition of 10 mM Ca^{2+} to the incubation medium (42), but there have been no attempts to measure Ca^{2+} levels surrounding aleurone and scutellar cells in intact, germinating grains. However, a Ca^{2+} content of 1.2 μg per isolated starchy endosperm in ungerminated barley has been determined by neutron activation analysis (235). Assuming a starchy endosperm volume of 25 μl and uniform distribution, the concentration of Ca^{2+} may be calculated at approximately 1 mM, a value that at least approaches Ca^{2+} levels required for an effect in vitro.

PASSAGE THROUGH CELL WALLS   Following secretion of hydrolytic enzymes across the plasma membranes of aleurone or scutellar epithelial cells, the enzymes are confronted with two potential barriers that may severely limit their access to substrates in the starchy endosperm. These barriers are the walls of the secretory aleurone or scutellar cells themselves, and those of the starchy endosperm cells. The fine structures of cereal arabinoxylans and $(1\rightarrow3, 1\rightarrow4)$-$\beta$-glucans have been defined in detail, and their molecular organization in the cell wall has been likened to a cellulose-reinforced multicomponent gel (69). The reinforced gel model satisfies a number of functional requirements for the cell walls: It would provide a skeletal framework and intercellular cohesive forces necessary for the maintenance of tissue integrity; it would provide a physical barrier, albeit weak, against penetration by insects and microorganisms; it would allow free diffusion of water, phytohormones, and low-molecular-weight metabolites; and it would help prevent desiccation of the grain during development or germination through the ability of constituent polysaccharides to bind water (69).

However, the gel-like matrix would not be freely permeable to the hydrolases secreted by the aleurone and scutellar epithelial cells. It has been estimated that globular proteins of molecular weights between 20,000 and 60,000 might penetrate the pores of primary plant cell walls (44, 248), but penetration into the wall cannot be equated with diffusion across the entire wall (69). Elegant immunocytochemical studies with gold-labeled antibodies (83) show that α-amylase is released from barley aleurone cells via channels formed in the outer wall layer (254). The digestion of the outer wall channels is presumably mediated by endoxylanases and other enzymes that participate in arabinoxylan depolymerization, since arabinoxylans appear to be major components of the outer wall. However, this suggestion has to be reconciled with the observation that endoxylanases are detected in the medium surrounding isolated barley aleurone layers somewhat later than other hydrolytic enzymes (243). A possible explanation is that the xylanases are tightly bound to the aleurone wall during the formation of channels and are only released when degradation of the outer wall layer is complete. This is consistent with the detection of arabinoxylan degradation products from barley aleurone layers some 16 hr before the release of xylanase itself; however, no evidence for wall-bound enzyme has been found (54), and the apparent discrepancy in the timing of endoxylanase appearance has yet to be satisfactorily explained.

It is not yet clear how secreted hydrolytic enzymes penetrate the thin, inner layer of the aleurone wall. This layer appears to remain intact until the secretory process is complete (83, 245); it presumably persists to provide continuing structural support for the active aleurone protoplast. The molecular organization of the inner layer may differ from that of other primary walls and allow enzyme diffusion without prior modification of the wall layer. More probably, a specific component of the inner wall layer, such as $(1\rightarrow3, 1\rightarrow4)$-$\beta$-glucan, is removed in the early stages of germination, leaving a resistant polymeric framework, possibly proteinaceous in nature, through which enzymes can move freely. The inner resistant layer of barley aleurone walls is continuous with tubes of similarly resistant material around plasmodesmata, and these plasmodesmal wall tubes could provide a network to facilitate the distribution of wall-degrading enzymes through the outer wall layer (83). Ferulic acid is not involved in the resistance of the inner wall layer to digestion, because it is distributed through both layers and aleurone wall hydrolysis is accompanied by its removal (82).

The outer wall layer of the scutellar epithelium is also digested early in the germination process, whereas the inner layer persists for many days (184, 242). The route of hydrolytic enzyme secretion through the scutellar wall has not been investigated.

SECRETION OF PEPTIDASES   Among the hydrolytic enzymes secreted from the aleurone and scutellum of germinating cereals are endo- and exo-peptidases, many of which might be expected to pose problems for the parent cells through their potential to nonspecifically degrade cellular proteins. In mammalian exocrine glands, intracellular proteins can be protected from the action of secreted peptidases by synthesis of the peptidases in inactive, zymogen forms, which are activated by proteolysis after secretion. Barley scutellar carboxypeptidase I is synthesized as a precursor polypeptide, and it has been suggested that the precursor is processed only after its secretion into the starchy endosperm (56). Although there is no evidence that the precursor form does in fact represent an inactive procarboxypeptidase I zymogen (56), its existence raises questions about whether cereal peptidases might be synthesized in zymogen form and about the locations and characteristics of proteolytic enzymes that mediate in their maturation.

REGULATION OF GENE EXPRESSION

In germinating cereals, different genes are progressively activated in the various tissues of the grain. Within individual cells, some genes are expressed in the early stages of cellular activity while others are activated later. Our understanding of the coordination of gene expression in the intact grain is rudimentary, but progress is being made in specific tissues. Once again the experimental convenience of isolated wheat and barley aleurone layers has resulted in their use as model systems for the examination of regulatory mechanisms controlling plant gene expression, with a major emphasis on the role of GA in the control of α-amylase expression. The underlying aim of this work has been to determine how aleurone cells become dedicated to the prolonged production and secretion of a relatively small number of hydrolytic enzymes. While it is evident that the overriding control of expression is exerted at the transcriptional level and that most of our current knowledge for germinating cereals is related to this level of control, one might predict that posttranscriptional and translational controls play significant complementary roles in the totality of gene regulation. Evidence to support this notion is now emerging.

Transcriptional Control

When aleurone layers are treated with GA, levels of mRNA for α-amylase, (1→3, 1→4)-β-glucanase, and other unidentified proteins increase, as measured by the abundance of specific polypeptide products in mRNA translation systems or by the use of DNA probes to quantitate levels of mRNA transcripts themselves (22, 47, 177, 214). In nuclei isolated from GA-treated barley aleurone protoplasts, total RNA transcripts decrease in the presence of GA, but the abundance of α-amylase transcripts increases 5.5-fold over untreated controls (107).

Based on these results, it can be concluded that expression of genes encoding α-amylase and other hydrolases that participate in endosperm mobilization results from transcriptional activation. The availability of specific cDNA clones has enabled examination of the abundance of mRNA encoding specific α-amylase isoenzymes. Within a particular α-amylase gene family, individual gene expression is synchronous (20, 109, 215). In contrast, different families of α-amylase genes are apparently subject to differential regulation (105, 135, 213). Thus, low-pI α-amylase mRNA is present in relatively high levels in unstimulated barley aleurone layers and increases 20-fold after treatment with low levels of GA. High-pI α-amylase mRNAs are barely detectable in unstimulated aleurone layers, but they increase up to 100-fold over 16 hr in the presence of high concentrations of GA and then decrease (109, 213). The different levels of mRNA for the α-amylase gene families and their differential response to GA are reflected in the levels of the corresponding isoenzymes (109), and similar differences are observed in the expression of α-amylase gene families in wheat aleurone (20). In addition to the differential expression of α-amylase gene families in isolated aleurone layers, there is some evidence for tissue-specific differences of α-amylase and (1→3, 1→4)-β-glucanase gene expression in the aleurone and scutellum (149, 160, 236).

To understand the molecular basis for temporal, tissue-specific, and hormone-inducible gene expression we need, in the first instance, to define the structure of individual genes and to identify promoter and enhancer elements in their 5'- and 3'-flanking regions that are required for expression and regulation. Eukaryotic promoters are typically 100–200 base pairs long and are located immediately 5' to the transcription start point, where they are required for accurate and efficient initiation of transcription by specific interaction with RNA polymerases. Promoters usually include an AT-rich TATA box and additional upstream promoter elements of 8–12 nucleotide pairs. Enhancers, which increase the rate of transcription from promoters, also contain short, modular recognition elements that specifically interact with proteins but are either closely associated with the promoter or located at a distance from it. A common characteristic of enhancers is that they can exert their effect on *cis*-linked promoters in an orientation-independent fashion from distances of several kilobase pairs either 5' or 3' to the promoter. One model to explain the mechanism of enhancer action predicts the interaction of proteins bound to both the enhancer and the promoter, leading to the formation of a transcription complex and a loop of intervening DNA (156). Inducible enhancers respond to a variety of external stimuli, including steroid hormones, and tissue-specific *trans*-acting transcription factors (proteins) could control the expression of a number of genes transcribed in that tissue (156).

Genes encoding barley α-amylase (121, 261), (1→3, 1→4)-β-glucanase (B. Ahluwalia, N. Slakeski, and G. B. Fincher, unpublished), a wheat carboxypeptidase (21) and a wheat α-amylase (23) have now been isolated. A barley gene of the low-pI α-amylase family has three introns and one from the high-pI α-amylase family, two introns (121, 261). The detection of both direct and inverted repeats in the 5'-flanking regions of the two α-amylase genes (121, 261) and a putative peptidase gene (261) may be an important first step in identifying key sequence elements that participate in tissue-specific and hormone-regulated expression of genes encoding the hydrolytic enzymes responsible for endosperm mobilizaton in germinating cereals. Promoters from these GA-induced genes can now be attached to suitable reporter genes and, following reintroduction of the DNA into appropriate GA-sensitive cells, sequences involved in transcription activation may be identified. Excised promoters could also be used to isolate or identify protein factors involved in gene expression in the aleurone and scutellum.

mRNA Stability

Steady-state levels of a eukaryotic mRNA are determined by both the rate of transcription of the gene and the rate of decay of the mRNA in the cytoplasm. The rate of mRNA appearance depends on the rate of transcription, the efficiency of mRNA processing, and transport of the mRNA to the cytoplasm. Eukaryotic mRNAs have half-life values ranging from minutes to hours or even days, and the rate of decay of mRNAs is therefore an important determinant of the net rate of gene expression. Stability of mRNAs has been attributed to their ability to form secondary stem-loop structures and to the abundance of sequence motifs at their 5' and 3' ends that are recognized and degraded by endo- and exonucleases (30, 274).

In germinating cereals, stable mRNAs would offer energetic advantages to cells dedicated to the long-term production of hydrolytic enzymes, and indeed the half-life of α-amylase mRNAs from barley aleurone has been estimated to be longer than 100 hr (99). These mRNAs have the potential to form very stable hairpin loops at their 5' ends (213). Association of mRNAs with the ER also increases their stability (25), possibly by steric protection from nucleases. Differential mRNA stability rather than transcriptional control could account for the differential expression of α-amylase gene families (20, 213). The various structural properties or cellular events that influence mRNA stability in the aleurone and scutellum of germinating cereals await detailed investigation.

Translational Efficiency

In vitro translation of barley aleurone RNA preparations indicates that α-amylase mRNA constitutes approximately 20% of total translatable mRNA, whereas in vivo labeling suggests that α-amylase protein may account for more than 50% of total protein synthesis (109). Although caution must be exercised in the comparison of in vivo and in vitro experiments, it is possible that α-amylase mRNA is translated at a higher rate than other aleurone mRNAs and that this represents an additional control point aimed at maximizing the expression of a relatively small number of genes.

In both eukaryotic and prokaryotic systems translation is a non-uniform process. The rate-limiting step in the elongation cycle is the stochastic search for the cognate, ternary tRNA complex corresponding to a particular codon, and differences in rates of translation are linked to tRNA availability for various codons (41, 224). In highly expressed genes, codons recognized by abundant tRNAs are used more than those recognized by rare tRNAs (41). Codon usage in highly expressed genes is more biased than in genes for rare proteins (224). In yeast, highly expressed genes have a higher (G+C) content, largely because of increased (G+C) usage in the third position of codons (224).

Genes encoding α-amylases (47, 213) and (1→3, 1→4)-β-glucanases (67) from barley aleurone, (1→3)-β-glucanase from barley scutellum (P. B. Hoj, N. P. Doan, D. J. Hartman, and G. B. Fincher, unpublished), and two (1→3, 1→4)-β-glucanases from wheat aleurone (D. Lai and G. B. Fincher, unpublished) are all characterized by (G+C) contents in excess of 60% and a strong bias (80–90%) towards the use of G and C in the wobble base position of codons. In contrast, genes encoding cereal storage proteins and the barley scutellar carboxypeptidase I have a more balanced codon usage, with overall (G+C) contents of approximately 50% (29, 56). It is not yet known whether these dramatic biases in codon usage are correlated with the relative abundance of corresponding isoaccepting tRNAs in aleurone and scutellar cells of germinating grain. However, if this is so, it could explain the apparent differences in translational efficiencies of mRNAs in these cells. Again, faster translation of abundant, stable mRNAs in the aleurone or scutellum of germinating cereals would further increase the rate of expression of hydrolytic enzymes required for rapid endosperm mobilization.

MOBILIZATION OF STARCHY ENDOSPERM RESERVES

Endosperm dissolution in germinating cereals generally begins adjacent to the scutellum and progresses from the proximal to distal end of the grain as a front moving away from, but approximately parallel to, the face of the scutellum, and advancing faster adjacent to the aleurone layer. This pattern reflects the secretion of hydrolytic enzymes from the scutellum and the aleurone. In small grains, such as millet and sorghum, and also in rice, the scutellum appears to be the major source of α-amylase; but the aleurone cells are extensively modified during germination and may secrete stored minerals, phosphate, low levels of α-amylase, or hydrolytic enzymes other than α-amylase (5, 7, 9, 80). In larger grains, such as barley, wheat, rye, oats, and maize, cooperative secretion of hydrolases from both the scutellum and aleurone (36, 61, 77, 160, 186) may overcome problems of longer diffusion pathways between the scutellum and reserve polymers in the distal region of the grain. Thus, degradative enzymes secreted initially from the scutellar epithelial layer cause degradation of the starchy endosperm adjacent to the scutellum. After dissolution of the proximal starchy endosperm, the scutellum appears to relinquish its secretory role to the aleurone, which becomes the major source of hydrolytic enzymes (77, 160). The absorptive function of the scutellar epithelium is presumably active throughout.

The spatially and temporally coordinated pattern of endosperm dissolution is demonstrated by the pattern of expression of genes encoding the cell wall degrading $(1{\rightarrow}3, 1{\rightarrow}4)$-$\beta$-glucanases in intact, germinating barley grains (160; Figure 1). Hybridization histochemistry shows that $(1{\rightarrow}3, 1{\rightarrow}4)$-$\beta$-glucanase mRNA is confined to the scutellar epithelium one day after the initiation of germination; no expression is detected in the aleurone at this stage. After two days, however, expression in the scutellar epithelium begins to decline and finally ceases, while expression in the aleurone commences and progresses from the proximal to the distal end of the grain (160; Figure 1). The participation of both the scutellar epithelium and the aleurone in production and secretion of these enzymes is consistent with the pattern of wall modification in the endosperm (77). A major advantage of hybridization histochemistry is that it avoids the interpretative problems associated with tissue disruption and with defining the origin of secreted enzymes (160, 191). On the other hand, care must be taken to avoid cross-hybridization of the cDNA probe with related but distinct mRNA species (160). As additional cDNA clones become available, hybridization histochemistry will undoubtedly be refined through the use of specific oligonucleotides or by the manipulation of hybridization stringency, to define the location and developmental patterns of individual isoenzymes. This approach has been taken in germinating rice, where α-amylase isoenzymes expressed in the scutellar epithelium differ from those expressed in the aleurone (R. L. Rodriguez, personal communication).

During dissolution of the starchy endosperm, polymeric components of cell walls, starch granules, protein bodies, and residual nucleic acids are depolymerized by the concerted action of many enzymes. The pH of the starchy endosperm is actively maintained in the range 5.0–5.2 during germination, possibly by the secretion of malic acid by the aleurone (86, 165), and the hydrolytic enzymes usually exhibit pH optima in the same range. I discuss the role of these enzymes in endosperm mobilization in the following sections, together with emerging evidence for a possible role for secreted hydrolases in protecting germinating grain against pathogen attack.

Cell Walls

During germination of barley and wheat, walls of the starchy endosperm appear to be completely degraded, although final remnants of the walls disappear relatively slowly after the initial front of wall-degrading enzymes has passed (68, 223). The framework of cell walls of sorghum endosperm appears visually unchanged after germination (80), but limited modification of the wall matrix has probably occurred to allow enzyme penetration.

ARABINOXYLAN DEGRADATION   Enzymes capable of degrading arabinoxylans to their constituent monosaccharides have been detected in germinating cereals. These include α-arabinofuranosidase, endoxylanase, exoxylanase, and xylobiase (or β-xylopyranosidase) (202, 243). Progressive product analyses indicate that the initial action of arabinofuranosidases to remove arabinosyl side chains opens the way for the release of xylobiose by an exoxylanase; the products of endoxylanases are not detected until much later (202). Three endoxylanases of M_r 41,000 and pI 5.2 have been purified from barley grain five days after the initiation of germination (A. M. Slade, P. B. Hoj, and G. B. Fincher, unpublished). As reported in early work (202), their levels during the initial stages of germination are very low. One might question, therefore, whether endoxylanases play an important role in the

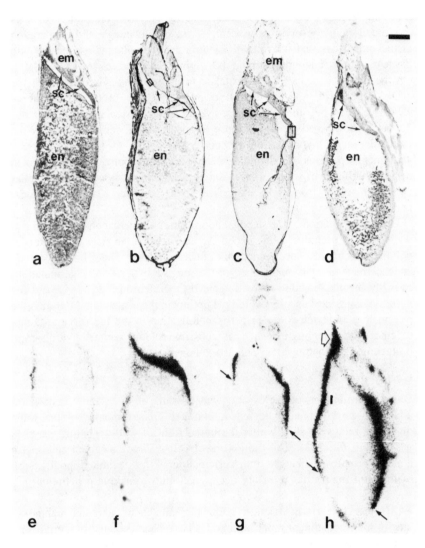

Figure 1 Localization of mRNAs encoding (1→3, 1→4)-β-glucanases in germinating barley grains by hybridization histochemistry (160). Near median longitudinal sections (*top panel*) were taken from barley grains 0, 1, 2, and 4 days after the initiation of germination and hybridized with a ^{32}P-labeled cDNA probe encoding (1→3, 1→4)-β-glucanase. Bound probe was detected by X-ray film autoradiography of the whole section (*bottom panel*). Control sections are shown elsewhere (160). Scale bar = 1.0 mm. *a*: After 0 days germination the embryo (em) is rehydrated. The scutellum (sc) can be detected at the interface of the embryo and the starchy endosperm (en). *e*: The probe binds diffusely over the entire section at low levels. *b*: After 1 day of germination, the starchy endosperm (en) is swollen. *f*: A high level of probe is bound to the scutellum (sc). *c*: After 2 days, endosperm dissolution has commenced adjacent to the scutellum.

initial degradation of walls in the starchy endosperm, at least in barley. Alternatively, they may simply remain bound to cell walls initially, making their extraction difficult. Supporting the latter view, depolymerization of wall arabinoxylans in germinating wheat proceeds without any dramatic, initial changes in xylose: arabinose ratios, suggesting that endoxylanases are active throughout (68).

In isolated barley aleurone layers, GA enchances the activity of endoxyla-nase, arabinofuranosidase, and xylopyranosidase (243); but again arabinofuranosidase and exoxylanases were released prior to endoxylanase (54, 243). A xylanase of M_r 34,000 and pI 4.6 has been purified 29-fold from the secreted proteins of barley aleurone layers (26), but its precise substrate specificity remains uncertain. In view of the diversity of the enzymes that participate in arabinoxylan degradation in germinating cereals, it is crucial to define in detail the structure of the substrate used, the structure of oligosaccharides released, and the time course of appearance of individual oligosaccharide products before a particular enzyme can be correctly identified and classified.

$(1{\rightarrow}3,\ 1{\rightarrow}4)$-$\beta$-GLUCAN DEGRADATION   Three major groups of β-glucan endohydrolases have the potential to degrade cell wall $(1{\rightarrow}3,\ 1{\rightarrow}4)$-$\beta$-glucans in germinating cereals. These include $(1{\rightarrow}3,\ 1{\rightarrow}4)$-$\beta$-glucan 4-glucanohydrolase (EC 3.2.1.73) (269, 270), $(1{\rightarrow}4)$-β-glucan 4-glucanohydrolase (cellulase, EC 3.2.1.4) (103, 158, 275), and, if adjacent $(1{\rightarrow}3)$-linkages are present, $(1{\rightarrow}3)$-β-glucan 3-glucanohydrolase (EC 3.2.1.39) (19, 100). Levels of $(1{\rightarrow}4)$-β-glucanase (cellulase), in germinating barley at least, are low and are mostly contributed by commensal microflora associated with the grain (103, 158). $(1{\rightarrow}3)$-β-Glucanase activity can be relatively high in germinating barley and rye (16, 17), and although the enzyme has been implicated in the initial release of cell wall $(1{\rightarrow}3,\ 1{\rightarrow}4)$-$\beta$-glucan by the hydrolysis of regions containing contiguous $(1{\rightarrow}3)$-linkages (19), no conclusive evidence for the widespread occurrence of blocks of adjacent $(1{\rightarrow}3)$-linkages in cereal $(1{\rightarrow}3,\ 1{\rightarrow}4)$-$\beta$-glucans is available (271). A role for $(1{\rightarrow}3)$-β-glucanases in cell wall degradation remains to be demonstrated (100).

←——

g: High levels of probe bind to the proximal regions of the aleurone layer and binding is more advanced, distally, on the dorsal side (arrows). There is less binding of probe to the scutellum (sc) in 2-day grain than in 1-day grain (compare g and f). d: After 4 days, extensive endosperm (en) dissolution is evident. h: The probe binds at high levels to the proximal and medial aleurone layer on the ventral side and more distally on the dorsal side (black arrows) and also to the monolayer of aleurone surrounding the scutellum (open arrow). Probe binding in the scutellum (sc) is minimal at 4 days.

Photograph prepared by Dr. Geoff McFadden, Plant Cell Biology Research Centre, University of Melbourne.

It appears, therefore, that (1→3, 1→4)-β-glucanases are primarily responsible for the degradation of cell wall (1→3, 1→4)-β-glucan during starchy endosperm mobilization. These enzymes are widely distributed in germinating cereals (6, 15, 237). Two (1→3, 1→4)-β-glucanases from barley have been characterized (269, 270), and the complete primary structure of isoenzyme II has been defined (67). Both isoenzymes exhibit similar specificities and action patterns on cell wall (1→3, 1→4)-β-glucan but differ in their primary structure and degree of glycosylation (269, 270, 272). The two genes encoding barley (1→3, 1→4)-β-glucanases probably originated by duplication of a common ancestral gene (272) and a subsequent translocation event, since the genes are now located on chromosomes 1 and 5 (144). In experiments with excised tissue fragments, isoenzyme I is secreted predominantly from isolated scutella, together with an immunologically related but as yet uncharacterized protein, designated isoenzyme III, which is not detected in whole grain (236). Isoenzyme II is secreted only from aleurone layers (236). A single isolated barley scutellum can secrete approximately 40% of the (1→3, 1→4)-β-glucanase activity secreted by a single aleurone layer (236), suggesting that the scutellum is a significant source of the enzyme, particularly in the early stages of germination (160). The relatively low abundance of translatable (1→3, 1→4)-β-glucanase mRNA in isolated scutella (176) is probably explained by the dilution of mRNA from the single layer of epithelial cells with mRNAs extracted simultaneously from metabolically active parenchymatous and vascular tissues of the scutellum (160, 176).

A β-glucanase detected in germinating sorghum does not hydrolyze barley (1→3, 1→4)-β-glucan (6) and may have a specificity tailored to the sorghum (1→3, 1→4)-β-glucan, which differs from barley and other cereal β-glucans in its relatively high content of (1→3)-linkages (60%, compared to about 30% in most cereal β-glucans) (273).

Exo-β-glucanases capable of degrading (1→3, 1→4)-β-glucan have been reported in germinating barley (201). An exo-β-glucanase recently purified from barley has an apparent M_r of 67,000 and is capable of rapidly degrading both (1→3, 1→4)- and (1→3)-β-glucans to glucose (P. B. Hoj, G. B. Fincher, unpublished), but its functional role in the germinating grain is unknown. Oligo-β-glucosides released from cell walls by endo-(1→3, 1→4)-β-glucanases can be degraded to glucose by β-glucosidases (157). The combined action of these enzymes leads to the complete conversion of wall (1→3, 1→4)-β-glucan to glucose, which represents an important energy source for the developing seedling. Up to 18.5% of total glucose released during endosperm mobilization in barley has been attributed to (1→3, 1→4)-β-glucan degradation (171).

Reserve Proteins

Mobilization of cereal storage proteins is mediated by the action of endo- and exopeptidases (163, 207). Rye, oats, barley, and wheat develop high carboxypeptidase activities relative to endopeptidases during germination, whereas rice, sorghum, and maize have much lower carboxypeptidase activity (266). Despite the widespread occurrence of endopeptidases in germinated cereal grain, relatively few have been purified and characterized. The predominant endopeptidases released from GA-treated barley aleurone layers are thiol endopeptidases (87, 238), which are synthesized de novo (87) and can be resolved into three isoenzymic forms of M_r 37,000 (123). Thiol endopeptidases from wheat, maize, and rice have M_r values in the range 20,000–25,000 (2, 58, 226), while an acid peptidase from germinated sorghum has an M_r of about 80,000 (75).

Acid carboxypeptidases are the major exopeptidases involved in storage protein mobilization in barley (163). The starchy endosperm is reportedly devoid of aminopeptidases, although these are found in the aleurone and embryo (164). Five carboxypeptidases with different but complementary substrate specificities have been identified in germinating wheat (166) and barley (162). Of these, barley carboxypeptidases I, II, and III have been purified (31–33, 256). Enzymes I and II exist as dimers of M_r approximately 110,000; each subunit consists of two polypeptide chains linked by disulfide linkages. The A-chains of enzymes I and II have 266 and 260 residues, and the B-chains 148 and 159 residues, respectively; but the amino acid sequence similarity is only 40% for the A-chains and 30% for the B-chains (232, 233). Sequence analysis of a cDNA encoding barley carboxypeptidase I shows that both chains are translated from a single mRNA. The A-chain is located at the NH_2-terminal end of the primary translation product and is separated from the B-chain by a 55-residue linker peptide (56). The linker peptide is rich in proline, arginine, and lysine residues, has a deduced pI of 11.9, and appears to be excised by endoproteolytic cleavage of peptide bonds on the COOH-terminal side of serine residues (56). In contrast, barley carboxypeptidase III is a monomer with a molecular weight of approximately 48,000 (32). Its amino acid sequence is very similar to that deduced from a GA-induced carboxypeptidase gene from wheat aleurone (21; K. Breddam, personal communication).

Wheat and barley carboxypeptidases I and III are absent from resting grains but increase to high levels during germination. Carboxypeptidases IV and V also increase, while carboxypeptidase II is abundant in ungerminated grain and decreases during germination (163, 166). A similar pattern may occur in germinating rice (57, 166). The different developmental patterns suggest that carboxypeptidases I and III differ from carboxypeptidase II in their physiological function.

Overall, very little information is available on the development and functional significance of individual endo- and exopeptidases in germinating grain or their response to GA, and this is partly attributable to difficulties in assaying specific enzymes in crude tissue extracts or secretions. Both the aleurone and scutellum appear to be involved in the secretion of peptidases (186), but zymogen forms of such enzymes might also exist in the starchy endosperm of ungerminated grain (57). In germinating barley, carboxypeptidase I is detected in the endosperm, but the main site of synthesis is in the scutellum (56, 176). Other endopeptidases and carboxypeptidases are secreted from isolated aleurone layers, where synthesis of endopeptidases (but not carboxypeptidases) is enhanced by GA (87). In contrast, expression of a carboxypeptidase gene in wheat aleurone (21) and a putative endopeptidase from barley aleurone is enhanced by GA (214).

Acting in concert, the endo- and exopeptidases secreted into (or preexisting in) the starchy endosperm of germinating cereals are capable of depolymerizing storage proteins to a mixture of amino acids and small peptides. Endopeptidases may be particularly important in the initial solubilization of reserve prolamins. The carboxypeptidases act rapidly on large peptides but relatively slowly on di- and tripeptides (167). However, these small peptides can be transported rapidly and actively into the scutellum (192, 231) where they are hydrolyzed to free amino acids before their transfer to the growing seedling or their entry into metabolic pathways (163). The peptidases in germinating grains may also release and/or activate preexisting zymogens such as β-amylase (145) and possibly carboxypeptidases or endopeptidases themselves (56).

Starch

Insoluble starch granules can be converted to glucose by the combined action of α-amylase, β-amylase, limit dextrinase, and α-glucosidase during cereal grain germination, but the physical mechanism of attack on the granules and the relative roles of individual enzymes in this attack are not clear.

In the following sections, I compare characteristics of the major groups of starch-degrading enzymes with respect to their tissue origin and physiological importance. The activities of each group in the mobilization of the starchy endosperm are related to aleurone and scutellar function.

α-AMYLASE   Two groups of α-amylases in germinating barley, wheat, rye, and triticale have been classified on the basis of their isoelectric points (150). In accordance with IUPAC-IUB recommendations, the group with the lower pI values (4.5–5.0) is referred to as the α-amylase 1 group and the higher-pI group (5.9–6.4) as the α-amylase 2 group. However, inconsistencies in nomenclature are evident in the literature; I therefore use "low-pI" and high-pI" here to specify α-amylase isoenzyme groups. Within each group, several isoforms can be identified. Some may represent true isoenzymes, being derived from different genes. Others may be generated by alternative splicing of mRNA derived from a single gene or may differ only by virtue of differences in covalent modification during the maturation of a single primary translation product; the latter would not be classified as isoenzymes. In addition, α-amylase is a Ca^{2+}-dependent enzyme, and its pI could be altered if Ca^{2+}-binding sites were not completely occupied. Nevertheless, wheat and barley α-amylases are the products of two related but distinct gene families (47, 109, 135, 215), which probably arose by duplication of an ancestral gene and transposition of new copies to a different chromosome, followed by further duplications and separate evolutionary divergence of individual genes. The low- and high-pI α-amylase gene families are located on chromosomes 1 and 6, respectively, in barley (37, 180). Directly determined amino acid sequences and those deduced from cDNA nucleotide sequences indicate that two members of the low- and high-pI α-amylase groups of barley contain 414 and 403 amino acid residues, respectively; have molecular weights of approximately 45,000; and exhibit 22% sequence divergence (47, 109, 213, 240). The barley α-amylase groups differ in their sensitivity to sulfhydryl reagents, metal ions, elevated temperatures, and low pH (109); and purified low-pI α-amylase degrades starch granules more rapidly than does the high-pI isoenzyme (151).

Four to seven days after the initiation of germination, the high-pI α-amylase group is far more abundant than the low-pI group in barley, wheat, rye, and triticale (150); but the relative abundance of the two groups earlier in the germination process is not known. Multiple forms of α-amylases synthesized in germinating oats, rice, millet, sorghum, and maize are predominantly of the low-pI group (150). Secretion of the high-pI α-amylase group from barley aleurone cells is dependent on added Ca^{2+} (43) and high levels of GA (109). There are also indications that the low-pI α-amylase is the major isoenzyme group synthesized in the barley embryo (149), while the high-pI α-amylase predominates in isolated aleurone layers (110). The significance of these findings in intact, germinating grain is unclear. It may be concluded that barley α-amylase isoenzymes, and probably those from other cereals as well, are subject to differential regulation, follow separate secretory pathways in aleurone cells, and perform specialized, yet complementary, roles in starch mobilization.

Two α-amylase gene families in wheat are also located on different chromosomes (20), and multiple forms of wheat α-amylases can be grouped according to their pI values (109, 159). A third multigene family, carried on a different chromosome from the low- and high-pI α-amylase genes, has been identified in wheat; but members of this family are expressed during grain maturation rather than during germination (23).

β-AMYLASE   In ungerminated cereal grains, β-amylase may constitute 1% of total protein. Some of the enzyme is bound to the periphery of starch granules during the final desiccation phase of grain maturation (89, 126, 133). A portion of the enzyme can be extracted with aqueous salt solutions, while "bound" enzyme, which is attached to proteins such as protein Z through disulfide linkages (4, 89, 94), can be released by reducing agents or papain. Multiple forms of β-amylases have been identified, although some of these are heterodimers that arise by disulfide cross-linking to other proteins (94, 128, 131). The chromosomal locations of two small gene families have been defined (127). Four major forms of β-amylase purified from ungerminated barley in the presence of thiol reagents are single polypeptide chains of M_r 54,000–60,000 and pI values of 5.2–5.7, but are probably generated by limited proteolysis of the COOH-terminal region of a single gene product (145). Sequence analysis of a cDNA encoding barley β-amylase shows a short repetitive domain consisting of 4 glycine-rich, 11-residue sequences close to the COOH-terminus; this repetitive domain could be involved in cross-linking β-amylase to protein Z or to other proteins at the periphery of starch granules (89, 126).

During germination, insoluble, latent β-amylase is released and activated by the action of proteolytic enzymes. Barley β-amylase has little or no action on intact starch granules, but synergistic hydrolysis of granules occurs in the presence of α-amylase (154). Although the peptidases mediating in the solubilization process have not been identified, they are presumably secreted by the scutellum and the aleurone. Whether the peptidases are specific for β-amylase or whether they are simply components of the battery of proteolytic enzymes attacking storage proteins is not known.

LIMIT DEXTRINASE   Low levels of limit dextrinase in ungerminated cereal grains can be extracted with reducing agents or papain, and probably arise from residual enzyme involved in grain development (129). During germination, limit dextrinase activity increases, although prolonged periods are required before maximum levels are obtained (137). The enzyme exists in a number of forms of pI 4.7–5.0, and molecular weights of 80,000 to 100,000 have been reported (146). The GA-induced, de novo synthesis of the enzyme in embryoless barley half-grains indicates that the aleurone participates in its production during germination (90); involvement of the scutellum has not yet been established.

α-GLUCOSIDASE   The maltose-degrading enzyme, α-glucosidase, is found in many ungerminated cereal grains, predominantly in the pericarp (148). After germination is initiated, α-glucosidase increases rapidly (148), and both the aleurone (90) and the scutellum (50) are involved in its synthesis.

Nucleic Acids

Although the starchy endosperm cells of mature grains are nonliving, they contain significant levels of residual RNA and DNA (160). Cereals have evolved an enzyme system for the recovery of these nucleic acids during germination. Barley nuclease I, a multifunctional enzyme that hydrolyzes RNA, DNA, and the 3'-phosphoester linkage of nucleoside 3'-monophosphates, has been purified and characterized (38, 39). The enzyme is a glycoprotein of apparent M_r 35,000, it has a pH optimum of 6.0, and its de novo synthesis and secretion from aleurone layers are increased by GA (38, 39). A similar enzyme has been purified from germinating wheat (88). In both species, additional ribonucleases are also present (38).

Nuclease I releases oligonucleotides and eventually 5'-mononucleotides from nucleic acids. The latter are converted to nucleosides by nucleotidases (138), which may correspond to acid phosphatase isoenzymes secreted from aleurone layers under the influence of GA (11, 74, 102). Glycosidic linkages in nucleosides are hydrolyzed by a group of nucleosidases to yield ribose or deoxyribose and free purine or pyrimidine bases (139). A purified adenosine nucleosidase from malted barley has a molecular weight of 120,000 and a pH optimum of 5.2 (139). The tissue origin of the barley nucleosidases is uncertain, but their activity increases significantly during germination and one would anticipate that they are synthesized in both the aleurone and the scutellum. In addition to the degradation of nucleic acids in the starchy endosperm, DNA in aleurone cells may also be depolymerized during aleurone senescence.

Protection from Pathogen Attack

Accumulated amino acids, fermentable sugars, nitrogenous bases, and other products in the germinating grain constitute an attractive nutrient medium for invading microorganisms. Some protection against microbial penetration of the grain is afforded by the physical barrier of highly lignified cell wall remnants in the husk, pericarp, and testa that surround the embryo and the endosperm. However, there is increasing evidence, albeit indirect, that cereal grains invoke additional strategies to counter pathogenic invasion.

Plants possess constitutive and inducible defense mechanisms that enable them to resist potentially pathogenic organisms through a variety of complex interactions involving both the plant and the pathogen (205). Components of plant-pathogen interactive systems can be recognized in both quiescent and germinating cereal grains, although their relationships with infection processes have not usually been established. I consider here the potential role of hydrolytic enzymes and specific inhibitors in the protection of germinating cereal grains from microbial attack. In addition to these, abundant cysteine-rich peptide toxins, referred to as thionins, are widely distributed in higher plants, including cereal endosperm (96, 200). The thionins exhibit pronounced antifungal and antibacterial activity, and their synthesis can be triggered by pathogens (27).

ENZYMES $(1\rightarrow3)$-β-Glucan endohydrolases in ungerminated barley increase markedly during germination (17), but endogenous $(1\rightarrow3)$-β-glucan in the grain is limited to small, extracellular deposits scattered through the starchy endosperm (13, 73). The low abundance of this material does not appear to justify the production of high levels of $(1\rightarrow3)$-β-glucanase in germinating grain. Branched $(1\rightarrow3, 1\rightarrow6)$-$\beta$-glucans do, however, constitute major wall components of common fungal pathogens of the Basidiomycetes, Ascomycetes, and Oomycetes (260). $(1\rightarrow3)$-β-Glucanases hydrolyze these fungal cell wall polysaccharides, degradation products can act as elicitors of other plant defense reactions (53, 218), and $(1\rightarrow3)$-β-glucanases have been identified among the "pathogenesis-related" (PR) proteins that accumulate in plants challenged with viruses, viroids, fungi, or bacteria (118, 257). If barley $(1\rightarrow3)$-β-glucanases do function as part of a general, nonspecific defense strategy, their constitutive expression in sterile, germinating grains indicates that pathogen invasion is not required for their synthesis (100). Two $(1\rightarrow3)$-β-glucanase isoenzymes have now been purified from germinating barley (100; D. J. Hartman, P. B. Hoj, and G. B. Fincher, unpublished). Their synthesis is probably mediated by both the aleurone and the scutellum, since isolated aleurone layers secrete $(1\rightarrow3)$-β-glucanases in response to GA (244) and a $(1\rightarrow3)$-β-glucanase cDNA has been isolated from a library prepared from scutellar mRNA (N. P. Doan, and G. B. Fincher, unpublished). Amino acid sequence similarities suggest that the barley $(1\rightarrow3)$-β-glucanases may have evolved from the cell wall–degrading $(1\rightarrow3, 1\rightarrow4)$-$\beta$-glucanases (100).

Endochitinases identified in the embryo and endosperm of ungerminated cereal grains (136, 211) may also participate in nonspecific protection against pathogen infection. Chitin is not present in cereal grains but is an important wall constituent of many pathogenic fungi (260). Chitinases have also been identified as PR proteins (140, 257) and are known both to be potent inhibitors of fungal growth and to degrade bacterial walls (211, 222). Fungal wall lysis may be synergistically increased by the simultaneous action of coordinately induced $(1\rightarrow3)$-β-glucanases and chitinases (28). The development of the barley chitinase during germination or in response to pathogen attack has not been examined.

INHIBITORS   Microbial infection, insect attack, and mechanical wounding lead to greatly increased levels of enzyme inhibitors in plants. Cereal endosperms and embryos contain several families of closely related water- and salt-soluble proteins that inhibit hydrolytic enzymes from a wide spectrum of plants, animals, and microorganisms (1, 10, 155, 185, 193, 239, 247, 259, 264, 265). Functionally, the cereal inhibitors can be classified into two groups: those that inhibit endogenous cereal hydrolases and hence participate in the regulation of grain development or germination, and those that act as inhibitors of exogenous enzymes and thereby provide a degree of protection against invading organisms. The second group includes inhibitors of fungal and insect α-amylases and peptidases (18, 120, 134, 209) and the barley protein synthesis inhibitor, which also has antifungal activity (10, 210). Inhibitors of exogenous enzymes might be expected to increase in abundance during germination or in response to stress or pathogen attack, either through de novo synthesis in the aleurone and scutellum or by activation of inert forms of the inhibitors. No evidence for this is so far available, but it remains a distinct possibility that inhibitors of exogenous wall-, starch-, or protein-degrading enzymes are elaborated by the aleurone or scutellum in germinating cereals, either constitutively (GA-inducible?) or in response to pathogen invasion (ABA-inducible?).

CONCLUDING REMARKS

Despite our best efforts, there remain serious gaps in our understanding of molecular and cellular events associated with endosperm mobilization in germinating cereals, as shown by the absence of even the most fundamental information on the molecular mechanisms of hormone perception and the stimulus-response pathway leading to gene activation. The significance of posttranscriptional control of gene expression in the aleurone and scutellum also awaits detailed investigation. Isolated aleurone layers have provided a vast amount of information and will doubtless continue to do so, but the tissue disruption necessary for their isolation imposes interpretative limitations on some results, particularly in relation to the coordination of gene regulation throughout the germinating grain. Recently developed immunocytochemical and hybridization histochemical techniques will find increasing application for high resolution experimentation with intact grains. Improved methods for labeling, purifying and characterizing low-abundance DNA-binding proteins will also play an important role in future advances in defining the coordinated regulation of gene expression in germinating cereal grains. This will provide the basis for the genetic manipulation of commercially important cereals; genes for hydrolytic enzymes expressed in germinating grains have been isolated and will be among the targets for these genetic manipulations. Our challenges are not restricted to the areas of molecular biology. Significant recent advances in our knowledge of the cellular processes leading to enzyme secretion can now be extended to address such questions as the secretory pathways for individual isoenzymes, the mechanism of action of Ca^{2+} in secretion, and the posttranslational modification of secreted enzymes. Detailed compositional analyses of cellular components, including cell walls, is necessary to further strengthen the fundamental information on which a thorough understanding of the germination process depends. In particular, the protein species in aleurone and scutellar protein bodies of the quiescent grain have not yet been defined, but may be especially important in view of their apparently superior nutritional quality. Increasing interest in the molecular events involved in plant-pathogen interactions and in plant responses to other environmental stresses will undoubtedly be reflected in an increased effort to understand these processes in germinating cereal grains.

ACKNOWLEDGMENTS

I thank Professor Bruce Stone and Drs. Peter Hoj, Jake Jacobsen, and Gideon Polya for critical review of the manuscript, and Emilia van Selm for her skilled editorial assistance. The support of the Australian Research Council and the Barley Research Council of Australia is gratefully acknowledged.

Literature Cited

1. Abe, K., Emori, Y., Kondo, H., Suzuki, K., Arai, S. 1987. Molecular cloning of a cysteine proteinase inhibitor of rice (Oryzacystatin). Homology with animal cystatins and transient expression in the ripening process of rice seeds. *J. Biol. Chem.* 262:16793–97
2. Abe, M., Arai, S., Fujimaki, M. 1978. Substrate specificity of a sulfhydryl protease purified from germinating corn. *Agric. Biol. Chem.* 42:1813–17
3. Adams, C. A., Novellie, L. 1975. Acid hydrolases and autolytic properties of protein bodies and spherosomes isolated from ungerminated seeds of *Sorghum bicolor* (Linn.) Moench. *Plant Physiol.* 55:7–11
4. Adams, C. A., Watson, T. G., Novellie, L. 1975. Lytic bodies from cereals hydrolysing maltose and starch. *Phytochemistry* 14:953–56
5. Aisien, A. O., Palmer, G. H. 1983. The sorghum embryo in relation to the hydrolysis of the endosperm during germination and seedling growth. *J. Sci. Food Agric.* 34:113–21
6. Aisien, A. O., Palmer, G. H., Stark, J. R. 1983. The development of enzymes during germination and seedling growth in Nigerian sorghum. *Starch/Stärke* 35:316–20
7. Aisien, A. O., Palmer, G. H., Stark, J. R. 1986. The ultrastructure of germinating sorghum and millet grains. *J. Inst. Brew.* 92:162–67
8. Akazawa, T., Hara-Nishimura, I. 1985. Topographic aspects of biosynthesis, extracellular secretion, and intracellular storage of proteins in plant cells. *Annu. Rev. Plant Physiol.* 36:441–72
9. Akazawa, T., Mitsui, T., Hayashi, M. 1988. Recent progress in α-amylase biosynthesis. In *The Biochemistry of Plants: A Comprehensive Treatise*, ed. J. Preiss, 14:465–92. New York: Academic

10. Asano, K., Svensson, B., Poulsen, F. M., Nygard, O., Nilsson, L. 1986. Influence of a protein synthesis inhibitor from barley seeds upon different steps of animal cell-free protein synthesis. *Carlsberg Res. Commun.* 51:75–81

11. Ashford, A. E., Jacobsen, J. V. 1974. Cytochemical localization of phosphatase in barley aleurone cells: the pathway of gibberellic-acid-induced enzyme release. *Planta* 120:81–105

12. Atzorn, R., Weiler, E. W. 1983. The role of endogenous gibberellins in the formation of α-amylase by aleurone layers of germinating barley caryopses. *Planta* 159:289–99

13. Bacic, A., Stone, B. A. 1981. Isolation and ultrastructure of aleurone cell walls from wheat and barley. *Aust. J. Plant Physiol.* 8:453–74

14. Bacic, A., Stone, B. A. 1981. Chemistry and organization of aleurone cell wall components from wheat and barley. *Aust. J. Plant Physiol.* 8:475–95

15. Ballance, G. M., Manners, D. J. 1975. The development of carbohydrases in germinating rye. *Biochem. Soc. Trans.* 3:989–91

16. Ballance, G. M., Manners, D. J. 1978. Partial purification and properties of an endo-1, 3-β-D-glucanase from germinated rye. *Phytochemistry* 17:1539–43

17. Ballance, G. M., Meredith, W. O. S., Laberge, D. E. 1976. Distribution and development of endo-β-glucanase activities in barley tissues during germination. *Can. J. Plant Sci.* 56:459–66

18. Barber, D., Sanchez-Monge, R., Mendez, E., Lazaro, A., Garcia-Olmedo, F., Salcedo, G. 1986. New α-amylase and trypsin inhibitors among the CM-proteins of barley *(Hordeum vulgare)*. *Biochim. Biophys. Acta* 869:115–18

19. Bathgate, G. N., Palmer, G. H., Wilson, G. 1974. The action of endo-β-1, 3-glucanases on barley and malt β-glucans. *J. Inst. Brew.* 80:278–85

20. Baulcombe, D., Lazarus, C., Martienssen, R. 1984. Gibberellins and gene control in cereal aleurone cells. *J. Embryol. Exp. Morphol. Suppl.* 83:119–35

21. Baulcombe, D. C., Barker, R. F., Jarvis, M. G. 1987. A gibberellin responsive wheat gene has homology to yeast carboxypeptidase Y. *J. Biol. Chem.* 262:13726–35

22. Baulcombe, D. C., Buffard, D. 1983. Gibberellic-acid-regulated expression of α-amylase and six other genes in wheat aleurone layers. *Planta* 157:493–501

23. Baulcombe, D. C., Huttly, A. K., Martienssen, R. A., Barker, R. F., Jarvis,

24. Bechtel, D. B., Pomeranz, Y. 1981. Ultrastructure and cytochemistry of mature oat *(Avena sativa L.)* endosperm. The aleurone layer and starchy endosperm. *Cereal Chem.* 58:61–69

25. Belanger, F. C., Brodl, M. R., Ho, T.-H.D. 1986. Heat shock causes destabilization of specific mRNAs and destruction of endoplasmic reticulum in barley aleurone cells. *Proc. Natl. Acad. Sci. USA* 83:1354–58

26. Benjavongkulchai, E., Spencer, M. S. 1986. Purification and characterization of barley-aleurone xylanase. *Planta* 169:415–19

27. Bohlmann, H., Clausen, S., Behnke, S., Giese, H., Hiller, C., et al. 1988. Leaf-specific thionins of barley—a novel class of cell wall proteins toxic to plant-pathogenic fungi and possibly involved in the defence mechanism of plants. *EMBO J.* 7:1559–65

28. Boller, T. 1987. Hydrolytic enzymes in plant disease resistance. In *Plant-Microbe Interactions. Molecular and Genetic Perspectives*, ed. T. Kosuge, E. W. Nester, 2:385–413. New York: Macmillan

29. Brandt, A., Montembault, A., Cameron-Mills, V., Rasmussen, S. K. 1985. Primary structure of a B1 hordein gene from barley. *Carlsberg Res. Commun.* 50:333–45

30. Brawerman, G. 1987. Determinants of messenger RNA stability. *Cell* 48:5–6

31. Breddam, K. 1985. Enzymatic properties of malt carboxypeptidase II in hydrolysis and aminolysis reactions. *Carlsberg Res. Commun.* 50:309–23

32. Breddam, K., Sorensen, S. B. 1987. Isolation of carboxypeptidase-III from malted barley by affinity chromatography. *Carlsberg Res. Commun.* 52:275–83

33. Breddam, K., Sorensen, S. B., Ottesen, M. 1983. Isolation of a carboxypeptidase from malted barley by affinity chromotography. *Carlsberg Res. Commun.* 48:217–30

34. Breddam, K., Sorensen, S. B., Svendsen, I. 1987. Primary structure and enzymatic-properties of carboxypeptidase-II from wheat bran. *Carlsberg Res. Commun.* 52:297–311

35. Briggs, D. E., Hough, J. S., Stevens, R., Young, T. W. 1981. In *Malting and Brewing Science*. Vol. I, *Malt and Sweet Wort*, p. 77. London: Chapman & Hall

36. Briggs, D. E., MacDonald, J. 1983.

M. G. 1987. A novel wheat α-amylase gene (α-Amy3). *Mol. Gen. Genet.* 209:33–40

Patterns of modification in malting barley. *J. Inst. Brew.* 89:260–73

37. Brown, A.H.D., Jacobsen, J. V. 1982. Genetic basis and natural variation of α-amylase isozymes in barley. *Genet. Res.* 40:315–24

38. Brown, P. H., Ho, T-H.D. 1986. Barley aleurone layers secrete a nuclease in response to gibberellic acid. *Plant Physiol.* 82:801–6

39. Brown, P. H., Ho, T-H.D. 1987. Biochemical properties and hormonal regulation of barley nuclease. *Eur. J. Biochem.* 168:357–64

40. Buckhout, T. J., Gripshover, B. M., Morré, D. J. 1981. Endoplasmic reticulum formation during germination of wheat seeds. *Plant Physiol.* 68:1319–22

41. Bulmer, M. 1987. Coevolution of codon usage and transfer RNA abundance. *Nature* 325:728–30

42. Bush, D. S., Biswas, A. K., Jones, R. L. 1988. Measurement of cytoplasmic Ca^{2+} and H^+ in barley aleurone protoplasts. *Plant Cell Tissue Organ Cult.* 12:159–62

43. Bush, D. S., Cornejo, M.-J., Huang, C.-N., Jones, R. L. 1986. Ca^{2+}-stimulated secretion of α-amylase during development in barley aleurone protoplasts. *Plant Physiol.* 82:566–74

44. Carpita, N., Sabularse, D., Montezinos, D., Delmer, D. P. 1979. Determination of the pore size of cell walls of living plant cells. *Science* 205:1144–47

45. Cassab, G. I., Varner, J. E. 1988. Cell wall proteins. *Annu. Rev. Plant Physiol. Plant Mol. Biol.* 39:321–53

46. Chandler, P. M., Ariffin, Z., Huiet, L., Jacobsen, J. V., Zwar, J. 1987. Molecular biology of expression of alpha-amylase and other genes following grain germination. In *4th Int. Symp. Pre-Harvest Sprouting in Cereals*, ed. D. J. Mares, pp. 295–303. Colo: Westview

47. Chandler, P. M., Zwar, J. A., Jacobsen, J. V., Higgins, T.J.V., Inglis, A. S. 1984. The effects of gibberellic acid and abscisic acid on α-amylase mRNA levels in barley aleurone layers studies using an α-amylase cDNA clone. *Plant Mol. Biol.* 3:407–18

48. Ching, T. M., Lin, T.-P., Metzger, R. J. 1987. Purification and properties of acid phosphatase from plump and shriveled seeds of triticale. *Plant Physiol.* 84:789–95

49. Chrispeels, M. J., Varner, J. E. 1967. Gibberellic acid-enhanced synthesis and release of α-amylase and ribonuclease by isolated barley aleurone layers. *Plant Physiol.* 42:398–406

50. Clutterbuck, V. J., Briggs, D. E. 1973.

Enzyme formation and release by isolated barley aleurone layers. *Phytochemistry* 12:537–46

51. Cornejo, M. J., Platt-Aloia, K. A., Thomson, W. W., Jones, R. L. 1989. A freeze-fracture study of barley aleurone protoplasts: Effects of GA_3 and Ca^{2+}. *Protoplasma.* In press

52. Daiber, K. H., Novellie, L. 1968. Kaffircorn malting and brewing studies. XIX. Gibberellic acid and amylase formation in kaffircorn. *J. Sci. Food Agric.* 19:87–90

53. Darvill, A. G., Albersheim, P. 1984. Phytoalexins and their elicitors—a defense against microbial infection in plants. *Annu. Rev. Plant Physiol.* 35:243–75

54. Dashek, W. V., Chrispeels, M. J. 1977. Gibberellic-acid-induced synthesis and release of cell-wall degrading endoxylanase by isolated aleurone layers of barley. *Planta* 134:251–56

55. Deikman, J., Jones, R. L. 1986. Regulation of the accumulation of mRNA for α-amylase isoenzymes in barley aleurone. *Plant Physiol.* 80:672–75

56. Doan, N. P., Fincher, G. B. 1988. The A- and B-chains of carboxypeptidase I from germinated barley originate from a single precursor polypeptide. *J. Biol. Chem.* 263:11106–10

57. Doi, E., Komori, N., Matoba, T., Morita, Y. 1980. Purification and some properties of a carboxypeptidase in rice bran. *Agric. Biol. Chem.* 44:85–92

58. Doi, E., Shibata, D., Matoba, T., Yonezawa, D. 1980. Evidence for the presence of two types of acid proteinases in germinating seeds of rice. *Agric. Biol. Chem.* 44:435–36

59. Donhowe, E. T., Peterson, D. M. 1983. Isolation and characterization of oat aleurone and starchy endosperm protein bodies. *Plant Physiol.* 71:519–23

60. Dubois, M., Geddes, W. F., Smith, F. 1960. The carbohydrates of the gramineae. X. A quantitative study of the carbohydrates of wheat germ. *Cereal Chem.* 37:557–68

61. Dure, L. S. 1960. Site of origin and extent of activity of amylases in maize germination. *Plant Physiol.* 35:925–34

62. Elliott, D. C., Kokke, Y. S. 1987. Partial-purification and properties of a protein kinase-C type enzyme from plants. *Phytochemistry* 26:2929–35

63. Fernandez, D. E., Qu, R., Huang, A.H.C., Staehelin, L. A. 1988. Immunogold localization of the L3 protein of maize lipid bodies during germination and seedling growth. *Plant Physiol.* 86:270–74

64. Fernandez, D. E., Staehelin, L. A. 1985. Structural organization of ultrarapidly frozen barley aleurone cells actively involved in protein secretion. *Planta* 165:455–68

65. Fernandez, D. E., Staehelin, L. A. 1987. Does gibberellic acid induce the transfer of lipase from protein bodies to lipid bodies in barley aleurone cells? *Plant Physiol.* 85:487–96

66. Festenstein, G. N., Hay, F. C., Shewry, P. R. 1987. Immunochemical relationships of the prolamin storage proteins of barley, wheat, rye and oats. *Biochim. Biophys. Acta* 912:371–83

67. Fincher, G. B., Lock, P. A., Morgan, M. M., Lingelbach, K., Wettenhall, R.E.H., et al. 1986. Primary structure of the (1→3, 1→4)-β-D-glucan 4-glucanohydrolase from barley aleurone. *Proc. Natl. Acad. Sci. USA* 83:2081–85

68. Fincher, G. B., Stone, B. A. 1974. Some chemical and morphological changes induced by gibberellic acid in embryo-free wheat grain. *Aust. J. Plant Physiol.* 1:297–311

69. Fincher, G. B., Stone, B. A. 1986. Cell walls and their components in cereal grain technology. *Adv. Cereal Sci. Technol.* 8:207–95

70. Freedman, R. B. 1984. Native disulphide bond formation in protein biosynthesis: evidence for the role of protein disulphide isomerase. *Trends Biochem. Sci.* 9:438–41

71. Fulcher, R. G., O'Brien, T. P., Lee, J. W. 1972. Studies on the aleurone layer. I. Conventional and fluorescence microscopy of the cell wall with emphasis on phenol-carbohydrate complexes in wheat. *Aust. J. Biol. Sci.* 25:23–34

72. Fulcher, R. G., O'Brien, T. P., Simmonds, D. H. 1972. Localization of arginine-rich proteins in mature seeds of some members of the gramineae. *Aust. J. Biol. Sci.* 25:487–97

73. Fulcher, R. G., Setterfield, G., McCully, M. E., Wood, P. J. 1972. Observations on the aleurone layer. II. Fluorescence microscopy of the aleurone-subaleurone junction with emphasis on possible β-1, 3-glucan deposits in barley. *Aust. J. Plant. Physiol.* 4:917–28

74. Gabard, K. A., Jones, R. L. 1986. Localization of phytase and acid phosphatase isoenzymes in aleurone layers of barley. *Physiol. Plant.* 67:182–92

75. Garg, G. K., Virupaksha, T. K. 1970. Acid protease from germinated sorghum. 1. Purification and characterization of the enzyme. *Eur. J. Biochem.* 17:4–12

76. Geraghty, D., Peifer, M. A., Ruben-stein, I., Messing, J. 1981. The primary structure of a plant storage protein: zein. *Nucleic Acids Res.* 9:5163–74

77. Gibbons, G. C. 1981. On the relative role of the scutellum and aleurone in the production of hydrolases during germination of barley. *Carlsberg Res. Commun.* 46:215–25

78. Giese, H., Hejgaard, J. 1984. Synthesis of salt-soluble proteins in barley. Pulse-labeling study of grain filling in liquid-cultured detached spikes. *Planta* 161:172–77

79. Gilmour, S. J., MacMillan, J. 1984. Effect of inhibitors of gibberellin biosynthesis on the induction of α-amylase in embryoless caryopses of *Hordeum vulgare* cv. Himalaya. *Planta* 162:89–90

80. Glennie, C. W., Harris, J., Liebenberg, N.V.D.W. 1983. Endosperm modification in germinating sorghum grain. *Cereal Chem.* 60:27–31

81. Gram, N. H. 1982. The ultrastructure of germinating barley seeds. I. Changes in the scutellum and the aleurone layer in Nordal barley. *Carlsberg Res. Commun.* 47:143–62

82. Gubler, F., Ashford, A. E., Bacic, A., Blakeney, A. B., Stone, B. A. 1985. Release of ferulic acid esters from barley aleurone. II. Characterization of the feruloyl compounds released in response to GA₃. *Aust. J. Plant Physiol.* 12:307–17

83. Gubler, F., Ashford, A. E., Jacobsen, J. V. 1987. The release of α-amylase through gibberellin-treated barley aleurone cell walls. *Planta* 172:155–61

84. Gubler, F., Jacobsen, J. V., Ashford, A. E. 1986. Involvement of the Golgi apparatus in the secretion of α-amylase from gibberellin-treated barley aleurone cells. *Planta* 168:447–52

85. Deleted in proof

86. Hamabata, A., Garcia-Maya, M., Romero, T., Bernal-Lugo, I. 1988. Kinetics of the acidification capacity of aleurone layer and its effect upon solubilization of reserve substances from starchy endosperm of wheat. *Plant Physiol.* 86:643–44

87. Hammerton, R. W., Ho, T.-H. D. 1986. Hormonal regulation of the development of protease and carboxypeptidase activities in barley aleurone layers. *Plant Physiol.* 80:692–97

88. Hanson, D. M., Fairley, J. L. 1969. Enzymes of nucleic acid metabolism from wheat seedlings. I. Purification and general properties of associated deoxyribonuclease, ribonuclease, and 3'-

340   FINCHER

nucleotidase activities. *J. Biol. Chem.*
244:2440–49

89. Hara-Nishimura, I., Nishimura, M., Daussant, J. 1986. Conversion of free β-amylase to bound β-amylase on starch granules in the barley endosperm during desiccation phase of seed development. *Protoplasma* 134:149–53

90. Hardie, D. G. 1975. Control of carbohydrase formation by gibberellic acid in barley endosperm. *Phytochemistry* 14:1719–22

91. Deleted in proof

92. Harvey, B.M.R., Oaks, A. 1974. Characteristics of an acid protease from maize endosperm. *Plant Physiol.* 53:449–52

93. Heidecker, G., Messing, J. 1986. Structural analysis of plant genes. *Annu. Rev. Plant Physiol.* 37:439–66

94. Hejgaard, J. 1978. "Free" and "bound" β-amylases during malting of barley. Characterization by two-dimensional immunoelectrophoresis. *J. Inst. Brew.* 84:43–46

95. Hepler, P. K., Wayne, R. O. 1985. Calcium and plant development. *Annu. Rev. Plant Physiol.* 36:397–439

96. Hernandez-Lucas, C., Royo, J., Paz-Ares, J., Ponz, F., Garcia-Olmedo, F., Carbonero, P. 1986. Polyadenylation site heterogeneity in mRNA encoding the precursor of the barley toxin β-hordothionin. *FEBS Lett.* 200:103–6

97. Heupke, H.-J., Robinson, D. G. 1985. Intracellular transport of α-amylase in barley aleurone cells: evidence for the participation of the Golgi apparatus. *Eur. J. Cell Biol.* 39:265–72

98. Higgins, T.J.V. 1984. Synthesis and regulation of major proteins in seeds. *Annu. Rev. Plant Physiol.* 35:191–221

99. Ho, T.-H. D., Nolan, R. C., Lin, L.-S., Brodl, M. R., Brown, P. H. 1987. Regulation of gene expression in barley aleurone layers. In *Molecular Biology of Plant Growth Control—UCLA Symp. Mol. Cell. Biol.* (NS), ed. J. E. Fox, M. Jacobs, 44:35–49. New York: Liss

100. Hoj, P. B., Slade, A. M., Wettenhall, R.E.H., Fincher, G. B. 1988. Isolation and characterization of a (1→3)-β-glucan endohydrolase from germinating barley *(Hordeum vulgare)*: amino-acid sequence similarity with barley (1→3, 1→4)-β-glucanases. *FEBS Lett.* 230:67–71

101. Hood, E. E., Shen, Q. X., Varner, J. E. 1988. A developmentally regulated hydroxyproline-rich glycoprotein in maize pericarp cell walls. *Plant Physiol.* 87:138–42

102. Hooley, R. 1984. Gibberellic acid controls specific acid-phosphatase isozymes in aleurone cells and protoplasts of *Avena fatua* L. *Planta* 161:355–60

103. Hoy, J. L., Macauley, B. J., Fincher, G. B. 1981. Cellulases of plant and microbial origin in germinating barley. *J. Inst. Brew.* 87:77–80

104. Huang, A.H.C. 1985. Lipid bodies. In *Modern Methods of Plant Analysis*, ed. H. F. Linskens, J. F. Jackson, 1:145–51. Berlin: Springer-Verlag

105. Huang, J.-K., Swegle, M., Dandekar, A. M., Muthukrishnan, S. 1984. Expression and regulation of α-amylase gene family in barley aleurones. *J. Mol. Appl. Genet.* 2:579–88

106. Huttner, W. B. 1987. Protein tyrosine sulfation. *Trends Biochem. Sci.* 12:361–63

107. Jacobsen, J. V., Beach, L. R. 1985. Control of transcription of α-amylase and rRNA genes in barley aleurone protoplasts by gibberellin and abscisic acid. *Nature* 316:275–77

108. Jacobsen, J. V., Bush, D. S., Sticher, L., Jones, R. L. 1989. Evidence for precursor forms of the low-pI α-amylase isozymes secreted by barley aleurone cells. *Plant Physiol.* In press

109. Jacobsen, J. V., Chandler, P. M. 1987. Gibberellin and abscisic acid in germinating cereals. In *Plant Hormones and Their Role in Plant Growth and Development*, ed. P. J. Davies, pp. 164–93. Dordrecht: Martinus Nijhoff

110. Jacobsen, J. V., Higgins, T.J.V. 1982. Characterization of the α-amylases synthesized by aleurone layers of Himalaya barley in response to gibberellic acid. *Plant Physiol.* 70:1647–53

111. Jacobsen, J. V., Higgins, T.J.V., Zwar, J. A. 1979. Hormonal control of endosperm function during germination. In *The Plant Seed*, ed. I. Rubenstein, R. L. Phillips, C. E. Green, B. G. Gengenbach, pp. 241–62. New York: Academic

112. Jelsema, C. L., Morré, D. J., Ruddat, M., Turner, C. 1977. Isolation and characterization of the lipid reserve bodies, spherosomes, from aleurone layers of wheat. *Bot. Gaz.* 138:138–49

113. Jones, R. L. 1980. Quantitative and qualitative changes in the endoplasmic reticulum of barley aleurone layers. *Planta* 150:70–81

114. Jones, R. L. 1985. Protein synthesis and secretion by the barley aleurone: a perspective. *Isr. J. Bot.* 34:377–95

115. Jones, R. L. 1987. Localization of ATPase in the endoplasmic reticulum and Golgi apparatus of barley aleurone. *Protoplasma* 138:73–88

116. Jones, R. L., Bush, D. S., Sticher, L., Simon, P., Jacobsen, J. V. 1987. Intracellular transport and secretion of barley aleurone α-amylase. In *Plant Membranes: Structure, Function, Biogenesis*, ed. C. Leaver, H. Sze, pp. 325–40. New York: Liss

117. Jones, R. L., Jacobsen, J. V. 1983. Calcium regulation of the secretion of α-amylase isoenzymes and other proteins from barley aleurone layers. *Planta* 158:1–9

118. Kauffmann, S., Legrand, M., Geoffroy, P., Fritig, B. 1987. Biological function of "pathogenesis-related" proteins: four PR proteins of tobacco have 1,3-β-glucanase activity. *EMBO J.* 6:3209–12

119. King, R. W. 1976. Abscisic acid in developing wheat grains and its relationship to grain growth and maturation. *Planta* 132:43–51

120. Kirsi, M., Mikola, J. 1971. Occurrence of proteolytic inhibitors in various tissues of barley. *Planta* 96:281–91

121. Knox, C.A.P., Sonthayanon, B., Chandra, G. R., Muthukrishnan, S. 1987. Structure and organization of two divergent α-amylase genes from barley. *Plant Mol. Biol.* 9:3–17

122. Kodicek, E., Wilson, P. W. 1960. The isolation of niacytin, the bound form of nicotinic acid. *Biochem. J.* 76:27P–28P

123. Koehler, S., Ho, T.-H.D. 1988. Purification and characterization of gibberellic acid-induced cysteine endoproteases in barley aleurone layers. *Plant Physiol.* 87:95–103

124. Koivu, J., Myllylä, R. 1987. Interchain disulfide bond formation in types-I and-II procollagen. Evidence for a protein disulfide isomerase catalyzing bond formation. *J. Biol. Chem.* 262:6159–64

125. Kreis, M., Forde, B. G., Rahman, S., Miflin, B. J., Shewry, P. R. 1985. Molecular evolution of the seed storage proteins of barley, rye and wheat. *J. Mol. Biol.* 183:499–502

126. Kreis, M., Williamson, M., Buxton, B., Pywell, J., Hejgaard, J., Svendsen, I. 1987. Primary structure and differential expression of β-amylase in normal and mutant barleys. *Eur. J. Biochem.* 169:517–25

127. Kreis, M., Williamson, M. S., Shewry, P. R., Sharp, P., Gale, M. 1988. Identification of a second locus encoding β-amylase on chromosome 2 of barley. *Genet. Res.* 51:13–16

128. Kruger, J. E. 1979. Modification of wheat β-amylase by proteolytic enzymes. *Cereal Chem.* 56:298–302

129. Kruger, J. E., Marchylo, B. 1978. Note on the presence of debranching enzymes in immature wheat kernels. *Cereal Chem.* 55:529–33

130. Kuo, T. M., VanMiddlesworth, J. F., Wolf, W. J. 1988. Content of raffinose oligosaccharides and sucrose in various plant seeds. *J. Agric. Food Chem.* 36:32–36

131. LaBerge, D. E., Marchylo, B. A. 1983. Heterogeneity of the beta-amylase enzymes of barley. *J. Am. Soc. Brew. Chem.* 41:120–24

132. Lang, K., Schmid, F. X., Fischer, G. 1987. Catalysis of protein folding by prolyl isomerase. *Nature* 329:268–70

133. Lauriere, C., Lauriere, M., Daussant, J. 1986. Immunohistochemical localization of β-amylase in resting barley seeds. *Physiol. Plant.* 67:383–88

134. Lazaro, A., Sanchez-Monge, R., Salcedo, G., Paz-Ares, J., Carbonero, P., García-Olmedo, F. 1988. A dimeric inhibitor of insect α-amylase from barley. Cloning of the cDNA and identification of the protein. *Eur. J. Biochem.* 172:129–34

135. Lazarus, C. M., Baulcombe, D. C., Martienssen, R. A. 1985. α-Amylase genes of wheat are two multigene families which are differently expressed. *Plant Mol. Biol.* 5:13–24

136. Leah, R., Mikkelsen, J. D., Mundy, J., Svendsen, I. 1987. Identification of a 28,000 dalton endochitinase in barley endosperm. *Carlsberg Res. Commun.* 52:31–37

137. Lee, W. J., Pyler, R. E. 1984. Barley malt limit dextrinase: varietal, environmental and malting effects. *J. Am. Soc. Brew. Chem.* 42:11–17

138. Lee, W. J., Pyler, R. E. 1985. Nucleic acid degrading enzymes of barley malt. I. Nucleases and phosphatases. *J. Am. Soc. Brew. Chem.* 43:1–6

139. Lee, W. J., Pyler, R. E. 1986. Nucleic acid degrading enzymes of barley malt. III. Adenosine nucleosidase from malted barley. *J. Am. Soc. Brew. Chem.* 44:86–90

140. Legrand, M., Kauffmann, S., Geoffroy, P., Fritig, B. 1987. Biological function of pathogenesis-related proteins: Four tobacco pathogenesis-related proteins are chitinases. *Proc. Natl. Acad. Sci. USA* 84:6750–54

141. Lending, C. R., Kriz, A. L., Larkins, B. A., Bracker, C. E. 1988. Structure of maize protein bodies and immunocytochemical localization of zeins. *Protoplasma* 143:51–62

142. Lin, L.-S., Ho, T.-H. D. 1986. Mode of action of abscisic acid in barley aleurone layers. *Plant Physiol.* 82:289–97

143. Deleted in proof

144. Loi, L., Ahluwalia, B., Fincher, G. B. 1988. Chromosomal location of genes encoding barley (1→3, 1→4)-β-glucan 4-glucanohydrolases. *Plant Physiol.* 87: 300–2

145. Lundgard, R., Svensson, B. 1987. The four major forms of barley β-amylase. Purification, characterization and structural relationship. *Carlsberg Res. Commun.* 52:313–26

146. MacGregor, A. W. 1987. α-Amylase, limit dextrinase, and α-glucosidase enzymes in barley and malt. *CRC Crit. Rev. Biotechnol.* 5:117–28

147. Deleted in proof

148. MacGregor, A. W., Lenoir, C. 1987. Studies on α-glucosidase in barley and malt. *J. Inst. Brew.* 93:334–37

149. MacGregor, A. W., Marchylo, B. A. 1986. α-Amylase components in excised, incubated barley embryos. *J. Inst. Brew.* 92:159–61

150. MacGregor, A. W., Marchylo, B. A., Kruger, J. E. 1988. Multiple α-amylase components in germinated cereal grains determined by isoelectric focusing and chromatofocusing. *Cereal Chem.* 65:326–33

151. MacGregor, A. W., Morgan, J. E. 1986. Hydrolysis of barley starch granules by alpha-amylases from barley malt. *Cereal Foods World* 31:688–93

152. MacLeod, A. M., Palmer, G. H. 1967. Gibberellin from barley embryos. *Nature* 216:1342–43

153. MacMasters, M. M., Hinton, J.J.C., Bradbury, D. 1971. Microscopic structure and composition of the wheat kernel. In *Wheat: Chemistry and Technology*, ed. Y. Pomeranz, 3:51–113. St. Paul, Minn: Am. Soc. Cereal Chem.

154. Maeda, I., Kiribuchi, S., Nakamura, M. 1978. Digestion of barley starch granules by the combined action of α- and β-amylases purified from barley and barley malt. *Agric. Biol. Chem.* 42:259–67

155. Maeda, K. 1986. The complete amino-acid sequence of the endogenous α-amylase inhibitor in wheat. *Biochim. Biophys. Acta* 871:250–56

156. Maniatis, T., Goodbourn, S., Fischer, J. A. 1987. Regulation of inducible and tissue-specific gene expression. *Science* 236:1237–45

157. Manners, D. J., Marshall, J. J. 1969. Studies on carbohydrate metabolizing enzymes. XXII. The β-glucanase system of malted barley. *J. Inst. Brew.* 75:550–61

158. Manners, D. J., Seiler, A., Sturgeon, R. J. 1982. Observations on the endo-

(1→4)-β-D-glucanase activity of extracts of barley. *Carbohydr. Res.* 100:435–40

159. Marchylo, B. A., Kruger, J. E., MacGregor, A. W. 1984. Production of multiple forms of alpha-amylase in germinated, incubated, whole, de-embryonated wheat kernels. *Cereal Chem.* 61:305–10

160. McFadden, G. I., Ahluwalia, B., Clarke, A. E., Fincher, G. B. 1988. Expression sites and developmental regulation of genes encoding (1→3, 1→4)-β-glucanases in germinated barley. *Planta* 173:500–8

161. Melroy, D., Jones, R. L. 1986. The effect of monensin on intracellular transport and secretion of α-amylase isoenzymes in barley aleurone. *Planta* 167:252–59

162. Mikola, J. 1983. Proteinases, peptidases, and inhibitors of endogenous proteinases in germinating seeds. In *Seed Proteins*, ed. J. Daussant, J. Mossé, J. Vaughan, pp. 35–52. London: Academic

163. Mikola, J. 1987. Proteinases and peptidases in germinating cereal grains. See Ref. 46, pp. 463–73.

164. Mikola, J., Kolehmainen, L. 1972. Localization and activity of various peptidases in germinating barley. *Planta* 104:167–77

165. Mikola, J., Virtanen, M. 1980. Secretion of L-malic acid by barley aleurone layers. *Plant Physiol.* 65:S-142

166. Mikola, L. 1986. Acid carboxypeptidases in grains and leaves of wheat, *Triticum aestivum* L. *Plant Physiol.* 81:823–29

167. Mikola, L., Mikola, J. 1980. Mobilization of proline in the starchy endosperm of germinating barley grain. *Planta* 149:149–54

168. Mitsui, T., Christeller, J. T., Hara-Nishimura, I., Akazawa, T. 1984. Possible roles of calcium and calmodulin in the biosynthesis and secretion of α-amylase in rice seed scutellar epithelium. *Plant Physiol.* 75:21–25

169. Miyata, S., Akazawa, T. 1982. α-Amylase biosynthesis: signal sequence prevents normal conversion of the unprocessed precursor molecule to the biologically active form. *Proc. Natl. Acad. Sci. USA* 79:7792–95

170. Miyata, S., Okamoto, K., Watanabe, A., Akazawa, T. 1981. Enzymic mechanism of starch breakdown in germinating rice seed. 10. In vivo and in vitro synthesis of α-amylase in rice seed scutellum. *Plant Physiol.* 68:1314–18

171. Morrall, P., Briggs, D. E. 1978.

Changes in cell wall polysaccharides of germinating barley grains. *Phytochemistry* 17:1495–1502

172. Morrison, I. N., Kuo, J., O'Brien, T. P. 1975. Histochemistry and fine structure of developing wheat aleurone cells. *Planta* 123:105–16

173. Motojima, K., Sakaguchi, K. 1982. Part of the lysyl residues in wheat α-amylase is methylated as N-ε-trimethyl lysine. *Plant Cell Physiol.* 23:709–12

174. Mozer, T. J. 1980. Control of protein synthesis in barley aleurone layers by the plant hormones gibberellic acid and abscisic acid. *Cell* 20:479–85

175. Mundy, J. 1984. Hormonal regulation of α-amylase inhibitor synthesis in germinating barley. *Carlsberg Res. Commun.* 49:439–44

176. Mundy, J., Brandt, A., Fincher, G. B. 1985. Messenger RNAs from the scutellum and aleurone of germinating barley encode (1→3, 1→4)-β-D-glucanase, α-amylase and carboxypeptidase. *Plant Physiol.* 79:867–71

177. Mundy, J., Fincher, G. B. 1986. Effects of gibberellic acid and abscisic acid on levels of translatable mRNA for (1→3, 1→4)-β-D-glucanase in barley aleurone. *FEBS Lett.* 198:349–52

178. Mundy, J., Hejgaard, J., Svendsen, I. 1984. Characterization of a bifunctional wheat inhibitor of endogenous α-amylase and subtilisin. *FEBS Lett.* 167:210–14

179. Mundy, J., Rogers, J. C. 1986. Selective expression of a probable amylase/protease inhibitor in barley aleurone cells: comparison to the barley amylase/subtilisin inhibitor. *Planta* 169:51–63

180. Muthukrishnan, S., Gill, B. S., Swegle, M., Chandra, G. R. 1984. Structural genes for α-amylases are located on barley chromosomes-1 and -6. *J. Biol. Chem.* 259:13637–39

181. Negbi, M., Sargent, J. A. 1986. The scutellum of *Avena:* a structure to maximize exploitation of endosperm reserves. *Bot. J. Linn. Soc.* 93:247–58

182. Nicholls, P. B., MacGregor, A. W., Marchylo, B. A. 1986. Production of α-amylase isozymes in barley caryopses in the absence of embryos and exogenous gibberellic acid. *Aust. J. Plant Physiol.* 13:239–47

183. Nieuwdorp, P. J. 1963. Electron microscopic structure of the epithelial cells of the scutellum of barley. The structure of the epithelial cells before germination. *Acta Bot. Neerl.* 12:295–301

184. Nieuwdorp, P. J., Buys, M. C., 1964.

Electron microscopic structure of the epithelial cells of the scutellum of barley. II. Cytology of the cells during germination. *Acta Bot. Neerl.* 13:559–65

185. Odani, S., Koide, T., Ono, T. 1986. Wheat germ trypsin inhibitors. Isolation and structural characterization of single-headed and double-headed inhibitors of the Bowman-Birk type. *J. Biochem.* 100:975–83

186. Okamoto, K., Kitano, H., Akazawa, T. 1980. Biosynthesis and excretion of hydrolases in germinating cereal seeds. *Plant Cell Physiol.* 21:201–4

187. Osborne, T. B. 1907. In *The Proteins of the Wheat Kernel.* Washington DC: Carnegie Inst. Wash. Publ. No. 84

188. Paleg, L. G. 1960. Physiological effects of gibberellic acid. II. On starch hydrolyzing enzymes of barley endosperm. *Plant Physiol.* 35:902–6

189. Paliyath, G., Poovaiah, B. W. 1988. Promotion of β-glucan synthase activity in corn microsomal membranes by calcium and protein phosphorylation. *Plant Cell Physiol.* 29:67–73

190. Palmer, G. H. 1970. Response of cereal grains to gibberellic acid. *J. Inst. Brew.* 76:378–80

191. Palmer, G. H., Duffus, J. H. 1986. Aleurone or scutellar hydrolytic enzymes in malting. *J. Inst. Brew.* 92:512–13

192. Payne, J. W., Walker-Smith, D. J. 1987. Isolation and identification of proteins from the peptide-transport carrier in the scutellum of germinating barley (*Hordeum vulgare* L.) embryos. *Planta* 170:263–71

193. Paz-Ares, J., Ponz, F., Rodriguez-Palenzuela, P., Lazaro, A., Hernandez-Lucas, C., et al. 1986. Characterization of cDNA clones of the family of trypsin/α-amylase inhibitors (CM-proteins) in barley (*Hordeum vulgare* L.). *Theor. Appl. Genet.* 71:842–46

194. Pedersen, K., Devereux, J., Wilson, D. R., Sheldon, E., Larkins, B. A. 1982. Cloning and sequence analysis reveal structural variation among related zein genes in maize. *Cell* 29:1015–26

195. Peterson, D. M., Saigo, R. H., Holy, J. 1985. Development of oat aleurone cells and their protein bodies. *Cereal Chem.* 62:366–71

196. Peumans, W. J., Stinissen, H. M. 1983. Gramineae lectins: occurrence, molecular biology and physiological function. In *Chemical Taxonomy, Molecular Biology, and Functions of Plant Lectins,* ed. I. J. Goldstein, M. E. Etzler, pp. 99–116. New York: Liss

197. Pharis, R. P., King, R. W. 1985. Gibberellins and reproductive development in seed plants. *Annu. Rev. Plant Physiol.* 36:517–68

198. Polya, G. M., Haritou, M. 1988. Purification and characterization of two wheat-embryo protein phosphatases. *Biochem. J.* 251:357–63

199. Polya, G. M., Klucis, E., Haritou, M. 1987. Resolution and characterization of two soluble calcium-dependent protein kinases from silver beet leaves. *Biochim. Biophys. Acta* 931:68–77

200. Ponz, F., Paz-Ares, J., Hernandez-Lucas, C., García-Olmedo, F., Carbonero, P. 1986. Cloning and nucleotide sequence of a cDNA encoding the precursor of the barley toxin α-hordothionin. *Eur. J. Biochem.* 156: 131–35

201. Preece, I. A., Hoggan, J. 1957. Carbohydrate modification during malting. *Eur. Brew. Conv. Proc. Congr. 6th Copenhagen,* pp. 72–83

202. Preece, I. A., MacDougall, M. 1958. Enzymic degradation of cereal hemicelluloses. II. Pattern of pentosan degradation. *J. Inst. Brew.* 64:489–500

203. Quatrano, R. S. 1986. Regulation of gene expression by abscisic acid during angiosperm embryo development. *Oxford Surv. Plant Mol. Cell Biol.* 3:467–77

204. Radley, M. 1967. Site of production of gibberellin-like substances in germinating barley embryos. *Planta* 75:164–71

205. Ralton, J. E., Smart, M. G., Clarke, A. E. 1987. Recognition and infection processes in plant pathogen interactions. See Ref. 28, pp. 217–52

206. Ranjeva, R., Boudet, A. M. 1987. Phosphorylation of proteins in plants: regulatory effects and potential involvement in stimulus/response coupling. *Annu. Rev. Plant. Physiol.* 38:73–93

207. Rastogi, V., Oaks, A. 1986. Hydrolysis of storage proteins in barley endosperms. Analysis of soluble products. *Plant Physiol.* 81:901–6

208. Raynes, J. G., Briggs, D. E. 1985. Genotype and the production of α-amylase in barley grains germinated in the presence and absence of gibberellic acid. *J. Cereal Sci.* 3:55–65

209. Richardson, M., Valdes-Rodriguez, S., Blanco-Labra, A. 1987. A possible function for thaumatin and a TMV-induced protein suggested by homology to a maize inhibitor. *Nature* 327:432–34

210. Roberts, W. K., Selitrennikoff, C. P. 1986. Isolation and partial characterization of two antifungal proteins from barley. *Biochim. Biophys. Acta* 880:161–70

211. Roberts, W. K., Selitrennikoff, C. P. 1988. Plant and bacterial chitinases differ in antifungal activity. *J. Gen. Microbiol.* 134:169–76

212. Rodaway, S. J. 1978. Composition of α-amylase secreted by aleurone layers of grains of Himalaya barley. *Phytochemistry* 17:385–89

213. Rogers, J. C. 1985. Two barley α-amylase gene families are regulated differently in aleurone cells. *J. Biol. Chem.* 260:3731–38

214. Rogers, J. C., Dean, D., Heck, G. R. 1985. Aleurain: a barley thiol protease closely related to mammalian cathepsin H. *Proc. Natl. Acad. Sci. USA* 82:6512–16

215. Rogers, J. C., Milliman, C. 1984. Coordinate increase in major transcripts from the high pI α-amylase multigene family in barley aleurone cells stimulated with gibberellic acid. *J. Biol. Chem.* 259: 12234–40

216. Roth, R. A., Pierce, S. B. 1987. In vivo cross-linking of protein disulfide isomerase to immunoglobulins. *Biochemistry* 26:4179–82

217. Deleted in proof

218. Ryan, C. A. 1987. Oligosaccharide signalling in plants. *Annu. Rev. Cell Biol.* 3:295–317

219. Sakai-Wada, A., Nakata, M. 1987. Effect of gibberellic acid on the ultrastructure and α-amylase activity of aleurone cells of *Avena sativa* L. *Plant Cell Physiol.* 28:1465–76

220. Saluja, D., Bansal, A., Sachar, R. C. 1987. Regulation of protein kinase through de novo enzyme synthesis in germinating embryos of wheat: enzyme purification and its autophosphorylation. *Plant Sci.* 50:37–48

221. Saluja, D., Berry, M., Sachar, R. C. 1987. Inorganic phosphate mimics the specific action of gibberellic acid in regulating the activity of monophenolase in embryo-less half-seeds of wheat. *Phytochemistry* 26:611–14

222. Schlumbaum, A., Mauch, F., Vögeli, U., Boller, T. 1986. Plant chitinases are potent inhibitors of fungal growth. *Nature* 324:365–67

223. Selvig, A., Aarnes, H., Lie, S. 1986. Cell wall degradation in endosperm of barley during germination. *J. Inst. Brew.* 92:185–87

224. Sharp, P. M., Tuohy, T.M.F., Mosurski, K. R. 1986. Codon usage in yeast: cluster analysis clearly differentiates highly and lowly expressed genes. *Nucleic Acids Res.* 14:5125–43

225. Shewry, P. R., Miflin, B. J. 1985. Seed storage proteins of economically impor-

tant cereals. *Adv. Cereal Sci. Technol.* 7:1–83
226. Shutov, A. D., Beltei, N. K., Vaintraub, I. A. 1985. A cysteine proteinase from germinating wheat seeds: partial purification and hydrolysis of gluten. *Biokhimika* 49:1004–10
227. Skerritt, J. H., Smith, R. A., Wrigley, C. W., Underwood, P. A. 1984. Monoclonal antibodies to gliadin proteins used to examine cereal grain protein homologies. *J. Cereal Sci.* 2:215–24
228. Smart, M. G., O'Brien, T. P. 1979. Observations on the scutellum. II. Histochemistry and autofluorescence of the cell wall in mature grain and during germination of wheat, barley, oats and ryegrass. *Aust. J. Bot.* 27:403–11
229. Smart, M. G., O'Brien, T. P. 1979. Observations on the scutellum. III. Ferulic acid as a component of the cell wall in wheat and barley. *Aust. J. Plant Physiol.* 6:485–91
230. Smith, M. M., Hartley, R. D. 1983. Occurrence and nature of ferulic acid substitution of cell-wall polysaccharides in graminaceous plants. *Carbohydr. Res.* 118:65–80
231. Sopanen, T., Burston, D., Taylor, E., Matthews, D. M. 1978. Uptake of glycylglycine by the scutellum of germinating barley grain. *Plant Physiol.* 61:630–33
232. Sorensen, S. B., Breddam, K., Svendsen, I. 1986. Primary structure of carboxypeptidase-I from malted barley. *Carlsberg Res. Commun.* 51:475–85
233. Sorensen, S. B., Svendsen, I., Breddam, K. 1987. Primary structure of carboxypeptidase-II from malted barley. *Carlsberg Res. Commun.* 52:285–95
234. Stevens, D. J., McDermott, E. E., Page, J. 1963. Isolation of endosperm proteins and aleurone cell contents from wheat, and determination of their amino acid composition. *J. Sci. Food Agric.* 14:284–87
235. Stewart, A., Nield, H., Lott, J.N.A. 1988. An investigation of the mineral content of barley grains and seedlings. *Plant Physiol.* 86:93–97
236. Stuart, I. M., Loi, L., Fincher, G. B. 1986. Development of (1→3, 1→4)-β-D-glucan endohydrolase isoenzymes in isolated scutella and aleurone layers of barley *(Hordeum vulgare)*. *Plant Physiol.* 80:310–14
237. Stuart, I. M., Loi, L., Fincher, G. B. 1987. Immunological comparison of (1→3, 1→4)-β-glucan endohydrolases in germinating cereals. *J. Cereal Sci.* 6:45–52

238. Sundblom, N.-O., Mikola, J. 1972. On the nature of the proteinases secreted by the aleurone layer of barley grain. *Physiol. Plant.* 27:281–84
239. Svendsen, I., Hejgaard, J., Mundy, J. 1986. Complete amino-acid sequence of the α-amylase/subtilisin inhibitor from barley. *Carlsberg Res. Commun.* 51:43–50
240. Svensson, B., Mundy, J., Gibson, R. M., Svendsen, I. B. 1985. Partial amino acid sequences of α-amylase isozymes from barley malt. *Carlsberg Res. Commun.* 50:15–22
241. Swift, J. G., O'Brien, T. P. 1972. The fine-structure of the wheat scutellum before germination. *Aust. J. Biol. Sci.* 25:9–22
242. Swift, J. G., O'Brien, T. P. 1972. The fine-structure of the wheat scutellum during germination. *Aust. J. Biol. Sci.* 25:469–86
243. Taiz, L., Honigman, W. A. 1976. Production of cell wall hydrolyzing enzymes by barley aleurone layers in response to gibberellic acid. *Plant Physiol.* 58:380–86
244. Taiz, L., Jones, R. L. 1970. Gibberellic acid, β-1, 3-glucanase and the cell walls of barley aleurone layers. *Planta* 92:73–84
245. Taiz, L., Jones, R. L. 1973. Plasmodesmata and an associated cell wall component in barley aleurone tissue. *Am. J. Bot.* 60:67–75
246. Tanaka, K., Yoshida, T., Kasai, Z. 1976. Phosphorylation of myo-inositol by isolated aleurone particles of rice. *Agric. Biol. Chem.* 40:1319–25
247. Tashiro, M., Hashino, K., Shiozaki, M., Ibuki, F., Maki, Z. 1987. The complete amino-acid sequence of rice bran trypsin-inhibitor. *J. Biochem.* 102:297–306
248. Tepfer, M., Taylor, I.E.P. 1981. The permeability of plant cell walls as measured by gel filtration chromatography. *Science* 213:761–63
249. Tkachuk, R., Kruger, J. E. 1974. Wheat α-amylases. II. Physical characterization. *Cereal Chem.* 51:508–29
250. Trewavas, A. J. 1987. Sensitivity and sensory adaptation in growth substance responses. In *Hormone Action in Plant Development—A Critical Appraisal*, ed. G. V. Hoad, J. R. Lenton, M. B. Jackson, R. K. Atkin, pp. 19–38. London: Butterworths
251. Trewavas, A. J. 1982. Growth substance sensitivity: the limiting factor in plant development. *Physiol. Plant.* 55:60–72
252. Triplett, B. A., Quatrano, R. S. 1982.

Timing, localization, and control of wheat germ agglutinin synthesis in developing wheat embryos. *Dev. Biol.* 91:491–96

253. Vakharia, D. N., Brearley, C. A., Wilkinson, M. C., Galliard, T., Laidman, D. L. 1987. Gibberellin modulation of phosphatidyl-choline turnover in wheat aleurone tissue. *Planta* 172:502–7

254. Van der Eb, A. A., Nieuwdorp, P. J. 1967. Electron microscopic structure of the aleuron cells of barley during germination. *Acta Bot. Neerl.* 15:690–99

255. Vance, V. B., Huang, A.H.C. 1988. Expression of lipid body protein gene during maize seed development. Spatial, temporal, and hormonal regulation. *J. Biol. Chem.* 263:1476–81

256. Visuri, K., Mikola, J., Enari, T.-M. 1969. Isolation and partial characterization of a carboxypeptidase from barley. *Eur. J. Biochem.* 7:193–99

257. Vögeli-Lange, R., Hansen-Gehri, A., Boller, T., Meins, F. 1988. Induction of the defense-related glucanohydrolases, β-1, 3-glucanase and chitinase, by tobacco mosaic virus infection of tobacco leaves. *Plant Sci.* 54:171–76

258. Wang, S-M., Huang, A.H.C. 1987. Biosynthesis of lipase in the scutellum of maize kernel. *J. Biol. Chem.* 262:2270–74

259. Weselake, R. J., Macgregor, A. W., Hill, R. D. 1985. Endogenous alpha-amylase inhibitor in various cereals. *Cereal Chem.* 62:120–23

260. Wessels, J.G.H., Sietsma, J. H. 1981. Fungal cell walls: a survey. In *Encyclopedia of Plant Physiology. Plant Carbohydrates II,* (NS), ed. W. Tanner, F. A. Loewus, 13B: 352–94. Berlin: Springer-Verlag

261. Whittier, R. F., Dean, D. A., Rogers, J. C. 1987. Nucleotide sequence analysis of alpha-amylase and thiol protease genes that are hormonally regulated in barley aleurone cells. *Nucleic Acids Res.* 15:2515–35

262. Wilkinson, M. C., Laidman, D. L., Galliard, T. 1984. Two sites of phosphatidylcholine synthesis in the wheat aleurone cell. *Plant Sci. Lett.* 35:195–99

263. Williamson, J. D., Quatrano, R. S. 1988. ABA-regulation of two classes of embryo-specific sequences in mature wheat embryos. *Plant Physiol.* 86:208–15

264. Williamson, M. S., Forde, J., Buxton, B., Kreis, M. 1987. Nucleotide sequence of barley chymotrypsin inhibitor-2 (CI-2) and its expression in normal and high-lysine barley. *Eur. J. Biochem.* 165:99–106

265. Williamson, M. S., Forde, J., Kreis, M. 1988. Molecular cloning of two isoinhibitor forms of chymotrypsin inhibitor 1 (CI-1) from barley endosperm and their expression in normal and mutant barleys. *Plant Mol. Biol.* 10:521–35

266. Winspear, M. J., Preston, K. R., Rastogi, V., Oaks, A. 1984. Comparisons of peptide hydrolase activities in cereals. *Plant Physiol.* 75:480–82

267. Wood, A., Paleg, L. G. 1974. Alteration of liposomal membrane fluidity by gibberellic acid. *Aust. J. Plant Physiol.* 1:31–40

268. Wood, P. J., Fulcher, R. G., Stone, B. A. 1983. Studies on the specificity of interaction of cereal cell wall components with Congo Red and Calcofluor. Specific detection and histochemistry of (1→3), (1→4) -β-D-glucan. *J. Cereal Sci.* 1:95–110

269. Woodward, J. R., Fincher, G. B. 1982. Purification and chemical properties of two 1,3;1,4-β-glucan endohydrolases from germinating barley. *Eur. J. Biochem.* 121:663–69

270. Woodward, J. R., Fincher, G. B. 1982. Substrate specificities and kinetic properties of two (1→3), (1→4)-β-D-glucan endo-hydrolases from germinating barley *(Hordeum vulgare). Carbohydr. Res.* 106:111–22

271. Woodward, J. R., Fincher, G. B., Stone, B. A. 1983. Water-soluble (1→3), (1→4)-β-D-glucans from barley *(Hordeum vulgare)* endosperm. II. Fine structure. *Carbohydr. Polym.* 3:207–25

272. Woodward, J. R., Morgan, F. J., Fincher, G. B. 1982. Amino acid sequence homology in two 1,3;1,4-β-glucan endohydrolases from germinating barley *(Hordeum vulgare). FEBS Lett.* 138: 198–200

273. Woolard, G. R., Rathbone, E. B., Novellie, L. 1976. A hemicellulosic β-D-glucan from the endosperm of sorghum grain. *Carbohydr. Res.* 51:249–52

274. Wreschner, D. H., Rechavi, G. 1988. Differential mRNA stability to reticulocyte ribonucleases correlates with 3' non-coding $(U)_n A$ sequences. *Eur. J. Biochem.* 172:333–40

275. Yamashita, H., Uehara, H., Tsumura, Y., Hayase, F., Kato, H. 1987. Precipitate-forming reaction of β-(1→4)-D-glucanase (I) in malt. *Agric. Biol. Chem.* 51:655–64

276. Yomo, H. 1960. Studies on the amylase-activating substance. IV. Amylase-activating activity of gibberellin. *Hakko Kyokai Shi* 18:600–3

Annu. Rev. Plant Physiol. Plant Mol. Biol. 1989. 40:347–69

PHYSIOLOGY AND MOLECULAR BIOLOGY OF PHENYLPROPANOID METABOLISM[1]

Klaus Hahlbrock and Dierk Scheel

Max-Planck-Institut für Züchtungsforschung, Abteilung Biochemie, D-5000 Köln 30, FRG

CONTENTS

[1]Abbreviations used: BMT, SAM : bergaptol O-methyltransferase; CAD, cinnamyl alcohol dehydrogenase; CHI, chalcone-flavanone isomerase; CHS, chalcone synthase; C4H, cinnamate 4-hydroxylase (monooxygenase); 4CL, 4-coumarate : CoA ligase; PAL, phenylalanine ammonia-lyase; *Pi, Phytophthora infestans; Pmg, Phytophthora megasperma* f. sp. *glycinea;* SAM, S-adenosyl-L-methionine; XMT, SAM : xanthotoxol O-methyltransferase

347

1040-2519/89/0601-0347$02.00

INTRODUCTION

The many plant-specific phenylpropanoid branch pathways and the corresponding functional diversity of their products have long attracted attention in plant physiology, as has the specific, differential inducibility of these pathways at the transcriptional level. Asked to summarize recent developments in this area, we concentrate on major advances in elucidating the structural organization, mode of expression, and functional relationships of genes encoding enzymes of phenylpropanoid metabolism in dicot plants. Where appropriate, we include the identification and some properties of phenolic constituents in the systems described.

The functions of phenylpropanoid derivatives are as diverse as their structural variations. Phenylpropanoids serve as low-molecular-weight flower pigments, antibiotics (phytoalexins), UV protectants, insect repellents, and signal molecules in plant-microbe interactions; they also function as complex, polymeric constituents of surface and support structures, such as suberin, lignin, and other cell-wall components. In accordance with the large structural and functional diversity of these compounds, the temporal patterns of phenylpropanoid pathway activities in development and the spatial distribution throughout individual plant organs and cell types vary greatly.

The individual branch pathways derive their basic phenylpropanoid building unit from the core reactions of general phenylpropanoid metabolism (Figure 1). Although the precise branch point is not known in all cases, the involvement of the last step of the core reactions, the formation of coenzyme A esters of cinnamate derivatives by 4-coumarate:CoA ligase, has been demonstrated for various ubiquitous pathways, such as those leading to the various flavonoid and lignin-like end products (38).

In the following, we first discuss in some detail three selected systems—parsley, bean, and potato plants and cell cultures—used for studies of the molecular genetics, mechanisms of activation, and functional connections between general phenylpropanoid metabolism and certain branch pathways. We then summarize some additional, related results obtained with other species.

THE PARSLEY SYSTEM

Special Features

The original purpose of studying phenylpropanoid derivatives in parsley *(Petroselinum crispum)* was the occurrence of flavonoid glycosides containing a branched-chain sugar, apiose, whose biosynthesis and mode of incorporation into flavonoids were first investigated (34). It soon turned out that cultured parsley cells were easy to propagate in simple, synthetic media and were a suitable system for elucidating the enzymology and signal-specific

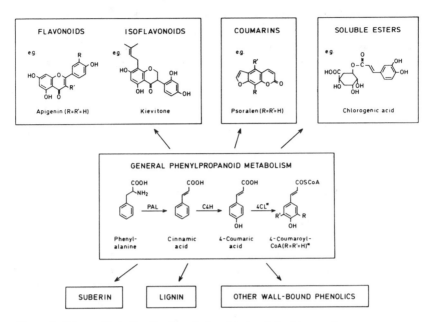

Figure 1 Scheme illustrating the flux of phenylalanine-derived intermediates from the core reactions of general phenylpropanoid metabolism to some of the major branch pathways mentioned in the text. Asterisks indicate the possible introduction of structural variations, predominantly in the 3 and 5 positions (R, R'), of 4-coumaroyl-CoA, catalyzed by enzymatic reactions either preceding or following CoA thiol ester formation by 4CL. Assignment of these reactions to individual pathways is open (38). Products shown are major phenolic constituents of parsley (apigenin, psoralen), French bean (kievitone), and potato (chlorogenic acid).

induction mechanisms of general phenylpropanoid metabolism as well as the flavone and flavonol glycoside and the furanocoumarin pathways (36, 38, 77). Of particular advantage was the selective inducibility of UV-protective flavonoids by UV light and of antibiotically active furanocoumarins (phytoalexins) by pathogen-derived elicitors, whose biosynthesis overlaps in the core reactions of general phenylpropanoid metabolism (36, 77). This differential response to signals causally related to specific biosynthetic end products and retention of the responsiveness by protoplasts (15) greatly facilitated some of the molecular analyses described below. Furthermore, several of the molecular probes used to localize pathway activities in whole plant tissue could not have been generated easily from starting material other than more or less uniformly stimulated cell cultures.

General Phenylpropanoid Metabolism

Although all three core reactions (Figure 1) have been investigated in parsley (38), studies of the genomic structure and mechanisms of gene activation have so far been confined to the first and the last enzymes, phenylalanine ammonia-lyase (PAL) and 4-coumarate : CoA ligase (4CL). Each occurs in cultured parsley cells in the form of heterogeneous protein populations and is encoded by multiple gene copies. PAL is a tetrameric enzyme (104) and has a complex structure with respect to both the catalytically active protein(s) and the corresponding genes (R. Lois, K. Hahlbrock, and W. Schulz, unpublished results). In accordance with the role of general phenylpropanoid metabolism as a supplier of substrates to various branch pathways, PAL is present constitutively, together with C4H and 4CL, in varying activities throughout all tested stages of plant development or cell culture growth (38, 54, 56).

In stem and leaf sections of young parsley seedlings PAL was detected immunohistochemically in all cell types. It was most abundant in epidermal and oil-duct epithelial cells, where flavonoids and furanocoumarins, respectively, are synthesized at high rates (52). PAL also accumulates in the tissue surrounding hypersensitive cell death at fungal infection sites. High enzyme levels are found in both the undamaged epidermis and an apparent, redifferentiating cell layer replacing the destroyed epidermis around the necrotic spot (51). In situ hybridization with labeled antisense RNA has demonstrated that transient PAL gene activation and increased transcript levels precede the local accumulation of PAL around infection sites (E. Schmelzer and K. Hahlbrock, unpublished results).

Unlike PAL, 4CL is a monomeric enzyme that occurs in two isoforms in parsley, each of which is encoded by a single-copy gene (23, 63). The two genes are very similar in exon-intron structure and nucleotide sequence, including major portions of the promoter region (23). All corresponding areas of exons and introns are 97–99% homologous. Two notable exceptions are a small (54-bp) and a larger (660-bp) insertion in the second intron of one 4CL gene relative to the other. This difference was used to demonstrate that both genes are expressed at similar relative rates, in unstimulated, UV-irradiated, or elicitor-stimulated cells (23). Thus, both 4CL genes are activated by either stimulus.

The two genes encode 4CL isoenzymes that differ in three amino acid residues, one of which results in a net charge difference. Separation on this basis yielded isoenzyme preparations with virtually identical catalytic properties (63). Hence, a differential role in phenylpropanoid branch pathways, as proposed for 4CL isoenzymes from several other plant species (38), is an unlikely explanation for the occurrence of two 4CL genes in parsley. Concomitant with PAL mRNA, 4CL mRNA is rapidly and transiently accumulated in small, confined areas around fungal penetration sites in infected parsley leaves (E. Schmelzer and K. Hahlbrock, unpublished results).

Flavone and Flavonol Glycoside Pathways

In light-grown leaves or cell suspension cultures of parsley, the structural relationships among more than 20 different flavone and flavonol glycosides and the enzymes involved in their biosynthesis have been described (38, 45, 57). The first committed step in the biosynthesis of all of these compounds is catalyzed by chalcone synthase (CHS), the key enzyme of the various flavonoid pathways (38, 45). Substrates of the CHS reaction are the major product of general phenylpropanoid metabolism, 4-coumaroyl-CoA, and the product of the acetyl-CoA carboxylase reaction, malonyl-CoA. Three acetate units from malonyl-CoA are converted to a new aromatic ring in the condensation product, naringenin chalcone. CHS has been studied extensively as the major representative of the flavonoid glycoside pathways in parsley, not only because of its pivotal role, but also because it is by far the most abundant of all biosynthetically related proteins in this system (46, 58, 59, 83). The abundance of this enzyme and its mRNA (relative to all the other biosynthetically related proteins and mRNAs) may be explicable as a compensation at the transcriptional level for poor catalytic efficiency. In CHS preparations of various degrees of purity, the specific catalytic activity in vitro is comparatively low, and the relative amount of catalytically inactive enzyme molecules is high (83).

The expression of flavonoid biosynthesis in parsley is dependent on light, both in cultured cells (8) and whole plant tissue (78). The wavelength requirements were determined by measuring the rate of CHS transcription in cell suspension cultures treated with different qualities and quantities of light (8, 71). UV light (applied at $\lambda_{max} = 350$ nm) is most efficient, although blue light ($\lambda_{max} = 436$ nm) in the absence of UV has some effect (approximately 10% of the UV effect). In addition, blue light has modulating effects when given prior to UV. These effects result in both the accumulation of increased absolute amounts of CHS mRNA and the elimination of a characteristic lag phase preceding the UV-stimulated increase in mRNA formation. While the signal for CHS mRNA induction provided by UV or blue light alone is comparatively short-lived and apparently converted immediately, the modulating effects of blue light on UV induction are stable. The triggering signal in blue pre-irradiated cells is stored in the system for at least 20 hr without detectable loss and is used up only upon UV irradiation (71). Comparatively small, but detectable, modulating as well as direct effects of red light (applied at $\lambda_{max} = 775$ nm; acting via phytochrome) were also observed (8, 71).

In intact parsley leaves, the entire light-dependent sequence of biosynthetic events, from CHS gene activation to enzyme and flavonoid glycoside accumulation, have been localized in situ in the epidermis (78). The results support the notion of a causal relationship between the triggering signal and the role of the accumulated products as protective agents against potential damage caused by UV irradiation in the remainder of the tissue. A direct flux of the information provided by the inducing agent, from transient gene activation through increased mRNA and enzyme activities to the accumulation of the biosynthetic end products, is indicated by a simple mathematical relationship (9).

In parsley, the CHS gene occurs in only one copy, whose two alleles are easily distinguished in heterozygous plants or cell cultures. They differ in size by a 927-bp transposon-like insertion in the promoter region of one allele relative to the other (position −538). Homozygous plants and cultured cell lines were used to demonstrate that both alleles are inducible by UV light and expressed during plant development (48). The CHS gene is located on the long arm of one of a small group of submetacentric chromosomes (50), a location different from that of the two UV-responsive 4CL genes (P.-L. Huang, K. Hahlbrock, and I. Somssich, unpublished results). Its occurrence as a single-copy gene and extremely efficient activation by UV light made this CHS gene suitable for "in vivo footprinting" as a means of identifying light-responsive, cis-acting elements. Several such elements were detected by this method and their functional necessity verified by in vitro mutagenesis and transient expression in transformed parsley protoplasts (84). One of these elements is significantly conserved in several other light-activated genes and in somewhat degenerated form in CHS genes from different plant species (84), suggesting the widespread occurrence of similar transcription activation mechanisms.

Furanocoumarin Pathway

The linear furanocoumarins, marmesin and psoralen, their coumarin precursor, umbelliferone, and the methoxylated psoralen derivatives, xanthotoxin, bergapten, and isopimpinellin, have been identified in the culture fluid of parsley cells treated with fungal elicitor (77, 92). Essentially the same mixture of compounds accumulates in infection droplets of parsley leaves inoculated with spores of the fungus *Phytophthora megasperma* f. sp. *glycinea (Pmg)* (51, 77). The dihydropyranocoumarin, graveolone, originally identified (92) as a major component in a different parsley cell line, was not detected in parsley plants (77). All of the furanocoumarins mentioned are antibiotically active and are considered to be potent phytoalexins in parsley and some related species (4, 77). With the exception of umbelliferone and marmesin, the same compounds are also present in uninfected, healthy leaves (54). This constitutive occurrence of furanocoumarins is confined to the lumen of oil ducts (51, 52). Incubation of methanolic extracts from elicitor-treated cells with β-glucosidases yielded all furanocoumarins previously identified from the culture medium of the same cells (D. Scheel, unpublished results). Therefore, the biosynthetic steps of furanocoumarin biosynthesis in parsley appear to include the intermediate formation of glucosides or glucose esters. These may be cleaved by cell wall– or plasma membrane–associated glucosidases as part of the excretion mechanism. Such a sequence of reactions would provide an efficient mechanism of protection against self-poisoning by these compounds.

In contrast to the flavonoid pathways mentioned above, not all of the enzymatic steps involved in the biosynthesis of the various furanocoumarins have been elucidated. In particular, the link between general phenylpropanoid metabolism and the furanocoumarin pathway proper is still obscure. It requires both *ortho*-hydroxylation of 4-coumarate and *cis-trans* isomerization of its side chain. Since the isomerization takes place in dark-grown parsley cells, it cannot be light dependent, as recently suggested (94). In analogy to chlorogenic acid synthesis (47), a hydroxycinnamate glycosyl or thiol ester may be the substrate for hydroxylation. The involvement of CoA esters has not been demonstrated unequivocally, and glucose esters may alternatively be considered as substrates for this branch pathway. Elicitor treatment of cultured parsley cells, but not irradiation, rapidly stimulates UDP-glucose:cinnamate, UDP-glucose:4-coumarate, and UDP-glucose:ferulate O-glucosyl transferases (D. Scheel, unpublished results).

The enzymes involved in formation of the furan ring are membrane associated. Dimethylallyl diphosphate:umbelliferone dimethylallyl transferase catalyzes the prenylation of umbelliferone to give demethylsuberosin (93), which is converted to (+)marmesin by marmesin synthase (40). Marmesin then serves as substrate for psoralen synthase (102). The activity of all three enzymes transiently increases in elicitor-treated parsley cells. Psoralen is the putative precursor for all linear furanocoumarins (7, 44). The hydroxylation reactions of the central aromatic ring have not been reported in vitro. SAM:xanthotoxol O-methyltransferase (XMT) and SAM:bergaptol O-methyltransferase (BMT), specifically methylating these hydroxyl groups, are also induced by elicitor in cultured parsley cells (44).

These furanocoumarin-specific methyltransferases have been selected for comparative studies of transcriptional activation patterns of this pathogen defense–related pathway and others (37) investigated in the parsley system. Specific antisera for each of the two methyltransferases were used to demonstrate the sequential induction of the corresponding mRNAs and enzymes (76). In the case of BMT, transcriptional activation was demonstrated by run-off transcription in nuclei from elicitor-treated cell cultures (K.-D. Hauffe, D. Scheel, and K. Hahlbrock, unpublished results) and by in situ RNA hybridization in *Pmg*-infected leaves (E. Schmelzer and K. Hahlbrock, unpublished results). Infection sites in leaves were examined by both immunohistochemical localization of BMT (51) and in situ identification of BMT mRNA (E. Schmelzer and K. Hahlbrock, unpublished results). The relatively late timing of transient mRNA and enzyme accumulation in infected tissue is consistent with the slow kinetics of furanocoumarin accumulation in infection droplets (77). The timing of BMT gene activation at infection sites is particularly slow when compared with the timing of transcript accumulation from the rapidly responding genes, such as pathogenesis-related protein 1 (88), in close agreement with similar observations made in elicitor-treated cell cultures (76, 87).

The chain of events leading from recognition of fungal elicitors to the activation of genes related to phenylpropanoid pathways has not been elucidated. However, it is noteworthy in this connection that different components of a crude *Pmg* elicitor preparation are active in soybean and parsley (72). Soybean plasma membranes contain a putative receptor specifically recognizing glucan fragments from *Pmg* (11, 25, 26, 79), which do not act as elicitors in parsley. In contrast, a proteinaceous component of the same fungal elicitor preparation stimulates phytoalexin synthesis in parsley, but not in soybean (72). Since parsley protoplasts completely retain elicitor responsiveness (15), the cell wall is apparently not involved in the recognition process. However, the plant cell wall serves as a source of a different, endogenous elicitor (18). Preliminary results indicate that protein phosphorylation (A. Dietrich and K. Hahlbrock, unpublished results) and Ca^{2+} influx (C. Colling, K. Hahlbrock, and D. Scheel, unpublished results) are associated with the elicitation of furanocoumarin production by cultured parsley cells.

Other Phenylpropanoid Derivatives

Histochemical staining of infected parsley leaves (51) and elicitor-treated cell cultures (W. Jahnen and K. Hahlbrock, unpublished results) indicated rapid incorporation of phenolic compounds into the cell wall as part of an early defense response. Elicitor treatment also stimulated the activity of SAM:caffeic acid O-methyltransferase (44), but did not affect feruloyl-CoA:NADP oxidoreductase (65), the enzyme catalyzing the formation of cinnamyl alcohols, the immediate precursors of lignin (35, 38). Improved methods of lignin determination revealed that there is no increase in lignification under these conditions (65; D. Scheel, unpublished results). However, alkaline hydrolysis of ethanol-extracted cell debris from elicitor-treated parsley cells released cinnamate, 4-coumarate, and ferulate, indicating esterification of these compounds to the cell wall. The activities of UDP-glucose:cinnamate, UDP-glucose:4-coumarate, and UDP-glucose:ferulate glucosyl transferases increased rapidly upon treatment of the cells with elicitor, but not with light (D. Scheel, unpublished results). Thus, impregnation of the cell wall with these esters appears to be an early component of the pathogen defense response in parsley. The coumarin esculetin is also believed to be incorporated into cell walls of elicitor-treated parsley cells (65). Its formation from caffeic acid is catalyzed by a polyphenol oxidase. However, esculetin was not detectable in extracts from the parsley cells. Finally, elicitor stimulates the formation of *trans*-5-O-caffeoylshikimate from 4-coumaroyl-

CoA and shikimate, successively catalyzed by 4-coumaroyl-CoA:shikimate coumaroyltransferase and 5-0-(4-coumaroyl)shikimate 3'-O-hydroxylase. Both enzymes increase in activity upon elicitor treatment. The reaction product may be polymerized into polyphenolic materials involved in the browning that is associated with hypersensitive cell death following microbial invasion (47).

THE BEAN SYSTEM

Special Features

The French bean (*Phaseolus vulgaris* L.) system has also greatly benefited from extensive comparative studies of intact plants and suspension-cultured cells. With regard to phenylpropanoid metabolism, one major difference between the two systems is the occurrence of isoflavonoid phytoalexins in bean but not in parsley. Flavonoid biosynthesis in bean thus diverges further into isoflavonoid and other flavonoid branch pathways, a metabolic specialization observed in many members of the Leguminoseae (3, 20, 21, 25, 26, 61). Individual cultivars of dwarf French bean have been studied in race/cultivar-specific interactions with physiological races of *Colletotrichum lindemuthianum,* the causal agent of anthracnose (2).

General Phenylpropanoid Metabolism

The first enzyme of phenylpropanoid biosynthesis, PAL, has been studied extensively in bean. The enzyme is rapidly synthesized de novo and accumulated in elicitor-stimulated bean cell cultures (13). Multiple tetrameric forms, differing in K_m as well as size and pI of the subunits, have been separated from partially purified PAL preparations (6). Each isoform contained subunits of M_r = 77,000, 70,000, and 53,000, the relative proportions of which depended on the conditions of enzyme purification and on the bean cultivar from which the cell culture used as a PAL source was derived. Interconversion of the subunits, in vivo as well as in vitro, was deduced from the results of pulse-chase, peptide-mapping, and in vitro translation experiments. The authors concluded that the native PAL subunit of M_r = 77,000 is inherently unstable and thus gives rise to the lower-M_r partial degradation products. An interesting observation is the rapid degradation of PAL activity in stimulated cells following exogenous addition of the product of the PAL reaction, *trans*-cinnamate (85). Whether this is a specific effect on PAL has not been unequivocally demonstrated and should probably be regarded with caution in view of nonspecific effects of *trans*-cinnamate on cultured parsley cells (101).

PAL is encoded in bean by a family of divergent classes of genes (C. J. Lamb, personal communication). Several immunoprecipitable PAL polypeptides were observed on a two-dimensional gel following in vitro translation of mRNA isolated from elicited bean cell cultures (6). In contrast, about 10–11 isoforms of the $M_r = 77,000$ subunit were detected by in vivo labeling of equivalent cells. These results suggest extensive posttranslational modification as a cause of at least part of the observed polymorphism (6). The PAL genes appear to be regulated differentially in roots, shoots, leaves, and flowers, as well as upon induction by mechanical wounding, fungal infection, or illumination. Antisense RNA was used for inhibition studies in vitro to identify specific isoforms encoded by transcripts of the different types of PAL gene (X. Liang and C. J. Lamb, personal communication).

Flavonoid Pathways

In bean and other legumes, the key enzyme of flavonoid biosynthesis, CHS, serves a multiple function. CHS is involved, on the one hand, in different branch pathways leading to flavone and isoflavone derivatives, and on the other hand, in the generation of 5'-deoxy as well as the more common 5'-oxy flavonoids (21, 45). In bean, differential de novo synthesis of up to 9 CHS isopeptides is observed in elicitor-treated cell cultures and wounded hypocotyls, and hypocotyls inoculated with spores of virulent or avirulent races of C. lindemuthianum (75). All of these treatments cause the accumulation of isoflavonoid phytoalexins following transient increases in CHS activity.

In agreement with the polymorphism observed at the polypeptide level, CHS in bean is encoded by a large, partly clustered gene family consisting of 6–8 members (75). Most nonlegume species investigated to date, including parsley (see above), contain only one or two CHS genes per haploid genome (48). Several CHS genes in bean are transiently activated by the treatments mentioned above, whereas only a small set is expressed in illuminated, previously etiolated hypocotyls. This genetic polymorphism in bean, as well as another legume, soybean (R. Wingender-Drissen, personal communication), has been proposed (75) to allow "a set of CHS genes encoding identical or very similar polypeptides to be positioned in different regulatory networks . . . to provide precise yet flexible adaptive and protective responses to a set of diverse environmental stresses." Since transcript-specific hybridization probes derived from 3' untranslated regions of elicitor-induced CHS mRNAs are available (75), it should be possible to test this hypothesis. Recent studies involving transformation of soybean protoplasts with chimeric gene constructs and identification of DNAse I hypersensitive sites suggest the presence of elicitor-regulated, cis-acting activator and silencer elements in the 5'-flanking region of a bean CHS gene (24).

The second committed step in flavonoid biosynthesis is catalyzed by chalcone-flavanone isomerase (CHI), which in bean is a monomeric protein of $M_r = 27,000$ (74). CHI mRNA is coordinately induced with PAL and CHS mRNAs in elicitor-treated cell cultures, in wounded hypocotyls, and in compatible or incompatible interactions of hypocotyls with *C. lindemu-thianum*. In contrast to the multigene families encoding several PAL and CHS isopolypeptides, respectively, CHI in bean appears as a single protein on a two-dimensional gel and is encoded by a single gene (66).

The timing of coordinate, transient PAL, CHS, and CHI induction differs greatly between compatible and incompatible interactions of bean hypocotyls with *C. lindemuthianum*. The activation of this and other, nonphenolic defense responses (for review, see 61) is much more rapid in incompatible than compatible interactions. This result supports the notion that rapid gene activation in the incompatible interaction is an important feature in disease resistance. Similar results, though on greatly reduced time scales, have been obtained with other systems—e.g. infected potato leaves or soybean roots (see below).

Lignin Pathway

Another phenylpropanoid branch pathway of great importance in plant development and environmental protection is that generating precursors for the deposition of lignin and related wall-bound phenolics (35, 49). The monomeric precursors are hydroxylated and methoxylated cinnamyl alcohols synthesized in two steps from the corresponding coenzyme A esters by cinnamoyl-CoA reductase and cinnamyl alcohol dehydrogenase (CAD) (35). The second of these two enzymes, CAD, was shown to be accumulated very rapidly in elicitor-treated bean cell cultures. The corresponding, transient increase in CAD mRNA was even more rapid than the coordinated increases in PAL, CHS, and CHI mRNAs, with maximal levels of hybridizable mRNA occurring about 1.5 and 4 hr after elicitor treatment, respectively (33, 100).

An interesting speculation is based on the observation that certain lignin precursors, dehydrodiconiferyl glucosides, exhibit cytokinin-like activity in plant cells (5, 64). It has been suggested (100) that the rapid elicitor stimulation of CAD may play a role in the generation of secondary signals involved in the induction of the pathogen defense response. In this connection, studies of both the temporal and spacial patterns of CAD induction in compatible and incompatible interactions of bean hypocotyls with *C. lindemuthianum*, as well as attempts to elucidate its biochemical function at infection sites (deposition of cell-wall components, generation of signals, or both; other functions?) are likely to give valuable further insight into the mechanism(s) of disease resistance. Another interesting observation is the occurrence of only one class of CAD gene in the bean genome. Thus, CAD and CHI, two enzymes catalyzing near-equilibrium second reactions of specific branch pathways, are encoded by genes of relatively simple organization, in contrast to the branch-point enzymes, PAL and CHS (100).

THE POTATO SYSTEM

Special Features

In potato *(Solanum tuberosum)*, as in most or all other plants, various phenylpropanoid pathways are stimulated by wounding, illumination, or pathogen attack (12, 62, 86). Unlike parsley, bean, and some of the systems discussed below, potato has frequently been studied with regard to general phenylpropanoid metabolism without detailed knowledge of the subsequent branch pathways or their biosynthetic end products. Sliced tubers were used in most of the earlier studies, and measurement of combined responses to wounding, illumination, or infection by pathogens, superimposed on senescence reactions, often complicated interpretation of the results. More recent investigations (see below) have attempted to avoid such complications by analyzing cell cultures and leaf tissue. As described above, this allows generation of specific probes for in situ measurements of gene activity.

General Phenylpropanoid Metabolism

Of the three enzymes of general phenylpropanoid metabolism, only PAL has been investigated thoroughly in potato at the protein level (43). The enzyme, its heterogeneous appearance on two-dimensional gels, and the size of its mRNA, are similar in potato (32), parsley, and bean (see above). However, the genomic organization of PAL in potato is particularly complex. Approximately 20 cloned genes have been distinguished on the basis of their restriction fragmentation patterns, and genomic DNA blots suggest that the total number of PAL genes is even higher (H. Fritzemeier, I. Häuser, H.-J. Joos, and K. Hahlbrock, unpublished results).

The timing of PAL gene activation in response to fungal infection or elicitor treatment of mature potato leaves, or to elicitor treatment of potato cell cultures, has been investigated (32). In all three cases, transient, large, and rapid increases in the rate of PAL transcription were observed within 1–2 hr. In fungal infections, a second wave of induction occurs a few hours later, though with different intensity and timing in compatible and incompatible interactions with *Phytophthora infestans*. This differential response is particularly obvious in young leaves, where hypersensitive cell death occurs within 3 hr. In the incompatible interaction, in situ hybridization using labeled antisense RNA has demonstrated rapid, transient accumulation of PAL mRNA in a sharply confined area around the fungal penetration site. The same cultivar responded in the compatible interaction by accumulating PAL mRNA more slowly and in a more diffuse and further spreading area (14).

Uptake of a solution containing L-2-aminooxy-3-phenylpropionic acid (AOPP), an inhibitor of PAL activity (1), prior to infection of leaves from this cultivar with an avirulent race of *P. infestans* completely abolished the typical resistance response (D. Scheel, unpublished results). However, the fungus was still unable to sporulate and the interaction was therefore not fully converted from incompatible to compatible. A case of full conversion by PAL inhibition in vivo is discussed below for the soybean-*Pmg* interaction. In both the potato and the soybean system, these results suggest that phenylpropanoid metabolism plays an essential role in the resistance response to fungal infections.

In contrast to the large complexity of the genomic organization of PAL in potato, 4CL occurs only in two copies. In every known respect, the two 4CL genes in potato are very similar to those in parsley, including the exon-intron structure (except that size differences are not observed for introns in the two potato 4CL genes) and a high degree of homology between the two genes (M. Becker-Andre and K. Hahlbrock, unpublished results).

Soluble Phenylpropanoid Derivatives

Various soluble phenylpropanoid derivatives have been described in potato (30, 31). One of the most abundant and extensively studied is chlorogenic acid (3-O-(caffeoyl)quinate), the levels of which increase greatly in potato tubers upon illumination (62), infection (86), or wounding (12). Transient increases were observed within approximately 40 hr of illumination or infection, whereas a relatively slow accumulation occurs within 12 days during wound healing. Increased activities of PAL and hydroxycinnamoyl-CoA : quinate hydroxycinnamoyltransferase have been correlated with chlorogenic acid accumulation (60). However, the biosynthesis of chlorogenic acid in potato is not completely understood. The acid moiety of 4-coumaroyl-CoA is probably transferred to quinate prior to hydroxylation in the 3' position (73). It should be noted in this connection that a different biosynthetic pathway has been proposed for sweet potato roots (97), where cinnamoyl-D-glucose is first formed from cinnamate and UDP-glucose. This ester is hydroxylated twice to give caffeoyl-D-glucose, which then serves as substrate for hydroxycinnamoyl-D-glucose : quinate hydroxycinnamoyltransferase in the transesterification to chlorogenic acid. Despite its abundance and widespread occurrence, the function(s) of chlorogenic acid in potato or other species is unknown.

Wall-Bound Phenolics

Wounding and pathogen attack drastically change the composition of cell walls in potato tubers (12, 31, 41, 42). Following either treatment, nitrobenzene or cupric oxide oxidation releases 4-hydroxybenzaldehyde and vanillin, but no syringaldehyde, from cell wall preparations (12, 41, 42). The absence of syringaldehyde, a characteristic product of oxidative lignin degradation, indicates that the release products originate from suberin rather

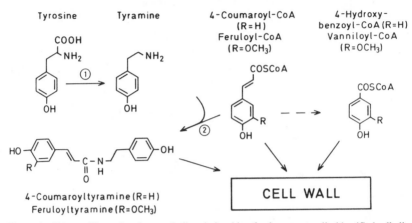

Figure 2   Scheme illustrating the metabolic relationship of a few structurally identified, alkali-labile phenolic constituents of potato cell walls. Two known enzymatic steps are catalyzed by 1. tyrosine decarboxylase, 2. hydroxycinnamoyl-CoA : tyramine hydroxycinnamoyltransferase. R indicates nature and position of major structural variations.

than lignin. Wound-induced deposition of suberin phenolics is slow, reaching maximum levels after 5–10 days, whereas suberization caused by pathogen attack occurs rapidly within 1 day (42). Alkaline hydrolysis of cell walls from infected tuber or leaf tissue releases two other classes of phenolic compounds. One class consists of hydroxybenzoic and hydroxycinnamic acids, such as 4-hydroxybenzoic, 4-coumaric and ferulic acids; the other consists of hydroxycinnamic amides, primarily 4-coumaroyltyramine and feruloyltyramine (10, 31). Accumulation of these compounds in the affected tissue is preceded by coordinate increases in the activities of PAL and tyrosine decarboxylase (H. Keller, K. Hahlbrock, and D. Scheel, unpublished results). Soluble and wall-bound amides of hydroxycinnamic acids appear to be involved in resistance reactions and developmental processes in several solanaceous plants (29, 91). The present view of these metabolic interconnections in potato is illustrated in Figure 2.

OTHER SYSTEMS

Space does not allow extensive discussion of more than three selected systems. However, some fundamental aspects are being studied in other systems. We mention these briefly, particularly as far as new developments in the molecular biology of phenylpropanoid metabolism are concerned. With respect to the large contributions made by the analysis of transposable elements in various plant systems, the reader is referred to relevant review articles (e.g., 22).

Disease Resistance

Besides bean, several legumes, including soybean (*Glycine max* L.) and pea (*Pisum sativum* L.), respond to pathogen attack and various other types of stress with the accumulation of isoflavonoid phytoalexins. In soybean, the biosynthesis of glyceollin and related isoflavonoid derivatives, as well as their accumulation pattern in race/cultivar-specific interactions with *Pmg,* have been studied extensively (25, 26). Cultured soybean cells and cut soybean cotyledons have been used to study the effects of *Pmg*-derived glucan fragments in the elicitation of glyceollin production (11, 79).

Recently, specific high-affinity binding sites for these fragments were described in plasma membrane fractions from soybean roots, hypocotyls, cotyledons, and cell suspension cultures. In these experiments, the ability of chemically related compounds to compete for binding of the glucan elicitor correlated well with their efficiency in stimulating glyceollin production (79). The mechanism of elicitation again involves gene activation and is associated with Ca^{2+} influx into elicitor-treated cells (89), whereas neither cAMP (39) nor phosphoinositol (90) concentration changes were observed.

PAL activity is inhibited in soybean roots by treatment with L-2-aminooxy-3-phenylpropionic acid (AOPP) or *R*-(1-amino-2-phenyl-ethyl)phosphonic acid (APEP) prior to infection with an avirulent race of *Pmg.* This inhibition completely abolishes glyceollin induction and changes the phenotype of the interaction from incompatible to compatible, suggesting that phenylpropanoid derivatives play a decisive role in the resistance of soybean to *Pmg* (69, 99). However, it remains open whether these compounds are phytoalexins or other phenolics. So far, the only established case of a causal relationship between phytoalexin levels and resistance of a plant to a fungal pathogen is that of the pea isoflavonoid phytoalexin pisatin and the fungus *Nectria haematococca.* Resistance of pisatin-containing pea leaves is overcome by those fungal strains possessing an inducible pisatin demethylase (53).

Evolution of Phenylpropanoid Pathways

An interesting question concerns the evolutionary origin and possible relationships of phenylpropanoid branch pathways. A first approach to an answer was the comparison of CHS genes from different sources. Among some notable, common features is the presence of introns in identical positions, including one intron inserted in a highly conserved cysteine codon, possibly indicating an important function of this amino acid in CHS activity as well as a common evolutionary origin of the various CHS genes (70).

Recently, the genes encoding the key enzymes of flavonoid and stilbenoid biosynthesis, chalcone and resveratrol synthases, have also been compared. Several stilbenoid derivatives serve as phytoalexins, *e.g.* resveratrol and some related compounds in peanut (*Arachis hypogaea* L.). Their synthesis involves two condensation steps of acetate units from malonyl-CoA with 4-coumaroyl-CoA, catalyzed by resveratrol synthase (80), as compared with three such condensation reactions catalyzed by CHS. Resveratrol synthase is readily induced by treatment of cultured peanut cells with elicitor or UV light (98). Comparison of resveratrol synthase and CHS at the level of cDNA and genomic clones revealed a high degree of homology for corresponding protein-encoding regions, including analogous positions of the intron splitting a cysteine codon in resveratrol synthase as well as CHS (82). These results may be the beginning of extensive studies on evolutionary relationships of related phenylpropanoid pathways within and between species.

Petunia hybrida

One of the classical systems for genetic and biochemical studies of flavonoid biosynthesis and its molecular basis is *Petunia hybrida* (81, 103). Recent advances include analyses of the genomic organizations of CHS, a particularly large multigene family in a nonleguminous plant, consisting of at least seven complete members (55), and CHI, which is encoded by one or two nonallelic genes (96). As expected from their close functional relationship, CHS and CHI in *P. hybrida* are expressed, at least to a large extent, in a coordinated manner (96). A new development with a high potential for further, more detailed investigations was the generation of *P. hybrida* mutants with impaired flower pigment synthesis through the expression of antisense CHS RNA (95).

A different approach, likewise exploiting the excellent genetic background in *P. hybrida* and opening up new experimental realms, was the construction of transgenic *P. hybrida* plants expressing a new flower color (67). The recipient white-flower mutant, which is unable to convert colorless dihydroflavonol precursors to anthocyanin (flavonoid) pigments, was transformed with the A_1 gene of *Zea mays,* encoding dihydroquercetin 4-reductase. This enzyme catalyzes the formation of a red pigment from the accumulated precursor and hence converts the white- to a red-flower mutant. The mutant produces an anthocyanin derivative (pelargonidin) slightly differing in the hydroxylation pattern, and therefore in color, from the normally occurring pigments. This results in an easily screenable phenotype that may serve as a useful marker for gene expression studies (67).

Arabidopsis thaliana

Finally, an increasing number of investigators are recognizing *Arabidopsis thaliana* as a system whose unusually low genome complexity well suits it for studies of plant genome organization and mechanisms of gene expression (27, 68). Initial work on phenylpropanoid metabolism has demonstrated that the core reactions (Figure 1) as well as CHS are induced by treatments similar to those described for other plants. High-intensity irradiation results in the accumulation of anthocyanins, which is preceded by sequential, transient increases in CHS transcription rates, mRNA amount, and enzyme activity (28). A. *thaliana* possesses a single CHS gene with extensive homology to CHS genes from other plants (28), again including the occurrence of the only intron within the conserved cysteine codon (see above). Treatment of cultured A. *thaliana* cells with various types of elicitor leads to elevated levels of PAL and 4CL activities, following transient increases in PAL mRNA (17; K. Davis, personal communication). So far, enzymes and products of phenylpropanoid branch pathways have not been identified, with the above-mentioned exception of CHS and anthocyanins. However, all theoretical and practical advantages of this system, including the small plant size and rapid growth, suggest that major contributions to our understanding of the molecular mechanisms regulating phenylpropanoid metabolism will soon be made using A. *thaliana* as an experimental tool.

CONCLUSIONS AND OUTLOOK

This article deals with an area of research that is expanding rapidly. A similar review article ten years ago (38) found the tools of molecular biology just beginning to provide insight into previously invisible realms. Two of the major types of question we asked then concerned (*a*) the extent to which changes in gene activity contribute to the fine tuning of phenylpropanoid pathways in cellular metabolism and development, and (*b*) the nature of the involved signal molecules and signal-transduction mechanisms.

The first question has been answered, at least in principle, in a number of cases. Changes in gene expression rates play a major role in the regulation of various phenylpropanoid pathways. As mentioned above, flavonoid biosynthesis in UV-irradiated parsley cells is an example where an appreciable contribution of other mechanisms could be excluded on the basis of mathematical calculations (9). However, it might be worth emphasizing that regulation of protein synthesis is not always exerted exclusively or predominantly at the gene level.

The second type of question, regarding signals and their trigger mechanisms, has remained largely unanswered. In fact, more or less the same questions are still being asked, and it has become evident that they are increasing in complexity. Typical examples are related to the topics discussed above. The light-triggered expression of flavonoid biosynthesis in parsley might serve as the first example. Until recently, it was held that UV light is essential for the induction, whereas blue and red light have only modulating effects (8). Although this has now been largely confirmed at the level of CHS gene expression, a more detailed analysis has revealed that blue and even red light, when given alone, do have small but measurable effects (71). Three different photoreceptors have been postulated to participate in the perception of UV, blue, and red light in this system (8, 71). Only one of these, the red light–responsive phytochrome, has been described in molecular terms. With this exception, all links in the signal-transduction chain are presently unknown.

The other example differs in detail, but is analogous in principle. Many plants, perhaps all, accumulate phytoalexins as part of their defense response to pathogens (3, 20, 25). Some of the responsible genes have been shown to be activated either as a result of a true plant-pathogen interaction or upon treatment of appropriate plant tissue with a pathogen- or plant-derived "biotic", or with an "abiotic" elicitor (16, 20, 25, 26). Again, a considerable biological complexity of the trigger mechanism became apparent recently, this time by comparison of two different systems, parsley and soybean. The same crude mixture of fungal cell-wall fragments acts as elicitor of phytoalexin synthesis in both systems. However, the active components are protein(s) in parsley and oligomeric glucan(s) in soybean (72). In both systems, the crude pathogen-derived elicitor acts more efficiently in a synergistic interaction with plant cell-wall fragments ("endogenous" oligoglucuronide elicitor), although either elicitor by itself is capable of eliciting phytoalexin synthesis (18). Thus, the complex situation resembles that of UV and/or blue or red light–triggered flavonoid synthesis, even though the products are different, and consequently the putative "second messengers" and/or *trans*-acting factors involved in gene activation may be different as well. Moreover, it is unknown whether any of the isolated compounds with elicitor activity occurs as such in a true plant-pathogen interaction. Again, questions about the chemical nature and complexity of all links in the signal-transduction chains have so far remained unanswered, despite some notable progress in the structural analysis of elicitors, including first indications of race specificity (19).

What then are the major recent advances in this field, and what is to be expected in the near future? The structural and functional analysis of genes, classed under the new term molecular genetics, is proceeding rapidly and on a broadening scale. The results will soon form a solid basis for new approaches in the unraveling of trigger mechanisms in signal-specific gene activation. The speed at which our understanding of events between signal perception and the mechanisms of gene activation increases will naturally depend on the complexity of the intervening steps. However, the goal is clearly defined, and suitable tools have become available. For example, monoclonal antibodies can be used to study the nature and location of both effector and effector-binding molecules, and new, powerful methods have been developed to identify those signal-specific *cis*-elements and *trans*-acting factors that are responsible for the control of gene expression rates. Some of the systems discussed above provide the necessary prerequisites for such studies.

ACKNOWLEDGMENTS

We thank the colleagues mentioned in the text who provided unpublished information. We are also grateful to Dr. J. Dangl for valuable comments on the manuscript.

Literature Cited

1. Amrhein, N., Gödeke, K.-H. 1977. α-Aminooxy-β-phenyl-propionic acid—a potent inhibitor of L-phenylalanine ammonia-lyase *in vitro* and *in vivo*. *Plant Sci. Lett.* 8:313–17
2. Bailey, J. A. 1982. Physiological and biochemical events associated with the expression of resistance to disease. In *Active Defense Mechanisms in Plants,* ed. R. K. S. Wood, pp. 39–65. New York: Plenum
3. Bailey, J. A., Mansfield, J. W., ed. 1982. *Phytoalexins.* London: Blackie & Son. 334 pp.
4. Beier, R. C., Oertli, E. H. 1983. Psoralen and other linear furocoumarins as phytoalexins in celery. *Phytochemistry* 22:2595–97
5. Binns, A. N., Chen, R. H., Wood, H. N., Lynn, D. G. 1987. Cell division promoting activity of naturally occurring dehydrodiconiferyl glucosides: Do cell wall components control cell division? *Proc. Natl. Acad. Sci. USA* 84:980–84
6. Bolwell, G. P., Bell, J. N., Cramer, C. L., Schuch, W., Lamb, C. J., Dixon, R. A. 1985. L-Phenylalanine ammonia-lyase from *Phaseolus vulgaris.* Characterisation and differential induction of multiple forms. *Eur. J. Biochem.* 149:411–19
7. Brown, S. A. 1985. Recent advances in the biosynthesis of coumarins. In *The Biochemistry of Plant Phenolics,* ed. C. F. van Sumere, P. J. Lea, 25:257–70. Oxford: Clarendon
8. Bruns, B., Hahlbrock, K., Schäfer, E. 1986. Fluence dependence of the ultraviolet-light-induced accumulation of chalcone synthase mRNA and effects of blue and far-red light in cultured parsley cells. *Planta* 169:393–98
9. Chappell, J., Hahlbrock, K. 1984. Transcription of plant defence genes in response to UV light or fungal elicitor. *Nature* 311:76–78
10. Clarke, D. D. 1982. The accumulation of cinnamic acid amides in the cell walls

of potato tissue as an early response to fungal attack. In *Active Defense Mechanisms in Plants,* ed. R. K. S. Wood, pp. 321–22. New York: Plenum
11. Cosio, E. G., Pöpperl, H., Schmidt, W. E., Ebel, J. 1988. High-affinity binding of fungal β-glucan fragments to soybean (*Glycine max* L.) microsomal fractions and protoplasts. *Eur. J. Biochem.* 175:309–15
12. Cottle, W., Kolattukudy, P. E. 1982. Biosynthesis, deposition, and partial characterization of potato suberin phenolics. *Plant Physiol* 69:393–99
13. Cramer, C. L., Bell, J. N., Ryder, T. B., Bailey, J. A., Schuch, W., et al. 1985. Co-ordinated synthesis of phytoalexin biosynthetic enzymes in biologically-stressed cells of bean (*Phaseolus vulgaris* L.). *EMBO J.* 4:285–90
14. Cuypers, B., Schmelzer, E., Hahlbrock, K. 1988. *In situ* localization of rapidly accumulated phenylalanine ammonia-lyase mRNA around penetration sites of *Phytophthora infestans* in potato leaves. *Mol. Plant-Microbe Interact.* 1:157–60
15. Dangl, J. L., Hauffe, K.-D., Lipphardt, S., Hahlbrock, K., Scheel, D. 1987. Parsley protoplasts retain differential responsiveness to uv-light and fungal elicitor. *EMBO J.* 6:2551–56
16. Darvill, A. G., Albersheim, P. 1984. Phytoalexins and their elicitors—a defense against microbial infection in plants. *Annu. Rev. Plant Physiol.* 35:243–75
17. Davis, K. R., Ausubel, F. M. 1988. Characterization of elicitor and pathogen induced gene expression in *Arabidopsis thaliana. J. Cell. Biochem. Suppl.* 12C:264
18. Davis, K. R., Hahlbrock, K. 1987. Induction of defense responses in cultured parsley cells by plant cell wall fragments. *Plant Physiol.* 85:1286–90
19. De Wit, P. J. G. M., Hofman, A. E., Velthuis, G. C. M., Kuc, J. A. 1985. Isolation and characterization of an elici-

tor of necrosis isolated from intercellular fluids of compatible interactions of *Cladosporium fulvum* (Syn. *Fulvia fulva*) and tomato. *Plant Physiol.* 77:642–47

20. Dixon, R. A. 1986. The phytoalexin response: elicitation, signalling and control of host gene expression. *Biol. Rev.* 61:239–91

21. Dixon, R. A., Dey, P. M., Lawton, M. A., Lamb, C. J. 1983. Phytoalexin induction in French bean. Intercellular transmission of elicitation in cell suspension cultures and hypocotyl sections of *Phaseolus vulgaris*. *Plant Physiol.* 71: 251–56

22. Döring, H.-P., Starlinger, P. 1986. Molecular genetics of transposable elements in plants. *Annu. Rev. Genet.* 20: 175–200

23. Douglas, C., Hoffmann, H., Schulz, W., Hahlbrock, K. 1987. Structure and elicitor or uv-light-stimulated expression of two 4-coumarate:CoA ligase genes in parsley. *EMBO J.* 6:1189–95

24. Dron, M., Clouse, S. D., Dixon, R. A., Lawton, M. A., Lamb, C. J. 1988. Glutathione and fungal elicitor regulation of a plant defense gene promoter in electroporated protoplasts. *Proc. Natl. Acad. Sci. USA* 85:6738–42

25. Ebel, J. 1986. Phytoalexin synthesis: The biochemical analysis of the induction process. *Annu. Rev. Phytopathol.* 24:235–64

26. Ebel, J., Grisebach, H. 1988. Defense strategies of soybean against the fungus *Phytophthora megasperma* f. sp. *glycinea*: a molecular analysis. *Trends Biochem. Sci.* 13:23–27

27. Estelle, M. A., Somerville, C. R. 1986. The mutants of *Arabidopsis*. *Trends Genet.* 2:89–93

28. Feinbaum, R. L., Ausubel, F. M. 1988. Transcriptional regulation of the *Arabidopsis thaliana* chalcone synthase gene. *Mol. Cell. Biol.* 8:1985–92

29. Filner, P. 1987. Candidate for a sexual reproduction specific pathway in plants: putrescine to GABA *via* hydroxycinnamic amide intermediates. In *Plant Molecular Biology*, ed. D. von Wettstein, N.-H. Chua, pp. 243–51. New York: Plenum

30. Friend, J. 1981. Plant phenolics, lignification and plant disease. *Progr. Phytochem.* 7:197–261

31. Friend, J. 1985. Phenolic substances and plant disease. In *The Biochemistry of Plant Phenolics*, ed. C. F. van Sumere, P. J. Lea, pp. 367–92. Oxford: Clarendon

32. Fritzemeier, K.-H., Cretin, C., Kombrink, E., Rohwer, F., Taylor, J., et al.

1987. Transient induction of phenylalanine ammonia-lyase and 4-coumarate:CoA ligase mRNAs in potato leaves infected with virulent or avirulent races of *Phytophthora infestans*. *Plant Physiol.* 85:34–41

33. Grand, C., Sarni, F., Lamb, C. J. 1987. Rapid induction by fungal elicitor of the synthesis of cinnamyl-alcohol dehydrogenase, a specific enzyme of lignin synthesis. *Eur. J. Biochem.* 169:73–77

34. Grisebach, H. 1980. Branched-chain sugars: occurrence and biosynthesis. In *The Biochemistry of Plants*, ed. J. Preiss, 3:171–97. New York: Academic

35. Grisebach, H. 1981. Lignins. In *The Biochemistry of Plants*, ed. E. E. Conn, 7:457–78. New York: Academic

36. Hahlbrock, K., Boudet, A. M., Chappell, J., Kreuzaler, F., Kuhn, D. N., Ragg, H. 1983. Differential induction of mRNAs by light and elicitor in cultured plant cells. In *NATO Advanced Studies Institute on "Structure and Function of Plant Genomes,"* ed. O. Ciferri, L. Dure III, pp. 15–23. London: Chapman & Hall

37. Hahlbrock, K., Cretin, C., Cuypers, B., Fritzemeier, K.-H., Hauffe, K.-D., et al. 1987. Tissue specificity and dynamics of disease resistance responses in plants. See Ref. 29, pp. 399–406.

38. Hahlbrock, K., Grisebach, H. 1979. Enzymic controls in the biosynthesis of lignin and flavonoids. *Annu. Rev. Plant Physiol.* 30:105–30

39. Hahn, M. G., Grisebach, H. 1983. Cyclic AMP is not involved as a second messenger in the response of soybean to infection by *Phytophthora megasperma* f. sp. *glycinea*. *Z. Naturforsch. Teil C* 38:578–82

40. Hamerski, D., Matern, U. 1988. Elicitor-induced biosynthesis of psoralens in *Ammi majus* L. suspension cultures. Microsomal conversion of demethylsuberosin into (+)marmesin and psoralen. *Eur. J. Biochem.* 171:369–75

41. Hammerschmidt, R. 1984. Rapid deposition of lignin in potato tuber tissue as a response to fungi non-pathogenic on potato. *Physiol. Plant Pathol.* 24:33–42

42. Hammerschmidt, R. 1985. Determination of natural and wound-induced potato tuber suberin phenolics by thioglycolic acid derivatization and cupric oxide oxidation. *Potato Res.* 28:123–27

43. Hanson, K. R., Havir, E. A. 1981. Phenylalanine ammonia-lyase. See Ref. 35, pp. 577–625

44. Hauffe, K.-D., Hahlbrock, K., Scheel, D. 1986. Elicitor-stimulated furanocoumarin biosynthesis in cultured pars-

ley cells: S-adenosyl-ʟ-methionine: bergaptol and S-adenosyl-ʟ-methionine: xanthotoxol O-methyltransferases. *Z. Naturforsch. Teil C* 41:228–39

45. Heller, W., Forkmann, G. 1988. Biosynthesis. In *The Flavonoids*, ed. J. B. Harborne, pp. 399–425. London: Chapman & Hall

46. Heller, W., Hahlbrock, K. 1980. Highly purified "flavanone synthase" from parsley catalyzes the formation of naringenin chalcone. *Arch. Biochem. Biophys.* 200: 617–19

47. Heller, W., Kühnl, T. 1985. Elicitor induction of a microsomal 5-O-(4-coumaroyl)shikimate 3'-hydroxylase in parsley cell suspension cultures. *Arch. Biochem. Biophys.* 241:453–60

48. Herrmann, A., Schulz, W., Hahlbrock, K. 1988. Two alleles of the single-copy chalcone synthase gene in parsley differ by a transposon-like element. *Mol. Gen. Genet.* 212:93–98

49. Higuchi, T. 1985. Biosynthesis and biodegradation of wood components. New York: Academic

50. Huang, P.-L., Hahlbrock, K., Somssich, I. 1988. Detection of a single-copy gene on plant chromosomes by in situ hybridization. *Mol. Gen. Genet.* 211: 143–17

51. Jahnen, W., Hahlbrock, K. 1988. Cellular localization of nonhost resistance reactions of parsley *(Petroselinum crispum)* to fungal infection. *Planta* 173: 197–204

52. Jahnen, W., Hahlbrock, K. 1988. Differential regulation and tissue-specific distribution of enzymes related to phenylpropanoid branch pathways in developing parsley seedlings. *Planta* 173: 453–58

53. Kistler, H. C., VanEtten, H. D. 1984. Regulation of pisatin demethylation in *Nectria haematococca* and its influence on pisatin tolerance and virulence. *J. Gen. Microbiol.* 130:2605–13

54. Knogge, W., Kombrink, E., Schmelzer, E., Hahlbrock, K. 1987. Occurrence of phytoalexins and other putative defense-related substances in uninfected parsley plants. *Planta* 171:279–87

55. Koes, R. E., Spelt, C. E., Mol, J. N. M., Gerats, A. G. M. 1987. The chalcone synthase multigene family of *Petunia hybrida* (V30): sequence homology, chromosomal localization and evolutionary aspects. *Plant Mol. Biol.* 10:375–85

56. Kombrink, E., Hahlbrock, K. 1986. Responses of cultured parsley cells to elicitors from phytopathogenic fungi. *Plant Physiol.* 81:216–21

57. Kreuzaler, F., Hahlbrock, K. 1973. Flavonoid glycosides from illuminated cell suspension cultures of *Petroselinum hortense. Phytochemistry* 12:1149–52

58. Kreuzaler, F., Light, R. J., Hahlbrock, K. 1978. Flavanone synthase catalyzes CO_2 exchange and decarboxylation of malonyl-CoA. *FEBS Lett.* 94:175–78

59. Kreuzaler, F., Ragg, H., Heller, W., Tesch, R., Witt, I., et al. 1979. Flavanone synthase from *Petroselinum hortense.* Molecular weight, subunit composition, size of messenger RNA, and absence of pantetheinyl residue. *Eur. J. Biochem.* 99:89–96

60. Lamb, C. J. 1977. *Trans*-cinnamic acid as a mediator of the light-stimulated increase in hydroxycinnamoyl-CoA: quinate hydroxycinnamoyl transferase. *FEBS Lett.* 75:37–40

61. Lamb, C. J., Bell, J. N., Cramer, C. C., Dildine, S. L., Grand, C., et al. 1986. Molecular response of plants to infection. In *Beltsville Agric. Res. Cent. Symp.*, Vol. 10. *Biotechnology for Solving Agricultural Problems*, pp. 237–51.

62. Lamb, C. J., Rubery, P. H. 1976. Photocontrol of chlorogenic acid biosynthesis in potato tuber discs. *Phytochemistry* 15:665–68

63. Lozoya, E., Hoffmann, H., Douglas, C., Schulz, W., Scheel, D., Hahlbrock, K. 1988. Primary structures and catalytic properties of isoenzymes encoded by the two 4-coumarate:CoA ligase genes in parsley. *Eur. J. Biochem.* 176:661–67

64. Lynn, D. G., Chen, R. H., Manning, K. S., Wood, H. N. 1987. The structural characterization of endogenous factors from *Vinca rosea* crown gall tumors that promote cell division of tobacco cells. *Proc. Natl. Acad. Sci. USA* 84:615–19

65. Matern, U., Kneusel, R. E. 1988. Phenolic compounds in plant disease resistance. *Phytoparasitica* 16:153–70

66. Mehdy, M. C., Lamb, C. J. 1987. Chalcone isomerase cDNA cloning and mRNA induction by fungal elicitor, wounding and infection. *EMBO J.* 6: 1527–33

67. Meyer, P., Heidmann, I., Forkmann, G., Saedler, H. 1987. A new petunia flower color generated by transformation of a mutant with a maize gene. *Nature* 330:677–78

68. Meyerowitz, E. M., Pruitt, R. E. 1985. *Arabidopsis thaliana* and plant molecular genetics. *Science* 229:1214–18

69. Moesta, P., Grisebach, H. 1982. ʟ-2-Aminooxy-3-phenylpropionic acid inhibits phytoalexin accumulation in soybean with concomitant loss of resistance against *Phytophthora megasperma* f. sp.

glycinea. Physiol. Plant Pathol. 21:65–70
70. Niesbach-Klösgen, U., Barzen, E., Bernhardt, J., Rohde, W., Schwarz-Sommer, Z., et al. 1987. Chalcone synthase genes in plants: a tool to study evolutionary relationships. *J. Mol. Evol.* 26:213–25
71. Ohl, S., Hahlbrock, K., Schäfer, E. 1989. A stable blue light–derived signal modulates UV light-induced chalcone synthase gene activation in cultured parsley cells. *Planta.* In press
72. Parker, J. E., Hahlbrock, K., Scheel, D. 1988. Different cell-wall components from *Phytophthora megasperma* f. sp. *glycinea* elicit phytoalexin production in soybean and parsley. *Planta* 176:75–82
73. Rhodes, J. M., Wooltorton, L. S. C. 1978. The biosynthesis of phenolic compounds in wounded plant storage tissues. In *Biochemistry of Wounded Plant Tissues,* ed. G. Kahl, pp. 243–86. Berlin: de Gruyter
74. Robbins, M. P., Dixon, R. A. 1984. Induction of chalcone isomerase in elicitor-treated bean cells. Comparison of rates of synthesis and appearance of immunodetectable enzyme. *Eur. J. Biochem.* 145:195–202
75. Ryder, T. B., Hedrick, S. A., Bell, J. N., Liang, X., Clouse, S. D., Lamb, C. J. 1987. Organization and differential activation of a gene family encoding the plant defense enzyme chalcone synthase in *Phaseolus vulgaris. Mol. Gen. Genet.* 210:219–33
76. Scheel, D., Dangl, J. L., Douglas, C., Hauffe, K.-D., Herrmann, A., et al. 1987. Stimulation of phenylpropanoid pathways by environmental factors. See Ref. 29, pp. 315–26
77. Scheel, D., Hauffe, K.-D., Jahnen, W., Hahlbrock, K. 1986. Stimulation of phytoalexin formation in fungus-infected plants and elicitor-treated cell cultures of parsley. In *Recognition in Microbe-Plant Symbiotic and Pathogenic Interactions,* ed. B. Lugtenberg, pp. 325–31. Berlin: Springer-Verlag
78. Schmelzer, E., Jahnen, W., Hahlbrock, K. 1988. *In situ* localization of light-induced chalcone synthase mRNA, chalcone synthase, and flavonoid end products in epidermal cells of parsley leaves. *Proc. Natl. Acad. Sci. USA* 85:2989–93
79. Schmidt, W. E., Ebel, J. 1987. Specific binding of a fungal glucan phytoalexin elicitor to membrane fractions from soybean *Glycine max. Proc. Natl. Acad. Sci. USA* 84:4117–21
80. Schöppner, A., Kindl, H. 1984.

Purification and properties of a stilbene synthase from induced cell suspension cultures of peanut. *J. Biol. Chem.* 259:6806–11
81. Schram, A. W., Jonsson, L. M. V., Bennink, G. J. H. 1984. Biochemistry of flavonoid synthesis in *Petunia hybrida.* In *Taxonomy. Monographs on Theoretical and Applied Genetics: Petunia,* ed. K. C. Sink, pp. 68–75. Berlin: Springer-Verlag
82. Schröder, G., Brown, J. W. S., Schröder, J. 1988. Molecular analysis of resveratrol synthase; cDNA, genomic clones and relationship with chalcone synthase. *Eur. J. Biochem.* 172:161–69
83. Schröder, J., Schäfer, E. 1980. Radioiodinated antibodies, a tool in studies on the presence and role of inactive enzyme forms: regulation of chalcone synthase in parsley cell suspension cultures. *Arch. Biochem. Biophys.* 203:800–8
84. Schulze-Lefert, P., Dangl, J. L., Becker-Andre, M., Hahlbrock, K., Schulz, W. 1989. Inducible *in vivo* DNA footprints define sequences necessary for light activation of the parsley chalcone synthase gene. *EMBO J.* In press
85. Shields, S. E., Wingate, V. P., Lamb, C. J. 1982. Dual control of phenylalanine ammonia-lyase production and removal by its product cinnamic acid. *Eur. J. Biochem.* 123:389–95
86. Smith, B. G., Rubery, P. H. 1981. The effects of infection by *Phytophthora infestans* on the control of phenylpropanoid metabolism in wounded potato tissue. *Planta* 151:665–68
87. Somssich, I. E., Schmelzer, E., Bollmann, J., Hahlbrock, K. 1986. Rapid activation by fungal elicitor of genes encoding "pathogenesis-related" proteins in cultured parsley cells. *Proc. Natl. Acad. Sci. USA* 83:2427–30
88. Somssich, I. E., Schmelzer, E., Kawalleck, P., Hahlbrock, K. 1988. Gene structure and in situ transcript localization of pathogenesis-related protein 1 in parsley. *Mol. Gen. Genet.* 213:93–98
89. Stäb, M. R., Ebel, J. 1987. Effects of Ca^{2+} on phytoalexin induction by fungal elicitor in soybean cells. *Arch. Biochem. Biophys.* 257:416–23
90. Strasser, H., Hoffmann, C., Grisebach, H., Matern, U. 1986. Are polyphosphoinositides involved in signal transduction of elicitor-induced phytoalexin synthesis in cultured plant cells? *Z. Naturforsch. Teil C* 41:717–24
91. Tanguy, J. M., Negrel, J., Paynot, M., Martin, C. 1987. Hydroxycinnamic acid amides, hypersensitivity, flowering

and sexual organogenesis in plants. See Ref. 29, pp. 253–63

92. Tietjen, K. G., Hunkler, D., Matern, U. 1983. Differential response of cultured parsley cells to elicitors from two non-pathogenic strains of fungi. 1. Identification of induced products as coumarin derivatives. *Eur. J. Biochem.* 131:401–7

93. Tietjen, K. G., Matern, U. 1983. Differential response of cultured parsley cells to elicitors from two non-pathogenic strains of fungi. 2. Effects on enzyme activities. *Eur. J. Biochem.* 131:409–13

94. Towers, G. H. N., Yamamoto, E. 1985. Interactions of cinnamic acid and its derivatives with light. In *The Biochemistry of Plant Phenolics,* ed. C. F. van Sumere, P. J. Lea, pp. 271–88. Oxford: Clarendon

95. van der Krol, A. R., Lenting, P. E., Veenstra, J., van der Meer, I. M., Koes, R. E., et al. 1988. An anti-sense chalcone synthase gene in transgenic plants inhibits flower pigmentation. *Nature* 333:866–69

96. van Tunen, A. J., Koes, R. E., Spelt, C. E., van der Krol, A. R., Stuitje, A. R., Mol, J. N. M. 1988. Cloning of the two chalcone flavanone isomerase genes from *Petunia hybrida:* coordinate, light-regulated and differential expression of flavonoid genes. *EMBO J.* 7:1257–63

97. Villegas, R. J. A., Kojima, M. 1986. Purification and characterization of hydroxycinnamoyl D-glucose. Quinate hydroxycinnamoyl transferase in the root of sweet potato, *Ipomoea batatas* Lam. *J. Biol. Chem.* 261:8729–33

98. Vornam, B., Schön, H., Kindl, H. 1988. Control of gene expression during induction of cultured peanut cells: mRNA levels, protein synthesis and enzyme activity of stilbene synthase. *Plant Mol. Biol.* 10:235–43

99. Waldmüller, T., Grisebach, H. 1987. Effects of R-(1-amino-2-phenylethyl)phosphonic acid on glyceollin accumulation and expression of resistance to *Phytophthora megasperma* f. sp. *glycinea* in soybean. *Planta* 172:424–30

100. Walter, M. H., Grima-Pettenati, J., Grand, C., Boudet, A. M., Lamb, C. J. 1988. Cinnamyl alcohol dehydrogenase, a molecular marker specific for lignin synthesis: cDNA cloning and mRNA induction by fungal elicitor. *Proc. Natl. Acad. Sci. USA* 85:5546–50

101. Walter, M. H., Hahlbrock, K. 1984. Cinnamic acid induces PAL mRNA and inhibits mRNA translation in cultured parsley cells. *Plant Physiol.* (Suppl.) 75:155

102. Wendorff, H., Matern, U. 1986. Differential response of cultured parsley cells to elicitors from two non-pathogenic strains of fungi. Microsomal conversion of (+)marmesin into psoralen. *Eur. J. Biochem.* 161:391–98

103. Wiernig, H., de Vlaming, P. 1984. Inheritance and biochemistry of pigments. In *Taxonomy. Monographs on Theoretical and Applied Genetics: Petunia,* ed. K. C. Sink, pp. 49–65. Berlin: Springer-Verlag

104. Zimmermann, A., Hahlbrock, K. 1975. Light-induced changes of enzyme-activities in parsley cell-suspension cultures. Purification and some properties of phenylalanine ammonia-lyase (EC 4.3.1.5). *Arch. Biochem. Biophys.* 166:47–53

Annu. Rev. Plant Physiol. Plant Mol. Biol. 1989. 40:371–414

THE UNIQUENESS OF PLANT MITOCHONDRIA

Roland Douce and Michel Neuburger

Laboratoire de Physiologie Cellulaire Végétale, Centre d'Etudes Nucléaires et Université Joseph Fourier, 85X F38041 GRENOBLE-CEDEX, France

CONTENTS

INTRODUCTION

Mitochondria from all organisms provide ATP as the principal energy source for the cell and deliver numerous substrates via specific carriers for biosynthetic reactions in the cytoplasm (61). According to supply and demand of metabolites the relative importance of these roles in the overall cell metabolism will thus vary with the particular tissue, development, and

371

1040-2519/89/0601-0371$02.00

environmental factors. Many basic features of mitochondrial structure and function, developed at an early stage of evolution, have been highly conserved between animals and plants despite a billion years of divergent evolution. Thus (a) the morphology of plant mitochondria closely resembles that of their animal counterparts (92); (b) the patterns of phospholipids in membranes from plant and animal mitochondria are virtually identical (99); (c) the sequence of electron carriers that mediates the flow in plants of electrons from NADH and succinate to O_2 via cytochrome oxidase (i.e. the cyanide-sensitive electron pathway) and the fundamental structure of the plant phosphorylation system (ATPase complex) appear very similar to those found in mitochondria from yeast and animals (156, 180, 220); (d) there are similarities in energy conservation (H^+ ejection mechanisms) between plant and mammalian mitochondria (156); (e) plant mitochondria possess transport systems for numerous anions ($H_2PO_4^-$; HPO_4^{2-}; ATP^{4-};ADP^{3-}; pyruvate$^-$ etc) similar to those of other organisms (52, 235); (f) tricarboxylic acid–cycle functioning in plant mitochondria resembles in many respects that in animal mitochondria (236, 237); and (g) the outer membrane of plant mitochondria possesses a 31-kD channel-forming protein called mitochondrial porine (or voltage-dependent anion selective channel) similar to that of mammalian and yeast mitochondria and responsible for the high permeability of the outer membrane to small molecules (138).

However, there are distinct differences between plant and animal mitochondria in the nature of the electron transport system, in the presence of specific dehydrogenases and anion carriers, and in the size and complexity of their DNA (61). In addition, the rate of O_2 consumption on a protein basis is much higher in plant than in animal mitochondria (63, 108), while fatty acid oxidation is either very low (221, 222) or not detectable in plant mitochondria (the bulk of fatty acid oxidation in the plant cell being confined to microbodies) (88). [A carnitine acyltransferase exhibiting a broad specificity has been well characterized in plant mitochondria, however (87).]

At one time these differences were felt to be artifacts caused mainly by obvious problems associated with the preparation of higher plant mitochondria (61, 164). For example, many workers have thought the rapid oxidation of NADH by plant mitochondria resulted from the rapid permeation of NADH across the inner membrane and its subsequent oxidation by complex I. This apparent "leakiness" was thought to arise from inner mitochondrial membrane damage induced by the rigorous grinding used when isolating the organelles (129, 177). Fortunately, this view is no longer widely held. Purification of mitochondria from various plant tissues on silica sol gradients results in the preparation of intact mitochondria with high rates of O_2 consumption in state 3 and with very low contamination by other organelles (61, 164).

In addition, the use of protoplasts has greatly extended the range of tissues from which active mitochondria exhibiting high respiratory control ratios can be obtained (172). Under these conditions, the mechanical force needed to rupture the protoplasts is much less than that required to disrupt a tissue, and any interfering substances present in vascular tissues or cell walls are removed during preparation of protoplasts. Furthermore, intactness (intact mitochondria are those that have retained the outer membrane) can be determined rapidly by any assay in which an electron donor or acceptor such as reduced or oxidized cytochrome c does not penetrate the outer membrane so that the reaction (reduced cytochrome c–dependent O_2 consumption or cytochrome c reduction rates) is latent in intact mitochondria (63, 164). For example, membrane intactness studies, as determined by KCN-sensitive ascorbate-cytochrome c–dependent O_2 consumption or succinate:cytochrome c oxidoreductase activity, reveal that mitochondria purified using new isolation and purification procedures possess a mean percentage of intactness of approximately 95% (63, 164). Thus the unique features of plant mitochondria, which one assumes reflect their functioning in autotrophic metabolism, are maintained during the purification process. They are not attributable to experimental artifacts associated with their preparation, as originally stated by several authors (61).

ORGANIZATION OF THE RESPIRATORY CHAIN

The plant respiratory chain, like its more extensively studied counterpart in animal and yeast mitochondria, consists of only four protein complexes: complex I, complex II, complex III (usually called the cytochrome b–c_1 complex), and complex IV (cytochrome oxidase). Ubiquinone in the lipid bilayer of the mitochondrial membrane behaves as a homogeneous pool connecting the respiratory-chain enzymes in a substrate-like fashion. Except for cytochrome c, these complexes are very hydrophobic and are soluble in the "fluid" lipid bilayer medium of the mitochondrial inner membrane, more generally known as the coupling membrane (147). The overall process of mitochondrial electron transport includes diffusion and random collisions of the redox components, which represent the basis for the random-collision model of electron transport (93).

However, distinct differences between plant and animal mitochondria include the cyanide- and antimycin A–insensitive electron pathway, which is also encountered in the mitochondria of microogranisms (131); and the respiration-linked oxidation of external NAD(P)H and rotenone-insensitive oxidation of internal NADH (44, 122, 125, 150, 180, 183, 211, 213).

Respiration-Linked Oxidation of External NAD(P)H

Lehninger (127) found that NADH added to mammalian mitochondria was not oxidized. If the mitochondria were gently disrupted by hypotonic swelling, however, oxidation of NADH was considerably enhanced via complex I.

In mammals the reducing equivalents enter the mitochondria indirectly by means of coordinated reactions catalyzed by cytosolic and mitochondrial isoenzymes. The most prevalent of these mechanisms appears to be the malate-aspartate shuttle, although in some tissues the glycerol-phosphate shuttle may be significant (124).

An important feature of mitochondria from plant tissues and fungi is their ability to oxidize added NADH at a high rate in the absence of added cytochrome c (150, 180). One of the most striking example of NADH oxidation occurs in the thermogenic spadix of aroids such as *Arum maculatum*, where NADH oxidation rates may reach 7 μmol/min per mg protein (121). This oxidation, which does not require NADH translocase, is attributable to the presence of a specific NADH dehydrogenase situated on the outer surface of the inner membrane (64). This dehydrogenase is specific for the β-4 hydrogen of NADH and feeds electrons directly to complex III, bypassing complex I and the first site of H^+ translocation (183). Consequently, NADH oxidation by this external dehydrogenase is sensitive to antimycin A, is insensitive to rotenone or piericidin, and has an ADP/O ratio similar to that of succinate. In other words, electrons from the external NADH dehydrogenase have a common pathway with electrons from endogenous NADH at the level of ubiquinone.

The mechanism whereby this external dehydrogenase transfers reducing equivalents to ubiquinone (or complex III) remains to be elucidated. According to several groups (18, 47, 68), plant mitochondria preferentially oxidize endogenous NADH when confronted with a mixture of NADH and NAD^+-linked TCA-cycle substrates. It seems that complex III in plant mitochondria has an absolute preference for electrons generated from complex I. The mechanism of this interaction remains unresolved. Taking into account quinone mobility, one can postulate that diffusion distances between complex III and either external NADH dehydrogenase or complex I are not identical.

Platanetine, a 3,5,7,8-tetrahydroxy-6-isoprenyl flavone isolated from the bud scales of the plane tree *(Platanus acerifolia)*, behaves as a potent inhibitor of the exogenous NADH dehydrogenase with one of the best efficiencies ever mentioned for this complex. A 50% inhibition of the NADH oxidation rate is obtained at a micromolar concentration (196). The oxidation of exogenous NADH by plant mitochondria is dependent on micromolar concentrations of Ca^{2+}, which is often found bound to the mitochondrial membranes in sufficient amounts to ensure maximal activity and which does not appear to involve calmodulin (36, 42, 150, 151, 153). Exogenous NADH oxidation is therefore inhibited by Ca^{2+} chelators such as EGTA (151). Møller & Palmer (151) have suggested that the oxidation of NADH somehow makes Ca^{2+} less accessible to the chelators because EGTA inhibited NADH oxidation more when added before the NADH than after. This may result from a conformational change of the dehydrogenase that locks Ca^{2+} into the active site (149). It is interesting that the apparent K_m for exogenous NADH varied 6-fold in response to changes in the surface potential, suggesting that the approach of the negatively charged NADH to the active site is hampered by the negative surface potential (148).

Plant mitochondria also oxidize exogenous NADPH apparently via a Ca^{2+}-dependent dehydrogenase located on the outer surface of the inner membrane (8, 117). There are similarities between the characteristics of the oxidations of external NADH and NADPH, including ADP/O ratios below 2, and rotenone insensitivity (181). Edman et al (73) showed conclusively, however, that the pH optimum for NADPH oxidation is lower than that for NADH oxidation. In addition, the responses of external NADH and NADPH oxidations to both chelators and mersalyl are quite different (151). Neither a phosphatase converting NADPH to NADH nor a nicotinamide transhydrogenase was involved in the oxidation of NADPH by plant mitochondria (152). These observations seem to support the idea that there are two separate dehydrogenases on the outer surface of the inner mitochondrial membrane, one specific for NADH and the other for NADPH.

To date, relatively little information is available on either the subunit or the polypeptide composition of these external NAD(P)H dehydrogenases. There have been several papers, however, on its solubilization and partial purification from cauliflower buds (116) and *Arum maculatum* spadixes (38, 39, 41). The rotenone-insensitive NADH dehydrogenase appears to be a flavoprotein (the enzyme probably contains FAD) and does not contain any electron spin resonance–detectable iron-sulfur center. However, it was not possible to rule out the involvement of an iron-sulfur center that was lost during the purification. If the external NADH dehydrogenase is lacking an iron-sulfur center, its mechanism of interaction with either complex III or the ubiquinone pool would be of considerable interest. Unfortunately, little evidence was presented on the degree of contamination by the other NADH dehydrogenases. In addition, the determination of the minimum polypeptide composition of this enzyme and the elucidation of the mechanism of Ca^{2+} activation and interaction with complex III will only be possible if the final purification step can be considerably improved.

The metabolic significance of the respiration-linked inner-membranal external dehydrogenase capable of oxidizing cytosolic NADH very rapidly and present in all the plant mitochondria isolated so far is unknown. This dehydrogenase exhibiting a low K_m value for NADH is likely to serve as an efficient redox shuttle from cytoplasmic NADH sources (glycolytic pathway; β-oxidation in peroxisomes, etc) to the respiratory chain; it therefore will favor the forward direction of glycolysis via glyceraldehyde-3-phosphate dehydrogenase and of β-oxidation via 3-hydroxyacyl-CoA dehydrogenase.

Permeability measurements revealed that peroxisomes are permeable to small solutes including NAD^+ owing to the presence in the membrane of a non-selective pore-forming protein (225). This dehydrogenase could also play an important role in avoiding an accumulation of lactate under conditions of relative hypoxia in the plant tissue.

Unfortunately, the evidence available indicates that plant mitochondria preferentially oxidize endogenous NADH produced in the matrix space during the course of TCA-cycle substrate oxidations, suggesting that direct oxidation of cytosolic NADH by the mitochondria is strongly modulated by substrate delivery (malate; pyruvate, etc) to the mitochondria. Since the regulation of glyceraldehyde-3-phosphate dehydrogenase can be readily accounted for by equilibrium effect alone (224), an increase in the free cytosolic $NADH/NAD^+$ ratio will strongly decrease the rate of respiratory substrate production (malate? pyruvate?) to the point where cytosolic NADH oxidation by mitochondria is engaged and vice versa. In other words the respiration-linked inner-membranal, external dehydrogenase could play an important role whenever the glycolytic input of pyruvate or malate is slowed down.

The rate of NADPH reoxidation by the mitochondria could strongly increase the rate of the cytosolic pentose phosphate pathway, because the most effective control of this pathway appears to be the concentration of cytosolic NADPH (4). The high concentration of glucose-6-P measured in the cytosolic compartment by ^{31}P-NMR in all the plant cells studied so far (199, 200) strongly suggests that the cytosolic NADPH/NADP ratio is considerable. Likewise, NADPH oxidation could favor the rate of citrate metabolism via conversion of the citrate to α-ketoglutarate owing to the presence of powerful cytosolic $NADP^+$-isocitrate dehydrogenase and aconitase (28, 193) (Figure 1). Under these conditions, the rapid reoxidation of NADPH by the mitochondria could play a significant role during active cell growth, because the supply of carbon skeletons (pentose-P; erythrose-4-P; α-ketoglutarate, etc) for biosynthetic purposes may be more important than the supply of NADPH.

It seems reasonable to expect, therefore, that the electron flux through this external NADPH dehydrogenase is strongly regulated. Since this dehydrogenase is Ca^{2+}-dependent, in vivo control by modulation of the cytosolic free Ca^{2+} is possible. For example, cycling of Ca^{2+} through cell organelles (mitochondria? vacuole? endoplasmic reticulum?) via independent influx and efflux pathway as demonstrated in the case of mammalian mitochondria (78) could be an efficient system that may provide a fine control over the cytosolic free Ca^{2+} concentration and may therefore be involved in the regulation of Ca^{2+}-sensitive cellular activities such as cytosolic NAD(P)H oxidation. Parenthetically, there is no evidence that plant mitochondria transport Ca^{2+} by two different Ca^{2+} transport systems, one responsible for the energy-dependent influx of Ca^{2+} and the other mediating Ca^{2+} efflux as shown in mammalian mitochondria (52, 154). Obviously a comprehensive view of Ca^{2+} homeostasis in higher plant cells is required, especially if one considers that intact vacuoles isolated from *Acer pseudoplatanus* cell suspension cultures release part of their Ca^{2+} content in the presence of inositol triphosphate (194), and that illumination of the Characean alga *Nitellopsis* with white light elicits a decrease in cytosolic Ca^{2+} that is reversed by darkness (145).

Rotenone-Insensitive Oxidation of Internal NADH Dehydrogenase

The oxidation of endogenous NADH in plant mitochondria appears to be more complex than its counterpart in mammalian mitochondria. The most obvious indication of this complexity is that inhibitors such as piericidin A or rotenone, which strongly inhibit the oxidation of endogenous NADH in animal mitochondria, cause only a partial and sometimes an imperceptible inhibition in the plant mitochondrial system (109, 180, 183). Plant mitochondria seem to possess two internal NADH dehydrogenases on the inner surface of the inner membrane. One of these internal dehydrogenases, similar to the complex I characterized in mammalian mitochondria, readily oxidizes NADH in a rotenone-sensitive manner (183). This dehydrogenase (apparent K_m for NADH: 8 μM) functions as the "first coupling site" carrying reversible electron flux from NADH to ubiquinone coupled to the generation of $\Delta\mu_H^+$. The biochemical characterization of complex I in plant mitochondria has not been undertaken so far. We believe that complex I, which operates in close relationship with all the NAD^+-linked TCA-cycle dehydrogenases, NAD-malic enzyme, and glycine decarboxylase (in mitochondria from leaf tissues having the C_3 pathway of photosynthesis), utilizes a common pool of NAD^+ present in the matrix space (Figure 1)

The second dehydrogenase connected to the respiratory chain via the ubiquinone pool is insensitive to inhibition by rotenone (109, 150, 183). It has been shown that the passage of a pair of electrons from NADH to ubiquinone via this rotenone-insensitive internal dehydrogenase does not result in the translocation of H^+ ions from the matrix into the medium and is therefore not coupled to the generation of $\Delta\mu_H^+$ (Figure 1). This dehydrogenase, in contrast with complex I, exhibits a low affinity for internal NADH (150, 183) and differs from the rotenone-resistant NADH dehydrogenase associated with the outer face of the inner membrane inasmuch as it is not sensitive to EGTA or Ca^{2+} (150, 183). NADH oxidation via the rotenone-insensitive pathway is therefore dependent on the intramitochondrial concentration of NAD+ and the kinetic parameters $(K_{m_{NAD+}};\ K_{i_{NADH}})$ of all the NAD^+-linked dehydrogenases present in the matrix space. This path is accessible to almost all matrix dehydrogenases, insofar as they can maintain a sufficient concentra-

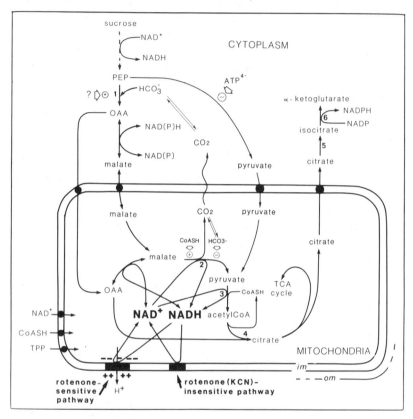

Figure 1   Schematic representation of malate oxidation in plant mitochondria and citrate delivery to the cytoplasm. This scheme emphasizes the great flexibility of plant mitochondria and indicates that there exists a concerted action of malate dehydrogenase, NAD^+-linked malic enzyme (label 2), and pyruvate dehydrogenase (label 3) to provide citrate in the anaplerotic function of the TCA cycle. The scheme also indicates that matrix NADH produced by the dehydrogenases, including NAD^+-linked malic enzyme, can be oxidized equally well either by the respiratory chain or by oxaloacetate, owing to the malate dehydrogenase working in the reverse direction.

One is faced with another choice at the level of the respiratory chain, since electron transport to O_2 can be either phosphorylating via the rotenone-sensitive pathway or nonphosphorylating via the rotenone KCN-insensitive pathway. The latter mechanism may be important when TCA-cycle intermediates such as citrate leave the mitochondrion for use elsewhere in the cell (see text). Note that carbon input to the TCA cycle could occur in the form of cytosolic oxaloacetate and malate [produced by the combined operation of PEP-carboxylase (label 1) and malate dehydrogenase in the cytosol], pyruvate being provided either by the action of pyruvate kinase in the cytosol or by operation of malic enzyme in the matrix, utilizing malate generated either in the matrix or the cytosol. Finally, this scheme emphasizes the fact that isolated intact plant mitochondria actively accumulate NAD^+, thiamine pyrophosphate (TPP), and coenzyme A (CoA) from the

tion of NADH. In contrast, malate dehydrogenase with its unfavorable equilibrium constant cannot trigger the rotenone-insensitive pathway (169). Unfortunately, no information is available to date on the structural and thermodynamic properties of the rotenone-insensitive dehydrogenase. Furthermore, its interaction with the respiratory chain is unknown.

The physiological significance of the rotenone-resistant internal NADH dehydrogenase is not understood. It has been suggested that complex I may be associated with the normal cyanide-sensitive respiratory pathway, whereas the nonphosphorylating internal NADH dehydrogenase is associated with the cyanide-resistant electron pathway, providing a totally nonphosphorylating pathway for the oxidation of endogenous NADH (150, 183). Knowledge of the mechanism whereby the rotenone-insensitive pathway is engaged and of the extent to which it operates is of the utmost importance in understanding the physiological role of this dehydrogenase. Nonetheless, the mechanism of the rotenone-insensitive pathway in plant mitochondria remains obscure. Perhaps it is unmasked by rotenone and has, therefore, little physiological significance. For example, it is possible that in the case of plant mitochondria the binding of rotenone to complex I leads to a potent inhibition of proton translocation without stopping electron transfer. In support of this suggestion, Ragan (192) has indicated that exogenous quinones are reduced by complex I by two pathways, one rotenone-sensitive and the other rotenone-insensitive. Furthermore, correlation among rotenone sensitivity, H+ extrusion, and $\Delta\mu_H^+$ rise indicated that only the rotenone-sensitive site is related to the activation of the H+ pump, as shown by Di Virgilio & Azzone (57) in rat liver mitochondria.

Cyanide-Insensitive Electron Pathway

Practically all the plant mitochondria isolated so far show a residual respiration in the presence of CO, N^{3-}, or CN^- (102, 122, 125, 211). The cyanide-resistant electron transport system consists of a branch point from the conventional electron transport system beginning with ubiquinone and terminating with a specific oxidase (alternative oxidase) distinct from cytochrome oxidase (16, 219). The cyanide-resistant electron pathway has

← _____

external medium. Such accumulation leads to a substantial increase in the matrix concentration of cofactors and stimulates TCA-cycle dehydrogenases, NAD-malic enzyme, and rotenone (KCN)-insensitive pathway exhibiting a low affinity for NADH. Other important reactions are cytosolic aconitase (label 5); cytosolic α-ketoglutarate dehydrogenase (label 6); and citrate synthase (label 4). (The first step in the TCA cycle catalyzed by citrate synthase is an aldol condensation of acetyl-CoA and oxaloacetate to give an enzyme-bound thioester intermediate. Hydrolysis of this thioester gives the citrate ion and coenzyme A. The equilibrium of this reaction lies heavily on the side of the products, because the free energy of hydrolysis of the thioester drives the reaction to completion.)

been reported not only in higher plants, but also in a wide range of fungi, including a large number of yeasts, *Neurospora*, and various microorganisms (131). Substituted benzohydroxamic acids such as salicylhydroxamic acid (SHAM) (208), the antioxidant n-propylgallate (3,4,5-trihydroxybenzoic acid propyl ester) (212), and disulfiram (tetraethylthiuram disulfide) (90) are potent inhibitors of the alternative pathway.

Cyanide-resistant respiration has been the subject of a number of interesting reviews in recent years, and these reviews should be consulted for detailed discussions of the cyanide-resistant pathway in plant mitochondria (122, 125, 211). In short, electrons seem to be partitioned between cytochrome oxidase and the alternative pathway according to the rate constants associated with the reactions between ubiquinol and either the alternative oxidase or complex III, respectively. In other words, the endogenous substrate for both pathways is the reduced form of ubiquinone-10 (211, 219), and this homogeneous pool of ubiquinone is kept reduced via substrate oxidation by complexes I and II.

Huq & Palmer (106) found that partial extraction of ubiquinone from *Arum maculatum* mitochondria by pentane treatment resulted in a preferential loss of the alternative pathway relative to the main pathway. A similar result was obtained with partial extraction of quinones in aged potato tuber mitochondria by Triton X-100 treatment (59). One possible explanation for these observations is that the affinity of complex III for ubiquinol is higher than that of the alternative oxidase. In support of this suggestion, several groups (for review see 120) have shown that electrons from TCA-cycle substrates are diverted to the alternative pathway only when the cytochrome pathway approaches saturation, either by inhibition (including state 4) or by flooding with electrons (13, 14, 120).

Numerous studies have provided convincing evidence that the ubiquinone molecules are free to diffuse laterally and independently of each other in the mid-plane of the mitochondrial membrane (118, 207) (Q-pool behavior), with the quinone ring oscillating within the hydrophobic core towards the membrane surfaces (128) although the diffusion path could be obstructed by the integral proteins. [The idea that ubiquinone can be reduced through direct interaction between the primary dehydrogenase and complex III via protein-bound semiquinone has also been suggested (240).] It should be pointed out, however, that plant mitochondria provide examples of deviations from Q-pool behavior as originally elucidated by Kroger & Klingenberg (118), most notably when comparing the ability of electrons from exogenous NADH and endogenous substrates to gain access to the alternative pathway (211).

The redox state of the quinone pool in plant mitochondria has been monitored voltametrically using a quinone-sensitive electrode (155). It was concluded that the engagement of the alternative oxidase was dependent on the redox state of the quinone pool and was not simply an overflow pathway.

Moore et al (155) also concluded that deviations from the simple kinetic model of Kroger & Klingenberg (118) could be well accommodated within the framework of a homogeneous pool without invoking the existence of multiple pools. In other words, the extent of deviation from simple first-order behaviour is dependent upon the equilibrium constant for the oxidation of quinol by the alternative oxidase.

A specific protein called "engaging factor" can be envisioned as being essential for coupling electron flow between the ubiquinone pool and the alternative pathway (216), because the expression or engagement of cyanide-resistant respiration is highly dependent upon developmental stage, tissue type, and physiological status. According to this hypothesis, the level of engaging factor present may serve as the rate-limiting component associated with the extent of cyanide resistance in any given mitochondria (216). For example, it is possible that some plant responses to cytokinin (60, 146, 160) involve the disengagement of the cyanide-resistant alternative respiratory pathway (159). Obviously, details of the mechanism by which electrons branch onto the alternative pathway are still unclear. A better understanding of how the quinol pool and the protein-bound semiquinone species interact is needed before we can fully understand how electrons are shunted between the main and alternative pathways. In addition, the alternative oxidase reaction may occur on either side of the inner membrane, but its location is presently unknown.

Because the alternative oxidase is indistinct both in its electron paramagnetic resonance and in its spectrophotometric parameters, the nature of the alternative oxidase has remained elusive (211). Consequently, several laboratories (for review see 122, 205) have expressed doubts as to the existence of this oxidase and have instead suggested lipoxygenase, or even fatty acid peroxidation as the cause of cyanide-insensitive O_2 consumption. The suggestion that the alternative oxidase might be a quinol oxidase led Huq & Palmer (105) and Rich (198) to use para-quinols (menaquinol, ubiquinol, and duroquinol) as artificial electron donors to assay for the alternative oxidase in isolated spadix mitochondria. (It is curious that mitochondria from any other plant source show almost no cyanide-resistant O_2 uptake with these quinols.) These quinols seem to donate electrons to a point that is at or very close to the alternative O_2-consuming step and offer an invaluable tool for the further investigation of the oxidase itself. These researchers showed that the enzyme could be solubilized from the inner membrane and still retain partial activity.

A major limitation in attempts to purify the alternative oxidase has been the extreme lability shown by the solubilized preparation. Nevertheless, two important articles have appeared in which efforts to purify the alternative oxidase further were made. One was by Bonner et al (21), in which some degree of purification from *Arum maculatum* mitochondria was achieved using potassium deoxycholate and lauryl maltoside to solubilize the preparation. Metal analyses of the preparation showed that only iron was closely correlated with the oxidase activity. By analogy with laccase and cytochrome oxidase, catalyzing the four-electron reduction of O_2 to water, it is likely that the alternative oxidase contains several metal centers for binding O_2. The second article was from Elthon & McIntosh (76), who reported on the identification of the proteins responsible for alternative oxidase activity in *Sauromatum guttatum* mitochondria. A 166-fold purification was achieved using N,N-bis-(3-D-glucoamidopropyl)-deoxycholamide (BigCHAP) to solubilize the inner mitochondrial membrane and a combination of cation-exchange and hydrophobic-interaction chromatography.

It is interesting that polyclonal antibodies raised to the fraction thus obtained readily immunoprecipitated at least three of the proteins (37, 36, and 35 kDa) that copurify with the activity. In addition, binding of anti-36-kDa protein antibodies to total mitochondrial protein blots of several plant species including *Vigna radiata* and *Arum italicum* indicated that similar proteins were always present when alternative pathway activity was observed. It is further interesting that in *Arum* similar proteins were more highly expressed in appendix tissues where the alternative pathway is high than in female floral tissues where the activity is low. Strong evidence supporting the role of the 35- and 36-kDa proteins in alternative oxidase activity is the fact that induction of alternative pathway activity in *Sauromatum guttatum* by calorigen (see below) is correlated with expression of these proteins. A radiation-inactivation analysis (or target theory) allowing a determination of the molecular mass of a membrane-associated activity in situ, indicated a functional molecular mass for the alternative oxidase of approximately 28,000 Da for skunk cabbage and *Sauromatum guttatum* mitochondria (18a). The alternative oxidase, although definitively a protein, has yet to be well characterized.

Although the cyanide-resistant alternative respiration was first described 60 years ago (86), its physiological significance remains unknown. When electrons from the quinol pool flow through the alternative pathway, energy is not conserved in the form of an electrochemical gradient, and no ATP is formed (156). Hence, the potential energy of the system is lost as heat. Lambers (120) has suggested that in tissues with low levels of the alternative pathway during situations in which excess carbohydrate is produced (i.e. carbohydrate that cannot be readily stored or used in growth or reproduction), the nonphosphorylating alternative pathway may contribute significantly to total respiration. In other words, the cyanide-resistant respiration may act as an overflow to drain off the energy of carbohydrates supplied in excess of demand. Under these conditions, this pathway may be considered energetically wasteful in terms of whole-plant carbon budgets, and the whole picture of respiration in plant cells relates at least as much to energy dissipation as it does to energy conservation (see also 119, 161, 162). The concept of wasteful oxidation of sucrose has been criticized judiciously by Bryce & ap Rees (29). According to these authors, sucrose in excess does not significantly increase the rate of respiration of pea roots above the characteristic rate of normally growing roots. Furthermore, male-sterile and fertile soybean tissues showed similar responses to KCN (172a). On the other hand, according to Palmer (180), the presence of a nonphosphorylating pathway in plants would permit the continued functioning of the TCA cycle whenever traffic through the cytochrome pathway is constrained by the energy charge. As a result, the alternative pathway can function as a mechanism for providing numerous substrates, such as citrate, for cytoplasmic synthetic purposes (Figure 1). Obviously this situation is quite distinct from the concept of the overflow pathway that responds simply to excess substrate.

The mechanism whereby the alternative pathway is engaged in nonthermogenic plants and the extent to which it operates in vivo are of overriding importance to an understanding of the physiological role of the cyanide-resistant respiration. Slow progress in understanding the physiological role of alternative respiration may in part be the result of obvious problems with its assay. Current methods rely upon titration with SHAM in the presence and absence of KCN. However, the ability of plant mitochondria to switch electrons between the two oxidase pathways (233) casts doubt on the inhibitor titrations. In addition, Sesay et al (209b) indicated the stimulation of O_2 uptake by intact tissues of mature soybean leaves and cotyledons in the presence of SHAM and KCN. They concluded that this stimulation of respiration by KCN and SHAM is not explained by characteristics of mitochondria. Of some interest, Guy et al (94) reported that discrimination against [18]O by the alternative oxidase was substantially greater than by cytochrome oxidase. It is clear that this differential discrimination would form the basis of a new non-invasive technique to estimate in vivo the partitioning of electron flow between the two mitochondrial electron transport paths.

Finally, it is very difficult to discriminate in vivo the reoxidation of NADH in the matrix space via the alternative oxidase (whose rate seldom exceeds the rate of state 4 respiration) from its reoxidation via the cytochrome oxidase pathway owing to a great increase in the proton permeability of the mitochondrial inner membrane at high values of "proton motive force" (156). Thus, it is now well established that in the resting state (state 4), control of the rate of O_2 consumption in isolated mitochondria is exerted by the leak of protons through the inner membrane (a high-proton-conductance pathway occurs at high $\Delta\mu_H^+$), whereas in more active phosphorylating states (up to state 3) control is distributed among a number of steps, including the proton leak, the nucleotide carrier, and cytochrome oxidase (26). In other words, the high proton conductance observed at high $\Delta\mu_H^+$ would permit the continued functioning of the TCA cycle without ATP formation.

In tissues with high levels of the alternative pathway (in the male reproductive structure of cycads and in the flowers or inflorescences of some species belonging to the families Annonaceae, Araceae, Aristolochiaceae, Cyclanthaceae, and Nympheaceae) the functional significance of this pathway is best understood (122, 211). In Araceae, for example, heat produced by way of electron flow through the alternative pathway is often used to volatilize insect attractants (amines and indoles exhibiting a putrescent odor), facilitating insect pollinization (140). On the day of flowering, the alternative pathway becomes operational, and starch that has been accumulated up to this stage is burned up in a few hours, as a consequence of prodigiously high rates of glycolysis (6) and respiration (122). At its climax, this phenomenon can lead to an increase in temperature up to 15°C above ambient in *Arum maculatum* (122, 140).

During flowering of *Sauromatum,* a water-soluble substance called "calorigen" is released from the male (staminate) flower primordia just below the appendix and migrates into the appendix region where it triggers thermogenesis (140). For more than 50 years, the identity of "calorigen" remained obscure. However, recently, Raskin et al (195) have obtained a highly purified calorigen preparation from male flowers of *Sauromatum guttatum.* Mass spectrometry analysis showed that the sample of purified calorigen contained 2-hydroxybenzoic (salicylic) acid. Furthermore, they demonstrated that both the calorigen extract and salicylic acid caused warming of the appendix tissue. Thus, at least in some arum lilies, salicylic acid functions as an endogenous regulator of heat production. As pointed out by Raskin et al (195), the stimulatory effect of salicylic acid on heat-generating, cyanide-insensitive respiration is interesting because its analog salicylhydroxamic acid and some other aromatic hydroxamates are potent inhibitors of the alternative oxidation. According to the work of Elthon & McIntosh (76) it is possible that salicylic acid behaves as a transcription factor for efficient alternative oxidase gene(s) expression. Again much remains to be done in order to understand the whole cascade of events via salicylic acid synthesis that leads to heat production in thermogenesis.

Energetics of Electron Transport and Oxidative Phosphorylation

The transfer of electrons from substrate to O_2 via the cytochrome pathway is coupled to an electrogenic translocation of protons across the inner mitochondrial membrane (156). Consequently, electrogenic proton translocation without cotransport of anions generates both a proton gradient (ΔpH) and a membrane potential ($\Delta\Psi$; negatively charged side: matrix side; positively charged side:cytoplasmic side) (Figure 1). The two parameters are additive, contributing to a "proton-motive force" differential across the membrane. Experimental determinations of $\Delta\Psi$ yield a considerably higher value (approximately 220 mV for state 4 respiration) in plant mitochondria (55, 56, 137) than in mammalian mitochondria (150–180 mV). Such a high value is likely attributable to the presence in the inner mitochondrial membrane of a powerful K+/H+ antiporter which partially collapses the Δ pH, thereby increasing $\Delta\Psi$ (98). In support of this observation, Nigericin alone (+K$^+$), in contrast with valinomycin (+K$^+$) does not uncouple plant mitochondria (103). Furthermore, experimental determinations of ΔpH (quantitated by measuring the distribution of a permeable weak acid) have yielded very low values (169).

This considerable transmembrane potential is a potent driving force for an electrophoretic asymetric nucleotide exchange (ATP^{4-}/ADP^{3-}). It drives ADP^{3-} inside the mitochondria and ATP^{4-} outside. Accordingly, at equilibrium the cytosolic ATP/ADP ratio should be high. In support of this assumption ^{31}P-NMR spectra obtained from plant cells or tissues clearly show that in vivo the cytosolic ATP/ADP ratio is considerable (199, 200). In our laboratory we have also shown that in situ plant leaf mitochondria resemble mitochondria from nongreen plant cells in that they are highly energized and, under light or dark conditions, continuously maintain a high cytosolic ATP/ADP quotient (20). Since the cytosolic mobile ADP concentration as measured by ^{31}P-NMR is extremely low, it is likely that the rate of O$_2$ consumption by a plant cell depends upon the rate of ADP delivery to the mitochondria during the course of metabolism. Mitochondria in cells are often described as being in a state intermediate between states 4 and 3 since respiration can be increased (to state 3) with uncouplers or decreased (to state 4) by inhibiting ATP production with inhibitors such as oligomycin (61, 67). The problems inherent in providing an appropriate diffusion pathway of ADP within the cytosol containing a high protein concentration can be partially overcome by movement of mitochondria within the cytosol. For example, the cytosolic territory exhibiting high ATP consumption might attract mitochondria through microtubules attached to the outer mitochondrial membrane.

METABOLIC EXCHANGE BETWEEN THE MITOCHONDRION AND THE CYTOSOL

For mitochondria to function as integral components of the plant cell there must be movement of numerous anions including phosphate (H$_2$PO$_4^-$, HPO$_4^{2-}$), ADP^{3-}, ATP^{4-}, various TCA-cycle intermediates, and amino acids between the matrix and the remainder of the cytoplasm. Transport of metabolites in plant mitochondria has recently been reviewed (52, 97). In addition to possessing carriers similar to those found in mitochondria from other sources (these include the mono-, di-, and tricarboxylate carriers as well as the phosphate, the α-ketoglutarate, the pyruvate, and the adenine nucleotide carriers) (24, 52, 227), plant mitochondria also possess specific carriers that make interactions with their surroundings very flexible. Nonetheless, not all plant mitochondria are alike in their transport functions, and distinctions based on species, tissues, developmental stage, and environment are frequently evident. Unfortunately, critical comparisons are not often made. In this section we concentrate on the specific transport systems of plant mitochondria and their importance in plant cell carbon metabolism.

Oxaloacetate Carrier

In mammalian cells it has been assumed that the inner membrane is impermeable to oxaloacetate under normal physiological conditions, although reports have shown that it can be slowly transported in gluconeogenic organs such as the liver or kidney (185). In marked contrast, oxaloacetate has been found to traverse the inner membrane rapidly in all of the plant mitochondria isolated so far (62). For example, at an extramitochondrial oxaloacetate concentration of 50 μM the influx of oxaloacetate is so severe that NAD^+-linked TCA-cycle substrate-dependent O_2 consumption stopped because of the competition for NADH by malate dehydrogenase (the equilibrium of the malate dehydrogenase reaction lies far towards malate formation; Keq 3.10^{-5}) (Figure 1). Alleviation of respiratory inhibition subsequently occurred as the oxaloacetate was reduced.

In plant mitochondria malate transport was sensitive to 2-N-butylmalonate, an inhibitor of the dicarboxylate carrier, while that of oxaloacetate apparently was not (52). Conversely, phthalonate (141) had little effect on malate transport but severely restricted oxaloacetate transport (50, 51). These results strongly suggest that malate efflux and oxaloacetate influx occur on separate carriers (52, 71). In support of this suggestion, the uptake of oxaloacetate was not inhibited by a 1000-fold excess of malate (174). Cauliflower bud mitochondria swell spontaneously when suspended in high concentrations of ammonium oxaloacetate, implying exchange for OH^- (or cotransport of a proton along with the anion) (52). On the other hand, pea leaf mitochondria did not swell when suspended in isotonic solutions of NH_4^+ or K^+-oxaloacetate until valinomycin was added, suggesting that oxaloacetate is taken up by electrogenic uniport facilated by the valinomycin-mediated uptake of the compensating cation (241).

For the present, the details of oxaloacetate transport in plant mitochondria remain a mystery, and more work is needed to confirm a mechanism. The question whether both malate and oxaloacetate are transported by a single transport protein or by two different ones cannot be answered at present, although Zoglowek et al (241) provided evidence that a malate-oxaloacetate shuttle across the inner mitochondrial membrane is catalyzed by an electrogenic uniport of malate and one of oxaloacetate linked to a counterexchange.

In some circumstances, intact mitochondria can export oxaloacetate. For example, during the course of malate oxidation catalyzed by matrix malate dehydrogenase, plant mitochondria in state 3 excrete oxaloacetate via the oxaloacetate carrier insofar as malic enzyme does not operate (169). Likewise, during the course of succinate oxidation, plant mitochondria in state 3 also excrete large amounts of oxaloacetate (66). As malate oxidation proceeds, the concentration of oxaloacetate in the medium increases slowly up to an equilibrium concentration. When this is achieved the efflux of oxaloacetate is stopped and the matrix malate dehydrogenase is reversed. Consequently, it is the concentration of oxaloacetate on both sides of the inner mitochondrial membrane that seems to govern the efflux or influx of oxaloacetate. Irrespective of its role, the low K_m (2 μM) of the oxaloacetate carrier for its substrate should allow it to compete successfully with cytosolic or matrix malate dehydrogenase (71, 174). ↗

The physiological role of this unique oxaloacetate carrier has yet to be determined. The extraordinarily low half-saturation of oxaloacetate transport would make it possible for a very active malate-oxaloacete shuttle to occur between the mitochondria and the cytosolic compartment under physiological conditions. For example according to Woo & Osmond (238) and Ebbighausen et al (72) the NADH generated during glycine oxidation by mitochondria isolated from green leaves (C_3-plants) may be reoxidized via a malate-oxaloacetate shuttle (Figure 2) that may be linked in vivo to hydroxypyruvate reduction in the peroxisomes. The recent findings of Ebbighausen et al (72) show clearly that a leaf cell has the means to shuttle the NADH formed during glycine oxidation to the peroxisomes for the reduction of hydroxypyruvate (see, however, 69). Since the equilibrium of hydroxypyruvate reduction is shifted towards glycerate formation (Keq approximately 3.10^{-5}), this reaction is expected to act as a drain, sequestering the redox equivalents from the mitochondria (72). In addition, the compartmentation of hydroxypyruvate reductase and malate dehydrogenase in the peroxisomes confers a higher efficiency in the supply of NADH for hydroxypyruvate reduction (239).

Since all of the plant mitochondria isolated so far possess a unique powerful oxaloacetate carrier, carbon input to the TCA cycle could also occur in the form of cytosolic oxaloacetate, thus disrupting the conventional operation of the TCA cycle (Figure 1). This anion can derive either from cytosolic aspartate amino transferase (69) and malate dehydrogenase functioning and/or from β-carboxylation of phosphoenolpyruvate (PEP). The latter reaction is catalyzed by a cytosolic PEP-carboxylase present in all the plant cells examined so far (3). Unfortunately, specific information about the exact role of

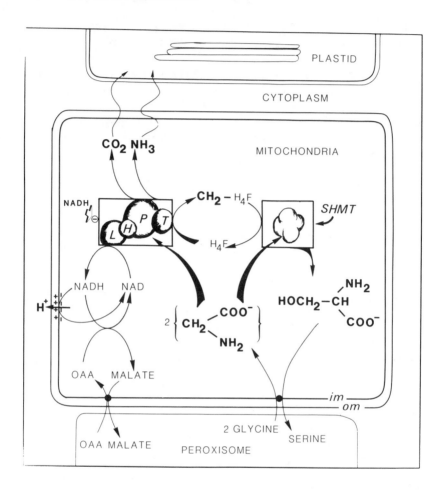

Figure 2   Schematic representation of glycine oxidation in green leaf mitochondria. During photorespiration glycine is cleaved in the matrix space by the glycine cleavage system (containing four protein components tentatively named P-protein, H-protein, T-protein, and L-protein) to CO_2, NH_3, and 5,10-methylenetetrahydropteroyl-L-glutamic acid ($5,10\text{-}CH_2\text{-}H_4F$). The latter compound reacts with a second mole of glycine to form serine and tetrahydropteroyl-L-glutamic acid (H_4F) in a reaction catalyzed by serine hydroxymethyltransferase (SHMT). NADH produced during the course of glycine oxidation is oxidized either by the respiratory chain or by oxaloacetate, owing to the malate dehydrogenase located in the matrix space working in the reverse direction. A rapid malate-oxaloacetate transport shuttle appears to play an important role in the photorespiratory cycle in catalyzing the transfer of reducing equivalents generated in the mitochondria during glycine oxidation to the peroxisomal compartment for the reduction of β-hydroxypyruvate. Note the unusual stoichiometry of two glycine molecules entering the mitochondrial matrix in exchange for one serine leaving.

PEP-carboxylase in the replenishment of the TCA-cycle intermediates (in the form of oxaloacetate or malate) remains sparse.

Glutamate and Aspartate Transport

Aspartate transport from mitochondrial matrixes to the cell cytosol in mammalian mitochondria is a necessary step in metabolic pathways such as gluconeogenesis and the transport of reducing equivalents from cell cytosol to the mitochondria (124). Aspartate is transported when its specific carrier catalyzes an exchange of glutamate for aspartate across the membrane. It is interesting that when a source of energy is available, the exchange is virtually unidirectional—the glutamate entering the mitochondrial matrix space, the aspartate leaving. Since a proton is cotransported on the carrier with glutamate, the process can be described as the exchange of neutral glutamic acid for the aspartate anion and is, therefore, electrogenic. In other words, in mammalian mitochondria entry of aspartate in exchange for glutamate is very slow and does not, in fact, occur in energized mitochondria (124).

Although a more detailed knowledge of glutamate and aspartate transport in plant mitochondria is desirable, it is obvious that in contrast with the situation observed in mammalian mitochondria there is no evidence for an electrogenic glutamate-aspartate transporter for plant mitochondria (52, 97). That plant mitochondria do not exhibit an electrogenic glutamate-aspartate antiporter driving glutamate inside the mitochondria and aspartate outside can be explained by the existence of an NADH dehydrogenase located on the outside of the inner membrane. Consequently, they do not require complex exchange systems for the transfer of reducing equivalents from the cytosol to the matrix. Instead fluxes of glutamate and aspartate will be determined by movement of their corresponding ketoacids and also by the local concentrations of each throughout the cells (110). For example, for NAD-malic enzyme C_4 species, aspartate formed from oxaloacetate in mesophyll cells is transported to the mitochondria of bundle sheath cells (via plasmodesmata?) where it is metabolized to CO_2 and pyruvate via an α-ketoglutarate/glutamate cycle (100) and by the combined action of aspartate aminotransferase and malate dehydrogenase. [In this system the operations of malic enzyme and malate dehydrogenase are coupled by an NAD/NADH cycle (100).]

Cofactor Uptake by Plant Mitochondria

Intact and well-coupled plant mitochondria appear capable of a net uptake of several important coenzymes. In this way, metabolic pathways with coenzyme-dependent enzymes localized in the matrix may be subject to modulation.

NET IMPORT OF ADENINE NUCLEOTIDES   The adenine nucleotide content (ATP + ADP + AMP) of freshly isolated animal mitochondria is usually quite constant in the range of 11–13 nmol per mg protein, and rapid leakage across the inner mitochondrial membrane does not usually occur. Consequently, addition of ADP to the medium triggers the full rate of nucleotide exchange. In contrast, very often in fully intact plant mitochondria—especially those isolated from dormant tissues such as potato (226) or Jerusalem artichoke (184) tubers—the total quantity of nucleotides is very low. Under these conditions, the initial rates of substrate oxidation by plant mitochondria in the presence of ADP are limited by the internal concentration of nucleotides impeding the nucleotide carrier. Abou-Khalil & Hanson (1, 2) have demonstrated the existence in plant mitochondria of a mechanism for the net uptake of ADP or ATP which is insensitive to carboxyatractyloside, a potent and specific inhibitor of the nucleotide carrier (226). This net uptake was strongly accelerated if an electrochemical gradient of protons across the membrane was established. Conversely, this net uptake was strongly inhibited by uncouplers. The results of Abou-Khalil & Hanson (1, 2) strongly suggest that this one-way adenine nucleotide transport (influx and efflux), distinct from the one-for-one exchange of ATP and ADP via the adenine nucleotide carrier, may be the means by which the total matrix adenine nucleotide pool is normally maintained (especially during mitochondrial proliferation) and by which the matrix adenine nucleotide concentration is altered in response to physiological stimuli. A similar mechanism for the net uptake of adenine nucleotides by mitochondria in vitro has also been characterized in liver mitochondria from newborn rat (7).

NET IMPORT OF NAD^+ Complex I, the segment of the respiratory chain responsible for electron transfer from NADH to ubiquinone, operates in close relationship with all of the NAD^+-linked TCA-cycle dehydrogenases that utilize a common pool of NAD^+ (6–7 nmoles per mg protein). Having been reduced by the TCA-cycle dehydrogenases, the pyridine nucleotide molecules diffuse to the inner membrane where they are oxidized. There exists, therefore, a competitive interaction between all the dehydrogenases for a common matrix NAD^+ pool. Considering matrix viscosity (0.4 g protein/ml), which may limit rates of NADH diffusion, there may be a steep downward gradient of NADH with higher concentrations in the center of the matrix and much lower concentrations near the periphery. (It is likely that the NADH produced by the dehydrogenases located near the periphery is channeled directly to complex I without equilibrating with the bulk solution.) Under these conditions, the steric placement of the NAD^+-linked dehydrogenases within the matrix (for example organized in the environment of NADH:ubiquinone oxidoreductase) as well as their kinetic parameters ($K_{m_{NAD+}}$; $K_{i_{NADH}}$) may be relevant in the outcome of the competition. For example in the case of pyruvate dehydrogenase, K_i values for NADH are smaller than K_m values. Consequently, increasing the ratio of NADH to NAD^+ in the matrix space resulted in a logarithmic increase in inhibition (142).

Stimulation of respiration in isolated plant mitochondria by exogenous NAD^+ is well known (180) and is seen with all NAD^+-linked substrates, especially with malate. Not all plant mitochondria, however, respond to added NAD^+; and those that do not, generally have high endogenous NAD^+ contents and rapid rates of respiration (223). Very often, the amount of NAD^+ present in plant mitochondria is sufficient to allow state 3 rates of approximately 50% of the maximal rate to be sustained whereas the rate of O_2 uptake in the presence of rotenone is almost completely dependent on NAD^+ (168, 170). Since the rotenone-insensitive internal NADH dehydrogenase has a much lower affinity for NADH than complex I it is clear that plant mitochondria need to maintain a sufficient internal NAD^+ pool to satisfy the requirements of the internal rotenone-insensitive pathway.

Although earlier results have suggested that NAD^+ slowly penetrates swollen mammalian mitochondria (104) it is generally accepted that NAD^+ cannot penetrate the inner membrane of intact mitochondria (127). However, Neuburger & Douce (170) have shown that isolated intact plant mitochondria actively accumulate NAD^+ from the external medium, leading to a marked increase in the matrix concentration of the cofactor and stimulating matrix dehydrogenases and electron transport activities, especially the rotenone-resistant respiration. Plant mitochondria apparently possess a specific NAD^+ carrier, since NAD^+ uptake is concentration dependent and exhibits Michaelis-Menten kinetics and the V value of the carrier is strongly affected by the initial concentration of NAD^+ present in the matrix space. Furthermore, the rate of NAD^+ transport is strongly temperature dependent, and the analogue N-4-azido-2-nitrophenyl-4-aminobutyryl-3'-NAD^+ (NAP4–NAD) almost completely inhibits NAD^+ import (168).

Neuburger & Douce (170) have also found that NAD^+ slowly effluxes from intact isolated mitochondria. As the level of NAD(H) falls progressively in the mitochondria it becomes too low to act as a substrate for NAD^+-linked matrix dehydrogenases exhibiting a low affinity for NAD^+, such as NAD-malic enzyme, and for rotenone-resistant respiration. The fact that intact purified mitochondria progressively lose their NAD^+ content and that this leads to a dramatic decrease in the O_2 uptake rates strongly suggests that most of the NAD^+ is not bound to the inner membrane or to the various dehydrogenases. The rate of NAD^+ efflux from the matrix space is dependent upon the intramitochondrial NAD^+ concentration and is inhibited by the analogue inhibitor of NAD^+ transport, indicating that a protein is required for net flux in either direction.

The physiological role of this NAD^+ carrier remains uncertain. Cycling of NAD^+ through mitochondrial influx and efflux pathways is an efficient system that may provide for fine control over the mitochondrial free NAD^+ concentration and, therefore, be involved in the regulation of NAD^+-sensitive mitochondrial activities such as NAD-malic enzyme and internal rotenone-insensitive NADH dehydrogenase activities. This transport system could also play an important role in the coarse control of metabolism, particularly during transition from a dormant stage (low NAD^+ matrix concentration) to a stage of active growth (high matrix NAD^+ concentration) (168, 210). Finally, if NAD^+ enters the matrix space without a simultaneous requirement for ATP then the concentration of NADH will rise. This may increase respiration via the rotenone-insensitive pathway without the need for gross changes in the cytoplasmic phosphorylation status. Proposals such as this must be tested in intact cells before a full appreciation of the role that mitochondrial NAD^+ transport plays in the regulation of mitochondrial activities is achieved.

NET IMPORT OF THIAMINE PYROPHOSPHATE   Thiamine pyrophosphate (TPP) can enter isolated plant mitochondria, since this cofactor is an essential addition for the oxidation of α-ketoglutarate and pyruvate (25, 123). Apparently, very often, the isolated plant organelles, in contrast with mammalian mitochondria, are depleted of endogenous TPP (presumably during their isolation, although this has yet to be demonstrated) but rapidly accumulate it when it is provided externally. The net movement of TPP into the matrix is concentration dependent in the physiological range ($K_m = 50$ μM), and the extent of TPP accumulation (V = 0.8 nmol/min per mg protein) is unaffected by the addition of NAP4-NAD to the incubation medium, indicating that NAD^+ and TPP do not share a common carrier (M. Neuburger and R. Douce, in preparation). The fact that intact plant mitochondria very rapidly lose their TPP content demonstrates that this coenzyme is not bound to the pyruvate and α-ketoglutarate dehydrogenase enzymes. It is interesting that highly purified preparations of α-ketoglutarate dehydrogenase (187) and pyruvate dehydrogenase (204) from plant mitochondria are completely de-pendent on added TPP, but similar preparations of the mammalian enzymes show little or no response to exogenous TPP. This transport process may be the means by which the total matrix TPP is normally maintained. Whether or not the matrix TPP concentration is altered in response to physiological stimuli leading to a coarse control of the multienzyme complexes that catalyze the TPP-mediated oxidative decarboxylation of pyruvate and α-ketoglutarate remains to be demonstrated.

NET IMPORT OF COENZYME A   Although a large number of synthetic and degradative reactions in all tissues depend on coenzyme A (CoASH), little is known about the regulation of CoASH levels in the cell and even less about the intracellular distribution of CoASH between the cytosol and mitochondrial matrix.

Plant mitochondria isolated from a number of tissues also have a relatively low endogenous CoASH content [the matrix CoASH pool (approximately 0.2 nmol per mg protein) is significantly smaller than the NAD^+ pool (approximately 1–2 nmol per mg protein)], and these mitochondria are capable of actively accumulating CoASH in a manner sensitive to uncouplers and low temperatures (167). This net uptake is catalyzed by a specific transport system distinct from the NAD^+ carrier and leads to an increase in the CoASH content of the matrix. This CoASH uptake followed saturation kinetics with an apparent K_m of 0.2 mM and a V of 4–6.5 nmol/min per mg protein. The physiological function of a mitochondrial system for CoASH transport would be to move the intact CoASH molecule from the cytosol where it is synthesized to the mitochondrial matrix where it is used in the entry of all major fuel substrates (pyruvate, α-ketoglutarate, and malate) into the TCA cycle. Parenthetically, within plant mitochondria, pyruvate dehydrogenase and α-ketoglutarate dehydrogenase interact via a common CoASH pool (70).

MATRIX ENZYMES

Except concerning the very large multienzyme complex that catalyzes the oxidative decarboxylation of pyruvate (30, 142–144), investigations of the TCA-cycle enzymes in plant mitochondria are incomplete, especially when compared to the extensive investigations of animal mitochondria. Nonetheless, sufficient investigation has been made to show that the basic principles of the operation and regulation of the TCA cycle are common among all eukaryotes (236, 237). Plant mitochondria do differ, however, in malate and glycine oxidations. These oxidations appear to be linked to autotrophic metabolism, although they are not necessarily manifested, or equally developed, in all species and tissues.

Malate Oxidation

Plant mitochondria, in contrast to mammalian mitochondria, readily oxidize malate without requiring the experimenter to remove oxaloacetate. (The unfavorable equilibrium of the reaction catalyzed by malate dehydrogenase necessitates the rapid removal of the product, oxaloacetate, in order for malate oxidation to proceed.) Pyruvate and/or oxaloacetate are the major products formed during the course of malate oxidation when pyruvate dehydrogenase is not operating (180). In the absence of TPP, O_2 uptake with malate as the substrate is attributed solely to malate dehydrogenase and/or NAD-malic enzyme [L-malate-NAD^+ oxidoreductase (decarboxylating); EC

1.1.1.39]. In the presence of TPP, pyruvate (the product of the malic enzyme) can be converted to acetyl-CoA by pyruvate dehydrogenase, thus enabling citrate synthase to remove oxaloacetate by condensing it with acetyl-CoA to form citrate (Figure 1).

NAD-malic enzyme was discovered in plant mitochondria by Macrae & Moorhouse (136). This enzyme, which is compartmentalized in the mitochondrial matrix (see, for example, 5), is specific for L-malate, has an absolute requirement for Mn^{2+}, and is distinguished by its inability to decarboxylate oxaloacetate and its low substrate affinity (11, 135). In addition, bicarbonate inhibits NAD-malic enzyme (the lower pH optimum may be the result of inhibition by HCO_3^- and Mn^{2+} precipitation as the pH increases) (33, 169), and this inhibition is relieved by CoASH and Mn^{2+} (CoASH and its analogues are the most potent activators yet found for the NAD-malic enzyme) (45, 101).

The ratio of the products oxaloacetate and pyruvate, formed during the course of malate oxidation, reflects the balance of the two malate-oxidizing enzymes (169, 182). When the activity of the NAD-malic enzyme is weakened [high HCO_3^- concentration (alkaline pH), low matrix CoASH and Mn^{2+} concentrations], oxaloacetate is preferentially excreted under state 3 conditions, and this process causes a decrease in the rate of O_2 consumption as the reaction proceeds owing to the accumulation of oxaloacetate. The activity of the matrix malate dehydrogenase is favored by high pH and is basically controlled by the oxaloacetate concentration and the redox state of pyridine nucleotides. Under these conditions, addition of either phthalonate (a potent inhibitor of oxaloacetate uptake and efflux) or rotenone to plant mitochondria respiring malate at alkaline pH (i.e. when NAD-malic enzyme is inhibited) induces a marked inhibition of O_2 uptake. When ADP is exhausted a very slow rate of O_2 consumption ensues and oxaloacetate production is stopped. On the other hand, when the activity of the NAD-malic enzyme is powerful [low HCO_3^- concentration (acidic pH); high matrix CoASH concentration] there is a marked increase in pyruvate production under state 3 conditions, accompanied by a decrease in oxaloacetate formation. When ADP is exhausted or after addition of rotenone, oxaloacetate production is also stopped; the rate of O_2 consumption is initially very slow but increases rapidly to reach a linear rate (biphasic state curves; 123). As shown by Neuburger & Douce (169) and Palmer et al (182), when the production of oxaloacetate is stopped in the matrix space (i.e. at high internal NADH concentration) excreted oxaloacetate reenters the matrix space and is rapidly converted to malate at the expense of NADH generated by the NAD-malic enzyme. When the oxaloacetate concentration becomes very low, the inhibition of O_2 consumption is released. It is clear, therefore, that malate dehydrogenase and NAD-malic enzyme are competing at the level of the pyridine nucleotide pool and that the regulation in vivo of malate dehydrogenase can be readily accounted for by equilibrium effects alone (166, 182) (Figure 1).

The effects of exogenous NAD^+ on malate oxidation by isolated plant mitochondria have been vigorously debated for the past decade. Coleman & Palmer (37) were the first to note that external NAD^+ stimulated the rate of malate oxidation and reduced its sensitivity to rotenone. It was consequently proposed that either some NAD-malic enzyme was localized between the inner and outer mitochondrial membranes (180) or a transmembrane transhydrogenase was capable of tranferring reducing equivalents from matrix NADH to external NAD^+ (49). Unfortunately, both of these hypotheses were formulated on the assumptions that the observed rotenone-insensitive O_2 uptake was catalyzed by the external NADH dehydrogenase and that the inner membrane of the mitochondrion was impermeable to exogenous NAD^+. However it is now known that NAD^+ can be accumulated by plant mitochondria from the external medium and that plant mitochondria possess an internal pathway for rotenone-insensitive NADH oxidation. Although this rate of NAD^+ uptake is slow, it can lead to very substantial increases in the matrix concentration of NAD^+. This, in turn, can considerably stimulate internal NAD-malic enzyme and dramatically increase the rate of electron transport via the internal rotenone-insensitive bypass, thus accounting for all of the early observations (48).

Rustin et al (206) have suggested that under certain circumstances, NAD-malic enzyme is specifically linked to the rotenone-insensitive and the cyanide-insensitive pathways. For example, in some CAM plant mitochondria and in bundle sheath mitochondria of NAD-malic enzyme type C_4 species, malate is actually decarboxylated via a NAD-malic enzyme system that is preferentially associated with the cyanide-insensitive pathway and operates independently of the TCA cycle (85). This allows CO_2 to be provided from malate to the Benson-Calvin cycle without any constraints that might occur by coupling malate oxidation to the oxidative phosphorylation.

Evidence has been provided for the exceptionally high C_4 acid-decarboxylating capacity of isolated mitochondria from bundle sheath cells of NAD-malic enzyme-type C_4 species (100). It is interesting that the PEP-carboxykinase-type C_4 species commonly show activities of the mitochondrial NAD-malic enzyme that fall between those in NAD-malic enzyme-type C_4 species and the low levels found in C_3 plants (100). In this case, phosphorylation linked to malate oxidation in mitochondria is the major or sole source of ATP for PEP-carboxykinase (31, 32). Recently we have observed that NAD-malic enzyme has an intrinsically lower sensitivity for NADH accumulation than the other matrix dehydrogenases, including glycine decarboxylase ($K_i = 450$ μM vs approximately 15 μM) (N. Pascal and R. Douce, unpublished observations). Under these conditions, once fully activated, NAD-malic enzyme can readily engage the rotenone-insensitive pathway and therefore the alternative oxidase, bypassing the electrogenic proton translocation at the level of complex I (206). Consequently there is no need to associate NAD-malic enzyme to a specific electron transport (122).

Plant NAD-malic enzyme differs from mammalian malic enzyme in several respects. One important difference is based on its subunit composition. Structural studies on the malic enzyme from mammalian tissues show that it is a tetrameric protein composed of a monomeric unit, which may range in mass from 49 to 67 kD (for example, see 163). In contrast, SDS gel electrophoresis of the enzyme from plant mitochondria reveals two main bands of equal proportions with molecular weights of 61,000 and 58,000. Peptide mapping of the NAD-malic enzyme subunits from *Crassula argentea* (a CAM plant) and *Solanum tuberosum* (a C_3 plant) indicates that the low-mobility subunit was not a proteolytic artifact or an isoenzyme (231). NAD-malic enzyme can exist as a dimer ($\alpha\beta$, the least active form), a tetramer ($\alpha_2\beta_2$, the most active form), and an octamer ($\alpha_4\beta_4$). Malate, CoASH, and perhaps bicarbonate regulate NAD-malic enzyme by controlling its state of oligomerization (91, 230, 231). Furthermore, a major change in the enzyme takes place as protonation nears the pH optimum. This change is recorded as a shift in the enzyme's intrinsic affinity for malate (232).

The physiological function of mitochondrial NAD-malic enzyme is unclear. This enzyme is present in all of the plant mitochondria isolated so far and serves as a branch-point for the metabolism of malate in plant cells. Obviously, NAD-malic enzyme allows for the continual turnover of the TCA cycle without the necessity of supplying pyruvate from glycolysis (180). This may be very important in vivo, since NAD-malic enzyme can easily engage the nonphosphorylating rotenone-insensitive NADH dehydrogenase leading to the anaplerotic (literally "filling up") functioning of the TCA cycle (unless pyruvate is utilized elsewhere in the cell during photosynthesis—e.g. in some CAM and C_4 plants where NAD-malic enzyme serves to release CO_2 for fixation in the Benson-Calvin cycle) (85, 100). In other words, the reaction catalyzed by NAD-malic enzyme compensates for the drain on the TCA cycle that occurs when intermediates that lead to citrate, oxaloacetate, and α-ketoglutarate are removed from the cycle for biosynthesis.

Beyond the simple elucidation of the pathway of malate oxidation, there are many additional questions to be resolved. For example, what factors influence aggregation states of NAD-malic enzyme in vivo? Similarly, considering the rather alkaline pH (7.5) of the cytoplasmic compartment as determined by ^{31}P-NMR (199), NAD-malic enzyme should be strongly inhibited in vivo [with a few exceptions, however (11)], unless free Mn^{2+} and a rather large pool of CoASH are present in the matrix space to overcome this inhibition. An intriguing question is how does PEP-carboxylase function in vivo? If the cytosolic PEP is utilized primarily in glycolysis by pyruvate kinase, are the differences in K_m (PEP) between pyruvate kinase and PEP-carboxylase sufficiently large to alleviate the diversion of glycolytic PEP to four-carbon dicarboxylic acids by cytosolic PEP-carboxylase? How might metabolite modulation of PEP-carboxylase permit it to sense events occurring in the cell cytosol and mitochondria and allow coordination of the glycolytic pathway with the NAD-malic enzyme during the anaplerotic function of the TCA cycle? Unfortunately, the extent to which this occurs in plants in general is not known, except in the thermogenesis club of *Arum maculatum* (5). The answers to all these questions will therefore prove very enlightening.

Glycine Oxidation

The oxidation of glycine by mitochondria represents an important step in the metabolic pathway of photorespiration in leaf tissue (134, 175). Glycine is oxidized in the matrix space (197) by the glycine cleavage system to produce CO_2, NH_3, NADH, and 5,10-methylenetetrahydropteroyl-L-glutamic acid (5,10-CH_2-H_4F) (40). The latter compound reacts with a second mole of glycine to form serine in a reaction catalyzed by serine hydroxymethyltransferase (SHMT) (Figure 2). NADH generated during the course of glycine oxidation must be reoxidized if the photorespiratory cycle is to continue. Reoxidation may occur equally well either via the mitochondrial respiratory chain or via a reversal of malate dehydrogenase in the presence of oxaloacetate (65, 110, 130, 238) (Figure 2).

Glycine decarboxylase activity has been shown to be present in mitochondria from leaves of several C_3 species (83, 84) and CAM plants (10, 85). In addition, the mitochondria from bundle sheath cells of some C_4 species (NAD-malic enzyme type) oxidize glycine whereas those from mesophyll cells do not (173, 186), probably because mesophyll cells of C_4 plants lack ribulose-1,5-bisphosphatecarboxylase (74). Immunogold labeling with monospecific antibodies used by Rawsthorne et al (197) to investigate glycine decarboxylase in C_3-C_4 *Moricandia* species strongly suggests that the sole, or at least by far the major, site of release of photorespiratory CO_2 in these leaves is in the mitochondria on the inner wall of the bundle sheath cells. Such a localization greatly enhances the potential of CO_2 recapture by the Benson-Calvin cycle because in these cells mitochondria are in close association with overlying chloroplasts through which the CO_2 must pass (197).

The recent data of Rawsthorne et al (197a) indicate that while mesophyll cells of *Moricandia arvensis* have the capacity to synthesize glycine during photorespiration they have only a low capacity to metabolize it. In fact, release of ammonia during photorespiration occurs almost exclusively in the bundle sheath mitochondria. Almost no glycine oxidation activity is present in mitochondria from nongreen or etiolated tissues (46, 83, 85). Likewise, glycine oxidation observed in mammalian mitochondria is almost negligible (96). The glycine decarboxylase is present in low amounts in etiolated pea leaves but increases dramatically upon exposure to light (9, 46, 197b, 229). Day et al (46) and Bourguignon et al (22) have shown that after centrifugation on a percoll gradient, mitochondria from etiolated leaves were located in a band higher in the tube than the mitochondria from mature leaves containing the glycine cleavage system, a result suggesting that the mitochondria from etiolated leaves are lighter—glycine decarboxylase represents a large proportion of the matrix protein in green leaf mitochondria (approximately 30–50% of the total amount of matrix proteins). This observation explains the relatively lower lipid-to-protein and cytochrome-to-protein ratios in the leaf mitochondria compared to those from nonphotosynthetic tissues (84).

The glycine cleavage system has been purified from plant (22, 165, 214, 228) and animal (115) mitochondria and contains four proteins that have tentatively been named P-protein (a pyridoxal–phosphate-containing protein: (2 × 94 kDa), H-protein (a lipoic acid–containing protein; 15 kDa), T-protein (a protein catalyzing the tetrahydrofolate-dependent step of the reaction; 41 kDa), and L-protein (a lipoamide dehydrogenase; 2 × 60 kDa) (Figure 2). Molecular weight similarities between the purified lipoamide dehydrogenase of glycine decarboxylase and pyruvate dehydrogenase raise the question whether a unique protein is involved in the reoxidation of the dihydrolipoyl moieties of both lipoamide dehydrogenase–containing enzyme complexes or whether isoenzymes are involved (22, 228). High-molecular-mass proteins from pea mitochondrial matrix retained on an XM-300 diaflo membrane (matrix extract) exhibited high rates of glycine oxidation in the presence of NAD^+ and tetrahydropteroyl-L-glutamic acid (H_4F) as long as the medium exhibited a low ionic strength (165). NADH acting directly on L-protein competitively inhibited glycine oxidation when NAD^+ was the varied substrate at saturating concentrations of glycine (22). In green leaf mitochondria glycine cleavage is, therefore, regulated by $NADH/NAD^+$ molar ratios and behaves like almost all the NAD^+-linked matrix dehydrogenases. In other words, increasing the ratio of NADH to NAD^+ in the matrix space resulted in a logarithmic increase in inhibition.

The results presented by Bourguignon et al (22) also suggest that a soluble pool of H_4F located in the matrix constitutes a biochemical link between the glycine cleavage system and SHMT. In vivo, during the course of glycine oxidation a large pool of $5,10\text{-}CH_2\text{-}H_4F$ is maintained in the matrix space owing to a large excess of glycine cleavage and H_4F pushing SHMT towards serine production (SHMT catalyzes a freely reversible reaction) (Figure 2). Such a situation triggers rapid O_2 consumption. This is in contrast with mammalian mitochondria where glycine added to intact mitochondria is unable to trigger rapid O_2 consumption because glycine cleavage represents a minute fraction of the total matrix protein. In contrast with the pyruvate dehydrogenase complex, all the protein components of the glycine cleavage system dissociate very easily and behave as separable nonassociated proteins as the concentration of ions and pH in the medium increase (165) or following acetone extraction (214, 228). However at low ionic strength all of the proteins aggregated to form a fully active labile complex (165). It should also be mentioned that glycine cleavage activity is readily solubilized by freezing and thawing the mitochondria (165).

Finally, it appears that the electron transport chain of pea leaf mitochondria has an absolute preference for the NADH generated from glycine oxidation (47, 68, 85). In contrast, the very low rate of glycine oxidation observed in rat liver mitochondria is severely restricted in the presence of other respiratory substrates (96). In fact, the preferential oxidation of glycine observed by several groups utilizing green leaf mitochondria (68, 85) is achieved by a dominance of complex I over both complex II and the external NADH dehydrogenase of the respiratory chain, by the ability of glycine de-carboxylase to compete favorably at the level of NAD^+ (165), and by the high concentration of glycine decarboxylase present in the matrix space. There is, therefore, no need to propose a specific association of glycine cleavage with a portion of the respiratory chain (see for example 237). However, it is likely that green cells in vivo will generally be respiring in states between 4 and 3, especially under light conditions, because the cytosolic ATP/ADP ratio is considerable. That is, the rate at which a green cell respires is limited by the availability of ADP for oxidative phosphorylation. Such a situation will lead to an increase in the matrix $NADH/NAD^+$ ratio impeding the functioning of glycine cleavage. Since NADH strongly inhibits glycine decarboxylase, we are forced to imagine that in vivo NADH produced during the course of glycine oxidation is reoxidized very rapidly by oxaloacetate owing to the matrix malate dehydrogenase working in the reverse direction (72).

Our present knowledge of the relationship between protein components of the glycine cleavage system has not improved much, and many challenging problems still confront the enzymologist and the protein chemist. Are these enzymes present as separable, nonassociated proteins? How do the proteins communicate? In what way does protein interaction affect enzymatic cataly-sis? An almost endless number of questions can be raised, and the complexi-ties involved defy simple answers. However, we do believe that all the proteins of glycine cleavage are associated in vivo in a fixed stoichiometry that permits the complex to maintain a more favorable surface-to-volume ratio, to limit the diffusion pathway of each intermediary substrate, and to achieve structural symmetry. The self-recognition of proteins in the associa-tion process is under investigation in our laboratory.

PLANT MITOCHONDRIAL GENOME: MOLECULAR
ORGANIZATION AND EXPRESSION

The mammalian mitochondrial genome is extraordinarily tightly packed and shows an extreme economy in gene arrangement. In this case, coding sequences for rRNAs, tRNAs, and proteins immediately abut one another, with few or no noncoding nucleotides separating individual genes (12). Furthermore, it appears there is only one major promoter, or initiation site, for RNA synthesis. The primary product of transcription is, therefore, a full-length RNA copy of each strand (12). The copy is then cleaved to yield the final RNAs.

Mitochondrial genetic systems in higher plants differ from those of other organisms in a number of important respects.

HETEROGENEITY   The mitochondrial DNA (mtDNA) of higher plants is much more heterogeneous in organization and size than that of fungi, yeast, and mammals. The complexity of plant mtDNA ranges from 218 kbp in turnip to about 2500 kpb (half the size of the *Escherichia coli* chromosome) in some cucurbits (17, 191). The heterogeneity appears to arise from recombination events between directly repeated sequences in the mitochondrial genome. For example, the genome of *Brassica campestris* (178) is organized as three physically distinct circular molecules. The largest and most fragile circle, 218 kb in size, bears the entire sequence of the genome (including two copies of an approximately 2-kb element, present as direct repeats separated by 135 and 83 kb), which constitutes a "master" chromosome; while two smaller circles contain distinct 135- and 83-kb subsets of the master molecule and one copy each of the 2-kb repeat element. These three chromosomes appear to interconvert via reciprocal recombination across the major repeat element (2 kb) present on all three molecules. In other words, an active recombinational system will result in sequence "flip-flop" between inverted repeats, or sequence "loop-out" from the interaction of direct repeats (126a, 158, 191). The larger the number of direct and inverted repeats, the more complex the genome organization becomes (190). Under these conditions, multiple circular size classes and long linear DNAs are observed by electron microscopy (191). The complexity of the plant genome organization is reflected in the complexity of the restriction endonuclease profiles (126a, 191). Nonetheless the plant mitochondrial genome appears to be stable, despite this complex multipartite structure. These subgenomic circles and minicircles presumably have coordinated replication to insure a balanced copy number within the mitochondria. Unfortunately, molecular analyses of plant mtDNA synthesis have been difficult owing to the large size and complexity of the genome.

An equally striking finding was that mtDNA of all plants tested contain plastid DNA sequences integrated in numerous sites on the master chromosomes (218), including altered versions of the chloroplast genes for 16S rRNA (217), several tRNAs (107, 111, 139), and the large subunit of ribulose-1,5-bisphosphatecarboxylase/oxygenase (133) when either total chloroplast DNA or cloned chloroplast DNA restriction fragments are used as hybridization probes. These promiscuous DNA-chloroplast genes (75) inside plant mitochondria seem unimportant for mitochondrial functioning except perhaps for mitochondrial tRNA genes, which are identical to the chloroplast tRNA (139).

Although plant mitochondria synthesize polypeptides not synthesized by mitochondria from fungi and animals (such as the α-subunit of mitochondrial F1-ATPase) (23, 95), the coding capacity of plant mitochondrial genomes appears to be very low in relation to their size (126). By comparing the in organello synthesis product of maize mitochondria purified from various organs, Newton & Walbot (171) have found several examples of organ-specific and developmental stage–specific patterns of polypeptide synthesis. Such a result offers evidence for the existence of elaborate mechanisms for regulating the synthesis of this set of proteins.

Unfortunately, although our knowledge of the physical structure and information content of plant mtDNA has increased greatly in recent years, relatively little is known about the mechanisms through which this information is expressed (77). According to Lonsdale (132) the functional genes in maize are dispersed around the genome and do not account for more than 15% of the sequence complexity. However, no correlation was observed between the size of a plant mitochondrial genome and the amount of chloroplast DNA–homologous sequences it contains (218). From the limited data available it is not yet clear whether "itinerant" DNA from nucleus and chloroplasts, more genes, optional introns, and/or DNA serving a sequence-independent function are responsible for either the large size of the plant mitochondrial genome or the variability among species.

It is likely that the unexpectedly large size of plant mitochondrial genomes results at least in part from the accumulation of nonessential DNA, as suggested by Grayburn & Bendich (89). The mode of transfer of nucleic acid information between the organelles is still unclear, as no direct evidence is available on the transport of nucleic acid into or out of plastids and mitochondria. However, according to Schuster & Brennicke (209a) transfer of nucleic acid sequence information between different cell compartments might have occurred through transfer of RNA followed by local reverse transcription and subsequent integration of this newly synthesized DNA into the organelle genomes. Of course, the ultimate answer will come when we are able to delete large amounts of putative noncoding DNA and compare plants carrying the reduced and normal-sized genomes.

PLASMIDS   In addition to the main high-molecular-weight genome (master chromosome + subgenomic circular chromosomes), the mitochondria of many higher plants contain smaller DNA molecules known as mitochondrial plasmids (27, 112, 179, 189). These molecules, whose function remains unknown, are present at a high stoichiometry relative to the main genome and can be either circular or linear in form (17, 190). For example, maize mitochondria have various combinations of minicircular DNAs ranging in size from 1.4 to 1.9 kb and minilinear DNAs ranging in size from 2.1 to 7.4 kb (190). Minicircular plasmids have been described in a large number of divergent plant species. Most of them have no sequence homology to the principle genome and are therefore dispensable. The minicircular DNAs appear to be replicated and maintained as a stable part of the mitochondrial genotype. The origin of most of the minicircles (nuclear? extracellular?) found in plant mitochondria is unknown, with the exception of those that are excised from the mitochondrial genome.

The linear mitochondrial plasmids are resistant to digestion by lambda exonuclease and presumably have $5'$-terminal proteins (114). For example, the mitochondria of all S-type cytoplasmic male sterile (cms) maize lines, a maternally inherited trait that results in the lack of pollen production (126, 126a, 132), contain two double-stranded, linear, plasmid-like DNAs of 6.4 and 5.4 kb (S_1 and S_2, respectively) (189). Free S_1 and S_2 plasmids recombine with specific sites in the mitochondrial genome. Likewise, linear, plasmid-like DNAs of 5.8 and 5.4 kb (N_1 and N_2, respectively) are associated with the mitochondrial genome of the IS1112C male sterile cytoplasm of *Sorghum bicolor* (58, 188). At this time it is not possible to draw a conclusion regarding the role of these linear plasmids in cms. The lack of homology between the N_1 and N_2 DNAs and the *Sorghum* chloroplast, nuclear, and mitochondrial genomes (34) indicates that these double-stranded linear plasmids have an exogenous origin. These plasmids might have been introduced into higher plant cells by symbionts or pathogens. In support of this suggestion, the double-stranded linear plasmid with $5'$-associated proteins is reminiscent of the structure of several viral genomes.

RECOMBINATORY CAPACITY   All the data presented so far emphasize the great flexibility of the plant mitochondrial genome. This complexity is compounded by the fact that culturing plant cells is sometimes accompanied by changes in the mtDNA both in the form of new low-molecular-weight plasmids and in altered restriction profiles (113). Likewise, cytoplasmic hybridization (somatic hybrids and cybrids) appears to promote mitochondrial genomic mixing and recombination (15, 35, 82, 126a, 157, 176, 201, 202). Thus, somatic hybrid mitochondrial genomes analyzed have been found to contain mixtures of restriction fragments characteristic of both parents as well as novel, nonparental fragments. The mechanisms leading to the formation of these new mitochondrial genotypes are as yet poorly understood.

Finally, mitochondrial recombinations may relate to cms (126a, 132) by creating novel open reading frames or altering mtDNA replication (158, 203, 234). The prevailing evidence indicates that cms is coded for in the mitochondrion and not in the chloroplast (126). Thus, a 13-kDa polypeptide is synthesized only by cms-T (Texas) maize mitochondria (79, 80, 126) and probably represents the translation product of one of these novel reading frames. Mitochondria isolated from cms-T maize are specifically sensitive to a toxin (Bm-T-toxin) produced by the fungal pathogen *Bipolaris maydis,* and the 13-kD polypeptide confers toxin sensitivity to mitochondria of cms-T maize (54). Recently, Dewey et al (53) have found a transcribed open reading frame, T-URF13, that may encode this 13-kD polypeptide, and Rottmann et al (203) provided direct evidence that homologous recombination involving a 127-bp repeated sequence leads to the deletion of this open reading frame and that this deletion is central to the fertility event. [The 13-kDa protein is characterized by a dramatically reduced abundance in cms-T plants that are restored to fertility by unique dominant nuclear restorer genes *Rf1* and *Rf2)* (79).] Sequence analyses suggest that this gene of chimeric origin (T-URF13) has arisen by recombinations among the coding and/or flanking regions of the maize mitochondrial *26S* ribosomal gene and *atp6* gene. Its transcription in the T cytoplasm is presumably under the control of the *atp6* promoter (215). Parenthetically, the ATPase subunit 6 protein is highly variable among distantly related species, and the polypeptides predicted from the tobacco and maize ATPase coding region reveal a hydrophobic presequence that is not found in the fungal or animal subunits (19).

Obviously, the plant mitochondrial genome has great capacity for recombination. It is apparent, therefore, that the basis for providing a physiological explanation of the plant mitochondrial genome plasticity should be available once the mechanism and control of mtDNA recombination are understood. Finally, the employment of a unique genetic code (81) and the presence of a 5S rRNA species (43) constitute further differences between the mitochondrial systems of plants and other organisms.

FINAL COMMENTS

Plant mitochondria have specific transport systems for metabolites and electrons that reflect their functioning in autotrophic metabolism. We now know a great deal about plant mitochondria, but there remain major gaps to be filled. Thus, it is important to understand how the unique properties of the plant mitochondrion, which make this cell organelle biochemically very flexible, are integrated into the general metabolism of various tissues such as leaves and roots. The art of isolating plant mitochondria in a physiologically active condition has advanced greatly in recent years but has only been achieved for a few species of higher plant. Techniques for the isolation of active, intact mitochondria from any tissues of the same plant would, for example, be of use in studies attempting to relate the specific functions of each tissue to intracellular biochemistry. (Mitochondria vary considerably in morphology in various plant tissues and clearly have a variety of physiological roles in green and various nongreen tissues.) In addition, only a few plant mitochondrial proteins have been reasonably well characterized. Clearly, the soluble enzymes of the stroma space should receive intensive investigation since they are likely involved in organ-specific functions.

Finally, considering the large size of the plant mitochondrial genome, analyses of the unidentified mitochondrial translation products in plant cells and of their relationships present a formidable challenge for the future. Such studies may reveal unexpected facets of the expression and regulation of the mitochondrial genome, of the respiratory chain assembly, and of the differentiation of the organelle. These studies constitute a natural prelude to our ultimate understanding of the central developmental problem of how cellular organelle number and function are determined in response both to the varying metabolic energy demands in differentiated cells and to environmental factors.

ACKNOWLEDGMENTS

Research work in this laboratory has been supported by research grants from the Centre National de la Recherche Scientifique (CNRS) and the Commissariat à l'Energie Atomique (CEA).

Literature Cited

1. Abou-Khalil, S., Hanson, J. B. 1979. Energy-linked adenosine diphosphate accumulation by corn mitochondria. I. General characteristics and effect of inhibitors. *Plant Physiol.* 64:276–80
2. Abou-Khalil, S., Hanson, J. B. 1979. Energy-linked adenosine diphosphate accumulation by corn mitochondria. II. Phosphate and divalent cation requirement. *Plant Physiol.* 64:281–84
3. Andreo, C. S., Gonzalez, D. H., Iglesias, A. A. 1987. Higher plant phosphoenolpyruvate carboxylase. *FEBS Lett.* 213:1–8
4. ap Rees, T. 1985. The organization of glycolysis and the oxidative pentose phosphate pathway in plants. See Ref. 63a, pp. 391–417
5. ap Rees, T., Bryce, J. L., Wilson, P. M., Green, J. H. 1983. Role and loca-

tion of NAD-malic enzyme in thermogenic tissues of *Araceae*. *Arch. Biochem. Biophys.* 227:511–21
6. ap Rees, T., Fuller, W. A., Wright, B. W. 1977. Measurements of glycolytic intermediates during the onset of thermogenesis in the spadix of *Arum maculatum*. *Biochim. Biophys. Acta* 461:274–82
7. Aprille, J. R., Austin, J. 1981. Regularisation of the mitochondrial adenine nucleotide pool size. *Arch. Biochem. Biophys.* 212:689–99
8. Arron, G. P., Edwards, G. E. 1979. Oxidation of reduced nicotinamide adenine dinucleotide phosphate by plant mitochondria. *Can. J. Biochem.* 57:1392–99
9. Arron, G. P., Edwards, G. E. 1980. Light induced development of glycine

oxidation by mitochondria from sunflower cotyledons. *Plant Sci. Lett.* 18: 229–35

10. Arron, G. P., Spalding, M. H., Edwards, G. E. 1979. Isolation and oxidative properties of intact mitochondria from the leaves of *Sedum praealtum*. A crassulacean acid metabolism plant. *Plant Physiol.* 64:18–26

11. Artus, N. N., Edwards, G. E. 1985. NAD-malic enzyme from plants. *FEBS Lett.* 182:225–33

12. Attardi, G. 1981. Organisation and expression of the mammalian mitochondrial genome: a lesson in economy. *Trends Biochem. Sci.* 6:86–89

13. Azcon-Bieto, J., Day, D. A., Lambers, H. 1983. The effect of photosynthesis and carbohydrate status on respiratory rates and the involvement of the alternative path in leaf respiration. *Plant Physiol.* 72:598–603

14. Azcon-Bieto, J., Lambers, H., Day, D. A. 1983. Respiratory properties of developing bean and pea leaves. *Aust. J. Plant Physiol.* 10:237–45

15. Belliard, G., Vedel, F., Pelletier, G. 1979. Mitochondrial recombination in cytoplasmic hybrids of *Nicotiana tabacum* by protoplast fusion. *Nature* 281:401–3

16. Bendall, D. S., Bonner, W. D. 1971. Cyanide-insensitive respiration in plant mitochondria. *Plant Physiol.* 47:236–45

17. Bendich, A. J. 1985. *Plant Mitochondrial DNA: Unusual Variation on a Common Theme in Genetic Flux in Plants*, ed. B. Hohn, E. S. Dennis, pp. 111–38. Berlin: Springer-Verlag

18. Bergman, A., Ericson, I. 1983. Effects of NADH, succinate and malate on the oxidation of glycine in spinach leaf mitochondria. *Physiol. Plant.* 59:421–27

18a. Berthold, D. A., Fluke, D. J., Siedow, J. N. 1988. Determination of molecular mass of the Aroid alternative oxidase by radiation-inactivation analysis. *Biochem. J.* 252:73–77

19. Bland, M. M., Levings, C. S. III, Matzinger, C. F. 1987. The ATPase subunit 6 gene of tobacco mitochondria contains an unusual sequence. *Curr. Genet.* 12:475–81

20. Bligny, R., Roby, C., Douce, R. 1988. Phosphorus-31 nuclear magnetic resonance studies in higher plant cells. In *Nuclear Magnetic Resonance in Agriculture*, ed. P. E. Pfeffer, W. V. Gerasimowicz. Boca Raton: CRC Press. In press

21. Bonner, W. D. Jr., Clarke, S. D., Rich, P. R. 1986. Partial purification and characterization of the quinol oxidase activity of *Arum maculatum* mitochondria. *Plant Physiol.* 80:838–42

22. Bourguignon, J., Neuburger, M., Douce, R. 1988. Resolution and characterization of the glycine cleavage reaction in pea leaf mitochondria. Properties of the forward reaction catalyzed by glycine decarboxylase and serine hydroxymethyltransferase. *Biochem. J.* 255: 169–78

23. Boutry, M., Briquet, M., Goffeau, A. 1982. The α- subunit of a plant mitochondrial F_1-ATPase is translated in mitochondria. *J. Biol. Chem.* 238:8524–26

24. Brailsford, M. A., Thompson, A. G., Kaderbhai, N., Beechey, R. B. 1986. The extraction and reconstitution of the α-cyanocinnamate-sensitive pyruvate transporter from castor bean mitochondria. *Biochem. Biophys. Res. Commun.* 140:1036–42

25. Brailsford, M. A., Thompson, A. G., Kaderbhai, N., Beechey, R. B. 1986. Pyruvate metabolism in castor bean mitochondria. *Biochem. J.* 239:355–61

26. Brand, M. D., Murphy, M. P. 1987. Control of electron flux through the respiratory chain in mitochondria and cells. *Biol. Rev.* 62:141–93

27. Brennicke, A., Blanz, P. 1982. Circular mitochondrial DNA species from *Oenothera* with unique sequences. *Mol. Gen. Genet.* 187:461–67

28. Brouquisse, R., Nishimura, M., Gaillard, J., Douce, R. 1987. Characterization of a cytosolic aconitase in higher plant cells. *Plant Physiol.* 84:1402–7

29. Bryce, J. H., ap Rees, T. 1985. Effects of sucrose on the rate of respiration of the roots of *Pisum sativum*. *J. Plant Physiol.* 120:363–67

30. Budde, R. J. A., Randall, D. D. 1987. Regulation of pea mitochondrial pyruvate dehydrogenase complex activity: inhibition of ATP-dependent inactivation. *Arch. Biochem. Biophys.* 258:600–6

31. Burnell, J. N., Hatch, M. D. 1988. Photosynthesis in phosphoenolpyruvate carboxykinase-type C_4 plants: photosynthetic activities of isolated bundle sheath cells from *Urochloa panicoides*. *Arch. Biochem. Biophys.* 260:177–86

32. Burnell, J. N., Hatch, M. D. 1988. Photosynthesis in phosphoenolpyruvate carboxykinase-type C_4 acid decarboxylation in bundle sheath cells of *Urochloa panicoides*. *Arch. Biochem. Biophys.* 260:187–99

33. Chapman, K. S. R., Hatch, M. D. 1977. Regulation of mitochondrial NAD^+-

malic enzyme involved in C_4 pathway photosynthesis. *Arch. Biochem. Biophys.* 184:298–306

34. Chase, C. D., Pring, D. R. 1986. Properties of the linear N1 and N2 plasmid-like DNAs from mitochondria of cytoplasmic male sterile *Sorghum bicolor. Plant Mol. Biol.* 6:53–64

35. Chetrit, P., Mathieu, C., Vedel, F., Pelletier, G., Primard, C. 1985. Mitochondrial DNA polymorphism induced by protoplast fusion in *Cruciferae. Theor. Appl. Genet.* 69:361–66

36. Coleman, J. O. D., Palmer, J. M. 1971. Role of Ca^{2+} in the oxidation of exogenous NADH by plant mitochondria. *FEBS. Lett.* 17:203–8

37. Coleman, J. O. D., Palmer, J. M. 1972. The oxidation of malate by isolated plant mitochondria. *Eur. J. Biochem.* 26:499–509

38. Cook, N. D., Cammack, R. 1984. Purification and characterization of the rotenone-insensitive NADH dehydrogenase of mitochondria from *Arum maculatum. Eur. J. Biochem.* 141:573–77

39. Cook, N. D., Cammack, R. 1985. Reactivation of the NADH-Q1 reductase activity of the isolated rotenone-insensitive NADH dehydrogenase of *Arum maculatum* mitochondria. Activation of quinone reductase activity by a hydrophobic component extractable from *Arum* mitochondrial membranes. *Plant Sci.* 42:73–76

40. Cossins, E. A. 1987. Folate biochemistry and the metabolism of one-carbon units. In *The Biochemistry of Plants: A Comprehensive Treatise. Biochemistry of Metabolism,* ed. D. D. Davies, 11:317–53. New York: Academic

41. Cottingham, I. R., Moore, A. L. 1984. Partial purification and properties of the external NADH dehydrogenase from cuckoo-pint *(Arum maculatum)* mitochondria. *Biochem. J.* 224:171–79

42. Cowley, R. C., Palmer, J. M. 1978. The interaction of citrate and calcium in regulating the oxidation of exogenous NADH in plant mitochondria. *Plant Sci. Lett.* 11:345–50

43. Cunningham, R. S., Bonen, L., Doolittle, W. F., Gray, M. W. 1976. Unique species of 5S, 18S, and 26S ribosomal RNA in wheat mitochondria. *FEBS Lett.* 69:116–22

43a. Davies, D. D., ed. 1980. *The Biochemistry of Plants: A Comprehensive Treatise, Metabolism, and Respiration,* Vol. 2. New York: Academic

44. Day, D. A., Arron, G. P., Laties, G. G. 1980. Nature and control of respiratory

pathways in plants: the interaction of cyanide-resistant respiration with the cyanide-sensitive pathway. See Ref. 43a, pp. 197–241

45. Day, D. A., Neuburger, M., Douce, R. 1984. Activation of NAD-linked malic enzyme in intact plant mitochondria by exogenous coenzyme A. *Arch. Biochem. Biophys.* 231:233–42

46. Day, D. A., Neuburger, M., Douce, R. 1985. Biochemical characterization of chlorophyll-free mitochondria from pea leaves. *Aust. J. Plant Physiol.* 12:219–28

47. Day, D. A., Neuburger, M., Douce, R. 1985. Interactions between glycine decarboxylase, the tricarboxylic acid cycle and the respiratory chain in pea leaf mitochondria. *Aust. J. Plant Physiol.* 12:119–30

48. Day, D. A., Neuburger, M., Douce, R., Wiskich, J. T. 1983. Exogenous NAD^+ effects on plant mitochondria: a reinvestigation of the transhydrogenase hypothesis. *Plant Physiol.* 73:1024–27

49. Day, D. A., Wiskich, J. T. 1974. The effect of exogenous nicotinamide adenine dinucleotide on the oxidation of nicotinamide adenine dinucleotide-linked substrates by isolated plant mitochondria. *Plant Physiol.* 54:360–63

50. Day, D. A., Wiskich, J. T. 1981. Effect of phthalonic acid on respiration and metabolite transport in higher plant mitochondria. *Arch. Biochem. Biophys.* 211:100–7

51. Day, D. A., Wiskich, J. T. 1981. Glycine metabolism and oxaloacetate transport by pea leaf mitochondria. *Plant Physiol.* 68:425–29

52. Day, D. A., Wiskich, J. T. 1984. Transport processes of isolated plant mitochondria. *Physiol. Veg.* 22:241–61

53. Dewey, R. E., Levings, C. S. III, Timothy, D. H. 1986. Novel recombinations in the maize mitochondrial genome produce a unique transcriptional unit in the Texas male-sterile cytoplasm. *Cell* 4:439–49

54. Dewey, R. E., Siedow, J. N., Timothy, D. H., Levings, C. S. III. 1988. A 13-kilodalton maize mitochondrial protein in *E. coli* confers sensitivity to *Bipolaris maydis* toxin. *Science* 239:293–95

55. Diolez, P., Moreau, F. 1985. Correlations between ATP synthesis, membrane potential and oxidation rate in plant mitochondria. *Biochim. Biophys. Acta* 806:56–63

56. Diolez, P., Moreau, F. 1987. Relationships between membrane potential and oxidation rate in potato mitochondria. In *Plant Mitochondria,* ed. A. L.

Moore, R. B. Beechey, pp. 17–25. New York/London: Plenum

57. Di Virgilio, F., Azzone, G. F. 1982. Activation of site I Redox-driven H$^+$ pump by exogenous quinones in intact mitochondria. *J. Biol. Chem.* 257:4106–13

58. Dixon, L. K., Leaver, C. J. 1982. Mitochondrial gene expression and cytoplasmic male sterility in sorghum. *Plant Mol. Biol.* 1:89–102

59. Dizengremel, P. 1983. Effect of Triton X-100 on the cyanide-resistant pathway in plant mitochondria. *Physiol. Veg.* 21:743–52

60. Dizengremel, P., Chauveau, M., Roussaux, J. 1982. Inhibition by adenine derivatives of the cyanide-insensitive electron pathway of plant mitochondria. *Plant Physiol.* 70:585–89

61. Douce, R. 1985. *Mitochondria in Higher Plants. Structure, Function and Biogenesis.* New York: Academic

62. Douce, R., Bonner, W. D. Jr. 1972. Oxaloacetate control of Krebs cycle oxidation in purified plant mitochondria. *Biochem. Biophys. Res. Commun.* 47:619–24

63. Douce, R., Christensen, E. L., Bonner, W. D. 1972. Preparation of intact plant mitochondria. *Biochim. Biophys. Acta* 275:148–60

63a. Douce, R., Day, D. A., eds. 1985. *Encyclopedia of Plant Physiology*, Vol. 18. Berlin: Springer-Verlag

64. Douce, R., Mannella, C. A., Bonner, W. D. 1973. The external NADH dehydrogenases of intact plant mitochondria. *Biochim. Biophys. Acta* 292:105–16

65. Douce, R., Moore, A. L., Neuburger, M. 1977. Isolation and oxidative properties of intact mitochondria isolated from spinach leaves. *Plant Physiol.* 60:625–28

66. Douce, R., Neuburger, M., Givan, C. J. 1986. Regulation of succinate oxidation by NAD$^+$ in mitochondria purified from potato tubers. *Biochim. Biophys. Acta* 850:64–71

67. Dry, I. B., Bryce, J. H., Wiskich, J. T. 1987. Regulation of mitochondrial respiration. In *The Biochemistry of Plants: A Comprehensive Treatise. Biochemistry of Metabolism*, ed. D. D. Davies, 11:213–52. New York: Academic

68. Dry, I. B., Day, D. A., Wiskich, J. T. 1983. Preferential oxidation of glycine by the respiratory chain of pea leaf mitochondria. *FEBS Lett.* 158:154–58

69. Dry, I. B., Dimitriadis, E., Ward, A. D., Wiskich, J. T. 1987. The photore-

spiratory hydrogen shuttle. Synthesis of phthalonic acid: its use in the characterization of the malate/aspartate shuttle in pea *(Pisum sativum)* leaf mitochondria. *Biochem. J.* 245:669–75

70. Dry, I. B., Wiskich, J. T. 1987. 2-Oxoglutarate dehydrogenase and pyruvate dehydrogenase activities in plant mitochondria. Interaction via a commun coenzyme A pool. *Arch. Biochem. Biophys.* 257:92–99

71. Ebbighausen, H., Chen, J., Heldt, H. W. 1985. Oxaloacetate translocator in plant mitochondria. *Biochim. Biophys. Acta* 810:184–99

72. Ebbighausen, H., Hatch, M. D., Lilley, R. McC., Kromer, S. 1987. On the function of malate-oxaloacetate shuttles in a plant cell. In *Plant Mitochondria*, ed. A. L. Moore, R. B. Beechey, pp. 171–80. New York/London: Plenum

73. Edman, K., Ericson, I., Moller, I. M. 1985. The regulation of exogenous NAD(P)H oxidation in spinach leaf mitochondria by pH and cations. *Biochem. J.* 232:471–77

74. Edwards, G., Walker, D. A. 1983. C_3, C_4: *Mechanisms and Cellular and Environmental Regulation of Photosynthesis.* Oxford: Blackwell

75. Ellis, J. 1982. Promiscuous DNA-chloroplast genes inside plant mitochondria. *Nature* 299:678–79

76. Elthon, T. E., McIntosh, L. 1987. Identification of the alternative terminal oxidase of higher plant mitochondria. *Proc. Natl. Acad. Sci. USA* 84:8399–403

77. Finnegan, P. M., Brown, G. G. 1987. *In organello* transcription in maize mitochondria and its sensitivity to inhibitors of RNA synthesis. *Plant Physiol.* 85:304–9

78. Fiskum, G., Lehninger, A. L. 1980. The mechanisms and regulation of mitochondrial Ca^{2+} transport. *Fed. Proc.* 39:2432–36

79. Forde, B. G., Leaver, C. J. 1980. Nuclear and cytoplasmic genes controlling synthesis of variant mitochondrial polypeptides in male-sterile maize. *Proc. Natl. Acad. Sci. USA* 77:418–22

80. Forde, B. G., Oliver, R. J. C., Leaver, C. J. 1978. Variation in mitochondrial translation products associated with male-sterile cytoplasms in maize. *Proc. Natl. Acad. Sci. USA* 75:3841–45

81. Fox, T. D., Leaver, C. J. 1981. The *Zea mays* mitochondrial gene coding cytochrome oxidase subunit II has an intervening sequence and does not contain TGA codons. *Cell* 26:315–23

82. Galun, E., Arzee-Gonen, P., Fluhr, R.,

Edelman, M., Aviv, D. 1982. Cytoplasmic hybridization in *Nicotiana:* mitochondrial DNA analysis in progenies resulting from fusion between protoplasts having different organelle constitutions. *Mol. Gen. Genet.* 186:50–56

83. Gardestrom, P., Bergman, A., Ericson, I. 1980. Oxidation of glycine via the respiratory chain in mitochondria prepared from different parts of spinach. *Plant Physiol.* 65:389–91

84. Gardestrom, P., Bergman, A., Sahlström, S., Edman, K-A., Ericson, I. 1983. A comparison of the membrane composition of mitochondria isolated from spinach leaves and leaf petioles. *Plant Sci. Lett.* 31:173–80

85. Gardestrom, P., Edwards, G. E. 1985. Leaf mitochondria (C$_3$ + C$_4$ + CAM). See Ref. 63a, pp. 314–46

86. Genevois, M. L. 1929. Sur la fermentation et sur la respiration chez les végétaux chlorophylliens. *Rev. Gen. Bot.* 41:252–71

87. Gerbling, H., Gerhardt, B. 1988. Carnitine-acyltransferase activity of mitochondria from nung-bean hypocotyls. *Planta* 174:90–93

88. Gerhardt, B. 1986. Basic metabolic function of the higher plant peroxisome. *Physiol. Veg.* 24:397–410

89. Grayburn, W. S., Bendich, A. J. 1987. Variable abundance of a mitochondrial DNA fragment in cultured tobacco cells. *Curr. Genet.* 12:257–61

90. Grover, S. D., Laties, G. G. 1981. Disulfiram inhibition of the alternative respiratory pathway in plant mitochondria. *Plant Physiol.* 68:393–400

91. Grover, S. D., Wedding, R. T. 1984. Modulation of the activity of NAD malic enzyme from *Solanum tuberosum* by changes in oligomeric state. *Arch. Biochem. Biophys.* 234:418–25

92. Gunning, B. E. S., Steer, M. W. 1975. *Ultrastructure and the Biology of Plant Cells.* London: Edward Arnold

93. Gupte, S. S., Wu, E. S., Hoechli, L., Hoechli, M., Jacobson, K., et al. 1984. Relationship between lateral diffusion, collision frequency, and electron transfer of mitochondrial inner membrane oxidation-reduction components. *Proc. Natl. Acad. Sci. USA* 81:2606–10

94. Guy, R. D., Berry, J. A., Fogel, M. L. 1988. Differential fractionation of oxygen isotopes by cyanide-resistant and cyanide-sensitive respiration in plants. *Planta.* In press

95. Hack, E., Leaver, C. J. 1983. The α-subunit of the maize F$_1$-ATPase is synthesized in the mitochondrion. *EMBO J.* 2:1783–89

96. Hampson, R. K., Barron, L. L., Olson, M. S. 1983. Regulation of the glycine cleavage system in isolated rat liver mitochondria. *J. Biol. Chem.* 258:2993–99

97. Hanson, J. B. 1985. Membrane transport systems of plant mitochondria. See Ref. 63a, pp. 248–80

98. Hanson, J. B., Koeppe, D. E. 1975. *Mitochondria in Ion Transport in Plant Cells and Tissues,* ed. D. A. Baker, J. L. Hall, pp. 79–99. Amsterdam: Elsevier

99. Harwood, J. L. 1985. Plant mitochondrial lipids: structure, function and biosynthesis. See Ref. 63a, pp. 37–71

100. Hatch, M. D. 1988. C$_4$ photosynthesis: a unique blend of modified biochemistry, anatomy and ultrastructure. *Biochim. Biophys. Acta.* In press

101. Hatch, M. D., Kagawa, T. 1974. Activity, location and role of NAD-malic enzyme in leaves of C$_4$ pathway photosynthesis. *Aust. J. Plant Physiol.* 1:357–69

102. Henry, M. F., Nyns, E. J. 1975. Cyanide-insensitive respiration. An alternative mitochondrial pathway. *Sub-Cell. Biochem.* 4:1–65

103. Hensley, J. R., Hanson, J. B. 1975. The action of valinomycin in uncoupling corn mitochondria. *Plant Physiol.* 56:13–18

104. Hunter, F. E., Malison, R., Bridgers, W. F., Schutz, B., Atchison, A. 1959. Reincorporation of diphosphopyridine nucleotide into mitochondrial enzyme systems. *J. Biol. Chem.* 234:693–99

105. Huq, S., Palmer, J. M. 1978. Oxidation of durohydroquinone via the cyanide-insensitive respiratory pathway in higher plant mitochondria. *FEBS Lett.* 92:317–20

106. Huq, S., Palmer, J. M. 1978. The involvement and possible role of quinone in cyanide-resistant respiration. In *Plant Mitochondria,* ed. G. Ducet, C. Lance, pp. 225–32. Amsterdam: Elsevier

107. Iams, K. P., Heckman, J. E., Sinclair, J. H. 1985. Sequence of histidyl tRNA present as a chloroplast insert in mt DNA of *Zea mays. Plant Mol. Biol.* 4:225–32

108. Ikuma, H. 1972. Electron transport in plant respiration. *Annu. Rev. Plant Physiol.* 23:419–36

109. Ikuma, H., Bonner, W. D. 1967. Properties of higher plant mitochondria. III. Effects of respiratory chain inhibitors. *Plant Physiol.* 42:1535–44

110. Journet, E. P., Neuburger, M., Douce, R. 1981. Role of glutamate-oxaloacetate transaminase and malate dehydrogenase

in the regeneration of NAD+ for glycine oxidation by spinach leaf mitochondria. *Plant Physiol.* 67:467–69

111. Joyce, P. B. M., Spencer, D. F., Bonen, L., Gray, M. W. 1988. Genes for tRNA^{Asp}, tRNA^{Pro}, tRNA^{Tyr}, and two tRNA^{Ser} in wheat mitochondrial DNA. *Plant Mol. Biol.* 10:251–62

112. Kemble, R. J., Bedbrook, J. R. 1980. Low molecular weight circular and linear DNA in mitochondria from normal and male sterile *Zea mays* cytoplasm. *Nature* 284:565–66

113. Kemble, R. J., Shepard, J. F. 1984. Cytoplasmic DNA variation in a potato protoclonal population. *Theor. Appl. Genet.* 69:211–16

114. Kemble, R. J., Thompson, R. D. 1982. S$_1$ and S$_2$, the linear mitochondrial DNAs present in a male sterile line of maize, possess terminally attached proteins. *Nucleic Acids Res.* 10:8181–90

115. Kikuchi, G., Hiraga, K. 1982. The mitochondrial glycine cleavage system. Unique features of the glycine decarboxylation. *Mol. Cell Biochem.* 45:137–49

116. Klein, R. R., Burke, J. J. 1984. Separation procedure and partial characterization of two NAD(P)H dehydrogenases from cauliflower mitochondria. *Plant Physiol.* 76:436–41

117. Koeppe, D. E., Miller, R. J. 1972. Oxidation of reduced nicotinamide adenine dinucleotide, phosphate by isolated corn mitochondria. *Plant Physiol.* 49:353–57

118. Kroger, A., Klingenberg, M. 1973. The kinetics of the redox reactions of ubiquinone related to the electron-transport activity in the respiratory chain. *Eur. J. Biochem.* 34:358–68

119. Kuiper, D. 1983. Genetic differentiation in *Plantago major:* growth and root respiration and their role in phenotypic adaptation. *Physiol. Plant.* 57:222–30

120. Lambers, H. 1985. Respiration in intact plants and tissues: its regulation and dependence on environmental factors, metabolism and invaded organism. See Ref. 63a, pp. 418–73

121. Lance, C., Chauveau, M. 1975. Evolution des activités oxydatives et phosphorylantes des mitochondries de l'*Arum maculatum* L. au cours du développement de l'inflorescence. *Physiol. Vég.* 13:83–94

122. Lance, C., Chauveau, M., Dizengremel, P. 1985. The cyanide-resistant pathway of plant mitochondria. See Ref. 63a, pp. 202–47

123. Lance, C., Hobson, G. E., Young, R. E., Biale, J. B. 1967. Metabolic processes in the cytoplasmic particles from the avocado fruit. IX. The oxidation of pyruvate and malate during the climacteric cycle. *Plant Physiol.* 42: 471–78

124. LaNoue, K. F., Schoolwerth, A. C. 1979. Metabolite transport in mitochondria. *Annu. Rev. Biochem.* 48:871–922

125. Laties, G. G. 1982. The cyanide-resistant, alternative path in higher plant respiration. *Annu. Rev. Plant Physiol.* 33:519–55

126. Leaver, C. J., Gray, M. W. 1982. Mitochondrial genome organization and expression in higher plants. *Annu. Rev. Plant Physiol.* 33:373–402

126a. Leaver, C. J., Lonsdale, D. M. 1988. *Mitochondrial Biogenesis.* London: The Royal Society

127. Lehninger, A. L. 1964. *The Mitochondrion.* New York/Amsterdam: Benjamin

128. Lenaz, G., Fato, R. 1986. Ubiquinone diffusion rate-limiting for electron-transfer. *J. Bioenerg. Biomembr.* 18: 369–401

129. Lieberman, M., Baker, J. E. 1965. Respiratory electron transport. *Annu. Rev. Plant Physiol.* 16:343–82

130. Lilley, R. M. C., Ebbighausen, H., Heldt, H. W. 1987. The simultaneous determination of carbon dioxide release and oxygen uptake in suspensions of plant leaf mitochondria oxidizing glycine. *Plant Physiol.* 83:349–53

131. Lloyd, D. 1974. *The Mitochondria of Microorganisms.* London/New York: Academic

132. Lonsdale, D. M. 1987. The molecular biology and genetic manipulation of the cytoplasm of higher plants. *Genet. Eng.* 6:50–104

133. Lonsdale, D. M., Hodge, T. P., Howe, C. J., Stern, D. B. 1983. Maize mitochondrial DNA containing sequence homologous to the ribulose-1,5-bisphosphate carboxylase large subunit gene of chloroplast DNA. *Cell* 34:1007–14

134. Lorimer, G. H., Andrews, T. J. 1980. The C$_2$ chemo- and photorespiratory carbon oxidation cycle. In *The Biochemistry of Plants: A Comprehensive Treatise.* Photosynthesis, ed. M. D. Hatch, N. K. Boardman, 8:329–74. New York: Academic

135. Macrae, A. R. 1971. Isolation and properties of a "malic" enzyme from cauliflower bud mitochondria. *Biochem. J.* 122:495–501

136. Macrae, A. R., Moorhouse, R. 1970. The oxidation of malate by mitochondria

isolated from cauliflower buds. *Eur. J. Biochem.* 16:96–102

137. Mandolino, G., De Santis, A., Melandri, B. A. 1983. Localized coupling in oxidative phosphorylation by mitochondria from Jerusalem artichoke. *Biochim. Biophys. Acta* 723:428–39

138. Mannella, C. A., Tedeschi, H. 1987. Importance of the mitochondrial membrane channel as a model biological channel. *J. Bioenerg. Biomembr.* 19:305–58

139. Marechal, L., Runeberg-Rolos, P., Grienenberger, J. M., Colin, J., Weil, J. H., et al. 1987. Homology in the region containing a tRNA^{Trp} gene and a (complete or partial)) tRNA^{Pro} gene in wheat mitochondrial and chloroplast genomes. *Curr. Genet.* 12:91–98

140. Meeuse, B. J. D. 1975. Thermogenic respiration in aroids. *Annu. Rev. Plant Physiol.* 26:117–26

141. Meijer, A. J., Van Woerkom, G. M., Eggelte, T. A. 1976. Phthalonic acid an inhibitor of α-oxoglutarate transport in mitochondria. *Biochim. Biophys. Acta* 430:53–61

142. Miernyk, J. A., Camp, P. J., Randall, D. D. 1985. Regulation of plant pyruvate dehydrogenase complexes. *Curr. Top. Plant Biochem. Physiol.* 4:175–90

143. Miernyk, J. A., Fang, T. K., Randall, D. D. 1985. Calmodulin antagonists inhibit the mitochondrial pyruvate dehydrogenase complex. *J. Biol. Chem.* 262:15338–40

144. Miernyk, J. A., Randall, D. D. 1987. Some properties of pea mitochondrial phospho-pyruvate dehydrogenase phosphatase. *Plant Physiol.* 83:311–15

145. Miller, A. J., Sanders, D. 1987. Depletion of cytosolic free calcium induced by photosynthesis. *Nature* 326:397–400

146. Miller, C. O. 1980. Cytokinin inhibition of respiration in mitochondria from six plant species. *Proc. Natl. Acad. Sci. USA* 77:4731–35

147. Mitchell, P. 1966. Chemiosmotic coupling in oxidative and photosynthetic phosphorylation. *Biol. Rev.* 41:445–502

148. Møller, I. M., Kay, C. J., Palmer, J. M. 1984. Electrostatic screening stimulates rate-limiting steps in mitochondrial electron transport. *Biochem. J.* 223:761–67

149. Møller, I. M., Kay, C. J., Palmer, J. M. 1986. Chlortetracycline and the transmembrane potential of the inner membrane of plant mitochondria. *Biochem. J.* 237:765–71

150. Møller, I. M., Lin, W. 1986. Membrane-bound NAD(P)H dehydrogenases in higher plant cells. *Annu. Rev. Plant Physiol.* 37:309–34

151. Møller, I. M., Palmer, J. M. 1981. The inhibition of exogenous NAD(P)H oxidation in plant mitochondria by chelators and mersalyl as a function of pH. *Physiol. Plant.* 53:413–20

152. Møller, I. M., Palmer, J. M. 1981. Properties of the oxidation of exogenous NADH and NADPH by plant mitochondria. Evidence against a phosphatase or a nicotinamide nucleotide transhydrogenase being responsible for NADPH oxidation. *Biochim. Biophys. Acta* 638:225–33

153. Moore, A. L., Akerman, K. E. G. 1982. Ca^{2+} stimulation of the external NADH dehydrogenase in Jerusalem artichoke (*Helianthus tuberosus*) mitochondria. *Biochem. Biophys. Res. Commun.* 109:513–17

154. Moore, A. L., Bonner, W. D. Jr. 1977. The effect of calcium on the respiratory responses of mung bean mitochondria. *Biochim. Biophys. Acta* 460:455–66

155. Moore, A. L., Fricaud, A. C., Dry, I. B., Wiskich, J. T., Day, D. A. 1988. The role of ubiquinone in the regulation of electron transport in a branched respiratory chain. *5th Eur. Bioenerg. Conf. Abstr. Book*, p. 51

156. Moore, A. L., Rich, P. R. 1985. Organization of the respiratory chain and oxidative phosphorylation. See Ref. 63a, pp. 134–72

157. Morgan, A., Maliga, P. 1987. Rapid chloroplast segregation and recombination of mitochondrial DNA in *Brassica* cybrids. *Mol. Gen. Genet.* 209:240–46

158. Mulligan, R. M., Walbot, V. 1986. Gene expression and recombination in plant mitochondrial genomes. *Trends Genet.* 2:263–66

159. Musgrave, M. E., Miller, C. O., Siedow, J. N. 1987. Do some plant responses to cytokinin involve the cyanide-resistant respiratory pathway? *Planta* 172:330–35

160. Musgrave, M. E., Siedow, J. N. 1985. A relationship between cyanide-resistant respiration and plant responses to cytokinins. *Physiol. Plant.* 64:161–66

161. Musgrave, M. E., Antonovics, J., Siedow, J. N. 1987. Is male-sterility in plants related to lack of cyanide-resistant respiration in tissues. *Plant Sci.* 44:7–11

162. Musgrave, M. E., Strain, B. R., Siedow, J. N. 1986. Response of two hybrids to CO_2 enrichment: a test of the energy overflow hypothesis for alternative respiration. *Proc. Natl. Acad. Sci. USA* 83:8157–61

163. Nagel, W. O., Sauer, L. A. 1982. Mitochondrial malic enzymes. Purification and properties of the NAD(P)-

dependent malic enzyme from canine small intestinal mucosa. *J. Biol. Chem.* 257:12405–11

164. Neuburger, M. 1985. Preparation of plant mitochondria, criteria for assessement of mitochondrial integrity and purity, survival *in vitro*. See Ref. 63a, pp. 7–24

165. Neuburger, M., Bourguignon, J., Douce, R. 1986. Isolation of a large complex from the matrix of pea leaf mitochondria involved in the rapid transformation of glycine into serine. *FEBS Lett.* 207:18–22

166. Neuburger, M., Day, D. A., Douce, R. 1984. The regulation of malate oxidation in plant mitochondria by the redox state of endogenous pyridine nucleotides. *Physiol. Veg.* 22:571–80

167. Neuburger, M., Day, D. A., Douce, R. 1984. Transport of coenzyme A in plant mitochondria. *Arch. Biochem. Biophys.* 229:253–58

168. Neuburger, M., Day, D. A., Douce, R. 1985. Transport of NAD^+ in percoll-purified potato tuber mitochondria. Inhibition of NAD^+ influx and efflux by N-4-azido-1-nitrophenyl-4 - aminobutyryl-3'-NAD^+. *Plant Physiol.* 78:405–10

169. Neuburger, M., Douce, R. 1980. Effect of bicarbonate and oxaloacetate on malate oxidation by spinach leaf mitochondria. *Biochim. Biophys. Acta* 589:176–89

170. Neuburger, M., Douce, R. 1983. Slow passive diffusion of NAD^+ between intact isolated plant mitochondria and suspending medium. *Biochem. J.* 216:443–50

171. Newton, K. J., Walbot, V. 1985. Maize mitochondria synthesize organ-specific polypeptides. *Proc. Natl. Acad. Sci. USA* 82:6879–83

172. Nishimura, M., Douce, R., Akazawa, T. 1982. Isolation and characterization of metabolically competent mitochondria from spinach leaf protoplasts. *Plant Physiol.* 69:916–20

172a. Obenland, D., Hiser, C., McIntosh, L., Shibles, R., Stewart, C. R. 1988. Occurrence of alternative respiratory capacity in soybean and pea. *Plant Physiol.* 88:528–31

173. Ohnishi, J., Kanai, R. 1983. Differentiation of photorespiratory activity between mesophyll and bundle sheath cells of C_4 plants. I. Glycine oxidation by mitochondria. *Plant Cell Physiol.* 24:1411–20

174. Oliver, D. J., Walker, G. H. 1984. Characterization of the transport of oxaloacetate by pea leaf mitochondria. *Plant Physiol.* 76:409–13

175. Osmond, C. B. 1981. Photorespiration and photoinhibition. Some implications for the energetics of photosynthesis. *Biochim. Biophys. Acta* 639:77–98

176. Ozias-Akins, P., Pring, D. R., Vasil, I. K. 1987. Rearrangements in the mitochondrial genome of somatic hybrid cell-lines of *Pennisetum americanum* (L.) K. Schum. + *Panicum maximum*. Jacq. *Theor. Appl. Genet.* 74:15–20

177. Packer, L., Murakami, S., Mehard, C. W. 1970. Ion transport in chloroplasts and plant mitochondria. *Annu. Rev. Plant Physiol.* 21:271–304

178. Palmer, J. D., Shields, C. R. 1984. Tripartite structure of the *Brassica campestris* mitochondrial genome. *Nature* 307:437–40

179. Palmer, J. D., Shields, C. R., Cohen, D. B., Orton, T. J. 1983. An unusual mitochondrial DNA plasmid in the genus *Brassica*. *Nature* 301:725–27

180. Palmer, J. M. 1976. The organization and regulation of electron transport in plant mitochondria. *Annu. Rev. Plant Physiol.* 27:133–57

181. Palmer, J. M., Møller, I. M. 1982. Regulation of NAD(P)H dehydrogenases in plant mitochondria. *Trends Biochem. Sci.* 7:258–61

182. Palmer, J. M., Schwitzguebel, J. P., Møller, I. M. 1982. Regulation of malate oxidation in plant mitochondria. Response to rotenone and exogenous NAD^+. *Biochem. J.* 208:703–11

183. Palmer, J. M., Ward, J. A. 1985. The oxidation of NADH by plant mitochondria. See Ref. 63a, pp. 173–201

184. Passam, H. C., Souveryn, J. H. M., Kemp, A. Jr. 1973. Adenine nucleotide translocation in Jerusalem artichoke mitochondria. *Biochim. Biophys. Acta* 305:88–94

185. Passarella, S., Atlante, A., Quagliariello, E. 1985. Oxaloacetate permeation in rat kidney mitochondria: pyruvate/oxaloacetate and malate/oxaloacetate translocators. *Biochem. Biophys. Res. Commun.* 129:1–10

186. Petit, P., Cantrel, C. 1986. Mitochondria from *Zea mays* leaf tissues: differentiation of carbon assimilation and photorespiratory activity between mesophyll and bundle sheath cells. *Physiol. Plant.* 67:442–46

187. Poulsen, L. L., Wedding, R. T. 1970. Purification and properties of the α-ketoglutarate dehydrogenase complex of cauliflower mitochondria. *J. Biol. Chem.* 245:5709–17

188. Pring, D. R., Conde, M. F., Schertz, K. F., Levings, C. S. III. 1982. Plasmid-like DNAs associated with mitochondria

of cytoplasmic male-sterile *Sorghum*. *Mol. Gen. Genet.* 186:180–84

189. Pring, D. R., Levings, C. S. III, Hu, W. W. L., Timothy, D. H. 1977. Unique DNA associated with mitochondria in the "S"-type cytoplasm of male sterile maize. *Proc. Natl. Acad. Sci. USA* 74:2904–8

190. Pring, D. R., Lonsdale, D. M. 1985. Molecular biology of higher plant mitochondrial DNA. *Int. Rev. Cytol.* 97:1–46

191. Quetier, F., Lejeune, B., Delorme, S., Falconet, D. 1985. Molecular organization and expression of the mitochondrial genome of higher plants. See Ref. 63a, pp. 25–36

192. Ragan, C. I. 1978. The role of phospholipids in the reduction of ubiquinone analogues by the mitochondrial reduced nicotinamide-adenine dinucleotide-ubiquinone oxidoreductase complex. *Biochem. J.* 172:539–47

193. Randall, D. D., Givan, C. V. 1981. Subcellular location of NADP$^+$-isocitrate dehydrogenase in *Pisum sativum* leaves. *Plant Physiol.* 68:70–73

194. Ranjeva, R., Carrasco, A., Boudet, A. M. 1988. Inositol triphosphate stimulates the release of calcium from intact vacuoles isolated from *Acer* cells. *FEBS Lett.* 230:137–41

195. Raskin, I., Ehmann, A., Melander, W. R., Meeuse, B. J. D. 1987. Salicylic acid: a natural inducer of heat production in *Arum* lilies. *Science* 237:1601–2

196. Ravanel, P., Tissut, M., Douce, R. 1986. Platanetin: a potent natural uncoupler and inhibitor of the exogenous NADH dehydrogenase in intact plant mitochondria. *Plant Physiol.* 80:500–5

197. Rawsthorne, S., Hylton, C. M., Smith, A., Woolhouse, H. 1988. Photorespiratory metabolism and immunogold localization of photorespiratory enzymes in leaves of C$_4$ and C$_3$-C$_4$ intermediate species of *Moricandia*. *Planta* 173:298–308

197a. Rawsthorne, S., Hylton, C. M., Smith, A., Woolhouse, H. 1988. Distribution of photorespiratory enzymes between bundle sheath and mesophyll cells in leaves of the C$_3$-C$_4$ intermediate species *Moricandia arvensis* (L)DC. *Planta*. In press

197b. Remy, R., Ambard-Bretteville, F., Colas des Francs, C. 1987. Analysis by two-dimensional electrophoresis of the polypeptide composition of pea mitochondria isolated from different tissues. *Electrophoresis* 8:528–32

198. Rich, P. R. 1978. Quinol oxidation in *Arum maculatum* mitochondria and its

application to the assay, solubilisation and partial purification of the alternative oxidase. *FEBS Lett.* 96:252–56

199. Roberts, J. K. M. 1984. Study of plant metabolism *in vivo* using NMR spectroscopy. *Annu. Rev. Plant Physiol.* 35: 375–86

200. Roby, C., Martin, J. B., Bligny, R., Douce, R. 1987. Biochemical changes during sucrose deprivation in higher plant cells. Phosphorus-31 nuclear magnetic resonance studies. *J. Biol. Chem.* 262:5000–7

201. Rothenberg, M., Boeshore, M. L., Hanson, M. R., Izhar, S. 1985. Intergenomic recombination of mitochondrial genomes in a somatic hybrid plant. *Curr. Genet.* 9:615–18

202. Rothenberg, M., Hanson, M. R. 1987. Recombination between parental mitochondrial DNA following protoplast fusion can occur in a region which normally does not undergo intragenomic recombination in parental plants. *Curr. Genet.* 12:235–40

203. Rottmann, W. H., Brears, T., Hodge, T. P., Lonsdale, D. M. 1987. A mitochondrial gene is lost via homologous recombination during reversion of CMS T-maize to fertility. *EMBO J* 6:1541–46

204. Rubin, P. M., Randall, D. D. 1977. Regulation of plant pyruvate dehydrogenase complex by phosphorylation. *Plant Physiol.* 60:34–39

205. Rustin, P. 1987. The nature of the terminal oxidation step of the alternative electron transport pathway. In *Plant Mitochondria*, ed. A. L. Moore, R. B. Beechey, pp. 37–46. New York/London: Plenum

206. Rustin, P., Moreau, F., Lance, C. 1980. Malate oxidation in plant mitochondria via malic enzyme and the cyanide-insensitive electron transport pathway. *Plant Physiol.* 66:457–62

207. Schneider, H., Lemasters, J. J., Hackenbrock, C. R. 1982. Lateral diffusion of ubiquinone in mitochondrial electron transfer. In *Function of Quinones in Energy Conserving Systems*, ed. B. L. Trumpower, pp. 125–39. New York: Academic

208. Schonbaum, G. R., Bonner, W. D., Storey, B. T., Bahr, J. T. 1971. Specific inhibition of the cyanide-insensitive respiratory pathway in plant mitochondria by hydroxamic acids. *Plant Physiol.* 47:124–28

209a. Schuster, W., Brennicke, A. 1987. Plastid, nuclear and reverse transcriptase sequences in the mitochondrial genome of *Oenothera* is genetic information transferred between organelles via RNA? *EMBO J.* 6:2857–63

209b. Sesay, A., Stewart, C. R., Shibles, R. 1988. Cyanide-resistant respiration in light- and dark-grown soybean cotyledons. *Plant Physiol.* 87:655–59

210. Shibasaka, M., Tsuji, H. 1988. Respiratory properties of mitochondria from rice seedlings germinated under water and their changes during air adaptation. *Plant Physiol.* 86:1008–12

211. Siedow, J. N. 1982. The nature of cyanide-resistant pathway in plant mitochondria. *Rec. Adv. Phytochem.* 16:47–84

212. Siedow, J. N., Girvin, M. E. 1980. Alternative respiratory pathway. Its role in seed respiration and its inhibition by propyl gallate. *Plant Physiol.* 65:669–74

213. Solomos, T. 1977. Cyanide-resistant respiration in higher plants. *Annu. Rev. Plant Physiol.* 28:279–97

214. Sorojini, G., Oliver, D. J. 1983. Extraction and partial characterization of the glycine decarboxylase multienzyme complex from pea leaf mitochondria. *Plant Physiol.* 72:194–99

215. Stamper, S. E., Dewey, R. E., Bland, M. M., Levings, C. S. III. 1987. Characterization of the gene *urf 13-T* and unidentified reading frame, ORF25, in maize and tobacco mitochondria. *Curr. Genet.* 12:457–63

216. Stegink, S. J., Siedow, J. N. 1986. Binding to butylgallate to plant mitochondria II. Relationship to the presence or absence of the alternative pathway. *Plant Physiol.* 80:196–201

217. Stern, D. B., Lonsdale, D. M. 1982. Mitochondrial and chloroplast genomes of maize have a 12-kilobase DNA sequence in common. *Nature* 299:698–702

218. Stern, D. B., Palmer, J. D., Thompson, W. F., Lonsdale, D. M. 1983. Mitochondrial DNA sequence evolution and homology to chloroplast DNA in angiosperms. *Plant Mol. Biol.* 1:467–77

219. Storey, B. T. 1976. Respiratory chain of plant mitochondria. XVIII. Point of interaction of the alternate oxidase with the respiratory chain. *Plant Physiol.* 58:521–25

220. Storey, B. T. 1980. Electron transport and energy coupling in plant mitochondria. See Ref. 43a, pp. 125–95

221. Thomas, D. R., Wood, C. 1986. The two β-oxidation sites in pea cotyledons. *Planta* 168:261–66

222. Thomas, D. R., Wood, C., Masterson, C. 1988. Long-chain acyl CoA synthetase, carnitine and β-oxidation in the pea-seed mitochondrion. *Planta* 173:263–66

223. Tobin, A., Djerdjour, B., Journet, E., Neuburger, M., Douce, R. 1986. Effect of NAD$^+$ on malate oxidation in intact plant mitochondria. *Plant Physiol.* 66:225–29

224. Turner, J. F., Turner, D. H. 1980. The regulation of glycolysis and the pentose phosphate pathway. See Ref. 43a, pp. 279–316

225. Van Veldhoven, P. P., Just, W. W., Mannaerts, G. P. 1987. Permeability of the peroxisomal membrane to cofactors of β-oxidation. Evidence for the presence of a pore-forming protein. *J. Biol. Chem.* 262:4310–18

226. Vignais, P. V., Douce, R., Lauquin, G. J. M., Vignais, P. M. 1976. Binding of radioactively labeled carboxyatractyloside, atractyloside and bongkrekic acid to the ADP translocator of potato mitochondria. *Biochim. Biophys. Acta* 440:688–96

227. Vivekananda, J., Beck, C. F., Oliver, D. J. 1988. Monoclonal antibodies as tools in membrane biochemistry. Identification and partial characterization of the dicarboxylate transporter from pea leaf mitochondria. *J. Biol. Chem.* 263:4782–88

228. Walker, J. L., Oliver, D. J. 1986. Glycine decarboxylase multienzyme complex. Purification and partial characterization from pea leaf mitochondria. *J. Biol. Chem.* 261:2214–21

229. Walker, J. L., Oliver, D. J. 1986. Light-induced increases in the glycine decarboxylase multienzyme complex from pea leaf mitochondria. *Arch. Biochem. Biophys.* 248:626–38

230. Willeford, K. O., Wedding, R. T. 1986. Regulation of the NAD malic enzyme from *Crassula. Plant Physiol.* 80:792–95

231. Willeford, K. O., Wedding, R. T. 1987. Evidence for a multiple subunit composition of plant NAD malic enzyme. *J. Biol. Chem.* 262:8423–29

232. Willeford, K. O., Wedding, R. T. 1987. pH Effects on the activity and regulation of the NAD malic enzyme. *Plant Physiol.* 84:1084–87

233. Wilson, S. B. 1988. The switching of electron flux from the cyanide-insensitive oxidase to the cytochrome pathway in mung-bean (*Phaseolus aureus* L.) mitochondria. *Biochem. J.* 249:301–3

234. Wise, R. P., Pring, D. R., Gengenbach, B. G. 1987. Mutation to male fertility and toxin insensitivity in Texas (T)-cytoplasm maize is associated with a

frameshift in a mitochondrial open reading frame. *Proc. Natl. Acad. Sci. USA* 84:2858–62
235. Wiskich, J. T. 1977. Mitochondrial metabolite transport. *Annu. Rev. Plant Physiol.* 28:45–69
236. Wiskich, J. T. 1980. Control of the Krebs cycle. See Ref. 43a, pp. 243–78
237. Wiskich, J. T., Dry, I. B. 1985. The tricarboxylic acid cycle in plant mitochondria: its operation and regulation. See Ref. 63a, pp. 281–313
238. Woo, K. C., Osmond, C. B. 1976. Glycine decarboxylation in mitochondria isolated from spinach leaves. *Aust. J. Plant Physiol.* 3:771–85

239. Yu, C., Huang, A. H. C. 1986. Conversion of serine to glycerate in intact spinach leaf peroxisomes: role of malate dehydrogenase. *Arch. Biochem. Biophys.* 245:125–33
240. Zhu, Q.-S., Beattie, D. S. 1988. Direct interaction between yeast NADH-ubiquinone oxidoreductase, succinate-ubiquinone oxidoreductase, and ubiquinol-cytochrome *c* oxidoreductase in the reduction of exogenous quinones. *J. Biol. Chem.* 263:193–99
241. Zoglowek, C., Krömer, S., Heldt, H. W. 1988. Oxaloacetate and malate transport of plant mitochondria. *Plant Physiol.* 87:109–15

Annu. Rev. Plant Physiol. Plant Mol. Biol. 1989. 40:415–39

STRUCTURE, EVOLUTION, AND REGULATION OF *Rbcs* GENES IN HIGHER PLANTS

Caroline Dean,[1] Eran Pichersky,[2] and Pamela Dunsmuir[1]

[1]Advanced Genetic Sciences, Inc., 6701 San Pablo Avenue, Oakland, California 94608; and [2]Department of Biology, The University of Michigan, Ann Arbor, Michigan 48109-1048

CONTENTS

INTRODUCTION

Ribulose bisphosphate carboxylase is frequently cited as the most abundant protein on earth. It is definitely the most abundant protein in the leaves of light-grown plants. The enzyme RuBPCase catalyzes the primary step in carbon fixation. It is structurally a hetero-16-mer, composed of (*a*) eight small

1040-2519/89/0601-0415$02.00

subunits encoded by nuclear genes and (*b*) eight large subunits that encompass the active site and are encoded in the chloroplast genome (30).

As a result of the prevalence of this protein, and hence of the mRNAs encoding the subunits, the genes for the RbcS small subunits and the large RbcL subunits were the first nuclear (4) and chloroplast protein-coding genes (3) cloned in plants. Tremendous effort has since been devoted to the characterization of these genes in a wide range of plant species. Here we focus upon the *RbcS* genes in higher plants. These genes continue to serve as a model for the nuclear gene families that frequently exist in plant genomes. Research into their organization, structure, evolution, and expression will play an important role in elucidating the basic issues of plant biology.

ORGANIZATION AND STRUCTURE OF THE GENES ENCODING THE SMALL SUBUNIT OF RIBULOSE BISPHOSPHATE CARBOXYLASE

Dicotyledonous Plants

PETUNIA   Eight nuclear genes encode the RbcS polypeptide in *Petunia* (Mitchell) (14, 66). All eight members have been cloned and characterized, although their positions in the petunia genome have not yet been mapped. The genes are divided into three subfamilies or lineages based on nucleotide sequence homology. One subfamily contains six genes, five of which are known to be closely linked in the petunia genome (within 25 kb). The other two subfamilies contain single genes (16) (Table 1). There is 10.2% nucleotide-sequence divergence in the mature small subunit coding regions among the genes of the three different subfamilies; within a subfamily the nucleotide-sequence divergence ranges from 0% to 3%. The amino acid composition of the three different mature small subunits in petunia differ by a maximum of three amino acids. The amino acid differences are all located at, or very close to, the COOH-terminus of the mature RbcS polypeptide, and it is not known whether these substitutions produce functionally distinct proteins.

In general, the nucleotide sequence encoding the transit peptide region of the petunia RBCS precursor polypeptide shows a higher degree of divergence among subfamilies (21% divergence) than the sequence encoding the mature small subunit; one gene contains one additional codon in this region (58 rather than 57) (14).

The coding region of seven of the eight petunia *RbcS* genes is interrupted by two introns (15). The position of these introns is conserved in all the higher-plant *RbcS* genes studied to date. The first intron interrupts the mature small subunit between amino acid residues 2 and 3. The second intron is positioned between amino acid residues 47 and 48. One of the petunia *RbcS* genes, *SSU301*, contains an additional intron, which interrupts the mature

Table 1 *RbcS* gene nomenclature

Species	Locus	On chromosome	Number of genes	Gene designation (proper name)	Gene designation (trivial name[1])	References
Petunia	1	ND	1		*SSU611*	14, 15, 16, 66
	2	ND	1		*SSU301*	
	3	ND	5+1[2]		*SSU511*	
					SSU231	
					SSU112	
					SSU491	
					SSU211	
					SSU911	
Tomato	1	2	1	*RbcS1*		46, 60
	2	3	1	*RbcS2*		
	3	2	3	*RbcS3A*		
				RbcS3B		
				RbcS3C		
Potato[3]	1	ND	1	*RbcS1*	RbcS c	70
	2(1)	ND	1	*RbcS2*	RbcS 1	
	3(2)	ND	3	*RbcS3A*	RbcS 2a	
				RbcS3B	RbcS 2b	
				RbcS3C	RbcS 2c	
Pea	1	5	5		*RbcS-3A*	22, 12, 49, 62
					RbcS-3B	
					RbcS-3.6	
					RbcS-8.0	
					RbcS-E9	
Soybean	ND	ND	>6		*SRS1*	25
					SRS4	
Arabidopsis	1	ND	3		*ats1B*	33, 61
					ats2B	
					ats3B	
	2	ND	1		*ats1A*	
Lemna gibba	ND	ND	>6		*5A*	58, 69
					5B	
					40A	
					40B	
Wheat	ND	ND	>12			9, 57

[1] Trivial name is often derived from the designation of the phage or plasmid clone from which the gene was isolated, or from the size of the specific restriction fragment on which the gene resides.
[2] No linkage data for one gene (SSU491) but likely, from sequence homology, to belong to this locus.
[3] We have arranged the *RbcS* loci in Solanaceae to indicate orthology (i.e. all loci designated as "locus 1" are orthologous to each other, and similarly for "locus 2" and locus 3"). The authors of the potato report designated the potato *RbcS* genes in a manner similar to the tomato designation (with one exception, RbcS c), but without regard to orthology. We therefore chose to rename these loci (original designation in parentheses) and genes (new designations in the "proper name" column, authors' original designations in the "trivial name" column).

small subunit coding region within amino acid residue 65. A third identically positioned intron has now been identified in one of the *RbcS* genes in tomato (60), tobacco (40), and potato (70), all species within the Solanaceae.

All the petunia *RbcS* genes share regions of homology within the nucleotide sequence located 5' and 3' to the coding region (15). Immediately 3' to the coding region there is no homology between genes from different subfamilies. However, within a subfamily this region is highly conserved. All the petunia *RbcS* genes share a 60-bp region of nucleotide-sequence homology in the region preceding the poly-A addition sites. This location suggests the sequence may function in mRNA processing and poly-A addition. The petunia *RbcS* transcripts are polyadenylated at several different sites in vivo (13).

The nucleotide sequence 5' to the coding region also shows a high degree of nucleotide-sequence homology. All of the genes sequenced in this region differ in the untranslated leader regions but show a high degree of nucleotide-sequence homology consisting of large blocks of identical sequence 5' to the transcription start sites in the genes. This region extends to approximately −200 from the transcription start sites. Additional nucleotide sequence homology, located between −400 and −200 is found between the two most highly expressed petunia *RbcS* genes (15). This is shown in the *SSU301* gene in Figure 1.

TOMATO   The *RbcS* genes map to three loci, two in chromosome 2 and one on chromosome 3 (45, 46, 67). It has been shown (60) that locus 1 on chromosome 2 and locus 2 on chromosome 3 contain single genes, whereas locus 3 on chromosome 2 contains three tandemly repeated genes located within a 10-kb region. The tandemly linked genes are closely related and in the terminology used for the petunia genes would be classified as one subfamily (Table 1). The exon sequences of the linked genes diverge by at most only 4.7% in the transit peptide region and 1.6% in the mature small subunit. The *RbcS* genes at the two other loci show a much higher degree of nucleotide-sequence divergence, ranging from 10% to 14%. The three linked genes encode identical mature small subunits, while the mature small subunit proteins encoded by the unlinked genes differ from each other by three amino acid residues, and from the linked genes by four or one residue. A higher degree of sequence divergence occurs within the transit peptide regions.

Four tomato *RbcS* genes contain two introns that interrupt the coding region in the same positions as the first and second introns in the petunia *RbcS* genes. The fifth gene contains an additional intron, also located in the same position as the third intron of the petunia *RbcS* gene (60). The intron sequences of the three clustered *RbcS* genes are more highly related to each other than they are to those of the unlinked genes.

The nucleotide sequence 5' and 3' to the coding region of the five tomato *RbcS* genes shows some nucleotide-sequence homology but far less than that seen in the petunia genes. Manzara & Gruissem, in an excellent review (39), have compared the 5' sequences of all five tomato *RbcS* genes with those from the other sequenced *RbcS* genes. Very few blocks of nucleotide sequence occur in all five tomato genes. The conserved sequences fall into three

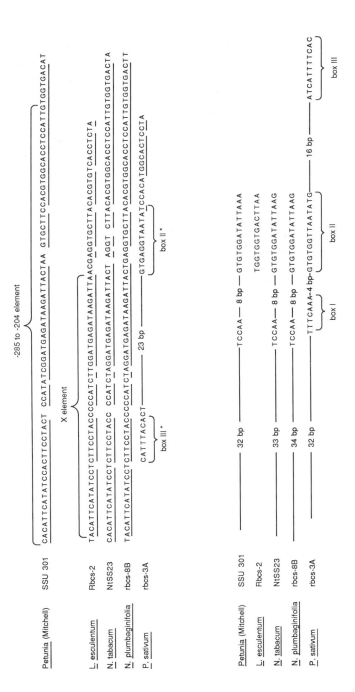

Figure 1 *RbcS* gene structure.

categories: (*a*) sequences present in all *RbcS* genes, (*b*) sequences conserved within one species or closely related species, and (*c*) sequences conserved only between members of a gene subfamily. The sequences that fall into category *a* include the "CAAT" and "TATA" sequences and the sequence 5'-GTGTGGTTAA/CTATG, termed box II, shown in Figure 1. The sequence with a consensus 5'-AAACCTTATCA, positioned upstream of the TATA box, is present in all of the tomato, petunia, and tobacco *RbcS* genes that have been sequenced and is an example of a category *b* sequence. Two of the tomato genes share several sequence elements that are also present in a subfraction of the petunia *RbcS* genes and the two genes sequences from the *Nicotiana* family. These sequences are shown in Figure 1 and represent an example of a sequence from category *c*.

POTATO   Wolter and colleagues (70) have recently analyzed the sequence of four unique *RbcS* genomic clones and one *RbcS* cDNA clone from potato, suggesting that there are at least five *RbcS* genes. The four genomic clones isolated represent two chromosomal loci (1 and 2) containing one and three *RbcS* genes, respectively. The three genes at locus 2 are arranged in a tandem array within a 10-kb region. They are positioned at least 12 kb from the gene at locus 1 (Table 1). The nucleotide sequence of the potato genes has not yet been published; however, the potato RbcS transit peptide sequences and intron number and length in the potato genes are available. The three genes at locus 2 have two introns located in the same position as the petunia and tomato introns. The gene at locus 1 contains three introns that occur in the same place as do the three introns of the petunia and tomato gene.

When the transit peptide regions of a potato and a tomato *RbcS* gene are compared, there is a remarkably high degree of conservation such that the transit peptides differ by only one amino acid (70). This high conservation is interpreted as indicative of an important and specific function for those particular transit peptide regions.

TOBACCO   The *RbcS* genes have been characterized from three *Nicotiana* species: *N. tabacum* (40, 44) (amphidiploid), *N. sylvestris* (48) (one of the progenitors of *N. tabacum*), and *N. plumbaginifolia* (51) (diploid).

Two unlinked *N. tabacum RbcS* genes have been cloned (40). Sequence analysis of one of these revealed a high level of homology to a cDNA clone isolated from *N. sylvestris,* from which it was concluded that the gene originated in the *N. sylvestris* parent of *N. tabacum*. The gene contains three intron sequences, with positions identical to those in the three-intron *RbcS* genes of petunia, tomato and potato. The 5' flanking sequence of the *N. tabacum RbcS* gene shows a high level of sequence homology to the petunia *RbcS* gene *SSU301* within the first 400 bp 5' to the transcription start site.

Additional *N. tabacum RbcS* genes have also been described (44). At least three *RbcS* genes differ from that characterized by Mazur & Chui (40) by virtue of differing restriction sites in the adjacent DNA. Only the 5' end of one of these genes *(TSSU 3.8)* has been cloned, but the sequence of this gene from the translation start site to the second intron is identical to that of *NtSS23*. Upstream of the coding region the sequence is 90% homologous to that of *NtSS23;* then, at approximately −400 from the transcription start site, the homology declines.

Genomic Southern analysis suggests there are at least seven *RbcS* genes in the nuclear DNA of *N. sylvestris* (48). Multiple cDNA clones isolated from *N. sylvestris* have been classified into two groups suggesting that only two families of genes are expressed in leaf tissue. Group I contains cDNA clones with a 73-nucleotide 5' leader sequence. Group II contains clones with a 60-nucleotide 5' leader sequence.

One *RbcS* gene has been characterized from *N. plumbaginifolia* (51). This gene shows a high degree of homology to the *N. tabacum NtSS23* gene and to the *N. sylvestris* cDNA clones. The coding region is interrupted by three introns, the first two sharing a high degree of homology with those of *NtSS23*. The only homology found within the third intron occurs at the intron/exon junctions. The 5' flanking region of the *N. plumbaginifolia RbcS* gene shares a high degree of homology to *RbcS* genes from petunia and tomato (39).

PEA   The *RbcS* gene family is composed of at least five members, which are clustered on chromosome 5 within a region of 4 map units (49) (Table 1). Two of the *RbcS* genes have been localized to a single genomic clone (11).

Unlike the *RbcS* genes of the Solanaceae, the pea *RbcS* genes cannot be divided into subfamilies based on nucleotide sequence homology. There is little nucleotide sequence divergence between the coding regions of the five pea *RbcS* genes, all encode an identical mature small subunit (22), and the only amino acid replacements occur in the transit peptide regions.

All of the pea *RbcS* genes have two introns interrupting the coding sequence, located in the same positions as those in the petunia and tomato genes (11, 12, 22). The second intron is well conserved in all five genes having only six nucleotide substitutions within its 86-bp length (22), while the first intron is more variable.

The 5' untranslated regions of the pea *RbcS* genes are relatively short, ranging from 9 to 22 nucleotides (11, 12, 22). Two of the genes are indistinguishable in the leader regions, and the leader regions of another two genes are homologous. Upstream of the transcription start site in the pea *RbcS* genes there are several blocks of nucleotide sequence homology that occur in all of the genes. These include the "CAAT" and "TATA" boxes, and five boxes I, II, III, II*, and III*, shown in Figure 1 (35). These regions have been found to significantly affect the expression of the *RbcS* genes, an issue that will be discussed later in the review.

Four of the pea *RbcS* genes show approximately 90% homology within the sequences immediately following the translation stop codon. The equivalent sequences from the fifth gene, however, show little sequence homology (22).

SOYBEAN   There are 6–10 genes that encode the RbcS polypeptide in soybean; two have been isolated and characterized (8, 25). *SRS1*, the first soybean *RbcS* gene to be analyzed, has two introns interrupting the coding region between residues 2–3 and 47–48 of the mature small subunit coding region. The two genes are highly related at the nucleotide sequence level (96%, 93%, and 96.5% homology in the three exons); the predicted amino acid composition of proteins encoded by the two genes thus differs at only one residue. The two introns are approximately 75% homologous, and the flanking regions are more than 85% homologous (700 bp at the 5' end and 300 bp at the 3' end). Hybridization studies demonstrated that the flanking homology extends at least 4 kb 5' and 1.1 kb 3' of the coding region. Based upon their extensive homology, it has been proposed that the two genes may be alloalleles or alleles of homeologous loci (25).

ARABIDOPSIS   The *Arabidopsis* genome contains four *RbcS* genes that are divided into two subfamilies based on linkage data and nucleotide sequence homology (33, 61) (Table 1). One subfamily contains three closely linked genes, which share greater than 95% nucleotide sequence and encode the same polypeptide. All four genes contain two introns. The three closely linked genes contain intron sequences that are similar in length and nucleotide sequence but share no homology to the unlinked gene. There is only limited homology in the 5' and 3' flanking regions of the four genes (33).

Monocotyledonous Plants

LEMNA   The *RbcS* gene family in *Lemna gibba* consists of approximately 13 genes (69). Five of these have been isolated and characterized and a sixth is represented by a cDNA clone (58). Two pairs of the five genes show linkage; one set is linked in an inverted orientation with 2.5 kb separating the 5' regions of the genes, and the other pair are in tandem within a 6.5-kb region (69) (Table 1).

It appears that the *RbcS* gene family is made up of several subfamilies in *Lemna*. One subfamily consists of two as yet unisolated genes, which contain sequences homologous to the 3' untranslated region of the isolated cDNA clone. The tandemly linked genes share a high degree of homology and can be considered another subfamily (69). The second, more tightly linked pair are not highly homologous, suggesting that they may be in different subfamilies. Each of the five isolated genes has been partially sequenced. They all contain only one intron, located in the position of intron 2, in the *RbcS* genes from dicotyledonous plants. The exon sequences of the five genes show a high degree of nucleotide-sequence homology such that the six characterized *Lemna* gene-coding sequences would produce an identical small subunit protein (65). There are, however, several differences in the amino acid sequences of the transit peptides.

WHEAT   The hexaploid genome of *Triticum aestivum* contains more than ten *RbcS* genes (57). One genomic clone and four different cDNA clones have been characterized (9, 57). The genomic clone contains only one intron, a feature presently characteristic of monocotyledonous *RbcS* genes. The 289-bp intron interrupts the mature small subunit coding region between amino acid residues 2 and 3. Comparison of the isolated clones indicates nucleotide and predicted amino acid sequence divergence occurs between the different wheat *RbcS* genes. However, since different wheat cultivars have been studied by different groups, it is not clear how much sequence divergence occurs among the genes within a cultivar.

MAIZE   One genomic *RbcS* clone and three distinct *RbcS* cDNA clones have been isolated from *Zea mays* L. (38, 54). Sequence analysis of the genomic clone showed that the coding sequence of the mature *RbcS* subunit is interrupted by a single 163-bp intron, located between amino acid residues 2 and 3. Although the sequences of the three cDNA clones was not determined, the 3' flanking regions of these cDNAs are clearly divergent, because they have been successfully used as gene-specific probes in the analysis of *RbcS* gene expression levels.

There are other species in which *RbcS* cDNA clones have been characterized but where analysis of the genomic clones has not yet been completed or published. These include spinach (63), cucumber (27), *Silene pratensis* (56), *Amaranthus hypochondriacus* (5), *Flaveria trinervia* (1), *Helianthus annuus* (68), and *Sinapis alba* (43).

SUMMARY OF GENE STRUCTURE AND EVOLUTION

Coding Regions for the Mature Peptides

In all of the different plant species analyzed, the mature small subunit coding sequence is more conserved than the transit peptide sequence. One region in the mature *RbcS* coding region that is absolutely conserved in all of the species so far examined is the hexadecapeptide YYDGRYWTMWKLPMFG, located in the petunia small subunit between amino acid residues 61 and 76.

Also relatively well conserved are the 16 amino acids located between positions 102 and 117. In view of the overall relatively high level of interspecific sequence divergence within other parts of the mature small subunit, the conservation of the hexadecapeptide suggests that this region of the polypeptide is important in the structure and/or function of the small subunit. Six of the residues are aromatic, suggesting that this part of the molecule may not be exposed at the surface of the RuBPCase holoenzyme. This sequence might therefore be involved in the binding of the small subunits to the central octameric core of large subunits. Three charged amino acids in the conserved sequence could participate in ionic interactions with charged residues in the large subunit polypeptides. Since the large subunit is relatively conserved among species one might expect that the region of the small subunit polypeptide that associates with the large subunit would also be relatively conserved.

Coding Regions for the Transit Peptides

Variation in both sequence and length of the transit peptide occurs among the species examined. In petunia (14), pea (22), tomato (45, 60), tobacco (40, 44), potato (70), and *Lemna* (69) most genes have transit sequences of 57 amino acids, although one of the genes in each of the Solanaceous species [petunia (14), tomato (45), and potato (70)] has one extra amino acid. In contrast, the soybean (25) genes have transit peptides of 55 amino acids, and in the wheat (9) and maize (38) genes they are either 46 or 47 amino acids.

Concerted Evolution of RbcS Genes within a Genome

In all species where more than one *RbcS* gene sequence has been obtained, data show little or no RbcS protein sequence heterogeneity within the cell. For example, the six *RbcS* genes from *Lemna* sequenced to date encode identical mature proteins (69). The five *RbcS* genes of tomato encode three different mature proteins which, however, differ from each other by at most four amino acids (45, 60). This situation is in contrast to the large differences in amino acid sequences of RbcS proteins among even closely related species. For example, the RbcS sequences of tomato and petunia, both in the Solanaceae family, diverge by as much as 18% (14). From the large number of silent base-pair substitutions among unlinked tomato *RbcS* genes, it has been inferred that selection for protein sequence homogeneity is the mechanism largely responsible for the conservation of protein coding information among such genes (45). This conclusion has been supported by similar analysis of the petunia (14) and potato (70) *RbcS* genes. It is not known at present why there is such a strong selection for RbcS sequence uniformity within the cell. This is especially intriguing in view of the large interspecific sequence variation.

Within a complex *RbcS* locus, mechanisms other than selection seem to be operating to bring about the concerted evolution of the genes. For example, the three genes in the tomato *Rbcs3* locus are very similar in nucleotide sequence as well as in coding information. In fact, *Rbcs-3A* and *Rbcs-3C* are identical throughout the three exons and the two introns (but not in the 5' and 3' noncoding regions), whereas *Rbcs-3B* differs from them in these regions only by a few nucleotides (60). Examination of the patterns of similarities in nucleotide sequences of the three genes in the exons, introns, and surrounding regions has led to the conclusion that the high degree of sequence homologies results from recent gene conversion events rather than recent gene duplications (60).

To further strengthen the suggestions that selection or gene conversion are involved in sequence conservation, identification and comparisons of orthologous and paralogous genes in closely related species are required. For example, it was demonstrated that the nucleotide sequence of the tomato *Rbcs2* gene shows the highest homology to the orthologous gene from *N. tabacum* yet it encodes a protein whose sequence is more homologous to other tomato RbcS proteins (45), thus supporting the conclusion that selection for RbcS protein sequence homogeneity within a genome is operating on the tomato *RbcS* genes. The determination that gene conversion events have recently occurred in the *Rbcs3* locus is based both on the high homology among the three genes (which, however, does not include the 5' and 3' noncoding regions) and on the observation that paralogous genes in this locus in tomato are more similar to one another than to orthologous genes in the petunia genome.

Intron Positions

All *RbcS* genes from dicots sequenced to date have at least two introns that are always found in the same positions (8, 11). In the *RbcS* genes in the monocot species *Lemna gibba,* the first of these two introns is missing (69); in two other monocot species, wheat and corn, the second intron is missing in at least one of their *RbcS* genes (9, 38). One *RbcS* gene in each of the species petunia, tomato, potato, and tobacco, all in the family Solanaceae, contains an additional intron downstream from the second intron. The two *RbcS* genes of the green alga *Chlamydomonas reinhardtii* also possess three introns, although none of them corresponds in exact location to any of the introns in the *RbcS* genes of higher plants (28). It has been suggested that three introns is the ancestral situation in the *RbcS* genes of angiosperms and that the lack of match of the intron positions with the *Chlamydomonas RbcS* genes results from intron "sliding" (70). One argument against this proposal is that *RbcS* genes with three introns have not been observed in plant species outside the Solanaceae, as Wolter and colleagues themselves point out (70).

A stronger argument for the origin of the third intron within Solanaceae or closely related taxa is based on phylogenetic analysis of the *RbcS* genes. The Solanaceae species examined to date possess three subfamilies of *RbcS* genes, which we term subfamilies I, II, and III. These subfamilies represent gene duplications at least as old as the Solanaceae family since the DNA sequences of coding and noncoding regions (although not the coding information; see the previous section) of genes from different species of the same subfamily are more similar to each other than to other *RbcS* genes from the same species (but from different subfamilies) (14, 45, 70). In tomato, the three subfamilies are found at distinct genetic loci termed *Rbcs1*, *Rbcs2*, and *Rbcs3*, respectively; and this is probably true in petunia as well.

The gene with the additional intron in the four Solanaceae species examined to date is the single gene of subfamily II. Sequence analysis of coding regions has shown that subfamily I diverged first from the rest of the *RbcS* genes and that subfamilies II and III represent a more recent split. Since *RbcS* genes in subfamilies I and III have two introns, it follows that the third intron in the subfamily II genes is a recent introduction. It thus appears that two is the ancestral number of introns in the *RbcS* genes of angiosperms and that both gain and loss of introns have occurred since the split between the monocots and dicots. Because both tomato and petunia, species representing the most divergent taxa in the Solanaceae, possess the three *RbcS* subfamilies, it is possible that these three lineages originated before the formation of the family. If so, the acquisition of a third intron in one lineage of *RbcS* genes may also have occurred before that time. These hypotheses can be tested by looking at the *RbcS* genes of species in families related to Solanaceae.

Number of Genes and Genetic Loci

The entire complement of *RbcS* genes has been isolated and fully sequenced only in tomato and petunia. Tomato has a total of five *RbcS* genes; petunia, however, has eight *RbcS* genes. Both petunia and tomato have two *RbcS* loci (subfamilies I and II) containing a single gene. The third locus in tomato (subfamily III) contains three genes (45, 60), whereas the orthologous locus in petunia contains six (14, 16). Thus, there is evidence in the Solanaceae for recent gene duplication or deletion (or both) within an *RbcS* locus. In pea, the only species outside the Solanaceae where gene cloning has been combined with formal genetic analysis, the five *RbcS* genes are found in a single locus (49). Thus, during angiosperm evolution, there has also been duplication or loss (or both) of *RbcS* loci as well as loss and gain of individual genes. The exact number of *RbcS* genes in other species has not been determined [except for *Arabidopsis* where there are four *RbcS* genes (33)], but estimates range from two to eight or nine for diploid species. It should be borne in mind when considering estimates of the number of *RbcS* genes in *N. tabacum*, soybean, wheat, and *Lemna* that these species are polyploids.

EXPRESSION OF THE *Rbcs* GENES IN THEIR ENDOGENOUS STATE AND AFTER TRANSFORMATION INTO HETEROLOGOUS SPECIES

Several techniques have been used to analyze the expression of *RbcS* genes. As a result, discrepancies have arisen in the assignment of relative expression levels to different genes as measured by different groups. The techniques used include Northern and slot blot hybridization with gene-specific probes or oligonucleotides; analysis of frequency of cloning in a cDNA library to approximate relative abundance of the transcript in the original RNA sample; primer extension analysis; and S1 or RNase protection experiments. The last two techniques are more accurate because the hybridization reactions are carried out in solution, and a large excess of probe can be added so that the hybridization reaction is driven to completion. Hence differences in the hybridization efficiencies of different probes do not complicate the analysis.

Dicotyledonous Plants

PETUNIA   The eight genes of *Petunia* (Mitchell) are all expressed in petunia leaf tissue albeit to very different levels. The relative expression of the genes, as assayed using gene-specific probes to slot blots and looking at the relative frequency of cloning in a cDNA library, was measured as: *SSU301*, 47.3%; *SSU611*, 23.2%; *SSU491*, 7%; *SSU112*, 5.4%; *SSU911*, < 0.5%; *SSU231* and *SSU511* (also known as 11A), 15.2%; *SSU211*, 1.9% (15). The ranking of the relative expression levels was confirmed using a primer extension analysis (13) after conflicting data were published (66). Using oligonucleotide probes to Northern blots, Tumer and colleagues (66) found that the *RbcS* gene *SSU11A* (equivalent to *SSU511*) was expressed at a 10-fold higher level than *SSU8* (equivalent to *SSU301*). Dean et al have speculated about the basis of the discrepancy (13).

The total *RbcS* mRNA levels were found to vary 500-fold in six organs of petunia: leaves, sepals, petals, stems, roots, and stigma/anthers (15). No significant differential expression was detected among the petunia *RbcS* genes in different organs, with *SSU301* showing the highest levels of steady-state mRNA in each of the organs examined. Stayton et al (57a) have recently shown that petunia *RbcS* steady-state mRNA levels do not vary diurnally. During this analysis they also showed that the *SSU301* and *SSU611* transcripts remained at a very high level for up to 3 days after transfer of the petunia plants to the dark. Expression of members of the 51 subfamily, however, declined much more rapidly after transfer of the plants to darkness.

The basis of the quantitative differences in the expression of the petunia *RbcS* genes has been investigated by Dean et al (12a, b). They have focused upon the most strongly expressed *(SSU301)* and the most weakly expressed

(SSU911) petunia *RbcS* genes and have shown by transfer of these genes to tobacco that the determinants of the quantitative differences in expression are intrinsic to the genes and their surrounding sequences (approximately 2 kb 5' and 2 kb 3').

They also demonstrated, by analyzing hybrid *SSU301* and *SSU911* genes in transgenic tobacco plants, that sequences both 5' and 3' to the translation initiation codon contribute to the quantitative differences in expression of the genes. They investigated which sequences 3' to the start of the coding region caused these effects by constructing a series of fusions that exchanged the sequences of *SSU301* (the three-intron *RbcS* gene) and *SSU911* (which has two introns) between each intron sequence and immediately following the end of the coding region. They also constructed an *SSU301* gene that completely lacked introns. The relative expression levels of the different fusions were assayed in transgenic tobacco plants, and the results indicated that sequences downstream of the coding region, sequences within introns, and also possibly an interaction between 3' flanking and intron sequences, contributed to the differential expression of the *RbcS* genes. Subsequently, nuclear run-on experiments indicated that the sequences downstream of the coding region effected transcription rate.

The sequences 5' to the coding region that contributed to the quantitative differences in expression of the *RbcS* genes have also been investigated (12b). The promoter regions of the two most abundantly expressed petunia *RbcS* genes, *SSU301* and *SSU611,* show sequence homology not present in the other petunia genes. Furthermore, this homology is also present in two of the tomato *RbcS* genes, *RbcS2* and *RbcS-3A* (60); the *N. tabacum NtSS23* gene (40), and the *N. plumbaginifolia RbcS-8B* gene (51) (see Figure 1). The significance of these and other sequences was investigated by adding specific regions of the *SSU301* promoter (the most strongly expressed gene) to equivalent regions in the *SSU911* promoter (the most weakly expressed gene). The expression of the constructs was then analyzed in transgenic tobacco plants. In this way, an *SSU301* promoter fragment (80 bp) was identified that acted as an "enhancer-like" element. This fragment contained the promoter sequences also present in the *SSU611* petunia gene and in the tomato and tobacco genes.

TOMATO   Gene-specific oligonucleotides complementary to the 3' flanking region of the genes were used to probe Northern blots to assess the relative expression of the five tomato genes (59, 60). All five genes are expressed in leaf tissue, and the relative transcript abundance of the different genes varies by less than 4-fold. *RbcS* mRNA was not detected in roots and ripe tomato fruit and was expressed in stem, immature fruit, and etiolated seedlings to only 3.2%, 6.5%, and 4.0%, respectively, of the level found in leaf tissue.

The tomato *RbcS* genes, in contrast to those in petunia, differ in their expression-level ranking in different tissues (59).

When tomato plants were transferred to the dark the transcript levels of *RbcS1, RbcS-3B,* and *RbcS-3C* decreased to undetectable levels within 12 hr, whereas transcript levels of *RbcS2* and *RbcS-3A* only decreased to 33% and 40%, respectively (39). In these experiments *RbcS1* showed an unusual pattern of expression. It was not expressed in the dark in mature leaves, but it was expressed in the dark in etiolated seedlings. This implies that *RbcS1* expression in the dark is under developmental and/or organ-specific control. Although *RbcS1* expression in independent of light in cotyledon tissue, it is strictly light-regulated in mature leaf tissue (39).

The tomato *RbcS* genes do show small fluctuations in their level of steady-state mRNA in tomato fruit during a diurnal cycle. Two other non-light-regulated genes also showed evidence for diurnal variation in their expression in fruit, indicating that whatever mechanism controls these cycles is not limited to light-regulated genes.

Manzara & Gruissem (39) tried to correlate the presence of certain promoter elements with different patterns of expression of the tomato *RbcS* genes. One clear correlation is the presence of the X-bp element (shown in Figure 1) in *RbcS2* and *RbcS-3A* with a very slow reduction in steady-state mRNA levels when the tomato plants were transferred to the dark. The same sequence element has been shown to contribute to quantitative differences in expression among the petunia genes in leaf tissue. The significance of this element in the light-regulated expression of the tomato *RbcS* genes is currently being investigated.

TOBACCO   The *RbcS-8B* gene of *N. plumbaginifolia* is expressed predominantly in leaves. An S1 analysis using 5' probes showed that mRNA levels of *RbcS-8B* in stems and roots were 50–100-fold lower than that in leaves (51). The expression of the gene in leaves decreased to a low level when plants were transferred to total darkness for three days.

The expression of an *RbcS-8B*:CAT fusion was examined in transgenic petunia, *N. tabacum,* and *N. plumbaginifolia* (51). The fusion contained *RbcS-8B* sequences from −1038 to +32 and 3' sequences from a pea *RbcS* gene *E9*. This fusion showed light-regulated, tissue-specific expression of CAT enzyme that followed the expression of the endogenous *RbcS-E9* gene. Deletion analysis of the −1038 to +32 5' fragment showed that it contained two enhancer-like elements. A proximal element, located between −312 and −102, confers organ-specific and light-inducible expression. A distal element, located between −1038 and −589, has enhancer-like characteristics if a basic promoter element is present on the fusion (in this case the −90 to −46 region of the cauliflower mosaic virus 35S promoter). The distal element confers organ specificity on the heterologous promoter, but the enhanced transcription in leaves is insensitive to light (50).

PEA S1 nuclease protection experiments have been used to examine the relative expression levels of the pea *RbcS* genes. Originally these were done using 5' probes (12), then subsequently repeated and extended using 3' probes (22). The *RbcS* transcripts from genes *3A, 3C,* and 8.0 represent more than 90% of the total *RbcS* expression in leaf tissue. The *RbcS-E9* and *3.6* genes are expressed at lower levels, but the maximum difference in expression between any of the two genes in leaf tissue is less than 10-fold. Individual genes showed slight differences in their relative expression patterns in different organs, but the relative ranking is the same in all the organs examined (22).

Fluhr & Chua (20) investigated the photoresponses of the pea *RbcS* genes *3A* and *3C.* Their results demonstrated that the photoresponses of two pea genes *RbcS-3A* and *3C* depend on the developmental stage of the pea leaves and also that a blue-light photoreceptor is required in addition to phytochrome. The blue-light effect could be overcome by a subsequent far-red pulse in mature pea leaves. This effect may well be at the transcriptional level, because a far-red pulse has been shown to rapidly decrease *RbcS* transcription in soybeans (7).

In order to characterize the *cis*-acting elements that control the expression of the pea *RbcS* genes, intact genes and gene fusions have been introduced into transgenic plants. Initially the *RbcS-E9* gene was introduced into petunia protoplasts, and its expression was assayed in petunia callus (10). The expression of the introduced gene was 50-fold higher in the light than in the dark, even though the expression of the gene in petunia calli was significantly lower than that in pea leaves. The first *RbcS* fusion analyzed in transgenic plant cells contained the 5' region of the pea *RbcS-3.6* gene fused to the coding region of chloramphenicol acetyltransferase. This was introduced into tobacco cells and also assayed in callus tissue (29). The expression of the fusion was light dependent and occurred only in chloroplast-containing transformed tissue.

Further characterization of the *cis*-acting sequences required for light induction was carried out by Morelli and colleagues (41) and Timko and coworkers (62). The two groups analyzed 5' deletion mutants of the pea *RbcS* genes, *RbcS-E9* (41), and *RbcS-3.6* (62), respectively in transformed petunia and tobacco calli. The former group (41) concluded that a 33-bp sequence close to the "TATA" box of the gene was sufficient to confer light inducibility. They also identified an upstream region (between −1052 to −352) needed for maximal expression. The second group (62) showed that a fragment extending from -973 to -90 from the pea $RbcS$-3.6 gene transcription start site contained sequences involved in the photoregulation of the gene. This fragment also conferred light-inducible expression to a heterologous promoter. The fragment functioned in either orientation 5' to the gene but did not work when it was positioned 3' to the coding region. Comparison of the two results suggested that elements responsible for light-inducible expression may be duplicated in the $RbcS$ promoters. The results also confirmed the observation that the light-regulation of expression of $RbcS$ expression is mediated primarily through transcriptional effects (23).

Transformed petunia or tobacco calli serve as convenient systems in which to analyze the expression of deletion mutants of $RbcS$ genes because many hundreds of independently transformed cells (minicalli) can be pooled for tissue analysis. This avoids the problem of between-transformant variability (31) in expression of the introduced gene, which compromises the quantitative description of gene transfer experiments. There are, however, many drawbacks to assaying $RbcS$ gene expression in callus tissue. First, the genes are expressed to only a fraction of the level achieved in leaves. Second, the light induction of $RbcS$ expression is much slower in dark-adapted callus than it is in dark-adapted leaves (5–10 days, compared to hours), and a pulse of red light has no effect on $RbcS$ mRNA levels in calli, precluding experiments on phytochrome-mediated responses (21). Third, the callus is grown on a carbon source with phytohormones, both of which can affect $RbcS$ expression. In experiments comparing the expression levels of deletion mutants of $RbcS$-$E9$, there were major discrepancies between results obtained with transformed calli and those obtained with transgenic plants. Nagy and coworkers (42) have analyzed the expression of the intact $RbcS$-$E9$ pea gene in regenerated petunia and tobacco plants and demonstrated that a deletion mutant of $RbcS$-$E9$ with only 352 bp of 5' upstream sequence shows photoinducible and leaf-specific expression. In callus, a 5' deletion from -352 and -35 of the $RbcS$-$E9$ gene had little effect on already low expression levels in transformed calli. In contrast, a similar deletion caused a large decrease in expression of both $RbcS$-$E9$ and $RbcS$-$3A$ in transgenic plants. For the expression of two other intact pea genes, $RbcS$-$3A$ and $3C$, in regenerated petunia plants it was shown that all the photoresponses of the two genes in pea were retained after transfer to petunia (20). This included the more complete far-red reversibility of $RbcS$-$3A$ than $RbcS$-$3C$ and the blue-light effects.

An enhancer-like element that confers light inducibility and tissue specificity upon a truncated 35S promoter from cauliflower mosaic virus is located between -327 and -48 in the pea $RbcS$-$3A$ promoter (21, 42). This element conferred all the phytochrome-mediated and blue-light responses. It has also been shown to confer the correct cell-type expression, namely high-level expression in mesophyll cells, expression only in guard cells in the epidermis, and expression in chlorenchyma and phloem parenchyma regions of the stem (2). Similarly an element from the pea *RbcS-3.6* gene has been shown to confer mesophyll cell-specific expression (55).

Further analysis of the promoter region of the pea *RbcS* gene *3A* has identified a 58-bp sequence, from −169 to −112 [containing boxes I, II and II (Figure 1)], that contains two regulatory elements that decrease transcription in the dark (35). Boxes II and III can function as transcritional silencers in the dark. Positive light responsive elements (LREs) were also found to overlap with these boxes, giving some insight into the complexity of the regulation of these genes. As removal of this 58-bp fragment from the pea *RbcS-3A* gene still leaves a fully functional, light-inducible promoter, Kuhlemeier and coworkers (35, 36) concluded that additional LREs are located both further upstream and downstream in the promoter. Two boxes homologous to boxes II and III, designated II* and III* (Figure 1) and located at approximately −220 from the transcription start site, may act as these additional LREs. The effects of mutations in boxes II and III could only be measured when sequences upstream of −170 were removed. Sequences upstream of −170 appear to be dispensable in mature leaves of a normal plant. In young, expanding leaves at the top of a plant as well as in seedlings, the upstream elements are also required to achieve high-level expression (34). In contrast to the petunia *RbcS* genes, it appears that sequences within the introns and 3' flanking region of the pea *RbcS* genes are not required for maximum expression (37).

The factors that interact with boxes II and III in the pea *RbcS-3A* promoter have been analyzed (26). Using gel retardation assays and DNAse I footprinting experiments, a factor has been identified, designated GT-1, that binds to boxes II and III and II* and III*. Two adjacent G residues in both boxes II and III were shown to be critical for GT-1 binding. Mutation of the GG dinucleotide in box II to CC eliminated the ability of box II to function as an enhancer element when adjacent to a heterologous promoter. Single G residues present in boxes III and III* have also been shown to be important for GT-1 binding. Since GT-1 is present in nuclear extracts from leaves of light-grown and dark-adapted pea plants, its regulatory role does not depend on de novo synthesis. It is possible that GT-1 binds differently in light and dark, causing the different expressions. Posttranslational modification—e.g. phosphorylation or interaction with another factor—may allow or prevent binding in vivo. A second protein factor (GBF) has been identified in nuclear extracts obtained from tomato and *Arabidopsis*. It binds to sequence motifs distinct from boxes II and III (24).

POTATO   Expression of four of the five potato genes has been examined in haploid plants from tissue culture (P. Schreier, personal communication). The potato *RbcS* gene that contains three introns, *RbcS1*, is the most highly expressed in all situations examined, including older leaves, young leaves before and after ethylene treatment, and in stem tissue. Two of the genes, *RbcS1* and *2A*, are induced by light more readily than *RbcS-2B* and *2C;* and a blue light receptor appears to be required in addition to phytochrome since only a combination of blue and red light results in maximal *RbcS* expression.

AMARANTH   Berry and colleagues (5, 6) have examined *RbcS* and *RbcL* expression upon changing illumination of amaranth cotyledons. When dark-grown cotyledons were transferred to light, synthesis of RbcS and RbcL polypeptides started rapidly, before any increase in levels of corresponding mRNAs. Similarly, when light-grown cotyledons were transferred to darkness, the synthesis of the polypeptides rapidly declined without concomitant changes in mRNA levels. Using in vivo pulse-chase experiments and in vitro translation experiments, these investigators showed that the changes in protein levels were not caused by alterations in the mRNA translatability in vitro or by changes in protein turnover. They concluded from their results that the expression of the *RbcS* genes (and *RbcL*) is controlled in vivo at both the transcriptional and translational levels.

Expression of a fusion between the soybean *RbcS* gene *SRS1* and the reporter gene neomycin phosphotransferase *(NPTII)* transferred to soybean callus appeared to show the expected light induction (18). Similar experiments using an *Arabidopsis RbcS* gene *Ats1-A* have shown that the 5' flanking region (1550 bp) can confer light-regulated and tissue-specific expression upon gene fusions (61). Investigations into the expression of mustard *RbcS* genes have demonstrated that there is a feedback loop from normal plastids necessary for *RbcS* expression (43). Experiments using norfluorazon, a herbicide that blocks carotenoid biosynthesis and thus indirectly arrests plastid development, have shown that normal plastid function plays a crucial role in the activation of pea *RbcS* expression (55).

Monocotyledonous Plants

LEMNA   The Northern blot technique employing gene-specific probes from the 3' untranslated regions of six of the *Lemna RbcS* genes was used to analyze the expression of the gene family in different parts of the plant and from plants exposed to different light conditions. The gene represented by the cDNA clone pLgSSU is the most abundantly expressed gene. *SSU40A* and *40B* are the least abundantly expressed of the six genes (65).

After five days in the dark, *RbcS* expression in *Lemna* is minimal; however, expression of the genes *LgSSU, SSU5A, SSU5B,* and *SSU26* increased rapidly in response to a 1-min exposure to red light. The expression of *SSU40A* and

40B could only be detected in plants given repeated red illumination (2 min to 8 hr). Although the expression of *SSU5B* and *SSU40A* was undetectable in root tissue at the level of steady-state mRNA levels, nuclear run-on transcription experiments suggested these genes were abundantly transcribed in roots. These data suggest that the organ specificity in expression, at least for these two genes, is controlled in part by posttranscriptional processes (64). Flores & Tobin previously proposed that posttranscriptional processes account for the increase in steady-state levels of *RbcS* mRNA after addition of cytokinin to dark-grown Lemna (19).

MAIZE   There is significant *RbcS* expression in dark-grown leaf material (53); however, the level of expression increases upon illumination of the plants. Sheen & Bogorad have investigated the expression of *RbcS* genes in bundle-sheath and mesophyll cells of maize (54). Transcripts of all three *RbcS* genes were detected in bundle sheath cells and mesophyll cells of etiolated maize leaves, but the levels of each transcript differ in etiolated mesophyll cells. Upon illumination, the levels of all three transcripts drop in mesophyll cells but follow a pattern of transitory rise and fall in bundle-sheath cells. Transcripts from two of the genes account for greater than 80% of the *RbcS* mRNA in 24-hr greening maize leaves. The third gene accounts for approximately 10%.

WHEAT   Expression of a wheat *RbcS* gene from its own promoter was not detected after transfer to tobacco. Expression of the gene from the CaMV 35S promoter resulted in the accumulation of several types of transcripts, both spliced and unspliced, all polyadenylated at multiple novel sites in the wheat 3' flanking region (32). This appears to be the result of inefficient splicing of monocotyledonous transcripts in dicotyledonous plants.

Summary

Many features of *RbcS* gene expression are common to genes from different species. Surprisingly, there are also many features that distinguish *RbcS* gene expression in different species.

All *RbcS* genes in one species are expressed, at least in leaf tissue. Even the *N. tabacum* gene *TSSU3.2*, which contains a translational stop codon in the transit peptide coding region, is an expressed gene (44). The *RbcS* gene families that have been characterized with respect to their expression in different light qualities [pea (27), tomato (39), and potato (70)] also show common characteristics, namely that the genes are regulated by phytochrome when at the seedling stage, either etiolated or young green, but that after the plants have matured, some if not all the genes are also regulated via the blue-light receptor. A subfraction of the *RbcS* genes from tomato (39) and petunia (57a) also show a very slow decline in mRNA level after the plants have been transferred to darkness. Most of the *RbcS* genes, from all the species examined, do not show a diurnal rhythm in expression in leaves at the steady-state mRNA level. The exception to this is the example of *RbcS* expression in tomato fruit, where the genes show only a small diurnal fluctuation (47).

The conservation of the sequence element, termed box II (35, 42), in the promoter regions of all the different *RbcS* genes indicates the possible importance of this sequence in the regulation of *RbcS* gene expression. It is surprising that *RbcS* genes do not share more sequence elements. *RbcS* gene expression in all species is highly coordinated with plastid development. Elucidation of the "plastidic" factor that affects *RbcS* expression will be an important step in fully understanding the regulation of *RbcS* expression.

The clear differences between the expressions of *RbcS* genes in different species include the different expression patterns in different organs. In petunia (15) and pea (22), the different genes show small differences in relative expression in different organs but the ranking of expression levels remains the same. In tomato, however, different genes show maximal expression in different organs and at different developmental stages (59). The other clear difference between species is the extent of quantitative differences in expression of the individual *RbcS* genes in leaf tissues. In petunia the maximum range in expression is 100-fold (15); it is approximately 4-fold in tomato (60) and 10-fold in pea (22). Significant differences also occur in *RbcS* expression in dark-grown tissue of different species. Dark-grown wheat and maize leaves show high levels of *RbcS* expression (53). This is also true for tomato (39) and amaranth (5), whereas dark-grown or dark-adapted (for at least 5 days) pea (22) and petunia (15) leaf tissues show very low levels of *RbcS* expression.

The location of sequences that contribute to the quantitative differences in steady-state mRNA levels also differ between species. In petunia, it is clear that sequences both 5' and 3' to the coding region affect transcription of the *SSU301* gene (12a, b). In pea, however, Kuhlemeier and coworkers have shown that quantitative differences in expression are determined exclusively by sequences 5' to the coding region (35).

ACKNOWLEDGMENTS

Many thanks to Eleanor Crump, Joyce Hayashi, Lily Kruger, and Diana Wyles for assistance in manuscript preparation.

Literature Cited

1. Adams, C. A., Babcock, M., Leung, F., Sun, S. M. 1987. Sequence of a ribulose-1,5-bisphosphate carboxylase/ oxygenase cDNA from the C_4 dicot *Flaveria trinervia*. *Nucleic Acids Res.* 15:1875–78

2. Aoyagi, K., Kuhlemeier, C., Chua, N.-H. 1988. The pea *rbcS-3A* enhancer-like element directs cell-specific expression in transgenic tobacco. *Mol. Gen. Genet.* In press

3. Bedbrook, J. R., Smith, S. M., Ellis, R. J. 1980. Molecular cloning and sequencing of cDNA encoding the precursor to the small subunit of chloroplast ribulose-1,5-bisphosphate carboxylase. *Nature* 287:692–97

4. Bedbrook, J. R., Coen, D. M., Beaton, A. R., Bogorad, L., Rich, A. 1979. Location of the single gene for the large subunit of ribulosebisphosphate carboxylase on the maize chloroplast chromosome. *J. Biol. Chem.* 254:905–10

5. Berry, J. O., Nikolau, B. J., Carr, J. P., Klessig, D. F. 1985. Transcriptional and post-transcriptional regulation of ribulose-1,5-bisphosphate carboxylase gene expression in light- and dark-grown amaranth cotyledons. *Mol. Cell Biol.* 5:2238–46

6. Berry, J. O., Nikolau, B. J., Carr, J. P., Klessig, D. F. 1986. Translational regulation of light-induced ribulose-1,5-bisphosphate carboxylase gene expression in amaranth. *Mol. Cell Biol.* 6:2347–53

7. Berry-Lowe, S. L., Meagher, R. B. 1985. Transcriptional regulation of a gene encoding the small subunit of ribulose-1,5-bisphosphate carboxylase in soybean tissue is linked to the phytochrome response. *Mol. Cell Biol.* 5:1910–17

8. Berry-Lowe, S. L., McKnight, T. D., Shaw, D. M., Meagher, R. B. 1982. The nucleotide sequence, expression and evolution of one member of a multigene family encoding the small subunit of ribulose-1,5-bisphosphate carboxylase in soybean. *J. Mol. Appl. Gen.* 2:483–98

9. Broglie, R., Coruzzi, G., Lamppa, G., Keith, B., Chua, N.-H. 1983. Structural analysis of nuclear genes coding for the precursor to the small subunit of wheat ribulose-1,5-bisphosphate carboxylase. *Bio/Technol.* 1:55–61

10. Broglie, R., Coruzzi, G., Fraley, R. T., Rogers, S. G., Horsch, R. B., et al. 1984. Light-regulated expression of a pea ribulose-1,5-bisphosphate carboxylase small subunit gene in transformed plant-cells. *Science* 224:838–43

11. Cashmore, A. R. 1983. Nuclear genes encoding the small subunit of ribulose-1,5-bisphosphate carboxylase. In *Genetic Engineering of Plants*, ed. M. Kosuge, A. Hollander, pp. 29–38. New York: Plenum

12. Coruzzi, G., Broglie, R., Edwards, C., Chua, N.-H. 1984. Tissue-specific and light-regulated expression of a pea nuclear gene encoding the small subunit of ribulose-1,5-bisphosphate carboxylase. *EMBO J.* 3:1671–79

12a. Dean, C., Favreau, M., Band-Nutter, D., Bedbrook, J., Dunsmuir, P. 1988. Sequences downstream of translation start effect expression of petunia *RbcS* genes. *Plant Cell* 1. In press

12b. Dean, C., Favreau, M., Bedbrook, J., Dunsmuir, P. 1988. Sequences 5' to the translation start regulate expression of petunia *RbcS* genes. *Plant Cell* 1. In press

13. Dean, C., Favreau, M., Dunsmuir, P., Bedbrook, J. 1987. Confirmation of the relative expression levels of the *Petunia* (Mitchell) *rbcS* genes. *Nucleic Acids Res.* 15:4655–68

14. Dean, C., van den Elzen, P., Tamaki, S., Black, M., Dunsmuir, P., Bedbrook, J. 1987. Molecular characterization of the *rbcS* multi-gene family of *Petunia* (Mitchell). *Mol. Gen. Genet.* 206:465–74

15. Dean, C., van den Elzen, P., Tamaki, S., Dunsmuir, P., Bedbrook, J. 1985. Differential expression of the eight genes of the petunia ribulose bisphosphate carboxylase small subunit multigene family. *EMBO J.* 4:3055–61

16. Dean, C., van den Elzen, P., Tamaki, S., Dunsmuir, P., Bedbrook, J. 1985. Linkage and homology analysis divides the eight genes for the small subunit of petunia ribulose 1,5-bisphosphate carboxylase into three gene families. *Proc. Natl. Acad. Sci. USA* 82:4964–68

17. Deleted in proof

18. Facciotti, D., O'Neal, J. K., Lee, S., Shewmaker, C. K. 1985. Light-inducible expression of a chimeric gene in soybean tissue transformed with *Agrobacterium*. *Bio/Technol.* 3:241–46

19. Flores, S., Tobin, E. M. 1986. Benzyladenine modulation of the expression of two genes for nuclear-encoded chloroplast proteins in *Lemna gibba*: ap-

parent post-transcriptional regulation. *Planta* 168:340–49

20. Fluhr, R., Chua, N.-H. 1986. Developmental regulation of two genes encoding ribulose-bisphosphate carboxylase small subunit in pea and transgenic petunia plants: phytochrome response and blue-light induction. *Proc. Natl. Acad. Sci. USA* 83:2358–62

21. Fluhr, R., Kuhlemeier, C., Nagy, F., Chua, N.-H. 1986. Organ-specific and light-induced expression of plant genes. *Sciences* 232:1106–12

22. Fluhr, R., Moses, P., Morelli, G., Coruzzi, G., Chua, N.-H. 1986. Expression dynamics of the pea *rbcS* multigene family and organ distribution of the transcripts. *EMBO J.* 5:2063–71

23. Gallagher, T. F., Ellis, R. J. 1982. Light-stimulated transcripts of genes for two chloroplast polypeptides in isolated pea leaf nuclei. *EMBO J.* 1:1493–98

24. Giuliano, G., Pichersky, E. Malik, V. S., Timko, M. P., Scolnik, P. A., Cashmore, A. 1988. An evolutionary conserved protein binding sequence upstream of a plant light-regulated gene. *Proc. Natl. Acad. Sci. USA* 85:7089–93

25. Grandbastien, M. A., Berry-Lowe, S., Shirley, B. W., Meagher, R. B. 1986. Two soybean ribulose-1,5-bisphosphate carboxylase small subunit genes share extensive homology even in distant flanking sequences. *Plant Mol. Biol.* 7:451–65

26. Green, P. J., Kay, S. A., Chua, N.-H. 1987. Sequence-specific interactions of a pea nuclear factor with light-responsive elements upstream of the *rbcS-3A* gene. *EMBO J.* 6:2543–49

27. Greenland, A. J., Thomas, M. V., Walden, R. M. 1987. Expression of two nuclear genes encoding chloroplast proteins during early development of cucumber seedlings. *Planta* 170:99–110

28. Goldschmidt-Clermont, M. 1986. The two genes for the small subunit of RuBP carboxylase/oxygenase are closely linked in *Chlamydomonas reinhardtii*. *Plant Mol. Biol.* 6:13–21

29. Herrera-Estrella, L., Van den Broeck, G., Maenhaut, R., Van Montagu, M., Schell, J. 1984. Light-inducible and chloroplast-associated expression of a chimaeric gene introduced into *Nicotiana-tabacum* using a Ti-plasmid vector. *Nature* 310:115–20

30. Jensen, R. G., Bahr, J. T. 1977. Ribulose 1,5-bisphosphate carboxylase-oxygenase. *Annu. Rev. Plant Physiol.* 28:379–400

31. Jones, J. D. G., Dunsmuir, P., Bedbrook, J. 1985. High-level expression

of introduced chimaeric genes in regenerated transformed plants. *EMBO J.* 4:2411–18

32. Keith, B., Chua, N.-H. 1986. Monocot and dicot pre-mRNAs are processed with different efficiencies in transgenic tobacco. *EMBO J.* 5:2419–25

33. Krebbers, E., Seurinck, J., Herdies, L., Cashmore, A. R., Timko, M. P. 1988. Four genes in the two diverged subfamilies encode the ribulose-1,5-bisphosphate carboxylase small subunit polypeptides of *Arabidopsis thaliana*. *Proc. Natl. Acad. Sci. USA*. In press

34. Kuhlemeier, C., Cuozzo, M., Green, P. J., Goyvaerts, E., Ward, K., Chua, N.-H. 1988. Localization and conditional redundancy of regulatory elements in the pea *rbcS-3A* gene. *Proc. Natl. Acad. Sci. USA* 85:4662–66

35. Kuhlemeier, C., Fluhr, R., Green, P. J., Chua, N.-H. 1987. Sequences in the pea *rbcS-3A* gene have homology to constitutive mammalian enhancers but function as negative regulatory elements. *Genes & Dev.* 1:247–55

36. Kuhlemeier, C., Green, P. J., Chua, N.-H. 1987. Regulation of gene expression in higher plants. *Annu. Rev. Plant Physiol.* 38:221–57

37. Kuhlemeier, C., Fluhr, R., Chua, N.-H. 1988. Upstream sequences determine the difference in transcript abundance of pea *rbcS* genes. *Mol. Gen. Genet.* 212:405–11

38. Lebrun, M., Waksman, G., Fressinet, G. 1987. Nucleotide sequence of a gene encoding corn ribulose-1,5-bisphosphate carboxylase/oxygenase small subunit (*rbcS*). *Nucleic Acids Res.* 15:4360–64

39. Manzara, T., Gruissem, W. 1988. Organization and expression of the genes encoding ribulose-1,5-bisphosphate carboxylase in higher plants. *Photosyn. Res.* 16:117–39

40. Mazur, B. J., Chui, C. 1985. Sequence of a genomic DNA clone for the small subunit of ribulose bisphosphate carboxylase-oxygenase from tobacco. *Nucleic Acids Res.* 13:2373–86

41. Morelli, G., Nagy, F., Fraley, R. T., Rogers, S. G., Chua, N.-H. 1985. A short conserved sequence is involved in the light-inducibility of a gene encoding ribulose 1,5-bisphosphate carboxylase small subunit of pea. *Nature* 315:200–4

42. Nagy, F., Morelli, G., Fraley, R. T., Rogers, S. G., Chua, N.-H. 1985. Photoregulated expression of a pea *rbcS* gene in leaves of transgenic plants. *EMBO J.* 4:3063–68

43. Oelmuller, R., Dietrich, G., Link, G., Mohr, H. 1986. Regulatory factors in-

volved in gene-expression (subunits of ribulose-1,5-bisphosphate carboxylase) in mustard (*Sinapis alba* L.) cotyledons. *Planta* 169:260–66

44. O'Neal, J. K., Pokalsky, A. R., Kiehne, K. L., Shewmaker, C. K. 1987. Isolation of tobacco SSU genes: characterization of a transcriptionally active pseudogene. *Nucleic Acids Res.* 15:8661–77

45. Pichersky, E., Bernatzky, R., Tanksley, S. D., Cashmore, A. R. 1986. Evidence for selection as a mechanism in the concerted evolution of *Lycopersicon esculentum* (tomato) genes encoding the small subunit of ribulose-1,5-bisphosphate carboxylase/oxygenase. *Proc. Natl. Acad. Sci. USA* 83:3880–84

46. Pichersky, E., Bernatzky, R., Tanksley, S. D., Malik, V. S., Cashmore, A. R. 1987. Genomic organization and evolution of the *rbcS* and *cab* gene families in tomato and other higher plants. In *Tomato Biotechnology*, ed. D. J. Nevins, R. A. Jones, pp. 229–38. NY: Alan Liss

47. Piechulla, B., Gruissem, W. 1987. Diurnal mRNA fluctuations of nuclear and plastid genes in developing tomato fruits. *EMBO J.* 6:3593–99

48. Pinck, M., Dore, J.-M., Guilley, E., Durr, A., Pinck, L., et al. 1986. A simple gene-expression system for the small subunit of ribulose bisphosphate carboxylase in leaves of *Nicotiana sylvestris*. *Plant Biol.* 7:301–9

49. Polans, N. O., Weeden, N. F., Thompson, W. F. 1985. Inheritance, organization, and mapping of *rbcS* and *cab* multigene families in pea. *Proc. Natl. Acad. Sci. USA* 82:5083–87

50. Poulsen, C., Chua, N.-H. 1988. Dissection of 5' upstream sequences for selective expression of the *Nicotiana plumbaginiflolia rbcS-8B* gene. *Mol. Gen. Genet.* In press

51. Poulsen, C., Fluhr, R., Kauffman, J. M., Boutry, M., Chua, N.-H. 1986. Characterization of an *rbcS* gene from *Nicotiana plumbaginifolia* and expression of an *rbcS-CAT* chimeric gene in homologous and heterologous nuclear background. *Mol. Gen. Genet.* 205: 193–200

52. Deleted in proof

53. Sheen, J.-Y., Bogorad, L. 1985. Differential expression of the ribulose bisphosphate carboxylase large subunit gene in bundle sheath and mesophyll cells of developing maize leaves is influenced by light. *Plant Physiol.* 79:1072–76

54. Sheen, J.-Y., Bogorad, L. 1986. Expression of the ribulose-1,5-bisphos-

phate carboxylase large subunit gene and three small subunit genes in two cell types of maize leaves. *EMBO J.* 5:3417–22

55. Simpson, J., Van Montagu, M., Herrera-Estrella, L. 1986. Photosynthesis-associated gene families: differences in response to tissue-specific and environmental factors. *Science* 233:34–38

56. Smeekens, S., Van Oosten, J., de Groot, M., Weisbeek, P. 1986. Silene cDNA clones for a divergent chlorophyll-*a/b*-binding protein and a small subunit of ribulose bisphosphate carboxylase. *Plant Mol. Biol.* 7:433–40

57. Smith, S. M., Bedbrook, J., Speirs, J. 1983. Characterization of three cDNA clones encoding different mRNA's for the precursor to the small subunit of wheat ribulose bisphosphate carboxylase. *Nucleic Acids Res.* 11:8719–34

57a. Stayton, M., Brosio, P., Dunsmuir, P. 1988. Circadian expression of photosynthetic genes. *Plant Physiol.* In press

58. Stiekema, W. J., Wimpee, C. F., Tobin, E. M. 1983. Nucleotide sequence encoding the precursor of the small subunit of ribulose-1,5-bisphosphate carboxylase from *Lemna gibba* L. G-3. *Nucleic Acids Res.* 11:8051–61

59. Sugita, M., Gruissem, W. 1987. Developmental, organ-specific, and light-dependent expression of the tomato ribulose-1,5-bisphosphate carboxylase small subunit gene family. *Proc. Natl. Acad. Sci. USA* 84:7104–8

60. Sugita, M., Manzara, T., Pichersky, E., Cashmore, A., Gruissem, W. 1987. Genomic organization, sequence analysis and expression of all five genes encoding the small subunit of ribulose-1,5-bisphosphate carboxylase/oxygenase from tomato. *Mol. Gen. Genet.* 209:247–56

61. Timko, M. P., Herdies, L., de Almeida, E., Cashmore, A. R., Leemans, J., Krebbers, E. 1988. Genetic engineering of nuclear-encoded components of the photosynthetic apparatus in *Arabidopsis*. In *The Impact of Chemistry on Biotechnology: Multidisciplinary Discussion*, ed. M. Phillips, S. Shoemaker, R. Middlekauff, R. Ottenbrite. Washington, DC: Am. Chem. Soc.

62. Timko, M. P., Kausch, A. P., Castresana, C., Fassler, J., Herrera-Estrella, L., et al. 1985. Light regulation of plant gene-expression by an upstream enhancer-like element. *Nature* 318:579–82

63. Tittgen, J., Hermans, J., Steppuhn, J., Jansen, T., Jansson, C., et al. 1986.

Isolation of cDNA clones for fourteen nuclear-encoded thylakoid membrane-proteins. *Mol. Gen. Genet.* 204:258–65

64. Tobin, E. M., Buzby, J., Karlin-Neumann, G. A., Kehoe, D. Naderi, M., et al. 1988. Expression and regulation of light-harvesting chlorophyll *A*/*B*-proteins and the small subunits of ribulose-1,5-bisphosphate carboxylase. *J. Cell Biochem.* S12C:144

65. Tobin, E. M., Wimpee, C. F., Karlin-Neumann, G. A., Silverthorne, J., Kohorn, B. D. 1985. Phytochrome regulation of nuclear gene expression. In *Molecular Biology of the Photosynthetic Apparatus,* ed. K. E. Steinback, S. Bonitz, C. J. Arntzen, L. Bogorad, pp. 373–80. New York: Cold Spring Harbor Lab.

66. Tumer, N. E., Clark, W. G., Tabor, G. J., Hironaka, C. M., Fraley, R. T., Shaw, P. M. 1986. The genes encoding the small subunit of ribulose-1,5-bisphosphate carboxylase are expressed differentially in petunia leaves. *Nucleic Acids Res.* 14:3325–42

67. Vallejos, C. E., Tanksley, S. D., Bernatzky, R. 1986. Localization in the tomato genome of DNA restriction fragments containing sequences homologous to the rRNA (45S), the major chlorophyll *a*/*b*-binding polypeptides and the ribulose 1,5-bisphosphate carboxylase genes. *Genetics* 112:93–105

68. Waksman, G., Freyssinet, G. 1987. Nucleotide sequence of a cDNA encoding the ribulose-1,5-bisphosphate carboxylase/oxygenase from sunflower *(Helianthus annus). Nucleic Acids Res.* 15:1328–34

69. Wimpee, C. F., Stiekema, W. J., Tobin, E. M. 1983. Sequence heterogeneity in the RuBP carboxylase small subunit gene family of *Lemna gibba.* In *Plant Molecular Biology,* pp. 391–401. New York: Liss

70. Wolter, F. P., Fritz, C. C., Willmitzer L., Schell, J. 1988. *rbcS* genes in *Solanum tuberosum:* conservation of transit peptide and exon shuffling during evolution. *Proc. Natl. Acad. Sci. USA* 85: 846–50.

Annu. Rev. Plant Physiol. Plant Mol. Biol. 1989. 40:441–470

THE DEVELOPMENT OF HERBICIDE RESISTANT CROPS

Barbara J. Mazur and S. Carl Falco

Agricultural Products Department, Experimental Station, E. I. du Pont de Nemours & Co., Wilmington, Delaware 19880–0402

CONTENTS

INTRODUCTION

Genetics in Agriculture

The impressive increases in crop productivity achieved over the last several decades have resulted from genetic improvements in crop cultivars and from advances in agricultural technology and management practices. The nongenetic improvements have included improved weed, disease, and insect control through the use of crop protection chemicals, as well as better mechanization, increased supplies of water and nitrogen, and optimization of planting densities.

The introduction of genetic improvements into crops through breeding, a part of agriculture for thousands of years, was more an art than a science until

441

1040-2519/89/0601-441$02.00

the work of Mendel revealed the rules of inheritance. Genetic engineering in agriculture, which in a broad sense refers to any practice that leads to the development of improved cultivars, began with the application of this knowledge. Genetic traits have been introduced into modern cultivars that improve their insect and disease resistance, harvesting and processing qualities, adaptation to particular environmental conditions, yield (through traits such as heterosis and lodging resistance), and nutritional qualities (through the development of low-glucosinilate and low–erucic acid lines of oilseed rape).

To accelerate and improve crop breeding processes further, a number of new genetic approaches that require increasing degrees of technological sophistication have been developed in the last decade. These methods have included improved seed and pollen mutagenesis techniques, mutagenesis of plant cells in culture, regeneration of plants from cultured cells, plant protoplast fusion, and plant transformation. Mutagenesis procedures have been used to generate new desirable traits, while protoplast fusion has provided a mechanism for moving preexisting desirable traits into crops across species barriers. These approaches have required that efficient screens or genetic selections be developed in order to identify the rare cells expressing the new trait.

With the advent of recombinant DNA technology, it has become possible to transfer specific and well-characterized traits across the broadest evolutionary boundaries. As a result, the term genetic engineering has been reserved for the isolation, amplification, and in vitro manipulation of genes. Transformation methodologies have enabled the subsequent reintroduction of the genes into living cells or organisms. Although a number of technological hurdles remain, plant transformation will clearly be a powerful and broadly applicable methodology that will make crop breeding faster, more predictable, and more far-reaching.

For a number of technical and practical reasons, resistance to herbicides was among the first traits to which these new genetic approaches were applied. In some cases, a specific target of herbicide action had been identified through physiological and biochemical studies. In other cases, genetic studies had shown that resistance to a herbicide was a dominant trait exhibiting the simple Mendelian inheritance pattern of a mutation in a single nuclear gene. Dominance makes genetic selection of herbicide-resistant mutants or transformants easier. The potential utility of herbicide-resistance genes as dominant selectable genetic markers for research in plants, in a manner analogous to that for antibiotic-resistance genes in bacteria, has provided an incentive for research. In addition to these technical considerations, the agronomic importance of herbicides has been a major driving force behind the development of herbicide-resistant plants.

Here we review recent work in the development of herbicide-resistant crops. A list of the herbicides for which these efforts have been initiated, along with their structures and their primary targets, is given in Table 1. We present examples of the introduction of herbicide resistance into crop plants by several different techniques, the status of development of herbicide-resistant cultivars, and future prospects for such work. A number of recent reviews have covered mechanisms of herbicide action (21, 61, 64) and the genetic engineering of herbicide-resistant plants (13, 89, 104).

Herbicides in Agriculture

Even those whose experience with weeds has been limited to hand cultivating a home garden can appreciate the usefulness and efficiency herbicides provide for commercial agriculture. Herbicide treatments are an integral part of modern agriculture because they provide cost-effective increases in agricultural productivity. Increased yields result from reduced weed competition for water, light, and nutrients. In addition, crop quality often improves in the absence of contaminating weed seeds, such as wild mustard seeds in harvested canola or wild garlic seeds in wheat. (Weed seed contamination can result in off-flavors following processing, and therefore in reduced remuneration for the grower.) Herbicides can also aid soil conservation efforts through no-till agricultural practices, wherein herbicides rather than tillage are used to reduce weed populations prior to planting.

Herbicides have traditionally been discovered by screening novel compounds in a series of increasingly specific tests. Compounds are first tested for activity against a spectrum of weeds and lack of activity against targeted crops. Promising compounds are further tested in more extensive greenhouse screens and finally in small-scale field trials. Development candidates must then run the gauntlet of acute and chronic toxicology tests. To be commercially successful, herbicides must have potent biological activity against a broad spectrum of weeds and at the same time be nontoxic to crop plants, mammals, and invertebrates; have relatively short soil residual properties; and have favorable production costs. It is becoming increasingly difficult to identify new compounds that meet these criteria and can compete with the many excellent existing products. In the 1950s about 1 in 2000 screened compounds were commercialized; by the 1970s, the rate had dropped to approximately 1 in 7000 compounds, while in the 1980s hardly 1 in 20,000 compounds emerged from these screens.

Selective toxicity of herbicides to weeds but not to crops is one of the most difficult properties to achieve, as might be expected from the biological relatedness of weeds and crops. Selectivity is a function of the physico-

Table 1 Herbicide structures & targets

Herbicide	Inhibited Pathway	Primary Target	Structure
Chlorsulfuron (Sulfonyl-urea)	Branched-chain amino acid biosynthesis	acetolactate synthase	
Imazapyr (Imidazo-linone)	Branched-chain amino acid biosynthesis	acetolactate synthase	
Glyphosate	Aromatic amino acid biosynthesis	5-enolpyruvyl-shikimate-3-phosphate synthase	
Phosphino thricin	Glutamine biosynthesis	glutamine syn-thase	
Atrazine (Triazine)	Photosynthesis	Q_β protein	
Bromoxynil	Photosynthesis	Q_β protein	

chemical properties of a compound, and of the biochemical interactions of the compound with the crop and the weeds. For example, herbicides that do not percolate beyond the top soil layer, and thus do not affect crop roots that extend below this layer, can provide selectivity. Environmental conditions such as climate, soil pH, and soil organic content influence these interactions. For some compounds, management practices (e. g. timing and/or site of the application) can be used to impart selectivity. Glyphosate, a nonselective herbicide, is used prior to planting as a substitute for tillage, thereby taking advantage of its rapid and wide-spectrum weed killing activity and short lifetime in the soil.

A number of important classes of herbicides (e.g. the triazines, sulfonyl-ureas, and imidazolinones) are more toxic to weeds than to specific crops. In these examples, selectivity results from a unique or enhanced metabolic detoxification of the herbicide by the crop plant but not by the weed. In other cases, herbicide selectivity results from the sequestering of the herbicide within an internal compartment of the crop plant. Alternatively, external barriers such as plant cuticles can prevent penetration of the herbicide. In some cases it has been possible to achieve selectivity by seed coat applications of a "safener," a second chemical that reduces the toxicity of the herbicide to the crop.

Genetic modification of crops to make them herbicide resistant could remove a major factor in determining the choice of herbicides available for use by farmers. It could allow for the wider use of more effective herbicides with broader weed-control spectrums. In addition, compounds that are effective at low application rates, have short lifetimes in the soil, and have more favorable toxicological properties might become more generally useful if the constraint of crop selectivity were removed. These possibilities would extend beyond major acreage crops to minor acreage crops that have not yet benefited from effective weed control compounds.

Genetic modification could also complement and enhance existing herbicide selectivity by increasing the margin of safety for selective compounds, particularly during periods of environmental stress when plant metabolism is reduced. It could also allow the grower to increase the application rates for selective herbicides, thus leading to improved and/or wider-spectrum weed control for these compounds. The introduction of herbicide resistance into crops could give the grower greater flexibility in choosing crops for rotations or double crop plantings. (Such choices are currently limited by the differential sensitivies of crops to particular herbicides.) Thus the combination of crop protection chemical technology and genetic technology will provide a new range of management options for more effective weed control.

PLANTS RESISTANT TO AMINO ACID BIOSYNTHESIS INHIBITORS

Sulfonylureas and Imidazolinones

MODE OF ACTION   The sulfonylureas are a broad class of compounds, many of which have herbicidal activity (Table 1). The sulfonylurea herbicides are notable for their low use rates (as low as 2 grams per hectare) and their low toxicity to animals. Thousands of analogues of the sulfonylureas have been synthesized and screened to find compounds selectively active against weeds. A number of such herbicides have been identified and commercialized. Selective toxicity in all of these cases results from the metabolic detoxification of the herbicide by a particular crop species. For example, in wheat the compound chlorsulfuron is hydroxylated by a cytochrome P-450 enzyme and then glycosylated (116). In soybeans, conjugation with homoglutathione plays a major role in detoxification of chlorimuron ethyl, as is the case for the nonsulfonylurea herbicides acifluorfen and metribuzin (42, 43).

Selective sulfonylurea herbicide toxicity has also been achieved by genetic modification of crops. The first example of a sulfonylurea herbicide–resistant mutant plant was obtained by selection of tobacco cells in tissue culture and the subsequent regeneration of mutant plants from the cell lines. Resistance was semi-dominant and segregated as a single nuclear gene (16). It has also been possible to isolate resistant mutants of other crop plants (e.g. soybeans) by seed mutagenesis, but considerable effort and time have been required (101). A third approach has been to engineer sulfonylurea-resistant crops through plant transformation. For this effort, the gene(s) responsible for the resistance trait, either the native genes in the metabolic detoxification pathway or the mutant genes identified by selection, had to be isolated. The use of microbial species as models has played a central role in the identification and isolation of the necessary plant genes.

Studies on the mode of action of the sulfonylurea herbicide chlorsulfuron showed that inhibition of cell division in plant tissue was an early response to treatment (91, 92). This result, along with the ability to isolate resistant mutants in cultured plant cells, suggested that the herbicide antagonized a single basic cellular function and encouraged the use of microbial models to investigate herbicide action. Physiological studies in *Salmonella typhimurium* suggested that the target of the sulfonylurea herbicide sulfometuron methyl was the enzyme acetolactate synthase [ALS (E.C.4.1.3.18); also known as acetohydroxy acid synthase, AHAS], which is required for the synthesis of isoleucine, leucine, and valine (see Figure 1; 65). Multiple ALS isozymes exist in the enterobacteria *S. typhimurium* and *Escherichia coli,* and it was shown that ALS II and ALS III, but not ALS I, are inhibited by sulfometuron methyl (66). In vitro analyses of ALS activity from yeast, pea, tobacco, and *Chlamydomonas* demonstrated that the eukaryotic enzymes are very sensitive to sulfometuron methyl (15, 37, 49, 93).

Proof that the sulfonylurea herbicides act by inhibition of ALS came from a combination of genetic and biochemical studies. Sulfonylurea-resistant mutants of *S. typhimurium, Saccharomyces cerevisiae, Nicotiana tabacum,* and *Arabidopsis thaliana* were isolated. Most of the mutants from each organism produced ALS activity insensitive to the herbicide, and the resistant enzyme activity cosegregated with cellular resistance in genetic crosses (15, 37, 51, 65). The bacterial and yeast mutations were mapped to the loci of the ALS structural genes *ilvG* and *ILV2,* respectively (37, 65). The identification of ALS as the target of the sulfonylurea herbicides provided at least a partial explanation of their low toxicity to animals: Animals lack this enzyme and must obtain the branched-chain amino acids from their diets.

ALS is also the target of two other structurally distinct classes of herbicides, the imidazolinones (Table 1; 83, 105) and the triazolopyrimidines or sulfonanilides (52, 63). Thus, ALS may be a particularly susceptible target for herbicides. It has been demonstrated that the toxicity of sulfometuron methyl to bacteria is enhanced by the accumulation of an ALS substrate, 2-ketobutyrate, which is itself toxic. It has therefore been suggested that the deficiency of branched-chain amino acids and the increase in concentration of the toxic intermediate combine to make ALS a particularly good target for herbicides (67). However, no evidence that 2-ketobutyrate is accumulated in and/or is toxic to plant cells has been reported.

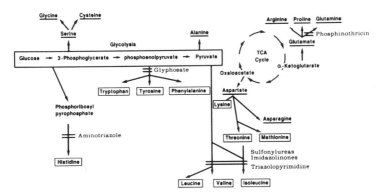

Figure 1   Herbicide targets in amino acid biosynthetic pathways. The figure shows the pathways for amino acid biosynthesis and the points in these pathways where herbicides or classes of herbicides can interrupt amino acid synthesis. Boxes indicate essential amino acids.

BIOCHEMICAL AND GENETIC CHARACTERIZATION OF ALS   Genetic and biochemical studies have provided strong evidence that ALS enzymes from enteric bacteria are tetramers containing two subunits. ALS I purified from *E. coli* is composed of two large 60 kD and two small 9.5 kD subunits (34). Molecular cloning and DNA sequencing have provided a physical characterization of the *ilvBN* operon that encodes these subunits (44, 123). Similarly, ALS II has been purified from *S. typhimurium* and shown to be composed of two large 59 kD and two small 9.7 kD subunits. DNA sequence analysis of the cloned *E. coli ilvGMEDA* operon, along with amino acid sequence analysis of purified ALS II, has demonstrated that the *ilvG* and *ilvM* genes encode these subunits (68, 100). Genetic studies first suggested that ALS III was specified by two genes, designated *ilvI* and *ilvH* (26). DNA sequence analysis again provided physical evidence for the existence of the two genes, organized as an operon (110). Purification of ALS III provided confirmation of an analogous subunit structure for this isozyme (6). Although the early genetic studies on ALS III suggested a role for the small subunit in controlling the sensitivity of the enzyme to valine feedback inhibition, more recent work has indicated an important role for the small subunit in enzyme activity. Mutational inactivation of the small subunit genes resulted in a 20- to greater than 100-fold reduction in ALS activity (35, 72, 111).

Comparison of the inferred amino acid sequences of the bacterial ALS enzymes has indicated a common evolutionary origin. This similarity is most obvious among the large-subunit polypeptides, which share about 40% amino acid sequence identity concentrated in three regions of the protein (see Figure 2). Amino acid sequence conservation, although still evident, is less extensive between the small-subunit genes (44, 110, 123).

The molecular cloning and DNA sequence analysis of the yeast ALS gene, designated *ILV2*, permitted a comparison of the ALS amino acid sequence from a eukaryotic source with those from *E. coli* (37, 38). Similarity between the 687-amino-acid-long *ILV2*-encoded polypeptide and each of the *E. coli* large-subunit polypeptides was as extensive as that between the *E. coli* polypeptides (38). The most striking structural difference between the deduced yeast and bacterial proteins was the presence of an approximately 90-amino-acid-long amino terminal sequence extension on the former that was absent from the latter. Because sub-cellular fractionation experiments (98) had shown yeast ALS to be localized in mitochondria and because the amino acid sequence in the region has characteristics common to other known mitochondrial transit sequences (31), it was suggested that this region might include a mitochondrial transit sequence (38).

HERBICIDE RESISTANCE MUTATIONS   Cloned yeast and bacterial ALS genes were used to investigate the molecular basis for resistance to the

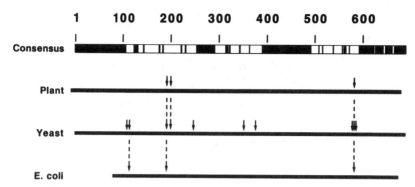

Figure 2 Comparison of ALS amino acid sequences. The top bar shows a schematic representation of homologies between *E. coli* ALS I, II, and III, yeast ALS, and ALS from the plants tobacco and *Arabidopsis*. The white boxes show stretches of amino acids that are conserved between all enzymes; the black boxes depict nonconserved regions. The arrows intersecting the horizontal lines show positions in these sequences where substitutions can produce a herbicide-insensitive enzyme. Dashed vertical lines indicate positions where such substitutions occur in more than one species.

sulfonylurea herbicides. Overexpression of the yeast ALS enzyme by about 4-fold occurred when the gene was present on a high-copy-number plasmid; this overexpression resulted in a 4–5-fold increase in the minimal concentration necessary to inhibit growth (37). An increase in the minimal inhibitory concentration of a sulfonylurea herbicide for *E. coli* was also observed when a functional *E. coli ilvG* gene was present on a high-copy-number plasmid (124). Mutations that resulted in the production of sulfonylurea-resistant ALS were isolated in the cloned yeast *ILV2* and *E. coli ilvG* genes by genetic selection. DNA sequencing showed that each mutant gene contained a single nucleotide change, resulting in a single amino acid substitution in the ALS protein. In yeast ALS, pro192 was substituted with ser, and in *E. coli* ALS II ala26 was changed to val (124).

Genetic engineering of yeast ALS was used to discover other mutations that resulted in sulfonylurea herbicide resistance. Spontaneous mutations in the yeast *ILV2* gene were isolated and characterized by DNA sequencing to determine the amino acid substitutions responsible for herbicide resistance (40). This analysis showed 24 different amino acid substitutions, at 10 different sites, ranging from the amino to the carboxy ends of the protein (Table 2, Figure 2).

The amino acid residues present in the wild-type yeast enzyme at these 10 sites are also present in the three wild-type *E. coli* ALS isozymes at most sites (40). One of the exceptions to this generalization occurs in *E. coli* ALS II, where a serine residue is present at the site analogous to that where the proline

Table 2   Herbicide resistant yeast ALS mutants

Wild type amino acid residue	Amino acid substitutions resulting in resistance
116G	S N
117A	P S T I L V N Q D E K R H W F Y M
192P	A S V Q E R W Y
200A	T V D E R W Y C
251K	P T N D E
354M	V K C
379D	G P S V N E W
583V	A N Y C
586W	G A S I L V N E K R H Y C
590F	G L N R C

to serine substitution had resulted in herbicide resistance in yeast. Since *E. coli* ALS II was known to be more resistant to the sulfonylurea herbicide sulfometuron methyl than were the yeast or plant enzymes (39), it was plausible that this serine residue contributed to the increased resistance. This hypothesis was tested by converting the serine codon in the ALS II gene of *E. coli* to a proline codon by site-directed mutagenesis; this change indeed increased the sensitivity of the enzyme to the herbicide. In another example, *E. coli* ALS I, which is much more resistant to the sulfonylurea herbicides than is ALS II, had a glutamine residue at one site analogous to that where substitutions for the wild-type tryptophan residue had resulted in herbicide-resistant yeast ALS, and a serine residue at a second site where substitutions for the wild-type alanine residue had resulted in herbicide resistance in both yeast ALS and *E. coli* ALS II. Conversion of the glutamine residue in *E. coli* ALS I to tryptophan also resulted in increased sensitivity to the herbicides (9). Thus these residues appear to contribute to inhibition by sulfonylureas in all ALS enzymes. The herbicide-resistant plant enzymes for which sequence information is available also have amino acid substitutions at these same sites (see below).

Site-directed mutagenesis was used to expand the spectrum of amino acid substitutions at the ten sites in yeast ALS (40). At some of these sites, such as ala117, pro192, and trp586, nearly any substitution for the wild-type amino acid resulted in a herbicide-resistant enzyme. At other sites, only a few substitutions had that result. A list of sulfonylurea herbicide–resistant yeast ALS mutants is shown in Table 2.

PLANT ALS GENES   Efforts to use these microbial genes to generate herbicide-resistant transgenic plants faced a number of difficulties and uncertainties. Two of these derived from the knowledge that plant ALS is localized in the chloroplast (58, 79) and that bacterial ALS is composed of two different subunits. Thus the generation of herbicide-resistant plants using bacterial genes might have required not only expression of the two protein subunits of the enzyme, but also their translocation into and assembly in plant chloroplasts. Since ALS had not been purified from yeast, neither the subunit structure of the enzyme nor the amino terminus of the mature protein present in the mitochondria was known, adding considerable uncertainty to any effort to express that enzyme in plants. For these reasons the isolation and use of plant ALS genes appeared to be a more attractive route for engineering sulfonylurea-resistant transgenic plants.

Even before the conservation of amino acid sequences between yeast and bacterial ALS enzymes had been discovered, hybridization between the yeast and *Salmonella ilvG* genes had been detected under low-stringency conditions (75). Together, these observations led to an attempt to detect ALS genes from other species, using heterologous DNA hybridization. A segment of the yeast ALS gene that spanned most of the coding region was used as a probe. Hybridization was detected between the yeast ALS gene and genomic DNA libraries from the cyanobacterium *Anabaena* 7120 and the higher plants *A. thaliana* and *N. tabacum;* ALS genes were isolated from all three species (76).

DNA sequence analysis of the *Arabidopsis* and *Nicotiana* ALS genes indicated that neither gene has introns, and that they code for proteins of 667 and 670 amino acids, respectively. The deduced ALS protein sequences are similar throughout most of their length; approximately 75% of the nucleotides and 85% of the encoded amino acids are identical (76). The 5' ends of the coding sequences, which are the only regions that are not highly conserved between the two plant ALS sequences, appear to encode chloroplast transit sequences. The nucleotide sequences of these regions are more similar than are the deduced amino acid sequences, suggesting that there are few constraints on the amino acid sequences in the transit peptides. Comparison of the deduced amino acid sequences of the plant ALS genes with those of the yeast and the three *E. coli* ALS genes showed that all share the same three conserved domains (Figure 2). At the ten sites where substitutions result in sulfonylurea-resistant yeast ALS, the amino acid residues present in the plant ALS enzymes are identical to those found in the wild-type yeast enzyme.

The number of ALS genes present in *N. tabacum* and *A. thaliana* was determined by Southern blot hybridization analyses. Homologous cloned ALS genes were used as probes. A single ALS gene hybridized to the probe in *Arabidopsis,* while two hybridized in *N. tabacum* (an allotetraploid) (76). The identification of two ALS genes in tobacco was consistent with genetic data, which had indicated that tobacco mutations that confer herbicide resistance define two loci (15, 16). Southern blot analyses of DNA from several other crop species has indicated that many, such as corn and soybean, carry multiple ALS genes (B. J. Mazur, unpublished observations).

The cloned plant ALS genes were used as hybridization probes to isolate genes carrying ALS mutations from herbicide-resistant plants. In tobacco, the Hra line, which is mutated at the *SURB* locus and which is 1000-fold more resistant to sulfonylureas than are wild-type lines (17), was used as one source of a mutant ALS gene. A second tobacco line, C3, which carries a mutation at the *SURA* locus (16), was used as a source of a second mutant ALS gene. Four genes, representing all of the ALS loci from the two mutant lines, were isolated and sequenced. The molecular characterization of mutant and wild-type genes from each plant line permitted the assignment of the genes to the appropriate genetic locus and the determination of the amino acid substitutions in the mutant enzymes. The *SURB-Hra* gene, which was isolated by two successive rounds of genetic selection, contained two mutations, which resulted in pro196 to ala and trp573 to leu substitutions. The *SURA-C3* gene carried a single mutation that resulted in a pro196 to gln substitution (70). Similarly, a gene that conferred sulfonylurea herbicide resistance was isolated from *Arabidopsis* and sequenced; this gene carried a single mutation that resulted in a pro197 to ser substitution (50). Both the pro196 to gln mutation in the tobacco gene and the analogous pro197 to ser mutation in the *Arabidopsis* gene conferred selective resistance to sulfonylurea herbicides but not to imidazolinone herbicides. The double tobacco mutant that carried the pro196 to ala and trp573 to leu substitutions was cross-resistant to both classes of herbicides (70).

HERBICIDE-RESISTANT TRANSGENIC PLANTS   The mutant plant ALS genes were introduced into *N. tabacum* by transformation and conferred useful levels of herbicide resistance in transgenic plants (50, 70). The tobacco *SURB-Hra* gene was also used to transform a number of heterologous species to sulfonylurea herbicide resistance at the cellular, and in some cases, whole plant level. These species include tomato, sugarbeet, oilseed rape, alfalfa, lettuce, and melon (9). In some of the heterologous transformants, such as tomato, expression of the resistant tobacco ALS gene was efficient, 20–60% of ALS activity being derived from the mutant gene (39, 77). In others, such as rape, only a low level of resistance was observed (9). Thus the effectiveness of heterologous genes must be evaluated on a case-by-case basis.

Additional herbicide-resistance mutations in plant ALS genes were generated by site-directed mutagenesis, based upon the mutations identified in the yeast ALS gene. The mutant genes were introduced into tobacco by transformation, and their ability to confer herbicide resistance was monitored

(48). In addition, a bacterial expression assay system for the *Arabidopsis* ALS gene was developed to permit the rapid isolation and characterization of new mutations. In this assay system, the plant gene, including its chloroplast transit sequence, was expressed in an *E. coli* auxotroph that produced no ALS. The wild-type *Arabidopsis* ALS gene promoted growth only in the absence of herbicides, while mutant *Arabidopsis* ALS genes promoted growth in both the presence and absence of herbicides. This selection thus allowed a facile assay for the efficacy of herbicide-resistance mutations in a plant ALS gene (107).

The *SURB-Hra* gene was introduced into a number of commercial lines of tobacco, and the transformed plants were tested to determine whether they could manifest useful levels of resistance in the field. Prior to field testing, the transformed plants were assayed for levels of sulfonylurea resistance by several methods. Resistance was measured by assaying leaf ALS activity in the presence of herbicide, by measuring secondary callus growth in the presence of increasing concentrations of herbicide, by monitoring the ability of progeny seeds to germinate and grow in the presence of increasing concentrations of herbicide, and by monitoring plant phytotoxicity after foliar spray applications of herbicides in greenhouse trials. The results of each of the tests were consistent but indicated the need for careful screening in order to identify those lines most suitable for crop breeding (39, 77, 106). For breeding purposes, a high level of herbicide resistance originating from a single genetic locus is preferred.

As a more critical measure of agronomically useful herbicide resistance, some of the tobacco transformants were treated with sulfonylurea herbicides in field tests and evaluated for phytotoxic symptoms. Foliar sprays were applied at rates corresponding to 0, 8, 16, and 32 grams of herbicide per hectare. Transformed plants showed no damage at the highest application rate tested, which was more than four times that of a typical field application rate. Wild-type plants showed damage at the 8g/hectare application rate (Figure 3; 77). Thus, expression of the *SURB-Hra* gene can provide an effective means of producing sulfonylurea herbicide–resistant crops.

ALTERNATIVE STRATEGIES FOR ACHIEVING HERBICIDE RESISTANCE   The metabolic detoxification of particular sulfonylureas by specific crop species, which is the basis for the selectivity of commercial herbicides, could provide another approach for genetically engineering sulfonylurea resistant crops. Native genes responsible for the detoxification pathways could be isolated from tolerant crops and transferred to sensitive ones. Again a microbial model has proved useful. In *Streptomyces griseolus* sulfonylurea herbicides such as chlorsulfuron, sulfometuron methyl, and chlorimuron ethyl are metabolized by a three-step process requiring an NADP-dependent reductase,

Figure 3   Field trial of herbicide-resistant transgenic tobacco plants. A gene encoding a sulfonylurea herbicide–insensitive form of tobacco ALS was transferred to elite cultivars of tobacco.
The resulting transgenic tobacco plants were tested for sulfonylurea resistance in a field trial.
Transgenic plants *(left and right)* were healthy, while nontransformed plants *(center)* were
severely stunted by the treatment.

an iron-sulfur protein, and either of two distinct, inducible cytochrome P-450
enzymes (88, 96). This bacterium could serve as a source for the isolation of
the genes encoding the metabolically active proteins. These genes could be
introduced into plants by transformation to confer herbicide resistance. Unfortunately, most of the same difficulties and uncertainties associated with the
expression of microbial genes encoding herbicide-resistant ALS in plants also
exist for the microbial detoxification genes. As an alternative, the bacterial
genes might serve as probes for the isolation of plant genes encoding detoxifying enzymes, in an approach analogous to that used with ALS.

   Because the most agronomically important crops were not amenable to
transformation, efforts were mounted to generate resistance to the sulfonylureas and the other ALS inhibitors using alternative methods. Imidazolinone-
resistant mutants of maize were isolated using genetic selection of cultured
maize cells. Fertile plants exhibiting a greater than 100-fold increase in
resistance to imidazolinones and cross-resistance to the sulfonylureas were
regenerated from one line. Homozygous progeny showed more than a 300-
fold increase in resistance to the herbicides. Resistance was inherited as a
single dominant nuclear gene, and ALS activity from the mutants was resis-

tant to the herbicides in vitro (3, 4). This mutant, which did not appear to have any associated growth or yield defects, is being developed by Pioneer Hi-Bred International, Inc. Although a significant breeding program has been required to introduce the resistance trait into elite lines, it is likely to be the first ALS-targeted herbicide-resistant crop to be commercialized, with release anticipated in the early 1990s. Imidazolinone-resistant mutants of maize that show no cross-resistance to the sulfonylureas have also been isolated. In one case the mutant showed tolerance to a much lower level of herbicide. Tolerance was inherited as a recessive trait, and no change in ALS activity was observed (3).

Sulfonylurea herbicide–tolerant mutants of soybean have been obtained by genetic selection from mutagenized soybean seeds. The first mutants to be described showed only a 5–10 fold increase in tolerance, were recessive, and did not affect the sensitivity of ALS to inhibition by sulfonylureas (101). Recently, sulfonylurea-resistant mutant lines of soybean with significantly higher levels of tolerance have been obtained. These mutants have an altered ALS that shows reduced sensitivity to inhibition by sulfonylureas (S. A. Sebastian, personal communication). There have been no reports yet of mutants selected for resistance to the triazolopyrimidine class of inhibitors.

Glyphosate

MODE OF ACTION   Glyphosate (Table 1) is an exceptionally reliable, phloem mobile, broad-spectrum herbicide with little residual soil activity. Because of these desirable properties, because it is toxic to all major crops, and because of its considerable commercial importance, extensive efforts have been aimed at the development of herbicide-resistant cultivars. A number of recent reviews have described this work in depth (21, 61, 89, 104).

The primary target of glyphosate is 5-enolpyruvyl-shikimate-3-phosphate synthase, or EPSPS, an enzyme in the aromatic amino biosynthetic pathway (see Figure 1). This pathway is found only in microbes and plants; glyphosate exhibits low acute toxicity to mammals. EPSPS was originally inferred to be the target of glyphosate action through identification of shikimic acid as an intermediate that accumulated following glyphosate treatment, through experiments showing that glyphosate activity could be suppressed by the addition of aromatic amino acids (57), and by biochemical studies of glyphosate inhibition of EPSPS (1, 55, 114). These findings were subsequently corroborated by the selection of glyphosate-resistant mutant strains of enteric bacteria. A resistant strain of *Salmonella* was identified following two cycles of chemical mutagenesis. The resistance mutation was shown to be in the *aroA* gene; the mutant gene was subsequently isolated, and was shown to confer resistance when transferred to *E. coli*. DNA sequencing indicated that two distinct types of mutations had led to the glyphosate-resistance phenotype.

The first round of mutagenesis had created a promoter mutation in the *aroA* gene, which conferred low levels of glyphosate tolerance by elevating the level of expression of the gene. The second cycle of mutagenesis had generated to a point mutation in the *aroA* structural gene, which caused a proline to serine substitution at residue 101 of the protein (20, 112). Overexpression of the *E. coli aroA* gene, as a consequence of its presence on a high-copy-number plasmid, could also confer glyphosate tolerance (95), providing additional evidence that EPSPS was the primary target of this compound.

HERBICIDE-RESISTANT TRANSGENIC PLANTS   Glyphosate-tolerant trans-genic plants were first generated by transferring the mutant *Salmonella aroA* gene, linked to either an octopine or mannopine synthase promoter for plant cell expression, to tobacco (19). The bacterial gene lacked a chloroplast transit sequence, and thus the herbicide-resistant EPSPS was expected to be localized in the cytoplasm; the plant EPSPS is predominantly localized in the chloroplast (28, 82). The transformed tobacco plants showed increased, but incomplete, tolerance to glyphosate. The same bacterial gene was also transferred to tomatoes. The gene again conferred glyphosate tolerance, but after foliar herbicide treatments the transformants were smaller than the unsprayed controls (41). Subsequent to these studies, mutant bacterial genes were fused to plant EPSPS chloroplast transit sequences (see below) and then transferred to plants. In one series of experiments, the petunia EPSPS gene transit sequence, along with the first 27 codons for the mature protein, was fused to a mutant *E. coli aroA* gene. The chimeric preprotein was imported into petunia chloroplasts and conferred glyphosate resistance in the transformed plants (29, 89).

Glyphosate resistance has also been imparted to plants through transformation with plant EPSPS genes. The plant genes were isolated by taking advantage of the finding that plant lines tolerant to glyphosate arose through overproduction of EPSPS (2, 85, 115). A glyphosate-tolerant petunia line was isolated by applying increasingly stringent stepwise selection conditions until a line was established in which EPSPS DNA, RNA, and protein levels were elevated approximately 20-fold (104, 115). This elevation in EPSPS-specific macromolecules facilitated the subsequent cloning of the EPSPS gene. EPSPS was purified from the line, and the N-terminal amino acid sequence of the protein was determined by microsequencing. Based on this sequence, three sets of potentially complementary oligonucleotide probes for the gene were synthesized and were used to screen messenger RNA populations from the amplified line. The set containing the oligonucleotide complementary to the gene was determined from Northern blot hybridizations, and was then used to identify a partial cDNA clone for EPSPS. The cDNA clone was in turn used to identify a genomic DNA clone; a complete cDNA clone was subsequently constructed (104).

The petunia EPSPS gene was shown to be a nuclear gene that spans 9 kb of DNA and is interrupted by 7 introns. The mature protein was predicted to have a molecular mass of approximately 48 kD (45). A comparison of the inferred petunia, *Arabidopsis,* and tomato EPSPS proteins showed that they are highly conserved, except in the region of the chloroplast transit peptide (45, 62). The petunia gene codes for a 72-amino-acid chloroplast transit sequence, while the tomato chloroplast transit sequence is more than twice as long, with 148 codons.

The petunia EPSPS gene was used to produce glyphosate-resistant transgenic plants. The petunia cDNA clone was linked to a Cauliflower Mosaic Virus 35S promoter, in order to obtain a high level of expression of the gene in plants. Expression of the wild-type gene from this promoter resulted in a 20-fold increase in EPSPS activity in transgenic petunia plants. The plants tolerated applications of glyphosate approximately four times greater than that needed to kill nontransformed plants (103). In order to provide additional herbicide tolerance in the transgenic plants, particularly in the meristematic regions where glyphosate accumulates (80), site-specific mutations were introduced into the wild-type petunia EPSPS gene. All of the substituted EPSPS proteins had reduced catalytic efficiencies (61, 89). The modified genes were again coupled to the Cauliflower Mosaic Virus 35S promoter, and were able to confer higher levels of resistance in transgenic plants. In a parallel series of experiments, mutated bacterial EPSPS genes were fused to plant EPSPS chloroplast transit sequences, and the chimeric genes introduced into plant cells (see above). These constructions also conferred herbicide resistance (61). Using such herbicide-resistance constructs, a wide variety of glyphosate-resistant transgenic plant species have been created. Some of these plants, including tomato and oilseed rape, have been tested in field trials during 1987 and 1988.

ALTERNATIVE STRATEGIES FOR GLYPHOSATE TOLERANCE   *Pseudomonas and Arthrobacter* species capable of growing on glyphosate as a sole carbon source have been identified (59, 90). These strains metabolize glyphosate to phosphate, glycine, and a one-carbon unit (56, 60, 90). In contrast, through soil organism metabolism, glyphosate is primarily degraded to aminomethylphosphonic acid (86, 97, 109). The bacterial genes that carry out these metabolic degradations could theoretically be isolated and expressed in plants, to produce glyphosate-tolerant plants. No glyphosate-tolerant plants have yet been reported in which expression of cloned genes enables metabolism of glyphosate to nontoxic products.

Phosphinothricin

MODE OF ACTION   Phosphinothricin (PPT) is an analogue of glutamine that inhibits the amino acid biosynthetic enzyme glutamine synthetase (GS) of bacteria and plants (see Figure 1) (8, 69). Bialaphos is a tripeptide precursor of PPT produced by some strains of *Streptomyces*, in which two alanine residues are linked to the PPT moiety; the active PPT moiety is released intracellularly by peptidase activity. Both PPT and bialaphos are marketed as broad-spectrum contact herbicides; they are not phloem mobile. Inhibition of GS by these compounds causes a rapid buildup of intracellular ammonia levels and an associated disruption of chloroplast structure, resulting in the inhibition of photosynthesis and plant cell death (117). Even though PPT inhibits GS from bacteria, plants, and mammals, its inability to cross the blood-brain barrier and its rapid clearance by the kidneys are the apparent reasons for its nontoxicity to mammals.

HERBICIDE-RESISTANT TRANSGENIC PLANTS   Several approaches have been used to produce plants tolerant to these herbicides. An alfalfa cell line tolerant to PPT was isolated following selection and was shown to produce elevated levels of GS (30). This line was used to facilitate the purification of GS. The amino acid sequences of two internal GS peptides were subsequently used to confirm the identification of an alfalfa cDNA clone (30), which had been isolated through hybridization with a *Phaseolus* GS cDNA clone (22). The corresponding genomic DNA clone for alfalfa GS was shown to be 4 kb long and to contain 11 introns. The encoded protein was predicted to be 356 amino acids long and to share several regions of sequence similarity with the *Anabaena* and Chinese hamster enzymes. The original tolerant line was demonstrated to have a 35-kb segment of DNA, carrying the coding region for GS, which had been amplified approximately 10-fold (120). In order to produce herbicide-resistant transgenic plants, the alfalfa GS cDNA clone was linked to the Cauliflower Mosaic Virus 35S promoter and introduced into tobacco plants via transformation. The resulting high level of expression of the alfalfa GS gene in the transgenic tobacco plants conferred a low level of tolerance to PPT (32).

In order to identify herbicide-resistance mutations that could occur in plant GS genes, and which might confer increased levels of tolerance to PPT, a bacterial expression assay system was developed. An alfalfa GS cDNA clone was linked to a bacterial promoter and transferred to a strain of *E. coli* devoid of GS activity; thus bacterial growth was dependent on the functional expression of the plant gene (24). Mutations that allowed bacterial growth in the presence of PPT were then selected. Such mutations, however, resulted from overexpression of the GS gene, as a consequence of gene amplification or promoter mutations. This problem was circumvented by fusing the 5' end of the plant GS gene to the 3' end of the *E. coli lacZ* gene, thus allowing a β-galactosidase assay to be used to screen for GS overproduction. Several point mutations in the alfalfa GS gene were identified by this method. DNA sequencing of these mutations predicted substitutions of serine, cysteine, or arginine for gly245 and lysine for arg332 (H. M. Goodman, unpublished).

A mutant alfalfa GS gene was then transferred to tobacco cells, and plants were regenerated. Although the mutant GS gene protected plants when herbicide was taken up through the roots, it did not protect against foliar applications of herbicide (H. M. Goodman, unpublished). One possible explanation for this finding derives from the existence of multiple nuclear GS genes in plants (118). In pea, for example, one GS gene encodes an isozyme found in the nodules, a second encodes a GS isozyme found in chloroplasts, and two others encode cytoplasmic forms of GS (119). Thus resistance mutations may have to be introduced into several GS genes to confer whole-plant resistance to this herbicide.

An alternative approach to achieving PPT resistance has been more successful. In this case, resistance was introduced into plants via a gene that produces a detoxifying enzyme. The gene, designated *bar,* is found in strains of *Streptomyces* that produce bialaphos, the tri-peptide precursor of PPT. The *bar* gene product protects these strains from the action of their own antibiotic by metabolizing PPT to an inactive, acetylated derivative (121). The *bar* gene was isolated from *Shygroscopicus* (84), placed under the control of the Cauliflower Mosaic Virus 35S promoter, and introduced into tobacco, potato, and tomato (25). Expression of the gene produced transferase levels that varied from 0.001% to 0.1% of the soluble protein in transgenic plants; protection against PPT occurred even in the plants producing the lowest levels of transferase (71). Field trials of transgenic tobacco, tomatoes, potatoes, oilseed rape, and poplars resistant to PPT have been or will be conducted (J. Leemans, personal communication).

PLANTS RESISTANT TO PHOTOSYNTHESIS INHIBITORS

Atrazine

MODE OF ACTION   The s-triazines (e.g. atrazine and simazine) and the ureas (e.g. diuron) (Table 1) are herbicidal because they inhibit photosynthesis. Atrazine was discovered in the 1950s and has become an important herbicide for corn and sorghum. These crops, in contrast to most other crops and weeds, have isozymes of glutathione-s-transferase that rapidly detoxify atrazine by conjugation. Because atrazine is slowly metabolized to inactive forms in the soil, residues from one growing season to the next limit the grower's opportunity to rotate atrazine-treated corn and sorghum with other crops such as soybeans or small grains.

The extensive agricultural use of atrazine has provided a strong genetic selection for the emergence of resistant weeds. The first atrazine-resistant mutant weed was reported in 1970 (98). About 38 resistant weed species have now been identified in Canada, Europe, and the United States (47). Atrazine resistance is maternally inherited, indicating that the gene responsible is located in the plastid or mitochondrion (108). Considerable advantage has been taken of the resistant weeds in research on the atrazine target and on the photosystem II reaction center. The protein site of action of atrazine and related herbicides was identified by a combination of physiological, biochemical, and genetic studies. Early research indicated that these herbicides were inhibitors of photosynthesis. Binding studies showed that atrazine displaced a plastoquinone from its binding site on a membrane protein in the photosystem II complex of chloroplasts. This protein has been variously designated D1, the 32-kD membrane protein, the herbicide-binding protein, and the Q_β protein (73).

The Q_β protein has been extensively studied in several plants. It undergoes a more rapid turnover than other chloroplast proteins (74, 94), and its degradation and synthesis are light regulated (87). The protein, encoded by the chloroplast gene *psbA* (12), is synthesized as a 34-kD precursor (33, 46, 94) and is posttranslationally processed and assembled into photosystem II complexes (14). The *psbA* gene has been cloned and sequenced from several higher plants and photosynthetic microorganisms. The deduced amino acid sequences indicate that the Q_β protein is a highly conserved, hydrophobic protein of about 350 amino acids (36).

The herbicide-resistant weeds, as well as resistant mutants selected in several microorganisms, such as *Chlamydomonas, Euglena,* and the cyanobacterium *Anacystis,* have provided a source of genes for studies of the molecular genetic basis for resistance. More than a dozen mutant *psbA* genes have been isolated and sequenced; at least five sites where amino acid substitutions result in herbicide resistance have been identified (54). More than two thirds of the mutations result in a substitution of ser264 with gly or ala. A list of herbicide-resistance mutations identified in *psbA* genes is given in Table 3. Crystallographic studies on the photosynthetic reaction center of purple bacteria (27), as well as correlations between the structural organization and herbicide binding sites of bacteria and higher-plant photosystems II, have led to a model for the topology of the Q_β protein in the thylakoid membrane and a proposed herbicide binding niche (122).

HERBICIDE-RESISTANT PLANTS   Atrazine-resistant weeds have been used to generate resistant crops by various means. The resistance trait has been

Table 3 Mutations in Q_β conferring herbicide resistance[1]

Amino acid change	Species	Relative resistance			
		Atrazine	Metribuzin	Diuron	Bromacil
219 Val → Ile	C. reinhardtii	2	15	1	1
251 Ala → Val	C. reinhardtii	25	1000	5	
255 Phe → Tyr	C. reinhardtii	15		0.5	1
264 Ser → Gly	A. hybridus	1000	200	1	
264 Ser → Gly	S. nigrum	1000		1	20
264 Ser → Gly	B. campestris	600	50	1	
264 Ser → Gly	P. paradoxa	>120	23	1.3	62
264 Ser → Ala	C. reinhardtii	100		10	
264 Ser → Ala	E. gracilis				
264 Ser → Ala	A. nidulans	17	5000	150	300
275 Leu → Phe	C. reinhardtii			5	4
264 Ser → Ala 255 Phe → Tyr	A. nidulans	360	200	300	330
264 Ser → Ala 255 Phe → Leu	A. nidulans	2.5	175	2650	35

[1] Taken from reference 54

transferred by sexual crosses from the weed *Brassica campestris* into several *Brassica* crops of commercial importance, such as oil seed rape, rutabaga, and Chinese cabbage (11). This work led to the release of the atrazine-resistant Canola cultivar "Triton," which was planted on about 250,000 acres in Canada in 1986, despite a yield penalty of about 20% associated with the resistant cytoplasm. In another such example, resistance from the weed *Setaria viridis* was crossed into the crop *Setaria italica* (23). Further such sexual crosses will be limited, however, because only a few resistant weeds are sexually compatible with crop plants. Protoplast fusion has also been used to transfer atrazine resistance between *Solanum* lines (5). In *Brassica,* transfer of atrazine-resistant cytoplasm into cytoplasmic male sterile lines by protoplast fusion was carried out to aid hybrid seed production (7, 10).

Even with this wealth of information and the availability of a number of resistance genes, genetic engineering of triazine herbicide–resistant crop plants via transformation has been difficult. A major problem has been the inability to introduce genes into chloroplasts reproducibly. To bypass this technical problem, transgenic tobacco plants were produced following nuclear transformation with a mutant *psbA* gene. To accomplish this, a chimeric gene was constructed in which a nuclear promoter and a chloroplast transit peptide-encoding sequence were attached to a mutant *psbA* structural gene. The transgenic plants showed an increased tolerance for atrazine (18).

As an alternative, engineering atrazine resistance through the introduction of genes encoding detoxifying enzymes has been attempted but has also proven difficult. Genes encoding atrazine and alachlor detoxifying glutathione-S-transferases (GSTI and GSTIII) have been isolated from maize (81, 103). Preliminary studies indicating some increased tolerance to atrazine of transgenic tobacco plants expressing an atrazine-metabolizing GST gene have been reported (53).

Bromoxynil

Bromoxynil is a benzonitrile compound that inhibits photosynthetic electron transport. This broad-leaf plant herbicide is used for weed control in cereal crops (Table 1). In order to engineer crops resistant to the effects of this herbicide, a search was initiated for bacteria able to metabolically detoxify the compound. A strain of the soil bacterium *Klebsiella ozaenae,* which could use bromoxynil as its sole nitrogen source, was isolated. This strain produces a bromoxynil-specific nitrilase protein that detoxifies the compound by conversion to 3,5-di-bromo 4-hydroxybenzoic acid, in a single step (78). The *bxn* gene that encodes this enzyme was shown to be plasmid borne and was subsequently cloned (113). The gene has been placed under the control of plant promoters and has been transferred to tobacco and tomato plants; it confers resistance to bromoxynil herbicide treatments in these plants (D. M. Stalker, personal communication).

CONCLUSION

In the last decade major research efforts on the genetic engineering of plants have been initiated in academic institutions and government agencies. In the private sector, a number of biotechnology companies devoted to this endeavor have been established; some have since failed or been partially or completely subsumed by larger concerns. Virtually all of the major agrichemical companies have established research and development programs in this area.

The field's early dreams, (e.g. of creating crops that could fix their own nitrogen) have been joined by more technically attainable goals. The generation of herbicide-resistant crops has provided a focal point for research, in part because it appears readily achievable. The targets of most herbicides are known, and many herbicide-resistant plants have been identified, either as weeds or as mutants selected in the laboratory. Furthermore, herbicide resistance clearly represents a valuable trait. From the standpoint of basic research, the availability of genetic determinants for herbicide resistance can provide a plant analog to the antibiotic-resistance genes that have proven so important for bacterial research. In agriculture, herbicide-resistant crops could lead to more cost effective weed control and could permit the wider use of more environmentally desirable herbicides.

Much of the work on herbicide resistance has centered on the isolation of the genes that could impart the herbicide-resistance trait and on the transfer of these genes into plants. The motivation for this work has come from the added flexibility and increased speed that gene transfer can provide. For example, gene transfer can allow another gene, which confers resistance to a second herbicide, to be linked on the introduced DNA. This second resistance trait could be useful either to prevent herbicide-resistant weeds from arising following prolonged use of a particular herbicide, or to allow use of multiple herbicides with differing attributes in order to achieve the broadest possible weed control. Gene transfer can also allow herbicide resistance to be introduced simultaneously into multiple crop species and into multiple cultivars of each of those species. Finally, because herbicide resistance is a selectable trait, genes whose phenotype cannot be readily scored can also be transferred if linked to the resistance trait. Examples of such genes would include disease resistance and nutritional quality traits.

What has been achieved so far? The first example of a herbicide-resistant crop, atrazine-resistant Canola, derived from crossing a resistant weed with a crop, has been released for use. Although there is a yield penalty associated with the use of this cultivar, it has gained acceptance because of the advantages it provides for weed control. Imidazolinone-resistant corn, selected from corn cells grown in tissue culture, is in the breeding and evaluation stage, with release of cultivars expected in the early 1990s. A sulfonylurea-resistant soybean line has been selected from mutagenized seed and is being introduced into soybean cultivars by standard crop breeding practices. Many examples of transgenic herbicide-resistant crops have been reported and are in various stages of evaluation and development. Not surprisingly, this approach has turned out to be a more complex undertaking than originally envisioned. For example, it has been difficult to obtain adequate levels of glyphosate resistance in some crops, using either mutant bacterial or plant EPSPS genes. Similarly, field-application-rate levels of resistance to phosphinothricin based on mutant plant GS genes have not yet been reported. It is encouraging that useful levels of sulfonylurea herbicide resistance have been obtained in transgenic plants using mutant plant ALS genes, demonstrating that a herbicide target-based engineering approach can work. Finally, resistance resulting from expression of herbicide detoxifying enzymes has been accomplished for phosphinothricin and bromoxynil. The levels of phosphinothricin resistance achieved through expression of the *bar* gene from *Streptomyces* in plants are suitable for agricultural use.

In addition to these practical achievements, much has been learned as a result of the efforts to engineer herbicide-resistant crops. Considerable basic biological knowledge has been generated, along with a tremendous increase in the development of the technologies necessary for sophisticated genetic manipulation of plants. Much progress has been made in the identification of DNA sequences that control the expression of genes in plant cells. This information derives from efforts to express bacterial genes that encode antibiotic and herbicide-resistance determinants, and from investigations of the expression of plant genes that encode herbicide-resistance determinants. Similarly, an increased understanding of the signals that control targeting of proteins to organelles has resulted from investigations employing herbicide-resistance genes. The study of mutant herbicide target proteins has also provided insight into protein-ligand interactions, enzyme functions, and the photosystem II reaction center. New information has been obtained in the areas of plant physiology and metabolism from investigations of the mode of action of herbicides and of plant detoxification pathways. Studies on herbicide-resistant mutants have demonstrated several instances of plant gene amplification, illuminating the dynamic nature of the plant genome. On the other hand, genes introduced into plants by transformation, inserted at many different and apparently random sites, exhibit stable inheritance.

Methods for propagating plant cells in culture and for regenerating whole, fertile plants from cells have been improved, modified, and extended to an ever-increasing number of species (41a). Among the plant transformation techniques that have been developed are *Agrobacterium*-mediated DNA transfer to plant protoplasts (co-cultivation) or to leaf disks; direct DNA uptake into protoplasts, with or without electroporation; and DNA injection into plant cells via micropipet or particle gun. Vectors of increasing utility have been constructed for the *Agrobacterium*-mediated transformation method, which remains the most reliable and widely used procedure for several plant species. It seems certain that transformation methods for all the major crop species will soon be available.

Along with these technical achievements, the genetic engineering of herbicide-resistant crops has provided a link between academic and industrial research efforts that should benefit both. It has also provided a test case for government regulatory agencies to use in the development of public policy. The agricultural biotechnology industry has used herbicide-resistant crops as an example to refine the evaluation process for product development. Thus, genetic engineering of herbicide-resistant plants has been not merely an end but also a means for exploring the potential of recombinant DNA technology in agriculture. The rapid progress already made will spur further research, innovation, and the development of commercial products.

ACKNOWLEDGMENTS

We thank our colleagues and collaborators at Du Pont, Advanced Genetic Sciences, and Michigan State University who have been involved in the generation of sulfonylurea herbicide–resistant crops. We also thank Jan Leemans at Plant Genetic Systems and Ganesh Kishore, Stephen Padgette, and Dilip Shah at Monsanto for providing us with manuscripts prior to publication. Finally, we thank John Goss for a critical reading of the manuscript from a non-genetic-engineering perspective.

Literature Cited

1. Amrhein, N., Deus, B., Gehrke, P., Steinrücken, H. C. 1980. The site of inhibition of the shikimate pathway by glyphosate. *Plant Physiol.* 66:830–34
2. Amrhein, N., Johänning, D., Schab, J., Schulz, A. 1983. Biochemical basis for glyphosate tolerance in a bacterium and a plant tissue culture. *FEBS Lett.* 157: 191–96
3. Anderson, P. C., Georgeson, M. 1986. Selection and characterization of imidazolinone tolerant mutants of maize. In *Biochemical Basis of Herbicide Action*, 27th Harden Conf. Prog. Abstr., Wye College, Ashford, UK
4. Anderson, P. C., Hibberd, K. A. 1985. Evidence for the interaction of an imidazolinone herbicide with leucine, valine, and isoleucine metabolism. *Weed Sci.* 33:479–83
5. Austin, S., Helgeson, J. P. 1987. Interspecific somatic fusions between *Solanum brevidens* and *S. tuberosum*. *Plant Mol. Biol.* 140:209–22
6. Barak, Z., Calvo, J. M., Schloss, J. V. 1988. Acetolactate synthase isozyme III *(Escherichia coli)*. *Methods Enzymol.* 166: In press
7. Barsby, T. L., Kemble, R. J., Yarrow, S. A. 1987. *Brassica* cybrids and their utility in plant breeding. *Plant Mol. Biol.* 140:223–34
8. Bayer, E., Gugel, K. H., Haegele, K., Hagenmaier, H., Jessipow, S., Koenig, W. A., Zaehner, Z. 1972. Stoffwechselprodukte von Microorganismen 98. Mitteilung: Phosphinothricin und Phosphinothricyl-Alanyl-Alanin. *Helv. Chim. Acta* 55:224–39
9. Bedbrook, J., Chaleff, R. S., Falco, S. C., Mazur, B. J., Yadav, N. 1988.
Nucleic acid fragment encoding herbicide resistant plant acetolactate synthase. Eur. Patent Appl., 0257993
10. Beversdorf, W. D., Erickson, L. R., Grant, I. 1985. Hybridization process utilizing a combination of cytoplasmic male sterility and herbicide tolerance. *US Patent No. 4517763*
11. Beversdorf, W. D., Weiss-Lerman, J., Erickson, L. R., Souza Machado, V. 1980. Transfer of cytoplasmically-inherited triazine resistance from bird's rape to cultivated *Brassica campestris* and *Brassica napus. Can. J. Genet. Cytol.* 22:167–72
12. Bogorad, L., Jolly, S. O., Link, G., McIntosh, L., Poulsen, C., Schwartz, Z., Steinmetz, A. 1980. Studies of the maize chloroplast chromosome. In *Biological Chemistry of Organelle Formation*, ed. T. Bucher, W. Sebald, H. Weiss, pp. 87–96. Berlin: Springer-Verlag
13. Botterman, J., Leemans, J. 1988. Engineering of herbicide resistance in plants. *Biotechnol. Genet. Eng. Rev.* 6: In press
14. Callahan, F. E., Edelman, M., Matoo, A. K., Autar, K. 1987. Posttranslational acylation and intra-thylakoid translocation of specific chloroplast proteins. In *Progress in Photosynthesis Research,* ed. J. Biggens, III: 799–802. Dordrecht: Martinus Nijhoff
15. Chaleff, R. S., Mauvais, C. J. 1984. Acetolactate synthase is the site of action of two sulfonylurea herbicides in higher plants. *Science* 224:1143–45
16. Chaleff, R. S., Ray, T. B. 1984. Herbicide-resistant mutants from tobacco cell cultures. *Science* 223:1148–51

17. Chaleff, R. S., Sebastian, S. A., Creason, G. L., Mazur, B. J., Falco, S. C., et al. 1987. Developing plant varieties resistant to sulfonylurea herbicides. In *Molecular Strategies for Crop Protection*, ed. C. J. Arntzen, C. Ryan. New York: A. R. Liss

18. Cheung, A. Y., Bogorad, L., Van Montagu, M., Schell, J. 1988. Relocating a gene for herbicide tolerance: a chloroplast gene is converted into a nuclear gene. *Proc. Natl. Acad. Sci. USA* 85: 391–95

19. Comai, L., Facciotti, D., Hiatt, W. R., Thompson, G., Rose, R. E., Stalker, D. M. 1985. Expression in plants of a mutant *aroA* gene from *Salmonella typhimurium* confers tolerance to glyphosate. *Nature* 317:741–44

20. Comai, L., Sen, L. C., Stalker, D. M. 1983. An altered *aroA* gene product confers resistance to the herbicide glyphosate. *Science* 221:370–71

21. Comai, L., Stalker, D. 1986. Mechanism of action of herbicides and their molecular manipulation. *Oxford Surv. Plant Mol. Cell Biol.* 3:167–95

22. Cullimore, J. V., Gebhardt, C., Saarelainen, R., Miflin, B. J., Idler, K. B., Barker, R. F. 1984. Glutamine synthetase of *Phaseolus vulgaris* L. Organspecific expression of a multigene family. *J. Mol. Appl. Genet.* 2:589–99

23. Darmency, H. M., Pernes, J. 1985. Use of wild *Setaria viridis (L.) Beauv.* to improve trazine resistance in cultivated *S. italica* by hybridization. *Weed Res.* 25:175–79

24. DasSarma, S., Tischer, E., Goodman, H. M. 1986. Plant glutamine synthetase complements a *glnA* mutation in *Escherichia coli. Science* 232:1242–44

25. DeBlock, M., Botterman, J., Vandewiele, M., Dockx, J., Thoen, C., et al. 1987. Engineering herbicide resistance in plants by expression of a detoxifying enzyme. *EMBO J.* 6:2513–18

26. DeFelice, M., Guardiola, J., Esposito, B., Iaccarino, M. 1974. Structural genes for a newly recognized acetolactate synthase in *Escherichia coli* K-12. *J. Bacteriol.* 120:1068–77

27. Deisenhofer, J., Epp, O., Miki, K., Huber, R., Michel, H. 1985. Structure of the protein subunits in the photosynthetic reaction center of *Rhodopseudomonas viridis* at 3A resolution. *Nature* 318: 618–24

28. Della-Cioppa, G., Bauer, S. C., Klein, B. K., Shah, D. M., Fraley, R. T., Kishore, G. M. 1986. Translocation of the precursor of 5-enolpyruvylshik-imate-3-phosphate synthase into chloroplasts of higher plants in vitro. *Proc. Natl. Acad. Sci. USA* 83:6873–77

29. Della-Cioppa, G., Bauer, S. C., Taylor, M. L., Rochester, D. E., Klein, B. K., et al. 1987. Targeting a herbicide-resistant enzyme from *Escherichia coli* to chloroplasts of higher plants. *Biol Technology* 5:579–84

30. Donn, G., Tischer, E., Smith, J. A., Goodman, H. M. 1984. Herbicide-resistant alfalfa cells: an example of gene amplification in plants. *J. Mol. Appl. Genet.* 2:621–35

31. Douglas, M. G., McCammon, M. T., Vassaroti, A. 1986. Targeting proteins into mitochondria. *Microbiol. Rev.* 50: 166–78

32. Eckes, P., Wengenmayer, F. 1987. Overproduction of glutamine synthetase in transgenic plants. In *Regulation of Plant Gene Expression,* 29th Harden Conf. Prog. Abstr., Wye College, Ashford, UK

33. Ellis, R. J. 1981. Chloroplast proteins: synthesis, transport and assembly. *Annu. Rev. Plant Physiol.* 32:111–37

34. Eoyang, L., Silverman, P. M. 1984. Purification and subunit composition of acetohydroxyacid synthase I from *Escherichia coli* K-12. *J. Bacteriol.* 157: 184–89

35. Eoyang, L., Silverman, P. M. 1986. Role of small subunit (IlvN polypeptide) of acetohydroxyacid synthase I from *Escherichia coli* K-12 in sensitivity of the enzyme to valine inhibition. *J. Bacteriol.* 166:901–4

36. Erickson, J. M., Rahire, M., Rochaix, J. D. 1985. Herbicide resistance and cross-resistance: changes at three distinct sites in the herbicide-binding protein. *Science* 228:204–7

37. Falco, S. C., Dumas, K. S. 1985. Genetic analysis of mutants of *Saccharomyces cerevisiae* resistant to the herbicide sulfometuron methyl. *Genetics* 109:21–35

38. Falco, S. C., Dumas, K. S., Livak, K. J. 1985. Nucleotide sequence of the yeast ILV2 gene which encodes acetolactate synthase. *Nucleic Acids Res.* 13: 4011–27

39. Falco, S. C., Knowlton, S., LaRossa, R. A., Smith, J. K., Mazur, B. J. 1987. Herbicides that inhibit amino acid biosynthesis: the sulfonylureas—a case study. In *1987 Brit. Crop Protection Conf.—Weeds,* pp. 149–58. Surrey, UK: BCPC Publications

40. Falco, S. C., McDevitt, R. E., Chui, C.-F., Hartnett, M. E., Knowlton, S., et al. 1988. Engineering herbicide resistant

acetolactate synthase. *Dev. Ind. Micro-biol.* In press
41. Fillatti, J. J., Kiser, J., Rose, R., Comai, L. 1987. Efficient transfer of a glyphosate tolerance gene into tomato using a binary *Agrobacterium tumefaciens* vector. *Bio/Technology* 5:726–30
41a. Fraley, R. T., Rogers, S. G., Horsch, R. B. 1986. Genetic transformation in higher plants. *CRC Crit. Rev. Plant Sci.* 4(1):1–46
42. Frear, D. S., Mansager, E. R., Swanson, H. R., Tanaka, F. S. 1983. Metribuzin metabolism in tomato: isolation and identification of N-glucoside conjugates. *Pest. Biochem. Physiol.* 19:270–81
43. Frear, D. S., Swanson, H. R., Mansager, E. R. 1985. Alternate pathways of metribuzin metabolism in soybean: formation of N-glucoside and homoglutathione conjugates. *Pest. Biochem. Physiol.* 23:56–65
44. Friden, P., Donegan, J., Mullen, J., Tsui, P., Freundlich, M., et al. 1985. The *ilvB* locus of *Escherichia coli* K-12 is an operon encoding both subunits of acetohydroxyacid synthase I. *Nucleic Acids Res.* 13:3979–93
45. Gasser, C. S., Winter, J. A., Hironaka, C. M., Shah, D. M. 1988. Structure, expression and evolution of the genes encoding 5-enolpyruvylshikimate-3-phosphate synthase of petunia and tomato. *J. Biol. Chem.* 263:4280–89
46. Grebanier, A. E., Coen, D. M., Rich, A., Bogorad, L. 1978. Membrane proteins synthesized but not processed by isolated maize chloroplasts. *J. Cell Biol.* 78:734–46
47. Gressel, J., Ben-Sinai, G. 1985. Low intra-specific competitive fitness in a trazine resistant, nearly isogenic line of *Brassica napus. Plant Sci.* 38:29–32
48. Hartnett, M. E., Chui, C.-F., Mauvais, C. J., McDevitt, R. E., Knowlton, S., et al. 1989. Herbicide resistant plants carrying mutated acetolactate synthase genes. In *ACS Symposium on Fundamental and Practical Approaches to Combating Resistance,* ed. H. LeBaron, W. Moberg, M. Green. Washington, DC: ACS Books. In press
49. Hartnett, M. E., Newcomb, J. R., Hodson, R. C. 1987. Mutations in *Chlamydomonas rheinhardtii* conferring resistance to the herbicide sulfometuron methyl. *Plant Physiol.* 85:898–901
50. Haughn, G. W., Smith, J., Mazur, B., Somerville, C. 1988. Transformation with a mutant *Arabidopsis* acetolactate synthase gene renders tobacco resistant to sulfonylurea herbicides. *Mol. Gen. Genet.* 211:266–71

51. Haughn, G. W., Somerville, C. R. 1986. Sulfonylurea-resistant mutants of *Arabidopsis thaliana. Mol. Gen. Genet.* 204:430–34
52. Hawkes, T. R., Howard, J. L., Pontin, S. E. 1989. Herbicidal Inhibition of Branched Chain Amino Acid Biosynthesis. In *Herbicides and Plant Metabolism (SEM Seminar Ser),* ed. E. D. Dodge. Cambridge: Academic. In press
53. Helmer, G. 1986. Genetic engineering of atrazine tolerance in plants. Presented at *Intl. Symp. Pest. Biotechnol.,* Michigan State Univ., Aug. 17–19
54. Hirschberg, J., Yehuda, A. B., Pecker, I., Ohad, N. 1987. Mutations resistant to photosystem II herbicides. *Plant Mol. Biol.* 140:357–66
55. Holländer, H., Amrhein, N. 1980. The site of the inhibition of the shikimate pathway by glyphosate. *Plant Physiol.* 66:823–29
56. Jacob, G. S., Schaefer, J., Stejskal, E. O., McKay, R. A. 1985. Solid-state NMR determination of glyphosate metabolism in a *Pseudomonas* sp. *J. Biol. Chem.* 260:5899–5905
57. Jaworski, E. G. 1972. Mode of action of N-phosphono-methylglycine: inhibition of aromatic amino acid biosynthesis. *J. Agric. Food. Chem.* 20:1195–98
58. Jones, A. V., Young, R. M., Leto, K. J. 1985. Subcellular localization and properties of acetolactate synthase, target site of the sulfonylurea herbicides. *Plant Physiol.* 77:S-293
59. Kent-Moore, J., Braymer, H. D., Larson, A. D. 1983. Isolation of a *Pseudomonas* sp. which utilizes the phosphonate herbicide glyphosate. *Appl. Environ. Microbiol.* 46:316–20
60. Kishore, G. M., Jacob, G. S. 1987. Degradation of glyphosate by *Pseudomonas* sp. PG2982 via a sarcosine intermediate. *J. Biol. Chem.* 262:12164–68
61. Kishore, G. M., Shah, D. 1988. Amino acid biosynthesis inhibitors as herbicides. *Annu. Rev. Biochem.* 57:627–63
62. Klee, H. J., Muskopf, Y. M., Gasser, C. S. 1988. Cloning of an *Arabidopsis thaliana* gene encoding 5-enolpyruvyl-shikimate-3-phosphate synthase: sequence analysis and manipulation to obtain glyphosate-tolerant plants. *Mol. Gen. Genet.* In press
63. Kleswick, W. A., Ehr, R. J., Gerwick, B. C., Monte, W. T., Pearson, N. R., et al. 1984. New 2-arylamino-sulphonyl-1,2,4-triazolo-(1,5-a)-pyrimidine(s) useful as selective herbicides and to suppress nitrificaton in soil. *Eur. Patent Appl. 0142152*

468   MAZUR & FALCO

64. LaRossa, R. A., Falco, S. C. 1984. Amino acid biosynthetic enzymes as targets of herbicide action. *Trends Biotechnol.* 2:158–61
65. LaRossa, R. A., Schloss, J. V. 1984. The sulfonylurea herbicide sulfometuron methyl is an extremely potent and selective inhibitor of acetolactate synthase in *Salmonella typhimurium. J. Biol. Chem.* 259:8753–57
66. LaRossa, R. A., Smulski, D. R. 1984. ilvB-encoded acetolactate synthase is resistant to the herbicide sulfometuron methyl. *J. Bacteriol.* 160:391–94
67. LaRossa, R. A., Van Dyk, T. K., Smulski, D. R. 1987. Toxic accumulation of 2-ketobutyrate caused by inhibition of the branched-chain amino acid biosynthetic enzyme acetolactate synthase in *Salmonella typhimurium. J. Bacteriol.* 169:1372–78
68. Lawther, R. P., Calhoun, D. H., Adams, C. W., Hauser, C. A., Gray, J., Hatfield, G. W. 1981. Molecular basis of valine resistance in *Escherichia coli* K-12. *Proc. Natl. Acad. Sci. USA* 78: 922–25
69. Leason, M., Dunliffe, D., Parkin, D., Lea, P. J., Miflin, B. J. 1982. Inhibition of pea leaf glutamine synthetase by methionine sulphoximine, phosphinothricin and other glutamate analogues. *Phytochemistry* 21:855–57
70. Lee, K. Y., Townsend, J., Tepperman, J., Black, M., Chui, C.-F., et al. 1988. The molecular basis of sulfonylurea herbicide resistance in higher plants. *EMBO J.* 7:1241–48
71. Leemans, J., De Block, M., D'Halluin, K., Botterman, J., De Greef, W. 1987. The use of glufosinate as a selective herbicide on genetically engineered resistant tobacco plants. In *1987 Proc. Brit. Crop Protection Conf.—Weeds,* pp. 867–70. Surrey, UK: BCPC Publications
72. Lu, M.-F., Umbarger, H. E. 1987. Effects of deletion and insertion mutations in the *ilvM* gene of *Escherichia coli. J. Bacteriol.* 169:600–4
73. Mathis, P. 1987. Primary reactions of photosynthesis: discussion of current issues. In *Progress in Photosynthesis Research, Proc. Int. Congr. Photosynth.,* 7th, ed. J. Biggens, 1:151–60. Dordrecht: Martinus Nijhoff
74. Mattoo, A. K., Hoffman-Falk, H., Marder, J. B., Edelman, M. 1984. Regulation of protein metabolism: coupling of photosynthetic electron transport to *in vivo* degradation of the rapidly metabolized 32-kilodalton protein of the chloroplast membranes. *Proc. Natl. Acad. Sci. USA* 81:1380–84
75. Mazur, B. J., Chui, C.-F., Falco, S. C., Mauvais, C. J., Chaleff, R. S. 1985. Cloning herbicide resistance genes into and out of plants. In *The World Biotech Report 1985.* New York: Online Int.
76. Mazur, B. J., Chui, C.-F., Smith, J. K. 1987. Isolation and characterization of plant genes coding for acetolactate synthase, the target enzyme for two classes of herbicides. *Plant Physiol.* 85:1110–17
77. Mazur, B. J., Falco, S. C., Knowlton, S., Smith, J. K. 1987. Acetolactate synthase, the target enzyme of the sulfonylurea herbicides. *Plant Mol. Biol.* 140:339–49
78. McBride, K. E., Kenny, J. W., Stalker, D. M. 1986. Metabolism of the herbicide bromoxynil by *Klebsiella pneumoniae* subsp. *ozaenae. Appl. Environ. Microbiol.* 52:325–30
79. Miflin, B. J. 1974. The location of nitrite reductase and other enzymes related to amino acid biosynthesis in the plastids of root and leaves. *Plant Physiol.* 54:550–55
80. Mollenhauer, C., Smart, C., Amrhein, N. 1987. Glyphosate toxicity in the shoot apical region of the tomato plant. *Pest. Biochem. Physiol.* 29:55–65
81. Moore, R. E., Davies, M. S., O'Connell, K. M., Harding, E. I., Wiegand, R. C., Tiemeier, D. C. 1986. Cloning and expression of a cDNA encoding a maize glutathione-S-transferase in *E. coli. Nucleic Acids Res.* 14:7227–35
82. Mousdale, D. M., Coggins, J. R. 1985. Subcellular localization of the common shikimate pathway enzymes in *Pisum sativum. L. Planta* 163:241–49
83. Muhitch, M. J., Shaner, D. L., Stidham, M. A. 1987. Imidazolinones and acetohydroxyacid synthase from higher plants. *Plant Physiol.* 83:451–56
84. Murakami, T., Anzai, H., Imai, S., Satoh, A., Nagaoka, K., Thompson, C. J. 1986. The bialaphos biosynthetic genes of *Streptomyces hygroscopicus:* molecular cloning and characterization of gene cluster. *Mol. Gen. Genet.* 205:42–50
85. Nafziger, E. D., Widholm, J. M., Steinrücken, H. C., Killmer, J. L. 1984. Selection and characterization of a carrot cell-line tolerant to glyphosate. *Plant Physiol.* 76:571–74
86. Nomura, N. S., Hilton, H. W. 1977. The adsorption and degradation of glyphosate in five Hawaiian sugarcane soils. *Weed Res.* 17:113–21

87. Ohad, I., Kyle, D. J., Hirschberg, J. 1985. Light-dependent degradation of the Q$_\beta$-protein in isolated pea thylakoids. *EMBO J.* 4:1655–59

88. O'Keefe, D. P., Romesser, J. A., Leto, K. J. 1988. Identification of constitutive and herbicide inducible cytochromes P-450 in *Streptomyces griseolus*. *Arch. Microbiol.* 149:406–12

89. Padgette, S. R., Della-Cioppa, G., Shah, D. M., Fraley, R. T., Kishore, G. M. 1988. Selective herbicide resistance through protein engineering. In *Cell Culture and Somatic Cell Genetics of Plants*, ed. J. Schell, I. Vasil, Vol. 6. New York: Academic. In press

90. Pipke, R., Amrhein, N., Jacob, G. S., Schaefer, J., Kishore, G. M. 1987. Metabolism of glyphosate in an *Arthrobacter* sp GLP-1. *Eur. J. Biochem.* 165:267–73

91. Ray, T. B. 1982. The mode of action of chlorsulfuron: a new herbicide for cereals. *Pest. Biochem. Physiol.* 17:10–17

92. Ray, T. B. 1982. The mode of action of chlorsulfuron: the basis of direct inhibition of plant DNA synthesis. *Pest. Biochem. Physiol.* 18:262–66

93. Ray, T. B. 1984. Site of action of chlorsulfuron. *Plant Physiol.* 75:827–31

94. Reisfeld, A., Mattoo, A. K., Edelman, M. 1982. Processing of a chloroplast-translated membrane protein *in vivo*. *Eur. J. Biochem.* 124:125–29

95. Rogers, S. G., Brand, L. A., Holder, S. B., Sharps, E. S., Brackin, M. J. 1983. Amplification of the *aroA* gene from *E. coli* results in tolerance to the herbicide glyphosate. *Appl. Environ. Microbiol.* 46:37–43

96. Romesser, J. A., O'Keefe, D. P. 1986. Induction of cytochrome P-450 dependent sulfonylurea metabolism in *Streptomyces griseolus*. *Biochem. Biophys. Res. Commun.* 140:650–59

97. Rueppel, M. L., Brightwell, B. B., Schaefer, J., Marvel, J. T. 1977. Metabolism and degradation of glyphosate in soil and water. *J. Agric Food Chem.* 25:517–28

98. Ryan, E. D., Kohlhaw, G. B. 1974. Subcellular localization of isoleucine-valine biosynthetic enzymes in yeast. *J. Bacteriol.* 120:631–37

99. Ryan, G. F. 1970. Resistance of common groundsel to simazine and atrazine. *Weed Sci.* 18:614–16

100. Schloss, J. V., Van Dyk, D. E., Vasta, J. F., Kutny, R. M. 1985. Purification and properties of *Salmonella typhimurium* acetolactate synthase isozyme II

from *E. coli* HB101/pDU9. *Biochemistry* 24:4952–59

101. Sebastian, S. A., Chaleff, R. S. 1987. Soybean mutants with increased tolerance for sulfonylurea herbicides. *Crop Sci.* 27:948–52

102. Shah, D. M., Gasser, C. S., Della-Cioppa, G., Kishore, G. M. 1988. Genetic engineering of herbicide resistance genes. In *Plant Gene Research, Temporal and Spatial Regulation of Plant Genes*, ed. D. P. S. Verma, R. B. Goldberg, 5:297–309. Berlin: Springer-Verlag

103. Shah, D. M., Hironaka, C. M., Wiegand, R. C., Harding, E. I., Krivi, G. G., Tiemeier, D. C. 1986. Structural analysis of a maize gene coding for glutathione S-transferase involved in herbicide detoxification. *Plant Mol. Biol.* 6:203–11

104. Shah, D. M., Horsch, R. B., Klee, H. J., Kishore, G. M., Winter, J. A., Tumer, N. E., et al. 1986. Engineering herbicide tolerance in transgenic plants. *Science* 233:478–81

105. Shaner, D. L., Anderson, P. C., Stidham, M. A. 1984. Imidazolinones (potent inhibitors of acetohydroxyacid synthase). *Plant Physiol.* 76:545–46

106. Smith, J. K., Mauvais, C. J., Knowlton, S., Mazur, B. J. 1988. Molecular biology of resistance to sulfonylurea herbicides. In *ACS Symposium on Biotechnology of Crop Protection*, 379:25–36. Washington, DC: ACS Books. In press

107. Smith, J. K., Schloss, J. V., Mazur, B. J. 1989. Functional expression of plant acetolactate synthase genes in *Escherichia coli*. *Proc. Natl. Acad. Sci. USA*. In press

108. Souza-Machado, V. 1982. Inheritance and breeding potential of triazine tolerance and resistance in plants. In *Herbicide Resistance in Plants*, ed. H. M. Le Baron, J. Gressel, pp. 257–74. New York: Wiley

109. Sprankle, P., Meggit, W. F., Penner, D. 1975. Adsorption, mobility, and microbial degradation of glyphosate in the soil. *Weed Sci.* 23:229–34

110. Squires, C. H., DeFelice, M., Devereux, J., Calvo, J. M. 1983. Molecular structure of *ilvIH* and its evolutionary relationship to *ilvG* in *Escherichia coli* K-12. *Nucleic Acids Res.* 11:5299–5313

111. Squires, C. H., DeFelice, M., Wessler, S. R., Calvo, J. M. 1981. Physical characterization of the *ilvHI* operon of *Escherichia coli* K-12. *J. Bacteriol.* 147:797–804

112. Stalker, D. M., Hiatt, W. R., Comai, L.

1985. A single amino acid substitution in the enzyme 5-enolpyruvylshikimate-3-phosphate synthase confers resistance to the herbicide glyphosate. *J. Biol. Chem.* 260:4724–28

113. Stalker, D. M., McBride, K. E. 1987. Cloning and expression in *Escherichia coli* of a *Klebsiella ozaenae* plasmid-borne gene encoding a nitrilase specific for the herbicide bromoxynil. *J. Bacteriol.* 169:955–60

114. Steinrucken, H. C., Amrhein, N. 1980. The herbicide glyphosate is a potent inhibitor of 5-enolpyruvyl-shikimic acid-3-phosphate synthase. *Biochem. Biophys. Res. Commun.* 94:1207–12

115. Steinrucken, H. C., Schulz, A., Amrhein, N., Porter, C. A., Fraley, R. T. 1986. Overproduction of a 5-enolpyruvylshikimate-3-phosphate synthase in a glyphosate-tolerant *Petunia hybrida* cell line. *Arch. Biochem. Biophys.* 244:169–78

116. Sweetser, P. B., Schow, G. S., Hutchinson, J. M. 1982. Metabolism of chlorsulfuron by plants: biological basis for selectivity of a new herbicide for cereals. *Pest. Biochem. Physiol.* 17:18–23

117. Tachibana, K., Watanabe, T., Sekizawa, T., Takematsu, T. 1986. Action mechanism of bialaphos. II. Accumulation of ammonia in plants treated with bialaphos. *J. Pest. Sci.* 11:33–37

118. Tingey, S. V., Coruzzi, G. M. 1987.

Glutamine synthetase of *Nicotiana plumbaginifolia*. Cloning and *in vivo* expression. *Plant Physiol.* 84:366–73

119. Tingey, S. V., Walker, E., Coruzzi, G. M. 1987. Glutamine synthetase genes of pea encode distinct polypeptides which are differentially expressed in leaves, roots and nodules. *EMBO J.* 6:1–9

120. Tischer, E., DasSarma, S., Goodman, H. M. 1986. Nucleotide sequence of an alfalfa glutamine synthetase gene. *Mol. Gen. Genet.* 203:221–29

121. Thompson, C., Movva, N., Tizard, R., Crameri, R., Davies, J., et al. 1987. Characterization of the herbicide-resistance gene *bar* from *Streptomyces hygroscopicus*. *EMBO J.* 6:2519–23

122. Trebst, A., Draber, W. 1978. Structure activity correlations of recent herbicides in photosynthetic reactions. In *Advances in Pesticide Science*, ed. H. Geissbuhler, 2:223–34. New York: Pergamon

123. Wek, R. C., Hauser, C. A., Hatfield, G. W. 1985. The nucleotide sequence of the *ilvBN* operon of *Escherichia coli*: sequence homologies of the acetohydroxyacid synthase isozymes. *Nucleic Acids Res.* 13:3995–4010

124. Yadav, N., McDevitt, R. E., Bernard, S., Falco, S. C. 1986. Single amino acid substitutions in the enzyme acetolactate synthase confer resistance to the herbicide sulfometuron methyl. *Proc. Natl. Acad. Sci. USA* 83:4418–22

Annu. Rev. Plant Physiol. Plant Mol. Biol. 1989. 40:471–501

CHLOROPLASTIC PRECURSORS AND THEIR TRANSPORT ACROSS THE ENVELOPE MEMBRANES

Kenneth Keegstra and Laura J. Olsen

Department of Botany, University of Wisconsin, Madison, Wisconsin 53706

Steven M. Theg

Department of Botany, University of California, Davis, California 95616

CONTENTS

INTRODUCTION

Most chloroplastic proteins are synthesized on cytosolic ribosomes as larger precursors containing an amino terminal extension called a transit peptide (22). Posttranslational transport of precursor proteins into chloroplasts was

471

1040-2519-89-0601-0471$02.00

first demonstrated with prSS[1] a decade ago (21, 59). Since that time considerable progress has been made in defining the signals that direct transport, and some progress has been made in understanding the transport process itself. The transport signals are contained in the transit peptide, which is necessary and sufficient for transport into chloroplasts. The overall process of protein import into chloroplasts can be separated into several steps. They are: 1. association of the precursor with the outer envelope membrane; 2. translocation of the polypeptide across the inner and outer envelope membranes, perhaps at contact sites; 3. proteolytic removal of the transit peptide by the stromal processing protease and, when necessary, 4. further sorting of modified precursor to other chloroplastic compartments, followed by further proteolytic processing and/or 5. association with other polypeptides to form multimeric protein complexes. Much of this information has been summarized in several reviews (21, 30, 40, 41, 75, 87, 95, 128, 135, 145, 161).

Here we focus on three aspects of protein transport into chloroplasts where recent advances have revealed new insights. The first is the rapid progress in determining amino acid sequences of transit peptides. This structural information is valuable in trying to understand the functions of transit peptides. The second area is the binding of precursor proteins to the surface of chloroplasts. This step, which requires a transit peptide, is thought to provide the specificity needed to ensure that only chloroplastic precursors are imported into chloroplasts. The third area is the energetics of the transport process. This topic has received renewed attention because of its relevance in evaluating possible mechanisms of protein translocation.

STRUCTURE OF TRANSIT PEPTIDES

The necessity of transit peptides has been documented by the demonstration that precursors lacking transit peptides cannot be imported into chloroplasts (3, 96). The more interesting observation has been that transit peptides are capable of transporting foreign proteins into chloroplasts (87). These observations raise interesting questions concerning the mechanism whereby a transit peptide causes the transport of an attached passenger protein. Addressing these questions requires knowledge about the primary and secondary structure of transit peptides. This structural information can then be used to investigate the structure-function relationships of transit peptides. Until recently, knowledge about the primary structure of transit peptides was limited to a relatively small number of precursors. Based on an analysis of this small sample set,

---

[1]Abbreviations: pr, precursor; SS, small subunit of Rubisco; CAB, chlorophyll a/b binding protein; FD, ferredoxin; PC, plastocyanin; PMF, protonmotive force; Rubisco, ribulose-1,5-bisphosphate carboxylase/oxygenase

both Schmidt & Mishkind (128) and Karlin-Neumann & Tobin (73) identified sequence similarities among the transit peptides. The latter authors proposed that the similar sequences represented a framework that was shared by all chloroplastic precursors. Both groups suggested these common sequences might be important in the various steps of the transport process. As described below, these common sequences are not found in all transit peptides, and new hypotheses now need to be considered.

Primary Structure of Transit Peptides

TRANSIT PEPTIDES OF RUBISCO SMALL-SUBUNIT PRECURSORS   The transit peptide from prSS of *Chlamydomonas reinhardtii* was the first to be sequenced (127). It still is the only precursor for which the transit peptide has been sequenced directly (Crpep in Figure 1). All other sequences have been deduced from the nucleotide sequence of precursor genes. The sequence of the prSS transit peptide has now been determined from 48 genes representing 22 plant species (Figure 1). As can be seen from inspection of Figure 1, the sequences from the various species exhibit striking similarities. To allow maximal alignment of conserved sequences, the prSS transit peptide has been divided into 4 regions. Region 1 (residues -37 to -57 of the consensus sequence) is the least conserved portion of the transit peptide. When prSS transit peptides vary in length, it is usually because of deletions or insertions in this region. For example, the monocot species have large deletions in region 1 compared to the consensus sequence. Region 2 (residues -18 to -36) has several features that are highly conserved; for example, the GLK sequence at -28 to -30, the FP sequence at -22 and -23, and the two positive charges at -18 and -19. Region 3 (residues -9 to -17) is generally highly conserved except in two plant groups that have completely different sequences in this region. One group is represented by *Flaveria, Lemna*, and sunflower *(Helianthus); the second group is the monocots. Finally, region 4 (residues -1 to -8) is very highly conserved in all species except *Chlamydomonas*. For example, the positive charge at -4 is found in all higher plant sequences.

It is tempting to speculate that highly conserved regions are important for transit peptide function, whereas poorly conserved regions are not. However, it has not been demonstrated that all the prSS transit peptides are equally effective in directing import. No one has yet conducted a quantitative analysis of prSS import comparing precursors with different transit peptides in an effort to exploit the set of alterations that nature has provided. Thus it is premature to draw firm conclusions concerning structure-function relationships from an analysis of the sequences in Figure 1. However, from these analyses it is possible to formulate specific hypotheses that can now be tested by specific alterations in transit sequences.

```
              -50        -40          -30       -20        -10
               |          |            |         |          |              REF
CONSENSUS  MASSMLSSAAV--ATRTNPAQAS  MVAPFTGLKSAASFPVSRK  QNLDITSIA  SNGGRVQC
At1A       .........TM----VAS....T  .....N....S.A..AT..  A.N.....T  ......N.   80
At1B       ............----V.S....T  ...........S.....T..  A.N.....T  ......S.   80
At2B       ........T..----V.S....T  ...........S.....T..  A.N.....T  ......S.   80
At3B       ............----V.S....T  ...........S.A...T..  T.K.....T  ......S.   80
Cuc        ....I......ASVNSAS.....  ...........S.G..IT..  N.V...TL.  ..A.K...   53
Ft         ...IPATV...--------.-.TN  ...........AN.A...TK.  V.G-FSTLP  ........    1
Ha         ...-IS..V.T--VS.TA....N  ...........N.A..TTK.  A.-.FSTLP  ........   158
Lg5A       .....MA.T.A--VA.AG...S.  ......N..R.SVA..AT..  A.N.LSTLP  ......S.   150
Lg5B       .....MA.T.A--VA.VG...TN  ......N..R.SVA..AT..  A.N.LSTLP  ......S.   150
Lg26       .....MA.T.A--VA.AG...SN  ......N..R.SVA..AT..  A.KNLSTLP  ....K.S.   150
LgSSU      ----.MV.T.A--VA.VR...TN  ..GA.N.CR.SVA..AT..  A.N.LSTLP  .S....S.   150
Mc         .....V.....ATVN..-.....  ..V..N....V.A...TK.  -.N....V.  T.......   31
Np         ....V.......--...S.V...N  ..................  .........  ........   111
Ns         ....V.......--...S.V...N  ..................  .........  ........   110
Nt         ....V.......--...S.V...N  ..................  .........  ........   91
Pea3A      ...-.I..S..TTVS.ASTV.SA  A....G....MTG...-K.  V.T.....T  ......K.   46
Pea3C      ...-.I..S..TTVS.ASRG.SA  A....G....MTG...-K.  V.T.....T  ......K.   46
Pea3D      ...-.I..S..TTVS.ASRG.SA  A....G....MTG...-K.  V.T.....T  ......K.   27
Pea3.6     ...-.I..S..TTVS.ASRG.SA  A....G....MTG...-K.  V.T.....T  ......K.   13
Pea8.0     ...-.I..S..TTVS.ASRV.SA  A....G....MTG...-K.  V.T.....T  ......K.   148
Pet511     ....VM.....--..S..A....  ..................  .........  ........   151
Pet112     ....VM..S.A-V..S..A....  ..................  .........  ........   28
Pet231     ....VM.....--..S..A....  ..................  .........  ........   28
Pet911     ....VM.....--..S..A....  ..................  .........  ........   28
Pet491     ....VM.....--..N..A....  ...........T..  .........  ........   28
Pet611     ....VI.....--..SS.AV...  ...........SA...TK.  N......L.  ......S.   28
Pet301     ....VI.....--.....V....  .....N.....V.......  .........  ........   28
PotC       ....IV.....--...S.V....  ...............TK.  NN.V....L.  ......S.   163
Pot1       ....VI.....-ATRTNVTQAG.  .I...........T......  .........  ......R.   163
Pot2A      ....VM.....--...G.G....  ..................  .........  ......R.   163
Pot2B      ....IV.....--...A.G....  ..G........T......  .........  ......R.   163
Pot2C      ....VM.....--...G.G....  ...........T......  .........  ......R.   163
Rape       ..Y........----V.S....T  ...........SSA...T..  A.N.....V  ......NS    7
Rs         ............----V.SQL..T  ...........S.A...T..  T.T.......  ......S.   55
Sp         ...-LM.N...--V.ASTA...N  .....S....TSA......  S.V....L.  ......N.   134
Soy1       .....I..P...--.TV.R.G.G  ...........M.G..-T..  T.N.......  ........    8
Soy4       .....I..P...---.TV.R.G.G  ...........M.GL.-T..  T.N.......  ........   52
Tom1       ....IV....A--...S.V....  ...............TK.  NN.V....L.  ......R.   143
Tom2A      ....VI.....--...S.VT...  ...........S.T...TK.  .........  ......S.   143
Tom3A      ....VM....:.--...G.G....  ...........T......  .........  ......S.   143
Tom3B      ....IV.....--...G.G....  ...........T......  .........  ......S.   143
Tom3C      ....VM.....--...G.G....  ...........T......  .........  ......S.   143
Rice       ..PSV-MASSAT----------  T....Q.---SSPPPAC.R  PPSELQLRR  QH...IR.   164
Wht9       ..PAV-MASSAT----------  T....Q....T.GL.I.CR  SGSTGLSSV  .....IR.   11
Wht4.3     ..PAV-MASSAT----------  T....Q....T.GL....R  S-RGSLGSV  .....IR.   11
Zm         ..PTVMMASSAT----------  A....Q....T..L..A.R  SSRSLGNV-  .....IR.   84


Cr1                             MAAVI AKSSVSAAVA RPARSSVRPM AALKPAVKAA PVAAPAQANQ   51
Cr2                             ..... .......... .......... .......... ..........   51
CrPep                           ..-.. .......... .......... .......... ......G..D  127
```

Figure 1 Amino acid sequences of the transit peptides from prSS. The sequences are shown using the standard single-letter code for the amino acids. The consensus sequence is the residue that appears at a particular position most frequently. A dot indicates the same amino acid as the consensus sequence. A dash indicates a space introduced to maximize alignment of the sequences. When the sequences are not related to the consensus, the dots refer to the top sequence in the group. The two-letter abbreviations represent the scientific name of the species as follows: At, *Arabidopsis thaliana;* Ft, *Flaveria trinervia;* Ha, *Helianthus annuus;* Lg, *Lemna gibba;*

Two groups have attempted to address the structure-function relations of prSS by generating and analyzing deletions of various regions of the transit peptide of the pea precursor (113, 145, 159). During the course of these studies all regions of the transit peptide were deleted in at least one construct. Very large deletions—e.g. deletion of residues 1–25 (113) or 5–51 (145)—abolished import. However, smaller deletions in any single region allowed import to occur. For example, deletions missing residues 1–5, 5–27, 26–35, 33–52, or 48–57 were capable of import, although some were imported poorly (113, 145, 159). Taken together, these results do not support the hypothesis that specific amino acid sequences have specific essential functions. They are more consistent with a hypothesis in which the essential feature of a transit peptide is some secondary structure (see below).

TRANSIT PEPTIDES OF CAB PRECURSORS   The second most intensively studied precursor with respect to sequence analysis is prCAB. Thirty-one different precursor genes representing 11 plant species have been sequenced (Figure 2). In Figure 2, the CAB precursors have been organized into four groups; two that serve as part of the antenna complex for photosystem II (108) and two that serve as part of the antenna complex for photosystem I (61, 107, 109). The transit peptides are highly conserved among members of each group, but little similarity is seen when comparisons are made among the four groups. As with the prSS transit peptides, it is easy to formulate hypotheses regarding the significance of the conserved features, but these hypotheses still need to be tested experimentally.

TRANSIT PEPTIDES OF OTHER STROMAL AND THYLAKOID MEMBRANE PRECURSORS   Sequences are now available for transit peptides of 13 additional precursors for other stromal proteins (Figure 3) and for five precursors of other thylakoid membrane proteins (Figure 4). Several of these precursors have been sequenced from more than one plant species. Inspection of these sequences reveals that they have little similarity to the sequences of prSS and prCAB transit peptides. Many of these transit peptides do not fit within the framework hypothesis of Karlin-Neumann & Tobin (73). For example, the transit peptide of nitrite reductase (4) has none of the features of the proposed framework. In several cases, similarities can be observed when comparing the same precursor from different species, as was true for prSS and prCAB. For example, glutamine synthase transit peptides from bean and pea are quite

---

Mc, *Mesembryanthemum crystallinum;* Np, *Nicotiana plumbaginifolia;* Ns, *Nicotiana sylvestris;* Nt, *Nicotiana tabacum;* Rs, *Raphanus sativus;* Sp, *Silene pratensis;* Zm, *Zea mays;* Cr, *Chlamydomonas reinhardtii.* The three-letter abbreviations represent the common names of plants as follows: Cuc, cucumber; Pet, petunia; Pot, potato; Soy, soybean; Tom, tomato; Wht, wheat.

```
                                                             REF
PSII-type I proteins:
CONSENSUS        MAAS--- TMALSSPSFA GKAVKLSPSS SEITGNGRVT
At165            ....--- .......A.. ....N...AA ..VL.S....    85
At180            ....--- .......A.. ....N...AA ..VL.S....    85
At140            ....--- .......A.. ........AA ..VL.S....    85
Bar2             ...A--- ......ST.. .....NLS.. ..VQ.DA..S    20
LgAB30           ....--- -.......LV ......A.AA ..VF.E...S    78
NpC              ....--- ......S... .......... .......K..    15
NpE              ....--- .T......-.. .......... ..V....K..   15
Pea80            ....SSS S......TL. ..QL..N... Q.-L.AA.F.    14
Pea66            ....SSS S......TL. ..QL..N... Q.-L.AA.F.   148
Pet13            ...A--- ......S... ....NV-... ....R..K..    33
Pet22L           ...A--- ......ST.. ..V....... .......KA.    33
Pet22R           ...T--- ......S... .......S.. .......K..    33
Pet25            ...A--- ...I..S... ....NV-... .Q.....KA.    33
Pet91R           ...A--- .......... .....F.... .......KA.    33
Tom1A            ...A--- A......... .Q.......A ..NS....I.   105
Tom1B            ...A--- .......... .Q.......A ...S....I.   105
Tom1C            ...A--- .......... .Q.......A ...S....I.   105
Tom3A            ....--- ......ST.. ..T...A... ........I.   105
Tom3B            ....--- ......ST.. .......... ...S....I.   105
Tom3C            ..T.--- ......ST.. .......... ..........   105
Wht              ...T--- ..S...S... .....NL... A-LI.DA..N    82
Zm1              ..S.--- ......TA.. .....-NV... .--F.EA..T    88
Zm1084           ....--- ...I..TAM. .TPI.-VG.- ---F.EG.IT   144


PSII-type II proteins:
Pet37            MATSAI QQSAFAGQTA LKSQNELVRK IGSFGGGRAT   140
Tom4             ...C.. .....V..AV G.....FI.. V.N..E..I.   108
Sp               ....T. .........L ..P....... V.-GN...VS   134
Lg19             ..A... .S........ ..QRD..... V.-VSD..FS    72


PSI-type I proteins:
Tom6A     MASN TLMSCGIPAV CPSFLSSTKS KFAAAMPVSV GATNSMSRFS?    61
Tom6B     .... .......... .......... .......... ..........?   107


PSI-type II proteins:
Pet15     M ASACASSTIA AVAFSSPSSQ KNGSIVGATK ASFLGGKRLR?   141
Tom7      . .......... .......R R......T.. ......R...?       109
```

Figure 2 Amino acid sequences of the transit peptides from prCAB. The sequences are shown using the same format as described in Figure 1 except that the residues are grouped in clusters of 10 amino acids rather than as conserved regions. The abbreviations are also the same, with the addition of Bar for barley. Although some disagreement exists concerning the location of the processing site for CAB, we have chosen to show it at the indicated location, which makes methionine the first residue of the mature protein. A question mark at the end of a sequence indicates that the location of the processing site has not been determined precisely. The sequence listed may contain some residues from the mature protein, or may lack a few residues of the transit peptide.

REF

Acyl Carrier Protein:
```
Barley I        MAHCLAAVS SFSPSAVRRR LSSQVANVVS SRSSVSFHSR QMSFVSISSR PSSLRFKICC   56
Spinach I       MASLSA TTTVRVQPSS SSLHKLSQGN GRCSSIVCLD WGKSSFPTLR TSRRRSFISA  124

Rape 28F10      MATT FSASVSMQAT SLATTTRISF QKPVLVSNHG RTNLSFNLSR ---TRLSISC  123
Rape 10H11      .... .......... .......... .......... .......... ---.......  123
Rape 04F05      .... .......... ..V....... .......... .......... ---.......  123
Rape 34C02      .... ......TL.. ....P..... ...A...--- ........R. SIP....V..  123
Rape 29C08      .S.. .CS....... ...A....... ...A...--- .......... SIP....V..  123
Turnip          .S.. .CS....... ...A....... ...A...--- T.......R. SIP..F....  121
```

Acetolactate Synthase:
```
Arabidopsis     MAAATTTTTT SSSISFSTKP SPSSSKSPLP ISRFSLPFSL NPNKSSSSSR RRGIKSSSPS
                SISAVLNTTT NVTTTPSPTK PTKPETFISR FAPDQPRKGA?   92
Tobacco         MAAAAPSPSS SAFSKTLSPS SSTSSTLLPR STFPFPHHPH KTTPPPLHLT HTHIHIHSQR
                RRFTISNVIS TNQKVSQTEK TETFVSRFAP DEPRKGSDVL?   92
```

EPSP Synthase: Petunia/Tomato/Arabidopsis
```
MAQINN MAQGIQTLNP NS--NFHKPQ VPKSSSFLVF GSKKL-KNSA NSMLVLKKDS -IFMQKFCSF RISASVATAQ  48a
....SS ........SL ..SNLSKTQK G.LV.NS.F. .....TQI.. K.LG.F.... LVRVVRKS.. .........E  48a
...VSR ICN.V.NPSL I-SNLSKSS. RKSPV.LKTQ QHPRAYPI.S SWGLKKSGMT L.GSELRPLK VM.-..S..E   76
```

Ferredoxin:
```
Silene          MA STLSTLSVSA SLLP-KQQPM -VASSLPTNM GQALFGLK-A GSRGRV-TAM  133
Pea             .. T.PALYGTAV .TSFLRT... PMSVTTTKAF SNGFL...TS LK..DLAV.. 32,a
```

Glutamine Synthetase:
```
Pea             MAQILAPS TQWQMRITKT SPCATPITSK MWSSLVMKQT KKVAHSAKFR VMAVNSENGT?  149
Bean            ........ ......F..S .RH.S....N T.,..L...N ..TS-..... .L..K.DGS.?   86
```

Glyceraldehyde-3P-dehydrogenase:
```
Corn            MASSML SATTVPLQQG GGLSEFSGLR SSASLPMRRN ATSDDFMSAV SFRTHAVGTS GGPRRAPTEA?  112
Tobacco A subunit    <NSSLQV SNKGFSEFSG LRTSSAIPFG RKTNDDLLSV VAFQTSVIGG GNSKRGVVEA?  129
Tobacco B subunit    <CLS KKFEVAEFAG LRSSGCVTFS NKESSFFDVV SAQLTPKTTR STPVKGERVA?  129
```

Heat shock protein:
```
Pea             MAQ SVSLSTIASP ILSQKPGSSV KSTPPCMASF PLRRQLPRLG LRNVRAQAGG?  154
```

Nitrite reductase:
```
Spinach         MA SLPVNKIIPS STTLLSSSNN NRRRNNSSIR    4
```

Phosphoribulokinase:
```
Spinach         M AVCTVYTIPT TTHLGSSFNQ NNKQVFFNYK RSSSSNNTLF TTRPSYVITC  116
```

Pyruvate Pi Dikinase: corn
```
M AASVSRAICV QKPGSKCTRD REATSFARRS VAAPRPPHAK AAGVIRSDSG AGRGQHCSPL RAVVDAAPIQ   89
```

Rubisco activase:
```
Spinach         MATAVSTVG AATRAPLNLN GSSAGA-SVP TSGFLGSSLK KHTNVRFPSS SRTTSMTVKA  162
Arabidobsis     ..A....... ..N....S.. ..GS..V.A. A.T...KKVV TVSRFAQSNK KSNG.FK.L.?    b
```

UDPGlc:starch glucosyl transferase: Corn-1st/Barley-2nd
```
MA ALATSQLVAT RAGLGVPDAS TFRRGAAQGL RGARASAAAD TLSMRTSARA APRHQQQARR GGRFPSLVVC   77
.. .......ATS GTV...T.RF RRPGFQGLRP .NPADA.LGM RTIGASA.PK QS.KAHRGS. RCLSVVVSAT?  117
```

Unknown:
```
Pea             MAVSSCQ SIMSNSMTNI SSRSRVNEFT NIPSVYIPTL RRNVSLKVRS?   79
```

Figure 3   Amino acid sequences of the transit peptides from precursors to stromal proteins. The sequences are shown using the same format as described in Figure 2. Generally, incomplete sequences of transit peptides are not shown. The only exceptions are the two subunits of glyceraldehyde-3-P dehydrogenase. The < indicates that the amino terminal residues are missing. Reference a: W. F. Thompson, personal communication; reference b: J. Werneke and W. L. Ogren, personal communication.

similar (Figure 3). In other cases few similarities are observed—e.g. between ferredoxin transit peptides from pea and *Silene* (Figure 3). The general picture that emerges is that sequence similarities generally exist among transit peptides of the same precursor derived from different plant species. In contrast, few similarities are found among different precursors, even when the precursors are derived from the same plant species (Figures 1–4).

TRANSIT PEPTIDES OF THYLAKOID LUMEN PRECURSORS   Cytosolically synthesized precursors for proteins that reside in the thylakoid lumen are a special group because these proteins must cross three membranes—i.e. the two envelope membranes plus the thylakoid membrane. Studies with plastocyanin provide evidence for two separate steps in the transport process (75, 131, 135). The first step, transport across the two envelope membranes, is similar to the import of other chloroplastic proteins. The second step, transport across the thylakoid membrane, is unique to lumenal proteins. The precursors of lumenal proteins have composite transit peptides with two domains that direct the two steps (75, 132, 135). The amino terminal domain is functionally analogous to the transit peptides of stromal precursors. It carries out the same function—i.e. transport across the envelope membranes—and consequently is similar in its gross structural features to other transit peptides. The second domain is very different in structure and function (132). It directs the transport of lumenal proteins across the thylakoid membrane (132, 161). Its structure is more analogous to that of a bacterial signal peptide (161) and contains a stretch of hydrophobic amino acids (underlined in Figure 5). Again, the sequence similarities between transit peptides from

```
                                                                              REF
CF1-delta subunit: Spinach
   MAALQNPVAL QSRTTTAVAA LSTSSTTSPP KPFSLSFSSS TATFNPLRLK ILTASKLTAK PRGGALGTRM   58

Ferredoxin/NADP reductase:
Pea        MAAAV TAAVSLPYSN STSLPIRTSI VA-PERLVFK KVSLN--NVS ISGRVGTIRA   97
Spinach    .TT.. .....F.STK T...SA.S.S .IS.DKISY. ..P.YYR... AT.KM.P...   66

PS I-Subunit II
Tomato     MAMA TQASLFTPPL S--VPKS--T TAPWKQSLVS FSTPKQLKST VSVTRPIRAM   62
Spinach    .... ...T..S.SS LSSAKPIDTR LTTSFKQPSA VTFASKPASR HHSIRAAAAA?   81

PS II-10kDa protein
Spinach    M ATSVMSSLSL KPSSFGVDTK SAVKGLPSLS RSSASFTVRA   83
Potato     . .ST....... ..-T.TLE-. TS.......A ...S..K.V.?   34,
                                                                              160
Rieske FeS: Spinach
   MIISIFNQ LHLTENSSLM ASFTLSSATP SQLCSSKNGM FAPSLALAKA GRVNVLISKE RIRGMKLTCQ   142
```

Figure 4   Amino acid sequences of transit peptides from precursors to thylakoid membrane proteins. The sequences are shown using the same format as described in Figure 2.

PROTEIN TRANSPORT INTO CHLOROPLASTS 479

Cytochrome C552:

```
                                                                                      REF

Chlamydomonas   MLQLANRS VRAKAARASQ SARSVSCAAA KRGADVAPLT SALAVTASIL LTTGAASASA    94
```

Oxygen evolving complex:

33 kDa proteins:

```
Spinach  MAAS LQASTTFLQP TKVASRNTLQ LRSTQNVCKA FGVESASSGG RLSLSLQSDL KELANKCVDA TKLAGLALAT SALIASGANA   152
Pea      .... ...AA.LM.. ..LR.-.... .K.N.S.S.. ..L.HY-..A KVTC.....F ....H...E. S.I.F..... ...VV...S.     a
Chlamydomonas                    M ALRAAQSAKA GVRAARPNRA TAVVCKAQDV GQAAAAALA TAMVRGSANA                   b
```

23 kDa proteins:

```
Spinach  MA STACFLHHHA AISSPAAGRG SAAQRYQAVS IKPNQIVCKA QKQDDN-EAN VLNSGVSRRL ALTVLIGAAA VGSKVSPADA    67
Pea      .. ..Q.....QY ..TT.T--.T LS-..-.V.T T...H..... .....VVD.V .-------.. ....S...... ..........     a
Chlamydomonas          MATALCN KAFAAAPVAR PASRRSAVVV RASGSDVSRR AALAGFAGAA ALVSSSPANA                   90
```

16 kDa protein:

```
Spinach  MAQ AMASMAGLRG ASQAVLEGSL QISGSNRLSG PTTSRVAVPK MGLNIRAQQV SAEAETSRRA MLGFVAAGLA SGSFVKAVLA    67
```

Plastocyanin:

```
Arabidopsis  MAA ITS-ATVTIP SFTGLKLAVS SKPKTLSTIS RSSSATRAPP KLALKSSLKD FGVIAVATAA SIVLAGNAMA   157
Pea          ..T V..-T..A.. ..S...TNAA T.VSAMA--K IPTSTSQS.- R.CVRA.... R........ .A...S..L.      a
Silene       ..T V..S.A.A.. ..A...ASST TRAATV---- --KV.VAT.- RMSI.A.... V..VVA.... AGI.......    132
Spinach      ..T VA.S.A.AV. ......ASG. I..T.A---K IIPTT.AV.- R.SV.A...N V.AAV..... AGL.......    122
```

Figure 5 Amino acid sequences of the transit peptides from precursors to thylakoid lumen proteins. The sequences are shown using the same format as described in Figure 2. The hydrophobic region of the second domain is underlined in each sequence. Reference a: J. C. Gray, personal communication; reference b: S. Mayfield, personal communication.

the same precursor are greater than the similarities between different precursors from the same species (Figure 5).

Common Features of Transit Peptides

GENERAL FEATURES OF THE PRIMARY STRUCTURE   Despite the lack of sequence similarities, the transit peptides of the various precursors (Figures 1–4) and the first domain of lumenal precursors (Figure 5) share several features. They are rich in the hydroxylated amino acids serine and threonine, with these two residues generally constituting 20–35% of the total. They are also rich in small hydrophobic amino acids such as alanine and valine. Transit peptides generally have a net positive charge, although they are not especially rich in basic amino acids. However, they are generally deficient in acidic amino acids, usually having only one or two.

The length of transit peptides varies from 29 residues in one CAB precursor from corn to nearly 100 residues in acetolactate synthase. The exact length of many transit peptides is not known because the precise location of the proteolytic processing site has not been established. In these cases, the transit peptides in Figures 2–4 have a question mark at the end of the sequence. This indicates that the listed sequence may include some of the mature protein or may not include all of the transit peptide.

The lack of sequence similarity among transit peptides from different precursors can be interpreted in at least two different ways. First is that the precursors interact with different receptors and that the differences in transit peptide sequences represent the requirements of the receptors. If this hypothesis were correct, one would predict that a limited number of receptors should exist, and it should be possible to identify groups of precursors that interact with a particular receptor. Although this hypothesis cannot be ruled out at this time, there is little evidence to support it. The second hypothesis is that the essential feature of transit peptides is not found in the amino acid sequence, but rather in some higher-order structure. This hypothesis draws support from analogies with signal sequences and mitochondrial presequences, where current evidence supports the idea that the essential features are particular secondary structures (119, 156). In the case of mitochondrial presequences (the most relevant analogy) it has been suggested that the essential secondary structure is an amphiphilic helix (118, 119, 155, 156). The evidence in favor of this hypothesis and its potential relevance to chloroplastic transit peptides are considered in more detail in the following paragraphs.

THE SECONDARY STRUCTURE OF TRANSIT PEPTIDES   Peptides that form an amphiphilic helix have polar residues and nonpolar residues interspersed in a periodic fashion so that when the peptide is folded into a helix, one face of the helix is nonpolar and the other face is polar. These structures are said to have a hydrophobic moment (39) and react spontaneously with the surfaces of biological membranes (71). Sequences capable of forming an amphiphilic helix can be identified by plotting the sequence of amino acids on a helical wheel (118, 125), or by using computer programs that can calculate the hydrophobic moment for various secondary structures (38, 39, 155). These methods have been used to predict that many mitochondrial presequences form amphiphilic helixes (118, 155).

In addition to the theoretical considerations predicting that mitochondrial presequences should form an amphiphilic helix, there are experimental data to support this conclusion. Synthetic peptides corresponding to the presequence from several precursors are water soluble but will spontaneously insert into a synthetic membrane (48, 65, 118, 120). Tamm has studied the interaction of one synthetic peptide with a membrane and demonstrated that it inserts with the axis of the helix parallel to the plane of the membrane (146). Allison & Schatz (2) used synthetic oligonucleotides to construct two types of artificial mitochondrial presequences. The first set contained only arginine, serine, and leucine in ratios to match that of basic, hydroxylated, and hydrophobic residues in natural presequences. These presequences could direct an attached protein into mitochondria in vivo and in vitro. The other set of artificial sequences contained only glutamine, arginine, and serine residues and were inactive in mitochondrial targeting. In a later study with additional artificial presequences, Roise et al (120) found that all import-active presequences were amphiphilic, whereas the inactive presequences were not amphiphilic. Most importantly, one of the active presequences was highly amphiphilic, but did not form an amphiphilic helix. They concluded that amphiphilicity is the key feature and that an amphiphilic helix is just one example of an amphiphilic structure that will provide biological activity. In this study, the active peptides were identified as being amphiphilic by establishing that they interacted with membranes (120). This point may be important in considering the essential features of chloroplastic transit peptides.

We have conducted preliminary analyses of selected chloroplastic transit peptides using both the helical wheel method (125) and computer programs to calculate the hydrophobic moment for various secondary structures (38, 39). We have not been able to identify regions of chloroplastic transit peptides that are predicted to form amphiphilic helixes. However, we cannot draw a firm conclusion from these results because the calculated hydrophobic moment is dependent upon the hydropathy values assigned to each amino acid (25, 42). Many different scales have been developed (25, 42). In several cases there is significant disagreement about the hydropathy values for the amino acids most prevalent in transit peptides. For example, Engelman et al (42) assign significantly more hydrophobic values to serine and threonine than those used by other authors. They point out that the hydropathy values for these amino acids should take into account whether or not the side-chain hydroxyl groups are involved in hydrogen bonds. In our analysis, we have used only one of the many possible hydropathy scales, and therefore our conclusions must be considered preliminary. We tentatively conclude that it is unlikely that an amphiphilic helix is an essential structure in all transit peptides. However, some transit peptides contain regions that are predicted to form amphiphilic beta structures (142, 152). Thus, the situation for chloroplastic transit peptides may be similar to that of mitochondrial presequences in that amphiphilicity is the essential feature and no one particular secondary structure is required.

This hypothesis needs to be evaluated experimentally in chloroplasts as has been done with mitochondrial presequences and synthetic peptide analogs. One factor that makes the use of synthetic peptides more difficult in chloroplast systems is the length of transit peptides. Most transit peptides contain 50 or more amino acids, and it is difficult and expensive to prepare synthetic peptides of this length. Thus, synthetic peptides corresponding to portions of a transit peptide must be used. Until more specific information is available regarding the essential portions of transit peptides, the regions to be used in preparing synthetic peptides will have to be chosen by trial and error.

The hypothesis that the essential feature of a transit peptide is amphiphilicity has interesting implications. Amphiphilic peptides interact directly with the lipid bilayer of biological membranes (71). They can cause alterations of bilayer structure (118, 120). These interactions may be important in the mechanism of protein translocation (74). In any event, the possibility of interactions between transit peptides and the lipids of the membrane forms an interesting complement to the more popular notion that precursor binding is mediated solely by receptor proteins.

PRECURSOR BINDING AND RECEPTOR IDENTIFICATION

The first step in the transport of precursor proteins into chloroplasts is a specific binding interaction between precursor molecules and the organelle surface. The binding step should be specific for chloroplastic precursors to account for the organelle specificity observed in plant cells (9). The components of the chloroplastic envelope involved in precursor binding and translocation have not been identified, but they most likely include both membrane lipids and intrinsic proteins.

Measurement of Binding

Unlike the early studies of precursor binding (40, 54, 104) in which a complex mixture of polyadenylated translation products was used, more recent investigations (24, 47, 99) have utilized radiochemically pure precursors. Binding assays are performed either in the cold (54, 47) or in the absence of high levels of ATP (24, 99) to inhibit transport of the bound precursors. Upon restoration of import conditions (by raising the temperature or adding exogenous ATP), 50–85% of bound prSS can subsequently be imported. Mature-sized SS binds poorly to intact chloroplasts (47). Therefore, the observed binding of prSS is specific and physiologically significant.

Although protein translocation into chloroplasts is known to require ATP hydrolysis (see next section), binding has generally been thought to be energy independent. Recently, however, work in our laboratory has shown that low levels of ATP (50–100 μM) stimulate high-affinity binding of several chloroplastic precursors (99).

Nature of Binding

EVIDENCE FOR RECEPTOR-MEDIATED BINDING   In 1979, Chua & Schmidt (22) predicted that "a specific envelope receptor or class of receptors facilitates post-translational transport" of precursors into chloroplasts. Definitive proof of the existence of such receptors is still not available today, though there is much indirect evidence supporting the involvement of envelope proteins in precursor binding.

One approach to showing membrane protein involvement in binding was taken by Cline et al (24) and Friedman & Keegstra (47). They pretreated intact chloroplasts with the protease thermolysin, which specifically destroys outer envelope polypeptides without affecting inner envelope proteins or envelope permeability properties (23). Pretreatment of intact chloroplasts with thermolysin prior to binding resulted in significant (though not total) inhibition of binding of both prSS and prCAB. This suggests that the protease destroyed or somehow inactivated a protein component of the chloroplast outer envelope that was necessary for high-affinity binding of precursors.

In the mitochondrial literature, protein transport is considered to be receptor mediated if it meets certain criteria (58, 114). These criteria can also be applied to protein transport into chloroplasts to support the idea of receptor involvement. First, a receptor should be exposed at the cytoplasmic face of the outer envelope so that it is accessible to newly synthesized precursors. The protease pretreatment experiments discussed above show that this requirement is met by chloroplasts. Second, precursor binding to chloroplasts should be rapid and productive; bound precursors can subsequently be transported into chloroplasts (see discussion above). Third, binding sites should be limiting and saturable. Friedman & Keegstra (47) determined that the extent of saturable binding of prSS varied between 1500 and 3500 binding sites per chloroplast. Pfisterer et al (104) calculated a comparable number of binding sites, using a mixture of precursor proteins. Fourth, precursor binding should be ligand specific; chloroplastic precursors lacking transit peptides do not bind to chloroplasts (47). (Precursors destined for other organelles have not been tested but should not bind to chloroplasts.) Finally, binding should be membrane specific. Chloroplastic precursors do not bind to mitochondria (A. L. Friedman, unpublished observations), plasma membranes (A. Yousif and K. Keegstra, unpublished observations), or erythrocyte membranes (40). They do, however, bind to thylakoids (40) and isolated inner envelope vesicles (A. Yousif and K. Keegstra, unpublished observations) as well as to outer envelopes. Taken together, these data present a strong case for receptor involvement in protein import into chloroplasts.

That energy in the form of ATP is required (99) provides new compelling evidence that binding of precursors to chloroplasts is a protein-mediated event. It is difficult to imagine a role for ATP in binding that does not involve a membrane protein in some way. There are many proteins in chloroplast envelopes that could be components of the transport apparatus, including protein kinases (138), chloroplast envelope ATPases (93), and several other proteins implicated as potential receptors (see discussion below).

Protein transport into chloroplasts and mitochondria is often considered to occur by similar mechanisms (153). Much evidence exists that mitochondrial import is receptor mediated (57, 102, 114, 165). Though most of the criteria for receptor involvement discussed above are met by import into mitochondria, the strongest evidence seems to be the existence of protease-sensitive components of the outer membrane that are required for high-affinity binding (114, 165) and subsequent translocation of mitochondrial precursors. Import into mitochondria also requires ATP (103), though its exact role is unclear. In fact, precursor-specific differences in energy requirements for import, as well as differences in protease sensitivity of membrane components (165), have been used to predict the existence of at least three receptor classes involved in mitochondrial import (102).

Gillespie and her coworkers showed that a synthetic mitochondrial peptide blocked the import of several mitochondrial precursors; the inhibition could be overcome by increasing the precursor concentrations (50). This synthetic peptide bound specifically to an outer membrane protein (49). Binding was saturable, reversible, and membrane specific—attributes characteristic of ligand-receptor interactions.

Buvinger et al (12) synthesized a peptide corresponding to the first 19 amino acids of *Arabidopsis* prCAB. Import of both prCAB and prSS is completely inhibited by 160 μM peptide. This synthetic peptide, which has been shown to have some correlation with chloroplastic import activity, may prove useful in attempting to identify receptor proteins on the surface of chloroplastic envelopes.

EVIDENCE FOR LIPID-MEDIATED BINDING   There is less evidence that precursor binding to chloroplasts is lipid mediated than there is for receptor-mediated binding. However, if the analogy to mitochondrial import is continued, some indirect evidence may be inferred. Like chloroplastic transit sequences (see discussion above), mitochondrial presequences are necessary and sufficient for targeting an attached passenger protein to its correct intramitochondrial location (63). No significant amino acid sequence similarities are seen in an analysis of nearly two dozen mitochondrial sequences (155). However, these sequences are generally thought to fold into similar secondary structures—i.e. amphiphilic helixes (118, 155). As discussed above, several groups (2, 118, 120) have synthesized artificial presequences for mitochondrial precursors. In each case, amphiphilic (but not necessarily helical) presequences were sufficient for import of the precursor, while the nonamphiphilic presequences were inactive. Thus, although it seems that chloroplastic transit peptides do not form amphiphilic helixes, an alternative amphiphilic secondary structure may be sufficient for direct interaction between the membrane lipids and the hydrophobic portion of the transit peptide.

It is also worth mentioning that some binding is seen even after chloroplasts have been pretreated with protease (24, 47). If the protease has actually digested most of the envelope surface proteins, it is possible that the residual binding observed is due to lipid-precursor interaction.

GENERAL MODEL INVOLVING BOTH LIPIDS AND MEMBRANE PROTEINS   Ultimately, it seems most likely that chloroplastic precursor binding involves both protein-protein and protein-lipid interactions. This very general hypothesis allows certain predictions to be made that can then be evaluated in an effort to focus the model.

To make the general model more specific and testable, the regions of the transit peptide that interact with membrane proteins or lipids will need to be identified, as well as the essential (or other) secondary structures involved. One possible approach is to examine the influence of synthetic peptides on binding and translocation. Kaiser & Kezdy have used synthetic peptides to test amphiphilic helical structure models for several peptide hormones (70, 71). They designed model peptides that optimized the amphiphilicity of the secondary structure, and observed stronger interactions and biological activity with the synthetic peptides than with the naturally occurring analogs (70). Synthetic peptides have also been used to test secondary structure predictions and structural requirements for function of mitochondrial presequences (see discussion above; 2, 118, 120, 155) and signal peptides involved in bacterial protein export (10, 18).

Two major roles of amphiphilic secondary structures that have been proposed for peptide hormones (71) may be considered as possible functions of transit peptides in precursor binding to chloroplasts. One is in positioning nonamphiphilic portions of the peptide (such as a specific recognition site) in the correct orientation for productive interactions with the receptor. Second, once a specific secondary structure of the peptide has been induced on a membrane surface near the receptor, the peptide can find its receptor more easily by diffusion in only two directions, rather than searching in three dimensions as would be necessary without the amphiphilic structure. Thus a transit peptide (lacking secondary structure in aqueous solutions) may be induced to form an amphiphilic secondary structure by a nonspecific binding event at the amphiphilic surface of the chloroplast envelope. This lipid-mediated binding may then facilitate more specific binding of the transit peptide to its receptor.

Identification of Receptor(s)

Several groups have attempted to identify putative receptors involved in protein import into chloroplasts and mitochondria. So far, only one specific protein has been implicated as a possible receptor for mitochondrial import (49). Several different approaches have been utilized in chloroplast studies and have led to the identification of at least two different proteins that may be components of the transport apparatus.

CROSS-LINKING STUDIES   Cornwell & Keegstra (26) used a heterobifunctional, photoactivatable cross-linking reagent to identify a chloroplast surface protein that was associated with prSS binding. The general strategy was to (a) modify the precursor by reaction with the cross-linker; (b) bind the modified precursor to intact chloroplasts, presumably to a specific receptor on the envelope surface; and (c) cross-link the modified precursor to the receptor by photoactivation of the cross-linking reagent. The resulting cross-linked conjugate was sensitive to protease digestion, and so is probably an envelope surface protein. The size of the protein identified in this study was calculated to be 66 kDa.

USE OF SYNTHETIC PEPTIDES   Two different groups have used chemically synthesized peptides as tools in the identification of a receptor for chloroplastic protein import. Pain et al (101) raised antibodies against antibodies that were directed against a synthetic peptide analog of the carboxyl-terminal 30 residues of the transit peptide of pea prSS. These anti-idiotypic antibodies mimic the transit peptide and presumably interact with the same binding site as the transit peptide on the putative import receptor. In support of this interpretation, the anti-idiotypic antibodies were able to block the import of prSS. The antibodies were also used for immunoblotting and were found to react with two different chloroplastic proteins. The first was the large subunit of Rubisco; the reason for this reaction remains unexplained. The second was a major 30-kDa protein of the envelope membranes. The major 30-kDa polypeptide in the inner envelope membrane is known to be the phosphate translocator (44, 68). However, Pain et al subsequently reported that their antibodies react with a 30-kDa protein different from the phosphate translocator (68). They conclude that this 30-kDa protein of the envelope membranes is the receptor for prSS (101).

Kaderbhai et al (69) also observed that a 30-kDa protein was labeled during their cross-linking experiments and concluded that they were labeling the same receptor identified by Pain et al (101). Kaderbhai et al (69) cross-linked a photoactivatable synthetic peptide corresponding to the first 24 residues of wheat prSS to isolated envelope membranes from pea chloroplasts. They observed that many proteins were labeled, with the largest quantity of label incorporated into the 52-kDa large subunit of Rubisco and the 30-kDa phosphate translocator. Unlike Pain et al (101), Kaderbhai et al (69) provided evidence that the 30-kDa protein they observed was the phosphate translocator. They speculated that it has an additional function in protein transport. However, no controls were reported to demonstrate the specificity of the cross-linking reaction, and it is possible that the observed labeling represents nonspecific reactions, the labeling of the most abundant proteins being most prominent.

Although the various approaches discussed above come to very different conclusions about the identity of the prSS receptor, they may all be correct. It is possible that a receptor has more than one polypeptide subunit and that the various approaches reveal this complexity. Alternatively these approaches may be complementary and may have identified different proteins involved in precursor transport. Further work will be needed to resolve which, if any, of these proteins is the prSS receptor.

THE ENERGY REQUIREMENT FOR PROTEIN IMPORT INTO CHLOROPLASTS

The translocation of proteins across biological membranes requires energy. ATP is the sole energy source driving protein transport (5, 6, 17, 64, 98) except in the export of proteins from Gram-negative bacteria and the import of proteins into mitochondria. While both still require ATP, the export of proteins from bacteria is facilitated by a protonmotive force (PMF) (16). Perhaps more relevant to the subject of chloroplast protein import is the fact that the import of proteins into mitochondria requires the presence of both ATP and an electric field across the inner membrane (19, 36, 103).

A PMF Is Not Required in Chloroplasts

The energy requirement for protein transport into chloroplasts was first addressed in 1980 by Grossman et al (54). Their experiments with inhibitors and ionophores led them to conclude that ATP alone is required for protein transport into the stroma. This conclusion was supported a few years later by Cline et al (24), who demonstrated that nigericin-induced inhibition of light-driven protein import could be overcome with exogenously added ATP. More recent experiments have shown that protein import cannot be supported by other nucleotide triphosphates (CTP, GTP, UTP), or by nonhydrolyzable ATP analogs (45, 100, 126).

The many similarities between protein import into mitochondria and chloroplasts raised the possibility that the energy requirement for the two processes might also be the same. In an attempt to discover any previously overlooked requirement for either an electric or pH component of a PMF, three independent research groups have recently reexamined the energy requirements for the translocation of prSS across the envelope membranes (45, 100, 126). Their experiments, which were performed with a larger array of ionophores than used previously, confirmed that a PMF is not involved, and that ATP is the sole energy source necessary for protein import into the chloroplast stroma. These experiments were subsequently expanded to include prFD and prPC as well as prSS, again with the same results (147).

It is noteworthy that the latter study included prPC (147). This is the only protein whose import energetics have been studied so far that crosses the thylakoid membrane in addition to the two envelope membranes. If the mitochondrial and chloroplastic energy requirements are in fact similar, one might expect that a PMF would be required for the translocation of only those proteins crossing the energy-transducing (PMF-bearing) membranes—i.e. the inner membrane in mitochondria and the thylakoid membrane in chloroplasts. Nonetheless, neither the electrical nor chemical components of a PMF are involved in the translocation of prPC across all three membranes in chloroplasts (147).

ATP Is Required for Precursor Binding

As described in previous sections, the first step in the import of proteins into chloroplasts is the binding of precursors to the external surface of the outer envelope membrane, presumably in association with a receptor. Until recently, this binding step was considered to be spontaneous. However, Olsen et al (99) have now demonstrated that the binding step is stimulated by the presence of low levels of ATP.

Though the binding energy requirement has not been characterized extensively, it appears to be different from the translocation energy requirement in a number of ways, indicating that ATP may be utilized differently. First, the ATP concentration required to support binding is 5–10-fold lower than needed to drive translocation [approximately 100 μM to support binding (99) vs approximately 1 mM to support translocation (45)]. Second, the specificity for nucleotide triphosphates appears to be broader for the binding reaction than for translocation; neither CTP nor GTP supports translocation (45, 100, 126), but both are able to substitute for ATP with less than a 50% drop in binding efficiency (99). On the other hand, neither the binding nor translocation reactions are supported by nonhydrolyzable ATP analogs (99, 100, 126), and both appear to require ATP within, rather than outside of, the chloroplasts (99, 100, 147; see below). This newly recognized energy requirement for binding should facilitate ongoing attempts to further subdivide the chloroplast protein import reaction into its component steps.

ATP is Utilized Within Chloroplasts

THE TRANSLOCATION REACTION   The site at which ATP is utilized during binding and subsequent translocation of chloroplastic precursors has a profound influence on the possible mechanisms one can envisage for the import process. This fundamental question concerning the site of ATP utilization has been addressed a number of times in the past few years. Working independently, three research groups (45, 100, 126) used similar techniques to manipulate separately the internal and external ATP concentrations, and examined the resultant effect on protein import. In one of the studies (45), the ATP levels in the two compartments were also measured under conditions similar to those used during the protein import reactions. Schindler et al (126) and Flugge & Hinz (45) concluded that the ATP supporting import was needed outside chloroplasts. In contrast, Pain & Blobel (100) concluded that the ATP was utilized in the stroma.

This issue was recently reexamined by Theg et al (147) using two independent approaches. In the first, the internal and external ATP concentrations were manipulated in the manner described above for the earlier studies. Where possible, the actual ATP levels in the internal and external phases were also measured in the same samples in which protein import was analyzed. These experiments revealed that the import of a number of different precursors occurred at high rates in the complete absence of external ATP, provided that ATP was present in the stroma.

In a second approach, Theg et al (147) monitored the kinetics of import of three different precursors under conditions where ATP was present inside the chloroplasts at the start of the import reaction, or where the internal ATP concentration rose from an initial low level via adenylate translocator activity (115). In the first case, import proceeded at a high rate from the moment the precursor was added to the chloroplast suspension. Under the latter conditions, import commenced only after a short lag phase. Presumably this lag phase represented the time required to accumulate enough internal ATP to support import. The most straightforward interpretation of these two sets of results is that ATP is utilized inside the chloroplasts for protein import.

One possible explanation for the different conclusions reached by the various groups may lie in the different methods used to deplete external ATP. Flugge & Hinz (45) relied on the use of alkaline phosphatase as an external ATP trap. This enzyme, which inhibits protein import, not only degrades ATP but also removes phosphorylated groups from proteins. When Flugge & Hinz (45) used an external trap specific for ATP (i.e. glucose/hexokinase), they observed no inhibition of protein import, in agreement with results obtained by Pain & Blobel (100) and by Theg et al (147). In light of new studies describing the phosphorylation of proteins in the outer envelope membrane (60; see below), it appears likely that the inhibition of protein import by alkaline phosphatase results from its dephosphorylation of protein moieties rather than its depletion of ATP from the external medium.

THE BINDING REACTION   Manipulations of internal and external ATP levels in chloroplasts were also used to study the precursor binding reaction (99). In these experiments, binding was stimulated by generating ATP within the chloroplasts while an ATP trap was present in the external medium. Binding was inhibited by increasing concentrations of an internal ATP trap in the presence of exogenous (external) ATP. The authors concluded from these results that the ATP utilized for the binding reaction is also located inside the chloroplasts and not in the external medium.

It is important to attempt to define further the internal site of ATP utilization during the binding (and translocation) step. The two possible locations for ATP utilization are (a) the stroma, which includes the inner surface of the inner envelope membrane, and (b) the intermembrane space, which includes the adjacent surfaces of the two envelope membranes. Since the outer envelope membrane is freely permeable to ATP (43), this question cannot be readily answered by manipulating the internal and extrachloroplastic ATP levels with ATP-consuming traps. However, the kinetic experiments (147) rule out the possibility that the translocation of proteins into chloroplasts is driven by ATP in the intermembrane space. In contrast, preliminary results from experiments currently in progress suggest that precursor binding may be mediated by ATP located in the intermembrane space (L. J. Olsen and K. Keegstra, unpublished; see below).

Differences Between Chloroplasts and Mitochondria

Three noteworthy differences in the energy requirements of the respective import mechanisms in mitochondria and chloroplasts can be described. First, protein import into mitochondria requires both ATP and a potential across the inner membrane, whereas chloroplastic protein import requires ATP alone (19, 24, 36, 45, 54, 100, 103, 126, 147). Second, chloroplasts appear to be unique among organelles in requiring ATP to bind precursors to their surfaces (99). While an energy requirement for the binding of precursors to mitochondria has been noted (103), the consensus now appears to be that this step is spontaneous (35, 36, 102). Finally, experiments with mitochondria have led to the conclusion that external, cytoplasmic ATP is utilized for protein import into that organelle (19, 36), whereas the evidence now favors an internal location for ATP utilization during import into chloroplasts (100, 147). The question remains whether these differences in energy requirements reflect subtle changes in a common mechanism for protein import, or whether they signify that two quite different mechanisms operate in the two organelles.

The Role of ATP in Protein Import into Chloroplasts

CHLOROPLASTIC PRECURSORS PROBABLY DO NOT UNDERGO ATP-DEPENDENT UNFOLDING IN THE CYTOPLASM   One hypothesis concerning the role of ATP during mitochondrial protein import posits that ATP is required to unfold precursors in the cytoplasm prior to their import (37). A similar unfolding mechanism has been suggested for protein import into chloroplasts (29). However, the experiments that demonstrate utilization of ATP inside the outer envelope membrane for both precursor binding (99) and translocation (147) (described in the previous section) do not support this proposed mechanism for chloroplastic protein import. If precursors targeted to the chloroplast must unfold prior to their import, they probably do not do so under the direction of cytosolic ATP-dependent unfoldases.

PHOSPHORYLATION OF PRECURSORS AND/OR COMPONENTS OF THE IMPORT APPARATUS   An alternative model for energy use during protein import into chloroplasts involves a phosphorylation/dephosphorylation cycle of components of the import machinery, or even of the precursor itself. There is ample precedent for the modulation of receptor activities through phosphorylation (cf 130). Such a phosphorylation hypothesis has been suggested for protein import into chloroplasts by a number of research groups (45, 100, 126, 138), in part to explain the observed inhibition of protein import by sodium fluoride and sodium molybdate (45, 126), both phosphatase inhibitors. Chloroplastic transit sequences are rich in potential phosphorylation sites (serines and threonines), and a number of protein kinases residing in both the inner and outer envelope membranes have been described (136, 138, 139). One of them has even been shown to accept SS as a phosphorylation substrate (138). However, it appears that none of the known envelope kinases (136, 139) possesses the same inhibitor and/or substrate specificity as the protein translocation reaction. Not enough is known about the ATP requirement for precursor binding to determine if one of the envelope kinases might function in this capacity.

Direct evidence for the participation of a phosphorylation event during chloroplastic precursor binding and import was recently reported by Hinz & Flugge (60). They demonstrated that a 51-kDa protein (termed P51), which may reside in the outer envelope membrane, is phosphorylated in a manner correlated with the import of prSS. Increasing inhibition of prSS import by incubation with pyridoxal 5'-phosphate resulted in a parallel increase in P51 phosphorylation. A similar increase in the phosphorylation of P51 was observed when chloroplasts were pre-incubated with high concentrations of a chimeric precursor under conditions where import was blocked by low ATP levels. Under these same conditions, phosphorylation of the chimeric precursor's transit peptide was occasionally observed. The phosphorylated group(s) on P51 were cleaved by the same low concentrations of thermolysin that inhibited the import of prSS, suggesting that P51 may be a component of the import machinery located in the outer envelope membrane.

Curiously, Hinz & Flugge (60) found a somewhat different pattern of phosphorylation of envelope proteins when the phosphorylation reaction was carried out in organello or in vitro with isolated membranes. Only in the latter case was the phosphorylation found to be reversible by a cold ATP chase, a feature to be expected of proteins undergoing a cycle of phosphorylation/ dephosphorylation. In addition, the phosphorylation of P51 occurred with an apparent K_m(ATP) of 5 μM, contrasting with the apparent K_m(ATP) of 0.9 mM determined for the import of prSS (45). This latter observation suggests that P51 may be a protein involved in the binding of precursors to chloroplast surfaces, rather than in the translocation step; the binding reaction also displays a relatively low ATP requirement compared to translocation (99). Regardless of the precise role of P51, this promising line of research will undoubtedly lead to an enhanced understanding of the overall import process.

CONCLUDING REMARKS

Amino acid sequences are now available from a large number of precursors, and analysis of the transit peptide sequences reveals both interesting conclusions and opportunities for further work. Transit peptides derived from the same precursor of different plant species show some striking similarities. Functional analysis of the similar sequences may provide valuable insight into the structure-function relationships of transit peptides. Consideration of the changes that have occurred during evolution may provide interesting insights into the evolutionary relationships among the various plants (106). Transit peptide sequences show few similarities when comparisons are made between different precursors, even if the precursors are derived from the same species. This makes it unlikely that the common amino acid framework proposed by Karlin-Neumann & Tobin (73) will be generally applicable. Rather, it now seems more likely that the common essential feature will be a secondary structure or a common property such as amphiphilicity. Further work is needed to define more precisely the common essential features.

It is attractive to consider that a specific region of each precursor interacts with a putative receptor protein, but such a region has not yet been identified. In searching for the region of a transit peptide that interacts with a putative receptor, it would be helpful to know whether a single receptor interacts with all precursors or whether different receptors interact with classes of precursors. Unfortunately such information is not yet available. Several groups have made preliminary progress in identifying potential candidates for the prSS receptor (26, 69, 101). However, more work will be required to determine whether any or all of these proteins function as receptors or whether some represent other proteins of the import apparatus.

A reexamination of the energetics of protein import has confirmed that ATP is the only source of energy needed, but has also led to some interesting new conclusions. Originally it was thought that ATP was needed only for protein translocation, but recent results demonstrate that ATP is also needed for efficient precursor binding (99). Although some disagreement still exists, the current evidence supports the hypothesis that ATP for translocation is needed inside the chloroplasts (100, 147). This is an important difference from the situation in mitochondria, where the most popular hypothesis regarding the mechanism of protein transport involves an ATP-dependent unfolding of precursor proteins outside the organelle (37). Although a similar unfolding mechanism has been proposed for transport into chloroplasts (29), alternative mechanisms should also be considered. One possible alternative described recently posits that precursors are transported through areas of the envelope membrane where lipids are temporarily rearranged from their typical bilayer configuration (74). In this scheme, the amphiphilicity of transit peptides plays an important role in triggering the rearrangement that allows translocation to occur. The phosphorylation of envelope proteins may also play a role in mediating or regulating these events. This alternative scheme is highly speculative and will certainly need to be modified as additional information regarding the mechanism of protein translocation becomes available. However, it does provide a valuable alternative model for interpreting data and designing new experiments.

ACKNOWLEDGMENTS

Work from our laboratory was supported by grants from the NSF and the Division of Energy Biosciences at DOE.

Literature Cited

1. Adams, C. A., Babcock, M., Leung, F., Sun, S. M. 1987. Sequence of a ribulose 1,5-bisphosphate carboxylase/oxygenase cDNA from the C4 dicot *Flaveria trinervia*. *Nucleic Acids Res.* 15:1875
2. Allison, D. S., Schatz, G. 1986. Artificial mitochondrial presequences. *Proc. Natl. Acad. Sci. USA* 83:9011–15
3. Anderson, S., Smith, S. M. 1986. Synthesis of the small subunit of ribulose-bisphosphate carboxylase from genes cloned into plasmids containing the SP6 promoter. *Biochem. J.* 240:709–15
4. Back, E., Burkhart, W., Moyer, M., Privalle, L., Rothstein, S. 1988. Isolation of cDNA clones coding for spinach nitrite reductase: Complete sequence and nitrate induction. *Mol. Gen. Genet.* 212:20–26
5. Balch, W. E., Elliott, M. M., Keller, D. S. 1986. ATP-coupled transport of vesicular stomatitis virus G protein between the endoplasmic reticulum and the Golgi. *J. Biol. Chem.* 261:14681–89
6. Balch, W. E., Keller, D. S. 1986. ATP-coupled transport of vesicular stomatitis virus G protein. Functional boundaries of secretory compartments. *J. Biol. Chem.* 261:14690–96
7. Baszczynski, C. L., Fallis, L., Bellemare, G. 1988. Nucleotide sequence of a full length cDNA clone of a *Brassica napus* ribulose bisphosphate carboxylase-oxygenase small subunit gene. *Nucleic Acids Res.* 16:4732
8. Berry-Lowe, S. L., McKnight, T. D., Shah, D. M., Meagher, R. B. 1982. The nucleotide sequence, expression, and evolution of one member of a multigene family encoding the small subunit of ribulose-1,5-bisphosphate carboxylase in soybean. *J. Mol. Appl. Genet.* 1:483–98
9. Boutry, M., Nagy, F., Poulsen, C., Aoyagi, K., Chua, N. H. 1987. Targeting of bacterial chloramphenicol acetyltransferase to mitochondria in transgenic plants. *Nature* 328:340–42
10. Briggs, M. S., Cornell, D. G., Dluhy, R. A., Gierasch, L. M. 1986. Conformations of signal peptides induced by lipids suggest initial steps in protein export. *Science* 233:206–8
11. Broglie, R., Coruzzi, G., Lamppa, G., Keith, B., Chua, N. H. 1983. Structural analysis of nuclear genes coding for the precursor to the small subunit of wheat ribulose-1,5-bisphosphate carboxylase. *Bio/Tech.* 1:55–61

12. Buvinger, W. E., Michel, H., Bennett, J. 1989. A truncated analog of a pre-LHC II transit peptide inhibits protein import into chloroplasts. *J. Biol. Chem.* 264:1195–1202
13. Cashmore, A. R. 1983. Nuclear genes encoding the small subunit of ribulose-1,5-bisphosphate carboxylase. In *Genetic Engineering of Plants*, ed. T. Kosuge, C. P. Meredith, A. Hollaender, pp. 29–38. New York: Plenum
14. Cashmore, A. R. 1984. Structure and expression of a pea nuclear gene encoding a chlorophyll a/b-binding polypeptide. *Proc. Natl. Acad. Sci. USA* 81:2960–64
15. Castresana, C., Staneloni, R., Malik, V. S., Cashmore, A. R. 1987. Molecular characterization of two clusters of genes encoding the Type I CAB polypeptides of PSII in *Nicotiana plumbaginifolia*. *Plant Mol. Biol.* 10:117–26
16. Chen, L., Tai, P. C. 1985. ATP is essential for protein translocation into *Escherichia coli* membrane vesicles. *Proc. Natl. Acad. Sci. USA* 82:4384–88
17. Chen, L., Tai, P. C. 1987. Evidence for the involvement of ATP in co-translational protein translocation. *Nature* 328:164–66
18. Chen, L., Tai, P. C., Briggs, M. S., Gierasch, L. M. 1987. Protein translocation into *Escherichia coli* membrane vesicles is inhibited by functional synthetic signal peptides. *J. Biol. Chem.* 262:1427–29
19. Chen, W. J., Douglas, M. G. 1987. Phosphodiester bond cleavage outside mitochondria is required for the completion of protein import into the mitochondrial matrix. *Cell* 49:651–58
20. Chitnis, P. R., Morishige, D. T., Mechushtai, R., Thornber, J. P. 1988. Assembly of the barley light-harvesting chlorophyll a/b proteins in barley etiochloroplasts involves processing of the precursor on thylakoids. *Plant Mol. Biol.* 11:95–107
21. Chua, N. H., Schmidt, G. W. 1978. Post-translational transport into intact chloroplasts of a precursor to the small subunit of ribulose-1,5-bisphosphate carboxylase. *Proc. Natl. Acad. Sci. USA* 75:6110–14
22. Chua, N. H., Schmidt, G. W. 1979. Transport of proteins into mitochondria and chloroplasts. *J. Cell Biol.* 81:461–83
23. Cline, K., Werner-Washburne, M., Andrews, J., Keegstra, K. 1984. Thermo-

lysin is a suitable protease for probing the surface of intact pea chloroplasts. *Plant Physiol.* 75:675–78

24. Cline, K., Werner-Washburne, M., Lubben, T. H., Keegstra, K. 1985. Precursors to two nuclear-encoded chloroplast proteins bind to the outer envelope membrane before being imported into chloroplasts. *J. Biol. Chem.* 260:3691–96

25. Cornette, J. L., Cease, K. B., Margalit, H., Spouge, J. L., Berzofsky, J. A., et al. 1987. Hydrophobicity scales and computational techniques for detecting amphipathic structures in proteins. *J. Mol. Biol.* 195:659–85

26. Cornwell, K. L., Keegstra, K. 1987. Evidence that a chloroplast surface protein is associated with a specific binding site for the precursor to the small subunit of ribulose-1,5-bisphosphate carboxylase. *Plant Physiol.* 85:780–85

27. Coruzzi, G., Broglie, R., Edwards, C., Chua, N. H. 1984. Tissue-specific and light-regulated expression of a pea nuclear gene encoding the small subunit of ribulose-1,5-bisphosphate carboxylase. *EMBO J.* 3:1671–79

28. Dean, C., van den Elzen, P., Tamaki, S., Black, M., Dunsmuir, P., et al. 1987. Molecular characterization of the *rbcS* multi-gene family of *Petunia* (Mitchell). *Mol. Gen. Genet.* 206:465–74

29. della-Cioppa, G., Kishore, G. M. 1988. Import of a precursor protein into chloroplasts is inhibited by the herbicide glyphosate. *EMBO J.* 7:1299–1305

30. della-Cioppa, G., Kishore, G. M., Beachy, R. N., Fraley, R. T. 1987. Protein trafficking in plant cells. *Plant Physiol.* 84:965–98

31. DeRocher, E. J., Ramage, R. T., Michalowski, C. B., Bohnert, H. J. 1987. Nucleotide sequence of a cDNA encoding *rbcS* from the desert plant *Mesembryanthemum crystallinum*. *Nucleic Acids Res.* 15:6301

32. Dobres, M. S., Elliott, R. C., Watson, J. C., Thompson, W. F. 1987. A phytochrome regulated pea transcript encodes ferredoxin I. *Plant Mol. Biol.* 8:53–59

33. Dunsmuir, P. 1985. The petunia chlorophyll a/b binding protein genes: a comparison of *Cab* genes from different gene families. *Nucleic Acids Res.* 13:2503–18

34. Eckes, P., Rosahl, S., Schell, J., Willmitzer, L. 1986. Isolation and characterization of a light-inducible, organspecific gene from potato and analysis of its expression after tagging and transfer into tobacco and potato shoots. *Mol. Gen. Genet.* 205:14–22

35. Eilers, M., Hwang, S., Schatz, G. 1988. Unfolding and refolding of a purified precursor protein during import into isolated mitochondria. *EMBO J.* 7:1139–45

36. Eilers, M., Oppliger, W., Schatz, G. 1987. Both ATP and an energized inner membrane are required to import a purified precursor protein into mitochondria. *EMBO J.* 6:1073–77

37. Eilers, M., Schatz, G. 1988. Protein unfolding and the energetics of protein translocation across biological membranes. *Cell* 52:481–83

38. Eisenberg, D., Schwarz, E., Komaromy, M., Wall, R. 1984. Analysis of membrane and surface protein sequences with the hydrophobic moment plot. *J. Mol. Biol.* 179:125–42

39. Eisenberg, D., Weiss, R. M., Terwilliger, T. C. 1984. The hydrophobic moment detects periodicity in protein hydrophobicity. *Proc. Natl. Acad. Sci. USA* 81:140–44

40. Ellis, R. J. 1983. Chloroplast protein synthesis: principles and problems. *Subcell. Biochem.* 9:237–61

41. Ellis, R. J., Robinson, C. 1987. Protein targeting. *Adv. Bot. Res.* 14:1–24

42. Engelman, D. M., Steitz, T. A., Goldman, A. 1986. Identifying nonpolar transbilayer helices in amino acid sequences of membrane proteins. *Annu. Rev. Biophys. Biophys. Chem.* 15:321–53

43. Flugge, U. I., Benz, R. 1984. Poreforming activity in the outer membrane of the chloroplast envelope. *FEBS Lett.* 169:85–89

44. Flugge, U. I., Heldt, H. W. 1984. The phosphate-triose phosphate-phosphoglycerate translocator of the chloroplast. *Trends Biochem. Sci.* 9:530–33

45. Flugge, U. I., Hinz, G. 1986. Energy dependence of protein translocation into chloroplasts. *Eur. J. Biochem.* 160:563–70

46. Fluhr, R., Moses, P., Morelli, G., Coruzzi, G., Chua, N. H. 1986. Expression dynamics of the pea *rbcS* multigene family and organ distribution of the transcripts. *EMBO J.* 5:2063–71

47. Friedman, A. L., Keegstra, K. 1989. Chloroplast protein import: quantitative analysis of receptor mediated binding. *Plant Physiol.* In press

48. Furuya, S., Okada, M., Ito, A., Aoyagi, H., Kanmera, T., et al. 1987. Synthetic partial extension peptides of P-450 (SCC) and adrenodoxin precursors:

effects on the import of mitochondrial enzyme precursors. *J. Biochem.* 102:821–32

48a. Gasser, C. S., Winter, S. A., Hironaka, C. M., Shah, D. M. 1988. Structure, expression, and evolution of the 5-enolpyruvylshikimate-3-phosphate synthase genes of petunia and tomato. *J. Biol. Chem.* 263:4280–89

49. Gillespie, L. L. 1987. Identification of an outer mitochondrial membrane protein that interacts with a synthetic signal peptide. *J. Biol. Chem.* 262:7939–42

50. Gillespie, L. L., Argan, C., Taneja, A. T., Hodges, R. S., Freeman, K. B., et al. 1985. A synthetic signal peptide blocks import of precursor proteins destined for the mitochondrial inner membrane or matrix. *J. Biol. Chem.* 260:16045–48

51. Goldschmidt-Clermont, M., Rahire, M. 1986. Sequence, evolution and differential expression of the two genes encoding variant small subunits of ribulose bisphosphate carboxylase/oxygenase in *Chlamydomonas reinhardtii. J. Mol. Biol.* 191:421–32

52. Grandbastien, M. A., Berry-Lowe, S., Shirley, B. W., Meagher, R. B. 1986. Two soybean ribulose-1,5-bisphosphate carboxylase small subunit genes share extensive homology even in distant flanking sequences. *Plant Mol. Biol.* 7:451–65

53. Greenland, A. J., Thomas, M. V., Walden, R. M. 1987. Expression of two nuclear genes encoding chloroplast proteins during early development of cucumber seedlings. *Planta* 170:99–110

54. Grossman, A., Bartlett, S., Chua, N. H. 1980. Energy-dependent uptake of cytoplasmically synthesized polypeptides by chloroplasts. *Nature* 285:625–28

55. Guidet, F., Fourcroy, P. 1988. Nucleotide sequence of a radish ribulose 1,5-bisphosphate carboxylase small subunit (*rbcS*) cDNA. *Nucleic Acids Res.* 16:2336

56. Hansen, L. 1987. Three cDNA clones for barley leaf acyl carrier proteins I and III. *Carlsberg Res. Commun.* 52:381–92

57. Hay, R., Bohni, P., Gasser, S. 1984. How mitochondria import proteins. *Biochim. Biophys. Acta* 779:65–87

58. Hermans, J., Rother, C., Bichler, J., Steppuhn, J., Herrmann, R. G. 1988. Nucleotide sequence of cDNA clones encoding the complete precursor for subunit delta of thylakoid-located ATP synthase from spinach. *Plant Mol. Biol.* 10:323–30

59. Highfield, P. E., Ellis, R. J. 1978. Synthesis and transport of the small subunit of chloroplast ribulose bisphosphate carboxylase. *Nature* 271:420–24

60. Hinz, G., Flugge, U. I. 1988. Phosphorylation of a 51-kDa envelope membrane polypeptide involved in protein translocation into chloroplasts. *Eur. J. Biochem.* 175:649–59

61. Hoffman, N. E., Pichersky, E., Malik, V. S., Castresana, C., Ko, K., et al. 1987. A cDNA clone encoding a photosystem-I protein with homology to photosystem-II chlorophyll a/b-binding polypeptides. *Proc. Natl. Acad. Sci. USA* 84:8844–48

62. Hoffman, N. E., Pichersky, E., Malik, V. S., Ko, K., Cashmore, A. R. 1988. Isolation and sequence of a tomato cDNA clone encoding subunit II of the photosystem I reaction center. *Plant Mol. Biol.* 10:435–45

63. Hurt, E. C., van Loon, A. P. G. M. 1986. How proteins find mitochondria and intramitochondrial compartments. *Trends Biochem. Sci.* 11:204–7

64. Imanaka, T., Small, G. M., Lazarow, P. B. 1987. Translocation of acyl-CoA oxidase into peroxisomes requires ATP hydrolysis but not a membrane potential. *J. Cell Biol.* 105:2915–22

65. Ito, A., Ogishima, T., Ou, W., Omura, T., Aoyagi, H., et al. 1985. Effects of synthetic model peptides resembling the extension peptides of mitochondrial enzyme precursors on import of the precursors into mitochondria. *J. Biochem.* 98:1571–82

66. Jansen, T., Reilander, H., Steppuhn, J., Herrmann, R. G. 1988. Analysis of cDNA clones encoding the entire precursor-polypeptide for ferredoxin: NADP⁺ oxidoreductase from spinach. *Curr. Genet.* 13:517–22

67. Jansen, T., Rother, C., Steppuhn, J., Reinke, H., Bayreuther, K., et al. 1987. Nucleotide sequence of cDNA clones encoding the complete "23 kDa" and "16 kDa" precursor proteins associated with the photosynthetic oxygen-evolving complex from spinach. *FEBS Lett.* 216:234–40

68. Joyard, J., Douce, R. 1988. Import receptor in chloroplast envelope. *Nature* 333:306–7

69. Kaderbhai, M. A., Pickering, T., Austen, B. M., Kaderbhai, N. 1988. A photoactivatable synthetic transit peptide labels 30 kDa and 52 kDa polypeptides of the chloroplast inner envelope membrane. *FEBS Lett.* 232:313–16

70. Kaiser, E. T., Kezdy, F. J. 1984. Amphiphilic secondary structure: design of peptide hormones. *Science* 223:249–55

71. Kaiser, E. T., Kezdy, F. J. 1987. Peptides with affinity for membranes. *Annu. Rev. Biophys. Biophys. Chem.* 16:561–81

72. Karlin-Neumann, G. A., Kohorn, B. D., Thornber, J. P., Tobin, E. M. 1985. A chlorophyll a/b-protein encoded by a gene containing an intron with characteristics of a transposable element. *J. Mol. Appl. Genet.* 3:45–61

73. Karlin-Neumann, G. A., Tobin, E. M. 1986. Transit peptides of nuclear-encoded chloroplast proteins share a common amino acid framework. *EMBO J.* 5:9–13

74. Keegstra, K. 1989. A new hypothesis for the mechanism of protein translocation into chloroplasts. In *Photosynthesis*, ed. W. Briggs. New York: Alan R. Liss. In press

75. Keegstra, K., Bauerle, C. 1988. Targeting of proteins into chloroplasts. *Bioessays* 9:15–19

76. Klee, H. J., Muskopf, Y. M., Gasser, C. S. 1987. Cloning of an *Arabidopsis thaliana* gene encoding 5-enolpyruvyl-shikimate-3-phosphate synthase: sequence analysis and manipulation to obtain glyphosate-tolerant plants. *Mol. Gen. Genet.* 210:437–42

77. Klosgen, R. B., Gierl, A., Schwarz-Sommer, Z., Saedler, H. 1986. Molecular analysis of the waxy locus of *Zea mays*. *Mol. Gen. Genet.* 203:237–44

78. Kohorn, B. D., Harel, E., Chitnis, P. R., Thornber, J. P., Tobin, E. M. 1986. Functional and mutational analysis of the light-harvesting chlorophyll a/b protein of thylakoid membranes. *J. Cell Biol.* 102:972–81

79. Kolanus, W., Scharnhorst, C., Kuhne, U., Herzfeld, F. 1987. The structure and light-dependent transient expression of a nuclear-encoded chloroplast protein gene from pea (*Pisum sativum* L.). *Mol. Gen. Genet.* 209:234–39

80. Krebbers, E., Seurinck, J., Herdies, L., Cashmore, A. R., Timko, M. P. 1988. Four genes in two diverged subfamilies encode the ribulose-1,5-bisphosphate carboxylase small subunit polypeptides of *Arabidopsis thaliana*. *Plant Mol. Biol.* In press

81. Lagoutte, B. 1988. Cloning and sequencing of spinach cDNA clones encoding the 20 kDa PS I polypeptide. *FEBS Lett.* 232:275–80

82. Lamppa, G. K., Morelli, G., Chua, N. H. 1985. Structure and developmental regulation of a wheat gene encoding the major chlorophyll a/b-binding polypeptide. *Mol. Cell. Biol.* 5:1370–78

83. Lautner, A., Klein, R., Ljungberg, U.,

84. Lebrun, M., Waksman, G., Freyssinet, G. 1987. Nucleotide sequence of a gene encoding corn ribulose-1,5-bisphosphate carboxylase/oxygenase small subunit (*rbcS*). *Nucleic Acids Res.* 15:4360

85. Leutwiler, L. S., Meyerowitz, E. M., Tobin, E. M. 1986. Structure and expression of three light-harvesting chlorophyll a/b-binding protein genes in *Arabidopsis thaliana*. *Nucleic Acids Res.* 10:4051–64

86. Lightfoot, D. A., Green, N. K., Cullimore, J. V. 1988. The chloroplast-located glutamine synthetase of *Phaseolus vulgaris* L: nucleotide sequence, expression in different organs and uptake into isolated chloroplasts. *Plant Mol. Biol.* 11:191–202

87. Lubben, T. H., Theg, S. M., Keegstra, K. 1988. Transport of proteins into chloroplasts. *Photosyn. Res.* 17:173–94

88. Matsuoka, M., Kano-Murakami, Y., Yamamoto, N. 1987. Nucleotide sequence of cDNA encoding the light-harvesting chlorophyll a/b binding protein from maize. *Nucleic Acids Res.* 15:6302

89. Matsuoka, M., Ozeki, Y., Yamamoto, N., Hirano, H., Kano-Murakami, Y., Tanaka, Y. 1988. Primary structure of maize pyruvate, orthophosphate dikinase as deduced from cDNA sequence. *J. Biol. Chem.* 263:11080–83

90. Mayfield, S. P., Rahire, M., Frank, G., Zuber, H., Rochaix, J. D. 1987. Expression of the nuclear gene encoding oxygen-evolving enhancer protein 2 is required for high levels of photosynthetic oxygen evolution in *Chlamydomonas reinhardtii*. *Proc. Natl. Acad. Sci. USA* 84:749–53

91. Mazur, B. J., Chui, C. F. 1985. Sequence of a genomic DNA clone for the small subunit of ribulose bis-phosphate carboxylase-oxygenase from tobacco. *Nucleic Acids Res.* 13:2373–86

92. Mazur, B. J., Chui, C. F., Smith, J. K. 1987. Isolation and characterization of plant genes coding for acetolactate synthase, the target enzyme for two classes of herbicides. *Plant Physiol.* 85:1110–17

93. McCarty, D. R., Selman, B. R. 1986. Properties of a partially purified nucleoside triphosphatase (NTPase) from the chloroplast of pea. *Plant Physiol.* 80:908–12

Reilander, H., Bartling, D., et al. 1988. Nucleotide sequence of cDNA clones encoding the complete precursor for the "10-kDa" polypeptide of photosystem II from spinach. *J. Biol. Chem.* 263: 10077–81

94. Merchant, S., Bogorad, L. 1987. The Cu(II)-repressible plastidic cytochrome c. Cloning and sequence of a complementary DNA for the pre-apoprotein. *J. Biol. Chem.* 262:9062–67
95. Mishkind, M. L., Scioli, S. E. 1988. Recent developments in chloroplast protein transport. *Photosyn. Res.* In press
96. Mishkind, M. L., Wessler, S. R., Schmidt, G. W. 1985. Functional determinants in transit sequences: import and partial maturation by vascular plant chloroplasts of the ribulose-1,5-bisphosphate carboxylase small subunit of *Chlamydomonas. J. Cell Biol.* 100: 226–34
97. Newman, B. J., Gray, J. C. 1988. Characterisation of a full-length cDNA clone for pea ferredoxin-NADP$^+$ reductase. *Plant Mol. Biol.* 10:511–20
98. Newmeyer, D. D., Forbes, D. J. 1988. Nuclear import can be separated into distinct steps in vitro: nuclear pore binding and translocation. *Cell* 52:641–53
99. Olsen, L. J., Theg, S. M., Selman, B. R., Keegstra, K. 1989. ATP is required for the binding of precursor proteins to chloroplasts. *J. Biol. Chem.* In press
100. Pain, D., Blobel, G. 1987. Protein import into chloroplasts requires a chloroplast ATPase. *Proc. Natl. Acad. Sci. USA* 84:3288–92
101. Pain, D., Kanwar, Y. S., Blobel, G. 1988. Identification of a receptor for protein import into chloroplasts and its localization to envelope contact zones. *Nature* 331:232–37
102. Pfanner, N., Hartl, F. U., Neupert, W. 1988. Import of proteins into mitochondria: a multi-step process. *Eur. J. Biochem.* 175:205–12
103. Pfanner, N., Tropschug, M., Neupert, W. 1987. Mitochondrial protein import: nucleoside triphosphates are involved in conferring import-competence to precursors. *Cell* 49:815–23
104. Pfisterer, J., Lachmann, P., Kloppstech, K. 1982. Transport of proteins into chloroplasts. Binding of nuclear-coded chloroplast proteins to the chloroplast envelope. *Eur. J. Biochem.* 126:143–48
105. Pichersky, E., Bernatzky, R., Tanksley, S. D., Breidenbach, R. B., Kausch, A. P., et al. 1985. Molecular characterization and genetic mapping of two clusters of genes encoding chlorophyll a/b-binding proteins in *Lycopersicon esculentum* (tomato). *Gene* 40:247–58
106. Pichersky, E., Bernatzky, R., Tanksley, S. D., Cashmore, A. R. 1986. Evidence for selection as a mechanism in the concerted evolution of *Lycopersicon esculentum* (tomato) genes encoding the

small subunit of ribulose-1,5-bisphosphate carboxylase/oxygenase. *Proc. Natl. Acad. Sci. USA* 83:3880–84
107. Pichersky, E., Hoffman, N. E., Bernatzky, R., Piechulla, B., Tanksley, S. D., et al. 1987. Molecular characterization and genetic mapping of DNA sequences encoding the Type I chlorophyll a/b-binding polypeptide photosystem I in *Lycopersicon esculentum* (tomato). *Plant Mol. Biol.* 9:205–16
108. Pichersky, E., Hoffman, N. E., Malik, V. S., Bernatzky, R., Tanksley, S. D., et al. 1987. The tomato *Cab-4* and *Cab-5* genes encode a second type of CAB polypeptide localized in photosystem II. *Plant Mol. Biol.* 9:109–20
109. Pichersky, E., Tanksley, S. D., Piechulla, B., Stayton, M. M., Dunsmuir, P. 1988. Nucleotide sequence and chromosomal location of *Cab-7*, the tomato gene encoding the type II chlorophyll a/b-binding polypeptide of photosystem I. *Plant Mol. Biol.* 11:69–71
110. Pinck, M., Guilley, E., Durr, A., Hoff, M., Pinck, L., et al. 1984. Complete sequence of one of the mRNAs coding for the small subunit of ribulose bisphosphate carboxylase of *Nicotiana sylvestris. Biochimie* 66:539–45
111. Poulsen, C., Fluhr, R., Kauffman, J. M., Boutry, M., Chua, N. H. 1986. Characterization of an *rbcS* gene from *Nicotiana plumbaginifolia* and expression of an *rbcS-CAT* chimeric gene in homologous and heterologous nuclear background. *Mol. Gen. Genet.* 205: 193–200
112. Quigley, F., Martin, W. F., Cerff, R. 1988. Intron conservation across the prokaryote-eukaryote boundary: structure of the nuclear gene for chloroplast glyceraldehyde-3-phosphate dehydrogenase from maize. *Proc. Natl. Acad. Sci. USA* 85:2672–76
113. Reiss, B., Wasmann, C. C., Bohnert, H. J. 1987. Regions in the transit peptide of SSU essential for transport into chloroplasts. *Mol. Gen. Genet.* 209: 116–21
114. Riezman, H., Hay, R., Witte, C., Nelson, N., Schatz, G. 1983. Yeast mitochondrial outer membrane specifically binds cytoplasmically-synthesized precursors of mitochondrial proteins. *EMBO J.* 2:1113–18
115. Robinson, S. P., Wiskich, J. T. 1976. Stimulation of carbon dioxide fixation in isolated pea chloroplasts by catalytic amounts of adenine nucleotides. *Plant Physiol.* 58:156–62
116. Roesler, K. R., Ogren, W. L. 1988. Nucleotide sequence of spinach cDNA

encoding phosphoribulokinase. *Nucleic Acids Res.* 16:7192

117. Rohde, W., Becker, D., Salamini, F. 1988. Structural analysis of the waxy locus from *Hordeum vulgare*. *Nucleic Acids Res.* 16:7185–86

118. Roise, D., Horvath, S. J., Tomich, J. M., Richards, J. H., Schatz, G. 1986. A chemically synthesized pre-sequence of an imported mitochondrial protein can form an amphiphilic helix and perturb natural and artificial phospholipid bilayers. *EMBO J.* 5:1327–34

119. Roise, D., Schatz, G. 1988. Mitochondrial presequences. *J. Biol. Chem.* 263: 4509–11

120. Roise, D., Theiler, F., Horvath, S. J., Tomich, J. M., Richards, J. H., et al. 1988. Amphiphilicity is essential for mitochondrial presequence function. *EMBO J.* 7:649–53

121. Rose, R. E., DeJesus, C. E., Moylan, S. L., Ridge, N. P., Scherer, D. E., et al. 1987. The nucleotide sequence of a cDNA clone encoding acyl carrier protein (ACP) from *Brassica campestris* seeds. *Nucleic Acids Res.* 15:7197

122. Rother, C., Jansen, T., Tyagi, A., Tittgen, J., Herrmann, R. G. 1986. Plastocyanin is encoded by an uninterrupted nuclear gene in spinach. *Curr. Genet.* 11:171–76

123. Safford, R., Windust, J. H. C., Lucas, C., DeSilva, J., James, C. M., et al. 1988. Plastid-localised seed acyl-carrier protein of *Brassica napus* is encoded by a distinct, nuclear multigene family. *Eur. J. Biochem.* 174:287–95

124. Scherer, D. E., Knauf, V. C. 1987. Isolation of a cDNA clone for the acyl carrier protein-I of spinach. *Plant Mol. Biol.* 9:127–34

125. Schiffer, M., Edmundson, A. B. 1967. Use of helical wheels to represent the structures of proteins and to identify segments with helical potential. *Biophys. J.* 7:121–35

126. Schindler, C., Hracky, R., Soll, J. 1987. Protein transport in chloroplasts: ATP is prerequisite. *Z. Naturforsch. Teil C* 42:103–8

127. Schmidt, G. W., Devillers-Thiery, A., Desruisseaux, H., Blobel, G., Chua, N. H. 1979. NH$_2$-terminal amino acid sequences of precursor and mature forms of the ribulose-1,5-bisphosphate carboxylase small subunit from *Chlamydomonas reinhardtii*. *J. Cell Biol.* 83:615–22

128. Schmidt, G. W., Mishkind, M. L. 1986. The transport of proteins into chloroplasts. *Annu. Rev. Biochem.* 55:879–912

129. Shih, M. C., Lazar, G., Goodman, H. M. 1986. Evidence in favor of the symbiotic origin of chloroplasts: primary structure and evolution of tobacco glyceraldehyde-3-phosphate dehydrogenases. *Cell* 47:73–80

130. Sibley, D. R., Benovic, J. L., Caron, M. G., Lefkowitz, R. J. 1987. Regulation of transmembrane signaling by receptor phosphorylation. *Cell* 48:913–22

131. Smeekens, S., Bauerle, C., Hageman, J., Keegstra, K., Weisbeek, P. 1986. The role of the transit peptide in the routing of precursors toward different chloroplast compartments. *Cell* 46:365–75

132. Smeekens, S., de Groot, M., van Binsbergen, J., Weisbeek, P. 1985. Sequence of the precursor of the chloroplast thylakoid lumen protein plastocyanin. *Nature* 317:456–58

133. Smeekens, S., van Binsbergen, J., Weisbeek, P. 1985. The plant ferredoxin precursor: nucleotide sequence of a full length cDNA clone. *Nucleic Acids Res.* 13:3179–94

134. Smeekens, S., van Oosten, J., de Groot, M., Weisbeek, P. 1986. *Silene* cDNA clones for a divergent chlorophyll-a/b-binding protein and a small subunit of ribulose bisphosphate carboxylase. *Plant Mol. Biol.* 7:433–40

135. Smeekens, S., Weisbeek, P. 1988. Protein transport towards the thylakoid lumen: posttranslational translocation in tandem. *Photosyn. Res.* 16:177–86

136. Soll, J. 1988. Purification and characterization of a chloroplast outer-envelope-bound, ATP-dependent protein kinase. *Plant Physiol.* 87:898–903

137. Deleted in proof

138. Soll, J., Buchanan, B. B. 1983. Phosphorylation of chloroplast ribulose bisphosphate carboxylase/oxygenase small subunit by an envelope-bound protein kinase *in situ*. *J. Biol. Chem.* 258:6686–89

139. Soll, J., Fischer, I., Keegstra, K. 1988. A GTP-dependent protein kinase is localized in the outer envelope membrane of pea chloroplasts. *Planta.* 176:488–96

140. Stayton, M. M., Black, M., Bedbrook, J., Dunsmuir, P. 1986. A novel chlorophyll a/b binding *(Cab)* protein gene from petunia which encodes the lower molecular weight Cab precursor protein. *Nucleic Acids Res.* 14:9781–96

141. Stayton, M. M., Brosio, P., Dunsmuir, P. 1987. Characterization of a full-length petunia cDNA encoding a polypeptide of the light-harvesting com-

plex associated with photosystem I. *Plant Mol. Biol.* 10:127–37

142. Steppuhn, J., Rother, C., Hermans, J., Jansen, T., Salnikow, J., et al. 1987. The complete amino-acid sequence of the Rieske FeS-precursor protein from spinach chloroplasts deduced from cDNA analysis. *Mol. Gen. Genet.* 210: 171–77

143. Sugita, M., Manzara, T., Pichersky, E., Cashmore, A., Gruissem, W. 1987. Genomic organization, sequence analysis and expression of all five genes encoding the small subunit of ribulose-1,5-bisphosphate carboxylase/oxygenase from tomato. *Mol. Gen. Genet.* 209:247–56

144. Sullivan, T. D., Christensen, A. H., Quail, P. H. 1989. Isolation and characterization of a maize chlorophyll a/b binding protein gene that produces high levels of mRNA in the dark. *Mol. Gen. Genet.* In press

145. Szabo, L. J., Cashmore, A. R. 1987. Targeting nuclear gene products into chloroplasts. In *Plant DNA Infectious Agents,* ed. T. Hohn, J. Schell, pp. 321–39. Wien/New York: Springer-Verlag

146. Tamm, L. K. 1986. Incorporation of a synthetic mitochondrial signal peptide into charged and uncharged phospholipid monolayers. *Biochemistry* 25:7470–76

147. Theg, S. M., Bauerle, C., Olsen, L. J., Selman, B. R., Keegstra, K. 1989. Internal ATP is the only energy requirement for the translocation of precursor proteins across chloroplastic membranes. *J. Biol. Chem.* In press

148. Timko, M. P., Kausch, A. P., Hand, J. M., Cashmore, A. R., Herrera-Estrella, L., et al. 1985. Structure and expression of nuclear genes encoding polypeptides of the photosynthetic apparatus. In *Molecular Biology of the Photosynthetic Apparatus,* ed. K. E. Steinback, S. Bonitz, C. J. Arntzen, L. Bogorad, pp. 381–96. Cold Spring Harbor: Cold Spring Harbor Lab.

149. Tingey, S. T., Tsai, F. Y., Edwards, J. W., Walker, E. L., Coruzzi, G. M. 1988. Chloroplast and cytosolic glutamine synthetase are encoded by homologous nuclear genes which are differentially expressed *in vivo. J. Biol. Chem.* 263:9651–57

150. Tobin, E. M., Wimpee, C. F., Karlin-Neumann, G. A., Silverthorne, J., Kohorn, B. D. 1985. Phytochrome regulation of nuclear gene expression. See Ref. 148, pp. 373–80

151. Tumer, N. E., Clark, W. G., Tabor, G.

J., Hironaka, C. M., Fraley, R. T., et al. 1986. The genes encoding the small subunit of ribulose-1,5-bisphosphate carboxylase are expressed differentially in petunia leaves. *Nucleic Acids Res.* 14:3325–42

152. Tyagi, A., Hermans, J., Steppuhn, J., Jansson, C., Vater, F., et al. 1987. Nucleotide sequence of cDNA clones encoding the complete "33 kDa" precursor protein associated with the photosynthetic oxygen-evolving complex from spinach. *Mol. Gen. Genet.* 207: 288–93

153. Verner, K., Schatz, G. 1988. Protein translocation across membranes. *Science* 241:1307–13

154. Vierling, E., Nagao, R. T., DeRocher, A. E., Harris, L. M. 1988. A heat shock protein localized to chloroplasts is a member of a eukaryotic superfamily of heat shock proteins. *EMBO J.* 7:575–81

155. von Heijne, G. 1986. Mitochondrial targeting sequences may form amphiphilic helices. *EMBO J.* 5:1335–42

156. von Heijne, G. 1988. Transcending the impenetrable; how proteins come to terms with membranes. *Biochim. Biophys. Acta* 947:307–33

157. Vorst, O., Oosterhoff-Teertstra, R., Vankan, P., Smeekens, S., Weisbeek, P. 1988. Plastocyanin of *Arabidopsis thaliana;* isolation and characterization of the gene and chloroplast import of the precursor protein. *Gene* 65:56–69

158. Waksman, G., Lebrun, M., Freyssinet, G. 1987. Nucleotide sequence of a gene encoding sunflower ribulose-1,5-bisphosphate carboxylase/oxygenase small subunit *(rbcS). Nucleic Acids Res.* 15:7181

159. Wasmann, C. C., Reiss, B., Bohnert, H. J. 1988. Complete processing of a small subunit of ribulose-1,5-bisphosphate carboxylase/oxygenase from pea requires the amino acid sequence Ile-Thr-Ser. *J. Biol. Chem* 263:617–19

160. Webber, A. N., Packman, L. C., Gray, J. C. 1989. A 10 kDa polypeptide associated with the oxygen-evolving complex of photosystem I has a putative C-terminal non-cleavable thylakoid transfer domain. *FEBS Lett.* In press

161. Weisbeek, P., Hageman, J., Smeekens, S., de Boer, D., Cremers, F. 1987. Chloroplast-specific import and routing of proteins. In *Plant Molecular Biology,* ed. D. von Wettstein, N. H. Chua, pp. 77–91. New York/London: Plenum

162. Werneke, J. M., Zielinski, R. E., Ogren, W. L. 1988. Structure and expression of spinach leaf cDNA encoding

ribulose bisphosphate carboxylase/oxygenase activase. *Proc. Natl. Acad. Sci. USA* 85:787–91

163. Wolter, F. P., Fritz, C. C., Willmitzer, L., Schell, J., Schreier, P. H. 1988. *rbcS* genes in *Solanum tuberosum:* conservation of transit peptide and exon shuffling during evolution. *Proc. Natl. Acad. Sci. USA* 85:846–50

164. Xie, Y., Wu, R. 1988. Nucleotide sequence of a ribulose-1,5-bisphosphate carboxylase/oxygenase small subunit gene *(rbcS)* in rice. *Nucleic Acids Res.* 16:7749

165. Zwizinski, C., Schleyer, M., Neupert, W. 1984. Proteinaceous receptors for the import of mitochondrial precursor proteins. *J. Biol. Chem.* 259:7850–56

Annu. Rev. Plant Physiol. Plant Mol. Biol. 1989. 40:503–37

CARBON ISOTOPE DISCRIMINATION AND PHOTOSYNTHESIS

G. D. Farquhar,[1] J. R. Ehleringer,[2] and K. T. Hubick[1]

[1]Research School of Biological Sciences, Australian National University, Canberra, ACT 2601 Australia

[2]Department of Biology, University of Utah, Salt Lake City, Utah 84112

CONTENTS

503

1040-2519/89/0601-503$02.00

INTRODUCTION

There are two naturally occurring stable isotopes of carbon, ^{12}C and ^{13}C. Most of the carbon is ^{12}C (98.9%), with 1.1% being ^{13}C. The isotopes are unevenly distributed among and within different compounds, and this isotopic distribution can reveal information about the physical, chemical, and metabolic processes involved in carbon transformations. The overall abundance of ^{13}C relative to ^{12}C in plant tissue is commonly less than in the carbon of atmospheric carbon dioxide, indicating that carbon isotope discrimination occurs in the incorporation of CO_2 into plant biomass. Because the isotopes are stable, the information inherent in the ratio of abundances of carbon isotopes, presented by convention as $^{13}C/^{12}C$, is invariant as long as carbon is not lost. Numerous contributions have been made to our understanding of carbon isotope discrimination in plants since this area was extensively reviewed by O'Leary (97). Here we discuss the physical and enzymatic bases of carbon isotope discrimination during photosynthesis, noting how knowledge of discrimination can be used to provide additional insight into photosynthetic metabolism and the environmental influences on that process.

ISOTOPE EFFECTS

Variation in the $^{13}C/^{12}C$ ratio is the consequence of "isotope effects," which are expressed during the formation and destruction of bonds involving a carbon atom, or because of other processes that are affected by mass, such as gaseous diffusion. Isotope effects are often classified as being either kinetic or thermodynamic, the distinction really being between nonequilibrium and equilibrium situations. One example of a kinetic effect is the difference between the binary diffusivity of $^{13}CO_2$ and that of $^{12}CO_2$ in air. Another example is the difference between the kinetic constants for the reaction of $^{12}CO_2$ and $^{13}CO_2$ with ribulose bisphosphate carboxylase-oxygenase (Rubisco). Both these examples are called "normal" kinetic effects in that the process discriminates against the heavier isotope. Thermodynamic effects represent the balance of two kinetic effects at chemical equilibrium and are therefore generally smaller than individual kinetic effects. An example of a thermodynamic effect is the unequal distribution of isotope species among phases in a system (e.g. in CO_2 in air versus in CO_2 in solution). Thermodynamic effects, like some kinetic ones, are temperature dependent.

Isotope effects, denoted by α, are also called fractionation factors because they result in fractionations of isotopes. They are here defined (as by some, but not all chemists) as the ratio of carbon isotope ratios in reactant and product

$$\alpha = \frac{R_r}{R_p}, \qquad\qquad 1.$$

where R_r is the $^{13}C/^{12}C$ molar ratio of reactant and R_p is that of the product. Defined in this way, a kinetic isotope effect can be thought of as the ratio of the rate constants for ^{12}C and ^{13}C containing substrates, k^{12} and k^{13}, respectively. Thus

$$\alpha_{kinetic} = \frac{k^{12}}{k^{13}}. \qquad\qquad 2.$$

A simple equilibrium isotope effect would be the ratio of the equilibrium constants for ^{12}C and ^{13}C containing compounds, K^{12} and K^{13}, respectively:

$$\alpha_{eqbm} = \frac{K^{12}}{K^{13}}. \qquad\qquad 3.$$

Diffusional effects belong to the category of kinetic effects, and the isotope effect is the ratio of the diffusivity of the ^{12}C compound to that of the ^{13}C compound. The above effects are discussed more fully in Part I of the Appendix. Isotope effects may occur at every reaction of a sequence, but the overall isotope effect will reflect only the isotope effects at steps where the reaction is partially reversible or where there are alternative possible fates for atoms, until an irreversible step is reached (97). Kinetic isotope effects of successive individual reactions are usually not additive, but the thermodynamic ones are. If all reactants are consumed and converted to product in an irreversible reaction, there is no fractionation. For example, plants grown in a closed system, where all CO_2 was fixed, showed no isotope effect (6).

ISOTOPIC COMPOSITION AND DISCRIMINATION

Definitions

Farquhar & Richards (39) proposed that whole plant processes should be analyzed in the same terms as chemical processes. From Equation 1 it is evident that this requires measurements of isotopic abundance of both source and product. For plants this means measuring R_a (isotopic abundance in the air) and R_p (isotopic abundance in the plant, where the plant can be considered the product referred to in Equation 1). For numerical convenience, instead of using the isotope effect ($\alpha = R_a/R_p$), Farquhar & Richards (39) proposed the use of Δ, the deviation of α from unity, as the measure of the carbon isotope discrimination by the plant:

$$\Delta = \alpha - 1 = \frac{R_a}{R_p} - 1. \qquad\qquad 4.$$

The absolute isotopic composition of a sample is not easy to measure directly. Rather, the mass spectrometer measures the deviation of the isotopic composition of the material from a standard,

$$\delta_p = \frac{R_p - R_s}{R_s} = \frac{R_p}{R_s} - 1, \qquad\qquad 5.$$

where R_s is the molar abundance ratio, $^{13}C/^{12}C$, of the standard. The reference material in determinations of carbon isotopic ratios has not normally been CO_2 in air but traditionally has been carbon in carbon dioxide generated from a fossil belemnite from the Pee Dee Formation, denoted PDB [for which $R = 0.01124$, (17)]. In this review all compositions that are denoted δ are with respect to PDB.

In contrast to δ, the discrimination, Δ, is independent of the isotopic composition of the standard used for measurement of R_p and R_a, and is also independent of R_a. Plants show a positive discrimination (Δ) against ^{13}C. Typically C_3 plants have a discrimination of $\sim 20 \times 10^{-3}$, which is normally presented in the literature as 20‰ ("per mil"). Consistent with this notation, we will use ‰ as equivalent to 10^{-3}. Note that "per mil" is not a unit, and is analogous to per cent; discrimination is therefore dimensionless. Equations involving the δ notation have been made unnecessarily complex by including the factor 1000 in the definition (i.e. $\delta_p = (R_p/R_s - 1)\cdot 1000$). We have opted for simplicity, but the reader should note that factors of 1000 in other treatments (including our own) should be deleted when comparing to the equations presented here. Other possible definitions of discrimination are discussed in Part III of the Appendix.

The value of Δ as defined above is obtained from δ_a and δ_p, where a and p refer to air and plant, respectively, using Equation 4, and the definitions of δ_a and δ_p ($R_a/R_s - 1$; $R_p/R_s - 1$, respectively):

$$\Delta = \frac{\delta_a - \delta_p}{1 + \delta_p}. \qquad\qquad 6.$$

On the PDB scale, free atmospheric CO_2 ($R_a \sim 0.01115$ in 1988) currently has a deviation, δ_a, of approximately -8‰, and typical C_3 material ($R_p \sim 0.01093$) a deviation, δ_p, of -27.6‰, which yields $\Delta = (-0.008 + 0.0276)/(1 - 0.0276) = 20.1$‰. O'Leary (97) pointed out that the simultaneous use of discrimination and δ is confusing for work with plants, since the discrimination values (Δ) are usually positive while those of δ are usually negative when PDB is the reference. Where possible, it is preferable to use molar abundance ratios (R) and compositional deviations (δ) only as intermediates in the calculation of final isotope effects (97).

Isotopic Composition of Source Air

The advantage of reporting Δ is that it directly expresses the consequences of biological processes, whereas composition, δ_p, is the result of both source isotopic composition and carbon isotope discrimination. This distinction is particularly important in the interpretation of some growth cabinet work where the isotopic composition of CO_2 can be affected by mixing of CO_2 derived from fossil fuel combustion with normal atmospheric CO_2. Of course, it is relevant for vegetation grown near vents outgassing the CO_2 produced from burning underground coal (for which $\delta_a = -32.5‰$) (46). Of wider relevance, the distinction between δ and Δ is important when interpreting results from canopies, if turbulent transfer is poor. In these conditions, there is a gradient, with height, in isotopic composition of CO_2 in the air, δ_a. This gradient occurs because of both canopy photosynthetic activity and soil respiration and litter decomposition. On the one hand, since photosynthetic processes discriminate against ^{13}C, the remaining CO_2 in air should be enriched in ^{13}C when CO_2 concentration is drawn down (32, 35). On the other hand, decomposition processes, which release CO_2 with an isotopic composition similar to that of the decaying vegetation, result in a much lower ^{13}C content of the soil CO_2 (1, 68, 116, 122, 123, 148). Francey et al (42) reported a CO_2 concentration of 20 ppm lower, 1 m above the ground, than outside the canopy in the daylight period in a dense (14 m) canopy of huon pine in Tasmania. The difference in δ_a between the top and bottom of the canopy was 0.8‰. In warm and dense tropical rainforests, the CO_2 concentration, c_a, is large near the forest floor, and δ_a is small [$c_a = 389$ ppm, $\delta_a = -11.4‰$ at 0.5 m (133); see also (88)]. The isotopic composition, δ_a, and CO_2 concentration, c_a, should be negatively related within a canopy (as in the above reports) so that for those field-grown plants where the gradients of c_a are found to be small, the gradient of δ_a is also likely to be small.

The isotopic composition of the free atmosphere also changes, slowly becoming depleted in ^{13}C (41, 45, 70, 92, 108). The progressive decrease in δ_a is caused by the anthropogenic burning of fossil fuels ($\delta \sim -26‰$). From 1956 to 1982, δ_a has decreased from $-6.7‰$ (at 314 ppm) to $-7.9‰$ (at 342 ppm) (70, 92).

There is also an annual cycle of 10 ppm in c_a, and 0.2‰ in δ_a, in the northern hemisphere, associated with seasonal changes in standing biomass; the amplitudes of changes in c_a and δ_a are much smaller in the southern hemisphere (92). In major metropolitan areas, δ_a may vary by as much as 2‰ both daily and annually, because of human activities (64, 65). Throughout this review when discussing studies where isotopic composition of plant material is presented without corresponding measurements of δ_a, we also provide an estimate of discrimination (Δ) using the assumption (for field-grown plants) of an atmospheric composition (δ_a) of $-8‰$.

"On-line" Measurement of Carbon Isotope Discrimination

In most studies, composition of CO_2 from combustion of plant material (δ_p) has been compared to that of the atmosphere in which the material was grown (δ_a) to yield an average discrimination over the period in which the carbon was fixed. A more direct and nondestructive means of measuring short-term carbon isotope discrimination is to measure the changes in the $^{13}C/^{12}C$ ratio of the CO_2 in air as it passes a leaf within a stirred cuvette, such as those commonly used for whole-leaf gas-exchange measurements (32, 36, 62, 125). If the reactions associated with photosynthetic CO_2 fixation discriminate against ^{13}C, the remaining CO_2 should be enriched in ^{13}C. Discrimination can be calculated from measurements of the concentration (c) and the isotopic composition (δ) of the CO_2 of the air entering (c_e and δ_e) and leaving (c_o and δ_o) the cuvette according to an equation derived by Evans et al (32),

$$\Delta = \frac{\xi(\delta_o - \delta_e)}{1 + \delta_o - \xi(\delta_o - \delta_e)} \qquad 7.$$

where $\xi = c_e/(c_e - c_o)$. Note that Evans et al (32) used the constant 1000 in the denominator rather than 1, because their values of δ had also been multiplied by 1000.

O'Leary et al (102) used a different "on-line" technique, where the plant was enclosed in a bell jar and allowed to deplete the CO_2. The continuing isotopic enrichment of the remaining CO_2 was monitored and discrimination calculated from a set of differential equations.

Estimates from these "on-line" methods are usually comparable to those from tissue combustion analyses (32, 62, 125). The clear advantage over tissue combustion of the "on-line" approaches is that they are nondestructive and rapid ($\sim$ 30 min), permitting studies of isotope discrimination as a function of time or of physiological and environmental conditions. The measurement of tissue is of course invaluable for longer-term integration, and for the ease with which small amounts of material can be collected, stored, and subsequently analyzed.

THEORY OF CARBON ISOTOPE DISCRIMINATION DURING PHOTOSYNTHESIS

Carbon isotope composition of plants was first used to indicate photosynthetic pathways in plants (2, 3, 89, 93, 106, 120, 127, 128, 130, 144, 145, 150, 151, 156, 159, 160, 163). This is because phospho*enol*pyruvate (PEP) carboxylase, the primary carboxylating enzyme in species having a C_4 metabolism, exhibits a different intrinsic kinetic isotope effect and utilizes a different species of inorganic carbon that has an isotopic composition at equilibrium different from that of Rubisco. Isotopic screening was a simple test for determining the photosynthetic pathway when it was unknown for a species. Over the past 15–20 years, the results of such surveys have provided a broad base of the distribution of photosynthetic pathways among different phylogenetic groups and ecological zones (97, 99, 106). Although major photosynthetic groups could clearly be distinguished by their isotopic composition, the results of these early studies also indicated that there was substantial variation in isotopic values at both the interspecific and intraspecific levels, as well as variation associated with different environmental growth conditions and with variation in dry-matter composition. Substantial theoretical and experimental progress has been made over the past ten years in understanding the biochemical, metabolic, and environmental factors contributing to the different isotopic compositions among plants. The major isotope effects of interest are listed in Table 1 and include kinetic discrimination factors associated with diffusion (denoted by a) and enzyme fractionation (denoted by b), as well as equilibrium discrimination factors (denoted by e). We refer to this table as we review the theory and supporting evidence.

C_3 *Photosynthesis*

HIGHER PLANTS Several models have been developed to describe the fractionation of carbon isotopes during C_3 photosynthesis (38, 69, 97, 109, 122, 149). The models are similar in structure, each assuming that the major components contributing to the overall fractionation are the differential diffusivities of CO_2 containing ^{12}C and ^{13}C across the stomatal pathway and the fractionation by Rubisco. Each of the models suggests additivity of fractionation factors weighted by the relative "limitation" or CO_2 partial pressure difference imposed by the step involved. Of the models, that of Farquhar et al (38) has been the most extensively developed and tested. Their expression for discrimination in leaves of C_3 plants in its simplest form is,

$$\Delta = a \frac{p_a - p_i}{p_a} + b \frac{p_i}{p_a} = a + (b - a) \frac{p_i}{p_a}, \qquad 8.$$

where a is the fractionation occurring due to diffusion in air (4.4‰, a theoretical value that has not been confirmed experimentally), b is the net fractionation caused by carboxylation (mainly b_3, discrimination by Rubisco; see Table 1 and also Part IV of the Appendix) and p_a and p_i are the ambient and intercellular partial pressures of CO_2, respectively. Equation 8 is derived in Part II of the Appendix; see also reference 5.

Table 1 Isotope effects of steps leading to CO_2 fixation in plants.

Process	Isotope effect (α)	Discrimination $(\permil)$	Symbol	Reference
Diffusion of CO_2 in air through the stomatal pore[a]	1.0044	4.4	a	Craig (16)
Diffusion of CO_2 in air through the boundary layer to the stomata[a]	1.0029	2.9	a_b	Farquhar (33)
Diffusion of dissolved CO_2 through water	1.0007	0.7	a_l	O'Leary (98)
Net C_3 fixation with respect to p_i	1.027	27	b	Farquhar & Richards (39)
Fixation of gaseous CO_2 by	1.030 (pH = 8)	30	b_3	Roeske & O'Leary (119)[b]
Rubisco from higher plants	1.029 (pH = 8.5)	20	b_3	Guy et al (50)
Fixation of HCO_3^- by PEP	1.0020	2.0	b_4^*	O'Leary et al (101)
carboxylase	1.0020	2.0		Reibach & Benedict (117)
Fixation of gaseous CO_2 (in equilibrium with HCO_3^- at 25°C) by PEP carboxylase	0.9943	−5.7	b_4	Farquhar (33)
Equilibrium hydration of CO_2 at 25°C	0.991	−9.0	e_b	Emrich et al (31)
	0.991	−9.0		Mook et al (91)
Equilibrium dissolution of CO_2 into water	1.0011	1.1	e_s	Mook et al (91)
	1.0011	1.1		O'Leary (98)

[a] Theoretical value
[b] Data corrected for dissolution of CO_2

The significance of Equation 8 is that when stomatal conductance is small in relation to the capacity for CO_2 fixation, p_i is small and Δ tends towards 4.4‰ (see Figure 1). Conversely, when conductance is comparatively large, p_i approaches p_a and Δ approaches b (perhaps 27–30‰; see Appendix Part IV). Nevertheless, it is a little dangerous to take the argument further and say that when p_i and Δ are small, stomata are necessarily limiting photosynthesis. That conclusion would only follow if the relationship between assimilation rate, A, and p_i remained linear beyond the operational point (40).

There are several cases where measurements of both Δ and p_i/p_a have been made in controlled conditions. Farquhar et al (35) found a significant correlation between Δ in dry matter and discrete measurements of p_i/p_a among different species over the range of p_i/p_a 0.3–0.85. The best fit, taking a as 4.4‰, was observed with a value for b of 27‰. The leaf with the lowest p_i/p_a was from an *Avicennia marina* plant, which showed discrimination of 11.8‰. Such low values of Δ had previously been considered to be in the range of C_4 plants. Downton et al (using spinach; 21) and Seemann & Critchley (using beans; 124) also observed significant correlations between Δ in dry matter and p_i/p_a, the best fit being obtained by setting b equal to 28.5‰ and 26.4‰,

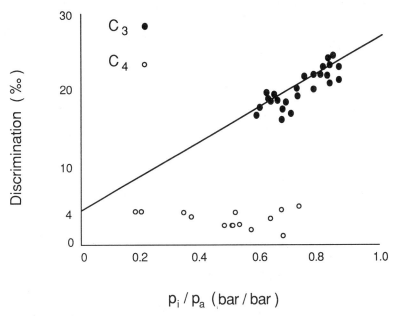

$$p_i / p_a \; (\text{bar / bar})$$

Figure 1   Carbon isotope discrimination, Δ, versus the ratio of intercellular and ambient partial pressures of CO_2, p_i/p_a, when all are measured simultaneously in a gas exchange system (36). The line drawn is Equation 8 with $a = 4.4‰$ and $b = 27‰$.

respectively. However, it should be noted that in none of the above studies was δ_a directly measured. Winter (155) showed that both Δ and p_i/p_a of leaves became smaller as *Cicer arietinum* plants were water stressed. Conversely, Bradford et al (9) showed that both were greater in a tomato mutant lacking abscisic acid (ABA) than in its isogenic parent. Phenotypic reversion of Δ and p_i/p_a occurred when the mutant was sprayed with ABA during its growth. Measurements of mistletoes and their hosts (25, 30) showed interspecific variation in both Δ and p_i/p_a. Guy et al (52) found that increased salinity decreased Δ in *Puccinellia* and p_i/p_a as expected from theory. Over the short term, Brugnoli et al (11) showed that the assimilation-weighted value of p_i/p_a and Δ of sugar produced by a leaf in a single day followed the predicted theoretical relationship (Equation 8) with a fitted value for a of $4.1‰$ and for b of 24–$25.5‰$. The overall discrimination to starch appeared to be slightly smaller. In all of the above cases, Δ, inferred from the carbon composition of leaf material, and p_i/p_a were positively correlated. The values of b that gave the best fit showed variation, which could have many causes (see Part IV of the Appendix for further elaboration).

NONVASCULAR PLANTS   Surveys of isotopic composition have been made on species of mosses, liverworts, and lichens. Isotope ratio variation in the range of $-21.3\%o$ to $-37.5\%o$ ($\Delta = 13.6$–$30.4\%o$) has been reported (121, 128, 135, 136).

For mosses, and some liverworts, the gametophytes are morphologically similar to higher plants but are restricted in size by their lack of vascular tissue. Their leaflike photosynthetic structures tend to be just one cell layer thick and do not have the specialized anatomy of higher plants. They do not consistently have an epidermis with impermeable cuticle and stomata, so we might not expect to observe variation in isotope discrimination arising from short-term variation in permeability to gases as with higher plants. It is possible, however, that permeability changes with water content. Even if this resistance remains constant, the gradient in partial pressure across it will change if the flux changes. For example, assimilation rate may change because of differing light levels, and this should increase the gradient and decrease Δ (32). For other liverworts with a thicker thallus and an epidermis, there may be pores that lead to air chambers, like stomata in higher plants, and we would expect to see variation in discrimination similar to that in higher plants.

In contrast to our explanations for variation in Δ in mosses and liverworts, Rundel et al (121) attributed the very negative values of δ_p in mosses in humid conditions to a large content of lipid in the tissue of those species [as lipids are depleted in ^{13}C compared to other plant compounds (97)]. Teeri (135) suggested that these differences may have arisen because of differences in the amount of carbon fixed by PEP carboxylase, but this is unlikely to differ from that in higher plants.

Among lichens, there are differences in carbon isotope discrimination that depend on the phycobionts in the symbiotic association (74, 76). Green algae as phycobionts are able to maintain positive photosynthetic rates when only misted, whereas when cyanobacteria are the phycobionts, surface liquid water is required for photosynthetic activity (75). This difference suggests that the CO_2 diffusion rate may be limiting when cyanobacteria are the phycobionts; correspondingly, the carbon isotope discrimination by lichens with cyanobacteria is 2–4$\%o$ less than that of lichens with green algae. Another possibility is that liquid water is needed for a bicarbonate transport system, which also has a characteristically smaller discrimination (see the section below on algae). A further complication is that discrimination by Rubisco, b_3, is 21$\%o$ in the only cyanobacterium measured compared to 29$\%o$ in higher plants (50). A great deal more work is required before an equation like Equation 8 could be applied with confidence to lichens.

C_4 Photosynthesis

Variation in isotopic composition among plants with the C_4 photosynthetic pathway is less than in C_3 plants, because the term b from Equation 8 (largely reflecting b_3, the discrimination by Rubisco, $\sim 30\%o$) is replaced by ($b_4 +$

$b_3\phi$) which is numerically much smaller than b_3. This is because b_4 (the effective discrimination by PEP carboxylase) is $\sim -5.7‰$ (see Table 1) and ϕ [the proportion of the carbon fixed by PEP carboxylation that subsequently leaks out of the bundle sheath, thereby allowing limited expression of Rubisco discrimination (b_3)] is necessarily less than unity (33). The bases for this model of discrimination are as follows: CO_2 diffuses through stomata to the mesophyll cells, where it dissolves (e_s) and is converted to HCO_3^- (e_b). At equilibrium, the heavier isotope concentrates in HCO_3^- compared to gaseous CO_2—i.e. the combined terms $e_s + e_b$ are negative (Table 1). In turn, PEP carboxylase discriminates against $H^{13}CO_3^-$—i.e. b_4^* is positive and normal for a kinetic effect. Thus if the gaseous intercellular CO_2 is in equilibrium with HCO_3^-, then the net discrimination from CO_2 to OAA is

$$b_4 = e_s + e_b + b_4^* \qquad\qquad 9.$$

which is negative because of the magnitude of e_b. Various transformations then occur, depending on C_4 subtype, but the net result in all cases is that CO_2 is released within the bundle sheath cells and refixed by Rubisco. There is little opportunity for discrimination in the release of CO_2 in the bundle sheath cells because of the lack of significant biochemical branches. No further discrimination would occur if the bundle sheath were gas tight (153). However, some quantities of CO_2 and HCO_3^- are likely to leak out of these cells and into the mesophyll cells, especially through the apoplastic portions of the bundle sheath cells, where they can then mix with other CO_2 that has diffused in through the stomata. The leak is a branch from the main path of carbon and allows some discrimination by Rubisco in the bundle sheath cells (b_3).

Various models (18, 33, 56, 110, 117) have addressed aspects of the ^{13}C discrimination during C_4 photosynthetic metabolism. Intrinsic to all of these models is the notion that variation in isotopic composition in C_4 plants is associated with leakage of CO_2 and/or HCO_3^-. The "leakiness" (ϕ) may also be regarded as a measure of the "overcycling" by PEP carboxylase that occurs in mesophyll cells, raising the partial pressure of CO_2 within the bundle sheath cells (33).

Farquhar (33) developed an expression for the discrimination occurring in C_4 photosynthesis, in which

$$\Delta = a\,\frac{p_a - p_i}{p_a} + (b_4 + b_3\phi)\,\frac{p_i}{p_a} = a + (b_4 + b_3\phi - a)\,\frac{p_i}{p_a}. \qquad 10.$$

Depending whether ($b_4 + b_3\phi - a$) is positive, zero, or negative, the dependence of Δ on p_i/p_a will be positive, zero, or negative. Experimental evidence suggests that the factor is often close to zero, with short-term discrimination responding little to variation in p_i/p_a (32, 36, and see Figure 1). From Table 1 it may be seen that the zero value is obtained with $\phi = 0.34$.

Farquhar (33) and Hattersley (56), using the Farquhar model, predicted that bundle sheath "leakiness" above $\phi = 0.37$ (the value differs from 0.34 because a smaller value was assumed for b_3) should result in a positive response of Δ to increasing p_i/p_a. Anatomical variations between C_4 types (55) may be associated with variations in ϕ. For example, Ehleringer & Pearcy (29) observed that quantum yields for CO_2 uptake are lower for all C_4 dicots and NAD-ME (malic enzyme) C_4 type grasses than for NADP-ME or PCK (phosphoenolpyruvate carboxykinase) types, which have bundle sheaths with suberized lamellae (12, 57). Diminished quantum yields are to be expected as a result of increased leakiness—i.e. increased overcycling within the mesophyll cells. The measured differences in carbon isotope discrimination by NAD-ME, NADP-ME, and PCK type C_4 grasses as deduced from isotopic composition (10, 56, 150, 163) and from "on-line" measurements (32, 36) are consistent with the expectation that leakage is greatest in the first type. Because ϕ is a measure of the overcycling as a proportion of the rate of PEP carboxylation, it is likely that Δ will increase whenever Rubisco activity is diminished more than PEP carboxylase activity by some treatment. Thus ϕ and Δ depend as much on coordination of mesophyll and bundle sheath activity as on anatomical features.

C_3-C_4 Intermediacy

Monson et al (90) measured isotopic composition of 6 C_3-C_4 intermediate species in *Flaveria* and reported Δ values of 9.6–22.6‰. They suggested that the isotopic variation resulted from differences in bundle sheath leakage (according to Equation 10). While this probably accounts for some of the variation, another biochemical factor may also be important. The C_3-C_4 "intermediate" species appear to have glycine decarboxylase confined to the bundle sheath cells (63, 115). The effect is that CO_2 released by photorespiration is released and partially refixed in the bundle sheath, so that discrimination by Rubisco can occur twice (S. von Caemmerer, unpublished). The modification to b, the C_3 carboxylation parameter from Equation 8, is thus the product of the proportion, A_s/A, of carbon fixed twice (where A_s is the rate of assimilation in the bundle sheath, and A that by the whole leaf), and ϕ, the proportion of the carbon supplied to the bundle sheath that leaks out. In the simplest form the equation becomes (G. D. Farquhar, unpublished)

$$\Delta = a \frac{p_a - p_i}{p_a} + b \left(1 + \frac{\phi A_s}{A} \right) \frac{p_i}{p_a} = a + \left[b \left(1 + \frac{\phi A_s}{A} \right) - a \right] \frac{p_i}{p_a}. \quad 11.$$

This theoretical prediction awaits experimental testing.

Crassulacean Acid Metabolism

The details of Crassulacean acid metabolism (CAM) that affect Δ have been recently reviewed by O'Leary (99). In this section, we present equations for Δ analogous to those discussed earlier for C_3 and C_4 carbon assimilation pathways. Plants in the full CAM mode take up CO_2 and synthesize oxaloacetate using PEP carboxylase, and the oxaloacetate (OAA) is then reduced and stored as malate (103). At dawn the plants close their stomata and decarboxylate the malate, refixing the released CO_2 using Rubisco. The malate that is stored at night will show the same discrimination as for C_4 species with zero leakage (33), i.e.

$$\Delta = a + (b_4 - a)\ \frac{p_i}{p_a}. \qquad\qquad 12.$$

Winter (154) reported that nocturnal values of p_i/p_a in *Kalanchoe pinnata* started at a C_4-like value (~ 0.4) and increased with time to a C_3-like value (~ 0.7) before dawn. On this basis, we could expect instantaneous Δ of the carbon being fixed to have decreased from ~ 0.4 to $-2.7\%o$ as the night progressed. This is consistent with observations that Δ of crystalline oxalate and of carbon-4 of malic acid were near to zero (58, 100).

If the stomata closed completely at dawn, the photosynthetic tissue would form a closed system and there would be no fractionation of the carbon between malate and the sugar products. However, consider the case where the stomata were not closed in the light while a CAM plant was enclosed in a cuvette with no external CO_2. In this case, there should be a discrimination in going from malate to the new C_3 carbon because we no longer have a closed system. The discrimination is given by $\phi(b_3 - a)$, where ϕ is the proportion of decarboxylated carbon that leaks out of the leaf. Nalborczyk (94) allowed CAM plants to fix CO_2 only at night and found that the overall discrimination was $\sim 3\%o$. This result implies that ϕ was about 0.05–0.15. However, when CAM plants are growing in normal air, evolution of CO_2 in the light is usually negligible.

Toward the end of the light period, after decarboxylation of all the stored malate, there is sometimes CO_2 uptake [denoted phase IV by Osmond (103)] via Rubisco, and possibly involving PEP carboxylase as well. Nalborczyk et al (95) allowed plants to fix carbon only in the light and observed a discrimination of $21\%o$, which is what one would expect with a typical C_3 value of p_i/p_a in Equation 8.

Therefore in the simplest case of C_4 fixation in the dark and C_3 fixation in the light, the average discrimination over a 24 hr period is

$$\Delta = a + \frac{\int^D A(b_4 - a)\,\frac{p_i}{p_a}\,dt + \int^L A(b - a)\,\frac{p_i}{p_a}\,dt}{\int^D A\,dt + \int^L A\,dt},$$

13.

where A is the assimilation rate, $\int^D dt$ denotes the time integral in the dark, and $\int^L dt$ that in the light, and b for the light period is the average of b_3 and b_4 weighted by the rates of RuBP and PEP carboxylation (if the latter occurs), respectively.

Aquatic Plants and Algae

Carbon isotope combinations measured in aquatic plants range between $-11‰$ and $-39‰$, potentially leading to the mistaken impression that both C_3 and C_4 photosynthetic pathways are present in aquatic plants (4, 22, 105, 113, 132). However, with limited exceptions (86, 147), C_4 plants are not known from aquatic habitats. When CO_2 fixation is via the normal C_3 pathway, Equation 8 applies, but with the parameter a modified to reflect diffusion in the aqueous phase ($e_s + a_l$) so that

$$\Delta = (e_s + a_l)\,\frac{p_a - p_c}{p_a} + b\,\frac{p_c}{p_a} = e_s + a_l + (b - e_s - a_l)\,\frac{p_c}{p_a},$$

14.

where the equivalent partial pressure of CO_2 at the site of carboxylation is denoted as p_c. Note that the discrimination during diffusion of CO_2 in water (a_l) is $0.7‰$ (98) and not $11‰$ as some authors have written. Much of the diffusion of inorganic carbon in the aqueous environment will be as bicarbonate rather than CO_2, but the discrimination here should also be small (38). Note further that the discrimination is with respect to gaseous CO_2 in equilibrium with the aqueous environment.

However, there is a widespread mechanism(s) among marine and freshwater organisms for raising the concentration of CO_2 at the site of carboxylation above that of the environment (5, 80). Farquhar (33) suggested that the equation for C_4 discrimination could be adapted to describe discrimination if the active species transported is bicarbonate as

$$\Delta = (e_s + a_l)\frac{p_a - p_c}{p_a} + (e_s + e_b + b_m + b_3\phi)\,\frac{p_c}{p_a}$$

$$= e_s + a_l + (e_b + b_m + b_3\phi - a_l)\,\frac{p_c}{p_a},$$

15.

where b_m is the fractionation during membrane transport. The value of b_m is unknown, but it has been cautiously assumed to be small, making ($e_s + e_b +$

b_m), which is the analog for b_4 from the C_4 model, close to $-7.9\%o$ (33). Note that in both Equations 14 and 15 the discrimination is again expressed in relation to a gaseous source. As with Equation 14, the discrimination in relation to *dissolved* CO_2 as the source, provided it were in equilibrium with the gas phase, would be found by subtracting e_s and with reference to bicarbonate (again, if in equilibrium) would be found by subtracting $e_s + e_b$. However, it is convenient to retain the same convention for source carbon as used for aerial plants (i.e. gaseous CO_2), especially when we have chosen gaseous CO_2 as our substrate for carboxylation by Rubisco (see definition of b_3 in Table 1). The latter choice is also reasonable in a mechanistic sense because the Rubisco site, with RuBP bound, probably reacts with gaseous substrates only.

The effects on Δ of induction of active carbon accumulation were elegantly demonstrated by Sharkey & Berry (125). The green alga *Chlamydomonas reinhardtii* was grown at 5% CO_2 and then transferred to normal air levels of CO_2. Before transfer, Δ was 27–29‰, and after 4 hr of induction Δ was 4‰. Sharkey & Berry (125) discussed their results in terms of slightly simplified versions of Equations 14 and 15. Berry (5) noted that measurements of Δ alone are insufficient to distinguish between a CO_2 concentrating mechanism (Equation 15) and a normal C_3 mechanism (Equation 14) with a large resistance to diffusion. In both cases, Δ is small because most of the CO_2 reaching Rubisco is fixed.

ENVIRONMENTAL EFFECTS ON CARBON ISOTOPE DISCRIMINATION

Goudriaan & van Laar (47), Körner et al (72), and Wong et al (161) were among the first to note a strong correlation between the photosynthetic rate and leaf conductance. This correlation was maintained over a wide variety of plant species and under a diversity of environmental treatments, implying some level of regulation between CO_2 demand by the chloroplasts and CO_2 supply by stomatal control. If in fact there were no deviations from the slope of the photosynthesis-versus-conductance relationship and if the intercept were zero (as was the case in the original papers), then the intercellular CO_2 pressure (p_i) of all plants would have been constant, dependent only on photosynthetic pathway. This constancy was mistakenly suggested in at least one early review (126). Although a number of studies that followed showed a significant tendency for photosynthesis and conductance to be correlated (161), many of these data sets exhibited some deviation from a linear relationship or a nonzero intercept (112, 152). It is unfortunate that in the search for general patterns the variance in p_i was, for a time, ignored. When it was recognized that there was a fundamental relationship between Δ or δ_p and p_i, more effort was put into documenting and understanding the isotopic variation at both the environmental and genetic (intra- and interspecific) levels. In the next sections, we describe what is known about the relationship between p_i (as measured by isotope discrimination) and environmental parameters.

Light

While some of the first experiments reported no consistent pattern between leaf isotopic composition, δ_p, and irradiance (129), later studies have indicated that δ_p increased with an increase in growth irradiance. Interpretation of carbon isotope composition of leaves experiencing different light levels has been somewhat controversial. The controversy lies in separating the effects of light on discrimination from correlated effects on δ_a (source air), both of which affect leaf carbon isotopic composition. In field studies, Vogel (148) was among the first to describe a consistent pattern of isotopic variation in leaves under canopy conditions where light levels varied substantially. He noted that δ_p within a canopy decreased by 3‰ between the top (19 m) and bottom (1 m) of the canopy. He further noted that the isotopic composition of soil CO_2 was approximately -19‰, while that of the atmosphere was only -7‰. He attributed all of the decrease in δ_p of leaves at lower layers to a recycling of soil CO_2 (a lighter source CO_2), although the isotopic composition of CO_2 within the canopy, δ_a, was not measured. He calculated that recycled CO_2 accounted for 15% of the carbon incorporated in lower leaf layers—assuming that the physiological discrimination was constant. Medina & Minchin (87) pursued these observations, reporting $\delta^{13}C$ gradients of 4.7 and 5.6‰ between upper and lower canopy leaves for two different tropical-forest types. Again the decrease in $\delta^{13}C$ of leaves at lower levels was attributed to a lighter source CO_2, with the implication that as much as 20% of the carbon fixed in lower leaf layers was derived from soil respiration. A third study by Schleser & Jayasekera (122) reports a similar pattern for forest beech and isolated lime trees. Again, they attributed this result to recycled soil CO_2.

Some recent studies have examined both δ_a and δ_p. In their study in a huon pine forest, Francey et al (42) observed that δ_p decreased with canopy depth, but without δ_a decreasing in a corresponding manner, which indicates a physiological effect. They found that leaves from lower in the canopy had greater p_i values than those from the upper canopy, suggesting, according to Equation 8, a greater discrimination in lower leaves. Ehleringer et al (27, 28) observed a similar pattern with ten shrub and tree species from a subtropical monsoon forest. Leaf δ_p decreased (i.e. became more negative) and p_i increased as observations were made deeper in the canopy. Furthermore, when only outer canopy leaves were measured on plants with differing degrees of canopy closure, δ_p was decreased with decreasing irradiance, consistent with the model of increasing p_i at lower light levels. These measurements were confirmed with gas exchange observations of the dependence of p_i on irradiance. While it is undoubtedly true that a fraction of the soil CO_2 is incorporated within leaves at the lower canopy level, much of the decrease in leaf isotopic composition is likely to be associated with stomatal and photosynthetic effects. Higher p_i values in understory leaves are likely to benefit plant performance when leaves are exposed to higher irradiances during sunflecks and when leaves are allowed to operate at higher quantum yields (71, 107). In the field, effects of irradiance on p_i are difficult to separate from those of leaf-to-air vapor pressure difference (vpd). The smaller vpd at the bottom of the canopy could also cause greater p_i, and greater Δ (another complication is discussed after Equation A13 in the Appendix).

Water

PHYSIOLOGICAL RESPONSE TO DROUGHT   When soil moisture levels are decreased, a common response is simultaneous decreases in photosynthesis, transpiration, and leaf conductance (40). If the "supply function" of photosynthesis (leaf conductance) decreases at a faster rate under stress than the "demand function" [photosynthetic dependence on p_i, *sensu* Farquhar & Sharkey (40)], then p_i will decrease. This effect should be measurable as either an increase in δ_p or correspondingly as a decrease in Δ. Over the short term when new growth has not occurred, the impact of stress can be detected in carbohydrate fractions within leaves (11, 163a, 81). Alternatively, the reduction in p_i/p_a can be measured using the "on-line" approach (62). In longer-term observations under both growth-chamber and field conditions, plants under water stress induced by low soil moisture availability produced leaves with lower p_i values as estimated by carbon isotopic composition (19, 23, 26, 39, 59–62, 131, 140, 155). Increasing the soil strength (physical resistance to root penetration), such as might occur in drier soils, induces a reduction of Δ, as observed with reduced soil moisture levels (84).

An increase in the leaf-to-air vapor pressure difference will also cause diminution of p_i and Δ in the short term (11) and long term (35, 39, 157).

PHOTOSYNTHETIC PATHWAY SWITCHING   In response to changes in leaf water status, a number of species show dramatic shifts in carbon isotope composition (up to 10–15‰) associated with changes in photosynthetic metabolism. Thus upon exposure to increased drought, some species can shift from C_3 to CAM photosynthesis (8, 54, 67, 78, 137–140, 146, 158). Correspondingly, there is an increase in δ_p (decrease in Δ). This shift in metabolism is reversible, dependent primarily on plant water status, and can occur in both annual and perennial leaf succulents of arid habitats. Other plants, notably "stranglers" of tropical habitats, exhibit CAM metabolism as epiphytic juveniles, but later switch to C_3 metabolism when roots reach the soil surface (111, 134, 143).

PHOTOSYNTHETIC TWIGS AND STEMS   In an interesting twist on the photo-synthetic-shift theme, at least two stem succulents native to southern Africa exhibit C_3 metabolism in the leaves (which are shed early in the drought period) and CAM in the stems (77, 142). In recent studies on green-twig plants from arid lands of North America, high rates of photosynthesis have been observed in twig tissues that are comparable to those observed in leaves (13, 24, 104, 131). Unlike the previous example, the twigs of these species all have C_3 photosynthesis. In all such species examined to date, p_i values as measured by gas exchange techniques are lower in twig than leaf tissues, leading to a significant difference in carbon isotopic composition of the two tissue types. Thus, in these cases, the decrease in Δ of the twigs is associated with increased diffusional constraints rather than with a change in metabolic pathway as described in the previous section.

Salinity

In nonhalophytic species, increased salinity has numerous metabolic effects (48). Stomatal closure is typically associated with increased salinity (20, 21, 79, 124). Thus it should not be surprising to note that in those species Δ decreased with increasing salinity, indicating a decrease in p_i with increasing stress (124). What is perhaps more intriguing is that halophytic species also exhibit a similar pattern whether in field or laboratory conditions (35, 51–53, 96, 163a).

Air Pollution

A long-term consequence of exposure to air pollutants (e.g. ozone, sulfur dioxide) at the leaf level is normally a decrease in both leaf conductance and photosynthesis (118). It is not clear, however, whether this decrease in gas exchange represents overall decline in metabolic activity or an increased diffusion limitation imposed by stomata. In each of the limited number of studies available that examine carbon isotope discrimination by leaves of plants exposed to pollutants, exposed plants exhibited lower Δ values, suggesting lower p_i (43, 49, 81). Changes in isotopic composition of leaf tissues from these studies of 1‰ or greater were common even under modest exposures to air pollution. Under long-term, chronic exposure to air pollut-ants, clear differences exist in the carbon isotope ratios of the wood of annual growth rings that are consistent with short-term, leaf-level observations (43, 44, 81).

WATER-USE EFFICIENCY OF C_3 SPECIES

Transpiration Efficiency and Carbon Isotope Discrimination

Measurements of Δ in C_3 species may usefully contribute to the selection for transpiration efficiency—i.e. the amount of carbon biomass produced per unit water transpired by the crop.

The instantaneous ratio of CO_2 assimilation rate of a leaf, A, to its transpiration rate, E, is given approximately by

$$\frac{A}{E} = \frac{p_a - p_i}{1.6\nu}, \qquad 16.$$

where ν is the water vapor pressure difference between the intercellular spaces and the atmosphere. The factor 1.6 arises because the binary diffusivity of water vapor and air is 1.6-fold greater than that of CO_2 and air. Equation 16 may be rewritten as

$$\frac{A}{E} = \frac{p_a\left(1 - \dfrac{p_i}{p_a}\right)}{1.6\nu}, \qquad 17.$$

to emphasize that a smaller value of p_i/p_a is equivalent to an increase in A/E, for a constant water vapor pressure difference, ν. Thus selecting for lower p_i/p_a should be, to a first approximation, a screen for greater A/E, which, in turn, is a component of transpiration efficiency. From Equation 8, Δ may be used as a surrogate measure of p_i/p_a in C_3 plants.

In all of the experiments relating gas exchange properties and short- and long-term discrimination (see the section above on C_3 photosynthesis) and where vapor pressure difference, ν, was maintained constant, the ratio of assimilation and transpiration rates, A/E, was negatively related to Δ, as expected from Equation 17. However, during whole-plant growth, losses of carbon and water occur that are not included in Equation 17. A proportion, ϕ_c, of the carbon fixed via the stomata during the day is lost from the shoot at night or from nonphotosynthetic organs such as the roots, during both the day and night. Further, some water is lost from the plant independently of CO_2 uptake. The stomata may not be completely closed at night, cuticular water loss occurs, and there is unavoidable evaporative loss from the pots in whole-plant experiments. If this "unproductive" water loss is a proportion, ϕ_w, of "productive" water loss, Equation 17 may be modified to describe the molar ratio, W, of carbon gain by a plant to water loss

$$W = \frac{p_a\left(1 - \dfrac{p_i}{p_a}\right)(1 - \phi_c)}{1.6\nu(1 + \phi_w)}, \qquad 18.$$

which, when combined with Equation 8, predicts a negative linear dependence of W on Δ (38, 60). By substitution, Equation 18 can be rewritten as

$$W = \frac{p_a \left(\dfrac{b - d - \Delta}{b - a} \right) (1 - \phi_c)}{1.6\nu(1 + \phi_w)},$$ 19.

where d is a correction related to assimilation rate (see Part III of the Appendix). The data from pot experiments using a combination of watering treatments and genotypes fit the theory reasonably well for a number of species—wheat (39, 84), peanuts (61, 62, 162), cotton (59), tomato (83), and barley (60). We suggest that future studies will provide better understanding of the relationships between W and Δ when account is taken of environmental and genetic effects on ϕ_c and ϕ_w.

Scaling from the Plant to the Canopy

Water-use efficiency is difficult to measure in the field. There have, however, been a few attempts to relate it to Δ, or at least to relate yield under water-limited conditions to Δ. Wright et al (162) measured total above-ground biomass yield and water use of eight peanut genotypes receiving adequate water (under a rain-excluding shelter). They obtained a negative relationship between W and leaf Δ.

There are several reasons why the negative relationship between W and Δ, given by Equation 19, might work well for individual plants in pots, or even for small plots in the field, but become inconsistent over larger areas. The uncontrolled loss of water is not an independent, fixed proportion (ϕ_w) of transpiration because, for example, soil evaporation tends to be negatively related to leaf area development. If ν fluctuates, then those genotypes that might grow more when ν is small will obtain a greater W for the same Δ.

Equation 19 also contains a simplification that becomes more problematic with increase of scale. The equation is written as if the vapor pressure difference, ν, were an independent variable. To some extent, however, it must vary as stomatal conductance, g_s, changes (as is the case for a single leaf). A reduction in g_s, and therefore in E, means more heat has to be lost by sensible heat transfer. The presence of a leaf boundary layer resistance to the transfer of heat translates this into an increase of leaf temperature and of ν and so the effect of decreased g_s on E is moderated. This moderating effect increases as the ratio of boundary layer resistance to stomatal resistance increases. With a sufficiently high ratio, the proportional reduction in E caused by partial stomatal closure is no greater than the associated proportional reduction in A. Farquhar et al (36) discussed the above problems and defined the conditions that would be necessary for A/E to become independent of stomatal conductance, p_i/p_a and Δ.

The problem is exacerbated in the field, where the aerodynamic resistance of the crop has to be taken into account. If the canopy and leaf boundary layer resistances to heat are very large, there is the possibility that a genotype with a greater stomatal conductance than another otherwise identical genotype will have a greater value of W (15), despite also having a greater Δ (36). This is more likely to occur at high temperatures. On the other hand, it is less likely to occur when crops have very small leaf area indexes, as would normally be the case under conditions where stress occurs early, and in crops sown in areas prone to severe, early water stress, because under these conditions the crop is more closely "coupled" to the atmosphere, like an isolated plant (15, 66). If the source of variation in Δ is the capacity for photosynthesis, the effects of boundary layers are unimportant (15). This appears to be the case for peanuts (62). Therefore at the crop level, identification of the causes underlying differences in Δ may become important—differences in conductance having different micrometeorological consequences from differences in photosynthetic capacity.

Carbon Isotope Discrimination and Plant Growth Characteristics

Hubick et al (62) found a negative relationship between dry matter production and Δ of peanut cultivars grown in field trials. On the other hand Condon et al (14) saw a positive relationship between yield and Δ for wheat cultivars in two years that included periods of greater than usual rainfall. The sign of the relationship under well-watered conditions is difficult to predict. It is clear that any associations between Δ and patterns of carbon partitioning will be important. The relative growth rate, r (sec^{-1}), of a plant depends on the assimilation rate per unit leaf area, A (mol C m^{-2} sec^{-1}), and the ratio of total plant carbon to leaf area, ρ (mol C m^{-2}), according to the following identity (84)

$$r = \frac{lA(1 - \phi_c)}{\rho}, \qquad\qquad 20.$$

where l is the photoperiod as a proportion of a day. Masle & Passioura (85) observed that wheat seedlings grew more slowly in soil of increased strength than in controls. Masle & Farquhar (84) showed that ρ increased with increasing soil strength. They also found that Δ decreased with increasing soil strength. Changing soil strength thus induced a negative relationship between ρ and Δ. They noted that a similar, negative, but genetic association between ρ and Δ would tend to cause a positive relationship between growth rate and Δ. A negative association between ρ and Δ has been observed among wheat and sunflower genotypes during early growth (J. Virgona, personal communication). If ν is low early in the life of a crop, then a positive association between Δ and relative growth rate among genotypes will confound the relationship between final W and Δ.

Genetic Control of Discrimination

Genetic studies of W, p_i/p_a, and Δ are in their infancy. These traits are most likely to be polygenic, since any gene that affects either assimilation rate per unit leaf area or stomatal conductance can have an effect. Despite the considerable genetic and environmental (nutrition, light intensity, etc) effects on the individual components A and g, separately, it is likely that the variation in the ratio A/g, and hence in p_i/p_a and Δ, is smaller, because of coordination between A and g (37). The coordination can lead to predictable genotypic differences in p_i/p_a and Δ as assessed from gas exchange (62), as well as in Δ assessed from δ_p.

The genetic control of Δ appears to be strong in wheat. Condon et al (14) showed that genotypic ranking was maintained at different sites and between plants grown in pots and in the field. The broad sense heritabilities [proportion of total variance of Δ that can be ascribed to genotype, rather than to environment or to interactions between the two ($G \times E$)] ranged between 60 and 90%. From analyses of Δ in 16 peanut genotypes grown at 10 sites in Queensland, Hubick et al (62) calculated an overall broad sense heritability of 81%. With *Phaseolus vulgaris* in Colombia, it was 71% (23). Hubick et al (62; and see earlier discussion in reference 36) examined the progeny of a cross between Tifton 8, a peanut genotype having a small value of Δ, and Chico, which has a large value of Δ. Statistical analyses of measurements of Δ and W in the F_2 generation gave estimates for the heritability of 53% for Δ and 34% for W. The phenotypic correlation between W and Δ was -0.78. As expected, the Δ values of F_2 plants were highly variable and there were several transgressive segregants with values of Δ lower than those of Tifton 8.

The Δ values of the F_1 generation of the Tifton 8 and Chico cross, while somewhat intermediate between the two parents, were very close to those of Tifton 8 in Δ and W. Martin & Thorstenson (83) examined the F_1 plants from a cross between *Lycopersicon pennellii*, a drought-tolerant species related to tomato, with tomato itself, *Lycopersicon esculentum*. *L. pennellii* had a lower Δ than *L. esculentum*, and again Δ of the F_1 was intermediate, but closer to the low-Δ parent. Both sets of data suggest some dominance of the low-Δ attribute.

Genetic analysis of a polygenic trait like Δ is obviously difficult, yet considerable progress has recently been made using modern techniques. Martin et al (82) reported that 70% of the variance for Δ in a variable tomato population derived from further generations of the above cross was associated with three restriction fragment length polymorphisms (RFLPs)—i.e. genetic markers identifying discrete DNA sequences within the genome. Additive gene action was observed in the three cases, and in one of them, there was also a significant nonadditive component. This kind of work may enable breeders to follow the results of backcrossing material with desirable Δ into commercial cultivars. However, in parallel with pursuing research on the genetic control of carbon isotope discrimination by the plant, it is important to establish what values of Δ are appropriate in a particular environment and for a particular species. This requires extensive physiological work at different scales, from the organelle to the canopy, and a much better understanding of the interactions among plants, canopies, and their microclimates.

CONCLUDING REMARKS

Carbon isotope discrimination has become a tool to help us understand photosynthesis and its coordination with water use in ecological and physiological studies of C_3 species. Future work will relate these more to growth characteristics and will differentiate between effects of photosynthetic capacity and stomatal conductance. The latter may perhaps be studied using observations of isotopic composition of organic oxygen and hydrogen (36). These compositions are affected by the ratio of ambient and intercellular humidities and should therefore reflect changes in the energy budgets of leaves, which are themselves influenced by stomatal conductance.

It is possible that measurements of Δ in C_4 species may aid in seeking changes in coordination between mesophyll and bundle sheath tissue during photosynthesis, perhaps revealing differences in quantum requirements.

Technological advances in combining gas chromatography and isotope ratio mass spectrometry should facilitate measurements of carbon isotope discrimination between and within organic compounds, thereby increasing our ability to identify origins of materials and to study the nature of the control of metabolic pathways following photosynthesis.

ACKNOWLEDGMENTS

We thank Drs. Joe Berry and Josette Masle for valuable discussions and comments on this manuscript.

APPENDIX

Part I. Definitions   Isotope effects (α) are here defined as the ratio of carbon isotope ratios in reactant and product (39)

$$\alpha = \frac{R_r}{R_p},$$ A1.

where R_r is the $^{13}C/^{12}C$ molar ratio of reactant and R_p is that of the product.

In a first order kinetic reaction, the definition is obvious, i.e.

$$\alpha = \frac{k^{12}}{k^{13}}, \qquad \text{A2.}$$

where k^{12} and k^{13} are the rate constants for reactions of the respective isotopic substances. Higher-order kinetic reactions including Michaelis-Menten ones (38) can be treated similarly (102), and k^{12} and k^{13} become pseudo-first-order rate constants. The isotope effect associated with diffusion is the ratio of the ^{12}C and ^{13}C diffusivities. The analogy with Equation A1 is the diffusion from a source (reactant) to a sink where the "product" is kept at a vanishingly small concentration. In an equilibrium, the "product" is the carbon-containing compound of interest on the right-hand side of an equilibrium reaction. So if the reaction of interest is

$$A \underset{k_{-1}}{\overset{k_1}{\rightleftharpoons}} B, \qquad \text{A3.}$$

where A and B might be CO_2 and HCO_3^-, for example, then application of this rule yields

$$\alpha = \frac{\dfrac{A^{13}}{A^{12}}}{\dfrac{B^{13}}{B^{12}}} = \frac{K^{12}}{K^{13}}, \qquad \text{A4.}$$

where K^{12} is the equilibrium constant,

$$K^{12} = \frac{k_1^{12}}{k_{-1}^{12}}, \qquad \text{A5.}$$

for the ^{12}C compounds and K^{13} is the analogous constant for ^{13}C compounds. Note that the equilibrium isotope effect, α, is the kinetic isotope effect for the forward reaction (α_1) divided by that of the reverse reaction (α_{-1})—i.e.

$$\alpha = \frac{K^{12}}{K^{13}} = \frac{\left[\dfrac{k_1^{12}}{k_{-1}^{12}}\right]}{\left[\dfrac{k_1^{13}}{k_{-1}^{13}}\right]} = \frac{\left[\dfrac{k_1^{12}}{k_1^{13}}\right]}{\left[\dfrac{k_{-1}^{12}}{k_{-1}^{13}}\right]} = \frac{\alpha_1}{\alpha_{-1}}. \qquad \text{A6.}$$

It is pleasing then that the forms of the isotope effect (α) for kinetic effects (k^{12}/k^{13}) and equilibrium effects (K^{12}/K^{13}) are superficially similar. We denote the discrimination for either effect as α minus one (39). In most cases discrimination associated with a kinetic effect will be positive, but there is no a priori reason why a thermodynamic discrimination should be positive.

Part II. Discrimination in a simple two stage model—diffusion followed by carboxylation   The carbon isotope ratio of CO_2 in air is R_a, and in the plant product is R_p. In turn R_p must be the same as the ratio of $^{13}CO_2$ assimilation rate, A^{13}, and $^{12}CO_2$ assimilation rate, A [no superscript is given here for a variable relating to the major isotope ^{12}C]—i.e.

$$R_p = \frac{A^{13}}{A}.$$ A7.

Further, if the isotope effect associated with carboxylation is $1 + b$, then we must have

$$\frac{R_i}{R_p} = 1 + b,$$ A8.

where R_i is the carbon isotope ratio of the intercellular CO_2.

In turn, R_i is simply found by relating A to g (conductance) and P (total pressure). Thus,

$$A = \frac{g(p_a - p_i)}{P}.$$ A9.

The kinetic isotope effect for diffusion is the ratio of the diffusivities of $^{12}CO_2$ and $^{13}CO_2$ in air. Thus,

$$1 + a = \frac{g}{g^{13}},$$ A10.

and so

$$A^{13} = \frac{g(R_a p_a - R_i p_i)}{(1 + a)P}.$$ A11.

Substituting Equations A9 and A11 in A7,

$$R_p = \frac{R_a p_a - R_i p_i}{(1 + a)(p_a - p_i)}.$$

Rearranging,

$$\frac{R_a}{R_p} = (1 + a)\frac{p_a - p_i}{p_a} + \frac{R_i}{R_p}\frac{p_i}{p_a}.$$

Thus, using the definition of discrimination and Equation A8

$$\alpha = 1 + \Delta = \frac{R_a}{R_p} = (1 + a)\frac{p_a - p_i}{p_a} + (1 + b)\frac{p_i}{p_a}.$$

Thus

$$\Delta = a\frac{p_a - p_i}{p_a} + b\frac{p_i}{p_a},$$

which is Equation 8 from the main text. Note that no assumption of linearity is made about the response of A to p_i in the derivation of this equation.

Part III. Alternative definitions of discrimination   There are other possible definitions of discrimination. For example one could write

$$\text{Discrimination*} = 1 - \frac{R_p}{R_a}.$$

This would correspond to $(1 - k^{13}/k^{12})$ for kinetic effects and to $(1 - K^{13}/K^{12})$ for equilibrium effects. The asterisk superscript is added to emphasize that the numerical values obtained differ from those made using Equation 4. On this basis

$$\Delta* = \frac{\delta_r - \delta_p}{1 + \delta_r}.$$

The numerical differences between this and our chosen definition of discrimination are usually less than 0.5‰. In the case of discrimination by ribulose bisphosphate carboxylase (Rubisco), the two definitions differ by ~ 0.9‰, which is significant. However, formulation of discrimination as $\Delta*$ rather than as $(R_a/R_p - 1)$, would make derivation of the theory much more complicated. This may be seen by repeating the derivation in Part II using $a* = 1 - g^{13}/g$ and $b* = 1 - R_p/R_i$.

Although it may seem odd to have the abundance ratio of the source, R_a, in the numerator of our chosen definition (Equation A1), we note that R_a/R_p may equally be thought of as S_p/S_a, where S is the molar ratio $^{12}C/^{13}C$.

Yet another notation is to use $R_p/R_a - 1$, $k^{13}/k^{12} - 1$, and $K^{13}/K^{12} - 1$, which leads to negative values of discrimination.

Part IV. Complications to the use of $\Delta = a + (b - a)p_i/p_a$ Farquhar (34) showed that the appropriate value of p_i in Equation 8 is the assimilation-rate-weighted value of p_i, whereas normal gas exchange gives a conductance-weighted value of p_i. These two estimates will differ if there is heterogeneity of stomatal opening (73, 141) and restricted lateral diffusion within the leaf. Greater degrees of heterogeneity will therefore cause smaller best fit values for b. The simplest value of b would be the isotope discrimination factor of Rubisco carboxylation, taking gaseous CO_2 as the substrate (b_3). Roeske & O'Leary (119) measured the isotope effect as 1.029, but with respect to dissolved CO_2, so that the result must be multiplied by the isotope effect of the dissolution of CO_2 in water (1.0011) making b_3 approximately 30‰ (36). Guy et al (50) measured the effect directly with respect to the gas by monitoring continuing isotopic enrichment of CO_2 in a reaction vessel and calculated b_3 to be ~ 29‰ using an equation analogous to that for Rayleigh distillation (7, 97). However, Farquhar & Richards (39) suggested that the net discrimination in C_3 photosynthesis should be less than that in the Rubisco carboxylation, because even in C_3 species a portion, β, of CO_2 fixation is via PEP carboxylase. With b_4 being the net fractionation by PEP carboxylase with respect to gaseous CO_2 in equilibrium with HCO_3^-, ($-5.7‰$; see Table 1) they suggested a net discrimination value of

$$b = (1-\beta)b_3 + \beta b_4 = b_3 - \beta(b_3 - b_4).$$

The difference ($b_3 - b_4$) is ~ 36‰, so that b is sensitive to the proportion of β-carboxylation. The latter depends on the amount of aspartate to be formed—unlikely to vary much between plants—and the amount of HCO_3^- formed for pH balance. This latter factor may contribute to the greater discrimination shown by *Ricinus* plants grown with NH_4^+ as N source than when NO_3^- was the sole source, although the phenomenon was interpreted in terms of changed stomatal behavior (114). In an unpublished study by Melzer & O'Leary (personal communication), C_4 fixation was found to reduce carbon discrimination by no more than 1‰ in C_3 species. Assuming p_i/p_a was ~ 0.7, this means that b could be reduced from b_3 by 1.9‰.

Other effects are ignored in the simple model represented by Equation 8. These include the presence of resistance between the intercellular spaces and the sites of carboxylation, and effects of respiratory losses and translocation. Many of these effects are taken into account in a more detailed equation (32) for which Equation 8 is a simplification:

$$\Delta = a_b \frac{p_a - p_s}{p_a} + a \frac{p_s - p_i}{p_a} + (e_s + a_1) \frac{p_i - p_c}{p_a} + b \frac{p_c}{p_a} - \frac{\dfrac{eR_d}{k} + f\Gamma^*}{p_a}, \quad \text{A12.}$$

where p_s is the $p(CO_2)$ at the leaf surface, p_c is the equivalent $p(CO_2)$ at the sites of carboxylation, a_b is the fractionation occurring during diffusion in the boundary layer (2.9‰), e_s is the fractionation occurring as CO_2 enters solution [1.1‰ at 25°C; (149)] a_l is the fractionation due to diffusion in water [0.7‰; (98)], e and f are fractionations with respect to average carbon composition associated with "dark" respiration (R_d) and photorespiration, respectively, k is the carboxylation efficiency, and Γ^* is the CO_2 compensation point in the absence of R_d (32).

Equation 8 overestimates discrimination compared to Equation A12 by

$$d = [r_b(a - a_b) + r_w(b - e_s - a_l)] \frac{AP}{p_a} + \frac{\dfrac{eR_d}{k} + f\Gamma^*}{p_a}, \quad \text{A13.}$$

The resistances r_b and r_w (m^2 sec mol^{-1}) are those of the boundary layer, and between the intercellular spaces and the sites of carboxylation, respectively, and P is the atmospheric pressure. Thus Equation 8 should overestimate discrimination at a fixed p_i / p_a by an amount (d) that increases with increasing assimilation rate, as may tend to occur naturally with increasing light intensity (32).

Literature Cited

1. Amundson, R. G., Chadwick, O. A., Sowers, J. M., Doner, H. E. 1988. Relationship between climate and vegetation and the stable carbon isotope chemistry of soils in the eastern Mohave Desert, Nevada. *Quat. Res.* In press

2. Bender, M. M. 1968. Mass spectrometric studies of carbon-13 variations in corn and other grasses. *Radiocarbon* 10:468–72

3. Bender, M. M. 1971. Variation in the $^{13}C/^{12}C$ ratios of plants in relation to the pathway of photosynthetic carbon dioxide fixation. *Phytochemistry* 10:1339–44

4. Benedict, C. R., Wong, W. W. L., Wong, J. H. H. 1980. Fractionation of the stable isotopes of inorganic carbon by seagrasses. *Plant Physiol.* 65:512–17

5. Berry, J. A. 1988. Studies of mechanisms affecting the fractionation of carbon isotopes in photosynthesis. In *Stable Isotopes in Ecological Research*, ed. P. W. Rundel, J. R. Ehleringer, K. A. Nagy, pp. 82–94. New York: Springer-Verlag

6. Berry, J. A., Troughton, J. H. 1974. Carbon isotope fractionation by C_3 and C_4 plants in 'closed' and 'open' atmospheres. *Carnegie Inst. Wash. Yearb.* 73:785–90

7. Bigeleisen, J., Wolfsberg, M. 1958. Theoretical and experimental aspects of isotope effects in chemical kinetics. *Adv. Chem. Phys.* 1:15–76

8. Bloom, A. J., Troughton, J. H. 1979. High productivity and photosynthetic flexibility in a CAM plant. *Oecologia* 38:35–43

9. Bradford, K. J., Sharkey, T. D., Farquhar, G. D. 1983. Gas exchange, stomatal behavior, and $\delta^{13}C$ values of the *flacca* tomato mutant in relation to abscisic acid. *Plant Physiol.* 72:245–50

10. Brown, W. V. 1977. The Kranz syndrome and its subtypes in grass systematics. *Mem. Torr. Bot. Club* 23:1–97
11. Brugnoli, E., Hubick, K. T., von Caemmerer, S., Farquhar, G. D. 1988. Correlation between the carbon isotope discrimination in leaf starch and sugars of C_3 plants and the ratio of intercellular and atmospheric partial pressures of carbon dioxide. *Plant Physiol.* 88:1418–24
12. Carolin, R. C., Jacobs, S. W. L., Vesk, M. 1973. The structure of the cells of the mesophyll and parenchymatous bundle sheath of the Graminae. *Bot. J. Linn. Soc.* 66:259–75
13. Comstock, J. P., Ehleringer, J. R. 1988. Contrasting photosynthetic behavior in leaves and twigs of *Hymenoclea salsola,* a green-twigged, warm desert shrub. *Am. J. Bot.* 75:1360–70
14. Condon, A. G., Richards, R. A., Farquhar, G. D. 1987. Carbon isotope discrimination is positively correlated with grain yield and dry matter production in field-grown wheat. *Crop Sci.* 27:996–1001
15. Cowan, I. R. 1988. Stomatal physiology and gas exchange in the field. In *Flow and Transport in the Natural Environment,* ed. O. T. Denmead. New York: Springer-Verlag. In press
16. Craig, H. 1954. Carbon-13 in plants and the relationship between carbon-13 and carbon-14 variations in nature. *J. Geol.* 62:115–49
17. Craig, H. 1957. Isotopic standards for carbon and oxygen and correction factors for mass spectrometric analysis of carbon dioxide. *Geochim. Cosmochim. Acta* 12:133–49
18. Deleens, E., Ferhi, A., Queiroz, O. 1983. Carbon isotope fractionation by plants using the C_4 pathway. *Physiol. Veg.* 21:897–905
19. DeLucia, E. H., Schlesinger, W. H., Billings, W. D. 1988. Water relations and the maintenance of Sierran conifers on hydrothermally altered rock. *Ecology* 69:303–11
20. Downton, W. J. S. 1977. Photosynthesis in salt-stressed grapevines. *Aust. J. Plant Physiol.* 4:183–92
21. Downton, W. J. S., Grant, W. J. R., Robinson, S. P. 1985. Photosynthetic and stomatal responses of spinach leaves to salt stress. *Plant Physiol.* 77:85–88
22. Dunton, K. H., Schell, D. M. 1987. Dependence of consumers on macroalgal *(Laminaria solidungula)* carbon in an arctic kelp community: $\delta^{13}C$ evidence. *Marine Biol.* 93:615–25
23. Ehleringer, J. R. 1988. Correlations between carbon isotope ratio, water-use efficiency and yield. In *Research on Drought Tolerance in Common Bean,* ed. J. W. White, G. Hoogenboom, F. Ibarra, S. P. Singh, pp. 165–91. Cali, Columbia: CIAT
24. Ehleringer, J. R., Comstock, J. P., Cooper, T. A. 1987. Leaf-twig carbon isotope ratio differences in photosynthetic-twig desert shrubs. *Oecologia* 71:318–20
25. Ehleringer, J. R., Cook, C. S., Tieszen, L. L. 1986. Comparative water use and nitrogen relationships in a mistletoe and its host. *Oecologia* 68:279–84
26. Ehleringer, J. R., Cooper, T. A. 1988. Correlations between carbon isotope ratio and microhabitat in desert plants. *Oecologia* 76:562–66
27. Ehleringer, J. R., Field, C. B., Lin, Z. F., Kuo, C. Y. 1986. Leaf carbon isotope ratio and mineral composition in subtropical plants along an irradiance cline. *Oecologia* 70:520–26
28. Ehleringer, J. R., Lin, Z. F., Field, C. B., Kuo, C. Y. 1987. Leaf carbon isotope ratios of plants from a subtropical monsoon forest. *Oecologia* 72:109–14
29. Ehleringer, J. R., Pearcy, R. W. 1983. Variation in quantum yield for CO_2 uptake among C_3 and C_4 plants. *Plant Physiol.* 73:555–59
30. Ehleringer, J. R., Schulze, E.-D., Ziegler, H., Lange, O. L., Farquhar, G. D., Cowan, I. R. 1985. Xylem-tapping mistletoes: water or nutrient parasites? *Science* 227:1479–81
31. Emrich, K., Ehhalt, D. H., Vogel, J. C. 1970. Carbon isotope fractionation during the precipitation of calcium carbonate. *Earth Planet. Sci. Lett.* 8:363–71
32. Evans, J. R., Sharkey, T. D., Berry, J. A., Farquhar, G. D. 1986. Carbon isotope discrimination measured concurrently with gas exchange to investigate CO_2 diffusion in leaves of higher plants. *Aust. J. Plant Physiol.* 13:281–92
33. Farquhar, G. D. 1983. On the nature of carbon isotope discrimination in C_4 species. *Aust. J. Plant Physiol.* 10:205–26
34. Farquhar, G. D. 1988. Models of integrated photosynthesis of cells and leaves. *Philos. Trans. R. Soc. Ser. B.* In press
35. Farquhar, G. D., Ball, M. C., von Caemmerer, S., Roksandic, Z. 1982. Effects of salinity and humidity on $\delta^{13}C$ value of halophytes—evidence for diffusional isotope fractionation determined by the ratio of intercellular/

atmospheric partial pressure of CO_2 under different environmental conditions. *Oecologia* 52:121–24

36. Farquhar, G. D., Hubick, K. T., Condon, A. G., Richards, R. A. 1988. Carbon isotope fractionation and plant water-use efficiency. In *Stable Isotopes in Ecological Research*, ed. P. W. Rundel, J. R. Ehleringer, K. A. Nagy, pp. 21–40. New York: Springer-Verlag

37. Farquhar, G. D., Hubick, K. T., Terashima, I., Condon, A. G., Richards, R. A. 1987. Genetic variation in the relationship between photosynthetic CO_2 assimilation and stomatal conductance to water loss. In *Progress in Photosynthesis IV*, ed. J. Biggins, pp. 209–12. Dordrecht: Martinus Nijhoff

38. Farquhar, G. D., O'Leary, M. H., Berry, J. A. 1982. On the relationship between carbon isotope discrimination and intercellular carbon dioxide concentration in leaves. *Aust. J. Plant Physiol.* 9:121–37

39. Farquhar, G. D., Richards, R. A. 1984. Isotopic composition of plant carbon correlates with water-use efficiency of wheat genotypes. *Aust. J. Plant Physiol.* 11:539–52

40. Farquhar, G. D., Sharkey, T. D. 1982. Stomatal conductance and photosynthesis. *Annu. Rev. Plant Physiol.* 33:317–45

41. Francey, R. J. 1985. Cape Grim isotope measurements—a preliminary assessment. *J. Atmos. Chem.* 3:247–60

42. Francey, R. J., Gifford, R. M., Sharkey, T. D., Weir, B. 1985. Physiological influences on carbon isotope discrimination in huon pine *(Lagastrobos franklinii). Oecologia* 44:241–47

43. Freyer, H. D. 1979. On the $\delta^{13}C$ record in tree rings. Part II. Registration of microenvironmental CO_2 and anomalous pollution effect. *Tellus* 31:308–12

44. Freyer, H. D., Belacy, N. 1983. $^{13}C/^{12}C$ records in northern hemispheric trees during the past 500 years—anthropogenic impact and climatic superpositions. *J. Geophys. Res.* 88:6844–52

45. Friedli, H., Siegenthaler, U., Rauber, D., Oeschger, H. 1987. Measurement of concentration, $^{13}C/^{12}C$ and $^{18}O/^{16}O$ ratios of tropospheric carbon dioxide over Switzerland. *Tellus* 39B:80–88

46. Gleason, J. D., Kyser, T. K. 1984. Stable isotope compositions of gases and vegetation near naturally burning coal. *Nature* 307:254–57

47. Goudriaan, J., van Laar, H. H. 1978. Relations between leaf resistance, CO_2-concentration and K-assimilation in maize, beans, lalang grass and sunflower. *Photosynthetica* 12:241–49

48. Greenway, H., Munns, R. 1980. Mechanisms of salt tolerance in nonhalophytes. *Annu. Rev. Plant Physiol.* 31:149–90

49. Greitner, C. S., Winner, W. E. 1988. Increases in $\delta^{13}C$ values of radish and soybean plants caused by ozone. *New Phytol.* 108:489–94

50. Guy, R. D., Fogel, M. F., Berry, J. A., Hoering, T. C. 1987. Isotope fractionation during oxygen production and consumption by plants. In *Progress in Photosynthetic Research III*, ed. J. Biggins, pp. 597–600. Dordrecht: Martinus Nijhoff

51. Guy, R. D., Reid, D. M., Krouse, H. R. 1980. Shifts in carbon isotope ratios of two C_3 halophytes under natural and artificial conditions. *Oecologia* 44:241–47

52. Guy, R. D., Reid, D. M., Krouse, H. R. 1986. Factors affecting $^{13}C/^{12}C$ ratios of inland halophytes. I. Controlled studies on growth and isotopic composition of *Puccinellia nuttalliana. Can. J. Bot.* 64:2693–99

53. Guy, R. D., Reid, D. M., Krouse, H. R. 1986. Factors affecting $^{13}C/^{12}C$ ratios of inland halophytes. II. Ecophysiological interpretations of patterns in the field. *Can. J. Bot.* 64:2700–7

54. Hartsock, T. L., Nobel, P. S. 1976. Watering converts a CAM plant to daytime CO_2 uptake. *Nature* 262:574–76

55. Hatch, M. D., Osmond, C. B. 1976. Compartmentation and transport in C_4 photosynthesis. In *Transport in Plants III: Encyclopedia of Plant Physiology (New Ser.)*, ed. C. R. Stocking, U. Heber, 3:144–84. New York: Springer-Verlag

56. Hattersley, P. W. 1982. $\delta^{13}C$ values of C_4 types in grasses. *Aust. J. Plant Physiol.* 9:139–54

57. Hattersley, P. W., Browning, A. J. 1981. Occurrence of the suberized lamella in leaves of grasses of different photosynthetic type. I. In the parenchymatous bundle sheath and PCR ('Kranz') sheaths. *Protoplasma* 109: 371–401

58. Holtum, J. A. M., O'Leary, M. H., Osmond, C. B. 1983. Effect of varying CO_2 partial pressure on photosynthesis and on carbon isotope composition of carbon-4 of malate from the Crassulacean acid metabolism plant *Kalanchoe daigremontiana* Hamet et Perr. *Plant Physiol.* 71:602–9

59. Hubick, K. T., Farquhar, G. D. 1987. Carbon isotope discrimination—selecting for water-use efficiency. *Aust. Cotton Grower* 8:66–68

60. Hubick, K. T., Farquhar, G. D. 1989. Genetic variation of transpiration efficiency among barley genotypes is negatively correlated with carbon isotope discrimination. *Plant Cell Environ.* In press

61. Hubick, K. T., Farquhar, G. D., Shorter, R. 1986. Correlation between water-use efficiency and carbon isotope discrimination in diverse peanut *(Arachis)* germplasm. *Aust. J. Plant Physiol.* 13:803–16

62. Hubick, K. T., Shorter, R., Farquhar, G. D. 1988. Heritability and genotype × environment interactions of carbon isotope discrimination and transpiration efficiency in peanut. *Aust. J. Plant Physiol.* 15:799–813

63. Hylton, M. C., Rawsthorne, S., Smith, A. M., Jones, D. A., Woolhouse, H. W. 1988. Glycine decarboxylase is confined to the bundle sheath cells of leaves of C$_3$-C$_4$ intermediate species. *Planta* 175:452–59

64. Inoue, H., Sugimura, Y. 1984. Diurnal change in δ^{13}C of atmospheric CO$_2$ at Tsukuba, Japan. *Geochem. J.* 18:315–20

65. Inoue, H., Sugimura, Y. 1985. The carbon isotope ratio of atmospheric carbon dioxide at Tsukuba, Japan. *J. Atmos. Chem.* 2:331–44

66. Jarvis, P. G., McNaughton, K. G. 1986. Stomatal control of transpiration: scaling up from leaf to region. *Adv. Ecol. Res.* 15:1–49

67. Kalisz, S., Teeri, J. A. 1986. Population-level variation in photosynthetic metabolism and growth in *Sedum wrightii*. *Ecology* 67:20–26

68. Keeling, C. D. 1958. The concentration and isotopic abundances of atmospheric carbon dioxide in rural areas. *Geochim. Cosmochim. Acta* 13:322–34

69. Keeling, C. D. 1961. A mechanism for cyclic enrichment of carbon-12 by terrestrial plants. *Geochim. Cosmochim. Acta* 24:299–313

70. Keeling, C. D., Mook, W. M., Tans, P. P. 1979. Recent trends in the ^{13}C/^{12}C ratio of atmospheric carbon dioxide. *Nature* 277:121–23

71. Kirschbaum, M. U. F., Pearcy, R. W. 1988. Gas exchange analysis of the relative importance of stomatal and biochemical factors in photosynthetic induction in *Alocasia macrorrhiza*. *Plant Physiol.* 86:782–85

72. Körner, C., Scheel, J. A., Bauer, H. 1979. Maximum leaf diffusive conductance in vascular plants. *Photosynthetica* 13:45–82

73. Laisk, A. 1983. Calculation of photosynthetic parameters considering the statistical distribution of stomatal apertures. *J. Exp. Bot.* 34:1627–35

74. Lange, O. L., Green, T. G. A., Ziegler, H. 1988. Water status related photosynthesis and carbon isotope discrimination in species of the lichen genus *Pseudocyphellaria* with green or blue-green photobionts and in photosymbiodemes. *Oecologia* 75:494–501

75. Lange, O. L., Kilian, E., Ziegler, H. 1986. Water vapor uptake and photosynthesis of lichens: performance differences in species with green and blue-green algae as phycobionts. *Oecologia* 71:104–10

76. Lange, O. L., Ziegler, H. 1986. Different limiting processes of photosynthesis in lichens. In *Biological Control of Photosynthesis*, ed. R. Marcelle, H. Clijsters, M. van Poucke, pp. 147–61. Dordrecht: Martinus Nijhoff

77. Lange, O. L., Zuber, M. 1977. *Frerea indica*, a stem succulent CAM plant with deciduous C$_3$ leaves. *Oecologia* 31:67–72

78. Lerman, J. C., Queiroz, O. 1974. Carbon fixation and isotope discrimination by a Crassulacean plant: dependence on photoperiod. *Science* 183:1207–9

79. Longstreth, D. J., Nobel, P. S. 1977. Salinity effects on leaf anatomy. Consequences for photosynthesis. *Plant Physiol.* 63:700–3

80. Lucas, W. L., Berry, J. A. 1985. Inorganic carbon transport in aquatic photosynthetic organisms. *Physiol. Plant.* 65:539–43

81. Martin, B., Bytnerowicz, A., Thorstenson, Y. R. 1988. Effects of air pollutants on the composition of stable isotopes (δ^{13}C) of leaves and wood, and on leaf injury. *Plant Physiol.* 88:218–23

82. Martin, B., Nienhuis, J., King, G., Schaefer, A. 1989. Restriction fragment length polymorphisms associated with water-use efficiency in tomato. *Science.* In press

83. Martin, B., Thorstenson, Y. R. 1988. Stable carbon isotope composition (δ^{13}C), water use efficiency, and biomass productivity of *Lycopersicon esculentum*, *Lycopersicon pennellii*, and the F$_1$ hybrid. *Plant Physiol.* 88:213–17

84. Masle, J., Farquhar, G. D. 1988. Effects of soil strength on the relation of water-use efficiency and growth to car-

bon isotope discrimination in wheat seedlings. *Plant Physiol.* 86:32–38

85. Masle, J., Passioura, J. B. 1987. Effects of soil strength on the growth of wheat seedlings. *Aust. J. Plant Physiol.* 14:643–56

86. Medina, E., de Bifano, T., Delgado, M. 1976. *Paspalum repens* Berg., a truly aquatic C_4 plant. *Acta Cient. Venezolana* 27:258–60

87. Medina, E., Minchin, P. 1980. Stratification of $\delta^{13}C$ values of leaves in Amazonian rain forests. *Oecologia* 45:377–78

88. Medina, E., Montes, G., Cuevas, E., Roksandic, Z. 1986. Profiles of CO_2 concentration and $\delta^{13}C$ values in tropical rainforests of the upper Rio Negro Basin, Venezuela. *J. Trop. Ecol.* 2:207–17

89. Medina, E., Troughton, J. H. 1974. Dark CO_2 fixation and the carbon isotope ratio in Bromeliaceae. *Plant Sci. Lett.* 2:357–62

90. Monson, R. K., Teeri, J. A., Ku, M. S. B., Gurevitch, J., Mets, L. J., Dudley, S. 1988. Carbon-isotope discrimination by leaves of *Flaveria* species exhibiting different amounts of C_3- and C_4-cycle co-function. *Planta* 174:145–51

91. Mook, W. G., Bommerson, J. C., Staverman, W. H. 1974. Carbon isotope fractionations between dissolved bicarbonate and gaseous carbon dioxide. *Earth Planet Sci. Lett.* 22:169–76

92. Mook, W. G., Koopmans, M., Carter, A. F., Keeling, C. D. 1983. Seasonal, latitudinal, and secular variations in the abundance of isotopic ratios of atmospheric carbon dioxide. 1. Results from land stations. *J. Geophys. Res.* 88:10915–33

93. Mooney, H. A., Troughton, J. H., Berry, J. A. 1977. Carbon isotope ratio measurements of succulent plants in southern Africa. *Oecologia* 30:295–305

94. Nalborczyk, E. 1978. Dark carboxylation and its possible effect on the value of $\delta^{13}C$ in C_3 plants. *Acta Physiol. Planta* 1:53–58

95. Nalborczyk, E., LaCroix, L. J., Hill, R. D. 1975. Environmental influences on light and dark CO_2 fixation by *Kalanchoe daigremontiana*. *Can. J. Bot.* 53:1132–38

96. Neales, T. F., Fraser, M. S., Roksandic, Z. 1983. Carbon isotope composition of the halophyte *Disphyma clavellatum* (Haw.) Chinnock (Aizoaceae), as affected by salinity. *Aust. J. Plant Physiol.* 10:437–44

97. O'Leary, M. H. 1981. Carbon isotope fractionation in plants. *Phytochemistry* 20:553–67

98. O'Leary, M. H. 1984. Measurement of the isotopic fractionation associated with diffusion of carbon dioxide in aqueous solution. *J. Phys. Chem.* 88:823–25

99. O'Leary, M. H. 1988. Carbon isotopes in photosynthesis. *BioScience* 38:325–36

100. O'Leary, M. H., Osmond, C. B. 1980. Diffusional contribution to carbon isotope fractionation during dark CO_2 fixation in CAM plants. *Plant Physiol.* 66:931–34

101. O'Leary, M. H., Reife, J. E., Slater, J. D. 1981. Kinetic and isotope effect studies of maize phosphenolpyruvate carboxylase. *Biochemistry* 20:7308–14

102. O'Leary, M. H., Treichel, I., Rooney, M. 1986. Short-term measurement of carbon isotope fractionation in plants. *Plant Physiol.* 80:578–82

103. Osmond, C. B. 1978. Crassulacean acid metabolism: a curiosity in context. *Annu. Rev. Plant Physiol.* 29:379–414

104. Osmond, C. B., Smith, S. D., Ben, G. Y., Sharkey, T. D. 1987. Stem photosynthesis in a desert ephemeral, *Eriogonum inflatum:* characterization of leaf and stem CO_2 fixation and H_2O vapor exchange under controlled conditions. *Oecologia* 72:542–49

105. Osmond, C. B., Valane, N., Haslam, S. M., Uotila, P., Roksandic, Z. 1981. Comparisons of $\delta^{13}C$ values in leaves of aquatic macrophytes from different habitats in Britain and Finland: some implications for photosynthetic processes in aquatic plants. *Oecologia* 50:117–24

106. Osmond, C. B., Winter, K., Ziegler, H. 1982. Functional significance of different pathways of CO_2 fixation in photosynthesis. In *Physiological Plant Ecology II: Encyclopedia of Plant Physiology (New Ser.)*, ed. O. L. Lange, P. S. Nobel, C. B. Osmond, H. Ziegler, 12B:479–547. New York: Springer-Verlag

107. Pearcy, R. W. 1988. Photosynthetic utilization of lightflecks by understory plants. *Aust. J. Plant Physiol.* 15:223–38

108. Pearman, G. I., Hyson, P. 1986. Global transport and inter-reservoir exchange of carbon dioxide with particular reference to stable isotopic distributions. *J. Atmos. Chem.* 4:81–124

109. Peisker, M. 1982. The effect of CO_2 leakage from bundle sheath cells on car-

bon isotope discrimination in C_4 plants. *Photosynthetica* 16:533–41

110. Peisker, M. 1985. Modelling carbon metabolism in C_3-C_4 intermediate species. 2. Carbon isotope discrimination. *Photosynthetica* 19:300–11

111. Popp, M., Kramer, D., Lee, H., Diaz, M., Ziegler, H., Lüttge, U. 1988. Crassulacean acid metabolism in tropical dicotyledonous trees of the genus *Clusia*. *Trees*. In press

112. Ramos, C., Hall, A. E. 1982. Relationships between leaf conductance, intercellular CO_2 partial pressure and CO_2 uptake rate in two C_3 and C_4 plant species. *Photosynthetica* 16:343–55

113. Raven, J. A., Beardall, J., Griffiths, H. 1982. Inorganic C-sources for *Lemna*, *Cladophora*, and *Ranunculus* in a fast-flowing stream: measurements of gas exchange and of carbon isotope ratio and their ecological implications. *Oecologia* 53:68–78

114. Raven, J. A., Griffiths, H., Allen, S. 1984. N source, transpiration rate and stomatal aperture in *Ricinus*. In *Membrane Transport in Plants*, ed. W. J. Cram, K. Janascek, R. Rybova, K. Sigler, pp. 161–62. Praha: Academia Praha

115. Rawsthorne, S., Hylton, C. M., Smith, A. M., Woolhouse, H. W. 1988. Photorespiratory metabolism and immunogold localisation of photorespiratory enzymes of C_3 and C_3-C_4 intermediate species of *Moricandia*. *Planta* 173:298–308

116. Reardon, E. J., Allison, G. B., Fritz, P. 1979. Seasonal chemical and isotopic variations of soil CO_2 at Trout Creek, Ontario. *J. Hydrol.* 43:355–71

117. Reibach, P. H., Benedict, C. R. 1977. Fractionation of stable carbon isotopes by phosphoenolpyruvate carboxylase from C_4 plants. *Plant Physiol.* 59:564–68

118. Reich, P. B., Amundson, R. G. 1985. Ambient levels of ozone reduce net photosynthesis in tree and crop species. *Science* 230:566–70

119. Roeske, C. A., O'Leary, M. H. 1984. Carbon isotope effects on the enzyme-catalyzed carboxylation of ribulose bisphosphate. *Biochemistry* 23:6275–84

120. Rundel, P. W., Rundel, J. A., Ziegler, H., Stichler, W. 1979. Carbon isotope ratios of Central Mexican Carassulaceae in natural and greenhouse environments. *Oecologia* 38:45–50

121. Rundel, P. W., Stichler, W., Zander, R. H., Ziegler, H. 1979. Carbon and hydrogen isotopes of bryophytes from arid and humid regions. *Oecologia* 44: 91–4

122. Schleser, G. H., Jayasekera, R. 1985. δ^{13}C-Variations of leaves in forests as an indication of reassimilated CO_2 from the soil. *Oecologia* 65:536–42

123. Schönwitz, R., Stichler, W., Ziegler, H. 1986. δ^{13}C Values of CO_2 from soil respiration on sites with crops of C_3 and C_4 types of photosynthesis. *Oecologia* 69:305–8

124. Seemann, J. R., Critchley, C. 1985. Effects of salt stress on growth, ion content, stomatal behaviour and photosynthetic capacity of a salt-sensitive species, *Phaseolus vulgaris* L. *Planta* 164:151–62

125. Sharkey, T. D., Berry, J. A. 1985. Carbon isotope fractionation of algae as influenced by an inducible CO_2 concentrating mechanism. In *Inorganic Carbon Uptake by Aquatic Organisms*, ed. W. J. Lucas, J. A. Berry, pp. 389–401. Rockville: Am. Soc. Plant Physiol.

126. Sinclair, T. R., Tanner, C. B., Bennett, J. M. 1984. Water-use efficiency in crop production. *BioScience* 34:36–40

127. Smith, B. N., Brown, W. V. 1973. The Kranz syndrome in the Gramineae as indicated by carbon isotope ratios. *Am. J. Bot.* 60:505–13

128. Smith, B. N., Epstein, S. 1971. Two categories of $^{13}C/^{12}C$ ratios for higher plants. *Plant Physiol.* 47:380–84

129. Smith, B. N., Oliver, J., McMillan, C. 1976. Influence of carbon source, oxygen concentration, light intensity, and temperature on $^{13}C/^{12}C$ ratios in plant tissues. *Bot. Gaz.* 137:99–104

130. Smith, B. N., Turner, B. L. 1975. Distribution of Kranz syndrome among Asteraceae. *Am. J. Bot.* 62:541–45

131. Smith, S. D., Osmond, C. B. 1987. Stem photosynthesis in a desert ephemeral, *Eriogonum inflatum*. Morphology, stomatal conductance and water-use efficiency in field populations. *Oecologia* 72:533–41

132. Stephenson, R. L., Tan, F. C., Mann, K. H. 1984. Stable carbon isotope variability in marine macrophytes and its implications for food web studies. *Marine Biol.* 81:223–30

133. Sternberg, L. D. L., Mulkey, S. S., Wright, S. J. 1989. Ecological interpretation of leaf carbon isotope ratios: influence of respired carbon dioxide. *Ecology*. In press

134. Sternberg, L. D. L., Ting, I. P., Price, D., Hann, J. 1987. Photosynthesis in epiphytic and rooted *Clusia rosea* Jacq. *Oecologia* 72:457–60

135. Teeri, J. A. 1981. Stable carbon isotope analysis of mosses and lichens growing in xeric and moist habitats. *Bryologist* 84:82–84

136. Teeri, J. A. 1982. Carbon isotopes and the evolution of C_4 photosynthesis and Crassulacean acid metabolism. In *Biochemical Aspects of Evolutionary Biology*, ed. M. H. Nitecki, pp. 93–129. Chicago: Univ. Chicago Press

137. Teeri, J. A. 1984. Seasonal variation in Crassulacean acid metabolism in *Dudleya blochmanae* (Crassulaceae). *Oecologia* 64:68–73

138. Teeri, J. A., Gurevitch, J. 1984. Environmental and genetic control of Crassulacean acid metabolism in two Crassulacean species and an F_1 hybrid with differing biomass $\delta^{13}C$ values. *Plant Cell Environ.* 7:589–96

139. Teeri, J. A., Tonsor, S. J., Turner, M. 1981. Leaf thickness and carbon isotope composition in the Crassulaceae. *Oecologia* 50:367–69

140. Teeri, J. A., Turner, M., Gurevitch, J. 1986. The response of leaf water potential and Crassulacean acid metabolism to prolonged drought in *Sedum rubrotinctum*. *Plant Physiol.* 81:678–80

141. Terashima, I., Wong, S. C., Osmond, C. B., Farquhar, G. D. 1988. Characterisation of non-uniform photosynthesis induced by abscisic acid in leaves having different mesophyll anatomies. *Plant Cell Physiol.* 29:385–94

142. Ting, I. P., Sternberg, L. O., DeNiro, M. J. 1983. Variable photosynthetic metabolism in leaves and stems of *Cissus quadrangularis* L. *Plant Physiol.* 71:677–79

143. Ting, I. P., Sternberg, L. O., DeNiro, M. J. 1985. Crassulacean acid metabolism in the strangler *Clusia rosea* Jacq. *Science* 229:969–71

144. Troughton, J. H. 1979. $\delta^{13}C$ as an indicator of carboxylation reactions. In *Photosynthesis II: Photosynthetic Carbon Metabolism and Related Processes. Encyclopedia of Plant Physiology (New Ser.)*, ed. M. Gibbs E. Latzko, 6:140–49. New York: Springer-Verlag

145. Troughton, J. H., Card, K., Björkman, O. 1974. Temperature effects on the carbon isotope ratio of C_3, C_4 and CAM plants. *Carnegie Inst. Wash. Yearb.* 73:780–84

146. Troughton, J. H., Mooney, H. A., Berry, J. A., Verity, D. 1977. Variable carbon isotope ratios of *Dudleya* species growing in natural environments. *Oecologia* 30:307–11

147. Ueno, O., Samejima, M., Muto, S., Miyachi, S. 1988. Photosynthetic characteristics of an amphibious plant, *Eleocharis vivipara:* expression of C_4 and C_3 modes in contrasting environments. *Proc. Natl. Acad. Sci. USA* 85:6733–37

148. Vogel, J. C. 1978. Recycling of carbon in a forest environment. *Oecol. Plant.* 13:89–94

149. Vogel, J. C. 1980. *Fractionation of Carbon Isotopes During Photosynthesis.* Heidelberg: Springer-Verlag

150. Vogel, J. C., Fuls, A., Ellis, R. P. 1978. The geographical distribution of Kranz grasses in South Africa. *S. Afr. J. Sci.* 74:209–15

151. Webster, G. L., Brown, W. V., Smith, B. N. 1975. Systematics of photosynthetic carbon fixation pathways in *Euphorbia. Taxon* 24:27–33

152. Werk, K. S., Ehleringer, J., Forseth, I. N., Cook, C. S. 1983. Photosynthetic characteristics of Sonoran Desert winter annuals. *Oecologia* 59:101–5

153. Whelan, T., Sackett, W. M., Benedict, C. R. 1973. Enzymatic fractionation of carbon isotopes by phosphenolpyruvate carboxylase from C_4 plants. *Plant Physiol.* 51:1051–54

154. Winter, K. 1980. Carbon dioxide and water vapor exchange in the Crassulacean acid metabolism plant *Kalanchoe pinnata* during a prolonged light period. *Plant Physiol.* 66:917–21

155. Winter, K. 1981. CO_2 and water vapor exchange, malate content and $\delta^{13}C$ value in *Cicer arietinum* grown under two water regimes. *Z. Pflanzenphysiol.* 101:421–30

156. Winter, K. 1981. C_4 plants of high biomass in arid regions of Asia—occurrence of C_4 photosynthesis in Chenopodiaceae and Polygonaceae from the Middle East and USSR. *Oecologia* 48:100–6

157. Winter, K., Holtum, J. A. M., Edwards, G. E., O'Leary, M. H. 1982. Effect of low relative humidity on $\delta^{13}C$ value in two C_3 grasses and in *Panicum milioides*, a C_3-C_4 intermediate species. *J. Exp. Bot.* 33:88–91

158. Winter, K., Lüttge, U., Winter, E., Troughton, J. H. 1978. Seasonal shift from C_3 photosynthesis to Crassulacean acid metabolism in *Mesembryanthemum crystallinum* growing in its natural environment. *Oecologia* 34:225–37

159. Winter, K., Troughton, J. H. 1978. Photosynthetic pathways in plants of coastal and inland habitats of Israel and the Sinai. *Flora* 167:1–34

160. Winter, K., Troughton, J. H., Card, K. A. 1976. $\delta^{13}C$ Values of grass species collected in the Northern Sahara Desert. *Oecologia* 25:115–23

161. Wong, S. C., Cowan, I. R., Farquhar, G. D. 1979. Stomatal conductance correlates with photosynthetic capacity. *Nature* 282:424–26

162. Wright, G. C., Hubick, K. T., Farquhar, G. D. 1988. Discrimination in carbon isotopes of leaves correlates with water-use efficiency of field grown peanut cultivars. *Aust. J. Plant Physiol.* 15:815–25

163. Ziegler, H., Batanouny, K. H., Sankhla, N., Vyas, O. P., Stichler, W. 1981. The photosynthetic pathway types of some desert plants from India, Saudi Arabia, Egypt, and Iraq. *Oecologia* 48:93–99

163a. Guy, R. D., Wample, R. L. 1984. Stable carbon isotope ratios of flooded and nonflooded sunflowers (*Helianthus annuus*). *Can. J. Bot.* 62:1770–74

Annu. Rev. Plant Physiol. 1989. 40:539–69

THE PHYSIOLOGY OF ION CHANNELS AND ELECTROGENIC PUMPS IN HIGHER PLANTS

Rainer Hedrich

Pflanzenphysiologisches Institut, Universität Göttingen, 3400 Göttingen, West Germany

Julian I. Schroeder

UCLA School of Medicine, Jerry Lewis Neuromuscular Research Center, University of California, Los Angeles, California 90024

CONTENTS

1040-2519/89/0601-539$02.00

INTRODUCTION

Membranes of living cells allow the maintenance of ionic and metabolic gradients essential for growth, development, movements and signal transduction. These membranes are barriers to continuous solute flux and they control ion and metabolite movement. Such tasks are accomplished by transport proteins embedded in these membranes (Figure 1). When activated, these proteins become permeable to ions or metabolites, allowing solute or messenger flow across the membranes.

The development of new techniques and the refinement of cell and tissue preparations have contributed to our ability to study ion channels, active pumps, and carriers as molecular entities rather than simply as membrane processes.

The door to the study of ion transport at the molecular level was thrown open by the patch-clamp technique, a revolutionary electrophysiological technique for the recording of ion currents from biological membranes, developed

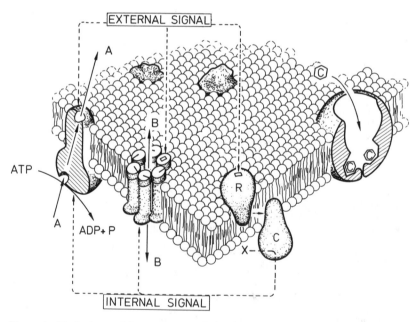

Figure 1 Mechanisms of ion transport and signal transduction in biological membranes. Three-dimensional diagram of the fluid mosaic model, showing the integral transport proteins embedded in the lipid bilayer. From left to right: (A) ion pump, (B) ion channel, (R, C) coupling proteins for signal perception and transduction and (C) carrier. Reprinted from Ref. 53, with permission from *Trends in Biochemical Sciences*.

by Erwin Neher and Bert Sakmann in 1976 at the Max-Planck-Institut für biophysikalische Chemie in Göttingen, West Germany. This highly sensitive technique was originally designed for the resolution of the current generated by ion flow through individual acetylcholine receptor channels (110).

Later, ion channels were found to be widely distributed in all animal tissues. They are required for the functioning of all kinds of nervous and nonnervous cell types, such as muscle cells, sperm, oocytes, blood cells, sensory cells, and epithelial cells (56). Ion channels have also been found in plants, yeast, bacteria, and primitive eukaryote protists (53, 56, 133). Thus, "these integral membrane proteins, which for many years were little more than an esoteric concept of biophysicists, have emerged from relative obscurity to become valued and highly visible members of the cell and molecular biology community" (cited in 84). In addition to its classical application for single-channel recordings, this technique has been applied to answer an enormous variety of open questions in the physiological sciences (152) during the past few years.

Patch clamping plant cells and organelles became possible because of refinements in protoplast and organelle isolation and purification techniques. The combined properties of these techniques circumvent problems associated with classical electrophysiology in plant cells.

Mature plant cells are characterized by a cell wall space (with a high ion-exchange capacity; 146), two membranes in series (plasma membrane and vacuolar membrane), separated by a thin layer of cytoplasm (Figure 2). Consequently the location of standard impalement microelectrodes and the composition of solutes in the various compartments are ill-defined and difficult to control (37, 89). Membrane potentials are not well determined if two membranes are in series (37). Furthermore, the rectifications of plasma membrane and vacuolar membrane are in opposition upon depolarization (24, 26, 50, 62, 76, 144). Therefore bipolar gating phenomena may be found when ion currents through both membranes are recorded in series (as reported for gap junctions; 113, 159). Moreover, the biochemical machinery of the cytoplasm may be disturbed by calcium and proton (and perhaps other) fluxes from the extracellular or vacuolar compartment through leaks along the inserted electrode.

Thus, electrophysiology on higher plants has contributed less to the understanding of membrane ion transport than corresponding investigations on giant algal or animal cells. The patch-clamp technique provides an alternative method that allows the study of small cells.

In this review we focus on new insights into molecular mechanisms of ion transport processes in higher plant cells and on some of the conceptual advances that have occurred since the initial application of the patch-clamp technique to plant cells in 1984 (104, 143). For information on related topics

| Microelectrode technique | Patch - Clamp technique |

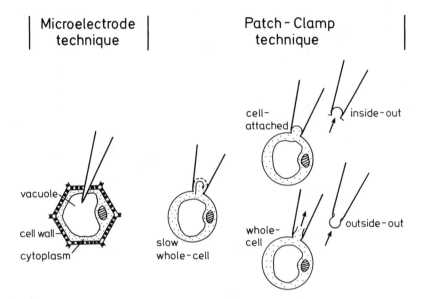

Figure 2   Schematic representations of various microelectrode configurations. *(Left)* Classical impalement technique applied to a plant cell. *(Right)* Patch-clamp technique applied to a plant protoplast. Modified from Ref. 53, with permission from *Trends in Biochemistry.*

the reader is referred to the following reviews on: the role of ion transport in algae (169); proton pumps in higher plant membranes (2, 112, 148, 149, 155, 158); and techniques for isolation of sealed membrane vesicles from plant tissues (164), isolation of protoplasts (117), isolation of vacuoles (91), membrane protein extraction-purification-reconstitution (101), genetic engineering (90), as well as common electrophysiological techniques (56, 154). Reviews on prospective applications of patch-clamp techniques to higher plant cells and microbes have appeared (53, 133, 136, 166).

PATCH-CLAMP TECHNIQUE AND SUITABLE PLANT MATERIAL

Patch-Clamp Technique

The patch-clamp technique was first used to obtain direct measurements of the elementary current passing through a single ion channel (110). Recent advances allow the patch-clamp technique to be used to record ionic currents from membrane patches and from entire small cells (47). This is accomplished by pressing a heat-polished glass pipette with a tip diameter on the order of 1 micrometer against a clean membrane surface. When suction is applied to the interior of the pipette, a seal forms between the pipette tip and the membrane (cell-attached configuration, Figure 2). The high seal resistance and small membrane area reduce background noise and ensure that currents passing ion channels in the membrane patch will flow into the pipette.

Both this cell-attached measuring configuration and excised-patch configurations (inside-out and outside-out patch, see Figure 2) allow the resolution of single-channel currents with amplitudes of less than one picoampere. Figure 3 shows current recordings from the vacuolar membrane of suspension-cultured *Chenopodium rubrum* cells. Single-channel currents consist of rectangular pulses of random duration reflecting conformational changes of a macromolecule. Each upward current step represents the opening, and each downward current step the closing, of a single ion channel. As long as the channel is open, ions pass through it driven by their electrochemical gradients. The current amplitude indicates the number of ions passing through the channel within a given time. The duration of the mean open and closed times can depend on the applied voltage and on a variety of chemical interactions with the channel protein (e.g. Ca^{2+}, pH, second messengers, phosphorylation; 56). A statistical analysis of open and closed time intervals gives insights into the molecular dynamics of the channel protein (25).

The mechanical stability of the pipette to membrane seal allows a membrane patch encircled by the pipette tip to be excised (inside-out or outside-out orientation), or to be ruptured without destroying the seal to provide access to the interior of the cell or organelle ("whole-cell or organelle" configuration, Figure 2). With these patch-clamp configurations, the membrane potential and the composition of the media on either side of the membrane are well defined and easy to control (97). Nevertheless, "wash-out" into the pipette of cellular factors essential for ion transport or for membrane-associated processes must be taken into account. On the other hand, "wash-in" (from the pipette interior) of essential compounds may prevent loss of activity ("rundown") or even reactivate processes under investigation (120), thus leading to an understanding of ion transport regulation. Complementary to the whole-

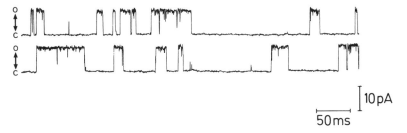

Figure 3   Single-channel currents of the SV type recorded with an isolated membrane patch of the vacuolar membrane of *Chenopodium rubrum*. Upward deflections indicate channel opening and downward deflections channel closure. Reprinted from Ref. 49, with permission from *Botanica Acta*.

cell, the "slow whole-cell" configuration (electrically equivalent to a cell impaled by a microelectrode; Figure 2) allows electrical measurements while the biochemistry remains effectively unperturbed (85). This was accomplished by agents perforating the membrane under the pipette (85, 141) so that electrolytes can be exchanged while proteins are retained (e.g. see the section on H^+ pump activation, below).

The whole-cell configuration enables experiments concerning overall current flow arising from populations of channels or other transporters distributed over the entire membrane surface. Using the whole-cell configuration, it is possible to study the properties of pumps or carriers, even though the current arising from a single pump protein is too small to be detected. Specific stimuli (such as pulses of substrate, light, or hormones) can elicit activation of these "low-turnover" transporters in the entire cell, resulting in a summed whole-cell current that can be readily measured and studied (3, 50, 141, 147). Whereas patch-clamp recordings of pump currents from several animal and plant systems have already been obtained (3, 33, 40, 50, 78, 108, 147), direct measurements of ion currents produced by carriers are still restricted to a few cases in animal systems (39).

Plant Material

The use of isolated cells or organelles in patch-clamp analysis avoids problems associated with plant tissues such as cell-cell junctions (plasmodesmata), diffusion gradients, and possible "wound effects" in dissected tissue. Simpler systems that normally consist of only one or a few cells such as guard cells, root-, salt-, or staminal hairs, cells of the aleurone layer, or suspension-cultured cells provide ideal material. However even in these systems a requirement for high-resolution patch-clamp recordings is the removal of the cell wall with lytic enzymes to obtain a clean membrane surface. It should be noted that it is generally much more difficult to seal patch pipettes on the plasma membrane of plant protoplasts than on the vacuolar membrane or on animal tissues. Procedures that may facilitate sealing of patch pipettes to guard cell protoplasts are described in detail elsewhere (124, 141, 143). Following these procedures, seals with various types of protoplasts could be obtained (e.g. from epidermal, mesophyll, coleoptile, and aleurone cells; 20, 30). After protoplast isolation, it is important to examine whether the protoplasts still maintain the biological functions of the intact cells in order to ensure the biological relevance of the transport systems investigated.

REMARKS ON NOMENCLATURE OF MEMBRANE ION TRANSPORT

Ion channels and carriers share many basic properties, such as allowing diffusion along a gradient and exhibiting both concentration-dependent and voltage-dependent saturation (11, 43, 77, 81). Channels and carriers have been distinguished from one another by Läuger (77, 81). Carrier proteins have a maximum turnover rate of 10^4–10^5 transported ions or molecules per second (77). An open ion channel, on the other hand, can pass more than 10^6 ions per second (see Figure 3). Since the electrical noise associated with carrier-mediated transmembrane currents (75) differs markedly from that of channel-mediated currents (70, 111), noise analysis allows distinction between channels and carriers without measurement of single-channel currents. Noise analysis has led to the discovery that some transport processes, previously ascribed to the activity of carriers, were mediated by ion channels (34).

The terms deactivation and inactivation apply to voltage-dependent channels and are occasionally confused. Deactivation refers to stepping the membrane to potentials at which ion channels are closed, such that relaxation ("tail") currents can be observed (see 56, 144). Inactivation (56, 141) refers to closing of ion channels in time, although the membrane is held to potentials that open ion channels (i.e. voltage-dependent "desensitization").

Ion specific currents through plant membranes have recently been referred to as "leaks." It became apparent at least 30 years ago that membrane components can be specifically permeable to various ions (58). In animal cells this was deduced from the pioneering research of Katz (69), Hodgkin & Huxley (58), and Hagiwara (46). In algae similar observations were made (22, 35, 37, 41, 61, 107). Physiologically speaking, the term leak is a misconception and should not be used in the context of ion-specific membrane currents. In biological systems, all ion channels, when activated, polarize the membrane potential to defined values and therefore control diffusion processes with accuracy. Leak implies noncontrolled passage.

There is increasing evidence that the basic structure of the outer plasma membrane of plant cells (plasmalemma) shares many essential features with the plasma membrane of fungi (among them yeast) and with animal plasma membranes. Therefore, we refer to the outer membrane of plant cells as the plasma membrane rather than the plasmalemma. Likewise the vacuolar membrane of plant cells (tonoplast) shares many properties with other lysosomal membranes (112), which we will therefore refer to as vacuolar membrane. Equality in the name for such basic membrane structures may help to recognize similarities as well as differences among different biological systems.

ION TRANSPORT THROUGH THE PLASMA MEMBRANE

Ion Channels in the Plasma Membrane

VARIOUS CHANNEL TYPES   Patch-clamp techniques allow the separate measurement of ionic currents through the plasma membrane and currents through the vacuolar membrane. The number of ion channels discovered in the plasma membrane of higher plant cells is currently increasing. Particular types of ion channels can be classified by the ion species able to permeate the open channel. Besides experimental evidence for K^+, Cl^-, and Ca^{2+} channels, transport of organic ions in higher plants (88) indicates that additional classes of channels for these ions may exist in the plasma membrane. Following the initial characterization of K^+ channels in guard cells (143), the presence of a variety of K^+ channels has been reported in the plasma membrane of various other higher plant tissues.

During development, transcellular K^+, H^+, and Ca^{2+} fluxes play an important role in polarized growth (54, 63, 64, 175). An increasing number of plant processes are being identified that are strongly influenced by external Ca^{2+} and cytoplasmic Ca^{2+}. Detailed reviews of such phenomena have been presented elsewhere (54, 170). Recent studies on carrot protoplasts found a correlation between high-affinity binding of Ca^{2+} channel blockers and inhibition of Ca^{2+} influx (44). Whether these binding sites represent molecular components of Ca^{2+} channels awaits elucidation.

K^+ Channels

K^+ transport across the plasma membrane of plant cells is closely linked to diverse tissue-specific and cell-specific actions related to plant growth and development, such as germination (65), leaf movements (135), stomatal action (122), ion uptake in roots, phloem transport, and nutrient storage (88).

K^+ channels have received attention due to their likely importance in the mediation of K^+ transport in plants. Specific tissues may have specialized types of K^+ channels appropriate to their physiological necessities. To date, two classes of K^+ channels have been characterized in guard cells (140, 141, 144). One type may allow K^+ efflux and the other K^+ uptake. Therefore, these channels have been designated "$I_{K^+,out}$ channels" for K^+ efflux and "$I_{K^+,in}$ channels" for K^+ uptake (141). Both $I_{K^+,out}$ and $I_{K^+,in}$ channels are strongly regulated by the membrane potential.

$I_{K^+,out}$ CHANNELS $I_{K^+,out}$ channels can allow K^+ efflux upon membrane depolarization (62, 105, 141, 142, 144). These K^+ channels become activated by depolarization of the membrane potential to values more positive than -40 mV. By averaging methods it was deduced that the outward rectifying K^+ current is carried by approximately 10^3 K^+ channels in the plasma membrane of a guard cell (144). Recently, $I_{K^+,out}$ channels have been characterized in a number of higher-plant cell types, including motor cells of pulvini (105), trap-lobe cells (62), *Asclepias tuberosa* suspension-culture cells (137), and aleurone layer cells (20). Furthermore, there is evidence for the existence of $I_{K^+,out}$ channels in mesophyll cells (104), corn suspension-culture cells (72), epidermal cells (J. I. Schroeder, unpublished), and callus protoplasts from *Arabidopsis* (R. Lew, personal communication). The $I_{K^+,out}$ channels found in various higher-plant protoplasts have properties very similar to the outward rectifying K^+ conductance in algal cells (5, 36, 169) and yeast (45). For instance, the voltage dependence and alkali-metal ion selectivity sequences are very similar in *Nitella* (156) and guard cells from *Vicia faba* (141) (permeability sequence: $P_{K^+} > P_{Rb^+} > P_{Na^+} > P_{Li^+} >> P_{Cs^+}$).

Furthermore, in recent patch-clamp studies using giant membrane vesicles (cytoplasmic droplets) of *Chara* and *Acetabularia*, voltage-dependent single K^+ channel currents were characterized (11, 59, 60, 82, 86). The kinetics of voltage-dependent single K^+ channels in cytoplasmic droplets have been investigated (12, 82). Further research will be required to determine whether the vesicle membrane originates from the vacuolar membrane, as proposed by Lühring (86) and Sakano & Tazawa (134), or from the plasma membrane (11, 59, 60).

Two physiologically significant roles of $I_{K^+,out}$ channels in plant cells have been proposed:

1. $I_{K^+,out}$ channels have been suggested to represent a predominant pathway for K^+ release from guard cells (143, 144), from pulvinus motor cells (136), and from algae (169). The magnitude of K^+ efflux through $I_{K^+,out}$ channels in guard cells can account for physiological K^+ fluxes of 0.7 $fmol \cdot s^{-1}$ per guard cell during stomatal closing (144). Properties of $I_{K^+,out}$ channels in guard cell protoplasts such as cation selectivity and lack of inactivation during sustained stimulation (141) agree with K^+ fluxes observed in guard cells embedded in their original environment of the epidermis (116). Moran, Satter, and colleagues have found that tetraethylammonium ions (TEA) in millimolar concentrations block $I_{K^+,out}$ channels in pulvini and reduce leaf movements, supporting the importance of these channels for turgor regulation (105). Pharmacological evidence for the importance of $I_{K^+,out}$ channels in K^+ release from cells other than motor cells remains to be found.

2. Apart from contributing to volume/turgor regulation the voltage-dependent outwardly rectifying K^+ conductance may play a role in the repolarization of plant action potentials in both algae and higher plant cells (6, 37, 41, 107, 151, 153). Therefore, $I_{K^+,out}$ channels in conjunction with Cl^- and Ca^{2+} permeabilities (4, 35, 176) may contribute to excitability in plants. Time course analysis of $I_{K^+,out}$ channels performed by whole-cell measurements leads to time constants very similar to repolarization times of plant action potentials (142), supporting the importance of these K^+ channels for repolarization. Ion transporters that can either transiently or continuously depolarize the membrane potential would be required to initiate K^+ efflux through $I_{K^+,out}$ channels. A clear understanding of depolarizing mechanisms in higher plant cells is, however, still lacking.

$I_{K^+,in}$ CHANNELS   Inwardly conducting plant K^+ channels were originally demonstrated in guard cells (144). These $I_{K^+,in}$ channels activate in a voltage- and time-dependent manner when the membrane is hyperpolarized to values more negative than -100 mV. In addition to the time-dependent inward K^+ currents, instantaneous inward rectifying K^+ currents have been found in algal cells (156). It has been proposed that the time-dependent inwardly conducting $I_{K^+,in}$ channels represent a pathway for K^+ uptake during stomatal opening (144). This was supported by the observation that Al^{3+} ions specifically block $I_{K^+,in}$ channels ($K_D \approx 15\mu M$) but not $I_{K^+,out}$ channels (141). These results correlate closely with the findings of Schnabl & Ziegler that Al^{3+} ions (1 mM) selectively block stomatal opening but not closing (139). It may be concluded that $I_{K^+,in}$ channels represent a major pathway for K^+ uptake in guard cells, and possibly in plant cells in general, as $I_{K^+,in}$ channels have also been recorded in aleurone layer cells (20), in corn suspension-culture cells (72), in coleoptile protoplasts (L. Taiz and J. I. Schroeder, unpublished), in tobacco tumor cells (S. Marx and R. Hedrich, unpublished), and in epidermal cells (J. I. Schroeder, unpublished).

H^+-Pump Activation

Patch-clamp studies allow investigation of pump dynamics by direct measurement of the electrogenic current (3, 50, 167). Furthermore, in whole-cell recordings the requirement of cytoplasmic substrates for pump activation can be studied effectively. Therefore, patch-clamp studies appear to be ideal for the investigation of the modulation of H^+ pumps. New insights into pump activation by blue light and red light (3, 141, 147) and by the hormone-like compound fusicoccin (32, 53, 92) have been obtained.

In guard cells, low fluence rates of blue light and high fluence rates of red light can activate plasma membrane H^+ pumps by different mechanisms (3, 147). Blue-light activation of pumps follows the initial exposure to light by a marked delay of approximately 30 sec (3). In contrast, exposure to red light activates H^+ pumps without a measurable delay (147). Both mechanisms require cytoplasmic substrates other than Mg-ATP for maximal pump activity (141, 147). By using nonperfused guard cells (slow-whole-cell, Figure 2) to prevent loss of cytoplasmic factors it was found that blue light–activated pump currents are increased 5–10 times by yet unidentified cytoplasmic substrates (141). Moreover, the magnitude of blue light–activated pump currents measured in nonperfused guard cells was found sufficient to activate $I_{K^+,in}$ channels and drive K^+ uptake required for stomatal opening (141). The current-voltage relationship of H^+ pumps in guard cells was calculated under the assumption that the H^+ pump inhibitor cyanide has no effect on channel conductances (13). However, more recent data suggest that cyanid also blocks K^+ channels (13a).

Biochemical studies provide evidence that activation of H^+ pumps by auxin and photonastic light stimuli may involve the pathway of phosphoinositol lipid metabolism (31, 106). Furthermore, a cytoplasmic acidification that preceded membrane hyperpolarization was observed in response to auxin (32).

Comparison Between K^+–H^+ Symporters and $I_{K^+,in}$ Channels

$I_{K^+,in}$ channels, as described in guard cells (141, 144), represent a mechanism for K^+ uptake in plant cells in addition to K^+-ATPases and coupled shuttles, which have been presumed on thermodynamic grounds (88, 118). K^+-ATPases have been cloned in *Escherichia coli* (55). To our knowledge, clear evidence for K^+-ATPases in plants has not yet been obtained.

Here we discuss an example of carrier and channel mediated K^+ uptake mechanisms. Recently investigations on *Neurospora crassa* as a model system for mechanisms of K^+ uptake were reported. These studies provided evidence for the hypothesis that active voltage-dependent K^+–H^+ carriers (symporters) represent a major mechanism for the uptake of K^+ into the cell (15, 129). Proposed for K^+–H^+ symport is an active mechanism in which the proton motive force (pH gradient) generated by the H^+ pump drives K^+ uptake through carriers. Return of one proton into the cell drives the uptake of one K^+ ion. This hypothesis is based on the following observations: 1. Uptake of K^+ into *Neurospora* saturates with an apparent K_m of approximately 10 μM (129). 2. Addition of extracellular K^+ leads to alkalization of the cytoplasmic pH by approximately 0.1–0.2 pH units (15). 3. A fixed-charge stoichiometry was found that shows two positive charges going into a cell per K^+ ion taken up (129); one charge was a K^+ ion, the other a proton.

In Figure 4 the current-voltage relationships of K^+–H^+ symporters in *Neurospora* and of $I_{K^+,in}$ channels in guard cells are superimposed. It is apparent that the two mechanisms have strikingly similar current-voltage curves. Furthermore, it should be noted that the current-voltage relationship for the proposed K^+–H^+ symport saturates at potentials more negative than -300 mV (15), and $I_{K^+,in}$ channels saturate at potentials of approximately -280 to -300 mV (J. I. Schroeder, unpublished). In the light of this similarity and of the findings of $I_{K^+,in}$ channels in various plant tissues (see above) an attempt can be made to decompose the proposed K^+–H^+ symport into H^+ pumps and $I_{K^+,in}$ channels:

1. When *Neurospora* cells were incubated in K^+-free solutions, membrane potentials in the range of -305 mV were measured (129). Addition of only 0.6 μM K^+ to the extracellular space would allow steady-state uptake of K^+ mediated via passive transport through K^+ channels, assuming 100 mM K^+ in the cytoplasm (129) and the operation of the H^+ pump. Therefore, passive K^+

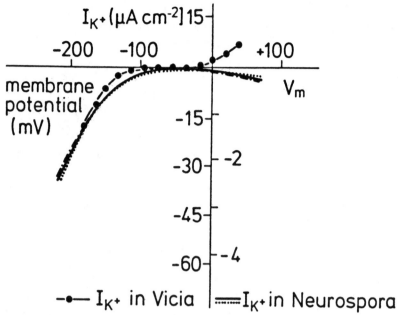

$$-\!\!\bullet\!\!-\ I_{K^+}\ \text{in Vicia}\ \ \Big|\ ====I_{K^+}\text{in Neurospora}$$

Figure 4   Comparison of current-voltage relationships of K^+–H^+ symporters *(continuous and dotted lines)* in *Neurospora crassa* (reproduced with permission from Figure 4 of Ref. 15) and inward K^+ currents in guard cell protoplasts of *Vicia faba (solid circles)* (reproduced with permission from Fig. 1, Ref. 144). The curves were superimposed after normalization with respect to the currents measured at -180 mV. The left scale of the ordinate shows the current density in the plasma membrane of guard cells and the right scale that of *Neurospora.* Inward K^+ currents of *V. faba* were recorded with K^+ glutamate solutions (as specified in Ref. 144). K^+-H^+ symport in *Neurospora* was measured after addition of 50 μM K^+ to K^+-free washed cells (as specified in Ref. 15). The current-voltage relationship of *Neurospora* was obtained by subtraction of curves before adding K^+, from curves after addition of K^+. Negative current corresponds to influx.

uptake could result at a K^+ concentration in the micromolar range in the cell wall.

2. The observed depolarization of the membrane potential from -305 mV to -220 mV (129) upon addition of 50 μM K^+ to the extracellular medium of *Neurospora* would be consistent with the presence of K^+ channels. The steep voltage-dependence of the H^+ pump from -300 to -200 mV (15) shows that the observed depolarization after K^+ addition leads to a marked stimulation of H^+ extrusion via the H^+ pump. This stimulation of the H^+ pump by K^+ may lead to the observed slow cytoplasmic alkalization.

3. On the other hand, the precise charge stoichiometry of one H^+ entering the cell per K^+ ion taken up supports the model of K^+–H^+ symporters as mechanisms of K^+ uptake.

Most plant cell walls have high cation binding capacities. At a physiological pH of 5.5, cell walls can release up to 30 mM K^+ kg^{-1} upon acidification by 0.1 pH units (146). Therefore it is conceivable that H^+ pump–mediated acidification of the cell-wall space may lead to elevated local K^+ concentrations that may play a role in non-steady-state uptake of K^+.

In conclusion, the possibility may not be excluded that the K^+ uptake phenomena in *Neurospora* may arise, in part, from K^+ flux through $I_{K^+,in}$ channels. To exclude the possibility of K^+ channel mediated K^+ uptake in *Neurospora* it may be necessary to test whether $I_{K^+,in}$ channels do not exist in these cells. Furthermore, questions such as the proton permeability of K^+ channels (56) deserve investigation.

Chemically Regulated K^+ Channels

K^+ fluxes in plant cells are under the control of various stimuli (88). Therefore it is likely that K^+ channels can be regulated chemically by external or cytoplasmic factors (53, 89). However, to date little evidence exists for chemical regulation of K^+ channels.

In cells of the alga *Eremosphaera viridis,* K^+ channels can be transiently activated by a number of stimuli such as external Ca^{2+}, Ba^{2+}, pH, light, and cytoplasmic substrates (74, 165). In naturally occurring endosperm protoplasts undergoing mitosis, a Ca^{2+}-activated, voltage-dependent, nonselective cation channel has been identified. When activated, this channel could lead to an outward current of K^+ (163).

A preliminary study of the action of fusicoccin in guard cells, using classical impalement electrodes, has led to the suggestion that this substance may block K^+ channels, in addition to the activation of H^+ pumps (14). Exposure of guard cells to 10 μM abscisic acid (ABA) increased the mean open time of K^+ channels from 1.25 ms to 2.05 ms when the cytoplasmic side of the plasma membrane was exposed to pH 5.8 and 2 mM Ca^{2+} (138). However, the increase in K^+ permeability under the experimental condition does not suffice to explain abscisic acid–induced K^+ release from guard cells (141). Further patch-clamp experiments concerning the interaction of pH, Ca^{2+}, and ABA (28) may in future elucidate the primary steps of stomatal action.

Comparison of K^+ Channels in Plant and Animal Cells

The most prominent difference between higher-plant and animal K^+ channels appears to lie in their physiological function. Plant K^+ channels play an important role in long-term K^+ transport into and out of cells. This role of higher-plant K^+ channels has been suggested and studied in detail in guard cells and pulvini (105, 141, 143, 144). In order to carry K^+ fluxes through these channels, mechanisms must exist that can shift the membrane potential from the K^+ equilibrium potential for prolonged durations (e.g. H^+ pumps drive K^+ uptake through $I_{K^+,in}$ channels). K^+ channels in animal cells, on the other hand, play a major role in resetting the membrane potential to the equilibrium potential for K^+, after the induction of short-term potential changes by other channels (58).

The enormous diversity of K^+ channel types and subtypes in animal cells, each of which serves a particular physiological purpose, makes a comparison with plant K^+ channels difficult. In general, animal K^+ channels can be classified by their mechanism of activation and their single-channel properties. The following classes of K^+ channels in animal cells have been found: various types of voltage-dependent K^+ channels, transmitter-activated K^+ channels, and second-messenger and ATP-regulated K^+ channels (56, 80, 130).

It is still too early to specify clearly differences in the biophysical properties of animal and plant K^+ channels. [Recently, some of the differences and similarities of biophysical properties have been addressed (62, 141, 144).] In general, most voltage-dependent K^+ channels in plant cells activate 10–100 times slower than those of animal cells (12, 144). The single-channel conductances of voltage-dependent K^+ channels in higher-plant cells and animal cells are within the same range (about 10–40 pS). However, single K^+ channel conductances in cytoplasmic droplets are much larger (100 pS range), resembling channel conductances of Ca^{2+}-activated vacuolar channels (49, 52) or of Ca^{2+}-activated "maxi-K^+" channels in animal cells (130).

The density of K^+ channels found in plant cells is on the order of one channel per μm^2 (62, 144). This density is similar to the density of K^+ channels in most animal cells (56). However, in specialized regions of nerve and muscle cells channel densities can be larger by up to several orders of magnitude (56).

Other Plasma Membrane Ion Channels

CL⁻ CHANNELS   Evidence for Cl^- channels was recently found in *Asclepias tuberosa* suspension-culture protoplasts (137). These channels recorded at very high cytoplasmic Ca^{2+} concentrations (2 mM) and pH 5.8 were characterized by a strong voltage dependence, large channel conductance, and inhibition by Zn^{2+} ions. Furthermore, single Cl^- channel currents were described in the plasma membrane of *Chara* (23, 169).

STRETCH-ACTIVATED CHANNELS   Channels activated by stretching of the plasma membrane have been found in protoplasts of cultured tobacco cells (29), barley aleurone cells (R. Hedrich and D. S. Bush, unpublished), guard cells (140), yeast, and the outer membrane of *E. coli* (93, 133). In all cases these channels were rather nonselective, with an approximate permeability ratio of 3:1 for anions over cations. It has been suggested that stretch-activated channels may function as both turgor sensors and mechanosensors (21, 29, 79, 93, 114, 131, 140, 141). In addition, stretch-activated channels, in conjunction with $I_{K^+,out}$ channels, may serve as down-regulators of osmotic potentials (141).

ION TRANSPORT THROUGH THE VACUOLAR MEMBRANE

Mature plant cells possess a large central vacuole. The storage of solutes in vacuoles and their subsequent release are important in cell metabolism and play a fundamental role in the balance of the osmotic pressure and the control of the electrical potential difference across the vacuolar membrane. The vacuolar membrane can generate and capitalize upon large concentration gradients of ions and metabolites, tending to force (for example) H^+, Ca^{2+}, or malate from the vacuolar compartment into the cytoplasm (42, 54, 67, 98). Regulation of solute transport across the vacuolar membrane, such as the diurnal accumulation and release of malate in vacuoles of photosynthetic tissue (42, 87), implicates changes in potential difference across the vacuolar membrane and in its permeability.

The intracellular location of this organelle has, until recently, complicated the study of the electrical properties of the vacuolar membrane from higher-plant cells by standard electrophysiology.

Improvement of methods for the isolation of stable vacuoles and the application of the patch-clamp technique made possible much more incisive studies on the electrophysiology of the vacuolar membrane (9, 24, 26, 50, 76). The fastest method for the isolation of small numbers of intact vacuoles directly from intact tissue is the one modified by Coyaud et al (26), from that of Klercker (73). In short, the surface of a freshly cut tissue slice is rinsed with buffer solution to wash the liberated vacuoles directly into the recording chamber. Thus, fresh vacuoles can be isolated for each experiment, and vacuole isolation and seal formation can be performed within 2–5 min.

A patch-clamp survey of the electrical properties of the vacuolar membrane from a large variety of plant material has demonstrated the presence of voltage-dependent ion channels and electrogenic pumps as a general feature of ion transport in higher-plant vacuoles (49). The characteristics of the ion channels and pumps detected by patch-clamp studies are reviewed in the following section.

Voltage-Dependent Channels

At high cytoplasmic Ca^{2+} (> 0.3 μM), the ionic conductance of the vacuolar membrane was found to be accounted for entirely by currents directed into the vacuole. These currents are activated at negative voltages (negative inside the vacuole) as well as at slightly positive potentials ($<$ $+20$ mV; 24, 26, 50, 76). The kinetics of activation of these currents are slow (τ = 100–200 ms) and were therefore termed slow-vacuolar (SV)-type currents (52).

Using excised patches, SV-type currents could be resolved at the single-channel level (49, 50). The reported single-channel conductance obtained under different patch-clamp configurations averaged 60–80 pS in 50–100 mM salt solutions (49).

In the range of 10–300 mM KCl, on either side of the membrane, the single-channel conductance of the sugar beet SV channel increased linearly with the salt concentration (R. Hedrich and J. Dunlop, unpublished). At the single-channel level, voltage dependence (inward rectification) was found to arise from an increased probability of opening at negative potentials (24, 50, 76). Voltage-dependence, time course of activation and deactivation (49, 50), as well as the permeability sequence (24) of the single channels were in agreement with measurements of whole-vacuolar currents, indicating that SV currents are carried by unitary SV-type channels.

SELECTIVITY   Whole-vacuole and single-channel recordings revealed that SV channels are permeable to both cations and anions, with variations in different types of tissue. The relative permeabilities in solutions containing various ion combinations and ionic gradients were characterized in sugar beet and red beet, barley leaf, *Acer* cells, and yeast vacuoles (24, 26, 50, 174; J. Alexandre et al, unpublished). SV channels of vacuoles isolated from various tissues are rather nonselective with respect to monovalent cations such as K^+, Na^+, Li^+ (24, 26, 174; J. Alexandre et al, unpublished). The permeability for anions was 2–10 times smaller (24, 50, 174; J. Alexandre et al, unpublished).

The reported anion-permeability sequence of the vacuolar membrane from sugar beet taproots was acetate $>$ nitrate $>$ malate $>$ chloride. This correlates with sequences obtained in tracer-flux experiments (94, 95), and indicates that SV channels may participate in vacuolar solute transport. It should be noted that a study on barley and *Catharanthus* vacuoles has revealed SV channels that seem to be exclusively permeable to cations (24, 76). Possibly the selectivity of these channels can vary with the cell type or during development. Future investigations may help to clarify the observed variance. Information concerning mechanisms of control of simultaneous uptake and retrieval of ions—e.g. malate and NO_3^- (16, 160)—through the described vacuolar channels is equally scant.

Gradients in ionic strength across the vacuolar membrane produce a shift in the activation threshold of the SV channels to positive potentials (50). This displacement could be caused by surface charges that become unshielded when the ionic strength of the extravacuolar solution is reduced (56). Because such ionic gradients across the vacuolar membrane occur under physiological conditions—e.g. during diurnal cycles in malate contents in mesophyll cells of C_3 and CAM plants (42, 87) or during annual cycles of citrate contents of *Hevea* latex (91)—shifts in the threshold potential of channel activation may serve physiological functions.

PHARMACOLOGY   In order to classify the SV channel and compare it to animal counterparts, SV vacuolar channels have been characterized pharmacologically by ion-channel inhibitors (51). In contrast to cation-transport inhibitors, anion-transport blockers strongly affected SV channels in sugar beet. Among them $ZnCl_2$ in micromolar concentrations effectively and reversibly blocked SV channels up to 95%. The stilbene derivatives DIDS and SITS, known to block anion channels in various animal cell types (56), caused an irreversible inhibition of ion movement through SV channels (51). As intravacuolar DIDS did not affect the channels, the inhibitor binding site (which in animal cells correlates with the anion binding site; 102) seems to be located on the cytoplasmic mouth of the SV channel. Through flux measurements, a DIDS-sensitive component of citrate transport was discovered recently in vesicles of the vacuolar membrane of tomato fruits (115).

The SV channel can also be blocked by protons. Low pH values at the vacuolar side reduced the single-channel conductance. At pH 3.8 the conductance was lower by about 30% with respect to the one found at 7.8. Acid pH values at the cytoplasmic face of the channel drastically reduced the mean open-time, whereas the single-channel conductance remained unaffected (J. Dunlop and R. Hedrich, unpublished).

Voltage-dependent K^+ channels were observed in membrane-bound protoplasmic or cytoplasmic droplets of *Chara australis*. The reported unitary conductance of these channels was about 100 pS with 80 mM KCl (60, 86). The membrane enclosing these protoplasmic or cytoplasmic droplets may have been of vacuolar origin, arranged inside-out (134), or may have been plasma membrane, or both (59, 60). Evidence for the occurrence of at least two types of vesicles (droplets) was shown by Lühring (86). One type was characterized by an acidic interior caused by an ATP-induced current; these were thus presumed to have been vacuole vesicles (10). These vesicles contained voltage-dependent potassium channels that were inhibited by sodium ions on either side of the membrane (10).

Channel-Regulation by Cytoplasmic Ca^{2+}

To date, two voltage-dependent channel types have been investigated in the vacuolar membrane of sugar beet taproots (52).

Both types are modulated by the cytoplasmic free Ca^{2+} concentration. One channel type is the above-described SV-type channel, which is activated by an increase in cytoplasmic Ca^{2+} (>0.3 μM). As in animal cells, the gating of the Ca^{2+}-dependent channel depends on both Ca^{2+} and voltage (96). The increase in Ca^{2+} shifted the threshold potential of SV channel activation towards less negative values inside the vacuole (52).

The other channel type is activated by a decrease in cytoplasmic Ca^{2+} (<0.3 μM) at both negative and positive voltages and was termed the fast-vacuolar (FV)-type (52) because it is characterized by "fast" kinetics ($\tau <$ 50 ms).

In sugar beet vacuoles both channel types allow cations and anions to pass the membrane, with permeability ratios: $P_{K^+}/P_{Cl^-} \approx 6$. When the cytoplasmic Ca^{2+} is low ($\approx 10^{-7}$ M) and the membrane potential becomes positive as a consequence of the activity of H^+-pumps (see the next section), only FV channels should be active.

FV channels may provide a pathway for anions to accumulate in the vacuole and potassium to equilibrate between the cytoplasm and the vacuole (94, 95).

Hyperpolarization of the vacuolar membrane and elevation of the Ca^{2+} level (from extracellular and/or intracellular stores) by physiological stimuli such as light-dark transitions (100) or by phosphoinositol metabolites (121, 145) would open SV channels for the release of anions and accompanying cation fluxes. The current through SV and FV channels at physiological membrane potentials and their density can account for ion movements (about 0.1 fmol vacuole^{-1}s^{-1}) measured in tracer-flux experiments (24, 94, 95).

Ca^{2+} efflux from the vacuole (Ca^{2+} channels) under the control of the phosphoinositol metabolism and the two channel types modulated by physiological changes in cytoplasmic Ca^{2+} may provide for a versatile regulation of ion and metabolite fluxes between the cytoplasm and the vacuole.

ATP- and PP$_i$-Dependent Pumps

Proton transport across biological membranes is a central process in many energy conversion reactions in the cell. In the vacuolar membrane protons are translocated by electrogenic ATPases and pyrophosphatases (PP$_i$-ases). The two phosphatases in the vacuolar membrane are physically distinct enzymes as confirmed by chromatographic separation and by their ionic specificities— e.g. activation by halides or cations (127).

With the development of techniques for the preparation of sealed membrane vesicles from various plant tissues in combination with pH- and voltage-sensing probes, several laboratories have demonstrated H^+-translocating phosphatases associated with fractions of the vacuolar membrane (164).

Sealed vesicles have been used successfully for the localization of H^+-

transport in cell membranes and identification of the energy source and the ion species involved in the transport process (for review see 112, 128). The application of the whole-vacuole configuration of the patch-clamp technique to isolated intact vacuoles further enables researchers to (*a*) control buffer capacities, ionic gradients, and electrical potentials across the vacuolar membrane; (*b*) directly measure potentials and currents generated by the proton pumps; and (*c*) determine the stoichiometry and thermodynamics of the pump reactions.

H^+ATPases   The "vacuolar type ATPase" is the most widespread H^+ ATPase in eukaryotic cells (112). It is present in lysosomes, vacuoles of plants and fungi, clathrin-coated vesicles, synaptic vesicles, and several secretory granules. All H^+ ATPases belonging to this group bear structural homologies and appear to be composed of three major polypeptides (112). Among them the 69-kD polypeptide revealed the highest degree of homology to the H^+ ATPase of archaebacteria (177).

A common characteristic of these enzymes is that they catalyze the same overall reaction,

$$nH^+_{cytoplasm} + ATP \rightarrow nH^+_{vacuole} + ADP + Pi$$

and utilize the free energy of this reaction for the transport of H^+ against an electrochemical gradient. The factor n signifies the number of H^+ translocated by the enzyme per one ATP or PP_i hydrolyzed. For the H^+ ATPase Bennett & Spanswick (8) deduced an ATP/H^+ ratio of 2 from kinetic measurements on ATP-hydrolysis and pH change using vacuole vesicles. A more detailed analysis of the pump stoichiometries of both the ATPase and PP_i-ase can be obtained by measuring the reversal potential of the pump currents. Under given phosphate potentials in the "cytoplasmic solution" the direction of the proton current through the pump can be recorded as a function of Δ pH increase across the vacuolar membrane.

Upon the application of Mg-ATP to the cytoplasmic face of a whole-vacuole, a rapid depolarization of 30–70 mV can be observed, indicating the presence of an inwardly directed electrogenic ATPase (24, 49–51). This membrane potential is immediately abolished by the H^+-ATPase inhibitor tributyltin (TBT; 50) or by the protonophore carbonylcyanid-dichlor-phenylhydrazon (CCCP, cf Figure 5, lower panel; 51a) indicating that the ATP-induced shift in membrane potential was generated by H^+ currents into the vacuole. When the ATP concentration was increased stepwise, the pump current also increased stepwise, reaching saturation at 5–10 mM Mg-ATP. K_m values with respect to ATP were about 0.6 mM for *Beta vulgaris* taproot (51a)

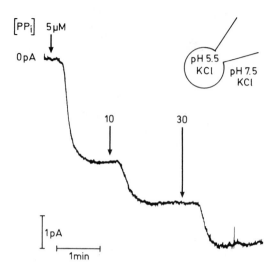

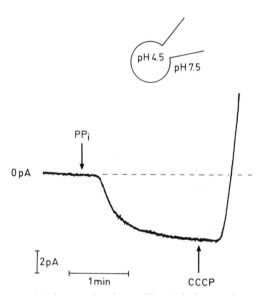

Figure 5   Proton translocating pyrophosphatase (PP$_i$-ase) in the vacuolar membrane from sugar beet taproots. Records of pump-currents resulting from the activity of H$^+$-translocating PP$_i$-ases in the entire vacuole ("whole-vacuole" configuration of the patch-clamp technique).

   (Upper panel) Increasing the pyrophosphate (PP$_i$) concentration in steps caused stepwise increase in H$^+$ current into the vacuole. *(Lower panel)* The protonophore carbonylcyanid-di-chlorophenylhydrazone (CCCP) abolished pump currents induced by PP$_i$ (inward current) and caused the release of H$^+$ from the vacuole (outward current upward deflection). Note that the vacuolar sap was clamped to pH 5.5 in upper panel and 4.5 in lower panel by buffer perfusion through patch pipettes. (From Hedrich & Kurkdjian, unpubl.)

0.5 mM for *Mesembryanthemum cristallinum* mesophyll, and 0.8 mM for *Hordeum vulgare* mesophyll (49, 50).

Under physiological conditions, such as 1 mM ATP (162) in the cytoplasm and a pH-gradient of 2–3 pH units across the vacuolar membrane (98), the H^+-ATPase will produce a H^+ current of about 50 pA/vacuole (1–5 $\mu A/cm^2$; 24, 49, 50). Assuming that anion fluxes shortcircuit the H^+ current, vacuoles (e.g. of barley mesophyll) can accumulate 0.1 fmol $malate^{2-}$ s^{-1} at the expense of 2 protons per malate—comparable to values found for CAM plants (87). This estimate demonstrates that the ATPase provides enough current to drive malate accumulation in the vacuole during photosynthesis (42).

The H^+-ATPase of the sugar beet taproot is able to pump against a 10^4-fold H^+-gradient at a membrane potential of 0 mV (pH 7.8 at the cytoplasmic side and pH 3.8 on the vacuolar side; R. Hedrich and A. Kurkdjian, unpublished). These experiments indicate that the proton binding sites at the cytoplasmic mouth of the proton channel of the ATPase complex must possess a pK above 7.8 and that the pK of H^+ release must have been below 3.8 at its vacuolar face. A proton binding site of pK < 3.8 may indicate the presence of carboxylates (167), because only these groups have a pK low enough to release protons at pH values < 4, a task that the vacuolar H^+-ATPase in the lemon fruit can certainly achieve (pH of lemon juice, the vacuolar sap, 2–2.2).

Under the imposed proton gradients, in symmetric KCl solutions and in the absence of ATP, the resting potential of e.g. sugarbeet vacuoles was zero mV, indicating the absence of a measurable proton conductance. In the presence of pH gradients, with the potential clamped to 0 mV, CCCP not only abolished the ATP-induced current but also mediated H^+ release through a CCCP-induced H^+ conductance (cf Figure 5, lower panel; 51a).

Other vacuolar ATPase properties, such as high preference to ATP over other nucleotides and NO_3^- inhibition established by vesicle experiments (164), could be confirmed (R. Hedrich and A. Kurkdjian, unpublished). As shown in pump current records, nitrate blocks the pump only when present at the cytoplasmic face of the vacuolar membrane (up to 200 mM NO_3 inside the vacuole did not block; 51a). No irreversible NO_3^--induced inhibition as deduced from vesicle studies (126) was found in patch-clamp investigations, since exchange of NO_3^- by Cl^- at the cytoplasmic face of the vacuole restored, at least partially, H^+-pump activity (51a).

H+ PYROPHOSPHATASE   This enzyme was first detected by Karlsson (68) in microsomal fractions of sugar beet taproots, several years before the transport function of the inorganic phosphatase in membranes of photosynthetic bacteria and plants was recognized. To a considerable extent, progress on PP_i-ase-related H^+-transport resulted from in vitro studies on vesicles derived from the vacuolar membrane (128).

While the electrogenic H^+-ATPase has been characterized in previous patch-clamp studies, an electrogenic proton current energized by PP_i-hydrolysis has been shown only recently on vacuoles of sugar beet (51). Upon the addition of PP_i to the cytoplasmic face of the vacuolar membrane, inwardly directed proton currents were activated, causing a depolarization of the vacuolar membrane (51). Saturation of pump current occurred at about 100 μM PP_i, with progressive inhibition at higher PP_i levels (Figure 5, upper panel; 51a). Comparative studies on PP_i-dependent activity of the pyrophosphate hydrolase (83), PP_i-dependent acidification of vacuolar vesicles (83), and PP_i-dependent pump currents of intact vacuoles (Figure 5, upper and lower panel), indicate that the observed changes in H^+ transport result from alterations in PP_i hydrolysis of the enzyme [as already suggested by Leigh & Pope (83)]. Like the ATPase, the PP_i-ase is able to pump H^+ against a 10^4-fold gradient (51a).

It was possible to induce separately and subsequently pump currents generated by the ATPase and by the PP_i-ase on the same vacuole (51). Both enzymes are located in the same membrane.

On the Role of Ion Channels and H^+-Pumps in ΔV and ΔpH Formation across the Vacuolar Membrane

In general, cell metabolism controls the energy status of the cytoplasm, thus determining the level of free energy available for proton pumps to form potential gradients, ΔV, and or pH gradients, Δ pH, across the vacuolar membrane. The ion permeability of this membrane (imposed by channels) in turn determines how the free energy is distributed among Δ V and Δ pH formation. If the membrane were permeable to anions only, anions would initially move into the organelle in response to Δ V, shortcircuiting the potential generated by the H^+-pumps creating a Δ pH (as expected from the chemiosmotic hypothesis; 103). An anion gradient will rapidly be established, creating a positive diffusion potential. Hence, the proton motive force will reach a limiting level and turn the H^+ pump off. Like the vacuolar membrane of plants (24, 50; J. Alexandre et al, personal communication), other lysosomal membranes are known to have high K^+ and anion conductance, which ensures that their internal pH will be maximally acidic (48). Opening of channels, modulating the ion permeability of the organelle membrane, could thus change the state of the organelle from one in which there is a membrane potential and small Δ pH to a state in which the organelle is quite acidic and the membrane potential is near zero, as well as a number of intermediate states (e.g. endosomes; 99).

The genetic information for the selectivity of vacuolar channels, stoichiometry of the pumps, and tissue- or plant-specific variations may determine whether a vacuole constitutes an acidic compartment in the sugar beet taproot (pH 6, storing sugar) or in the lemon fruit (pH 2, storing citric acid).

MEMBRANES OF OTHER ORGANELLES

Chloroplasts

Bulychev et al (18) reported small membrane potentials (+10 to +15 mV relative to the cytoplasm) across the chloroplast membranes when giant chloroplasts from leaves of *Peperomia metallica* were impaled by microelectrodes. Since the magnitude of these potentials depended on light, the tip of the electrode was likely inside the thylakoid (17–19, 173). pH measurements demonstrated that photosynthetic electron transport of the thylakoid generated a proton gradient that in turn drove the ATP-synthase (66). Surprisingly, the membrane potential only transiently exceeded +20 mV (173). Thus it was proposed that counter-ion fluxes across the thylakoid (e.g. chloride and magnesium) may have shortcircuited photosynthetic H^+ transport (57).

By osmotic swelling of giant *P. metallica* chloroplasts, Schönknecht et al (139a) gained access to the thylakoid with patch pipettes. Voltage-dependent chloride-selective channels with a conductance of 80–100 pS (in 100 mM KCl) were identified (Figure 6). These anion channels may be essential to balance the transthylakoid potential and to establish a pH gradient.

Recent patch-clamp studies on the inner mitochondrial membrane of giant mice mitochondria described voltage-dependent ion channels. These channels are characterized by 107 pS conductance (in 150 mM KCl) and poor selectivity of Cl^- over K^+ (157).

Ion Channel Reconstitution—Incorporation of Membrane Fractions into Bilayers and Giant Liposomes

Ion channel proteins, like other membrane proteins, are amphiphilic; they can thus be incorporated into lipid bilayers and liposomes. The reconstitution of membrane-related functions in artificial liposomes allows the association of complex biological processes to defined membrane proteins such as ion channels, pumps, and carriers (Figure 1). By simplifying the membrane system, reconstitution experiments also provide information about the minimal amount of components required for biological activity. One may realize that a membrane fraction of 99.9% purity (based on marker protein assays) still could contain enough foreign channel proteins, appearing like channels of the major membrane fraction when analyzed in bilayers or patch-clamp measurements on liposomes.

Single-channel recordings of reconstituted channel proteins have become possible by incorporation of channel-containing liposomes into bilayers (101) and direct patch clamping of giant liposomes (27, 71, 168).

Planar lipid bilayers have been successfully used by Flügge & Benz (38) to study the porin of the outer chloroplast envelope (720 pS conductance in 100 KCl).

Plasma membrane vesicles prepared from whole leaves of *Vicia faba* L.

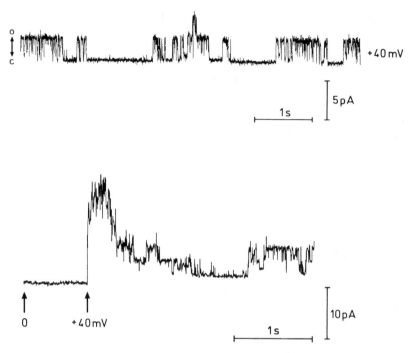

Figure 6   Chloride-channels in the thylakoid membrane of *Peperomia metallica*. Single Cl⁻
channel currents recorded from an excised membrane patch with its intrathylakoid side facing the
bath solution. *(Upper panel)* Single-channel activity at a holding potential of +40 mV inside the
thylakoid. *(Lower panel)* Voltage steps to +40 mV from a holding potential of 0 mV, simulating
polarization of the membrane during light flashes, caused additional transient channel openings.
From Ref. 139a.

were used recently to solubilize and purify a 70-kD protein capable of
mediating Rb^+ fluxes (K. Zeilinger and F. W. Weiles, unpublished). After
reconstitution into lipid bilayers this protein had properties resembling those
of a voltage-gated K^+-selective channel, showing a single-channel conductance of about 40 pS (in 200 mM KCl; K. Zeilinger and F. W. Weiles,
unpublished).

By patch clamping giant liposomes, investigators have studied the properties of reconstituted ion channels in various animal and plant membranes (27,
71). Plant membranes such as chloroplast envelopes, the vacuolar membrane,
and eye spots of the alga *Chlamydomonas reinhardtii* (B. Keller, M. Criado,
personal communication) are now accessible to single-channel analysis.

CONCLUSION AND OUTLOOK

Recent patch-clamp studies have led to the discovery that ion channels occur
in the membranes of higher-plant cells. The channels found in plasma mem-

brane and vacuolar membrane appear to be present in all higher-plant cells studied and may constitute a basic equipment of plant cells. Future investigations may lead to the characterization of tissue-specific subtypes of the discussed channel classes. Some channel types are postulated (e.g. Ca^{2+} channels) but their action has not yet been demonstrated directly, and evidence for their abundance and mode of activation is still lacking. A search for such evidence may become a focus for intensive future investigations. The patch-clamp method in combination with other techniques such as single-cell Ca^{2+}- and pH-fluorescence analysis (109, 171) should be useful for studies on the transduction of signals in plant cells.

Considerable efforts have been directed towards elucidating the genetic control of physiological processes linked to ion channels in animal systems by the use of mutants (90, 132). Especially, plants with a small genome, such as *Arabidopsis,* may lend themselves to the study of plant or tissue specificity and channel density. Moreover, the combination of molecular biology with the high resolution of the patch-clamp technique may become a suitable approach to the study of the relation between the structure and function of ion channels and ion pumps (1, 90, 119, 149, 150, 161, 172).

ACKNOWLEDGMENTS

We thank the groups of E. Neher and B. Sakmann at the Max-Planck-Institut and K. Raschke and U. I. Flügge at the University of Göttingen for stimulating discussions.

We also wish to thank our colleagues in the field of plant membrane ion transport for information concerning recent research prior to publication. We are grateful to S. Hagiwara (UCLA), H. A. Kolb (Konstanz), W. Lahr (University of Göttingen), P. Nobel (UCLA), G. Mathews (Stony Brook), R. Penner (MPI Göttingen), L. Taiz (UCSC), K. Takeda (University of Strasbourg), and J. Umbach (UCLA) for comments on the manuscript.

R. Hedrich was funded by the Max-Planck-Gesellschaft and a DFG grant to K. Raschke. J. I. Schroeder is a Feodor Lynen fellow from the Alexander von Humboldt Foundation and was supported by NIH Grant No. 5-T32-NSO-7101.

Literature Cited

1. Aaronson, L. R., Slayman, C. W., Slayman, C. L. 1988. Expression of the *Neurospora* plasma membrane H^+-ATPase in *Xenopus oocytes. Biophys. J.* 53:137a (Abstr.)
2. Al-Awqati, Q. 1986. Proton-translocating ATPases. *Annu. Rev. Phys. Chem.* 2:179–99
3. Assmann, S. M., Simoncini, L., Schroeder, J. I. 1985. Blue light activates electrogenic ion pumping in guard
cell protoplasts of *Vicia faba. Nature* 318:285–87
4. Beilby, M. J. 1982. Cl^- channels in *Chara. Philos. Trans. R. Soc. London Ser. B* 299:435–55
5. Beilby, M. J. 1984. Potassium channels and different states of plasmalemma. *J. Membr. Biol.* 89:241–46
6. Beilby, M. J., Coster, H. G. L. 1979. The action potential in *Chara corallina.* II. Two activation transients in voltage

clamps of the plasmalemma. *Aust. J. Plant Physiol.* 6:323–35

7. Bennett, A. B., O'Neill, S. D., Eilmann, M., Spanswick, R. M. 1985. H$^+$-ATPase activity from storage tissue of *Beta vulgaris. Plant Physiol.* 78:495–99

8. Bennett, A. B., Spanswick, R. M. 1984. H$^+$-ATPase activity from storage tissue of *Beta vulgaris.* II. H-ATP stoichiometry of an anion-sensitive H$^+$-ATPase. *Plant Physiol.* 74:545–48

9. Bentrup, F. W., Gogarten-Boekels, M., Hoffmann, B., Gogarten, J. P., Baumann, C. 1986. ATP-dependent acidification and tonoplast hyperpolarization in isolated vacuoles from green suspension cells of *Chenopodium rubrum* L.. *Proc. Natl. Acad. Sci. USA* 83:2431–33

10. Bertl, A. 1988. Current-voltage relationship of a sodium-sensitive potassium selective channel in the tonoplast of *Chara corallina. J. Membr. Biol.* Submitted

11. Bertl, A., Gradmann, D. 1987. Current-voltage relationship of potassium channels in the plasmamembrane of *Acetabularia. J. Membr. Biol.* 99:41–49

12. Bertl, A., Klieber, H. G., Gradmann, D. 1988. Slow kinetics of a potassium channel in *Acetabularia. J. Membr. Biol.* 102:141–52

13. Blatt, M. R. 1987. Electrical characteristics of stomatal guard cells: the contribution of ATP-dependent, "electrogenic" transport revealed by current-voltage and difference-current-voltage analysis. *J. Membr. Biol.* 98:257–74

13a. Blatt, M. R. 1988. Potassium-dependent bipolar gating of K$^+$ channels in guard cells. *J. Membr. Biol.* 102:235–46

14. Blatt, M. R. 1988. Mechanism of fusicoccin action: a dominant role for secondary transport in a higher-plant cell. *Planta* 174:187–200

15. Blatt, M. R., Slayman, C. L. 1987. Role of "active" potassium transport in the regulation of cytoplasmic pH by nonanimal cells. *Proc. Natl. Acad. Sci. USA* 84:2737–41

16. Boudet, A. M., Alibert, G., Marigo, G., Ranjeva, R. 1986. The vacuole: possible role in signal transduction versus cytoplasmic homeostasis. See Ref. 91

17. Bulychev, A. A., Andrianov, V. K., Kurella, G. A. 1980. Effect of dicyclohexylcarbodiimide on the proton conductance of thylakoid membranes in intact chloroplasts. *Biochim. Biophys. Acta* 590:300–8

18. Bulychev, A. A., Andrianov, V. K.,

Kurella, G. A., Litvin, F. F. 1972. Micro-electrode measurements of the transmembrane potential of chloroplasts and its photoinduced changes. *Nature* 236:175–77

19. Bulychev, A. A., Andrianov, V. K., Kurella, G. A., Litvin, F. F. 1976. Photoinduction kinetics of electrical potentials in a single chloroplast as studied with micro-electrode technique. *Biochim. Biophys. Acta* 420:336–51

20. Bush, D. S., Hedrich, R., Schroeder, J. I., Jones, R. L. 1988. Channel-mediated K$^+$ flux in barley aleurone protoplasts. *Planta* 176:368–377

21. Christensen, O. 1987. Mediation of cell volume regulation by Ca^{2+} influx through stretch-activated channels. *Nature* 330:66–68

22. Cole, K. S., Curtis, H. J. 1938. Electrical impedance of *Nitella* during activity. *J. Gen. Physiol.* 22:37–64

23. Coleman, H. A. 1986. Cl$^-$ currents in *Chara*—a patch clamp study. *J. Membr. Biol.* 93:55–61

24. Colombo, R., Cerana, R., Lado, P., Peres, A. 1988. Voltage-dependent channels permeable to K$^+$- and Na$^-$ in the membranes of *Acer pseudoplatanus. J. Membr. Biol.* 103:227–36

25. Colquhoun, D., Hawkes, A. G. 1981. On the stochastic properties of single ion channels. *Proc. R. Soc. London Ser. B* 211:205–35

26. Coyaud, L., Kurkdjian, A., Kado, R., Hedrich, R. 1987. Ion channels and ATP-driven pumps involved in ion transport across the tonoplast of sugarbeet vacuoles. *Biochim. Biophys. Acta* 902:263–68

27. Criado, M., Keller, B. U. 1987. A membrane fusion strategy for single-channel recordings of membranes usually non-accessible to patch-clamp pipette electrodes. *FEBS Lett.* 224:172–76

28. De Silva, D. L. R., Hetherington, A. M., Mansfield, T. A. 1985. Synergism between calcium ions and abscisic acid in preventing stomatal opening. *New Phytol.* 100:473–82

29. Edwards, K. L., Pickard, B. G. 1987. Detection and Transduction of physical stimuli in plants. In *The Cell Surface in Signal Transduction,* ed. E. Wagner, H. Greppin, B. Biller. NATO ASI Ser. H12, pp. 41–66. Heidelberg: Springer-Verlag

30. EMBO Experimental Course. 1986. Patch clamp studies on higher plant cells. Organized by Hedrich, R., Schroeder, J. I., Fernandez, J. M. in Göttingen, W. Germany

31. Ettlinger, C., Lehle, L. 1988. Auxin in-

duces rapid changes in phosphatidylinositol metabolites. *Nature* 331:176–78
32. Felle, H., Brummer, B., Bertl, A., Par-
    ish, R. W. 1986. Indole-3-acetic acid
    and fusicoccin cause cytosolic acidifica-
    tion of *corn* coleoptile cells. *Proc. Natl.
    Acad. Sci. USA* 83:8992–95
33. Fendler, K., Grell, E., Haubs, M.,
    Bamberg, E. 1985. Pump currents
    generated by the purified Na^+-K^+-
    ATPase from kidney on black lipid
    membranes. *EMBO J.* 4:3079–85
34. Fesenko, E. E., Kolesnikov, S. S.,
    Lyubarsky, A. L. 1985. Induction by
    cyclic GMP of cationic conductance in
    plasma membrane of retinal rod outer
    segment. *Nature* 313:310–13
35. Findlay, G. P. 1961. Voltage-clamp ex-
    periments with *Nitella*. *Nature* 191:812–
    14
36. Findlay, G. P., Coleman, H. A. 1983.
    Potassium channels in the membrane of
    Hydrodictyon africanum. *J. Membr.
    Biol.* 75:241–51
37. Findlay, G. P., Hope, A. B. 1976. Elec-
    trical properties of plant cells: methods
    & findings. In *Encyclopedia of Plant
    Physiology*, New Ser., Vol. 2, Part A.
    Transport in Plants, ed. U. Lüttge, M.
    G. Pitman, pp. 53–92. Berlin/
    Heidelberg/New York: Springer
38. Flügge, U. I., Benz, R. 1984. Pore-
    forming activity in the outer membrane
    of the chloroplast envelope. *FEBS Lett.*
    169:85–89
39. Fröhlich, O. 1988. The "tunneling"
    mode of biological carrier-mediated
    transport. *J. Membr. Biol.* 101:189–98
40. Gadsby, D. C., Kimura, J., Noma, A.
    1985. Voltage dependence of Na/K
    pump current in isolated heart cells. *Na-
    ture* 315:63–65
41. Gaffey, C. T., Mullins, L. J. 1958. Ion-
    ic fluxes during the action potential in
    Chara. *J. Physiol.* 144:505–24
42. Gerhardt, R., Heldt, H. W. 1987. Sub-
    cellular metabolite levels in spinach
    leaves. *Plant Physiol.* 83:399–407
43. Gradmann, D., Klieber, H. G., Hansen,
    U. P. 1987. Reaction kinetic parameters
    for ion transport from steady-state cur-
    rent-voltage curves. *Biophys. J.* 51:569–
    85
44. Graziana, A., Fosset, M., Ranjeva, R.,
    Hetherington, A. M., Lazdunski, M.
    1988. Ca^{2+} channel inhibitors that bind
    to plant cell membranes block Ca^{2+}-
    entry into protoplasts. *Biochem. J.*
    27:764–68
45. Gustin, M. C., Martinac, B., Saimi, Y.,
    Culbertson, M. R., Kung, C. 1986. Ion
    channels in yeast. *Science* 233:1195–97
46. Hagiwara, S., Tasaki, I. 1958. A study

on the mechanism of impulse transmission across the giant synapse of the
squid. *J. Physiol.* 143:114–37
47. Hamill, O. P., Marty, A., Neher, E.,
    Sakmann, B., Sigworth, F. J. 1981. Im-
    proved patch-clamp techniques for high-
    resolution current recording from cells
    and cell-free membrane patches. *Pflüg-
    ers Arch.* 391:85–100
48. Harikumar, P., Reeves, J. P. 1983. The
    lysosomal proton pump is electrogenic.
    J. Biol. Chem. 258:10403–10
49. Hedrich, R., Barbier-Brygoo, H., Felle,
    H., Flügge, U. I., Lüttge, U., et al.
    1988. General mechanisms for solute
    transport across the tonoplast of plant
    vacuoles. A patch-clamp survey of ion
    channels and proton pumps. *Botan. Acta*
    101:7–13
50. Hedrich, R., Flügge, U. I., Fernandez,
    J. M. 1986. Patch-clamp studies of ion
    transport in isolated plant vacuoles.
    FEBS Lett. 204:228–32
51. Hedrich, R., Kurkdjian, A. 1988.
    Characterization of an anion-permeable
    channel from sugar beet vacuoles: effect
    of inhibitors. *EMBO J.* 7:366–66
51a. Hedrich, R., Kurkdjian, A., Guern, J.,
    Flügge, U. I. 1989. Comparative studies
    on the electrical properties of the H^+
    translocating ATPase and pyrophospha-
    tase of the vacuolar-lysosomal compart-
    ment. *EMBO J.* Submitted
52. Hedrich, R., Neher, E. 1987.
    Cytoplasmic calcium regulates voltage-
    dependent ion channels in plant
    vacuoles. *Nature* 329:833–35
53. Hedrich, R., Schroeder, J. I., Fernan-
    dez, J. M. 1986. Patch-clamp studies on
    higher plant cells: a perspective. *Trends
    Biochem. Sci.* 12:49–52
54. Hepler, P. K., Wayne, R. O. 1985. Cal-
    cium and plant development. *Annu. Rev.
    Plant Physiol.* 36:397–439
55. Hesse, J. E., Wieczorek, L., Altendorf,
    K., Reicin, A. S., Dorus, E., Epstein,
    W. 1984. Sequence homology between
    two membrane-transport ATPases, the
    Kdp-ATPase of *Escherichia coli* and the
    Ca^{2+}-ATPase of sarcoplasmic reticu-
    lum. *Proc. Natl. Acad. Sci. USA*
    81:4746–50
56. Hille, B. 1984. *Ionic Channels of Excit-
    able Membranes*. Sunderland, Mass.:
    Sinauer Assoc.
57. Hind, G., Nakatani, H. Y., Izawa, S.
    1974. Light-dependent redistribution of
    ions in suspensions of chloroplast thyla-
    koid membranes. *Proc. Natl. Acad. Sci.
    USA* 71:1484–88
58. Hodgkin, A. L., Huxley, A. F. 1952. A
    quantitative description of membrane
    current and its application to conduction

and excitation in nerve. *J. Physiol.* 117:500–44

59. Homblé F. 1987. A tight-seal whole cell study of the voltage-dependent gating mechanism of K^+ channels of protoplasmic droplets of *Chara corallina*. *Plant Physiol.* 84:433–37

60. Homblé F., Ferrier, J. M., Dainty, J. 1987. Voltage-dependent K^+-channels in protoplasmic droplets of *Chara corallina*. *Plant Physiol.* 83:53–57

61. Hope, A. B., Walker, N. A. 1975. The physiology of Giant Algal Cells. New York: Cambridge Univ. Press

62. Iijima, T., Hagiwara, S. 1987. Voltage dependent K^+ channels in protoplasts of trap-lobe cells of *Dionea muscipula*. *J. Membr. Biol.* 100:73–81

63. Jaffe, L. F. 1966. Electrical currents through the developing *Fucus* egg. *Proc. Natl. Acad. Sci. USA* 56:1102–9

64. Jaffe, L. F., Nuccitelli, R. 1977. Electrical controls of development. *Annu. Rev. Biophys. Bioeng.* 6:445–76

65. Jones, R. L. 1973. Gibberelic acid and ion release from barley aleurone tissue. *Plant Physiol.* 52:303–8

66. Junge, W. 1982. Eletrogenic reactions and proton pumping in green plant photosynthesis. *Curr. Top. Membr. Transp.* 16:431–65

67. Kaiser, G., Martinoia, E., Wiemken, A. 1982. Rapid appearance of photosynthetic products in the vacuoles isolated from barley mesophyll protoplasts by a new fast method. *Z. Pflanzenphysiol.* 107:103–13

68. Karlsson, J. 1975. Membrane-bound potassium and magnesium ion-stimulated inorganic pyrophosphatase from roots and cotyledons of sugar beet *(Beta vulgaris L.)*. *Biochim. Biophys. Acta* 399:356–63

69. Katz, B. 1949. Les constants életriques de la membrane du muscle. *Arch. Sci. Physiol.* 2:285–99

70. Katz, B., Miledi, R. 1970. Membrane noise produced by acetylcholine. *Nature* 226:962–63

71. Keller, B., Hedrich, R., Vaz, W. L. C., Criado, M. 1988. Single channel recordings of reconstituted ion channel proteins: an improved technique. *Pflügers Arch.* 411:94–100

72. Ketchum, K. A., Poole, R. J. 1988. *Plant Physiol.* 86(4):82 (Abstr.)

73. Klercker, J. af. 1892. Eine Methode zur Isolierung lebender Protoplasten. *Öfvers. Kongl. Vetensk. Akad. Förandlingar*, Stockholm 9:463–74

74. Köhler, K., Steigner, W., Simonis, W., Urbach, W. 1986. Potassium channels

in *Eremosphaera viridis*. *Planta* 166:490–99

75. Kolb, H. A., Frehland, E. 1980. Noisecurrent generated by carrier mediated ion transport at non-equilibrium. *Biophys. Chem.* 12:21–34

76. Kolb, H. A., Köhler, K., Martinoia, E. 1987. Single potassium channels in membranes of isolated mesophyll barley vacuoles. *J. Membr. Biol.* 95:163–69

77. Läuger, P. 1979. Ion transport through lipid bilayer membranes. In *Membranes and Intercellular Communication*, ed. R. Balian, et al. Amsterdam: North/Holland

78. Lafaire, A. V., Schwarz, W. 1986. Voltage dependence of the rheogenic Na^+/K^+ ATPase in the membrane of oocytes of *Xenopus laevis*. *J. Membr. Biol.* 91:43–51

79. Lansman, J. B., Hallam, T. J., Rink, T. J. 1987. Single stretch-activated ion channels in vascular endothelial cells as mechanotransducers? *Nature* 325:811–13

80. Latorre, R., Miller, C. 1983. Conduction and selectivity in potassium channels. *J. Membr. Biol.* 71:11–30

81. Läuger, P. 1973. Ion transport through pores: a rate-theory analysis. *Biochim. Biophys. Acta* 311:423–41

82. Laver, D. R., Walker, N. A. 1987. Steady-state voltage dependent gating and conduction kinetics of single K^+ channels in the membrane of cytoplasmic drops of *Chara australis*. *J. Membr. Biol.* 100:31–42

83. Leigh, R. A., Pope, A. J. 1987. Understanding tonoplast function: Some emerging problems. See Ref. 91, pp. 223–228

84. Levitan, I. B. 1985. Phosphorylation of ion channels. *J. Membr. Biol.* 87:177–90

85. Lindau, M., Fernandez, J. M. 1986. IgE mediated degranulation of mast cells does not require opening of ion channels. *Nature* 319:150–53

86. Lühring, H. E. 1986. Recording of single K^+ channels in the membrane of cytoplasmic drop of *Chara australis*. *Protoplasma* 133:19–28

87. Lüttge, U. 1987. Carbon dioxide and water demand: crassulacean acid metabolism (CAM): a versatile ecological adaptation exemplifying the need for integration in ecophysiological work. *New Phytol.* 106:593–629

88. Lüttge, U., Pitman, M. G., eds. 1976. *Transport in Plants*, Vols. I, II, III. Berlin/Heidelberg/New York: Springer

89. MacRobbie, E. A. C. 1985. Ion channels in plant cells. *Nature* 313:529
90. Maelicke, A. 1988. Structural similarities between ion channel proteins. *Trends Biochem. Sci.* 13:199–202
91. Marin, B., ed. 1987. *Plant Vacuoles.* NATO ASI Ser. New York: Plenum
92. Marrè, E. 1979. Fusicoccin: a tool in plant physiology. *Annu. Rev. Plant Physiol.* 30:273–88
93. Martinac, B., Buechner, M., Delcour, A. H., Adler, J., Kung, C. 1987. Pressure-sensitive ion channel in *Escherichia coli. Proc. Natl. Acad. Sci. USA* 84:2297–301
94. Martinoia, E., Flügge, U. I., Kaiser, G., Heber, U., Heldt, H. W. 1985. Energy-dependent uptake of malate into vacuoles isolated from barley mesophyll protoplasts. *Biochim. Biophys. Acta* 806:311–19
95. Martinoia, E., Schramm, M. J., Kaiser, G., Kaiser, W. M., Heber, U. 1986. Transport of anions in isolated barley vacuoles. I. permeability to anions and evidence for a Cl-uptake system. *Plant Physiol.* 80:895–901
96. Marty, A. 1983. Ca^{2+}-dependent K^+-channels with large unitary conductance. *Trends Neurosci.* 6:262–65
97. Marty, A., Neher, E. 1983. Tight seal whole-cell recording. In *Single-Channel Recording,* ed. Sakmann, B., Neher, E., pp. 107–22. New York: Plenum
98. Matile, P. 1978. Biochemistry and function of vacuoles. *Annu. Rev. Plant Physiol.* 29:193–213
99. Maxfield, F. R. 1985. Calcium and pH in cytoplasmic organelles. *Trends Biochem. Sci.* 10:443–47
100. Miller, A. J., Sanders, D. 1987. Depletion of cytosolic free calcium induced by photosynthesis. *Nature* 326:397–400
101. Miller, C. 1986. *Ion Channel Reconstitution.* New York: Plenum
102. Miller, C., White, M. W. 1984. Dimeric structure of single chloride channels from *Torpedo* electroplax. *Proc. Natl. Acad. Sci. USA* 81:2772–75
103. Mitchell, P. 1967. Translocations through natural membranes. *Adv. Enzymol.* 29:33–87
104. Moran, N., Ehrenstein, G., Iwasa, K., Bare, C., Mischke, C. 1984. Ion channels in plasmalemma of wheat protoplast. *Science* 226:835–38
105. Moran, N., Ehrenstein, G., Iwasa, K., Mischke, C., Bare, C., Satter, R. L. 1988. Potassium channels in motor cells of *Samanea saman:* a patch clamp study. *Plant Physiol.* 88:643–48
106. Morse, M., Crain, R.C., Satter, R. L.

1987. Light stimulated phosphatidylinositol turnover in *Samanea saman* pulvini. *Proc. Natl. Acad. Sci. USA* 84:7075–78
107. Mullins, L. J. 1962. Efflux of chloride ions during the action potential of *Nitella. Nature* 196:986–87
108. Nakao, M., Gadsby, D. C. 1986. Voltage dependence of Na translocation by the Na/K pump. *Nature* 323:628–30
109. Neher, E., Almers, W. 1986. Fast calcium transients in rat peritoneal mast cells are not sufficient to trigger exocytosis. *EMBO J.* 5:193–214
110. Neher, E., Sakmann, B. 1976. Single-channel currents recorded from membrane of denervated frog muscle fibres. *Nature* 260:779–802
111. Neher, E., Stevens, C. F. 1977. Conductance fluctuations and ionic pores in membranes. *Annu. Rev. Biophys. Bioeng.* 6:345–81
112. Nelson, N. 1988. Structure, function, and evolution of proton-ATPases. *Plant Physiol.* 86:1–3
113. Neyton, J., Trautmann, A. 1985. Single-channel currents of an intercellular junction. *Nature* 317:331–35
114. Olesen, S. P., Clapham, D. E., Davies, P. F. 1988. Haemodynamic shear stress activates a K^+ current in vascular endothelial cells. *Nature* 331:168–70
115. Oleski, N., Mahdavi, P., Bennett, A. B. 1987. Transport properties of the tomato fruit tonoplast. II. Citrate transport. *Plant Physiol.* 84:997–1000
116. Outlaw, W. H. 1983. Current concepts on the role of potassium in stomatal movements. *Physiol. Plant.* 59:407–511
117. Pilet, P. E. 1985. *The Physiological Properties of Plant Protoplasts.* Berlin: Springer-Verlag
118. Poole, R. J. 1978. Energy coupling for membrane transport. *Annu. Rev. Plant Physiol.* 29:437–60
119. Portillo, F., Serrano, R. 1988. Dissection of functional domains of the yeast proton-pumping ATPase by directed mutagenesis. *EMBO J.* 7:1788–98
120. Pusch, M., Neher, E. 1988. Rates of diffusional exchange between small cells and a measuring patch pipette. *Pflügers Arch.* 411:204–11
121. Ranjeva, R., Carrasco, A., Boudet, A. M. 1988. Inositol triphosphate stimulates the release of calcium from intact vacuoles isolated from *Acer* cells. *FEBS Lett.* 230:137–41
122. Raschke, K. 1979. Movements of stomata. In *Encyclopedia of Plant Physiology,* New Ser., ed. W. Haupt, E. Feinleib, 7:383–441. Berlin: Springer

123. Raschke, K., Hedrich, R. 1987. Stomatal movement: ion transport and carbon metabolism. In *Model Building in Plant Physiology-Biochemistry,* ed. D. Newman, U. Wilson. Boca Raton: CRC

124. Raschke, K., Hedrich, R. 1989. Patch-clamp measurements on isolated guard-cell protoplasts and vacuoles. *Methods Enzymol.* In press

125. Raschke, K., Hedrich, R., Reckmann, U., Schroeder, J. I. 1988. Exploring biophysical and biochemical components of the osmotic motor that drives stomatal movement. *Botan. Acta.* 101: 283–94

126. Rea, P. A., Griffith, C. J., Manolson, M. F., Sanders, D. 1987. Irreversible inhibition of H^+-ATPase of higher-plant tonoplast by chaotropic anions: evidence for peripheral location of nucleotide-binding subunits. *Biochim. Biophys. Acta* 904:1–12

127. Rea, P. A., Poole, R. J. 1986. Chromatographic resolution of H^+-translocating ATPase of higher plant tonoplast. *Plant Physiol.* 81:126–29

128. Rea, P. A., Sanders, D. 1987. Tonoplast energization: two H^+ pumps, one membrane. *Physiol. Plant* 71:131–41

129. Rodriguez-Navarro, A., Blatt, M. R., Slayman, C. L. 1986. A potassium-proton symport in *Neurospora crassa*. *J. Gen. Physiol.* 87:649–74

130. Rudy, B. 1988. Diversity and ubiquity of K^+ channels. *Neuroscience* 25:729–49

131. Sachs, F. 1986. Mechanotransducing ion channels. In *Ionic Channels in Cells and Model Systems,* ed. R. Latorre. New York/London: Plenum

132. Saimi, Y., Kung, C. 1987. Behavioral genetics of *Paramecium*. *Annu. Rev. Genet.* 21:47–65

133. Saimi, Y., Martinac, B., Gustin, M. C., Culbertson, M. R., Adler, J., Kung, C. 1988. Ion channels in *Paramecium,* yeast and *Escherichia coli*. *Trends Biochem. Sci.* 13:304–9

134. Sakano, K., Tazwa, M. 1986. Tonoplast origin of the envelope membrane of cytoplasmic droplets prepared from *Chara* internodial cells. *Protoplasma* 131:247–49

135. Satter, R. L., Geballe, G. T., Applewhite, P. B., Galston, A. W. 1974. Potassium flux of leaf movement in *Samanea Saman*. I. Rhythmic movement. *J. Gen. Physiol.* 64:413–30

136. Satter, R. L., Moran, N. 1988. What's new in plant physiology: ionic channels in plant cell membranes. *Physiol. Plant.* 72:816–20

137. Schauf, C. L., Wilson, K. J. 1987. Properties of single K^+ and Cl^- channels in *Asclepias tuberosa* protoplasts. *Plant Physiol.* 85:413–18

138. Schauf, C. L., Wilson, K. J. 1987. Effect of abscisic acid on K^+ channels in *Vicia faba* guard cell protoplasts. *Biochem. Biophys. Res. Commun.* 145:284–90

139. Schnabl, H., Ziegler, H. 1975. Über die Wirkung von Aluminiumionen auf die Stomatabewegung von *Vicia faba* Epidermen. *Z. Pflanzenphysiol.* 74:394–403

139a. Schönknecht, G., Hedrich, R., Junge, W. Raschke, K. 1988. A voltage-dependent chloride channel in the photosynthetic membrane of a higher plant, Nature 336:589–92

140. Schroeder, J. I. 1987. *K^+-Kanäle in der Plasmamembran von Schliesszellen. Eine Patch-Clamp Untersuchung molekularer Mechanismen des K^+ Transportes in höheren Pflanzenzellen.* Thesis. Univ. Göttingen

141. Schroeder, J. I. 1988. K^+ transport properties of U^+ channels in the plasma membrane of *Vicia faba* guard cells. *J. Gen. Physiol.* 92:667–83

142. Schroeder, J. I. 1989. A quantitative analysis of outward rectifying K^+ channels in guard cell protoplasts from *Vicia faba*. *J. Membr. Biol.* 107:229–35

143. Schroeder, J. I., Hedrich, R., Fernandez, J. M. 1984. Potassium-selective single channels in guard cell protoplasts of *Vicia faba*. *Nature* 312:361–62

144. Schroeder, J. I., Raschke, K., Neher, E. 1987. Voltage dependence of K^+ channels in guard-cell protoplasts. *Proc. Natl. Acad. Sci. USA* 84:4108–12

145. Schumaker, K. S., Sze, H. 1986. Inositol 1,4,5-trisphosphate releases Ca^{2+} from vacuolar membrane vesicles of oat roots. *J. Biol. Chem.* 262:3944–46

146. Sentenac, H., Grignon, C. 1981. A model for predicting ion equilibrium concentrations in cell walls. *Plant Physiol.* 68:415–19

147. Serrano, E. E., Zeiger, E., Hagiwara, S. 1988. Red-light stimulates an electrogenic proton pump in *Vicia faba* guard-cell protoplasts. *Proc. Natl. Acad. Sci. USA* 85:436–40

148. Serrano, R. 1985. *Plasma Membrane ATPase of Plant and Fungi.* Boca Raton: CRC

149. Serrano, R. 1988. Structure and function of proton translocating ATPase in plasma membranes of plants and fungi. *Biochem. Biophys. Acta* 947:1–28

150. Serrano, R., Kielland-Brandt, M. C.,

Fink, G. R. 1986. Yeast plasma membrane ATPase is essential for growth and has homology with (Na$^+$ + K$^+$), K$^+$-, and Ca^{2+}-ATPases. *Nature* 319:689–93

151. Sibaoka, T. 1966. Action potentials in plant organs. *Symp. Soc. Exp. Biol.* 20:49–74

152. Sigworth, F. J. 1986. The patch clamp is more useful than anyone had expected. *Fed. Proc.* 45:2673–77

153. Simons, P. J. 1981. The role of electricity in plant movements. *New Phytol.* 87:11–37

154. Skulachev, V. 1988. *Membrane Biogenergetics*. New York: Springer-Verlag

155. Slayman, C. L. 1987. The plasma membrane ATPase of *Neurospora:* a proton pumping electroenzyme. *J. Bioenerg. Biomembr.* 19:1–20

156. Sokolik, A. I., Yurin, V. M. 1986. Potassium channels in plasmalemma of *Nitella* cells at rest. *J. Membr. Biol.* 89:9–22

157. Sorgato, M. C., Keller, B. U., Stühmer, W. 1987. Patch-clamping of the inner mitochondrial membrane reveals a voltage-dependent ion channel. *Nature* 330:498–500

158. Spanswick, R. M. 1981. Electrogenic ion pumps. *Annu. Rev. Plant Physiol.* 32:267–89

159. Spray, D. C., Harris, A. L., Bennett, M. V. L. 1981. Equilibrium properties of a voltage-dependent junctional conductance. *J. Gen. Physiol.* 77:77–93

160. Steingröver, E., Ratering, P., Siesling, J. 1986. Daily changes in uptake, reduction and storage of nitrate in spinach grown at low light intensity. *Physiol. Plant.* 66:550–55

161. Stevens, C. F. 1987. Channel families in the brain. *Nature* 328:198–99

162. Stitt, M., Lilley, R. M., Heldt, H. W. 1982. Adenine nucleotide levels in the cytosol, chloroplast, and mitochondria of wheat leaf protoplasts. *Plant Physiol.* 70:971–77

163. Stoeckel, H., Takeda, K. 1988. Calcium-activated, voltage dependent nonselective currents in endosperm plasma membrane from higher plants. *Proc. Roy. Soc. (London) Ser. B* Submitted

164. Sze, H. 1985. H$^+$-translocating ATPases: advances using membrane vesicles *Annu. Rev. Plant Physiol.* 36:175–208

165. Tahler, M., Steigner, W., Köhler, K., Simonis, W., Urbach, W. 1987. Release of repetitive transient potentials and opening of potassium channels by barium in *Eremosphaera viridis*. *FEBS Lett.* 219:351–54

166. Takeda, K., Kurkdjian, A. C., Kado, R. T. 1985. Ionic channels, ion transport and plant cell membranes: potential applications of the patch-clamp technique. *Protoplasma* 127:147–63

167. Tanford, C. 1983. Mechanisms of free energy coupling in active transport. *Annu. Rev. Biochem.* 52:379–409

168. Tank, D. W., Miller, C. 1983. Patch-clamped liposomes: reconstituted ion channels. In *Single-Channel Recording*, ed. B. Sakmann, E. Neher. New York: Plenum

169. Tazawa, M., Shimmen, T., Mimura, T. 1987. Membrane control in the Characeae. *Annu. Rev. Plant Physiol.* 38:95–117

170. Trewavas, A. J. 1986. *Molecular and Cellular Aspects of Calcium in Plant Development*. NATO Adv. Stud. Inst. Ser. A, Vol. 104. New York: Plenum

171. Tsien, R. Y., Poenie, M. 1986. Fluorescence ratio imaging: a new window into intracellular ionic signaling. *Trends Biochem. Sci.* 11:450–55

172. Unwin, N. 1986. Is there a common design for cell membrane channels? *Nature* 323:12–13

173. Vredenberg, W. J., Tonk, W. J. M. 1975. On the steady-state electrical potential difference across the thylakoid membranes of chloroplasts in illuminated plant cells. *Biochim. Biophys. Acta* 387:580–87

174. Wada, Y., Ohsumi, Y., Tanifuji, M., Kasai, M., Anraku, Y. 1987. Vacuolar ion channel of the yeast, *Saccharomyces cerevisiae*. *J. Biol. Chem.* 262:17260–63

175. Weissenseel, M. H., Kicherer, R. M. 1981. *Ionic Currents as Control Mechanisms in Cytomorphogenesis in Plants*, ed. O. Krermayer. New York: Springer-Verlag

176. Williamson, R. E., Ashley, C. C. 1982. Free Ca$^+$ and cytoplasmic streaming in the alga *Chara*. *Nature* 296:647–51

177. Zimniak, L., Dittrich, P., Gogarten, J. P., Kibak, H., Taiz, L. 1988. The cDNA sequence of the 69-kDa subunit of the carrot vacuolar H$^+$-ATPase. *J. Biol. Chem.* 263:9102–12

AUTHOR INDEX

SUBJECT INDEX

595

CUMULATIVE INDEXES

CONTRIBUTING AUTHORS, VOLUMES 32–40

CHAPTER TITLES, VOLUMES 32–40

ORGANELLES AND CELLS

Function

Annual Reviews Inc.

A NONPROFIT SCIENTIFIC PUBLISHER

4139 El Camino Way
P.O. Box 10139
Palo Alto, CA 94303-0897 • USA

ORDER FORM

**ORDER TOLL FREE
1-800-523-8635**
(except California)

Telex: 910-290-0275

Annual Reviews Inc. publications may be ordered directly from our office by mail, Telex, or use our Toll Free Telephone line (for orders paid by credit card or purchase order*, and customer service calls only); through booksellers and subscription agents, worldwide; and through participating professional societies. Prices subject to change without notice. ARI Federal I.D. #94-1156476

- **Individuals:** Prepayment required on new accounts by check or money order (in U.S. dollars, check drawn on U.S. bank) or charge to credit card—American Express, VISA, MasterCard.
- **Institutional buyers:** Please include purchase order number.
- **Students:** $10.00 discount from retail price, per volume. Prepayment required. Proof of student status must be provided (photocopy of student I.D. or signature of department secretary is acceptable). Students must send orders direct to Annual Reviews. Orders received through bookstores and institutions requesting student rates will be returned. You may order at the Student Rate for a maximum of 3 years.
- **Professional Society Members:** Members of professional societies that have a contractual arrangement with Annual Reviews may order books through their society at a reduced rate. Check with your society for information.
- **Toll Free Telephone orders:** Call 1-800-523-8635 (except from California) for orders paid by credit card or purchase order and customer service calls only. California customers and all other business calls use 415-493-4400 (not toll free). Hours: 8:00 AM to 4:00 PM, Monday-Friday, Pacific Time. *Written confirmation is required on purchase orders from universities before shipment.
- **Telex: 910-290-0275**

Regular orders: Please list the volumes you wish to order by volume number.
Standing orders: New volume in the series will be sent to you automatically each year upon publication. Cancellation may be made at any time. Please indicate volume number to begin standing order.
Prepublication orders: Volumes not yet published will be shipped in month and year indicated.
California orders: Add applicable sales tax.
Postage paid (4th class bookrate/surface mail) **by Annual Reviews Inc.** Airmail postage or UPS, extra.

ANNUAL REVIEWS SERIES		Prices Postpaid per volume USA & Canada/elsewhere	Regular Order Please send:	Standing Order Begin with:
Annual Review of **ANTHROPOLOGY**			Vol. number	Vol. number
Vols. 1-14	(1972-1985)................$27.00/$30.00			
Vols. 15-16	(1986-1987)................$31.00/$34.00			
Vol. 17	(1988) ....$35.00/$39.00			
Vol. 18	(avail. Oct. 1989) ....$35.00/$39.00		Vol(s). _____	Vol. _____
Annual Review of **ASTRONOMY AND ASTROPHYSICS**				
Vols. 1, 4-14, 16-20	(1963, 1966-1976, 1978-1982) ..$27.00/$30.00			
Vols. 21-25	(1983-1987)................$44.00/$47.00			
Vol. 26	(1988) ....$47.00/$51.00			
Vol. 27	(avail. Sept. 1989) ....$47.00/$51.00		Vol(s). _____	Vol. _____
Annual Review of **BIOCHEMISTRY**				
Vols. 30-34, 36-54	(1961-1965, 1967-1985).......$29.00/$32.00			
Vols. 55-56	(1986-1987)................$33.00/$36.00			
Vol. 57	(1988) ....$35.00/$39.00			
Vol. 58	(avail. July 1989)............$35.00/$39.00		Vol(s). _____	Vol. _____
Annual Review of **BIOPHYSICS AND BIOPHYSICAL CHEMISTRY**				
Vols. 1-11	(1972-1982)................$27.00/$30.00			
Vols. 12-16	(1983-1987)................$47.00/$50.00			
Vol. 17	(1988) ....$49.00/$53.00			
Vol. 18	(avail. June 1989) ....$49.00/$53.00		Vol(s). _____	Vol. _____
Annual Review of **CELL BIOLOGY**				
Vol. 1	(1985) ....$27.00/$30.00			
Vols. 2-3	(1986-1987)................$31.00/$34.00			
Vol. 4	(1988) ....$35.00/$39.00			
Vol. 5	(avail. Nov. 1989)...........$35.00/$39.00		Vol(s). _____	Vol. _____

ANNUAL REVIEWS SERIES	Prices Postpaid per volume USA & Canada/elsewhere	Regular Order Please send:	Standing Order Begin with:
		Vol. number	Vol. number

Annual Review of COMPUTER SCIENCE

Vols. 1-2	(1986-1987)..............$39.00/$42.00		
Vol. 3	(1988) ....$45.00/$49.00		
Vol. 4	(avail. Nov. 1989)............$45.00/$49.00	Vol(s). _____	Vol. _____

Annual Review of EARTH AND PLANETARY SCIENCES

Vols. 1-10	(1973-1982)..............$27.00/$30.00		
Vols. 11-15	(1983-1987)..............$44.00/$47.00		
Vol. 16	(1988) ....$49.00/$53.00		
Vol. 17	(avail. May 1989)............$49.00/$53.00	Vol(s). _____	Vol. _____

Annual Review of ECOLOGY AND SYSTEMATICS

Vols. 2-16	(1971-1985)..............$27.00/$30.00		
Vols. 17-18	(1986-1987)..............$31.00/$34.00		
Vol. 19	(1988) ....$34.00/$38.00		
Vol. 20	(avail. Nov. 1989)............$34.00/$38.00	Vol(s). _____	Vol. _____

Annual Review of ENERGY

Vols. 1-7	(1976-1982)..............$27.00/$30.00		
Vols. 8-12	(1983-1987)..............$56.00/$59.00		
Vol. 13	(1988) ....$58.00/$62.00		
Vol. 14	(avail. Oct. 1989)............$58.00/$62.00	Vol(s). _____	Vol. _____

Annual Review of ENTOMOLOGY

Vols. 10-16, 18	(1965-1971, 1973)		
20-30	(1975-1985)..............$27.00/$30.00		
Vols. 31-32	(1986-1987)..............$31.00/$34.00		
Vol. 33	(1988) ....$34.00/$38.00		
Vol. 34	(avail. Jan. 1989)............$34.00/$38.00	Vol(s). _____	Vol. _____

Annual Review of FLUID MECHANICS

Vols. 1-4, 7-17	(1969-1972, 1975-1985).......$28.00/$31.00		
Vols. 18-19	(1986-1987)..............$32.00/$35.00		
Vol. 20	(1988) ....$34.00/$38.00		
Vol. 21	(avail. Jan. 1989)............$34.00/$38.00	Vol(s). _____	Vol. _____

Annual Review of GENETICS

Vols. 1-19	(1967-1985)..............$27.00/$30.00		
Vols. 20-21	(1986-1987)..............$31.00/$34.00		
Vol. 22	(1988) ....$34.00/$38.00		
Vol. 23	(avail. Dec. 1989)............$34.00/$38.00	Vol(s). _____	Vol. _____

Annual Review of IMMUNOLOGY

Vols. 1-3	(1983-1985)..............$27.00/$30.00		
Vols. 4-5	(1986-1987)..............$31.00/$34.00		
Vol. 6	(1988) ....$34.00/$38.00		
Vol. 7	(avail. April 1989)...........$34.00/$38.00	Vol(s). _____	Vol. _____

Annual Review of MATERIALS SCIENCE

Vols. 1, 3-12	(1971, 1973-1982)...........$27.00/$30.00		
Vols. 13-17	(1983-1987)..............$64.00/$67.00		
Vol. 18	(1988) ....$66.00/$70.00		
Vol. 19	(avail. Aug. 1989)............$66.00/$70.00	Vol(s). _____	Vol. _____

Annual Review of MEDICINE

Vols. 9, 11-15	(1958, 1960-1964)		
17-36	(1966-1985)..............$27.00/$30.00		
Vols. 37-38	(1986-1987)..............$31.00/$34.00		
Vol. 39	(1988) ....$34.00/$38.00		
Vol. 40	(avail. April 1989) ....$34.00/$38.00	Vol(s). _____	Vol. _____